Engineering Mathematics

Engineering Mathematics

Vol. II

R. L. Garg
Nishu Gupta

PEARSON

Delhi • Chennai

Senior Editor—Acquisitions: Anita Yadav
Editor—Production: Vipin Kumar

ISBN 978-93-325-3633-3

First Impression

Published by Dorling Kindersley (India) Pvt. Ltd, licensees of Pearson Education in South Asia.

Head Office: 7th Floor, Knowledge Boulevard, A-8(A), Sector 62, Noida 201 309, UP, India.
Registered Office: 11 Community Centre, Panchsheel Park, New Delhi 110 017, India.

Compositor: SRS Technologies, Puducherry.
Printed in India by HT Media Ltd., Noida

Dedication

To My Wife

Late Smt. Shashi Kiran Garg

R. L. Garg

To My Parents

Smt. Adarsh Garg

and

Late Shri B.S.Garg

Nishu Gupta

Contents

Preface xiii

Acknowledgements xiv

About the Authors xv

Symbols, Basic Formulae and Useful Informations xvii

1. Functions of Complex Variables 1

1.1 Introduction — 1

1.2 Definition, Limit, Continuity and Differentiability of a Function of Complex Variable — 1

1.2.1 Definition 1

1.2.2 Limit of a Function 1

1.2.3 Continuity of a Function 4

1.2.4 Differentiability of a Function 4

1.3 Analytic Functions — 5

1.4 Cauchy–Riemann Equations — 5

1.4.1 Sufficient Conditions for a Function to be Analytic 6

1.4.2 Polar Form of Cauchy–Riemann Equations 7

1.5 Harmonic Functions — 21

1.5.1 Orthogonal System of Level Curves 22

1.5.2 Method to Find Conjugate Harmonic Function 23

1.5.3 Milne Thomson Method 23

1.6 Line Integral in the Complex Plane — 39

1.6.1 Continuous Curve or Arc 39

1.6.2 Smooth Curve or Smooth Arc 39

1.6.3 Piecewise Continuous Curve 39

1.6.4 Piecewise Smooth Curve 39

1.6.5 Contour 39

1.6.6 Line Integral 39

1.7 Cauchy Integral Theorem — 44

1.7.1 Simply Connected Domain 44

1.7.2 Multiply Connected Domain 44

1.7.3 Independence of Path 46

1.7.4 Integral Function 47

1.7.5 Fundamental Theorem of Integral Calculus 47

1.7.6 Extension of Cauchy–Goursat Theorem for Multiply Connected Domains 49

1.8 Cauchy Integral Formula — 50
1.8.1 Cauchy Integral Formula for Derivatives of Analytic Function 51
1.8.2 Morera's Theorem (Converse of Cauchy Integral Theorem) 53
1.8.3 Cauchy Inequality 53
1.8.4 Liouville's Theorem 54
1.8.5 Poisson's Integral Formula 54
1.9 Infinite Series of Complex Terms — 66
1.9.1 Power Series 67
1.10 Taylor Series — 68
1.11 Laurent's Series — 69
1.12 Zeros and Singularities of Complex Functions — 84
1.12.1 Zeros of an Analytic Function 84
1.12.2 Singularities of a Function 84
1.12.3 Method to Find Type of Isolated Singularity 85
1.13 Residue — 90
1.13.1 Residue at a Removable Singularity 90
1.13.2 Residue at a Simple Pole 90
1.13.3 Residue at Pole of Order m 91
1.13.4 Residue at an Isolated Essential Singularity 91
1.14 Evaluation of Contour Integrals using Residues — 97
1.15 Application of Cauchy Residue Theorem to Evaluate Real Integrals — 106
1.15.1 Integration Around the Unit Circle 106
1.15.2 Improper Real Integrals of the Form $\int_{-\infty}^{\infty} f(x)\,dx$ or $\int_{0}^{\infty} f(x)\,dx$ where f(z) has no Real Singularity 114
1.15.3 Some Special Improper Real Integrals 121
1.15.4 Improper Integrals with Singularities on Real Axis 122
1.16 Conformal Mapping — 132
1.17 Some Standard Mappings — 136
1.17.1 Translation Mapping 136
1.17.2 Magnification/Contraction and Rotation 136
1.17.3 Linear Transformation 137
1.17.4 Inverse Transformation (Inversion and Reflection) 139
1.17.5 Square Transformation 144
1.17.6 Bilinear Transformation (Mobius Transformation or Fractional Transformation) 149
1.17.7 Cross Ratio of Four Points 149

2. Laplace Transform 165

2.1 Introduction — 165
2.2 Definition of Laplace Transform and Inverse Laplace Transform — 165
2.2.1 Piecewise Continuous Function 166
2.2.2 Function of Exponential Order 166

2.3 Sufficient Conditions for Existence of Laplace Transform — 166

2.4 Properties of Laplace Transforms — 167

2.5 Laplace Transform of Elementary Functions — 168

2.6 Laplace Transforms of Derivatives and Integrals — 170

2.7 Differentiation and Integration of Laplace Transform — 172

2.8 Evaluation of Real Integrals using Laplace Transform — 187

2.9 Laplace Transform of Unit Step Function — 192

2.10 Laplace Transform of Unit Impulse Function (Dirac–delta function) — 199

2.11 Laplace Transform of Periodic Functions — 202

2.12 Inverse Laplace Transform — 209

2.13 Use of Partial Fractions to find Inverse Laplace Transform — 210

2.14 Convolution Theorem — 221

2.15 Applications of Laplace Transform to Solve Linear Differential Equations, Simultaneous Linear Differential Equations and Integral Equations — 227

2.16 Applications of Laplace Transform to Engineering Problems — 257

2.16.1 Problems Related to Electrical Circuits 257

2.16.2 Problem Related to Deflection of a Loaded Beam 265

2.16.3 Problems Related to Mechanical Systems 271

3. Fourier Series, Fourier Integrals and Fourier Transforms 281

3.1 Introduction — 281

3.1.1 Periodic Functions 281

3.1.2 Trigonometric Series 282

3.1.3 Orthogonality of Trigonometric System 282

3.1.4 Fourier Series 283

3.1.5 Euler Formulae for Fourier Coefficients 283

3.1.6 Dirichlet's Conditions for Convergence of Fourier Series of f(x) in [c, c + 2l] 286

3.1.7 Fourier Series of Even and Odd Functions 286

3.2 Fourier Half-range Series — 323

3.2.1 Convergence of Half-range Cosine Series 324

3.2.2 Convergence of Half-range Sine Series 324

3.3 Others Formulae — 341

3.3.1 Parseval's Formulae 341

3.3.2 Root Mean Square (R.M.S.) Value 343

3.3.3 Complex Form of Fourier Series 343

3.4 Harmonic Analysis — 352

3.5 Fourier Integrals and Fourier Transforms — 367

3.5.1 Fourier Series to Fourier Integrals 367

3.5.2 Fourier Cosine and Fourier Sine Integrals 368

3.5.3 Fourier Cosine and Sine Transforms 369

3.5.4 Complex Form of the Fourier Integral 370

3.5.5 Fourier Transform and Its Inverse 371

3.5.6 Spectrum 372

3.6 Properties of Fourier Transforms — 372

3.7 Convolution Theorem and Parseval's Identities — 381

3.7.1 Convolution 381

3.7.2 Convolution Theorem (or Faltung Theorem) for Fourier Transforms 381

3.7.3 Parseval's Identities (Energy Theorem) 382

3.7.4 Relation between Fourier and Laplace Transforms 384

3.8 Applications of Fourier Transforms — 416

4. Partial Differential Equations 433

4.1 Introduction — 433

4.2 Formation of Partial Differential Equations — 433

4.2.1 Elimination of Arbitrary Constants 433

4.2.2 Elimination of Arbitrary Functions 434

4.3 Definitions — 434

4.3.1 Linear and Non-Linear Partial Differential Equations 434

4.3.2 Homogenous and Non-Homogenous Partial Differential Equations 434

4.3.3 Partial Differential Equations Linear in Partial Derivatives 434

4.3.4 Linear Homogenous in their Order Partial Differential Equations 435

4.3.5 Solution of Partial Differential Equations 435

4.4 Direct Integration Method for Solutions — 444

4.5 Partial Differential Equations of the First Order — 448

4.5.1 Lagrange's Method 448

4.5.2 Geometrical Interpretation of Lagrange's Method 450

4.5.3 Charpit's Method 460

4.5.4 Standard Form $f(p,q) = 0$ 462

4.5.5 Standard Form $f(z, p, q) = 0$ 462

4.5.6 Standard Form $f(x,p) = \phi(y,q)$ 463

4.5.7 Clairut's Equation 463

4.6 Linear in Second Order Partial Derivatives Differential Equations: Monge's Method — 479

4.7 Partial Differential Equations Linear and Homogenous in Partial Derivatives with Constant Coefficients — 486

4.7.1 Superposition or Linearity Principle 487

4.7.2 Rules for Finding the Complementary Function 487

4.7.3 Inverse Operator 489

4.7.4 Operator Methods for Finding Particular Integrals 490

4.8 Linear Partial Differential Equations with Constant Coefficients, Non-Homogeneous in Partial Derivatives — 507

4.8.1 Rules for Finding Complementary Function 507

4.8.2 Operator Methods for Finding Particular Integral 509

4.9 Partial Differential Equations with Variable Coefficients Reducible to Partial Differential Equations with Constant Coefficients — 510

4.10 Applications of Partial Differential Equations — 520

4.11 Vibrations of a Stretched String (One Dimensional Wave Equation) — 527

4.11.1 Solution of the Wave Equation 528

4.11.2 D'Alembert's Method of Solving Wave Equation 530

4.12 One Dimensional Heat Flow — 546

4.12.1 Solution of the Heat Equation 547

4.13 Transmission Line Equations — 558

4.14 Two dimensional Heat Flow — 567

4.15 Solution of Two Dimensional Laplace equation — 569

4.16 Two Dimensional Wave Equation — 576

4.16.1 Solution of Two Dimensional Wave Equation 577

5. Numerical Methods in General and Linear Algebra 585

5.1 Introduction — 585

5.2 Errors in Numerical Computations — 585

5.3 Algebraic and Transcendental Equations — 586

5.3.1 Bisection Method or Bolzano Method or Halving Method 587

5.3.2 Direct Iteration Method 588

5.3.3 Secant and Regula-falsi Methods 594

5.3.4 Newton-Raphson Method (or Newton's Iteration Method or Method of Tangents) 600

5.4 System of Linear Equations — 615

5.4.1 Gauss Elimination Method 615

5.4.2 Gauss–Jordan Method 616

5.4.3 Triangularisation Method 623

5.4.4 Doolittle Method 625

5.4.5 Crout's Method 626

5.5 Iterative Methods for Solving System of Linear Equations — 634

5.5.1 Jacobi's Iterative Method 634

5.5.2 Gauss–Seidel Iteration Method 635

5.6 Algebraic Eigenvalue Problems — 642

5.6.1 Power Method 642

5.6.2 Modification of Power Method 644

5.7 Linear Operators — 656

5.7.1 Forward Differences 661

5.7.2 Backward Differences 662

5.7.3 Central Differences 662

5.7.4 Factorial Polynomials 663

5.7.5 Error Propagation 665

5.7.6 Missing Entries in the Data 667

5.8 Interpolation — 683

5.8.1 Lagrange's Interpolation Formula 683

5.8.2 Divided Differences 685

5.8.3 Newton's Divided Difference Interpolation Formula 687

5.8.4 Newton's Forward Difference Interpolation Formula 695

5.8.5 Newton's Backward Difference Interpolation Formula 696

5.8.6 Gauss Forward Interpolation Formula 701

5.8.7 Gauss Backward Interpolation Formula 702

5.8.8 Stirling's Formula 703

5.8.9 Bessel's Interpolation Formula 703

5.8.10 Laplace–Everett's Interpolation Formula 704

5.9 Inverse Interpolation — 712

5.9.1 Lagrange's Method for Inverse Interpolation 712

5.9.2 Inverse Interpolation using Newton's Forward Interpolation Formula 713

5.9.3 Inverse Interpolation using Everett's Formula 713

6. Numerical Methods for Differentiation, Integration and Ordinary Differential Equations 725

6.1 Introduction — 725

6.2 Numerical Differentiation — 725

6.2.1 Derivatives at Interior Points 725

6.2.2 Derivative at Grid Points 726

6.3 Numerical Quadrature — 739

6.3.1 General Quadrature Formula 739

6.3.2 Trapezoidal Rule 739

6.3.3 Simpson's One-third Rule 740

6.3.4 Simpson's Three-eight Rule 741

6.3.5 Weddle's Rule 742

6.3.6 Cote's Formulas 743

6.3.7 Error Term in Quadrature Formulae 745

6.4 Numerical Solutions of Ordinary Differential Equations — 756

6.4.1 Taylor-Series Method 756

6.4.2 Picard's Method 764

6.4.3 Euler's Method 772

6.4.4 Improved Euler's Method 773

6.4.5 Modified Euler's Method 774

6.4.6 Runge's Method 780

6.4.7 Runge–Kutta Method 781

6.4.8 Milne's Method 791

Index 801

Preface

The main aim of this book is to provide the readers with a thorough knowledge of the fundamental concepts and methods of applied mathematics used in all streams of engineering, technology and science in a very lucid and unambiguous style. For better understanding of students, all the concepts have been illustrated with sufficient number of solved examples, followed by a collection of graded practical exercises, their respective answers and their direct engineering applications. A large number of solved questions from different university examination papers form an integral part of this book.

The book is divided into two volumes: Volume I and II. Volume I contains nine chapters which cover topics like differential calculus, integral calculus, infinite series, linear algebra: matrices, vector calculus, ordinary differential equations, series solution and special functions. This is volume II of the book which contains six chapters and comprises topics like functions of complex variables, Fourier series, Fourier integrals, Fourier and Laplace transforms, partial differential equations, numerical methods in general and linear algebra and numerical methods for differentiation, integration and ordinary differential equations.

The topic complex numbers have been uploaded on the book's website that contains many solved examples on each topic discussed in the book. Multiple-choice solved questions, from various competitive examinations and related to the topics covered in the chapters with their exercises, have also been uploaded on the website.

Students of various institutes, engineering colleges and universities will definitely find this book to be more than handful apart from students preparing for competitive examinations like GATE, NET, MAT, etc.

Any suggestions for further improvement of this book will be gladly accepted.

R. L. Garg
Nishu Gupta

Acknowledgements

We express our sincere gratitude to Nand Kishore Garg, Hon'ble Chairman, Maharaja Agrasen Technical Education Society and M.L. Goyal, Director, Maharaja Agrasen Institute of Technology, Delhi, for their constant inspiration, encouragement and providing necessary facilities.

We acknowledge our innumerable thanks to our colleagues and friends, for their support and suggestions. No words can suffice to express our deep feelings for our family members, for their moral support and understanding.

We are specially thankful to Veena Rastogi (mother-in-law of Nishu Gupta), Sachin Rastogi (husband of Nishu Gupta), Mahesh Garg and Dinesh Garg (sons of R. L. Garg), Shalini Garg (daughter of R. L. Garg) and Geetima Rai (daughter-in-law of R. L. Garg), for their constant cooperation, motivation and encouragement. We would also like to appreciate the contribution of Nishu Gupta's daughter Sneha Rastogi and R. L. Garg's grandson Sparsh Goel, for their unending patience and understanding, without which completion of the book would have been impossible. It was a long and difficult journey for them.

Our special thanks to the reviewers, for their valuable comments. We are thankful to the Pearson Education's Editorial team, especially to Anita Yadav and Vipin Kumar, for their effective cooperation during various stages of work on this book.

Last but not the least, we apologize to all those who have been with us over the course of the years and whose names we have failed to mention.

R. L. Garg
Nishu Gupta

About the Authors

R.L. GARG

R. L. Garg received his Ph.D. in Mathematics in 1978 from the Kurukshetra University, Kurukshetra, India. He is a former Professor of Mathematics at the Department of Statistics and Operations Research, Kurukshetra University, and has been teaching Mathematics and Statistics for past 35 years. He had been a member of the Kurukshetra University Court. His eight research papers were published in some journals of international repute. He has successfully supervised three Ph.D. students. After his voluntary retirement from the Kurukshetra University in 2001, he started his own coaching classes to teach Mathematics for IIT-JEE, AIEEE and CEET examinations successfully for eight years. After that, he joined as a Professor of Mathematics at the Maharaja Agrasen Institute of Technology, Delhi. He has been on the examination panel of various universities and state service board's examinations.

NISHU GUPTA

Nishu Gupta is presently working as an Assistant Professor at the Maharaja Agrasen Institute of Technology, Delhi. She is a gold medalist in M.Sc. (Maths) from the C.C.S. University, Meerut (UP), India. She has been awarded with University Medal and Durga Trust Prize for scoring the highest marks in M.Phil. (Maths) from the University of Roorkee (presently IIT-Roorkee), India. She has qualified National Eligibility Test (JRF-NET) conducted by CSIR-UGC and was one of the toppers in the subject of Mathematical Sciences. Also, she qualified fellowship test for Ph.D. degree conducted by the National Board for higher Mathematics, Department of Atomic Energy. She has made contribution to the research in the field of Functional Analysis and was awarded with a Ph.D. degree from IIT-Roorkee in 2006. She has been teaching students at B.Sc., M.Sc. and B.Tech. levels, for the past 16 years.

Symbols, Basic Formulae and Useful Informations

1. Greek Letters

α alpha	ε epsilon	ζ zeta	λ lambda
β beta	ι iota	χ chi	Γ capital gamma
γ gamma	μ mu	π pi	Σ capital sigma
δ delta	ν nu	σ sigma	Δ capital delta
θ theta	ω omega	τ tau	Φ capital phi
ϕ phi	ξ xi	ρ rho	Ψ capital psi
ψ psi	η eta	κ kappa	Ω capital omega

2. Useful constants

$e = 2.71828$	$\log_e 3 = 1.0986$
$\pi = 3.14159$	$\log_e 2 = 0.6931$
$\sqrt{2} = 1.41421$	$\log_e 10 = 2.3026$
$\sqrt{3} = 1.73205$	$\log_{10} e = 0.4343$

3. Some Notations

$\in$ belongs to	iff if and only if
$\notin$ does not belong to	N Set of natural numbers
$\Rightarrow$ implies	I Set of all integers
$\Leftrightarrow$ implies and implied by	Q Set of all rational numbers
$\forall$ for all	R Set of all real numbers
$\cup$ union	C Set of all complex numbers
$\cap$ intersection	

4. Partial Fractions

A fraction of the form $\dfrac{a_0x^n + a_1x^{n-1} + a_2x^{n-2} + \cdots + a_n}{b_0x^m + b_1x^{m-1} + b_2x^{m-2} + \cdots + b_m}$

in which n and m are positive integers and $n < m$ is called a proper fraction. If $n \geq m$ then dividing numerator by denominator, remainder divided by denominator will be proper fraction. Only a proper fraction is resolved into partial fractions. We first factorize the denominator into real factors. These will be either linear or quadratic (we consider only such cases) and some factors repeated. Then proper fraction is resolved into a sum of partial fractions such that

(a) To a non-repeated linear factor $(x - a)$ in the denominator corresponds a partial fraction of the form $\dfrac{A}{x-a}$.

(b) To a repeated linear factor $(x-b)^r$ in the denominator corresponds the sum of r partial fractions of the form $\dfrac{A_1}{x-b}+\dfrac{A_2}{(x-b)^2}+\cdots+\dfrac{A_r}{(x-b)^r}$.

(c) To non-repeated quadratic factor (x^2+cx+d) in the denominator corresponds a partial fraction of the form $\dfrac{Bx+C}{x^2+cx+d}$.

(d) To a repeated quadratic factor $(x^2+ex+f)^k$ in the denominator corresponds the sum of k partial fractions of the form $\dfrac{B_1x+C_1}{x^2+ex+f}+\dfrac{B_2x+C_2}{(x^2+ex+f)^2}+\cdots+\dfrac{B_kx+C_k}{(x^2+ex+f)^k}$.

Then, we are to find the constants A, B, C, A_i, $B_j, C_j; i=1,2,\ldots,r,\ j=1,2,\ldots,k$ used in above partial fractions from (a) to (d).

Constants A and A_r in (a) and (b) are found by **Suppression method** explained below.

For finding A, put $x = a$, everywhere in the given proper fraction except in the factor $(x-a)$ itself, the value obtained will be value of A.

For finding A_r, put $x=b$, everywhere in the given proper fraction except in the factor $(x-b)^r$ itself, the value obtained will be value of A_r.

Other constants are obtained by writing the given proper fraction as sum of partial fractions as explained above from (a) to (d) and then multiply by the denominator of given proper fraction. Equate coefficients of various powers of x on both sides and solve the equations obtained for the constants.

5. Synthetic Division

It is used to divide a polynomial by a linear polynomial with leading coefficient unity. For this, we write the coefficients of dividend polynomial (writing zero for missing terms). Suppose, we are to divide the polynomial $a_0x^n+a_1x^{n-1}+a_2x^{n-2}+\cdots+a_{n-1}x+a_n$ by $(x-a)$. We write the coefficients $a_0, a_1, a_2, \ldots, a_{n-1}, a_n$ and a to the left as shown below. Then we write a_0 below the line as it is. Multiply a_0 by a and write the result below a_1 above the line and add it to a_1 and write below the line. We continue this process as follows

$a)$	a_0	a_1	a_2	$\cdots$	a_{n-1}	a_n
		aa_0	$aa_1+a^2a_0$	$\cdots$	$\cdots$	$\cdots$
	a_0	a_1+aa_0	$a_2+aa_1+a^2a_0$	$\cdots$	$\cdots$	$\cdots$

Then last term below the line will be remainder and before it there will be coefficients of Quotient polynomial. For example, we divide $5x^5+3x^3-7x+8$ by $x+2$ as follows

$$\begin{array}{r|rrrrrr} -2) & 5 & 0 & 3 & 0 & -7 & 8 \\ & & -10 & 20 & -46 & 92 & -170 \\ \hline & 5 & -10 & 23 & -46 & 85 & -162 \end{array}$$

$\therefore\ 5x^5+3x^3-7x+8=(x+2)(5x^4-10x^3+23x^2-46x+85)-162$

6. Trigonometric formulae

(i) Values of T-ratios (trigonometrical ratios) for some angles are

θ	0	$\pi/12$	$\pi/6$	$\pi/4$	$\pi/3$	$5\pi/12$	$\pi/2$
$\sin\theta$	0	$\frac{\sqrt{3}-1}{2\sqrt{2}}$	$\frac{1}{2}$	$\frac{1}{\sqrt{2}}$	$\frac{\sqrt{3}}{2}$	$\frac{\sqrt{3}+1}{2\sqrt{2}}$	1
$\cos\theta$	1	$\frac{\sqrt{3}+1}{2\sqrt{2}}$	$\frac{\sqrt{3}}{2}$	$\frac{1}{\sqrt{2}}$	$\frac{1}{2}$	$\frac{\sqrt{3}-1}{2\sqrt{2}}$	0
$\tan\theta$	0	$\frac{\sqrt{3}-1}{\sqrt{3}+1}=2-\sqrt{3}$	$\frac{1}{\sqrt{3}}$	1	$\sqrt{3}$	$\frac{\sqrt{3}+1}{\sqrt{3}-1}=2+\sqrt{3}$	∞

Note that $\tan\left(\frac{\pi}{2}-0\right)=\infty$, $\tan\left(\frac{\pi}{2}+0\right)=-\infty$, $\sin\frac{\pi}{5}=\frac{\sqrt{5}-1}{4}$, $\cos\frac{2\pi}{5}=\frac{\sqrt{5}+1}{4}$

$\sin(n\pi+\theta)=(-1)^n\sin\theta$, $\cos(n\pi+\theta)=(-1)^n\cos\theta$

(ii) Signs of T-ratios

All T-ratios are positive in first quadrant

$\sin\theta$ and $\operatorname{cosec}\theta$ are positive when θ is in second quadrant

$\tan\theta$ and $\cot\theta$ are positive when θ is in third quadrant

$\cos\theta$ and $\sec\theta$ are positive when θ is in fourth quadrant

T-ratios not mentioned are negative in that quadrant.

(iii) Formulae

(a) $\sin(A\pm B)=\sin A\cos B\pm\cos A\sin B$

(b) $\cos(A\pm B)=\cos A\cos B\mp\sin A\sin B$

(c) $\tan(A\pm B)=\dfrac{\tan A\pm\tan B}{1\mp\tan A\tan B}$

(d) $\sin 2A=2\sin A\cos A=\dfrac{2\tan A}{1+\tan^2 A}$

(e) $\cos 2A=\cos^2 A-\sin^2 A=2\cos^2 A-1=1-2\sin^2 A=\dfrac{1-\tan^2 A}{1+\tan^2 A}$

(f) $\tan 2A = \dfrac{2\tan A}{1-\tan^2 A}$

(g) $\sin 3A = 3\sin A - 4\sin^3 A$

(h) $\cos 3A = 4\cos^3 A - 3\cos A$

(i) $\tan 3A = \dfrac{3\tan A - \tan^3 A}{1-3\tan^2 A}$

(j) $\sin C + \sin D = 2\sin\dfrac{C+D}{2}\cos\dfrac{C-D}{2}$

(k) $\sin C - \sin D = 2\cos\dfrac{C+D}{2}\sin\dfrac{C-D}{2}$

(l) $\cos C + \cos D = 2\cos\dfrac{C+D}{2}\cos\dfrac{C-D}{2}$

(m) $\cos C - \cos D = 2\sin\dfrac{C+D}{2}\sin\dfrac{D-C}{2} = -2\sin\dfrac{C+D}{2}\sin\dfrac{C-D}{2}$

(n) $2\sin A\sin B = \cos(A-B) - \cos(A+B)$

(o) $2\sin A\cos B = \sin(A+B) + \sin(A-B)$

(p) $2\cos A\sin B = \sin(A+B) - \sin(A-B)$

(q) $2\cos A\cos B = \cos(A+B) + \cos(A-B)$

7. Differentiation

(a) $\dfrac{d}{dx}x^n = nx^{n-1}$

(b) $\dfrac{d}{dx}(ax+b)^n = na(ax+b)^{n-1}$

(c) $\dfrac{d}{dx}(f(x))^n = n(f(x))^{n-1}f'(x)$

(d) $\dfrac{d}{dx}a^x = a^x\log a\ ;\ a>0$

(e) $\dfrac{d}{dx}e^x = e^x$

(f) $\dfrac{d}{dx}\log_e x = \dfrac{1}{x}$

(g) $\dfrac{d}{dx}(f(x))^{g(x)} = g(x)(f(x))^{g(x)-1}f'(x) + (f(x))^{g(x)}\log f(x).g'(x)$

This formula is remembered as sum of derivatives considering [$g(x)$ as constant, $f(x)$ variable] and [$f(x)$ as constant, $g(x)$ variable]

(h) $\dfrac{d}{dx}\sin x = \cos x$

(i) $\dfrac{d}{dx}\cos x = -\sin x$

(j) $\dfrac{d}{dx}\tan x = \sec^2 x$

(k) $\dfrac{d}{dx}\text{cosec}\, x = -\text{cosec}\, x \cot x$

(l) $\dfrac{d}{dx}\sec x = \sec x \tan x$

(m) $\dfrac{d}{dx}\cot x = -\text{cosec}^2 x$

(n) $\dfrac{d}{dx}\sin^{-1} x = \dfrac{1}{\sqrt{1-x^2}}$

(o) $\dfrac{d}{dx}\cos^{-1} x = -\dfrac{1}{\sqrt{1-x^2}}$

(p) $\dfrac{d}{dx}\tan^{-1} x = \dfrac{1}{1+x^2}$

(q) $\dfrac{d}{dx}\text{cosec}^{-1} x = -\dfrac{1}{|x|\sqrt{x^2-1}}$

(r) $\dfrac{d}{dx}\sec^{-1} x = \dfrac{1}{|x|\sqrt{x^2-1}}$

(s) $\dfrac{d}{dx}\cot^{-1} x = -\dfrac{1}{1+x^2}$

(t) $\dfrac{d}{dx}\sinh x = \cosh x$

(u) $\dfrac{d}{dx}\cosh x = \sinh x$

8. Integration

In all below formulae, arbitrary constant should also be taken.

(a) $\int x^n dx = \dfrac{x^{n+1}}{n+1}$; $n \neq -1$

(b) $\int \dfrac{1}{x}\, dx = \log|x|$

(c) $\int (ax+b)^n dx = \dfrac{(ax+b)^{n+1}}{(n+1)a}$; $n \neq -1$

(d) $\int \dfrac{1}{ax+b}\, dx = \dfrac{1}{a}\log|ax+b|$

(e) $\int e^x dx = e^x$

(f) $\int a^x dx = \frac{a^x}{\log a}$; $a > 0,\ a \neq 1$

(g) $\int \sin x\, dx = -\cos x$

(h) $\int \cos x\, dx = \sin x$

(i) $\int \tan x\, dx = \log|\sec x|$

(j) $\int \cot x\, dx = \log|\sin x|$

(k) $\int \sec x\, dx = \log|\sec x + \tan x| = \log\left|\tan\left(\frac{\pi}{4} + \frac{x}{2}\right)\right|$

(l) $\int \operatorname{cosec} x\, dx = \log|\operatorname{cosec} x - \cot x| = \log\left|\tan\frac{x}{2}\right|$

(m) $\int \sec^2 x\, dx = \tan x$

(n) $\int \operatorname{cosec}^2 x\, dx = -\cot x$

(o) $\int \frac{dx}{a^2 + x^2} = \frac{1}{a}\tan^{-1}\frac{x}{a}$

(p) $\int \frac{dx}{a^2 - x^2} = \frac{1}{2a}\log\frac{a+x}{a-x}$

(q) $\int \frac{dx}{x^2 - a^2} = \frac{1}{2a}\log\frac{x-a}{x+a}$

(r) $\int \frac{dx}{\sqrt{a^2 - x^2}} = \sin^{-1}\frac{x}{a}$

(s) $\int \frac{dx}{\sqrt{a^2 + x^2}} = \log\left|x + \sqrt{a^2 + x^2}\right|$ or $\int \frac{dx}{\sqrt{a^2 + x^2}} = \sinh^{-1}\frac{x}{a}$

(t) $\int \frac{dx}{\sqrt{x^2 - a^2}} = \log\left|x + \sqrt{x^2 - a^2}\right|$ or $\int \frac{dx}{\sqrt{x^2 - a^2}} = \cosh^{-1}\frac{x}{a}$

(u) $\int \sqrt{a^2 + x^2}\, dx = \frac{x\sqrt{a^2 + x^2}}{2} + \frac{a^2}{2}\sinh^{-1}\frac{x}{a}$

(v) $\int \sqrt{x^2 - a^2}\, dx = \frac{x\sqrt{x^2 - a^2}}{2} - \frac{a^2}{2}\cosh^{-1}\frac{x}{a}$

(w) $\int \sqrt{a^2 - x^2}\, dx = \frac{x\sqrt{a^2 - x^2}}{2} + \frac{a^2}{2}\sin^{-1}\frac{x}{a}$

(x) $\int e^{ax}\sin(bx + c)\, dx = \frac{e^{ax}}{a^2 + b^2}\left[a\sin(bx + c) - b\cos(bx + c)\right]$

(y) $\int e^{ax}\cos(bx + c)\, dx = \frac{e^{ax}}{a^2 + b^2}\left[a\cos(bx + c) + b\sin(bx + c)\right]$

The terms within the brackets in above two formulae can be remembered as a (given trigonometric function)–(derivative of given trigonometric function)

(z) **Integration by parts**

Generalized formula for integration by parts is

$$\int u(x).v(x)\,dx = u(x)\,v_1(x) - u'(x)\,v_2(x) + u''(x)\,v_3(x) - \cdots (-1)^{n-1} u^{(n-1)}(x)\,v_n(x)$$
$$+(-1)^n \int u^{(n)}(x)\,v_n(x)\,dx$$

where $v_1(x) = \int v(x)\,dx$, $v_2(x) = \int v_1(x)\,dx$, $\cdots, v_n(x) = \int v_{n-1}(x)\,dx.$

If $u(x)$ is polynomial of $(n-1)$ degree in x then last term will be zero.

Functions of Complex Variables

1

1.1 INTRODUCTION

In this chapter, we study the concepts of limits, continuity and differentiability of a function of complex variable and define an analytic function and prove that it possesses all order derivatives, the concept which is not true in real variables. Integration of functions of complex variable is very much helpful in science and engineering. In case of real integral $\int_a^b f(x)\,dx$, path of integration is a real line from $x = a$ to $x = b$, but in case of complex variables, $\int_a^b f(z)\,dz$, it will depend upon the path of the variable z if $f(z)$ is not analytic and does not depend on the path if $f(z)$ is analytic and in that case fundamental theorem of integral calculus holds. Many definite integrals and improper integrals of real variables are evaluated with the help of complex integration using Cauchy residue theorem but these integrals are difficult or cannot be found with methods of integration in real variables. Finally, conformal and bilinear transformations are considered.

1.2 DEFINITION, LIMIT, CONTINUITY AND DIFFERENTIABILITY OF A FUNCTION OF COMPLEX VARIABLE

1.2.1 Definition

Function of a complex variable is defined in the similar way as the function of a real variable. A rule f which assigns to each $z = x + iy \in A$, a complex number $w = u(x, y) + iv(x, y) \in B$, is called a function from A to B. A is called domain of f, B is called codomain of f and w is called image of z. The set $\{f(z): z \in A\}$ is called range of f. If f assigns each z the unique w then f is a single-valued function and if f assigns to some z more than one w then f is multiple-valued function. Trigonometric and hyperbolic functions are single-valued functions, whereas square root and logarithmic functions are multi-valued functions. If we take principal value of multiple-valued function then it will become single-valued function.

1.2.2 Limit of a Function

Let $f(z)$ be a single-valued function defined on set S. Let z_0 be a complex number which may or may not be a member of S. If corresponding to $\varepsilon > 0$, however small, there exists $\delta > 0$ such that

$$|f(z) - l| < \varepsilon \text{ for all } 0 < |z - z_0| < \delta$$

for some complex number l and $0 < |z - z_0| < \delta$ is contained in S then if l is finite, it is called the limit of $f(z)$ as $z \to z_0$ and we write

$$\lim_{z \to z_0} f(z) = l$$

Theorem 1.1 If $\lim_{z\to z_0} f(z)$ exists, then it is unique.

Proof: If possible, let

$$\lim_{z\to z_0} f(z) = l_1,\ \lim_{z\to z_0} f(z) = l_2 \text{ where } l_1 \neq l_2$$

Then, by definition corresponding to $\varepsilon > 0$ there exist $\delta_1,\ \delta_2 > 0$ such that

$$|f(z)-l_1| < \frac{\varepsilon}{2};\ 0 < |z-z_0| < \delta_1$$

$$|f(z)-l_2| < \frac{\varepsilon}{2};\ 0 < |z-z_0| < \delta_2$$

Let $\delta = \min\ (\delta_1, \delta_2)$

$$\therefore \quad |f(z)-l_1| < \frac{\varepsilon}{2};\ 0 < |z-z_0| < \delta \tag{1.1}$$

$$|f(z)-l_2| < \frac{\varepsilon}{2};\ 0 < |z-z_0| < \delta \tag{1.2}$$

Now,

$$|l_1 - l_2| = |(f(z)-l_2)-(f(z)-l_1)| \leq |f(z)-l_2| + |f(z)-l_1|$$

$$< \frac{\varepsilon}{2} + \frac{\varepsilon}{2} = \varepsilon \text{ for } 0 < |z-z_0| < \delta \quad \text{by (1.1) and (1.2)}$$

But ε is arbitrary, so $|l_1 - l_2| = 0 \Rightarrow l_1 = l_2$.

Remark 1.1: In the definition given in 1.2.2, $z \to z_0$ along any path. If along two different paths limits are different then from Theorem 1.1 limit does not exist. As for example

For the $\lim_{z\to 0} \frac{z}{|z|}$

along the path positive real axis

$$\lim_{z\to 0} \frac{z}{|z|} = \lim_{x\to 0^+} \frac{x}{|x|} = \lim_{x\to 0^+} \frac{x}{x} = 1$$

and along the path negative real axis

$$\lim_{z\to 0} \frac{z}{|z|} = \lim_{x\to 0^-} \frac{x}{|x|} = \lim_{x\to 0^-} \frac{x}{-x} = -1$$

Both are different. Therefore, $\lim_{z\to 0} \frac{z}{|z|}$ does not exist.

Theorem 1.2 Let $f(z) = u(x, y) + iv(x, y);\ z = x + iy, z_0 = x_0 + iy_0$.

Then $\lim_{z\to z_0} f(z) = u_0 + iv_0$ iff $\lim_{\substack{x\to x_0 \\ y\to y_0}} u(x, y) = u_0$ and $\lim_{\substack{x\to x_0 \\ y\to y_0}} v(x, y) = v_0$

Proof: Let $\lim_{z\to z_0} f(z) = u_0 + iv_0$. Then corresponding to $\varepsilon > 0$ there exists $\delta > 0$ such that

$$|u(x,y)+iv(x,y)-(u_0+iv_0)| < \varepsilon \text{ when } 0<|z-z_0|<\delta$$

$$\therefore \quad |u(x,y)-u_0+i(v(x,y)-v_0)| < \varepsilon \text{ when } 0<|(x-x_0)+i(y-y_0)|<\delta$$

Now,

$$|u(x,y)-u_0| \le |u(x,y)-u_0+i(v(x,y)-v_0)| < \varepsilon \text{ when } 0<|(x-x_0)+i(y-y_0)|<\delta$$

$$|v(x,y)-v_0| \le |u(x,y)-u_0+i(v(x,y)-v_0)| < \varepsilon \text{ when } 0<|(x-x_0)+i(y-y_0)|<\delta$$

and $\quad 0<|x-x_0|<\dfrac{\delta}{2}, 0<|y-y_0|<\dfrac{\delta}{2}$

$$\Rightarrow \quad 0<|(x-x_0)+i(y-y_0)| \le |x-x_0|+|y-y_0| < \frac{\delta}{2}+\frac{\delta}{2}=\delta$$

$$\therefore \quad |u(x,y)-u_0|<\varepsilon \text{ when } 0<|x-x_0|<\frac{\delta}{2},\ 0<|y-y_0|<\frac{\delta}{2}$$

and $\quad |v(x,y)-v_0|<\varepsilon$ when $0<|x-x_0|<\dfrac{\delta}{2},\ 0<|y-y_0|<\dfrac{\delta}{2}$

Hence, $\quad \lim_{\substack{x\to x_0\\ y\to y_0}} u(x,y)=u_0$ and $\lim_{\substack{x\to x_0\\ y\to y_0}} v(x,y)=v_0$

Conversely, let $\lim_{\substack{x\to x_0\\ y\to y_0}} u(x,y)=u_0,\ \lim_{\substack{x\to x_0\\ y\to y_0}} v(x,y)=v_0$.

Then, corresponding to $\varepsilon>0$ there exist $\delta_1,\ \delta_2>0$ such that

$$|u(x,y)-u_0|<\frac{\varepsilon}{2} \text{ when } 0<|x-x_0|<\delta_1,\ 0<|y-y_0|<\delta_1$$

and $\quad |v(x,y)-v_0|<\dfrac{\varepsilon}{2}$ when $0<|x-x_0|<\delta_2,\ 0<|y-y_0|<\delta_2$.

Let $\quad \delta=\min(\delta_1,\ \delta_2)$

$$\therefore \quad |u(x,y)-u_0|<\frac{\varepsilon}{2} \text{ when } 0<|x-x_0|<\delta,\ 0<|y-y_0|<\delta$$

and $\quad |v(x,y)-v_0|<\dfrac{\varepsilon}{2}$ when $0<|x-x_0|<\delta,\ 0<|y-y_0|<\delta$.

$$\therefore \quad |u(x,y)+iv(x,y)-(u_0+iv_0)| \le |u(x,y)-u_0|+|v(x,y)-v_0| < \frac{\varepsilon}{2}+\frac{\varepsilon}{2}=\varepsilon$$

when $\quad 0<\left|x+iy-(x_0+iy_0)\right| \le \left|x-x_0\right|+\left|y-y_0\right| < \delta+\delta = 2\delta$

$\quad$ i.e., $0<\left|z-z_0\right|<2\delta$

$\therefore \quad \lim_{z\to z_0} f(z) = u_0 + iv_0$

1.2.3 Continuity of a Function

If a single-valued function $f(z)$ is defined in some neighbourhood of z_0 i.e., in $\left|z-z_0\right|<\delta$ for some δ, $\lim_{z\to z_0} f(z)$ exists and $\lim_{z\to z_0} f(z) = f(z_0)$, then $f(z)$ is called continuous at $z = z_0$.

If a function $f(z)$ is continuous at all points in some domain D, then $f(z)$ is called continuous in domain D.

1.2.4 Differentiability of a Function

If $f(z)$ is a single-valued function defined in some neighbourhood of z_0 i.e., $\left|z-z_0\right|<\delta$ for some δ then $f(z)$ is said to be differentiable at $z = z_0$ iff

$$\lim_{z\to z_0}\frac{f(z)-f(z_0)}{z-z_0} = \lim_{\Delta z\to 0}\frac{f(z_0+\Delta z)-f(z_0)}{\Delta z} \text{ exists.}$$

If this limit exists then it is the value of derivative at $z = z_0$ and is written as $f'(z_0)$ or $\frac{d}{dz}f(z_0)$.

Remark 1.2: In case of function of real variable there is concept of left-handed derivative and right-handed derivative, but in case of function of complex variable there is no such concept as the limit here is to taken along any path.

Theorem 1.3 If $f(z)$ is differentiable at $z = z_0$ then it is continuous at $z = z_0$ but converse is not true.

Proof: Let $f(z)$ be differentiable at $z = z_0$

then $\quad f'(z_0) = \lim_{z\to z_0}\frac{f(z)-f(z_0)}{z-z_0}$

Now, $\quad f(z) = \frac{f(z)-f(z_0)}{z-z_0}(z-z_0)+f(z_0);\ z\ne z_0$

$\therefore \quad \lim_{z\to z_0} f(z) = f'(z_0)\cdot(0)+f(z_0) = f(z_0)$

$\therefore \quad f(z)$ is continuous at $z = z_0$.

Now, $\quad f(z) = \bar{z} = x-iy$ (where $z = x+iy$) is continuous at $z = 0$.

But $\quad \lim_{z\to 0}\frac{f(z)-f(0)}{z-0} = \lim_{z\to 0}\frac{\bar{z}-0}{z} = \lim_{z\to 0}\frac{\bar{z}}{z}$

Along the path $y = 0$, $\lim\limits_{z\to 0} \dfrac{\overline{z}}{z} = \lim\limits_{x\to 0} \dfrac{x}{x} = 1$

Along the path $x = 0$, $\lim\limits_{z\to 0} \dfrac{\overline{z}}{z} = \lim\limits_{y\to 0} \dfrac{-iy}{iy} = -1$

These two are not equal.

$\therefore \qquad \lim\limits_{z\to 0} \dfrac{f(z)-f(0)}{z-0}$ does not exist.

$\therefore \qquad f(z)$ is not differentiable at $z = 0$.

Remark 1.3: It should be noted that if a function is not continuous at a point then it is not differentiable at that point.

1.3 ANALYTIC FUNCTIONS

A function $f(z)$ is said to be analytic or holomorphic or regular at $z = z_0$ iff there exists a neighbourhood $|z - z_0| < \delta$ of z_0 for some $\delta > 0$ such that $f(z)$ is differentiable at all points of this neighbourhood. A function $f(z)$ is called analytic in some domain D if it is analytic at all points of domain D. If $f(z)$ is not analytic at $z = z_0$ and there is a neighbourhood $|z - z_0| < \delta$ of z_0 for some δ in which $f(z)$ is differentiable at all points except $z = z_0$ then $z = z_0$ is called isolated singularity of $f(z)$.

If a function is differentiable for all z, then it is called an entire function.

1.4 CAUCHY–RIEMANN EQUATIONS

Theorem 1.4 Let $w = f(z) = u(x, y) + iv(x, y)$ be defined and continuous in some neighbourbood of a point $z = x + iy$ and differentiable at z itself. Then at this point, the first order partial derivatives of u and v exist and satisfy Cauchy–Riemann (C–R) equations:

$$u_x = v_y \text{ and } u_y = -v_x$$

Proof: $f(z)$ is differentiable at z

$$\therefore \qquad f'(z) = \lim_{\Delta z\to 0} \frac{f(z+\Delta z) - f(z)}{\Delta z}$$

exists and is unique along every path along which $\Delta z \to 0$.

$$f'(z) = \lim_{\substack{\Delta x\to 0\\ \Delta y\to 0}} \frac{[u(x+\Delta x, y+\Delta y) + iv(x+\Delta x, y+\Delta y)] - [u(x,y) + iv(x,y)]}{\Delta x + i\Delta y}$$

Along the path $\Delta x = 0$, i.e., parallel to y-axis

$$f'(z) = \lim_{\Delta y\to 0} \frac{[u(x, y+\Delta y) + iv(x, y+\Delta y)] - [u(x,y) + iv(x,y)]}{i\Delta y}$$

$$= \lim_{\Delta y \to 0} \left[\frac{u(x, y+\Delta y) - u(x, y)}{i\Delta y} + \frac{v(x, y+\Delta y) - v(x, y)}{\Delta y} \right]$$

$$= -i\frac{\partial u}{\partial y} + \frac{\partial v}{\partial y} = \frac{\partial v}{\partial y} - i\frac{\partial u}{\partial y} \tag{1.3}$$

Along the path $\Delta y = 0$, i.e., parallel to x-axis

$$f'(z) = \lim_{\Delta x \to 0} \left[\frac{u(x+\Delta x, y) - u(x, y)}{\Delta x} + i\frac{v(x+\Delta x, y) - v(x, y)}{\Delta x} \right]$$

$$= \frac{\partial u}{\partial x} + i\frac{\partial v}{\partial x} \tag{1.4}$$

Since $f(z)$ is differentiable, the two limits in equations (1.3) and (1.4) are equal:

$$\therefore \qquad \frac{\partial u}{\partial x} + i\frac{\partial v}{\partial x} = \frac{\partial v}{\partial y} - i\frac{\partial u}{\partial y}$$

Equate real and imaginary parts

$$\frac{\partial u}{\partial x} = \frac{\partial v}{\partial y}, \quad \frac{\partial v}{\partial x} = -\frac{\partial u}{\partial y}$$

or $\qquad u_x = v_y$ and $u_y = -v_x$

which are Cauchy–Riemann (C–R) equations

Remark 1.4: C–R equations are necessary conditions for a function to be differentiable or analytic at a point. Thus, a function not satisfying C–R equations at a point will neither be differentiable nor analytic at that point. These conditions are not sufficient. Thus, there exist functions which satisfy C–R equations at a point but not differentiable (and hence not analytic) at that point.

1.4.1 Sufficient Conditions for a Function to be Analytic

Theorem 1.5 The sufficient conditions for a function $w = f(z) = u(x, y) + iv(x, y)$ to be analytic at a point z are

(i) u, v, u_x, u_y, v_x, v_y are continuous functions of x and y in a certain neighbourhood of z.

(ii) C–R equations $u_x = v_y, u_y = -v_x$ are satisfied in neighbourhood of z.

Proof: Let $z + \Delta z$ lies in neighbourhood of z in which u, v, u_x, u_y, v_x, v_y are continuous

Then, by Taylor's expansion

$$u(x+\Delta x, y+\Delta y) - u(x, y) = \Delta x\, u_x + \Delta y\, u_y$$
$$+ \text{ terms of at least second order in } \Delta x \text{ and } \Delta y \tag{1.5}$$

$$v(x+\Delta x, y+\Delta y)-v(x,y)=\Delta x\ v_x+\Delta y\ v_y$$
$$+\text{ terms of at least second order in } \Delta x \text{ and } \Delta y \qquad (1.6)$$

In equations (1.5) and (1.6) use C–R equations $u_y = -v_x$ and $v_y = u_x$, respectively.

$$u(x+\Delta x, y+\Delta y)-u(x,y)=\Delta x\ u_x-\Delta y\ v_x$$
$$+\text{ terms of at least second order in } \Delta x \text{ and } \Delta y \qquad (1.7)$$

$$v(x+\Delta x, y+\Delta y)-v(x,y)=\Delta x\ v_x+\Delta y\ u_x$$
$$+\text{ terms of at least second order in } \Delta x \text{ and } \Delta y \qquad (1.8)$$

Now, $f(z+\Delta z)-f(z)=\left[u(x+\Delta x,y+\Delta y)+iv(x+\Delta x,y+\Delta y)\right]-\left[u(x,y)+iv(x,y)\right]$

$$=\left[u(x+\Delta x,y+\Delta y)-u(x,y)\right]+i\left[v(x+\Delta x,y+\Delta y)-v(x,y)\right]$$

Use equations (1.7) and (1.8)

$$f(z+\Delta z)-f(z)=\Delta x\ u_x-\Delta y\ v_x+i(\Delta x\ v_x+\Delta y\ u_x)$$
$$=(\Delta x+i\Delta y)\ u_x+i(\Delta x+i\Delta y)\ v_x$$
$$+\text{ terms of at least second order in } \Delta x \text{ and } \Delta y$$
$$=(u_x+iv_x)\ (\Delta x+i\Delta y)$$
$$+\text{ terms of at least second order in } \Delta x \text{ and } \Delta y$$

$\therefore$ $$\frac{f(z+\Delta z)-f(z)}{\Delta z}=u_x+iv_x+\text{ terms of at least first order in } \Delta x \text{ and } \Delta y$$

$\therefore$ $$\lim_{\Delta z\to 0}\frac{f(z+\Delta z)-f(z)}{\Delta z}=u_x+iv_x$$

$\therefore$ $f(z)$ is differentiable at z and $f'(z)=u_x+iv_x$

Similarly, $f(z)$ is differentiable in neighbourhood of z.

Hence, $f(z)$ is analytic at z.

Remark 1.5: If $f(z)$ is differentiable at z, then

$$f'(z)=u_x+iv_x=v_y-iu_y=u_x-iu_y=v_y+iv_x \qquad \text{(By C–R equations)}$$

1.4.2 Polar Form of Cauchy–Riemann Equations

Theorem 1.6 Polar form of Cauchy–Riemann equations are

$$\frac{\partial u}{\partial r}=\frac{1}{r}\frac{\partial v}{\partial \theta},\ \frac{\partial v}{\partial r}=-\frac{1}{r}\frac{\partial u}{\partial \theta}$$

and using these equations, we have

$$\frac{\partial^2 u}{\partial r^2}+\frac{1}{r}\frac{\partial u}{\partial r}+\frac{1}{r^2}\frac{\partial^2 u}{\partial \theta^2}=0$$

Proof:

$$z = x+iy = re^{i\theta}$$

$$\therefore \quad w = f(z) = f(re^{i\theta}) = u(r,\theta)+iv(r,\theta)$$

Differentiate partially w. r. t. r and θ

$$f'(z)\,\frac{\partial z}{\partial r}=\frac{\partial u}{\partial r}+i\frac{\partial v}{\partial r} \tag{1.9}$$

$$f'(z)\,\frac{\partial z}{\partial \theta}=\frac{\partial u}{\partial \theta}+i\frac{\partial v}{\partial \theta} \tag{1.10}$$

But $$\frac{\partial z}{\partial r}=e^{i\theta},\ \frac{\partial z}{\partial \theta}=ire^{i\theta} \qquad (\because z=re^{i\theta})$$

$\therefore$ from equations (1.9) and (1.10)

$$f'(z)=e^{-i\theta}\left(\frac{\partial u}{\partial r}+i\frac{\partial v}{\partial r}\right)=\frac{1}{ri}e^{-i\theta}\left(\frac{\partial u}{\partial \theta}+i\frac{\partial v}{\partial \theta}\right) \tag{1.11}$$

$$\therefore \quad \frac{\partial u}{\partial r}+i\frac{\partial v}{\partial r}=-\frac{i}{r}\frac{\partial u}{\partial \theta}+\frac{1}{r}\frac{\partial v}{\partial \theta}$$

Equate real and imaginary parts

$$\frac{\partial u}{\partial r}=\frac{1}{r}\frac{\partial v}{\partial \theta} \tag{1.12}$$

$$\frac{\partial v}{\partial r}=-\frac{1}{r}\frac{\partial u}{\partial \theta} \tag{1.13}$$

which are C–R equations in polar form.

To prove the other part, differentiate both sides of equation (1.12) partially w. r. t. r and both sides of equation (1.13) partially w. r. t. θ

$$\frac{\partial^2 u}{\partial r^2}=\frac{-1}{r^2}\frac{\partial v}{\partial \theta}+\frac{1}{r}\frac{\partial^2 v}{\partial r\,\partial \theta} \tag{1.14}$$

$$\frac{\partial^2 v}{\partial \theta \partial r}=\frac{-1}{r}\frac{\partial^2 u}{\partial \theta^2} \tag{1.15}$$

Multiply equation (1.15) by $\frac{1}{r}$ and add (1.14)

$$\frac{\partial^2 u}{\partial r^2}=-\frac{1}{r^2}\frac{\partial v}{\partial \theta}-\frac{1}{r^2}\frac{\partial^2 u}{\partial \theta^2} \qquad \left(\because \frac{\partial^2 u}{\partial r\,\partial \theta}=\frac{\partial^2 u}{\partial \theta\,\partial r}\right)$$

Using C–R equation (1.12), we have

$$\frac{\partial^2 u}{\partial r^2}+\frac{1}{r}\frac{\partial u}{\partial r}+\frac{1}{r^2}\frac{\partial^2 u}{\partial \theta^2}=0$$

Also, note that $f'(z)$ in polar form is given by equation (1.11).

Example 1.1: Find the following limits

(i) $\lim\limits_{z\to i} \dfrac{z^3-iz^2+z-i}{z-i}$ (ii) $\lim\limits_{z\to i} \left[x+\dfrac{i}{1-x}\right]e^{xy}$

Solution: (i) $\lim\limits_{z\to i} \dfrac{z^3-iz^2+z-i}{z-i}$ $\left[\text{Form } \dfrac{0}{0}\right]$

$$=\lim_{z\to i} \frac{\left(3z^2-2iz+1\right)}{1} \quad \text{[By } L' \text{ Hospital rule]}$$

$$= 3i^2 - 2i^2 + 1 = -3 + 2 + 1 = 0$$

or

$$\lim_{z\to i} \frac{z^3-iz^2+z-i}{z-i}=\lim_{z\to i} \frac{z^2(z-i)+(z-i)}{(z-i)}$$

$$=\lim_{z\to i} \frac{(z^2+1)(z-i)}{(z-i)}$$

$$=\lim_{z\to i} (z^2+1)=i^2+1=-1+1=0$$

(ii) $\lim\limits_{z\to i} \left[x+\dfrac{i}{1-x}\right]e^{xy}$ $(\because\ z=x+iy\ \therefore\ z\to i \Rightarrow x\to 0\ \&\ y\to 1)$

$$=\lim_{\substack{x\to 0\\ y\to 1}} \left[x+\frac{i}{1-x}\right]e^{xy}$$

$$=\left(0+\frac{i}{1-0}\right)e^0=i$$

Example 1.2: (i) Show that the function $f(z)=\overline{z}$ is continuous everywhere but not differentiable at any point in the complex plane.
(ii) Show that $f(z)=\overline{z}$ is not differentiable at $z=0$ and it is nowhere analytic.

Solution: (i) $f(z)=\overline{z};\ z=x+iy,\ \overline{z}=x-iy$

Let $z_0=x_0+iy_0$ be any complex number.

$\therefore\ \overline{z}_0=x_0-iy_0$

$$\lim_{z\to z_0} f(z) = \lim_{z\to z_0} \bar{z} = \lim_{\substack{x\to x_0\\ y\to y_0}} (x-iy) = x_0 - iy_0 = \bar{z}_0 = f(z_0)$$

$\therefore f(z)$ is continuous at $z = z_0$.

But z_0 is any complex number.

$\therefore f(z)$ is continuous at all points in complex plane.

Now, $$\lim_{z\to z_0} \frac{f(z)-f(z_0)}{z-z_0} = \lim_{\Delta z\to 0} \frac{f(z_0+\Delta z)-f(z_0)}{\Delta z}$$

$$= \lim_{\Delta z\to 0} \frac{\overline{z_0+\Delta z} - \bar{z}_0}{\Delta z}$$

$$= \lim_{\Delta z\to 0} \frac{\overline{\Delta z}}{\Delta z}$$

$$= \lim_{\substack{\Delta x\to 0\\ \Delta y\to 0}} \frac{\Delta x - i\Delta y}{\Delta x + i\Delta y}$$

Now, $$\lim_{\Delta x\to 0}\lim_{\Delta y\to 0} \frac{\Delta x - i\Delta y}{\Delta x + i\Delta y} = \lim_{\Delta x\to 0} \frac{\Delta x}{\Delta x} = 1$$

and $$\lim_{\Delta y\to 0}\lim_{\Delta x\to 0} \frac{\Delta x - i\Delta y}{\Delta x + i\Delta y} = \lim_{\Delta y\to 0} \frac{-i\Delta y}{i\Delta y} = -1$$

$\therefore$ $$\lim_{\Delta x\to 0}\lim_{\Delta y\to 0} \frac{\Delta x - i\Delta y}{\Delta x + i\Delta y} \neq \lim_{\Delta y\to 0}\lim_{\Delta x\to 0} \frac{\Delta x - i\Delta y}{\Delta x + i\Delta y}$$

Therefore, $\lim_{z\to z_0} \frac{f(z)-f(z_0)}{z-z_0}$ does not exist.

$\therefore f(z)$ is not differentiable at z_0.

But z_0 is any complex number.

$\therefore f(z)$ is not differentiable at any point in the complex plane.

(ii) By (i), $f(z) = \bar{z}$ is not differentiable at any point in the complex plane. Therefore, $f(z)$ is not differentiable at $z = 0$ and it is nowhere analytic.

Example 1.3: Show that the function $f(z) = |z|^2$ is continuous everywhere but differentiable only at the origin. Is it analytic at the origin?

Solution: $f(z) = u + iv = |z|^2 = x^2 + y^2$

$\therefore \quad u(x,y) = x^2 + y^2, v = 0$

$u(x,y)$ and $v(x,y)$ are continuous functions of x and y for all $x, y \in R$.

$\therefore$ $f(z)$ is continuous everywhere.

Now, $$\frac{\partial u}{\partial x}=2x,\ \frac{\partial u}{\partial y}=2y,\ \frac{\partial v}{\partial x}=0,\ \frac{\partial v}{\partial y}=0$$

These four partial derivatives are continuous everywhere and C–R equations $\frac{\partial u}{\partial x}=\frac{\partial v}{\partial y}$; $\frac{\partial u}{\partial y}=-\frac{\partial v}{\partial x}$ are satisfied only at $x = 0, y = 0$.

$\therefore$ $f(z)$ is differentiable only at $z = 0$.

Since there is no neighbourhood of $z = 0$ in which $f(z)$ is differentiable, therefore $f(z)$ is not analytic at $z = 0$.

Example 1.4: Show that the function $f(z) = u + iv$, where

$$f(z)=\begin{cases}\dfrac{x^3(1+i)-y^3(1-i)}{x^2+y^2}; & z\neq 0\\ 0 & ;\ z=0\end{cases}$$

is continuous everywhere. Also, show that Cauchy–Riemann equations are satisfied at origin, yet $f'(0)$ does not exist.

Solution: $$f(z)=u(x,y)+iv(x,y)=\begin{cases}\dfrac{x^3(1+i)-y^3(1-i)}{x^2+y^2}; & z\neq 0\\ 0 & ;\ z=0\end{cases}$$

$\therefore$ $$u(x,y)=\begin{cases}\dfrac{x^3-y^3}{x^2+y^2}; & (x,y)\neq(0,0)\\ 0\ ; & (x,y)=(0,0)\end{cases}$$

and $$v(x,y)=\begin{cases}\dfrac{x^3+y^3}{x^2+y^2}; & (x,y)\neq(0,0)\\ 0\ ; & (x,y)=(0,0)\end{cases}$$

At all points, $(x, y) \neq (0, 0)$,

u and v are rational functions and hence continuous at all points $(x, y) \neq (0, 0)$.

Now, we check the continuity of u at $(0, 0)$.

Let $\varepsilon > 0$. Then

$$|u(x,y)-u(0,0)|=\left|\frac{x^3-y^3}{x^2+y^2}\right|$$

$$\leq\ \frac{|x^3|+|y^3|}{x^2+y^2}=\frac{x^2|x|+y^2|y|}{x^2+y^2}$$

$$\leq |x|+|y| \qquad \left(\because \frac{x^2}{x^2+y^2} \leq 1 \ \ \& \ \ \frac{y^2}{x^2+y^2} \leq 1\right)$$

$$< \varepsilon \text{ when } |x| < \frac{\varepsilon}{2} \text{ and } |y| < \frac{\varepsilon}{2}$$

Therefore, there exists $\delta = \frac{\varepsilon}{2} > 0$ s.t.

$$|u(x,y) - u(0,0)| < \varepsilon \text{ when } |x| < \delta \text{ and } |y| < \delta$$

$\therefore$ $\lim\limits_{\substack{x \to 0 \\ y \to 0}} u(x,y) = u(0,0)$

$\Rightarrow$ $u(x, y)$ is continuous at (0, 0).

Similarly, we can prove that $v(x, y)$ is continuous at (0, 0).

Hence, $u(x, y)$ and $v(x, y)$ are continuous at all points.

Thus, $f(z) = u + iv$ is continuous everywhere.

At origin

$$\frac{\partial u}{\partial x} = \lim_{x \to 0} \frac{u\,(x,0) - u\,(0,0)}{x} = \lim_{x \to 0} \frac{\frac{x^3}{x^2} - 0}{x} = 1$$

$$\frac{\partial u}{\partial y} = \lim_{y \to 0} \frac{u\,(0,y) - u\,(0,0)}{y} = \lim_{y \to 0} \frac{\frac{-y^3}{y^2} - 0}{y} = -1$$

$$\frac{\partial v}{\partial x} = \lim_{x \to 0} \frac{v\,(x,0) - v\,(0,0)}{x} = \lim_{x \to 0} \frac{\frac{x^3}{x^2} - 0}{x} = 1$$

$$\frac{\partial v}{\partial y} = \lim_{y \to 0} \frac{v\,(0,y) - v\,(0,0)}{y} = \lim_{y \to 0} \frac{\frac{y^3}{y^2} - 0}{y} = 1$$

$\therefore$ C–R equations $\frac{\partial u}{\partial x} = \frac{\partial v}{\partial y}, \ \frac{\partial u}{\partial y} = -\frac{\partial v}{\partial x}$ are satisfied at the origin.

Along $y = 0$, $$\lim_{z \to 0} \frac{f(z) - f(0)}{z} = \lim_{x \to 0} \frac{\frac{x^3(1+i)}{x^2} - 0}{x} = 1 + i \qquad (1)$$

Along $y = x$, $$\lim_{z \to 0} \frac{f(z) - f(0)}{z} = \lim_{x \to 0} \frac{i\frac{2x^3}{2x^2} - 0}{(1+i)x} = \frac{i}{1+i} = \frac{1+i}{2} \qquad (2)$$

From (1) and (2)

$$\lim_{z\to 0}\frac{f(z)-f(0)}{z} \text{ does not exist.}$$

$\therefore f'(0)$ does not exist.

Example 1.5: Show that the function $f(z)=\sqrt{|xy|}$ is not regular at the origin, although C–R equations are satisfied at the point.

Solution: $f(z)=u(x,y)+iv(x,y)=\sqrt{|xy|}$

$\therefore u(x,y)=\sqrt{|xy|},\ v(x,y)=0$

At origin

$$\frac{\partial u}{\partial x}=\lim_{x\to 0}\frac{u(x,0)-u(0,0)}{x}=\lim_{x\to 0}\frac{0-0}{x}=0$$

$$\frac{\partial u}{\partial y}=\lim_{y\to 0}\frac{u(0,y)-u(0,0)}{y}=\lim_{y\to 0}\frac{0-0}{y}=0$$

$$\frac{\partial v}{\partial x}=\lim_{x\to 0}\frac{v(x,0)-v(0,0)}{x}=\lim_{x\to 0}\frac{0-0}{x}=0$$

$$\frac{\partial v}{\partial y}=\lim_{y\to 0}\frac{v(0,y)-v(0,0)}{y}=\lim_{y\to 0}\frac{0-0}{y}=0$$

$\therefore$ C–R equations $\dfrac{\partial u}{\partial x}=\dfrac{\partial v}{\partial y},\ \dfrac{\partial u}{\partial y}=-\dfrac{\partial v}{\partial x}$ are satisfied at the origin.

Along $y=x$ in the first quadrant

$$\lim_{z\to 0}\frac{f(z)-f(0)}{z}=\lim_{x\to 0^+}\frac{\sqrt{x^2}-0}{(1+i)x}=\lim_{x\to 0^+}=\frac{|x|}{(1+i)x}=\lim_{x\to 0^+}\frac{x}{(1+i)x}=\frac{1}{(1+i)} \quad (1)$$

and along $y=x$ in the third quadrant

$$\lim_{z\to 0}\frac{f(z)-f(0)}{z}=\lim_{x\to 0^-}\frac{|x|}{(1+i)x}=\lim_{x\to 0^-}\frac{-x}{(1+i)x}=-\frac{1}{(1+i)} \quad (2)$$

From (1) and (2) $\lim_{z\to 0}\dfrac{f(z)-f(0)}{z}$ does not exist.

$\therefore f(z)$ is not differentiable at $z=0$

$\therefore f(z)$ is not regular at $z=0$ although C–R equations are satisfied at $z=0$.

Example 1.6: Show that the function $f(z)=\begin{cases} e^{-z^{-4}}, & z\neq 0\\ 0, & z=0\end{cases}$ is not analytic at the origin although the Cauchy–Riemann equations are satisfied there.

Solution: $f(z) = u(x, y) + iv(x, y) = \begin{cases} e^{-z^{-4}}, & z \neq 0 \\ 0, & z = 0 \end{cases}$

$$\therefore u(x, 0) + iv(x, 0) = \begin{cases} e^{-x^{-4}}, & x \neq 0 \\ 0, & x = 0 \end{cases}$$

and $u(0, y) + iv(0, y) = e^{-(iy)^{-4}} = \begin{cases} e^{-y^{-4}}, & y \neq 0 \\ 0, & y = 0 \end{cases}$

$$\therefore u(x, 0) = \begin{cases} e^{-x^{-4}}, & x \neq 0 \\ 0, & x = 0 \end{cases}, \quad u(0, y) = \begin{cases} e^{-y^{-4}}, & y \neq 0 \\ 0, & y = 0 \end{cases}$$

$v(x, 0) = 0$ for all $x \in R$, $v(0, y) = 0$ for all $y \in R$

At origin

$$\begin{aligned} \frac{\partial u}{\partial x} &= \lim_{x \to 0} \frac{u(x,0) - u(0,0)}{x} \\ &= \lim_{x \to 0} \frac{e^{-x^{-4}}}{x} \\ &= \lim_{x \to 0} \frac{1}{x e^{1/x^4}} \\ &= \lim_{x \to 0} \frac{1}{x\left[1 + \frac{1}{x^4} + \frac{1}{2! x^8} + \frac{1}{3! x^{12}} + \cdots\right]} \\ &= \lim_{x \to 0} \frac{1}{x + \frac{1}{x^3} + \frac{1}{2! x^7} \frac{1}{3! x^{11}} + \cdots} = 0 \end{aligned} \qquad (1)$$

$$\begin{aligned} \frac{\partial u}{\partial y} &= \lim_{y \to 0} \frac{u(0,y) - u(0,0)}{y} \\ &= \lim_{y \to 0} \frac{e^{-y^{-4}}}{y} = 0 \quad \text{from (1)} \end{aligned}$$

$$\frac{\partial v}{\partial x} = \lim_{x \to 0} \frac{v(x,0) - v(0,0)}{x} = \lim_{x \to 0} \frac{0 - 0}{x} = 0$$

$$\frac{\partial v}{\partial y} = \lim_{y \to 0} \frac{v(0,y) - v(0,0)}{y} = \lim_{y \to 0} \frac{0 - 0}{y} = 0$$

Thus, C–R equations $\frac{\partial u}{\partial x} = \frac{\partial v}{\partial y}$, $\frac{\partial u}{\partial y} = -\frac{\partial v}{\partial x}$ are satisfied at origin.

Along $y = x$,

$$\lim_{z\to 0}\frac{f(z)-f(0)}{z} = \lim_{x\to 0}\frac{e^{-[(1+i)x]^{-4}}}{(1+i)x} = \lim_{x\to 0}\frac{e^{-\frac{1}{[(1+i)^2]^2 x^4}}}{(1+i)x}$$

$$= \lim_{x\to 0}\frac{e^{-\frac{1}{(2i)^2 x^4}}}{(1+i)x} = \frac{1}{(1+i)}\lim_{x\to 0}\frac{e^{\frac{1}{4x^4}}}{x}$$

which is infinite and hence does not exist.

$\therefore$ $f(z)$ is not analytic at origin although C–R equations are satisfied at origin.

Example 1.7: Show that the function $f(z) = \dfrac{x^2 y^5 (x+iy)}{x^4 + y^{10}}$, $z \neq 0$, $f(0) = 0$ satisfies C–R equations at origin yet $f'(0)$ does not exist.

Solution: $f(z) = u(x, y) + iv(x, y) = \begin{cases} \dfrac{x^3 y^5}{x^4+y^{10}} + i\dfrac{x^2 y^6}{x^4+y^{10}}; & z \neq 0 \\ 0 & ; z = 0 \end{cases}$

$$\therefore \qquad u(x,y) = \begin{cases} \dfrac{x^3 y^5}{x^4+y^{10}}; & (x,y) \neq (0,0) \\ 0 & ; (x,y) = (0,0) \end{cases}$$

$$v(x,y) = \begin{cases} \dfrac{x^2 y^6}{x^4+y^{10}}; & (x,y) \neq (0,0) \\ 0 & ; (x,y) = (0,0) \end{cases}$$

At origin

$$\frac{\partial u}{\partial x} = \lim_{x\to 0}\frac{u(x,0)-u(0,0)}{x} = \lim_{x\to 0}\frac{0-0}{x} = 0$$

$$\frac{\partial u}{\partial y} = \lim_{y\to 0}\frac{u(0,y)-u(0,0)}{y} = \lim_{y\to 0}\frac{0-0}{y} = 0$$

$$\frac{\partial v}{\partial x} = \lim_{x\to 0}\frac{v(x,0)-v(0,0)}{x} = \lim_{x\to 0}\frac{0-0}{x} = 0$$

$$\frac{\partial v}{\partial y} = \lim_{y\to 0}\frac{v(0,y)-v(0,0)}{y} = \lim_{y\to 0}\frac{0-0}{y} = 0$$

$\therefore$ C–R equations $\dfrac{\partial u}{\partial x} = \dfrac{\partial v}{\partial y}$; $\dfrac{\partial u}{\partial y} = -\dfrac{\partial v}{\partial x}$ are satisfied at origin.

Along $y = x$,

$$\lim_{z\to 0} \frac{f(z)-f(0)}{z} = \lim_{x\to 0} \frac{x^7}{x^4+x^{10}} = \lim_{x\to 0} \frac{x^3}{1+x^6} = 0$$

Along $y^5 = x^2$,

$$\lim_{z\to 0} \frac{f(z)-f(0)}{z} = \lim_{x\to 0} \frac{x^4}{x^4+x^4} = \frac{1}{2}$$

$\therefore$ $$\lim_{z\to 0} \frac{f(z)-f(0)}{z} \text{ does not exist.}$$

$\therefore$ $f(z)$ is not differentiable at $z = 0$.

$\therefore$ C–R equations are satisfied at origin yet $f'(0)$ does not exist.

Example 1.8: Use Cauchy–Riemann equations to show that $f(z) = z^3$ is analytic in the entire complex plane.

Solution: $f(z) = u(x, y) + iv(x, y) = (x + iy)^3$

$$= x^3 - 3xy^2 + i(3x^2y - y^3)$$

$\therefore$ $u(x, y) = x^3 - 3xy^2,\ v(x, y) = 3x^2y - y^3$

$$\frac{\partial u}{\partial x} = 3x^2 - 3y^2 = 3(x^2 - y^2),\quad \frac{\partial v}{\partial x} = 6xy$$

$$\frac{\partial u}{\partial y} = -6xy \quad \frac{\partial v}{\partial y} = 3x^2 - 3y^2 = 3(x^2 - y^2)$$

$\therefore$ For all $x, y \in R$

C–R equations $\dfrac{\partial u}{\partial x} = \dfrac{\partial v}{\partial y},\ \dfrac{\partial u}{\partial y} = \dfrac{-\partial v}{\partial x}$ are satisfied.

Also, $\dfrac{\partial u}{\partial x}, \dfrac{\partial u}{\partial y}, \dfrac{\partial v}{\partial x}, \dfrac{\partial v}{\partial y}$ being polynomial functions are continuous for all $x, y \in R$.

$\therefore$ $f(z)$ is analytic for all z in complex plane.

$\therefore$ $f(z) = z^3$ is analytic in the entire complex plane.

Example 1.9: Determine 'p' such that the function $f(z) = \dfrac{1}{2}\log\left(x^2+y^2\right) + i\tan^{-1}\dfrac{px}{y}$ is analytic.

Solution: $f(z) = u(x, y) + iv(x, y)$

$$= \frac{1}{2}\log(x^2 + y^2) + i\tan^{-1}\frac{px}{y};\ (x, y) \neq (0, 0)$$

$\therefore$ $$u(x, y) = \frac{1}{2}\log\left(x^2+y^2\right);\ (x, y) \neq (0, 0)$$

$$v(x,y) = \tan^{-1}\frac{px}{y};\ (x,y) \neq (0,0)$$

$$\frac{\partial u}{\partial x} = \frac{x}{x^2+y^2},\ \frac{\partial u}{\partial y} = \frac{y}{x^2+y^2}$$

$$\frac{\partial v}{\partial x} = \frac{1}{1+\frac{p^2x^2}{y^2}} \cdot \frac{p}{y} = \frac{py}{p^2x^2+y^2}$$

$$\frac{\partial v}{\partial y} = \frac{1}{1+\frac{p^2x^2}{y^2}} \cdot \left(\frac{-px}{y^2}\right) = \frac{-px}{p^2x^2+y^2}$$

Since $f(z)$ is analytic at $z \neq 0$

$\therefore$ C–R equations $\frac{\partial u}{\partial x} = \frac{\partial v}{\partial y},\ \frac{\partial u}{\partial y} = \frac{-\partial v}{\partial x}$ are satisfied.

$\therefore \quad \frac{x}{x^2+y^2} = \frac{-px}{p^2x^2+y^2}$ and $\frac{y}{x^2+y^2} = \frac{-py}{p^2x^2+y^2}$

These are satisfied only for $p = -1$.

$\therefore \quad p = -1.$

Example 1.10: Find the constants a, b, c such that the function $f(z)$, where
(i) $f(z) = x - 2ay + i(bx - cy)$
(ii) $f(z) = -x^2 + xy + y^2 + i(ax^2 + bxy + cy^2)$
is analytic. Express $f(z)$ in terms of z.

Solution: (i) $f(z) = x - 2ay + i(bx - cy) = u + iv$

$\therefore \quad u = x - 2ay,\ v = bx - cy$

$\therefore \quad \frac{\partial u}{\partial x} = 1,\ \frac{\partial u}{\partial y} = -2a,\ \frac{\partial v}{\partial x} = b,\ \frac{\partial v}{\partial y} = -c$

Since $f(z)$ is analytic, therefore C–R equations

$\frac{\partial u}{\partial x} = \frac{\partial v}{\partial y},\ \frac{\partial u}{\partial y} = -\frac{\partial v}{\partial x}$ are satisfied.

$\therefore \quad 1 = -c,\ -2a = -b$

$\therefore \quad c = -1,\ 2a = b$

$\therefore \quad a = k,\ b = 2k,\ c = -1$; where k is any real number.

$$f(z) = x - 2ky + i(2kx + y)$$

$= x + iy + 2ki(x + iy) = z + 2kiz = (1 + 2ki)z$; where k is any real number.

(ii) $f(z) = -x^2 + xy + y^2 + i(ax^2 + bxy + cy^2) = u + iv$

$\therefore \quad u = -x^2 + xy + y^2, \; v = ax^2 + bxy + cy^2$

$\therefore \quad \frac{\partial u}{\partial x} = -2x + y, \; \frac{\partial u}{\partial y} = x + 2y, \; \frac{\partial v}{\partial x} = 2ax + by, \; \frac{\partial v}{\partial y} = bx + 2cy$

Since $f(z)$ is analytic, therefore C–R equations

$$\frac{\partial u}{\partial x} = \frac{\partial v}{\partial y}, \; \frac{\partial u}{\partial y} = -\frac{\partial v}{\partial x} \text{ are satisfied.}$$

$\therefore \quad -2x + y = bx + 2cy, \; x + 2y = -2ax - by$

$\Rightarrow \quad b = -2, \; c = \frac{1}{2}, \; a = -\frac{1}{2}$

$$\therefore \quad f(z) = -x^2 + xy + y^2 + \frac{i}{2}(-x^2 - 4xy + y^2)$$

$$= -(x^2 - y^2 + 2ixy) - \frac{i}{2}(x^2 - y^2 + 2ixy)$$

$$= -\left(1 + \frac{i}{2}\right)(x + iy)^2$$

$$= -\frac{1}{2}(2 + i)z^2$$

Example 1.11: Show that the functions defined by

(i) $f(z) = z^3 + 1 - iz^2$ and (ii) e^z are analytic everywhere.

Solution: (i) $\lim\limits_{\Delta z \to 0} \frac{f(z + \Delta z) - f(z)}{\Delta z}$

$$= \lim_{\Delta z \to 0} \frac{\left[(z + \Delta z)^3 + 1 - i(z + \Delta z)^2\right] - \left[z^3 + 1 - iz^2\right]}{\Delta z}$$

$$= \lim_{\Delta z \to 0} \frac{3z^2 \Delta z + 3z(\Delta z)^2 + (\Delta z)^3 - i\left[2z\Delta z + (\Delta z)^2\right]}{\Delta z}$$

$$= \lim_{\Delta z \to 0} \left[3z^2 + 3z \, \Delta z + (\Delta z)^2 - 2iz - i\Delta z\right]$$

$$= 3z^2 - 2iz$$

$\therefore \quad f'(z) = 3z^2 - 2iz$ which exists for all z

$\therefore f(z)$ is analytic everywhere.

(ii) $f(z) = e^z = e^{x+iy} = e^x \cdot e^{iy} = e^x(\cos y + i \sin y) = u + iv$

$\therefore \quad u = e^x \cos y, \; v = e^x \sin y$

$$\therefore \qquad \frac{\partial u}{\partial x} = e^x \cos y, \ \frac{\partial u}{\partial y} = -e^x \sin y$$

$$\frac{\partial v}{\partial x} = e^x \sin y, \ \frac{\partial v}{\partial y} = e^x \cos y$$

$\therefore$ C–R equations $\frac{\partial u}{\partial x} = \frac{\partial v}{\partial y}, \ \frac{\partial u}{\partial y} = -\frac{\partial v}{\partial x}$ are satisfied for all $x, y \in R$.

Also exponential and trigonometrical functions are continuous.

$\therefore \qquad \frac{\partial u}{\partial x}, \frac{\partial u}{\partial y}, \frac{\partial v}{\partial x}, \frac{\partial v}{\partial y}$ are continuous functions for all $x, y \in R$.

$\therefore \qquad f(z)$ is analytic function for all z.

Example 1.12: Prove that the function $e^x(\cos y + i \sin y)$ is analytic and find its derivative in terms of z.

Solution: Let $f(z) = e^x(\cos y + i \sin y) = u + iv$

Now, $f(z)$ is analytic for all z (see above example (ii) part)

$$\therefore \qquad f'(z) = \frac{\partial u}{\partial x} + i\frac{\partial v}{\partial x} = e^x \cos y + ie^x \sin y$$

$$= e^x(\cos y + i \sin y) = e^x \cdot e^{iy}$$

$$= e^{x+iy} = e^z.$$

Example 1.13: Show that if $f(z)$ is analytic and

(i) $\operatorname{Re} f(z) =$ constant (ii) $\operatorname{Im} f(z) =$ constant or (iii) $|f(z)|$ is a non – zero constant, then $f(z)$ is a constant.

Solution: (i) Let $f(z) = u + iv$ be analytic where

$$\operatorname{Re} f(z) = u - \text{constant}$$

$$\therefore \qquad \frac{\partial u}{\partial x} = \frac{\partial u}{\partial y} = 0$$

$$\therefore \qquad f'(z) = \frac{\partial u}{\partial x} + i\frac{\partial v}{\partial x} = \frac{\partial u}{\partial x} - i\frac{\partial u}{\partial y} \quad \text{(By C – R equation)}$$

$$= 0 - i\,(0) = 0$$

$$\Rightarrow \qquad f(z) = \text{constant}$$

(ii) Let $\quad f(z) = u + iv$ be analytic where

$$\operatorname{Im} f(z) = v = \text{ constant}$$

$$\therefore \qquad \frac{\partial v}{\partial x} = \frac{\partial v}{\partial y} = 0$$

$$\therefore \qquad f'(z) = \frac{\partial u}{\partial x} + i\frac{\partial v}{\partial x} = \frac{\partial v}{\partial y} + i\frac{\partial v}{\partial x} \quad (\text{By C} - \text{R equation})$$

$$= 0 + i(0) = 0$$

$$\Rightarrow \qquad f(z) = \text{constant}$$

(iii) Let $f(z) = u + iv$ be analytic where

$$|f(z)| = \sqrt{u^2 + v^2} = \text{constant} \neq 0$$

$$\therefore \qquad u^2 + v^2 = c; \quad c \neq 0 \tag{1}$$

Differentiate partially w. r. t. x and y

$$2u\frac{\partial u}{\partial x} + 2v\frac{\partial v}{\partial x} = 0$$

$$2u\frac{\partial u}{\partial y} + 2v\frac{\partial v}{\partial y} = 0$$

$$\therefore \qquad u\frac{\partial u}{\partial x} + v\frac{\partial v}{\partial x} = 0 \tag{2}$$

$$u\frac{\partial u}{\partial y} + v\frac{\partial v}{\partial y} = 0 \tag{3}$$

Use C–R equations in (3)

$$-u\frac{\partial v}{\partial x} + v\frac{\partial u}{\partial x} = 0 \tag{4}$$

Square and add (2) and (4)

$$\left(u^2 + v^2\right)\left[\left(\frac{\partial u}{\partial x}\right)^2 + \left(\frac{\partial v}{\partial x}\right)^2\right] = 0$$

$$\Rightarrow \qquad \left(\frac{\partial u}{\partial x}\right)^2 + \left(\frac{\partial v}{\partial x}\right)^2 = 0 \qquad \left(\because \text{ from (1) } u^2 + v^2 = c \neq 0\right)$$

$$\Rightarrow \qquad \left|\frac{\partial u}{\partial x} + i\frac{\partial v}{\partial x}\right|^2 = 0$$

$$\Rightarrow \qquad |f'(z)|^2 = 0$$

$$\Rightarrow \qquad |f'(z)| = 0$$

$$\Rightarrow \qquad f'(z) = 0$$

$$\Rightarrow \qquad f(z) = \text{constant}$$

Example 1.14: If the function f is given by

$f(z)=\sqrt{z}=\sqrt{r}\left(\cos\frac{\theta}{2}+i\sin\frac{\theta}{2}\right)$ where $r>0$ and $0<\theta<2\pi$, then show that $f(z)$ is analytic and find $f'(z)$.

Solution: Let $f(z)=u+iv=\sqrt{r}\left(\cos\frac{\theta}{2}+i\sin\frac{\theta}{2}\right)$ (given)

$$\therefore \quad u=\sqrt{r}\cos\frac{\theta}{2},\ v=\sqrt{r}\sin\frac{\theta}{2}$$

$$\therefore \quad \frac{\partial u}{\partial r}=\frac{1}{2\sqrt{r}}\cos\frac{\theta}{2},\ \frac{\partial u}{\partial \theta}=-\frac{\sqrt{r}}{2}\sin\frac{\theta}{2}$$

$$\frac{\partial v}{\partial r}=\frac{1}{2\sqrt{r}}\sin\frac{\theta}{2},\ \frac{\partial v}{\partial \theta}=\frac{\sqrt{r}}{2}\cos\frac{\theta}{2}$$

$$\therefore \quad \frac{\partial u}{\partial r}=\frac{1}{r}\left(\frac{\sqrt{r}}{2}\cos\frac{\theta}{2}\right)=\frac{1}{r}\frac{\partial v}{\partial \theta} \text{ and } \frac{\partial v}{\partial r}=\frac{1}{r}\left(\frac{\sqrt{r}}{2}\sin\frac{\theta}{2}\right)=\frac{-1}{r}\frac{\partial u}{\partial \theta}$$

$\therefore$ C–R equations are satisfied.

Also, $\frac{\partial u}{\partial r},\frac{\partial u}{\partial \theta},\frac{\partial v}{\partial r},\frac{\partial v}{\partial \theta}$ are continuous functions for $r>0$.

$\therefore$ $f(z)$ is analytic.

Now, $$f(z)=f(re^{i\theta})=u(r,\theta)+iv(r,\theta)$$

$$\Rightarrow \quad f'(re^{i\theta})e^{i\theta}=\frac{\partial u}{\partial r}+i\frac{\partial v}{\partial r}$$

$$\therefore \quad f'(z)=e^{-i\theta}\left(\frac{\partial u}{\partial r}+i\frac{\partial v}{\partial r}\right)=\frac{e^{-i\theta}}{2\sqrt{r}}\left(\cos\frac{\theta}{2}+i\sin\frac{\theta}{2}\right)$$

$$=\frac{e^{-i\theta}.e^{i\theta/2}}{2\sqrt{r}}=\frac{1}{2\sqrt{re^{i\theta}}}=\frac{1}{2\sqrt{z}}$$

$$\therefore \quad f'(z)=\frac{1}{2\sqrt{z}}=\frac{1}{2\sqrt{r}}\left(\cos\frac{\theta}{2}-i\sin\frac{\theta}{2}\right).$$

1.5 HARMONIC FUNCTIONS

A real function $f(x, y)$ is called harmonic function in a domain D if it satisfies Laplace equation $\frac{\partial^2 f}{\partial x^2}+\frac{\partial^2 f}{\partial y^2}=0$ and all its second order partial derivatives are continuous in D.

We now prove the harmonic property of real and imaginary parts of an analytic function.

Theorem 1.7 If $f(z)=u(x,y)+iv(x,y)$ is analytic in some domain D, then $u(x, y)$ and $v(x, y)$ are harmonic functions in D.

Proof: $f(z) = u(x, y) + iv(x, y)$ is analytic in D.

$\therefore$ By C–R equations

$$u_x = v_y \tag{1.16}$$

$$u_y = -v_x \tag{1.17}$$

Differentiate both sides of equation (1.16) partially w. r. t. x and equation (1.17) partially w. r. t. y and add

$$u_{xx} + u_{yy} = v_{xy} - v_{yx} = 0 \qquad (\because v_{xy} = v_{yx})$$

$\therefore$ $u(x, y)$ satisfies Laplace equation.

Now differentiate both sides of equation (1.16) partially w. r. t. y and equation (1.17) partially w. r. t. x and subtract

$$u_{yx} - u_{xy} = v_{yy} + v_{xx} = 0 \qquad (\because u_{xy} = u_{yx})$$

or

$$v_{xx} + v_{yy} = 0$$

$\therefore$ $v(x, y)$ satisfies Laplace equation.

We shall prove in Theorem 1.20 that all order derivatives of an analytic function exist. Thus, all partial derivatives of second order of $u(x, y)$ and $v(x, y)$ are continuous functions of x and y. Hence, $u(x, y)$ and $v(x, y)$ are harmonic functions.

Remark 1.6: $u(x, y)$ and $v(x, y)$ are called conjugate harmonic of each other and conjugate harmonic of each of these is unique except for an additive constant when $f(z) = u(x, y) + iv(x, y)$ is analytic.

1.5.1 Orthogonal System of Level Curves

Two curves $f(x, y) = c_1$, $g(x, y) = c_2$ where c_1 and c_2 are constants are called system of level curves.

These two system of curves are orthogonal *iff* tangents to these curves at point of intersection are perpendicular.

Theorem 1.8 Level curves $u(x, y) = c_1$ and $v(x, y) = c_2$ of analytic function $f(z) = u(x, y) + iv(x, y)$ form an orthogonal system.

Proof: Let (x, y) be point of intersection of

$$u(x, y) = c_1 \tag{1.18}$$

$$v(x, y) = c_2 \tag{1.19}$$

For curve with equation (1.18)

$$m_1 = \frac{dy}{dx} = \frac{-u_x}{u_y} \tag{1.20}$$

For curve with equation (1.19)

$$m_2 = \frac{dy}{dx} = \frac{-v_x}{v_y} \tag{1.21}$$

But $f(z) = u(x, y) + iv(x, y)$ is analytic

$\therefore$ By C–R equations, $v_x = -u_y$, $v_y = u_x$.

$\therefore$ From equation (1.21)

$$m_2 = -\frac{(-u_y)}{u_x} = \frac{u_y}{u_x} \tag{1.22}$$

From equations (1.20) and (1.22)

$$m_1 m_2 = -1$$

$\therefore$ System of level curves $u(x, y) = c_1$ and $v(x, y) = c_2$ are orthogonal.

1.5.2 Method to Find Conjugate Harmonic Function

Let $f(z) = u(x, y) + iv(x, y)$ be analytic in domain D and $u(x, y)$ is given and we are to find $v(x, y)$.

From given $u(x, y)$ we find $\frac{\partial u}{\partial x}$ and $\frac{\partial u}{\partial y}$.

Now, by C–R equation

$$\frac{\partial v}{\partial x} = -\frac{\partial u}{\partial y}$$

$$v = \int_{y=\text{constant}} \left(-\frac{\partial u}{\partial y}\right) dx + k(y) \tag{1.23}$$

where $k(y)$ is real function of y.

$$\therefore \qquad \frac{\partial v}{\partial y} = \frac{\partial}{\partial y}\left\{\int_{y\,\text{constant}} \left(-\frac{\partial u}{\partial y}\right) dx\right\} + \frac{d}{dy}k(y) = \frac{\partial u}{\partial x} \qquad \text{(By C–R equation)}$$

Solving it we find $\frac{d}{dy}k(y)$ and its integral will provide $k(y)$.

Substituting its value in equation (1.23), we find v.

Similarly, if $v(x, y)$ is given we can find $u(x, y)$.

1.5.3 Milne Thomson Method

Theorem 1.9 If $f(z) = \phi(x, y) + i\psi(x, y)$ where $z = x + iy$

then $\qquad f(z) = \phi(z, 0) + i\psi(z, 0)$

Proof: We have $z = x + iy$, $\bar{z} = x - iy$

$$\therefore \qquad x = \frac{z + \bar{z}}{2}, y = \frac{z - \bar{z}}{2i}$$

Thus, $\qquad f(z) \equiv \phi(x, y) + i\psi(x, y)$

can be written as

$$f(z) \equiv \phi\left(\frac{z+\bar{z}}{2}, \frac{z-\bar{z}}{2i}\right) + i\psi\left(\frac{z+\bar{z}}{2}, \frac{z-\bar{z}}{2i}\right)$$

This relation can be considered as a formal identity in the variables z and $\bar{z}$.
In this identity replacing $\bar{z}$ by z we have $f(z) = \phi(z,0) + i\psi(z,0)$

Remark 1.7: (i) Milne–Thomson method can be applied in both cases when $f(z)$ is analytic or not analytic.
(ii) If $f(z) = f(x+iy) = \phi(x,y)$ then $f(z) = \phi(z,0)$.

Method to find analytic function $f(z) = u(x, y) + iv(x, y)$ as a function of z when one of $u(x, y)$ or $v(x, y)$ is given.

When $u(x, y)$ is given

$$f'(z) = u_x + iv_x = u_x - iu_y \qquad \text{(By C–R equation)}$$

$$f'(z) = u_x(x, y) - iu_y(x, y)$$

By Milne–Thomson method

$$f'(z) = u_x(z, 0) - iu_y(z, 0)$$

$$\therefore \qquad f(z) = \int \left[u_x(z, 0) - iu_y(z, 0)\right] dz + ic$$

where c is arbitrary real constant.
Similarly, when $v(x, y)$ is given

$$f(z) = \int \left[v_y(z, 0) + iv_x(z, 0)\right] dz + c$$

where c is arbitrary real constant.

Method to find $f(z) = u(x, y) + iv(x, y)$ as function of z when $u - v$ or $u + v$ is given.

We have $\qquad f(z) = u(x, y) + iv(x, y)$

$$\therefore \qquad if(z) = iu(x, y) - v(x, y)$$

$$\therefore \qquad (1+i)f(z) = \left[u(x,y) - v(x,y)\right] + i\left[u(x,y) + v(x,y)\right]$$

Let $\qquad F(z) = (1+i)\, f(z)$

$$U = u - v$$

$$V = u + v$$

If one of U or V is given then by Milne–Thomson method as explained above we can find $F(z)$ and then

$$f(z) = \frac{1}{1+i} F(z) \text{ can be found.}$$

Method to find conjugate harmonic function in polar form.

Let $f(z) = u(r, \theta) + iv(r, \theta)$ be analytic in a domain D and $u(r, \theta)$ is given and we are to find $v(r, \theta)$.

From given $u(r, \theta)$ we find $\frac{\partial u}{\partial r}$ and $\frac{\partial u}{\partial \theta}$.

Now, by C–R equation

$$\frac{\partial v}{\partial r} = -\frac{1}{r}\frac{\partial u}{\partial \theta}$$

$$v = \int_{\theta\ \text{constant}} \left(-\frac{1}{r}\frac{\partial u}{\partial \theta}\right) dr + k(\theta) \tag{1.24}$$

where $k(\theta)$ is real function of θ.

$$\therefore \qquad \frac{\partial v}{\partial \theta} = \frac{\partial}{\partial \theta}\left\{\int_{\theta\ \text{constant}} \left(-\frac{1}{r}\frac{\partial u}{\partial \theta}\right) dr\right\} + \frac{d}{d\theta}k(\theta) = r\frac{\partial u}{\partial r} \qquad \text{(By C–R equation)}$$

Solving it we find $\frac{d}{d\theta}k(\theta)$ and its integral will provide $k(\theta)$.

Substituting its value in equation (1.24), we find v.

Similarly, if $v(r, \theta)$ is given we can find $u(r, \theta)$.

Milne–Thomson Method (polar form)

For any function

$$f(z) = u(r, \theta) + iv(r, \theta)\ ;\ z = re^{i\theta}$$

We can write

$$f(z) = u(z, 0) + iv(z, 0)$$

Proof:

$$z = re^{i\theta},\ \bar{z} = re^{-i\theta}$$

$$\therefore \qquad r^2 = z\bar{z} \text{ and } e^{2i\theta} = \frac{z}{\bar{z}}$$

$$\therefore \qquad r = \sqrt{z\bar{z}},\ \theta = \frac{1}{2i}\text{Log}\left(\frac{z}{\bar{z}}\right)$$

$$\therefore \qquad f(z) = u\left(\sqrt{z\bar{z}}, \frac{1}{2i}\text{Log}\left(\frac{z}{\bar{z}}\right)\right) + iv\left(\sqrt{z\bar{z}}, \frac{1}{2i}\text{Log}\left(\frac{z}{\bar{z}}\right)\right)$$

This can formally be considered as an identity in z and $\bar{z}$.

Replacing $\bar{z}$ by z, we have

$$f(z) = u\,(z, 0) + iv\,(z, 0)$$

Thus, if $f(z) = f(re^{i\theta}) = \phi(r, \theta)$ then $f(z) = \phi(z, 0)$.

Method to find $f(z) = u(r, \theta) + iv(r, \theta)$ as a function of z when one of $u(r, \theta)$ or $v(r, \theta)$ is given.

Let $u(r, \theta)$ be given. Then

$$f'(z) = \frac{1}{iz}\left(\frac{\partial u}{\partial \theta} + i\frac{\partial v}{\partial \theta}\right) \quad \left(\text{from equation (1.11) where } e^{-i\theta} = \frac{r}{z}\right)$$

$$= \frac{1}{iz}\left(\frac{\partial u}{\partial \theta} + ir\frac{\partial u}{\partial r}\right) \quad \left(\text{by C} - \text{R equation } \frac{\partial u}{\partial r} = \frac{1}{r}\frac{\partial v}{\partial \theta}\right)$$

$$= \frac{r}{z}\frac{\partial u}{\partial r} + \frac{1}{iz}\frac{\partial u}{\partial \theta}$$

$$= \frac{1}{z}\left[ru_r(r,\theta) - iu_\theta(r,\theta)\right]$$

By Milne–Thomson method

$$f'(z) = \frac{1}{z}\left[zu_r(z,0) - iu_\theta(z,0)\right]$$

$$\therefore \qquad f(z) = \int\left\{u_r(z,0) - \frac{i}{z}u_\theta(z,0)\right\}dz + ic$$

where c is arbitrary real constant.

Similarly if $v(r, \theta)$ is given, we can find $f(z)$.

Example 1.15: If $f(z)$ is an analytic function of z show that

(i) $\left(\frac{\partial^2}{\partial x^2} + \frac{\partial^2}{\partial y^2}\right)|\text{Re}\, f(z)|^2 = 2\,|f'(z)|^2$

(ii) $\left(\frac{\partial^2}{\partial x^2} + \frac{\partial^2}{\partial y^2}\right)|f(z)|^2 = 4\,|f'(z)|^2$

Solution: Let $f(z) = u + iv$ be analytic.

$$\therefore \qquad |\text{Re}\, f(z)|^2 = u^2 \text{ and } |f(z)|^2 = u^2 + v^2$$

Now, $$\frac{\partial}{\partial x}u^2 = 2u\frac{\partial u}{\partial x} \Rightarrow \frac{\partial^2}{\partial x^2}u^2 = \frac{\partial}{\partial x}\left(2u\frac{\partial u}{\partial x}\right) = 2\left[u\frac{\partial^2 u}{\partial x^2} + \left(\frac{\partial u}{\partial x}\right)^2\right]$$

Similarly, $$\frac{\partial^2}{\partial y^2}u^2 = 2\left[u\frac{\partial^2 u}{\partial y^2} + \left(\frac{\partial u}{\partial y}\right)^2\right]$$

$$\therefore \qquad \left(\frac{\partial^2}{\partial x^2} + \frac{\partial^2}{\partial y^2}\right)u^2 = 2u\left(\frac{\partial^2 u}{\partial x^2} + \frac{\partial^2 u}{\partial y^2}\right) + 2\left[\left(\frac{\partial u}{\partial x}\right)^2 + \left(\frac{\partial u}{\partial y}\right)^2\right] \qquad (1)$$

Since $f(z)$ is analytic, therefore u is harmonic and hence satisfies Laplace equation

$$\frac{\partial^2 u}{\partial x^2} + \frac{\partial^2 u}{\partial y^2} = 0.$$

∴ from (1)

$$\left(\frac{\partial^2}{\partial x^2}+\frac{\partial^2}{\partial y^2}\right)|\text{Re}\, f(z)|^2 = 2\left[\left(\frac{\partial u}{\partial x}\right)^2+\left(\frac{\partial u}{\partial y}\right)^2\right]$$

$$= 2\left[\left(\frac{\partial u}{\partial x}\right)^2+\left(-\frac{\partial v}{\partial x}\right)^2\right] \qquad \text{(By C-R equation)}$$

$$= 2\left[\left(\frac{\partial u}{\partial x}\right)^2+\left(\frac{\partial v}{\partial x}\right)^2\right] = 2\left|\frac{\partial u}{\partial x}+i\frac{\partial v}{\partial x}\right|^2$$

$$= 2|f'(z)|^2 \qquad (2)$$

Similarly, we can prove that

$$\left(\frac{\partial^2}{\partial x^2}+\frac{\partial^2}{\partial y^2}\right)|\text{Im}\, f(z)|^2 = 2|f'(z)|^2 \qquad (3)$$

Add (2) and (3)

$$\left(\frac{\partial^2}{\partial x^2}+\frac{\partial^2}{\partial y^2}\right)\left(u^2+v^2\right)= 4\;|f'(z)|^2$$

$$\Rightarrow \qquad \left(\frac{\partial^2}{\partial x^2}+\frac{\partial^2}{\partial y^2}\right)|f(z)|^2 = 4\;|f'(z)|^2$$

Example 1.16: Verify that $u = x^2 - y^2 - y$ is harmonic in the whole complex plane and find conjugate harmonic v of u.

Solution:

$$u = x^2 - y^2 - y$$

$$\therefore \qquad \frac{\partial u}{\partial x} = 2x, \; \frac{\partial u}{\partial y} = -2y-1$$

$$\frac{\partial^2 u}{\partial x^2} = 2, \; \frac{\partial^2 u}{\partial y^2} = -2, \; \frac{\partial^2 u}{\partial x \partial y} = 0$$

$$\therefore \qquad \frac{\partial^2 u}{\partial x^2}+\frac{\partial^2 u}{\partial y^2} = 0$$

∴ u satisfies Laplace equation and all second order partial derivatives of u are constants and hence continuous functions.

∴ u is harmonic in the whole complex plane.

Now, we find its conjugate harmonic v.

By C–R equations, $\dfrac{\partial v}{\partial x} = \dfrac{-\partial u}{\partial y}, \; \dfrac{\partial v}{\partial y} = \dfrac{\partial u}{\partial x}$

$$\therefore \qquad \frac{\partial v}{\partial x} = 2y+1 \qquad (1)$$

$$\frac{\partial v}{\partial y} = 2x \tag{2}$$

From (1),

$$v = \int_{y \text{ constant}} (2y+1)\,dx + k(y) = x(2y+1) + k(y)$$

where $k(y)$ is real function of y.

$$\therefore \qquad \frac{\partial v}{\partial y} = 2x + \frac{d}{dy}k(y) \tag{3}$$

From (2) and (3),

$$\frac{d}{dy}k(y) = 0$$

$$\Rightarrow \qquad k(y) = \text{constant} = c$$

where c is arbitrary real constant.

$\therefore$ Conjugate harmonic of u is $v = x\,(2y+1) + c$ where c is arbitrary real constant.

Example 1.17: Show that the function $u = e^{-2xy} \sin(x^2 - y^2)$ is harmonic. Find the conjugate function v and express $u + iv$ as an analytic function.

Solution:

$$u = e^{-2xy} \sin(x^2 - y^2) \tag{1}$$

$$\frac{\partial u}{\partial x} = e^{-2xy}\left\{-2y \sin\left(x^2 - y^2\right) + 2x \cos\left(x^2 - y^2\right)\right\} \tag{2}$$

Interchanging x and y in R.H.S. of (1), we get $-u$

$\therefore$ From (2), we have

$$\frac{\partial u}{\partial y} = -e^{-2xy}\left\{-2x \sin\left(y^2 - x^2\right) + 2y \cos\left(y^2 - x^2\right)\right\}$$

$$= e^{-2xy}\left\{-2x \sin\left(x^2 - y^2\right) - 2y \cos\left(x^2 - y^2\right)\right\} \tag{3}$$

From (2)

$$\frac{\partial^2 u}{\partial x^2} = e^{-2xy}\Big[-2y\left\{-2y \sin\left(x^2 - y^2\right) + 2x \cos\left(x^2 - y^2\right)\right\}$$

$$-2y \cdot 2x \cos\left(x^2 - y^2\right) + 2\cos\left(x^2 - y^2\right) - 4x^2 \sin\left(x^2 - y^2\right)\Big]$$

$$= e^{-2xy}\left[4\left(y^2 - x^2\right) \sin\left(x^2 - y^2\right) + (2 - 8xy)\cos\left(x^2 - y^2\right)\right] \tag{4}$$

Interchanging x and y and changing the sign in R.H.S. of (4), we shall get $\dfrac{\partial^2 u}{\partial y^2}$

$$\therefore \qquad \frac{\partial^2 u}{\partial y^2} = -e^{-2xy}\left[4\left(x^2 - y^2\right) \sin\left(y^2 - x^2\right) + (2 - 8xy)\cos\left(y^2 - x^2\right)\right]$$

$$= -e^{-2xy}\left[4\left(y^2 - x^2\right) \sin\left(x^2 - y^2\right) + (2 - 8xy)\cos\left(x^2 - y^2\right)\right] \tag{5}$$

Add (4) and (5)

$$\frac{\partial^2 u}{\partial x^2}+\frac{\partial^2 u}{\partial y^2}=0$$

Also, all second order partial derivatives of u are continuous.

$\therefore$ u is harmonic function.

Now, we find its conjugate harmonic v.

By C–R equations

$$\frac{\partial v}{\partial x}=\frac{-\partial u}{\partial y},\ \frac{\partial v}{\partial y}=\frac{\partial u}{\partial x}$$

$$\therefore \quad \frac{\partial v}{\partial x}=-\frac{\partial u}{\partial y}=e^{-2xy}\left[2x\ \sin\left(x^2-y^2\right)+2y\ \cos\left(x^2-y^2\right)\right] \tag{6}$$

$$\text{and} \quad \frac{\partial v}{\partial y}=\frac{\partial u}{\partial x}=e^{-2xy}\left\{-2y\ \sin\left(x^2-y^2\right)+2x\ \cos\left(x^2-y^2\right)\right\} \tag{7}$$

From (6),

$$v=\int_{y\,\text{constant}} e^{-2xy}\left\{2x\sin\left(x^2-y^2\right)+2y\ \cos\left(x^2-y^2\right)\right\}dx+k(y)$$

where $k(y)$ is real function of y.

$$\therefore \quad v=-e^{-2xy}\cos\left(x^2-y^2\right)+k(y)$$

$$\therefore \quad \frac{\partial v}{\partial y}=e^{-2xy}\left[2x\cos\left(x^2-y^2\right)-2y\sin\left(x^2-y^2\right)\right]+\frac{d}{dy}k(y) \tag{8}$$

From (7) and (8),

$$\frac{d}{dy}k(y)=0$$

$\Rightarrow$ $k(y)$ = constant = c

where c is arbitrary real constant.

$\therefore$ conjugate harmonic of u is $v=-e^{-2xy}\cos\left(x^2-y^2\right)+c$ where c is arbitrary real constant.

$$\begin{aligned}f(z)&=u+iv=e^{-2xy}\sin\left(x^2-y^2\right)-ie^{-2xy}\cos\left(x^2-y^2\right)+ic\\&=-ie^{-2xy}\left[\cos\left(x^2-y^2\right)+i\ \sin\left(x^2-y^2\right)\right]+ic\\&=-ie^{-2xy}e^{i\left(x^2-y^2\right)}=-ie^{i\left(x^2-y^2+2i\,xy\right)}+ic\\&=-ie^{i(x+iy)^2}+ic=-ie^{iz^2}+ic\end{aligned}$$

where c is arbitrary real constant.

Example 1.18: If $\omega=\phi+i\psi$ represents the complex potential function form electric field and $\psi=x^2-y^2+\dfrac{x}{x^2+y^2}$, find the functions ϕ and ω.

Solution: $\omega = \phi + i\psi$ is an analytic function.

$$\psi = x^2 - y^2 + \frac{x}{x^2 + y^2}$$

$$\therefore \quad \frac{\partial \psi}{\partial x} = 2x + \frac{x^2 + y^2 - 2x^2}{\left(x^2 + y^2\right)^2} = 2x + \frac{y^2 - x^2}{\left(x^2 + y^2\right)^2}$$

$$\frac{\partial \psi}{\partial y} = -2y - \frac{2xy}{\left(x^2 + y^2\right)^2}$$

By C–R equations

$$\frac{\partial \phi}{\partial x} = \frac{\partial \psi}{\partial y} = -2y - \frac{2xy}{\left(x^2 + y^2\right)^2} \tag{1}$$

$$\frac{\partial \phi}{\partial y} = -\frac{\partial \psi}{\partial x} = -2x - \frac{\left(y^2 - x^2\right)}{\left(x^2 + y^2\right)^2} \tag{2}$$

$\therefore$ From (1),

$$\phi = \int_{y \text{ constant}} \left\{-2y - \frac{2xy}{\left(x^2 + y^2\right)^2}\right\} dx + k(y), \text{ where } k(y) \text{ is real function of } y.$$

$$\therefore \quad \phi = -2xy + \frac{y}{\left(x^2 + y^2\right)} + k(y)$$

$$\therefore \quad \frac{\partial \phi}{\partial y} = -2x + \frac{\left(x^2 + y^2\right) \cdot 1 - y \cdot 2y}{\left(x^2 + y^2\right)^2} + \frac{d}{dy} k(y)$$

$$= -2x - \frac{\left(y^2 - x^2\right)}{\left(x^2 + y^2\right)^2} + \frac{d}{dy} k(y) \tag{3}$$

From (2) and (3)

$$\frac{d}{dy} k(y) = 0$$

$$\Rightarrow \quad k(y) = \text{constant} = c$$

where c is arbitrary real constant.

$$\therefore \quad \phi = -2xy + \frac{y}{\left(x^2 + y^2\right)} + c$$

where c is arbitrary real constant.

$$\therefore \quad \omega = \phi + i\psi = -2xy + \frac{y}{\left(x^2 + y^2\right)} + i\left[x^2 - y^2 + \frac{x}{\left(x^2 + y^2\right)}\right] + c$$

$$= i\left(x^2 - y^2 + 2ixy\right) + i\frac{(x-iy)}{(x-iy)(x+iy)} + c$$

$$= i\left[(x+iy)^2 + \frac{1}{(x+iy)}\right] + c$$

$\therefore$ $\qquad \omega = i\left(z^2 + \frac{1}{z}\right) + c$ where c is arbitrary real constant.

Example 1.19: Determine the analytic function whose real part is $u = x^3 - 3xy^2 + 3x^2 - 3y^2 + 2x + 1$ Also, prove that the given function satisfies Laplace equation.

Solution: $\qquad u = x^3 - 3xy^2 + 3x^2 - 3y^2 + 2x + 1$

$\therefore$ $\qquad \frac{\partial u}{\partial x} = 3x^2 - 3y^2 + 6x + 2, \quad \frac{\partial u}{\partial y} = -6xy - 6y$

$\therefore$ $\qquad \frac{\partial^2 u}{\partial x^2} = 6x + 6, \quad \frac{\partial^2 u}{\partial y^2} = -6x - 6$

$\therefore$ $\qquad \frac{\partial^2 u}{\partial x^2} + \frac{\partial^2 u}{\partial y^2} = 0$

$\therefore$ u satisfies Laplace equation.

Let $f(z) = u + iv$ be analytic function.

$\therefore$ $\qquad f'(z) = \frac{\partial u}{\partial x} + i\frac{\partial v}{\partial x} = \frac{\partial u}{\partial x} - i\frac{\partial u}{\partial y} \qquad$ (By C – R equation)

$$= \left(3x^2 - 3y^2 + 6x + 2\right) + i(6xy + 6y)$$

By Milne–Thomson method, replace x by z and y by 0

$$f'(z) = 3z^2 + 6z + 2$$

$\therefore$ $\qquad f(z) = z^3 + 3z^2 + 2z + 1 + ic \qquad$ ($\because$ u contains constant 1)

where c is arbitrary real constant.

Example 1.20: Determine the analytic function $f(z) = u + iv$, if $v = \log(x^2 + y^2) + x - 2y$.

Solution:

$$v = \log(x^2 + y^2) + x - 2y, \quad (x, y) \neq (0, 0)$$

$$\frac{\partial v}{\partial x} = \frac{2x}{x^2 + y^2} + 1, \quad \frac{\partial v}{\partial y} = \frac{2y}{x^2 + y^2} - 2$$

If $f(z) = u + iv$ is an analytic function, then

$$f'(z) = \frac{\partial u}{\partial x} + i\frac{\partial v}{\partial x} = \frac{\partial v}{\partial y} + i\frac{\partial v}{\partial x} \qquad \text{(By C – R equation)}$$

$$= \left(\frac{2y}{x^2 + y^2} - 2\right) + i\left(\frac{2x}{x^2 + y^2} + 1\right)$$

By Milne–Thomson method, replace x by z and y by 0.

$$f'(z) = -2 + i\left(\frac{2z}{z^2+0} + 1\right) = -2 + \frac{2i}{z} + i$$

$$\therefore \quad f(z) = -2z + 2i\log z + iz + c$$

$$= (i-2)z + 2i\log z + c$$

where c is arbitrary real constant.

Example 1.21: Find the analytic function $f(z) = u + iv$, given $u + v = \dfrac{2\sin 2x}{e^{2y} + e^{-2y} - 2\cos 2x}$

Solution: $f(z) = u + iv$

$$\therefore \quad if(z) = -v + iu$$

$$\therefore \quad (1+i)f(z) = u - v + i(u+v)$$

$$\therefore \quad F(z) = U + iV \text{ where } F(z) = (1+i)f(z),\ U = u - v \text{ and } V = u + v$$

$f(z)$ is analytic $\Rightarrow$ $F(z)$ is analytic.

$$V = u + v = \frac{2\sin 2x}{e^{2y} + e^{-2y} - 2\cos 2x} = \frac{\sin 2x}{\cosh 2y - \cos 2x}$$

$$\frac{\partial V}{\partial x} = \frac{2\cos 2x(\cosh 2y - \cos 2x) - 2\sin^2 2x}{(\cosh 2y - \cos 2x)^2} = \frac{2(\cos 2x\ \cosh 2y - 1)}{(\cosh 2y - \cos 2x)^2}$$

$$\frac{\partial V}{\partial y} = \frac{-2\sin 2x\ \sinh 2y}{(\cosh 2y - \cos 2x)^2}$$

$$F'(z) = \frac{\partial U}{\partial x} + i\frac{\partial V}{\partial x} = \frac{\partial V}{\partial y} + i\frac{\partial V}{\partial x} \qquad \text{(By C – R equation)}$$

$$= \frac{-2\sin 2x\ \sinh 2y}{(\cosh 2y - \cos 2x)^2} + 2i\frac{(\cos 2x\ \cosh 2y - 1)}{(\cosh 2y - \cos 2x)^2}$$

By Milne–Thomson method, replace x by z and y by 0

$$F'(z) = \frac{2i(\cos 2z - 1)}{(1 - \cos 2z)^2} = \frac{2i}{(\cos 2z - 1)} = \frac{2i}{-2\sin^2 z} = -i\,\text{cosec}^2 z$$

$$\therefore \quad F(z) = i\cot z + c \text{ where } c \text{ is arbitrary real constant.}$$

$$\therefore \quad (1+i)\ f(z) = i\cot z + c$$

$$\therefore \quad f(z) = \frac{i}{(1+i)}\cot z + \frac{c}{(1+i)} = \frac{1+i}{2}\cot z + \frac{1-i}{2}c$$

$$\therefore \quad f(z) = \frac{1}{2}(1+i)\cot z + (1-i)C$$

where $C = \dfrac{c}{2}$ is arbitrary real constant.

Example 1.22: Determine the analytic function $f(z) = u + iv$, if

$$u - v = \frac{\cos x + \sin x - e^{-y}}{2(\cos x - \cosh y)} \quad \text{and} \quad f(\pi/2) = 0$$

Solution: $f(z) = u + iv$

$\therefore \quad i\,f(z) = -v + iu$

$\therefore \quad (1 + i)\,f(z) = u - v + i(u + v)$

$\therefore \quad F(z) = U + iV$ where $F(z) = (1 + i)\,f(z)$, $U = u - v$, $V = u + v$

$f(z)$ is analytic $\Rightarrow F(z)$ is analytic

$$U = u - v = \frac{\cos x + \sin x - e^{-y}}{2(\cos x - \cosh y)}$$

$$\frac{\partial U}{\partial x} = \frac{(\cos x - \cosh y)(-\sin x + \cos x) - (\cos x + \sin x - e^{-y})(-\sin x)}{2(\cos x - \cosh y)^2}$$

$$= \frac{1 + (\sin x - \cos x)\cosh y - e^{-y}\sin x}{2(\cos x - \cosh y)^2} \tag{1}$$

$$\frac{\partial U}{\partial y} = \frac{(\cos x - \cosh y)e^{-y} + (\cos x + \sin x - e^{-y})\sinh y}{2(\cos x - \cosh y)^2}$$

$$= \frac{e^{-y}(\cos x - \cosh y - \sinh y) + (\cos x + \sin x)\sinh y}{2(\cos x - \cosh y)^2} \tag{2}$$

$$F'(z) = \frac{\partial U}{\partial x} + i\frac{\partial V}{\partial x} = \frac{\partial U}{\partial x} - i\frac{\partial U}{\partial y} \qquad \text{(By C – R equation)}$$

Substitute values from (1) and (2) and then applying Milne–Thomson method (replacing x by z and y by 0), we have

$$F'(z) = \frac{1 + (\sin z - \cos z) - \sin z - i(\cos z - 1)}{2(\cos z - 1)^2} = \frac{(1+i)(1-\cos z)}{2(1-\cos z)^2},$$

$$= \frac{1+i}{2(1-\cos z)} = \frac{1+i}{4\sin^2\frac{z}{2}} = -\frac{1+i}{2}\left(-\frac{1}{2}\operatorname{cosec}^2\frac{z}{2}\right)$$

$\therefore \quad F(z) = -\dfrac{1+i}{2}\cot\dfrac{z}{2} + ic$ where c is arbitrary real constant.

$\therefore \quad (1+i)\,f(z) = -\dfrac{1+i}{2}\cot\dfrac{z}{2} + ic$

$$\therefore \quad f(z) = -\frac{1}{2}\cot\frac{z}{2} + \frac{i}{(1+i)}c \tag{3}$$

$$f\left(\frac{\pi}{2}\right) = -\frac{1}{2} + \frac{i}{(1+i)}c = 0 \qquad \text{(given)}$$

$\therefore$
$$\frac{i}{(1+i)}c = \frac{1}{2}$$

$\therefore$ From (3)
$$f(z) = \frac{1}{2}\left(1 - \cot\frac{z}{2}\right)$$

Example 1.23: Find analytic function $f(z) = u(r, \theta) + iv(r, \theta)$ such that $v(r, \theta) = r^2 \cos 2\theta - r\cos\theta + 2$. Also find its harmonic conjugate.

Solution:
$$v(r, \theta) = r^2 \cos 2\theta - r\cos\theta + 2$$

$\therefore$
$$\frac{\partial v}{\partial r} = 2r\cos 2\theta - \cos\theta,\quad \frac{\partial v}{\partial \theta} = -2r^2 \sin 2\theta + r\sin\theta$$

By C–R equations
$$\frac{\partial u}{\partial r} = \frac{1}{r}\frac{\partial v}{\partial \theta} \quad \text{and} \quad \frac{\partial u}{\partial \theta} = -r\frac{\partial v}{\partial r}$$

$$\frac{\partial u}{\partial r} = -2r\sin 2\theta + \sin\theta \tag{1}$$

$$\frac{\partial u}{\partial \theta} = -2r^2 \cos 2\theta + r\cos\theta \tag{2}$$

From (1),
$$u = \int_{\theta \text{ constant}} \left(-2r\sin 2\theta + \sin\theta\right) dr + k(\theta)$$

where $k(\theta)$ is real function of θ.

$\therefore$
$$u = -r^2 \sin 2\theta + r\sin\theta + k(\theta)$$

$\therefore$
$$\frac{\partial u}{\partial \theta} = -2r^2 \cos 2\theta + r\cos\theta + \frac{d}{d\theta}k(\theta) \tag{3}$$

From (2) and (3),
$$\frac{d}{d\theta}k(\theta) = 0$$

$\Rightarrow$
$$k(\theta) = \text{constant} = c$$

where c is arbitrary real constant.

conjugate harmonic of v is $u = -r^2 \sin 2\theta + r\sin\theta + c$

and
$$\begin{aligned} f(z) = u + iv &= -r^2 \sin 2\theta + r\sin\theta + c + i\left(r^2 \cos 2\theta - r\cos\theta + 2\right) \\ &= ir^2\left(\cos 2\theta + i\sin 2\theta\right) - ir\left(\cos\theta + i\sin\theta\right) + 2i + c \\ &= ir^2 e^{2i\theta} - ire^{i\theta} + 2i + c \\ &= i\left(z^2 - z\right) + 2i + c \end{aligned}$$

$\therefore$
$$f(z) = c + i\left(z^2 - z + 2\right)$$

where c is arbitrary real constant.

Example 1.24: If $u(r,\theta)=\left(r-\dfrac{1}{r}\right)\sin\theta, r\neq 0,$ find analytic function $f(z) = u + iv$.

Solution:
$$u(r,\theta)=\left(r-\frac{1}{r}\right)\sin\theta,\ r\neq 0,$$

$$\frac{\partial u}{\partial r}=\left(1+\frac{1}{r^2}\right)\sin\theta,\ \frac{\partial u}{\partial \theta}=\left(r-\frac{1}{r}\right)\cos\theta$$

$f(z) = f(re^{i\theta}) = u(r, \theta) + iv\ (r, \theta)$ is analytic function

$$\therefore\quad f'(re^{i\theta})e^{i\theta}=\frac{\partial u}{\partial r}+i\frac{\partial v}{\partial r}=\frac{\partial u}{\partial r}-i\frac{1}{r}\frac{\partial u}{\partial \theta}\qquad \text{(By C – R equation)}$$

$$\therefore\quad f'(z)=e^{-i\theta}\left(\frac{\partial u}{\partial r}-\frac{i}{r}\frac{\partial u}{\partial \theta}\right)$$

$$=e^{-i\theta}\left\{\left(1+\frac{1}{r^2}\right)\sin\theta-\frac{i}{r}\left(r-\frac{1}{r}\right)\cos\theta\right\}$$

By Milne–Thomson method for polar, replace r by z and θ by 0

$$\therefore\quad f'(z)=-\frac{i}{z}\left(z-\frac{1}{z}\right)=-i+\frac{i}{z^2}$$

$$\therefore\quad f(z)=-iz-\frac{i}{z}+ic$$

where c is arbitrary real constant.

i.e.,
$$f(z)=-i\left(z+\frac{1}{z}-c\right)$$

where c is arbitrary real constant.

Example 1.25: Find the orthogonal trajectories of the family of curves $r^2 \cos 2\theta = C_1$.
Solution: Let $u\ (r, \theta) = r^2 \cos 2\theta = r^2 (\cos^2 \theta - \sin^2 \theta)$.
Then the family of curves given by v = constant will be the required orthogonal trajectories if $f(z) = u + iv$ is analytic.

Now,
$$u(x, y) = x^2 - y^2 \quad (\because x = r\cos\theta, y = r\sin\theta)$$

$$\therefore\quad \frac{\partial u}{\partial x}=2x,\ \frac{\partial u}{\partial y}=-2y$$

By C–R equations,
$$\frac{\partial v}{\partial x}=-\frac{\partial u}{\partial y},\ \frac{\partial v}{\partial y}=\frac{\partial u}{\partial x}$$

$$\frac{\partial v}{\partial x}=2y \qquad (1)$$

$$\frac{\partial v}{\partial y}=2x \qquad (2)$$

From (1),

$$v = \int_{y \text{ constant}} 2y\,dx + k(y)$$

where $k(y)$ is real function of y.

$$\therefore \qquad v = 2xy + k(y)$$

$$\therefore \qquad \frac{\partial v}{\partial y} = 2x + \frac{d}{dy}k(y) \qquad (3)$$

From (2) and (3),

$$\frac{d}{dy}k(y) = 0$$

$$\Rightarrow \qquad k(y) = \text{constant}$$

$$\therefore \qquad v = 2xy + \text{constant}$$

$\therefore$ orthogonal trajectories v = constant is

$$2xy = \text{constant}$$

or

$$r^2\,(2\cos\theta \sin\theta) = \text{constant}$$

or

$$r^2 \sin 2\theta = \text{constant}$$

Note: We can also solve the above question directly in polar form.

EXERCISE 1.1

1. If $f(z) = \dfrac{(x+y)^2}{x^2+y^2}$ then show that $\lim\limits_{x\to 0}\left[\lim\limits_{y\to 0} f(z)\right] = 1$ and $\lim\limits_{y\to 0}\left[\lim\limits_{x\to 0} f(z)\right] = 1$ but $\lim\limits_{z\to 0} f(z)$ does not exist.
2. If $f(z) = \dfrac{x^2-y^2}{x^2+y^2}$, then show that $\lim\limits_{x\to 0}\left[\lim\limits_{y\to 0} f(z)\right] \neq \lim\limits_{y\to 0}\left[\lim\limits_{x\to 0} f(z)\right]$.
3. Show that $\lim\limits_{z\to 0} \dfrac{x^2 y}{x^4+y^2}$ does not exist even though this function approaches the same limit along every straight line through the origin.
4. Find the following limits (i) $\lim\limits_{z\to 0} \dfrac{z^2}{|z|}$ (ii) $\lim\limits_{z\to 1} \dfrac{z^2+z-2}{z-1}$.
5. Check the continuity of the function $f(z) = \dfrac{z}{|z|}$, $z \neq 0$ and $f(0) = 0$ at the point $z = 0$.
6. Show that the function $f(z) = xy + iy$ is continuous everywhere but is not analytic.
7. Show that the function $f(z) = Re\, z$ is continuous but not differentiable.
8. Prove that $\dfrac{d}{dz}\left(z^2\,\overline{z}\right)$ does not exist anywhere except at $z = 0$.

9. Show that the function $f(z) = z^2$ is differentiable for all z and has the derivative $f'(z) = 2z$.
10. Show that the function $f(z)=\frac{(\bar{z})^2}{z}$, $z \neq 0$, $f(0)=0$ satisfies the C–R equations at the origin yet $f'(0)$ does not exist.
11. Discuss the analyticity of the function $f(z) = z\bar{z}$.
12. Show that the function $z|z|$ is not analytic anywhere.
13. Show that the real and imaginary parts of the function $w = \text{Log}\, z$ satisfy the Cauchy–Riemann equations when z is not zero.
14. Find the point where the Cauchy–Riemann equations are satisfied for the function $f(z) = xy^2 + ix^2y$. When does $f'(z)$ exist? Where $f(z)$ is analytic?
15. Find the constants a and b so that the function $f(z) = a(x^2 - y^2) + ib\,xy + c$ is analytic at all points.
16. Show that the function $v(x, y) = e^x \sin y$ is harmonic. Find its conjugate harmonic function $u(x, y)$ and the corresponding analytic function $f(z)$.
17. Show that the function $u(x, y) = 2x + y^3 - 3x^2y$ is harmonic. Find its conjugate harmonic function $v(x, y)$ and the corresponding analytic function $f(z)$.
18. Let $f(z) = u + iv$ be an analytic function. If $u = 3x - 2xy$, then find v and express $f(z)$ in terms of z.
19. Prove that the function $u = 3x^2y + 2x^2 - y^3 - 2y^2$ is harmonic and find its harmonic conjugate. Hence find $f(z)$ as function of z.
20. Is the function $u(x, y) = 2xy + 3xy^2 - 2y^3$ harmonic?
21. Prove that $u = x^2 - y^2$ and $v = \frac{y}{x^2 + y^2}$ are harmonic functions of (x, y) but are not harmonic conjugates.
22. Show that $v(x, y) = -\sin x \sinh y$ is harmonic. Find the conjugate harmonic of v and corresponding analytic function.
23. Show that $u = e^{-x}(x \sin y - y \cos y)$ is harmonic and find the conjugate harmonic of u and corresponding analytic function.
24. If $f(z) = u + iv$ is an analytic function where $u = \frac{\sin 2x}{\cosh 2y - \cos 2x}$, then find $f(z)$.
25. Show that the function $u(x, y) = 4xy - 3x + 2$ is harmonic. Construct the corresponding analytic function $f(z) = u(x, y) + iv(x, y)$. Express $f(z)$ in terms of complex variable z.
26. Find an analytic function $w = u + iv$ given that $v = \frac{x}{x^2 + y^2} + \cosh x \cos y$
27. If $u = x^2 - y^2$, find a corresponding analytic function.
28. Find the analytic function whose real part is $e^{2x}(x \cos 2y - y \sin 2y)$.
29. Find the analytic function $f(z) = u + iv$, given that $v = e^x(x \sin y + y \cos y)$
30. If $f(z) = u + iv$ is an analytic function and $u - v = (x - y)(x^2 + 4xy + y^2)$, find $f(z)$.
31. If $f(z) = u + iv$ is an analytic function of z and $u + v = e^x(\cos y + \sin y)$, find $f(z)$ in terms of z.
32. If $f(z) = u + iv$ is an analytic function and $u - v = \frac{e^y - \cos x + \sin x}{\cosh y - \cos x}$ find $f(z)$ subject to the condition $f\left(\frac{\pi}{2}\right) = \frac{3-i}{2}$.
33. Show that the function $u(r, \theta) = r^2 \cos 2\theta$ is harmonic. Find its conjugate harmonic function and the corresponding analytic function $f(z)$.
34. Find p such that the function $f(z)$ expressed in polar coordinates as $f(z) = r^2 \cos 2\theta + ir^2 \sin p\theta$ is analytic.

35. If $f(z) = u(r, \theta) + iv(r, \theta)$ is an analytic function where $u = -r^3 \sin 3\theta$, then find $f(z)$ in terms of z.
36. For what values of z, the function w defined by $z = \log \rho + i\phi$ where $w = \rho(\cos\phi + i\sin\phi)$ ceases to be analytic?
37. If n is real, show that $r^n(\cos n\theta + i\sin n\theta)$ is analytic except possibly when $r = 0$, and that its derivative is $nr^{n-1}[\cos(n-1)\theta + i\sin(n-1)\theta]$.
38. Find the orthogonal trajectories of the family of curves $x^3y - xy^3 =$ constant.
39. Given $w = z^3$ where $w = u + iv$, show that the family of curves $u = c_1$ and $v = c_2$ (where c_1 and c_2 are constants) are orthogonal to each other.

ANSWERS 1.1

4. (i) 0 (ii) 3
5. Discontinuous at $z = 0$
11. $f(z)$ is analytic nowhere.
14. C–R equations are satisfied only at $z = 0$, $f'(z)$ exists only for $z = 0$ and $f(z)$ is analytic nowhere.
15. $a = k$, $b = 2k$ where k is any real number.

In all the answers below, c is an arbitrary real constant.

16. $u(x, y) = e^x \cos y + c$; $f(z) = e^z + c$
17. $v = x^3 + 2y - 3xy^2 + c$; $f(z) = 2z + iz^3 + ic$
18. $v = x^2 - y^2 + 3y + c$; $f(z) = iz^2 + 3z + ic$
19. $v = -x^3 + x(3y^2 + 4y) + c$; $f(z) = -iz^3 + 2z^2 + ic$
20. Not harmonic
22. $u = \cos x \cosh y$; $f(z) = \cos z + c$
23. $v = e^{-x}(x\cos y + y\sin y) + c$; $f(z) = ize^{-z} + ic$
24. $f(z) = \cot z + ic$
25. $f(z) = -2iz^2 - 3z + 2 + ic$
26. $f(z) = i\left(\frac{1}{z} + \cosh z\right) + c$
27. $f(z) = z^2 + ic$
28. $f(z) = ze^{2z} + ic$
29. $f(z) = ze^z + c$
30. $f(z) = -iz^3 + (1 + i)c$
31. $f(z) = e^z + (1 - i)c$
32. $f(z) = \cot\frac{z}{2} + \frac{1-i}{2}$
33. $v(r, \theta) = r^2 \sin 2\theta + c$; $f(z) = z^2 + ic$
34. $p = 2$
35. $f(z) = iz^3 + ic$
36. w ceases to be analytic nowhere.
38. $x^4 + y^4 - 6x^2y^2 =$ constant.

1.6 LINE INTEGRAL IN THE COMPLEX PLANE

We shall be requiring certain terms in the definition of line integral and thus we are defining them.

1.6.1 Continuous Curve or Arc

Any path C in complex plane joining the points $z(a)$ and $z(b)$ defined by parametric equation $z(t) = x(t) + iy(t)$; $a \le t \le b$ where $x(t)$, $y(t)$ are continuous functions of t is called a continuous curve or arc.

If the curve does not intersect itself then it is called a simple curve otherwise a multiple curve. If $z(a) = z(b)$, i.e., the starting and end points coincide then the curve is called a closed curve. A closed curve which does not intersect itself is called a simple closed curve.

1.6.2 Smooth Curve or Smooth Arc

A continuous curve $z(t) = x(t) + iy(t)$; $a \le t \le b$ is called smooth curve or smooth arc if $x'(t)$ and $y'(t)$ are continuous functions of t in $[a, b]$ and $z'(t) \neq 0$ for any t in $[a, b]$.

1.6.3 Piecewise Continuous Curve

If there exists a partition $a = t_0 < t_1 < t_2 \ldots < t_{n-1} < t_n = b$ of $[a, b]$ such that the curve $z(t) = x(t) + iy(t)$ is continuous curve in every subinterval (t_{k-1}, t_k); $k = 1, 2, 3, \ldots, n$ then the curve $z(t) = x(t) + iy(t)$ is called piecewise continuous curve in $[a, b]$.

1.6.4 Piecewise Smooth Curve

If there exists a partition $a = t_0 < t_1 < t_2 \ldots < t_{n-1} < t_n = b$ of $[a, b]$ such that the curve $z(t) = x(t) + iy(t)$ is smooth curve in every subinterval (t_{k-1}, t_k); $k = 1, 2, 3, \ldots, n$ then the curve $z(t) = x(t) + iy(t)$ is called piecewise smooth curve in $[a, b]$.

1.6.5 Contour

A piecewise continuous closed smooth curve is called a contour.

1.6.6 Line Integral

Let $z(t) = x(t) + iy(t)$; $a \le t \le b$ be a piecewise smooth curve C and $a = t_0 < t_1 < t_2 < \ldots < t_{n-1} < t_n = b$ be a partition of $[a, b]$ such that $z(t)$ is continuous on each subinterval (t_{k-1}, t_k).

Let $\xi_k \in (t_{k-1}, t_k)$, $k = 1, 2, \ldots, n$

If $\lim_{n\to\infty} \sum_{k=1}^{n} f(\xi_k)\Delta z_k$; $\Delta z_k = z_k - z_{k-1}$; exists when $|\Delta z_k| \to 0$ then this limit is called the line integral of $f(z)$ over C and is written as $\int_C f(z)\,dz$. If C is a closed curve, i.e., $z(a) = z(b)$ then $\int_C f(z)\,dz$ may be written as $\oint_C f(z)dz$. Here, C is path of integration.

In case of real variables, the path of integration of $\int_a^b f(x)\,dx$ is always along the real line from $x = a$ to $x = b$, but in case of complex function $f(z)$, the path of integration of definite internal $\int_a^b f(z)\,dz$ can be along any piecewise continuous simple curve from $z = a$ to $z = b$. Its value depends on the path of integration. In 1.7.3, we shall prove that integral is independent of path if $f(z)$ is analytic in certain domain containing C.

Example 1.26: Evaluate $\int_0^{1+i}\left(x-y+ix^2\right)dz$

(i) along the straight line from (0, 0) to (1, 1)
(ii) over the path along the lines $y = 0$ and $x = 1$
(iii) over the path along the lines $x = 0$ and $y = 1$
(iv) along the path $x = y^2$
(v) along the curve C ; $x = t$, $y = t^2$.

Solution: (i) Equation of straight line from (0, 0) to (1, 1) is $y = x$ and x varies from 0 to 1.

$$\therefore \qquad z = x + iy = (1+i)\,x \Rightarrow dz = (1+i)dx$$

$$\therefore \qquad \int_0^{1+i}\left(x-y+ix^2\right)dz = \int_0^1\left(x-x+ix^2\right)(1+i)\,dx = i(1+i)\left(\frac{x^3}{3}\right)_0^1 = \frac{1}{3}(i-1)$$

(ii) over the path along the lines $y = 0$ and $x = 1$, i.e., along the lines OA and AB (see Figure 1.1)

Along the line OA

$y = 0$ and so $z = x \Rightarrow dz = dx$ and x varies from 0 to 1.

Along the line AB

$x = 1$ and so $z = 1 + iy \Rightarrow dz = idy$ and y varies from 0 to 1.

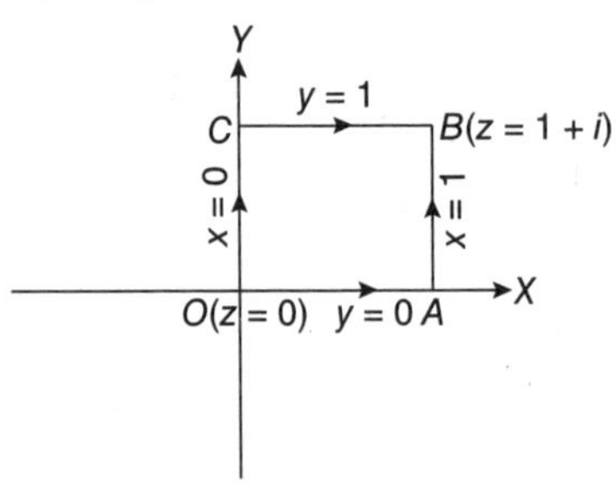

Figure 1.1

$$\therefore \qquad \int_0^{1+i}\left(x-y+ix^2\right)dz = \int_0^1\left(x+ix^2\right)dx + \int_0^1(1-y+i)\,idy$$

$$= \left(\frac{1}{2}x^2 + \frac{ix^3}{3}\right)_0^1 + i\left\{(1+i)\,y - \frac{y^2}{2}\right\}_0^1$$

$$= \frac{1}{2} + \frac{i}{3} + i\left(1 + i - \frac{1}{2}\right) = -\frac{1}{2} + \frac{5}{6}i$$

(iii) over the path along the lines $x = 0$ and $y = 1$, i.e., along the lines OC and CB (see Figure 1.1).
Along the line OC, $x = 0$ and so $z = iy \Rightarrow dz = idy$ and y varies from 0 to 1.
Along the line CB, $y = 1$ and so $z = x + i \Rightarrow dz = dx$ and x varies from 0 to 1.

$$\therefore \qquad \int_0^{1+i}\left(x-y+ix^2\right)dz = \int_0^1 -y(idy)+\int_0^1\left(x-1+ix^2\right)dx$$

$$= \left\{\frac{-iy^2}{2}\right\}_0^1 + \left\{\frac{(x-1)^2}{2}+\frac{ix^3}{3}\right\}_0^1$$

$$= \frac{-i}{2}-\frac{1}{2}+\frac{i}{3} = -\frac{1}{2}-\frac{i}{6}$$

(iv) Along the path $x = y^2$, $z = x + iy = y^2 + iy$

$$\therefore \qquad dz = (2y + i)\, dy$$

when $z \to 0$, $y \to 0$ and when $z \to 1 + i$, $y \to 1$

$$\therefore \qquad \int_0^{1+i}\left(x-y+ix^2\right)dz = \int_0^1\left(y^2-y+iy^4\right)(2y+i)\,dy$$

$$= \int_0^1\left[\left(2y^3-2y^2-y^4\right)+i\left(y^2-y+2y^5\right)\right]dy$$

$$= \left\{\frac{1}{2}y^4-\frac{2}{3}y^3-\frac{y^5}{5}+i\left(\frac{y^3}{3}-\frac{y^2}{2}+\frac{y^6}{3}\right)\right\}_0^1$$

$$= \frac{1}{2}-\frac{2}{3}-\frac{1}{5}+i\left(\frac{1}{3}-\frac{1}{2}+\frac{1}{3}\right) = -\frac{11}{30}+\frac{1}{6}i$$

(v) Along the curve $C: x = t, y = t^2$

$$z = x + iy = t + it^2$$

$$\therefore \qquad dz = (1 + 2it)\, dt$$

As $z \to 0$, $t \to 0$ and $z \to 1 + i$, $t \to 1$

$$\therefore \qquad \int_0^{1+i}\left(x-y+ix^2\right)dz = \int_0^1\left(t-t^2+it^2\right)(1+2it)\,dt$$

$$= \int_0^1\left[t-t^2-2t^3+i\left(t^2+2t^2-2t^3\right)\right]dt$$

$$= \left[\frac{t^2}{2}-\frac{t^3}{3}-\frac{t^4}{2}+i\left(t^3-\frac{1}{2}t^4\right)\right]_0^1$$

$$= \frac{1}{2}-\frac{1}{3}-\frac{1}{2}+i\left(1-\frac{1}{2}\right) = -\frac{1}{3}+\frac{1}{2}i$$

Example 1.27: Evaluate $\int_C (z-a)^n\, dz$ where C is the circle with center 'a' and radius r and n is an integer. Also, discuss the case when $n = -1$, that is, evaluate $\int_C \frac{dz}{(z-a)}$.

Solution: On C, $z-a=re^{i\theta}$; $0\le\theta<2\pi$

$\therefore$ $$dz = ire^{i\theta}\ d\theta$$

$\therefore$ $$\int_C (z-a)^n dz = \int_0^{2\pi} r^n e^{in\theta}\cdot ir\, e^{i\theta}\ d\theta = \left\{\frac{r^{n+1}\ e^{i(n+1)\theta}}{(n+1)}\right\}_0^{2\pi};\ n\neq -1$$

$$= \frac{r^{n+1}}{n+1}\left\{e^{2\pi i(n+1)}-1\right\} = \frac{r^{n+1}}{n+1}(1-1)=0;\ n\neq -1$$

When $n=-1$,

$$\int_C \frac{dz}{z-a} = \int_0^{2\pi}\frac{ire^{i\theta}d\theta}{re^{i\theta}} = i\int_0^{2\pi} d\theta = 2\pi i$$

Example 1.28: Evaluate $\int_C |z|\,dz$ where C is the contour

(i) the straight line from $z=-i$ to $z=i$

(ii) left half of the unit circle $|z|=1$ from $z=-i$ to $z=i$.

Solution: (i)

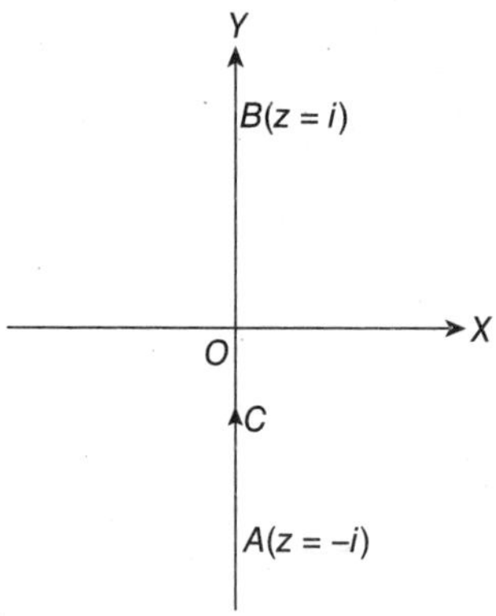

Figure 1.2 (a)

On straight line from $z=-i$ to $z=i$, $x=0$

$\therefore$ $$z=iy \Rightarrow dz = idy \text{ and } |z|=|y|$$

and y varies from -1 to 1.

$\therefore$ $$\int_C |z|dz = \int_{-1}^{1}|y|\,idy = 2i\int_0^1 y\,dy \qquad \text{(even function)}$$

$$= 2i\left(\frac{y^2}{2}\right)_0^1 = i$$

(ii)

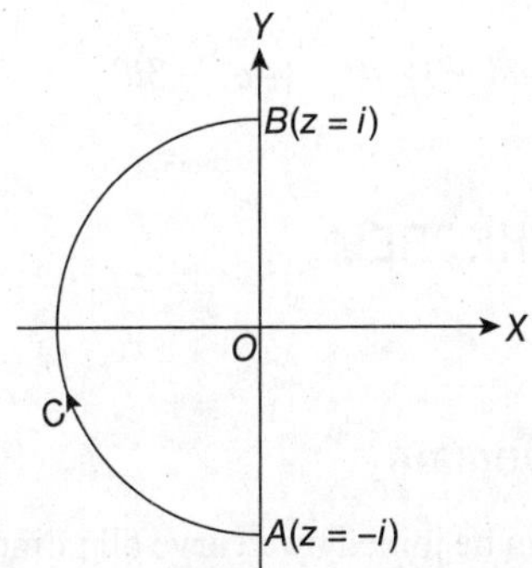

Figure 1.2 (b)

On left half of unit circle $|z| = 1$ from $-i$ to $z = i$,

$$z = e^{i\theta};\ \theta \text{ varies from } \frac{-\pi}{2} \text{ to } \frac{-3\pi}{2}$$

$$\therefore \quad |z| = 1 \text{ and } dz = ie^{i\theta}\, d\theta$$

$$\therefore \quad \int_C |z|\, dz = \int_{-\pi/2}^{-3\pi/2} ie^{i\theta}\, d\theta = \left(e^{i\theta}\right)_{-\pi/2}^{-3\pi/2} = e^{-\frac{3\pi}{2}i} - e^{-\frac{\pi}{2}i} = i - (-i) = 2i$$

Example 1.29: Evaluate the integral $\int_C \frac{2z+3}{z} dz$, where C is

(a) upper half of the circle $|z| = 2$ in the clockwise direction,

(b) lower half of the circle $|z| = 2$ in the anti-clockwise direction and

(c) the circle $|z| = 2$ in the anti-clockwise direction.

Solution: $|z| = 2 \Rightarrow z = 2e^{i\theta}$

$$\therefore \quad dz = 2ie^{i\theta}\, d\theta$$

$$\therefore \quad \frac{2z+3}{z} dz = \frac{4e^{i\theta}+3}{2e^{i\theta}} \cdot 2ie^{i\theta}\, d\theta = i\left(4e^{i\theta}+3\right) d\theta$$

(a) On the upper half of $|z| = 2$ in clockwise direction, θ varies from π to 0

$$\therefore \quad \int_C \frac{2z+3}{z} dz = \int_\pi^0 i\left(4e^{i\theta}+3\right) d\theta = \left(4e^{i\theta}+3i\theta\right)_\pi^0 = 4 - (-4 + 3i\pi) = 8 - 3\pi i$$

(b) On the lower half of $|z| = 2$ in anti-clockwise direction, θ varies from π to 2π

$$\therefore \quad \int_C \frac{2z+3}{z} dz = \int_\pi^{2\pi} i\left(4e^{i\theta}+3\right) d\theta = \left(4e^{i\theta}+3i\theta\right)_\pi^{2\pi} = (4+6\pi i) - (-4+3i\pi) = 8 + 3\pi i$$

(c) On circle $|z| = 2$ in anti-clockwise direction, θ varies from 0 to 2π

$$\therefore \qquad \int_C \frac{2z+3}{z}\,dz = \int_0^{2\pi} i\left(4e^{i\theta}+3\right)d\theta = \left(4e^{i\theta}+3i\theta\right)_0^{2\pi} = (4+6\pi i)-4 = 6\pi i$$

1.7 CAUCHY INTEGRAL THEOREM

We first define the following.

1.7.1 Simply Connected Domain

If every two points in a domain can be joined by a curve all points of which lie in domain then the domain is called a connected domain. A connected domain D bounded by a simple closed curve is called simply connected domain if every closed curve C_1 inside D encloses only points of D. Thus, in a simply connected domain if any closed curve inside it is shrunk then it will shrunk to a point inside it without leaving D. Figure 1.3 shows simply connected domain D by closed curve C in which C_1 is any curve with the stated properties.

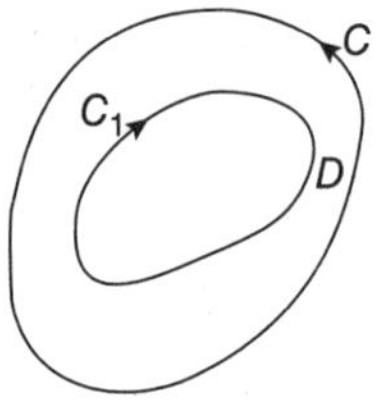

Figure 1.3

1.7.2 Multiply Connected Domain

A domain D bounded by two or more than two simple closed curves such that any two points lying inside it can be joined by a simple curve which totally lies inside the domain D is called multiply connected domain. Multiply connected domain cannot shrink to a point. The multiconnected domain has holes in it. If it has one hole then it is doubly connected and if it has two holes then it is triply connected. Let D_1 be removed from simply connected domain D then it will be doubly connected: $r < |z| < R$ is doubly connected. Figures 1.4 and 1.5 respectively show doubly connected and triply connected domains.

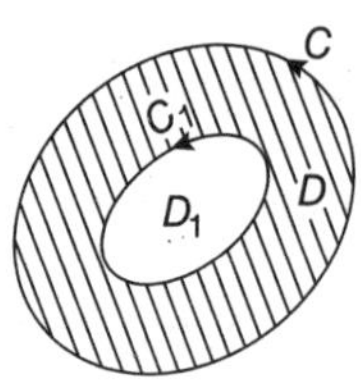

Figure 1.4

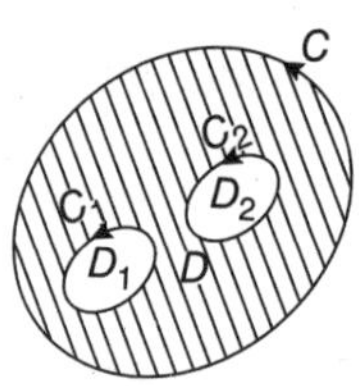

Figure 1.5

Any multiply connected domain can be converted into a simply connected domain by introducing cuts in the domain. In doubly connected domain in Figure 1.4 introduce cut AB as shown in Figure 1.6 and the new domain can be viewed as in Figure 1.7.

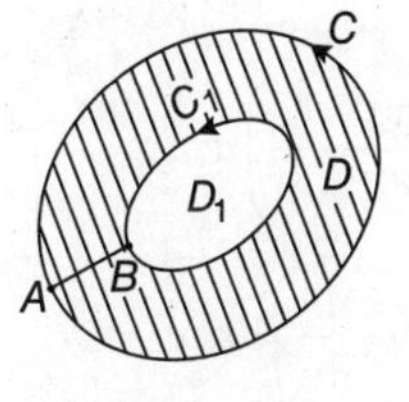

Figure 1.6

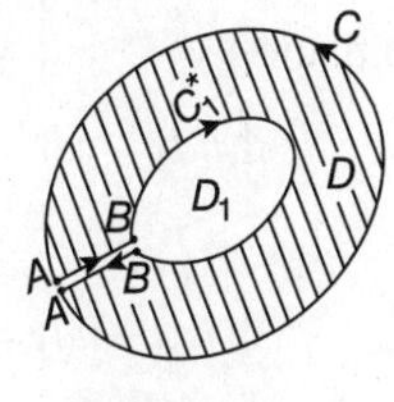

Figure 1.7

Note that here C_1^* is in opposite sense to C_1.

Similarly by introducing two cuts AB and PQ to triply connected domain in Figure 1.5, we get the simply connected domain as shown in Figure 1.8.

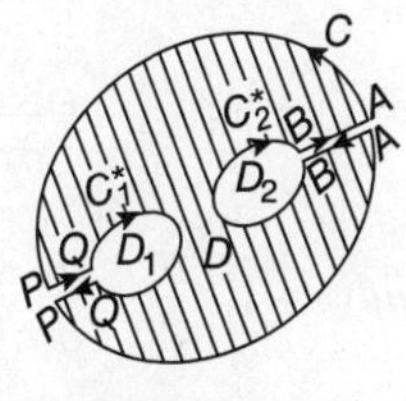

Figure 1.8

We now state and prove Cauchy integral theorem.

Theorem 1.10 (Cauchy integral theorem)

If a function $f(z)$ is analytic and its derivative $f'(z)$ is continuous in a simply connected domain D and C is any simple closed curve contained in D then

$$\oint_C f(z)\,dz = 0$$

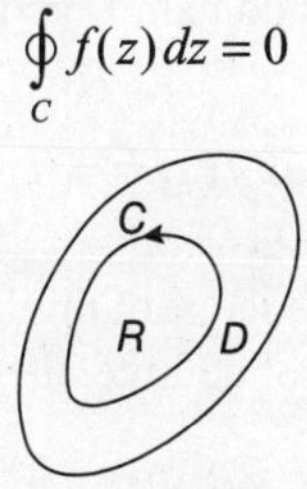

Figure 1.9

Proof: Let $f(z) = u(x,y) + iv(x,y)$. Then

$$\oint_C f(z)dz = \oint_C (u+iv)(dx+idy)$$

$$= \oint_C (udx - vdy) + i\oint_C (vdx + udy) \tag{1.25}$$

Let R be the region enclosed within C.

Now, $\frac{\partial u}{\partial x}, \frac{\partial u}{\partial y}, \frac{\partial v}{\partial x}, \frac{\partial v}{\partial y}$ are continuous in R.

$\therefore$ By Green's theorem in plane

$$\oint_C udx - vdy = \iint_R \left(\frac{-\partial v}{\partial x} - \frac{\partial u}{\partial y} \right) dxdy \tag{1.26}$$

and

$$\oint_C vdx + udy = \iint_R \left(\frac{\partial u}{\partial x} - \frac{\partial v}{\partial y} \right) dxdy \tag{1.27}$$

But $f(z)$ is analytic in R.

$\therefore$ By C–R equations

$$\frac{\partial u}{\partial x} = \frac{\partial v}{\partial y}, \frac{\partial u}{\partial y} = -\frac{\partial v}{\partial x}$$

$\therefore$ From equations (1.26) and (1.27)

$$\int_C (u\,dx - v\,dy) = \iint_R \left(-\frac{\partial v}{\partial x} + \frac{\partial v}{\partial x} \right) dx\,dy = 0$$

and

$$\int_C (vdx + udy) = \iint_R \left(\frac{\partial v}{\partial y} - \frac{\partial v}{\partial y} \right) dxdy = 0$$

$\therefore$ From equation (1.25)

$$\oint_C f(z)dz = 0 + i0 = 0$$

Remark 1.8: The condition that D is simply connected is necessary. But Cauchy and Goursat proved that the condition of $f'(z)$ is continuous can be relaxed.

Theorem 1.11 (Cauchy–Goursat theorem)

Let $f(z)$ be analytic in a simply connected domain D and C be any simple closed curve contained in D then

$$\oint_C f(z)dz = 0$$

The proof of this theorem is beyond the scope of this book. From this theorem we can prove the following result of independence of path of definite integral when integrand is analytic function.

1.7.3 Independence of Path

Theorem 1.12 Let $f(z)$ be analytic in a simply connected domain D and C be any path contained in D joining two points z_1 and z_2 in D then $\int_C f(z)dz$ is independent of the path C and depends only on z_1 and z_2.

Proof: Consider the simple paths C_1 and C_2 in D from z_1 to z_2. Let C_2^* denote the path C_2 with its orientation reversed then C_1 and C_2^* constitute a simple closed curve inside D. Then by Cauchy–Goursat theorem

$$\int_{C_1 \text{ and } C_2^*} f(z)dz = 0$$

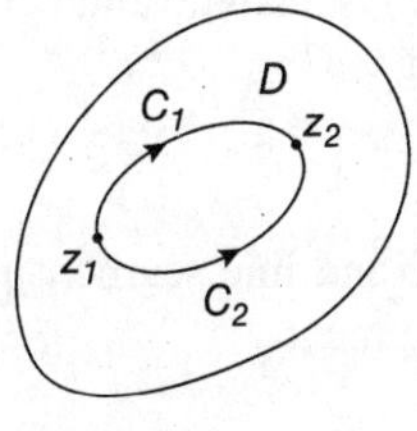

Figure 1.10

$$\therefore \quad \int_{C_1} f(z)dz + \int_{C_2^*} f(z)dz = 0$$

$$\therefore \quad \int_{C_1} f(z)dz - \int_{C_2} f(z)dz = 0 \quad (\because \text{ orientation of } C_2 \text{ and } C_2^* \text{ are reverse of each other})$$

$$\therefore \quad \int_{C_1} f(z)dz = \int_{C_2} f(z)dz$$

Thus, the integrals of $f(z)$ from z_1 to z_2 along paths C_1 and C_2 are equal. But the paths C_1 and C_2 joining z_1 and z_2 are arbitrary. Hence the integral of $f(z)$ is independent of path C joining z_1 and z_2 and depends only on z_1 and z_2 and thus the integral can be written as $\int_{z_1}^{z_2} f(z)dz$.

1.7.4 Integral Function

Let $f(z)$ be an analytic function in a simply connected domain D. Let z_0 be a fixed point in D and z be any point in D then $\int_{z_0}^{z} f(z)\,dz$ is independent of path and hence depends on z only.

$$F(z) = \int_{z_0}^{z} f(z)dz$$

is called integral function of $f(z)$.

1.7.5 Fundamental Theorem of Integral Calculus

Theorem 1.13 Let $f(z)$ be analytic function in a simply connected domain D, $F(z)$ is integral function of $f(z)$ then

$$\int_{z_0}^{z_1} f(z)dz = F(z_1) - F(z_0)$$

where z_0 and z_1 are in D.

Proof: $F(z)=\int_{z_0}^{z} f(z)dz$ is integral function of $f(z)$.

Let z be a fixed point in D. Now, $f(\xi)$ is analytic and hence continuous at z. Thus, corresponding to $\varepsilon > 0$ there exists $\delta > 0$ such that

$$|f(\xi)-f(z)|<\varepsilon \text{ when } |\xi - z|<\delta \tag{1.28}$$

Now, we take Δz such that $|\Delta z|<\delta$ and line segment joining z and $z+\Delta z$ lies inside D and take the path of integration this line segment.

Now

$$\frac{F(z+\Delta z)-F(z)}{\Delta z}-f(z)=\frac{1}{\Delta z}\left[\int_{z_0}^{z+\Delta z} f(\xi)d\xi-\int_{z_0}^{z} f(\xi)d\xi-\int_{z}^{z+\Delta z} f(z)d\xi\right]$$

$$\left(\because \int_{z}^{z+\Delta z} f(z)d\xi = f(z)\Delta z\right)$$

$$=\frac{1}{\Delta z}\int_{z}^{z+\Delta z}[f(\xi)-f(z)]d\xi$$

$$\therefore \quad \left|\frac{F(z+\Delta z)-F(z)}{\Delta z}-f(z)\right|=\frac{1}{|\Delta z|}\left|\int_{z}^{z+\Delta z}[f(\xi)-f(z)]d\xi\right|$$

$$\le \frac{1}{|\Delta z|}\int_{z}^{z+\Delta z}|f(\xi)-f(z)|\,|d\xi|$$

$$<\frac{1}{|\Delta z|}\varepsilon\,|\Delta z|=\varepsilon \qquad \text{(from 1.28)}$$

$$\therefore \quad \lim_{\Delta z\to 0}\left[\frac{F(z+\Delta z)-F(z)}{\Delta z}-f(z)\right]=0$$

$$\therefore \quad \lim_{\Delta z\to 0}\frac{F(z+\Delta z)-F(z)}{\Delta z}=f(z)$$

$$\therefore \quad F'(z)=f(z)$$

Since z is an arbitrary point in D

$\therefore$ $F(z)$ is analytic in D and $F'(z)=f(z)$.

Thus, $F(z)$ is indefinite integral of $f(z)$.

$\therefore\ F(z)=\int f(z)dz$

Let $H(z)$ be another definite integral of $f(z)$.

Then, $H'(z)=f(z)$

$\therefore \quad F'(z)=H'(z)$

$$\therefore \qquad F(z) = H(z) + C = \int_{z_0}^{z} f(z)dz \tag{1.29}$$

Take limit as $z \to z_0$

$$F(z_0) = H(z_0) + C = 0$$

$$\therefore \qquad C = -H(z_0)$$

$$\therefore \qquad F(z) = H(z) - H(z_0) = \int_{z_0}^{z} f(z)dz$$

Take limit as $z \to z_1$

$$H(z_1) - H(z_0) = \int_{z_0}^{z_1} f(z)dz$$

$$\therefore \qquad \int_{z_0}^{z_1} f(z)dz = H(z_1) - H(z_0) = \big(F(z_1) - C\big) - \big(F(z_0) - C\big) \quad \text{(from 1.29)}$$

$$= F(z_1) - F(z_0)$$

1.7.6 Extension of Cauchy–Goursat Theorem for Multiply Connected Domains

Theorem 1.14 Let D be doubly connected domain with outer boundary curve C and inner boundary curve C_1. If $f(z)$ is analytic in D and over C and C_1 then $\int_C f(z)dz = \int_{C_1} f(z)dz.$

Proof: Giving a cut AB, we have simply connected domain as shown in Figure 1.11. Here orientation of C_1^* is opposite to orientation of C_1 as C_1^* is clockwise and C_1 is anticlockwise.

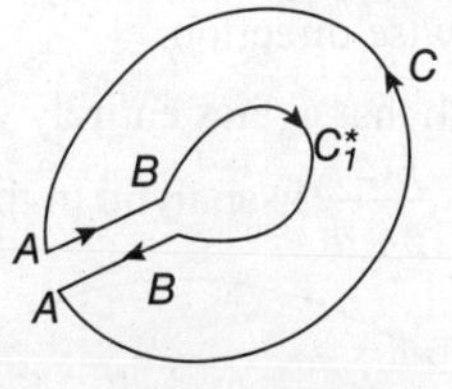

Figure 1.11

Thus, by Cauchy–Goursat theorem

$$\int_C f(z)dz + \int_{AB} f(z)dz + \int_{C_1^*} f(z)dz + \int_{BA} f(z)dz = 0$$

But $\qquad \int_{AB} f(z)dz = -\int_{BA} f(z)dz$

$$\therefore \qquad \int_C f(z)dz - \int_{C_1} f(z)dz = 0 \qquad (\because C_1^* = -C_1)$$

$$\therefore \qquad \int_C f(z)dz = \int_{C_1} f(z)dz$$

Remark 1.9: If domain D is multiply connected with outer boundary C and inner boundaries C_1, C_2, … C_n such that C_1, C_2, … C_n do not intersect and $f(z)$ is analytic in D and C, C_1, … C_n then

$$\int_C f(z)dz = \sum_{k=1}^{n} \int_{C_k} f(z)dz$$

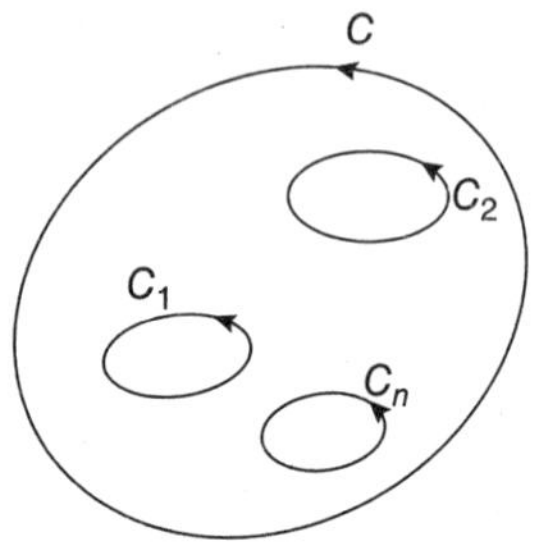

Figure 1.12

1.8 CAUCHY INTEGRAL FORMULA

Theorem 1.15 Let $f(z)$ be analytic in a simply connected domain D and 'a' is any point in D and C be any closed curve in D enclosing 'a' then

$$f(a) = \frac{1}{2\pi i}\int_C \frac{f(z)}{z-a}dz$$

where C is traversed in anti-clockwise direction.

Proof: Let C_1 be circle $|z-a| = r$ such that C_1 lies entirely within C. Then the region D_1 bounded by C and C_1 is doubly connected and $\dfrac{f(z)}{z-a}$ is analytic in this region and on C and C_1.

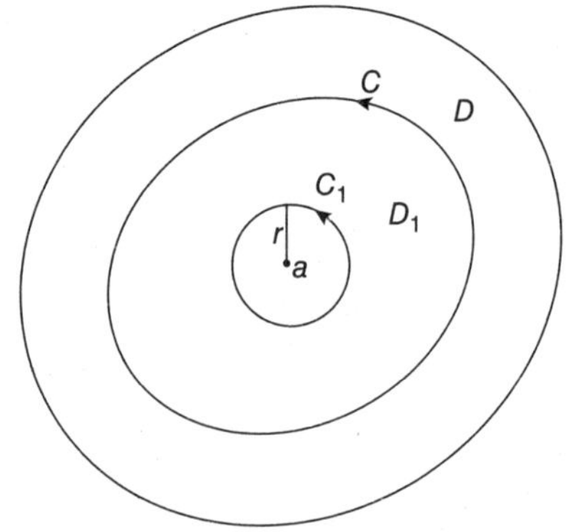

Figure 1.13

∴ By Cauchy–Goursat theorem for doubly connected regions

$$\int_C \frac{f(z)}{z-a}dz = \int_{C_1} \frac{f(z)}{z-a}dz = \int_{C_1} \frac{f(z)-f(a)}{z-a}dz + f(a)\int_{C_1} \frac{dz}{z-a}$$

$$= I_1 + f(a)\int_0^{2\pi} \frac{rie^{i\theta}d\theta}{re^{i\theta}} \qquad \left(\because \text{ On } C_1,\ z-a = re^{i\theta};\ 0 \le \theta < 2\pi\right)$$

$$= 2\pi i\ f(a) + I_1 \tag{1.30}$$

where $$I_1 = \int_{C_1} \frac{f(z)-f(a)}{z-a}dz$$

Now, $f(z)$ is analytic at $z = a$ and hence continuous at $z = a$.

∴ $$\lim_{z\to a} f(z) = f(a)$$

∴ Corresponding to $\varepsilon > 0$, however small there exist $\delta > 0$ such that $|f(z) - f(a)| < \varepsilon$ when $|z-a| < \delta$.

We take r such that $r < \delta$, then

$$|f(z) - f(a)| < \varepsilon \text{ when } |z-a| = r$$

∴ $$|I_1| \le \int_{C_1} \frac{|f(z)-f(a)|}{|z-a|}|dz| < \varepsilon \cdot \frac{1}{r} 2\pi r = 2\pi\varepsilon \qquad \left(\because \int_{C_1} |dz| = 2\pi r\right)$$

But ε is arbitrary, therefore $I_1 = 0$.

∴ From equation (1.30) $f(a) = \frac{1}{2\pi i}\int_C \frac{f(z)}{z-a}dz$

1.8.1 Cauchy Integral Formula for Derivatives of Analytic Function

Using Cauchy integral formula, we shall establish that if $f(z)$ is analytic in domain D, then its derivatives of all orders exist and are also analytic in D. Such a result does not exist for functions of real variables.

Theorem 1.16 Let $f(z)$ be analytic in a simply connected domain D. Let 'a' be any point in D and C be any simple closed curve in D enclosing point $z = a$. Then, $f(z)$ has derivatives of all order in D which are also analytic in D. Further,

$$f^{(n)}(a) = \frac{n!}{2\pi i}\int_C \frac{f(z)}{(z-a)^{n+1}}dz;\ n = 1, 2, 3, \cdots$$

Proof: $z = a$ lies inside C. For $|\Delta a|$ sufficiently small, $a + \Delta a$ will lie within C.

∴ By Cauchy integral formula

$$f(a) = \frac{1}{2\pi i}\int_C \frac{f(z)}{z-a}dz$$

and $$f(a+\Delta a) = \frac{1}{2\pi i}\int_C \frac{f(z)}{z-a-\Delta a}dz$$

$$\therefore \qquad \frac{f(a+\Delta a)-f(a)}{\Delta a} = \frac{1}{2\pi i\,\Delta a}\int_C \left(\frac{1}{z-a-\Delta a}-\frac{1}{z-a}\right) f(z)\,dz$$

$$= \frac{1}{2\pi i}\int_C \frac{f(z)}{(z-a)(z-a-\Delta a)}\,dz$$

$$= \frac{1}{2\pi i}\int_C \frac{(z-a-\Delta a)+\Delta a}{(z-a)^2(z-a-\Delta a)} f(z)\,dz$$

$$= \frac{1}{2\pi i}\int_C \left[\frac{1}{(z-a)^2}+\frac{\Delta a}{(z-a)^2(z-a-\Delta a)}\right] f(z)\,dz$$

$$= \frac{1}{2\pi i}\int_C \frac{f(z)}{(z-a)^2}\,dz + I_1 \qquad (1.31)$$

where $$I_1 = \frac{\Delta a}{2\pi i}\int_C \frac{f(z)}{(z-a)^2(z-a-\Delta a)}\,dz$$

Let minimum distance of any z on C from $z=a$ is d then

$$|z-a|^2 \geq d^2 \text{ and } |z-a-\Delta a| \geq |z-a|-|\Delta a| \geq d-|\Delta a|$$

and $f(z)$ is continuous at all points on C

$$\therefore \qquad |f(z)| \leq M \text{ for some } M$$

$$\therefore \qquad |I_1| = \frac{|\Delta a|}{2\pi}\left|\int_C \frac{f(z)}{(z-a)^2(z-a-\Delta a)}\,dz\right|$$

$$\leq \frac{|\Delta a|}{2\pi}\int_C \frac{|f(z)|}{|z-a|^2\,|z-a-\Delta a|}\,|dz|$$

$$\leq \frac{|\Delta a|}{2\pi}\frac{M}{d^2\left(d-|\Delta a|\right)}\int_C |dz| = \frac{|\Delta a|}{2\pi}\frac{M.L}{d^2\left(d-|\Delta a|\right)}$$

where L = length of closed curve C

$$\therefore \qquad \lim_{\Delta a\to 0} I_1 = 0$$

$\therefore$ from (1.31)

$$\lim_{\Delta a\to 0}\frac{f(a+\Delta a)-f(a)}{\Delta a} = \frac{1}{2\pi i}\int_C \frac{f(z)}{(z-a)^2}\,dz$$

$$\therefore \qquad f'(a) = \frac{1}{2\pi i}\int_C \frac{f(z)}{(z-a)^2}\,dz$$

But 'a' is any point in D

$\therefore f'(z)$ is analytic in D

$\therefore$ If $f(z)$ is analytic in D, then $f'(z)$ is analytic in D

and $$f'(a)=\frac{1}{2\pi i}\int_C \frac{f(z)}{(z-a)^2}dz \text{ for any } a \text{ in } D.$$

Now, replacing Cauchy integral formula for $f(a)$ by above expression for $f'(a)$ and repeating the above process, we shall have $f''(z)$ is analytic in D and

$$f''(a)=\frac{2!}{2\pi i}\int_C \frac{f(z)}{(z-a)^3}dz$$

Continuing this procedure, we shall have $f^{(n)}(z)$ is analytic in D for all $n\in N$ and

$$f^{(n)}(a)=\frac{n!}{2\pi i}\int_C \frac{f(z)}{(z-a)^{n+1}}dz$$

1.8.2 Morera's Theorem (Converse of Cauchy Integral Theorem)

Theorem 1.17 If $f(z)$ is continuous in a simply connected domain D and if $\int_C f(z)dz=0$ for every simple closed curve C in D then $f(z)$ is analytic in D.

Proof: Since $\int_C f(z)dz=0$ along any closed curve C in D and $f(z)$ is continuous at all point of C, therefore, the line integral of $f(z)$ from fixed point z_0 in D to any point z in D will be independent of path and hence will be single-valued function of z.

Let $$F(z)=\int_{z_0}^{z} f(z)dz.$$

After this, follow the proof of Theorem 1.13 and in the proof, we proved that $F(z)$ is analytic in D and $F'(z)=f(z)$.

But derivative of an analytic function is an analytic function and hence $f(z)$ is analytic in D.

1.8.3 Cauchy Inequality

If $f(z)$ is analytic within and on the circle $C:|z-a|=r$ then, $\left|f^{(n)}(a)\right|\le\frac{Mn!}{r^n}$ where $M=\max|f(z)|$ on C.

Proof: By Cauchy integral formula for nth order derivative

$$f^{(n)}(a)=\frac{n!}{2\pi i}\int_C \frac{f(z)}{(z-a)^{n+1}}dz$$

$$\therefore \quad \left|f^{(n)}(a)\right|=\frac{n!}{2\pi}\left|\int_C \frac{f(z)}{(z-a)^{n+1}}dz\right|$$

$$\therefore \qquad \left|f^{(n)}(a)\right| \le \frac{n!}{2\pi}\int_C \frac{|f(z)|}{|z-a|^{n+1}}|dz| \le \frac{Mn!}{2\pi r^{n+1}}\int_C |dz|$$

$$\therefore \qquad \left|f^{(n)}(a)\right| \le \frac{Mn!2\pi r}{2\pi r^{n+1}} = \frac{Mn!}{r^n}$$

$$\therefore \qquad \left|f^{(n)}(a)\right| \le \frac{Mn!}{r^n}$$

1.8.4 Liouville's Theorem

If an entire function $f(z)$ is bounded for all values of z, then $f(z)$ must be constant.

Proof: Since $f(z)$ is bounded for all z, so there exists M such that $|f(z)| \le M$ for all z.

By Cauchy integral formula for derivatives

$$f'(z) = \frac{1}{2\pi i}\int_C \frac{f(w)}{(w-z)^2}dw$$

where $C: |w-z| = r$

$$\therefore \qquad |f'(z)| = \frac{1}{2\pi}\left|\int_C \frac{f(w)}{(w-z)^2}dw\right| \le \frac{1}{2\pi}\int_C \frac{|f(w)|\,|dw|}{|w-z|^2}$$

$$\le \frac{1}{2\pi}\ \frac{M}{r^2}\int_C |dw| = \frac{1}{2\pi}\ \frac{M}{r^2}\cdot 2\pi r = \frac{M}{r}$$

$$\therefore \qquad |f'(z)| \le \frac{M}{r}$$

$r \to \infty$ gives $f'(z) = 0$ for all z

$\therefore \qquad f(z) = \text{constant}$

1.8.5 Poisson's Integral Formula

If $f(z)$ is analytic on and inside $C: |z| = R$ and $a = re^{i\theta}$ is any point within C then

$$f\left(re^{i\theta}\right) = \frac{1}{2\pi}\int_0^{2\pi} \frac{R^2 - r^2}{R^2 - 2Rr\cos(\theta - \phi) + r^2} f\left(\mathrm{R}e^{i\phi}\right)d\phi$$

Proof: By Cauchy integral formula

$$f(a) = \frac{1}{2\pi i}\int_C \frac{f(z)}{z-a}dz \qquad \text{where } z = \mathrm{R}e^{i\phi},\ a = re^{i\theta} \tag{1.32}$$

Now 'a' lies inside C, hence its inverse point w. r. t. circle C, i.e., $\dfrac{R^2}{\bar{a}}$ lies outside C.

∴ By Cauchy–Goursat theorem

$$0 = \frac{1}{2\pi i}\int_C \frac{f(z)}{z-\left(R^2/\bar{a}\right)}dz \tag{1.33}$$

Subtract (1.33) from (1.32)

$$f(a) = \frac{1}{2\pi i}\int_C \left(\frac{1}{z-a} - \frac{1}{z-\left(R^2/\bar{a}\right)}\right) f(z)\, dz$$

$$= \frac{1}{2\pi i}\int_C \left(\frac{1}{z-a} - \frac{\bar{a}}{z\bar{a}-R^2}\right) f(z)\,dz$$

$$= \frac{1}{2\pi i}\int_C \frac{a\bar{a}-R^2}{z^2\bar{a}-\left(a\bar{a}+R^2\right)z+R^2 a} f(z)\,dz$$

$$\therefore \quad f\left(re^{i\theta}\right) = \frac{1}{2\pi i}\int_0^{2\pi} \frac{\left(r^2-R^2\right) f\left(\mathrm{Re}^{i\phi}\right) \mathrm{R}\, i\, \mathrm{e}^{i\phi}\, d\phi}{R^2 e^{2i\phi} re^{-i\theta} - \left(r^2+R^2\right) Re^{i\phi} + R^2 re^{i\theta}} \qquad \left(\because\ z = \mathrm{Re}^{i\phi};\ 0 \le \theta < 2\pi \text{ on } C\right)$$

$$= \frac{1}{2\pi}\int_0^{2\pi} \frac{\left(r^2-R^2\right) f\left(\mathrm{Re}^{i\phi}\right) d\phi}{Rre^{i(\phi-\theta)} - \left(r^2+R^2\right) + Rre^{-i(\phi-\theta)}}$$

$$= \frac{1}{2\pi}\int_0^{2\pi} \frac{\left(R^2-r^2\right) f\left(\mathrm{Re}^{i\phi}\right) d\phi}{R^2 - 2rR\cos(\phi-\theta) + r^2}$$

Remark 1.10: If we write $f\left(re^{i\theta}\right) = u(r,\theta) + iv(r,\theta)$ and $f\left(\mathrm{Re}^{i\phi}\right) = U(R,\phi) + iV(R,\phi)$ and equate real and imaginary parts

$$u(r,\theta) = \frac{1}{2\pi}\int_0^{2\pi} \frac{\left(R^2-r^2\right) U(R,\phi)}{R^2 - 2rR\cos(\phi-\theta) + r^2} d\phi$$

and

$$v(r,\theta) = \frac{1}{2\pi}\int_0^{2\pi} \frac{\left(R^2-r^2\right) V(R,\phi)}{R^2 - 2rR\cos(\phi-\theta) + r^2} d\phi$$

Example 1.30: Evaluate the following integrals

(i) $\int_0^{3+i} z^2\, dz$ along the parabola $y^2 = x/3$.

(ii) $\int_C z^2\, dz$ where C is the arc of the circle $|z| = 2$ from $\theta = 0$ to $\theta = \frac{\pi}{3}$.

(iii) $\int_C \left(z - z^2\right) dz$ where C is the upper half of the circle $|z| = 1$.

(iv) $\int_C \left(3z^2 + 4z + 1\right) dz$ where C is the arc of the cycloid $x = a\,(\theta + \sin\theta)$, $y = a\,(1 - \cos\theta)$ between

(a) (0, 0) and (2 π*a*, 0) (b) (0, 0) and (π*a*, 2*a*).

Solution:

(i) Since z^2 is analytic everywhere, therefore $\int_0^{3+i} z^2 dz$ is independent of path and depends upon end points 0 and $3 + i$ only. Therefore, by fundamental theorem of integral calculus

$$\int_0^{3+i} z^2 dz = \left(\frac{z^3}{3}\right)_0^{3+i} = \frac{(3+i)^3}{3} = \frac{27+27i-9-i}{3} = 6 + \frac{26i}{3}$$

(ii) $|z| = 2 \Rightarrow z = 2e^{i\theta}$

$\therefore$ By fundamental theorem of integral calculus

$$\int_C z^2 dz = \int_2^{2e^{i\pi/3}} z^2 dz = \left(\frac{z^3}{3}\right)_2^{2e^{i\pi/3}} = \frac{8e^{i\pi}-8}{3} = \frac{-8-8}{3} = \frac{-16}{3}$$

(iii) Since $z - z^2$ is analytic everywhere, therefore by fundamental theorem of integral calculus

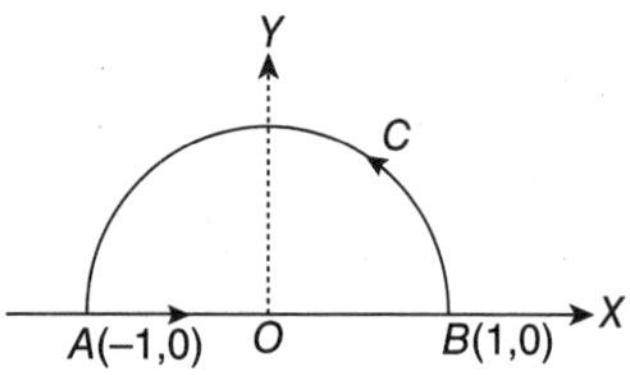

Figure 1.14

$$\int_C (z - z^2)\, dz = \int_1^{-1} (z - z^2)\, dz = \int_{-1}^{1} (z^2 - z)\, dz$$

$$= 2\int_0^1 z^2 dz \quad (\because z^2 \text{ is even and } z \text{ is odd})$$

$$= 2\left(\frac{z^3}{3}\right)_0^1 = \frac{2}{3}$$

(iv) Since $3z^2 + 4z + 1$ is analytic everywhere, therefore by fundamental theorem of integral calculus

(a) $$\int_C (3z^2 + 4z + 1)\, dz = \int_{0+i0}^{2\pi a + i0} (3z^2 + 4z + 1)\, dz$$

$$= \left(z^3 + 2z^2 + z\right)_0^{2\pi a} = \left\{z(z+1)^2\right\}_0^{2\pi a} = 2\pi a(2\pi a + 1)^2$$

(b) $$\int_C (3z^2 + 4z + 1)\, dz = \int_{0+i0}^{\pi a + 2ia} (3z^2 + 4z + 1)\, dz$$

$$= \left\{z(z+1)^2\right\}_0^{(\pi+2i)a} = (\pi + 2i)a\ \{(\pi + 2i)a + 1\}^2$$

Example 1.31: Prove that $\int_C \frac{dz}{z-a} = 2\pi i$ where C: $|z - a| = r$

Solution: Let $f(z) = 1$. Then $f(z)$ is analytic on and inside $C : |z - a| = r$ and $z = a$ lies inside C.

$\therefore$ By Cauchy integral formula $f(a) = \frac{1}{2\pi i}\int_C \frac{f(z)}{z-a}dz$

But
$$f(z) = f(a) = 1$$

$\therefore$
$$\int_C \frac{dz}{z-a} = 2\pi i$$

Example 1.32: Evaluate $\oint_C \frac{e^{-z}}{z+1}dz$ where C is (i) $|z| = 2$ (ii) $|z| = \frac{1}{2}$.

Solution: (i) $f(z) = e^{-z}$ is analytic on and inside $|z| = 2$ and $z = -1$ lies inside $|z| = 2$.

$\therefore$ By Cauchy integral formula

$$\oint_C \frac{e^{-z}}{z+1}dz = 2\pi i . f(-1) = 2\pi i e$$

(ii) $\frac{e^{-z}}{z+1}$ is analytic on and inside $|z| = \frac{1}{2}$. $\left(\because z = -1 \text{ lies outside} |z| = \frac{1}{2} \right)$

$\therefore$ By Cauchy–Goursat theorem $\oint_C \frac{e^{-z}}{(z+1)}dz = 0$

Example 1.33: State and prove Cauchy's integral formula and hence find the value of

(i) $F(3.5)$ (ii) $F(i)$ (iii) $F'(-1)$ and $F''(-i)$, if

$$F(a) = \int_C \frac{4z^2 + z + 5}{z-a}dz \text{ where } C \text{ is the ellipse } \left(\frac{x}{2}\right)^2 + \left(\frac{y}{3}\right)^2 = 1$$

Solution: For statement and proof of Cauchy integral formula see (1.8).

Let
$$f(z) = 4z^2 + z + 5$$

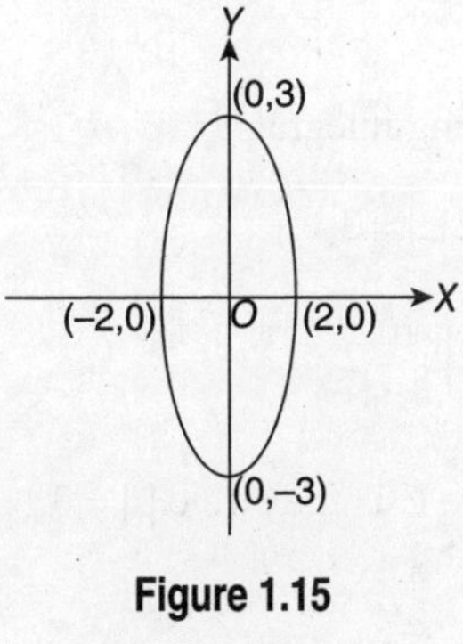

Figure 1.15

$f(z)$ is analytic on and inside C where C is the ellipse $\left(\frac{x}{2}\right)^2 + \left(\frac{y}{3}\right)^2 = 1$

Now, i, -1 and $-i$ lie within C and $z = 3.5$ lies outside C.

(i) $F(3.5)=\int_C \frac{4z^2+z+5}{z-3.5}$ is analytic on and inside C.

$\therefore$ By Cauchy–Goursat theorem $F(3.5)=0$

Now, when a lies inside C then $F(a)=2\pi i\, f(a)=\int_C \frac{f(z)}{(z-a)}dz$

$\therefore$ $$F(a)=2\pi i\left(4a^2+a+5\right)$$

$\therefore$ $$F'(a)=2\pi i(8a+1) \text{ and } F''(a)=2\pi i(8)=16\pi i$$

(ii) $F(i) = 2\pi i\left[4(i)^2+i+5\right]=2\pi i(1+i)=2\pi(i-1)$

(iii) $F'(-1)=2\pi i(-8+1)=-14\pi i$

(iv) $F''(-i)=16\pi i$

Example 1.34: Evaluate $f(2)$ and $f(3)$ if $f(a)=\oint_C \frac{2z^2-z-2}{z-a}dz$ where $C:|z|=2.5$.

Solution: $F(z)=2z^2-z-2$ is analytic on and inside $C:|z|=2.5$.

$\therefore$ if a lies within C then by Cauchy integral formula

$$\int_C \frac{F(z)}{z-a}dz=2\pi i\, F(a)=f(a)$$

Now, $z=2$ lies inside C.

$\therefore$ $$f(2)=2\pi i\, F(2)=2\pi i\,(8-2-2)=8\pi i$$

Now, $z=3$ lies outside C.

$\therefore$ $\frac{2z^2-z-2}{z-3}$ is analytic on and inside C.

$\therefore$ By Cauchy–Goursat theorem

$$f(3)=\oint_C \frac{2z^2-z-2}{z-3}dz=0$$

Example 1.35: Evaluate the following integrals:

(i) $\int_C \frac{e^{2z}}{\left(z^2-3z+2\right)}dz$ where $C:|z|=3$

(ii) $\int_C \frac{e^z}{(z-1)(z-4)}dz$ where $C:|z|=2$

(iii) $\int_C \frac{\sin\pi z^2+\cos\pi z^2}{(z-1)(z-2)}dz$ where C is the circle $|z|=3$.

Solution: (i) $\int_C \frac{e^{2z}}{z^2-3z+2}dz=\int_C \frac{e^{2z}}{(z-1)(z-2)}dz$

$$=\int_C \frac{e^{2z}}{z-2}dz-\int_C \frac{e^{2z}}{z-1}dz \qquad (1)$$

$f(z) = e^{2z}$ is analytic on and inside $C: |z| = 3$ and $z = 1, 2$ lie inside C.

$\therefore$ By Cauchy integral formula

$$\int_C \frac{e^{2z}}{(z-2)} dz = 2\pi i \ f(2) = 2\pi i e^4$$

$$\int_C \frac{e^{2z}}{(z-1)} dz = 2\pi i \ f(1) = 2\pi i e^2$$

$\therefore$ From (1)

$$\int_C \frac{e^{2z}}{z^2 - 3z + 2} dz = 2\pi i e^4 - 2\pi i e^2 = 2\pi i \left(e^2 - 1\right) e^2$$

(ii) $f(z) = \dfrac{e^z}{(z-4)}$ is analytic on and inside $C: |z| = 2$ ($\because$ $z = 4$ lies outside C) and $z = 1$ lies inside C.

$\therefore$ By Cauchy integral formula

$$\int_C \frac{e^z dz}{(z-1)(z-4)} = \int_C \frac{f(z)}{(z-1)} dz = 2\pi i \, f(1) = 2\pi i \frac{e}{1-4} = \frac{-2\pi i e}{3}$$

(iii)
$$\frac{1}{(z-1)(z-2)} = \frac{1}{z-2} - \frac{1}{z-1}$$

$$\therefore \quad \oint_C \frac{\sin \pi z^2 + \cos \pi z^2}{(z-1)(z-2)} dz = \oint_C \frac{\sin \pi z^2 + \cos \pi z^2}{z-2} dz - \oint_C \frac{\sin \pi z^2 + \cos \pi z^2}{z-1} dz$$

Now, $f(z) = \sin \pi z^2 + \cos \pi z^2$ is analytic on and inside $C: |z| = 3$.
and $z = 1, 2$ lie inside C.

$\therefore$ By Cauchy integral formula

$$\oint_C \frac{\sin \pi z^2 + \cos \pi z^2}{z-2} dz = 2\pi i \ f(2) = 2\pi i [\sin 4\pi + \cos 4\pi] = 2\pi i$$

$$\oint_C \frac{\sin \pi z^2 + \cos \pi z^2}{z-1} dz = 2\pi i \ f(1) = 2\pi i [\sin \pi + \cos \pi] = -2\pi i$$

$\therefore$ from (1)

$$\oint_C \frac{\sin \pi z^2 + \cos \pi z^2}{(z-1)(z-2)} dz = 2\pi i - (-2\pi i) = 4\pi i$$

Example 1.36: Evaluate $\displaystyle\oint_C \frac{(z-3)}{\left(z^2 + 2z + 5\right)} dz$ where

(i) $C: |z| = 1$ (ii) $C: |z + 1 + i| = 2$

Solution: $z^2 + 2z + 5 = 0$

iff $z = \frac{-2 \pm \sqrt{4-20}}{2} = -1 \pm 2i$

Now, $|-1 \pm 2i| > 1$

$\therefore$ $-1 \pm 2i$ lie outside $|z| = 1$

and $|-1 + 2i + 1 + i| = 3 > 2,$

$|-1 - 2i + 1 + i| = 1 < 2$

$\therefore$ $-1 + 2i$ lies outside $|z+1+i| = 2$ and $-1 - 2i$ lies inside $|z+1+i| = 2$

(i) $\frac{z-3}{z^2+2z+5}$ is analytic on and inside $C: |z|=1$

$\therefore$ By Cauchy–Goursat theorem $\int_C \frac{z-3}{z^2+2z+5} dz = 0$

(ii) $f(z) = \frac{z-3}{z-(-1+2i)}$ is analytic on and inside $C: |z+1+i| = 2$ and $-1-2i$ lies inside C.

$\therefore$ By Cauchy integral formula

$$\oint_C \frac{z-3}{z^2+2z+5} dz = \oint_C \frac{f(z)}{z-(-1-2i)} dz = 2\pi i f(-1-2i)$$

$$= 2\pi i \left\{ \frac{-1-2i-3}{-1-2i-(-1+2i)} \right\} = \frac{2\pi i(-4-2i)}{-4i}$$

$$= \pi(2+i)$$

Example 1.37: Using Cauchy's integral formula evaluate $\int_C \frac{z^4 dz}{(z+1)(z-i)^2}$ where C is the ellipse $9x^2 + 4y^2 = 36$.

Solution: C is ellipse $\left(\frac{x}{2}\right)^2 + \left(\frac{y}{3}\right)^2 = 1$

Both $z = -1, i$ lie inside C.

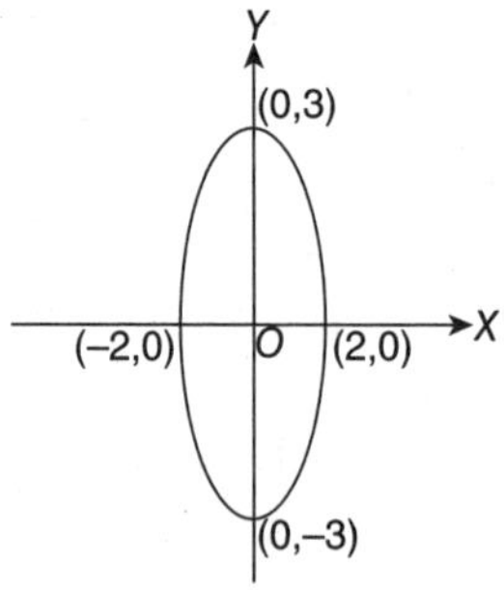

Figure 1.16

By suppression method

$$\frac{1}{(z+1)(z-i)^2} \equiv \frac{1}{(z+1)(-1-i)^2} + \frac{1}{(i+1)(z-i)^2} + \frac{A}{z-i}$$

$$= \frac{-\frac{i}{2}}{z+1} + \frac{\frac{1-i}{2}}{(z-i)^2} + \frac{A}{(z-i)}$$

$$\therefore \qquad 1 \equiv -\frac{1}{2}i\,(z-i)^2 + \frac{1-i}{2}(z+1) + A(z+1)(z-i)$$

Equate coefficient of z^2

$$0 = -\frac{1}{2}i + A \Rightarrow A = \frac{i}{2}$$

$$\therefore \qquad \frac{1}{(z+1)(z-i)^2} = \frac{-i}{2(z+1)} + \frac{i}{2(z-i)} + \frac{1-i}{2(z-i)^2}$$

Now by Cauchy integral formula

$$\int_C \frac{z^4}{z+1}dz = 2\pi i\,(-1)^4 = 2\pi i$$

$$\int_C \frac{z^4}{z-i}dz = 2\pi i\,(i)^4 = 2\pi i$$

and by Cauchy integral formula for derivatives

$$\int_C \frac{z^4}{(z-i)^2}dz = \frac{2\pi i}{1!}\left\{\frac{d}{dz}(z^4)\right\}_{z=i} = 2\pi i(4i^3) = 8\pi$$

$$\therefore \qquad \int_C \frac{z^4 dz}{(z+1)(z-i)^2} = \frac{-i}{2}\int_C \frac{z^4 dz}{(z+1)} + \frac{i}{2}\int_C \frac{z^4}{(z-i)}dz + \frac{(1-i)}{2}\int_C \frac{z^4}{(z-i)^2}dz$$

$$= \frac{-i}{2}(2\pi i) + \frac{i}{2}(2\pi i) + \frac{1-i}{2}(8\pi)$$

$$= 4\pi(1-i)$$

Example 1.38: Use Cauchy's integral formula to evaluate $\int_C \frac{e^{2z}}{(z+1)^4}dz$ where C is the circle $|z| = 2$.

Solution: $f(z) = e^{2z}$ is analytic on and inside $C: |z| = 2$ and $z = -1$ lies inside C.

$\therefore$ By Cauchy integral formula for derivatives

$$\oint_C \frac{e^{2z}}{(z+1)^4}dz = \frac{2\pi i}{3!}\left(\frac{d^3}{dz^3}e^{2z}\right)_{z=-1} = \frac{\pi i}{3}\left(8e^{2z}\right)_{z=-1} = \frac{8\pi i}{3e^2}$$

Example 1.39: Evaluate $\int_C \frac{e^z dz}{z(1-z)^3}$ where C is

(i) $|z| = \frac{1}{2}$ (ii) $|z-1| = \frac{1}{2}$ (iii) $|z| = 2$

Solution:

(i) $z = 0$ lies inside and $z = 1$ lies outside $C : |z| = \frac{1}{2}$ and $f(z) = \frac{e^z}{(1-z)^3}$ is analytic on and inside C.

$\therefore$ By Cauchy integral formula

$$\int_C \frac{e^z}{z(1-z)^3} dz = 2\pi i\, f(0) = 2\pi i$$

(ii) $z = 0$ lies outside $C : |z-1| = \frac{1}{2}$ and $z = 1$ lies inside C and $f(z) = \frac{-e^z}{z}$ is analytic on and inside C.

$\therefore$ By Cauchy integral formula for derivatives

$$\int_C \frac{e^z}{z(1-z)^3} dz = \int_C \frac{f(z)}{(z-1)^3} dz = \frac{2\pi i}{2!} \left\{ \frac{d^2}{dz^2} f(z) \right\}_{z=1}$$

$$= \pi i \left\{ \frac{d^2}{dz^2} \left(\frac{-e^z}{z} \right) \right\}_{z=1}$$

$$= -\pi i \left\{ \frac{d}{dz} \left(\frac{e^z}{z} - \frac{e^z}{z^2} \right) \right\}_{z=1}$$

$$= -\pi i \left\{ \frac{e^z}{z} - \frac{2e^z}{z^2} + \frac{2e^z}{z^3} \right\}_{z=1}$$

$$= -\pi i(e - 2e + 2e) = -\pi e i$$

(iii) Let $\frac{1}{z(z-1)^3} \equiv \frac{1}{(z-1)^3} - \frac{1}{z} + \frac{A}{z-1} + \frac{B}{(z-1)^2}$ (By suppression method) (1)

$\therefore$ $1 = z - (z-1)^3 + Az(z-1)^2 + Bz(z-1)$ (2)

Equate coefficient of z^3

$$0 = -1 + A \Rightarrow A = 1$$

Substitute $z = -1$ in (2)

$$1 = -1 + 8 - 4 + 2B \Rightarrow B = -1$$

$$\therefore \quad \frac{1}{z(z-1)^3} = \frac{1}{(z-1)} - \frac{1}{z} - \frac{1}{(z-1)^2} + \frac{1}{(z-1)^3}$$

Both $z = 0$ and $z = 1$ lie inside $|z| = 2$ and $f(z) = -e^z$ is analytic on and inside C

$\therefore$ By Cauchy integral formula and Cauchy integral formula for derivatives

$$\int_C \frac{e^z}{z(1-z)^3}\,dz = \int_C \frac{f(z)}{(z-1)}\,dz - \int_C \frac{f(z)}{z}\,dz - \int_C \frac{f(z)}{(z-1)^2}\,dz + \int_C \frac{f(z)}{(z-1)^3}\,dz$$

$$= 2\pi i\, f(1) - 2\pi i\, f(0) - \frac{2\pi i}{1!}\left(\frac{d}{dz} f(z)\right)_{z=1} + \frac{2\pi i}{2!}\left(\frac{d^2}{dz^2} f(z)\right)_{z=1}$$

$$= -2\pi i e + 2\pi i + 2\pi i e - \pi i e = \pi i(2-e)$$

Example 1.40: Evaluate $\int_C \frac{z}{(z^2+1)}\,dz$ where C is

(i) $C: \left|z+\frac{1}{z}\right| = 2$ (ii) $C: |z+i| = 1$

Solution: $z^2+1 = (z+i)(z-i)$

(i) for $z = i$, $\left|z+\frac{1}{z}\right| = |i-i| = 0 < 2$

for $z = -i$, $\left|z+\frac{1}{z}\right| = |-i+i| = 0 < 2$

$\therefore$ Both $z = i, -i$ lie inside $\left|z+\frac{1}{z}\right| = 2$

$$\frac{z}{z^2+1} = \frac{z}{(z+i)(z-i)} = \frac{1}{2}\left(\frac{1}{z-i} + \frac{1}{z+i}\right) \text{ (By suppression method)}$$

$\therefore$ By Cauchy integral formula

$$\int_C \frac{z}{z^2+1}\,dz = \frac{1}{2}\left[\int_C \frac{1}{z-i}\,dz + \int_C \frac{1}{z+i}\,dz\right] = \frac{1}{2}(2\pi i + 2\pi i) = 2\pi i$$

(ii) for $z = i$, $|z+i| = |2i| = 2 > 1$

and for $z = -i$, $|z+i| = |-i+i| = 0 < 1$

$\therefore$ $z = -i$, lies inside and $z = i$ outside $C: |z+i| = 1$

$$f(z) = \frac{z}{z-i} \text{ is analytic on and inside } C \text{ and } z = -i \text{ lies inside } C$$

$\therefore$ By Cauchy integral formula

$$\int_C \frac{z}{z^2+1}\,dz = \int_C \frac{f(z)}{z+i}\,dz = 2\pi i f(-i) = 2\pi i\left(\frac{-i}{-2i}\right) = \pi i$$

Example 1.41: Evaluate the following integral by Cauchy's integral formula

$\int_C \frac{\cos z}{z^{2n+1}}\,dz$ where $C: |z| = 1$

Solution: (i) $z=0$ lies inside $|z|=1$ and $f(z)=\cos z$ is analytic on and inside C.

$\therefore$ By Cauchy integral formula for derivatives

$$\int_C \frac{\cos z}{z^{2n+1}}dz = \frac{2\pi i}{(2n)!}\left(\frac{d^{2n}}{dz^{2n}}\cos z\right)_{z=0} = \frac{2\pi i}{(2n)!}\{(-1)^n \cos z\}_{z=0} = \frac{2\pi i(-1)^n}{(2n)!}$$

EXERCISE 1.2

1. Evaluate $\int_0^{1+i}(x^2+iy)dz$ along the path $y=x^2$.

2. If $f(z)=x^2+ixy$ evaluate $\int_A^B f(z)\,dz$ where A (1,1) and B (2,4) along
   (i) the straight line AB
   (ii) the curve $C: \quad x=t, \quad y=t^2$

3. Find the value of the integral $\int_C (x+y)dx+x^2y\,dy$
   (i) along $y=x^2$ having (0, 0), (3, 9) as end points.
   (ii) along $y=3x$ between the same points. Do the values depend upon path?

4. Evaluate $\int_{(0,3)}^{(2,4)}\left[(2y+x^2)dx+(3x-y)dy\right]$ along the parabola $x=2t, y=t^2+3$.

5. Evaluate $\int_{(0,0)}^{(1,1)}(3x^2+4xy+3y^2)dx+$

   $$2(x^2+3xy+4y^2)dy$$

   (i) along $y^2=x$ (ii) along $y=x^2$
   (iii) along $y=x$

6. Evaluate $\int_A^B z^2dz$ where $A=(1, 1)$, $B=(2, 4)$ along
   (i) the line segment AC parallel to x-axis and CB parallel to y-axis.
   (ii) the straight line AB joining the two points A and B.
   (iii) the curve $C \ : \ y=x^2$.

7. Evaluate $\int_0^{2+i}(\bar{z})^2\,dz$ along
   (i) the real axis to 2 and then vertically to $2+i$.
   (ii) along the line $2y=x$

8. Evaluate the integral $I=\int_C \text{Re}\left(z^2\right)dz$ from 0 to $2+4i$ along the
   (i) line segment joining the points (0, 0) and (2, 4),
   (ii) x-axis from 0 to 2, and then vertically to $2+4i$ and
   (iii) parabola $y=x^2$.

9. Evaluate $I=\oint_C \frac{dz}{z-2}$ around
   (i) circle $|z-2|=4$
   (ii) Rectangle with vertices at $3\pm 2i, -2\pm 2i$
   (iii) Triangle with vertices at (0, 0), (1, 0), (0, 1).

10. Evaluate $\oint_C |z|^2\,dz$ around the square with vertices at (0, 0), (1, 0), (1, 1) (0, 1).

11. Show that $\int_C \frac{dz}{z}=-\pi i$ or πi according as C is the semi-circular arc of $|z|=1$ above or below the x-axis.

12. Evaluate $\int_C \frac{1}{z^3}dz, C:|z|=1.$

13. Evaluate $\int_C z^2dz$ using Cauchy's integral theorem, where $C:|z|=1$.

14. Evaluate $\oint_C (5z^4 - z^3 + 2)\,dz$ around
   (i) unit circle $|z| = 1$
   (ii) square with vertices at (0, 0), (1, 0), (1, 1), (1, 0)
   (iii) curve consisting of the parabola $y = x^2$ from (0, 0) to (1, 1) and $y^2 = x$ from (1, 1) to (0, 0).
15. Can the Cauchy integral theorem be applied for evaluating the following integrals? Hence, evaluate these integrals.
   (i) $\oint_C e^{\sin z^2}\,dz;\ C : |z| = 1$
   (ii) $\oint_C \tan z\,dz,\ C : |z| = 1$
   (iii) $\oint_C \frac{e^z}{(z^2+9)}\,dz,\ C : |z| = 2$
   (iv) $\oint_C \frac{dz}{(z^3-1)}$, C is a triangle with vertices at 0, $\pm\frac{1}{4}+\frac{i}{2}$.
16. Evaluate the following integrals
   (i) $\int_1^{2-i} z\,dz$
   (ii) $\int_{-\pi i}^{\pi i} e^{2z}\,dz$
   (iii) $\int_0^{i} \sinh \pi z\,dz$
   (iv) $\int_0^{1+2i} ze^z\,dz$
   (v) $\int_0^1 z^2 e^{z^3}\,dz.$
17. Evaluate $\oint_C \frac{dz}{z^2+9}$ where C is
   (i) $|z-3i| = 4$ (ii) $|z+3i| = 2$
   (iii) $|z| = 5$
18. If $f(a) = \int_C \frac{3z^2+7z+1}{z-a}\,dz$, where C is the circle $x^2 + y^2 = 4$, find the values of $f(3)$, $f'(1-i)$ and $f''(1-i)$.
19. Determine $F(2)$, $F(4)$, $F(-3i)$, $F'(i)$, $F''(-2i)$ if $F(\alpha) = \oint_C \frac{5z^2-4z+3}{z-\alpha}\,dz$ where C is the ellipse $16x^2 + 9y^2 = 144$
20. Evaluate the following integrals by using Cauchy's integral formula
   (i) $\int_C \frac{dz}{z^2}$ where C is the circle $|z| = 1$
   (ii) $\int_C \left\{\frac{2z}{(z-2)^2}+\frac{3(z-1)}{(z-2)^3}\right\} dz$ where C is the circle $|z| = 3$
   (iii) $\int_C \frac{dz}{(z-a)^n}$ where n is any positive integer greater than 1 and C is the closed curve containing a.
   (iv) $\int_C \frac{z-1}{(z+1)^2(z-2)}\,dz$ where $C : |z-i| = 2$
   (v) $\int_C \frac{2z^2+z}{(z^2-1)}\,dz;\ C : |z-1| = 1$
   (vi) $\frac{1}{2\pi i}\int_C \frac{ze^z}{(z-a)^3}\,dz$ if the point a lies inside the simple closed curve C.
21. Evaluate by Cauchy's integral formula
   (i) $\int_C \frac{dz}{(z^2-1)}\,dz$ where C is the circle $x^2 + y^2 = 4$.
   (ii) $\int_C \frac{1}{z}\cos z\,dz;\ C : 9x^2 + 4y^2 = 1$
   (iii) $\frac{1}{2\pi i}\int_C \frac{z^2+7}{z-2}\,dz$ where $C : |z| = 5$
   (iv) $\int_C \frac{\sin 3z}{z+\pi/2}$ where $C : |z| = 5$
   (v) $\int_C \frac{z\,dz}{(9-z^2)(z+i)}$ where $C : |z| = 2$

(vi) $\int_C \frac{dz}{(z^2+4)}$ where $C: |z| = 4$

(vii) $\int_C \frac{\sin \pi z^2 + \cos \pi z^2}{(z+1)(z+2)} dz$ where $C: |z| = 3$

(viii) $\frac{1}{2\pi i} \int_C \frac{e^{zt}}{(z^2+1)} dz$ where $C: |z| = 3$ and $t > 0$

(ix) $\int_C \frac{\tan z}{(z-\pi/4)^2} dz$ where $C: |z| = 1$

ANSWERS 1.2

1. $(5i-1)/6$ 2. (i) $\frac{-29}{3} + 11i$ (ii) $\frac{-151}{15} + \frac{45}{4}i$

3. (i) $256\frac{1}{2}$ (ii) $200\frac{1}{4}$ 4. 33/2 5. (i) 26/3 (ii) 26/3 (iii) 26/3

6. (i) $\frac{-86}{3} - 6i$ (ii) $\frac{-86}{3} - 6i$ (iii) $\frac{-86}{3} - 6i$

7. (i) $\frac{14}{3} + \frac{11}{3}i$ (ii) $\frac{10}{3} - \frac{5i}{3}$

8. (i) $-8(1+2i)$ (ii) $\frac{8}{3}(1-2i)$ (iii) $\frac{-56}{15} - \frac{40}{3}i$ 9. (i) $2\pi i$ (ii) $2\pi i$ (iii) 0

10. $-1+i$ 12. 0 13. 0 14. (i), (ii), (iii) = 0

15. yes; (i), (ii), (iii), (iv) = 0

16. (i) $1-2i$ (ii) 0 (iii) $-2/\pi$ (iv) $2ie^{1+2i}+1$ (v) $\frac{1}{3}(e-1)$

17. (i) $\pi/3$ (ii) $-\pi/3$ (iii) 0 18. (i) 0 (ii) $2\pi(6+13i)$ (iii) $12\pi i$

19. $F(2) = 30\pi i$, $F(4) = 0$, $F(-3i) = -12\pi(2+7i)$, $F'(i) = -4\pi(5+2i)$, $F''(-2i) = 20\pi i$

20. (i) 0 (ii) $4\pi i$ (iii) 0 (iv) $-2\pi i/9$ (v) $3\pi i$ (vi) $e^a(1+a/2)$

21. (i) 0 (ii) $2\pi i$ (iii) 11 (iv) $2\pi i$ (v) $\pi/5$ (vi) 0 (vii) $-4\pi i$ (viii) $\sin t$ (ix) $4\pi i$

1.9 INFINITE SERIES OF COMPLEX TERMS

$\sum_{k=1}^{\infty}(a_k + ib_k)$; $a_k, b_k \in R$ is an infinite series of complex terms. If the series $\sum_{k=1}^{\infty} a_k$ and $\sum_{k=1}^{\infty} b_k$ converge to the sum A and B respectively then the series $\sum_{k=1}^{\infty}(a_k + ib_k)$ converges to the sum $A + iB$. Conversely if the series $\sum_{k=1}^{\infty}(a_k + ib_k)$ converges to $A + iB$ then $\sum_{k=1}^{\infty} a_k$ and $\sum_{k=1}^{\infty} b_k$ converge to A and B respectively. In (5.2) of vol 1, we have shown that if $\sum_{k=1}^{\infty} a_k$ and $\sum_{k=1}^{\infty} b_k$ converge then $\lim_{n\to\infty} a_n = 0$, $\lim_{n\to\infty} b_n = 0$. Thus, if series $\sum_{k=1}^{\infty}(a_k + ib_k)$ converges then $\lim_{n\to\infty}(a_n + ib_n) = 0$

The series $\sum_{k=1}^{\infty}(a_k + ib_k)$ is absolutely convergent iff $\sum_{k=1}^{\infty}|a_k + ib_k|$ is convergent.

Now, $|a_n| \le |a_n + ib_n|$ and $|b_n| \le |a_n + ib_n|$ for all $n \in N$.

$\therefore$ By comparision test in (5.3) of vol 1, if $\sum(a_n + ib_n)$ converges absolutely then $\sum a_n$ and $\sum b_n$ converge absolutely and thus $\sum a_n$ and $\sum b_n$ converge which implies $\sum(a_n + ib_n)$ converges.

Thus, an absolute convergent series of complex numbers is convergent.

Now, let the series $\sum_{k=1}^{\infty} u_k(z)$ converges to $s(z)$ and $\sum_{k=1}^{n} u_k(z) = s_n(z)$. If corresponding to $\varepsilon > 0$ there exists $m \in N$ depending on ε and not on z such that $|s_n(z) - s(z)| < \varepsilon$ for all $n > m$ then the series $\sum_{k=1}^{\infty} u_k(z)$ is said to be uniformly convergent.

A uniformly convergent series of continuous complex functions is itself continuous and can be differentiated or integrated term by term.

If a uniformly convergent series of continuous complex functions converges to $f(z)$ then $f(z)$ is an analytic function.

1.9.1 Power Series

An infinite series of the form

$$\sum a_n (z - z_0)^n = a_0 + a_1(z - z_0) + a_2(z - z_0)^2 + \cdots\cdots\cdots + a_n(z - z_0)^n + \cdots\cdots \tag{1.34}$$

is called a power series about $z = z_0$. The point z_0 is called the centre of the power series. If the power series converges for all z in $|z - z_0| < R$ and diverges for all z in $|z - z_0| > R$ then R is called radius of convergence of the power series.

The power series is always convergent at its centre $z = z_0$ since for $z = z_0$ series reduces to a_0.

If $\lambda = \lim_{n\to\infty}\left|\frac{a_{n+1}}{a_n}\right|$ then by De Alembert's ratio test proved in (5.4) of vol 1, the series converges for $|z - z_0|\lambda < 1$ and diverges for $|z - z_0|\lambda > 1$.

$\therefore$ Radius of convergence is $\frac{1}{\lambda}$

and series converges in $|z - z_0| < \frac{1}{\lambda}$

and diverges in $|z - z_0| > \frac{1}{\lambda}$.

Without proving, we state that the power series $\sum a_n (z-z_0)^n$ with non-zero radius of convergence R is uniformly convergent in the circle $|z-z_0| \le r < R$.

Thus, if the power series $\sum a_n (z-z_0)^n$ converges to function $f(z)$ then $f(z)$ is analytic function and also term by term differentiation and term by term integration are allowed within the circle of convergence.

1.10 TAYLOR SERIES

Theorem 1.18 If a function $f(z)$ is analytic inside a circle $C: |z-z_0| = R$ then for all z inside C, $f(z)$ can be represented by power series

$$f(z) = \sum_{n=1}^{\infty} a_n (z-z_0)^n \quad \text{where}$$

$$a_n = \frac{f^{(n)}(z_0)}{n!} = \frac{1}{2\pi i} \int_{C_r} \frac{f(w)}{(w-z_0)^{n+1}} dw$$

where C_r is circle $|w-z_0| = r < R$ containing z inside it.

Proof: Let z be an arbitrary point inside $C: |z-z_0| = R$.

Let C_r: $|w-z_0| = r < R$ be circle containing point z inside C_r and w is any point on C_r.

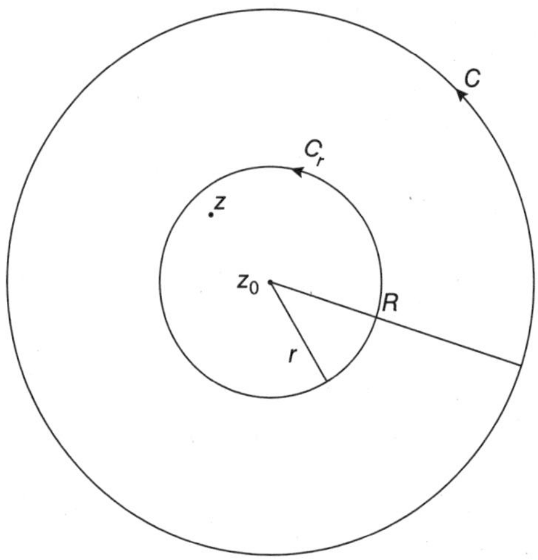

Figure 1.17

∴ By Cauchy integral formula

$$f(z) = \frac{1}{2\pi i} \oint_{C_r} \frac{f(w)}{w-z} dw \tag{1.35}$$

Now,
$$\frac{1}{w-z} = \frac{1}{(w-z_0)-(z-z_0)} = \frac{1}{(w-z_0)} \left[1 - \frac{z-z_0}{w-z_0}\right]^{-1} \tag{1.36}$$

Since $\left|\dfrac{z-z_0}{w-z_0}\right| = \dfrac{|z-z_0|}{|w-z_0|} < 1$

We can expand R.H.S. of equation (1.36) in binomial series

$$\therefore \qquad \frac{1}{w-z} = \frac{1}{(w-z_0)}\left[1+\frac{z-z_0}{w-z_0}+\left(\frac{z-z_0}{w-z_0}\right)^2+\left(\frac{z-z_0}{w-z_0}\right)^3+\cdots\cdots\right]$$

This series converges uniformly and hence multiplying by $f(w)$ and integrating term by term, we have

$$\frac{1}{2\pi i}\oint_{C_r}\frac{f(w)}{w-z}dw = \frac{1}{2\pi i}\left[\int_{C_r}\frac{f(w)}{(w-z_0)}dw+(z-z_0)\int_{C_r}\frac{f(w)}{(w-z_0)^2}dw+\cdots\cdots+(z-z_0)^n\int_{C_r}\frac{f(w)}{(w-z_0)^{n+1}}dw\right]$$

$\therefore$ Using Cauchy integral formula and Cauchy integral formula for derivatives, we have

$$f(z) = f(z_0)+(z-z_0)f'(z_0)+\frac{(z-z_0)^2}{2!}f''(z_0)+\cdots\cdots\cdots+\frac{(z-z_0)^n}{n!}f^n(z_0)+\cdots\cdots\cdots$$

or $\quad f(z) = a_0+a_1(z-z_0)+a_2(z-z_0)^2+\cdots\cdots\cdots+a_n(z-z_0)^n+\cdots\cdots\cdots$

where $a_n = \dfrac{f^{(n)}(z_0)}{n!} = \dfrac{1}{2\pi i}\displaystyle\int_{C_r}\frac{f(w)}{(w-z_0)^{n+1}}dw$

Remark 1.11: If we are to expand $f(z)$ about z_0 and $f(z)$ is not analytic at z_0 or in some neighbourhood of z_0 then Laurent's series is used.

1.11 LAURENT'S SERIES

Theorem 1.19 If a function $f(z)$ is analytic in the ring-shaped region R bounded by two concentric circles C and C_1 with radii r and r_1 $(r > r_1)$ and with centre at z_0 then for all z in R

$$f(z) = \sum_{n=0}^{\infty}a_n(z-z_0)^n+\sum_{n=1}^{\infty}\frac{a_{-n}}{(z-z_0)^n}$$

where
$$a_n = \frac{1}{2\pi i}\int_{C'}\frac{f(w)}{(w-z_0)^{n+1}}dw = \frac{f^{(n)}(z_0)}{n!};\ n = 1,2,3,\cdots\cdots\cdots$$

$$a_0 = \frac{1}{2\pi i}\int_{C'}\frac{f(w)}{(w-z_0)}dw = f(z_0)$$

$$a_{-n} = \frac{1}{2\pi i}\int_{C_1'}\frac{f(w)}{(w-z_0)^{-n+1}}dw;\ n = 1,2,3,\cdots\cdots\cdots$$

where C' and C_1' are circles in R containing z in the annular ring with inner circle C_1' and outer circle C'.

Proof: Let z be any point in R. Draw circles C' and C_1' such that z lies in annular ring bounded by inner circle C_1' and outer circle C' with centre z_0 and C_1', C' completely lie in R. Introduce cut AB such that AB does not pass through z as shown in Figure 1.18.

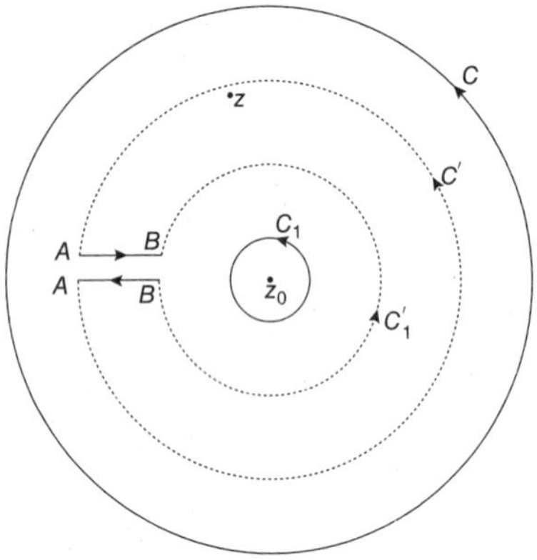

Figure 1.18

Then $f(z)$ is analytic in the simply connected domain bounded by AB, $C_1'^*$ (in clockwise direction), BA and C' including circles C' and C_1' where $C_1'^*$ is in opposite direction of C_1'.

∴ By Cauchy integral formula

$$f(z) = \frac{1}{2\pi i}\oint_{C'}\frac{f(w)}{w-z}dw - \frac{1}{2\pi i}\oint_{C_1'}\frac{f(w)}{w-z}dw \tag{1.37}$$

where integrals over AB and BA cancel and integral over $C_1'^*$ is negative of integral over C_1' as $C_1'^*$ is traversed clockwise and C_1' is traversed anticlockwise.

Now, when w lies on C'; $\left|\frac{z-z_0}{w-z_0}\right| < 1$

$$\therefore \qquad \frac{1}{w-z} = \frac{1}{w-z_0-(z-z_0)} = \frac{1}{(w-z_0)}\left(1-\frac{z-z_0}{w-z_0}\right)^{-1}$$

$$= \frac{1}{(w-z_0)}\sum_{n=0}^{\infty}\left(\frac{z-z_0}{w-z_0}\right)^n = \sum_{n=0}^{\infty}\frac{(z-z_0)^n}{(w-z_0)^{n+1}}$$

This series is uniformly convergent, hence after multiplying with $f(w)$ and term by term integration, we have

$$\oint_{C'}\frac{f(w)}{w-z}dw = \sum_{n=0}^{\infty}(z-z_0)^n\oint_{C'}\frac{f(w)}{(w-z_0)^{n+1}}dw \tag{1.38}$$

When w lies on C_1', $\left|\frac{w-z_0}{z-z_0}\right| < 1$

$$\therefore \qquad \frac{1}{w-z} = \frac{1}{w-z_0-(z-z_0)} = -\frac{1}{(z-z_0)}\left(1-\frac{w-z_0}{z-z_0}\right)^{-1}$$

$$= -\frac{1}{(z-z_0)}\sum_{n=0}^{\infty}\frac{(w-z_0)^n}{(z-z_0)^n} = -\sum_{n=0}^{\infty}\frac{(w-z_0)^n}{(z-z_0)^{n+1}}$$

This series is uniformly convergent, hence after multiplying with $f(w)$ and term by term integration, we have

$$\oint_{C_1'}\frac{f(w)}{w-z}dw = -\sum_{n=0}^{\infty}\frac{1}{(z-z_0)^{n+1}}\oint_{C_1'} f(w).(w-z_0)^n dw \tag{1.39}$$

Substituting from equations (1.38) and (1.39) in (1.37)

$$f(z) = \sum_{n=0}^{\infty}\frac{(z-z_0)^n}{2\pi i}\oint_{C'}\frac{f(w)}{(w-z_0)^{n+1}}dw + \sum_{n=0}^{\infty}\frac{1}{2\pi i(z-z_0)^{n+1}}\oint_{C_1'} f(w)(w-z_0)^n dw$$

$$= \sum_{n=0}^{\infty}a_n(z-z_0)^n + \sum_{n=1}^{\infty}\frac{a_{-n}}{(z-z_0)^n}$$

where $\quad a_n = \dfrac{1}{2\pi i}\displaystyle\int_{C'}\frac{f(w)}{(w-z_0)^{n+1}}dw = \frac{f^n(z_0)}{n!}; \quad n \in N$

$$a_0 = \frac{1}{2\pi i}\int_{C'}\frac{f(w)}{w-z_0}dw = f(z_0)$$

$$a_{-n} = \frac{1}{2\pi i}\int_{C_1'}\frac{f(w)}{(w-z_0)^{-n+1}}dw$$

Now, we shall give another proof of the theorem that every order derivative of an analytic function is an analytic function.

Theorem 1.20 If $f(z)$ is an analytic function at $z = a$ then mth order derivative of $f(z)$, i.e., $f^{(m)}(z)$ is analytic at $z = a$ for all $m \in N$.

Proof: $f(z)$ is analytic at $z = a$ and hence $f(z)$ can be expanded in Taylor series about $z = a$.

$$\therefore \qquad f(z) = \sum_{n=0}^{\infty}a_n(z-a)^n$$

This power series is uniformly convergent and circle of convergence is $|z-a| = \dfrac{1}{k}$

where $\qquad k = \lim\limits_{n\to\infty}\left|\dfrac{a_{n+1}}{a_n}\right|$

Let $m \in N$

$$\therefore \qquad f^{(m)}(z) = \sum_{n=m}^{\infty}a_n\frac{n!}{(n-m)!}(z-a)^{n-m}$$

This power series is also uniformly convergent and circle of convergence is $|z-a|=\frac{1}{k}$.
Hence, $f^{(m)}(z)$ is analytic at $z=a$.
But m is any natural number.
$\therefore f^{(m)}(z)$ is analytic at $z=a$ for all $m \in N$.

Example 1.42: Expand $f(z)=\dfrac{1}{z(z^2-3z+2)}$ in the region

(i) $0<|z|<1$ (ii) $1<|z|<2$ (iii) $|z|>2$

Solution: $f(z)=\dfrac{1}{z(z-1)(z-2)}=\dfrac{1}{2z}-\dfrac{1}{z-1}+\dfrac{1}{2(z-2)}$ (By suppression method)

(i) $$f(z)=\frac{1}{2z}+(1-z)^{-1}-\frac{1}{4}\left(1-\frac{z}{2}\right)^{-1}=\frac{1}{2z}+\sum_{n=0}^{\infty}z^n-\frac{1}{4}\sum_{n=0}^{\infty}\frac{z^n}{2^n} \qquad \left(\because |z|<1,\ \therefore \frac{|z|}{2}<1\right)$$

$$=\frac{1}{2z}+\sum_{n=0}^{\infty}\left(1-\frac{1}{2^{n+2}}\right)z^n;\ 0<|z|<1$$

(ii) $$f(z)=\frac{1}{2z}-\frac{1}{z}\left(1-\frac{1}{z}\right)^{-1}-\frac{1}{4}\left(1-\frac{z}{2}\right)^{-1}=\frac{1}{2z}-\frac{1}{z}\sum_{n=0}^{\infty}\frac{1}{z^n}-\frac{1}{4}\sum_{n=0}^{\infty}\frac{z^n}{2^n}$$

$$\left(\because 1<|z|<2,\ \therefore \frac{1}{|z|}<1, \frac{|z|}{2}<1\right)$$

$$=\frac{1}{2z}-\sum_{n=1}^{\infty}\frac{1}{z^{n+1}}-\sum_{n=0}^{\infty}\frac{z^n}{2^{n+2}}=\frac{1}{2z}-\sum_{n=2}^{\infty}\frac{1}{z^n}-\sum_{n=0}^{\infty}\frac{z^n}{2^{n+2}};\ 1<|z|<2$$

(iii) $$f(z)=\frac{1}{2z}-\frac{1}{z}\left(1-\frac{1}{z}\right)^{-1}+\frac{1}{2z}\left(1-\frac{2}{z}\right)^{-1}$$

$$=\frac{1}{2z}-\frac{1}{z}\sum_{n=0}^{\infty}\frac{1}{z^n}+\frac{1}{2z}\sum_{n=0}^{\infty}\frac{2^n}{z^n} \qquad \left(\because |z|>2,\ \therefore \frac{2}{|z|}<1, \frac{1}{|z|}<1\right)$$

$$=-\sum_{n=1}^{\infty}\frac{1}{z^{n+1}}+\sum_{n=1}^{\infty}\frac{2^{n-1}}{z^{n+1}}=\sum_{n=2}^{\infty}\frac{2^{n-1}-1}{z^{n+1}}=\sum_{n=3}^{\infty}\frac{2^{n-2}-1}{z^n};\ |z|>2$$

Example 1.43: Expand $f(z)=\dfrac{1}{(z+1)(z+3)}$ in Laurent series valid for

(i) $1<|z|<3$ (ii) $|z|>3$ (iii) $0<|z+1|<2$ (iv) $|z|<1$

Solution: $f(z)=\dfrac{1}{(z+1)(z+3)}=\dfrac{1}{2}\left(\dfrac{1}{z+1}-\dfrac{1}{z+3}\right)$ (By suppression method)

(i) $f(z)=\frac{1}{2}\left[\frac{1}{z}\left(1+\frac{1}{z}\right)^{-1}-\frac{1}{3}\left(1+\frac{z}{3}\right)^{-1}\right]$

$$=\frac{1}{2}\left[\frac{1}{z}\sum_{n=0}^{\infty}\frac{(-1)^n}{z^n}-\frac{1}{3}\sum_{n=0}^{\infty}\frac{(-1)^n z^n}{3^n}\right] \qquad \left(\because\ 1<|z|<3,\ \therefore \frac{1}{|z|}<1, \frac{|z|}{3}<1\right)$$

$$=\frac{1}{2}\left[\sum_{n=0}^{\infty}\frac{(-1)^n}{z^{n+1}}-\sum_{n=0}^{\infty}\frac{(-1)^n z^n}{3^{n+1}}\right]=\frac{1}{2}\left[\sum_{n=1}^{\infty}\frac{(-1)^{n-1}}{z^n}-\sum_{n=0}^{\infty}\frac{(-1)^n z^n}{3^{n+1}}\right];\ 1<|z|<3$$

(ii) $f(z)=\frac{1}{2}\left[\frac{1}{z}\left(1+\frac{1}{z}\right)^{-1}-\frac{1}{z}\left(1+\frac{3}{z}\right)^{-1}\right]$

$$=\frac{1}{2}\left[\frac{1}{z}\sum_{n=0}^{\infty}\frac{(-1)^n}{z^n}-\frac{1}{z}\sum_{n=0}^{\infty}\frac{(-1)^n 3^n}{z^n}\right] \qquad \left(\because\ |z|>3,\ \therefore \frac{3}{|z|}<1, \frac{1}{|z|}<1\right)$$

$$=\frac{-1}{2}\sum_{n=0}^{\infty}\frac{(-1)^n\left(3^n-1\right)}{z^{n+1}}=\frac{-1}{2}\sum_{n=1}^{\infty}\frac{(-1)^{n-1}\left(3^{n-1}-1\right)}{z^n};\ |z|>3$$

(iii) $f(z)=\frac{1}{2}\left[\frac{1}{z+1}-\frac{1}{z+1+2}\right]=\frac{1}{2}\left[\frac{1}{z+1}-\frac{1}{2}\left(1+\frac{z+1}{2}\right)^{-1}\right]$

$$=\frac{1}{2}\left[\frac{1}{z+1}-\frac{1}{2}\sum_{n=0}^{\infty}(-1)^n\frac{(z+1)^n}{2^n}\right] \qquad \left(\because\ 0<|z+1|<2,\ \therefore \frac{|z+1|}{2}<1\right)$$

$$=\frac{1}{2(z+1)}-\sum_{n=0}^{\infty}\frac{(-1)^n(z+1)^n}{2^{n+2}};\ 0<|z+1|<2$$

(iv) $f(z)=\frac{1}{2}\left[(1+z)^{-1}-\frac{1}{3}\left(1+\frac{z}{3}\right)^{-1}\right]$

$$=\frac{1}{2}\left[\sum_{n=0}^{\infty}(-1)^n z^n-\frac{1}{3}\sum_{n=0}^{\infty}\frac{(-1)^n z^n}{3^n}\right] \qquad \left(\because\ |z|<1,\ \therefore \frac{|z|}{3}<1\right)$$

$$=\frac{1}{2}\sum_{n=0}^{\infty}(-1)^n\left(1-\frac{1}{3^{n+1}}\right)z^n;\ |z|<1$$

Example 1.44: Find the first four terms of the Taylor series expansion of the complex variable function $f(z)=\frac{z+1}{(z-3)(z-4)}$ about $z=2$. Find the region of convergence.

Solution: $f(z)=\dfrac{z+1}{(z-3)(z-4)}=\dfrac{5}{z-4}-\dfrac{4}{z-3}=\dfrac{5}{z-2-2}-\dfrac{4}{z-2-1}$ (By suppression method)

$$=-\frac{5}{2}\left(1-\frac{z-2}{2}\right)^{-1}+4\left[1-(z-2)\right]^{-1}$$

$$=-\frac{5}{2}\sum_{n=0}^{\infty}\frac{(z-2)^n}{2^n}+4\sum_{n=0}^{\infty}(z-2)^n$$

$$=\sum_{n=0}^{\infty}\left(4-\frac{5}{2^{n+1}}\right)(z-2)^n$$

∴ First four terms are

$$\frac{3}{2},\frac{11}{4}(z-2),\frac{27}{8}(z-2)^2,\frac{59}{16}(z-2)^3$$

Region of convergence is the common region of

$$\left|\frac{z-2}{2}\right|<1 \text{ and } |z-2|<1$$

i.e., $$|z-2|<2 \text{ and } |z-2|<1$$

Common region is $|z-2|<1$

∴ Region of convergence is $|z-2|<1$.

Example 1.45: Find the expansion of $f(z)=\dfrac{1}{z-z^3}$ in the region $1<|z-1|<2$.

Solution:

$$f(z)=-\frac{1}{z(z-1)(z+1)}=\frac{1}{z-1}\left(\frac{1}{z+1}-\frac{1}{z}\right)$$

$$=\frac{1}{z-1}\left(\frac{1}{z-1+2}-\frac{1}{z-1+1}\right)$$

$$=\frac{1}{z-1}\left[\frac{1}{2}\left(1+\frac{z-1}{2}\right)^{-1}-\frac{1}{z-1}\left(1+\frac{1}{z-1}\right)^{-1}\right]$$

$$=\frac{1}{z-1}\left[\frac{1}{2}\sum_{n=0}^{\infty}\frac{(-1)^n(z-1)^n}{2^n}-\frac{1}{z-1}\sum_{n=0}^{\infty}\frac{(-1)^n}{(z-1)^n}\right]\quad\left(\because 1<|z-1|<2,\therefore \frac{|z-1|}{2}<1,\frac{1}{|z-1|}<1\right)$$

$$=\sum_{n=1}^{\infty}\frac{(-1)^n(z-1)^{n-1}}{2^{n+1}}+\frac{1}{2(z-1)}-\sum_{n=0}^{\infty}\frac{(-1)^n}{(z-1)^{n+2}}$$

$$=\sum_{n=0}^{\infty}\frac{(-1)^{n+1}(z-1)^n}{2^{n+2}}+\frac{1}{2(z-1)}-\sum_{n=2}^{\infty}\frac{(-1)^{n-2}}{(z-1)^n}\ ;\ 0<|z-1|<2$$

Example 1.46: When $|z+1|<1$, show that

$$\frac{1}{z^2}=1+\sum_{n=1}^{\infty}(n+1)(z+1)^n$$

Solution:
$$\frac{1}{z^2}=\frac{1}{(z+1-1)^2}=\left[1-(z+1)\right]^{-2}$$

$$=\sum_{n=0}^{\infty}(n+1)(z+1)^n=1+\sum_{n=1}^{\infty}(n+1)(z+1)^n$$

Example 1.47: Find all the possible Taylor's and Laurent series expansions of the function $f(z)$ about the point $z=1$ where $f(z)=\dfrac{1}{(z+1)(z+2)^2}$.

Solution: By suppression method

Let
$$\frac{1}{(z+1)(z+2)^2}\equiv\frac{1}{z+1}-\frac{1}{(z+2)^2}+\frac{A}{z+2}$$

$\therefore$
$$1\equiv(z+2)^2-(z+1)+A(z+1)(z+2)$$

Equate coefficient of z^2

$$0=1+A \qquad \therefore A=-1$$

$\therefore$
$$f(z)=\frac{1}{z+1}-\frac{1}{z+2}-\frac{1}{(z+2)^2}$$

$f(z)$ is not defined at $z=-1,-2$

Distances of $z=1$ from $z=-1,-2$ are respectively 2 and 3.

$\therefore$ We shall have Taylor expansion in $|z-1|<2$ and Laurent expansions in $2<|z-1|<3$ and $|z-1|>3$.

In the region $|z-1|<2$

$$f(z)=\frac{1}{z-1+2}-\frac{1}{z-1+3}-\frac{1}{(z-1+3)^2}$$

$$=\frac{1}{2}\left(1+\frac{z-1}{2}\right)^{-1}-\frac{1}{3}\left(1+\frac{z-1}{3}\right)^{-1}-\frac{1}{9}\left(1+\frac{z-1}{3}\right)^{-2}$$

$$=\frac{1}{2}\sum_{n=0}^{\infty}(-1)^n\frac{(z-1)^n}{2^n}-\frac{1}{3}\sum_{n=0}^{\infty}(-1)^n\frac{(z-1)^n}{3^n}-\frac{1}{3^2}\sum_{n=0}^{\infty}(-1)^n(n+1)\frac{(z-1)^n}{3^n}$$

$$=\sum_{n=0}^{\infty}(-1)^n\left[\frac{1}{2^{n+1}}-\frac{n+4}{3^{n+2}}\right](z-1)^n$$

In the region $2<|z-1|<3$

$$f(z)=\frac{1}{z-1}\left(1+\frac{2}{z-1}\right)^{-1}-\frac{1}{3}\left(1+\frac{z-1}{3}\right)^{-1}-\frac{1}{3^2}\left(1+\frac{z-1}{3}\right)^{-2}$$

$$=\frac{1}{z-1}\sum_{n=0}^{\infty}(-1)^n\frac{2^n}{(z-1)^n}-\frac{1}{3}\sum_{n=0}^{\infty}(-1)^n\frac{(z-1)^n}{3^n}-\frac{1}{3^2}\sum_{n=0}^{\infty}(-1)^n(n+1)\frac{(z-1)^n}{3^n}$$

$$=\sum_{n=0}^{\infty}\frac{(-1)^n 2^n}{(z-1)^{n+1}}-\sum_{n=0}^{\infty}(-1)^n\frac{(n+4)}{3^{n+2}}(z-1)^n$$

$$=\sum_{n=1}^{\infty}\frac{(-1)^{n-1}2^{n-1}}{(z-1)^n}-\sum_{n=0}^{\infty}(-1)^n\frac{(n+4)}{3^{n+2}}(z-1)^n$$

In the region $|z-1|>3$

$$f(z)=\frac{1}{z-1}\left(1+\frac{2}{z-1}\right)^{-1}-\frac{1}{z-1}\left(1+\frac{3}{z-1}\right)^{-1}-\frac{1}{(z-1)^2}\left(1+\frac{3}{z-1}\right)^{-2}$$

$$=\frac{1}{z-1}\sum_{n=0}^{\infty}(-1)^n\frac{2^n}{(z-1)^n}-\frac{1}{(z-1)}\sum_{n=0}^{\infty}(-1)^n\frac{3^n}{(z-1)^n}-\frac{1}{(z-1)^2}\sum_{n=0}^{\infty}(-1)^n\frac{(n+1)3^n}{(z-1)^n}$$

$$=\frac{1}{z-1}-\frac{1}{z-1}+\sum_{n=2}^{\infty}\left[(-1)^{n-1}2^{n-1}-(-1)^{n-1}3^{n-1}-(-1)^n(n-1)3^{n-2}\right]\frac{1}{(z-1)^n}$$

$$=\sum_{n=2}^{\infty}(-1)^{n-1}\frac{\left[2^{n-1}+(n-4)3^{n-2}\right]}{(z-1)^n}$$

Example 1.48: Write all possible Laurent series for the function $f(z)=\dfrac{1}{z(z+2)^3}$ about $z=-2$.

Solution: Singularities of $f(z)$ are at $z=0,-2$, their distances from $z=-2$, are respectively 2 and 0. Hence, Laurent series in regions $0<|z+2|<2$ and $|z+2|>2$.

In region $0<|z+2|<2$

$$f(z)=\frac{1}{(z+2)^3}\cdot\frac{1}{z+2-2}=-\frac{1}{2(z+2)^3}\left[1-\frac{z+2}{2}\right]^{-1}$$

$$=-\frac{1}{2(z+2)^3}\sum_{n=0}^{\infty}\frac{(z+2)^n}{2^n}=-\sum_{n=0}^{\infty}\frac{(z+2)^{n-3}}{2^{n+1}}$$

$$=-\frac{1}{2(z+2)^3}-\frac{1}{4(z+2)^2}-\frac{1}{8(z+2)}-\sum_{n=0}^{\infty}\frac{(z+2)^n}{2^{n+4}}$$

In region $|z+2|>2$

$$f(z)=\frac{1}{(z+2)^4}\left(1-\frac{2}{z+2}\right)^{-1}=\frac{1}{(z+2)^4}\sum_{n=0}^{\infty}\frac{2^n}{(z+2)^n}=\sum_{n=0}^{\infty}\frac{2^n}{(z+2)^{n+4}}=\sum_{n=4}^{\infty}\frac{2^{n-4}}{(z+2)^n}$$

Example 1.49: Find the Taylor's series and Laurent's series of $f(z)=\dfrac{-2z+3}{z^2-3z+2}$ with centre at the origin.

Solution: $f(z)=\dfrac{-2z+3}{(z-1)(z-2)}=-\dfrac{1}{z-1}-\dfrac{1}{z-2}$ (By suppression method)

Singularities of $f(z)$ are at $z = 1, 2$ which are at distance 1 and 2 respectively from origin.
Hence, Taylor expansion in $|z|<1$ and Laurent expansion in $1<|z|<2$, $|z|>2$.

In region $|z|<1$

$$f(z)=(1-z)^{-1}+\frac{1}{2}\left(1-\frac{z}{2}\right)^{-1}=\sum_{n=0}^{\infty}z^n+\frac{1}{2}\sum_{n=0}^{\infty}\frac{z^n}{2^n}=\sum_{n=0}^{\infty}\left(1+\frac{1}{2^{n+1}}\right)z^n$$

In region $1<|z|<2$

$$f(z)=-\frac{1}{z}\left(1-\frac{1}{z}\right)^{-1}+\frac{1}{2}\left(1-\frac{z}{2}\right)^{-1}$$

$$=-\frac{1}{z}\sum_{n=0}^{\infty}\frac{1}{z^n}+\frac{1}{2}\sum_{n=0}^{\infty}\frac{z^n}{2^n}$$

$$=\sum_{n=0}^{\infty}\frac{z^n}{2^{n+1}}-\sum_{n=0}^{\infty}\frac{1}{z^{n+1}}$$

$$=\sum_{n=0}^{\infty}\frac{z^n}{2^{n+1}}-\sum_{n=1}^{\infty}\frac{1}{z^n}$$

In region $|z|>2$

$$f(z)=-\frac{1}{z}\left(1-\frac{1}{z}\right)^{-1}-\frac{1}{z}\left(1-\frac{2}{z}\right)^{-1}=-\frac{1}{z}\sum_{n=0}^{\infty}\frac{1}{z^n}-\frac{1}{z}\sum_{n=0}^{\infty}\frac{2^n}{z^n}$$

$$=-\sum_{n=0}^{\infty}\frac{2^n+1}{z^{n+1}}=-\sum_{n=1}^{\infty}\frac{2^{n-1}+1}{z^n}$$

Example 1.50: Find the Taylor's expansion of function $f(z)=\dfrac{2z^3+1}{z^2+z}$ about the point $z=i$.

Solution: Singularities of $f(z)$ are at $z = 0, -1$ which are at distances 1 and $\sqrt{2}$ respectively from $z=i$.

Thus, Taylor expansion will be in region $|z-i|<1$.

By synthetic division

$$\begin{array}{r|rrr|r}
0) & 2 & 0 & 0 & 1 \\
 & & 0 & 0 & 0 \\
\hline
-1) & 2 & 0 & 0 & 1 \\
 & & -2 & 2 & \\
\hline
 & 2 & -2 & 2 &
\end{array}$$

$$\therefore \quad f(z) = 2z - 2 + \frac{2}{z+1} + \frac{1}{z(z+1)}$$

$$= 2z - 2 + \frac{2}{z+1} + \frac{1}{z} - \frac{1}{z+1}$$

$$= 2z - 2 + \frac{1}{z} + \frac{1}{z+1}$$

$$= 2(z-i) + 2i - 2 + \frac{1}{z-i+i} + \frac{1}{z-i+1+i}$$

$$= -2 + 2i + 2(z-i) + \frac{1}{i}\left(1 + \frac{z-i}{i}\right)^{-1} + \frac{1}{1+i}\left(1 + \frac{z-i}{1+i}\right)^{-1}$$

$$= -2 + 2i + 2(z-i) - i\left(1 - i(z-i)\right)^{-1} + \frac{1}{1+i}\left(1 + \frac{z-i}{1+i}\right)^{-1}$$

$$= -2 + 2i + 2(z-i) - i\sum_{n=0}^{\infty} i^n (z-i)^n + \frac{1}{1+i}\sum_{n=0}^{\infty} (-1)^n \frac{(z-i)^n}{(1+i)^n}$$

$$= -2 + 2i + 2(z-i) - i + (z-i) + \frac{1}{1+i} - \frac{1}{2i}(z-i) - i\sum_{n=2}^{\infty} i^n (z-i)^n + \sum_{n=2}^{\infty} (-1)^n \frac{(z-i)^n}{(1+i)^{n+1}}$$

$$= \left(-2 + 2i - i + \frac{1-i}{2}\right) + \left(2 + 1 - \frac{1}{2i}\right)(z-i) + \sum_{n=2}^{\infty}\left[i^{n-1} + \frac{(-1)^n}{(1+i)^{n+1}}\right](z-i)^n$$

$$= \frac{-3+i}{2} + \left(3 + \frac{i}{2}\right)(z-i) + \sum_{n=2}^{\infty} i^{n-1}\left[1 + \left(\frac{i}{1+i}\right)^{n+1}\right](z-i)^n$$

$$= \frac{-3+i}{2} + \left(3 + \frac{i}{2}\right)(z-i) + \sum_{n=2}^{\infty} i^{n-1}\left[1 + \left(\frac{1+i}{2}\right)^{n+1}\right](z-i)^n$$

Example 1.51: Expand $\dfrac{z}{(z^2-1)(z^2+4)}$ in $1 < |z| < 2$

Solution: $f(z) = \dfrac{z}{(z^2-1)(z^2+4)} = \dfrac{z}{5}\left(\dfrac{1}{z^2-1} - \dfrac{1}{z^2+4}\right)$ (By suppression method)

$$= \frac{z}{5}\left[\frac{1}{z^2}\left(1 - \frac{1}{z^2}\right)^{-1} - \frac{1}{4}\left(1 + \frac{z^2}{4}\right)^{-1}\right] \qquad \left(\because\ 1 < |z| < 2, \therefore \frac{1}{|z|} < 1, \frac{|z|^2}{4} < 1\right)$$

$$= \frac{1}{5z}\sum_{n=0}^{\infty} \frac{1}{z^{2n}} - \frac{z}{20}\sum_{n=0}^{\infty} (-1)^n \frac{z^{2n}}{4^n}$$

$$= \frac{1}{5}\sum_{n=0}^{\infty} \frac{1}{z^{2n+1}} - \frac{1}{10}\sum_{n=0}^{\infty} (-1)^n \left(\frac{z}{2}\right)^{2n+1}$$

Example 1.52: Expand the function $\dfrac{1-\cos z}{z^3}$ in Laurent series about the point $z = 0$.

Solution: $\cos z = 1 - \dfrac{z^2}{2!} + \dfrac{z^4}{4!} - \dfrac{z^6}{6!} + \cdots\cdots = \sum_{n=0}^{\infty} \dfrac{(-1)^n z^{2n}}{(2n)!}$

$$1 - \cos z = \sum_{n=1}^{\infty} \frac{(-1)^{n+1} z^{2n}}{(2n)!}$$

$$\therefore \quad \frac{1-\cos z}{z^3} = \sum_{n=1}^{\infty} \frac{(-1)^{n+1} z^{2n-3}}{(2n)!}; \; z \neq 0$$

$$= \frac{1}{2z} + \sum_{n=2}^{\infty} \frac{(-1)^{n+1} z^{2n-3}}{(2n)!}; \; z \neq 0$$

Example 1.53: Find the Laurent series of $z^2 e^{1/z}$ with centre as origin.

Solution:

$$e^{1/z} = \sum_{n=0}^{\infty} \frac{1}{n!\,z^n}; \; z \neq 0$$

$$\therefore \quad z^2 e^{1/z} = \sum_{n=0}^{\infty} \frac{1}{n!\,z^{n-2}}; \; z \neq 0$$

$$= z^2 + z + \frac{1}{2} + \sum_{n=3}^{\infty} \frac{1}{n!\,z^{n-2}}; \; z \neq 0$$

$$= z^2 + z + \frac{1}{2} + \sum_{n=1}^{\infty} \frac{1}{(n+2)!\,z^n}; \quad z \neq 0$$

Example 1.54: Expand $\cos z$ in a Taylor's series about $z = \dfrac{\pi}{4}$

Solution: Let $\quad z - \dfrac{\pi}{4} = w$

$$\therefore \quad z = w + \frac{\pi}{4}$$

$$\therefore \quad \cos z = \cos\left(w + \frac{\pi}{4}\right) = \cos\frac{\pi}{4}\cos w - \sin\frac{\pi}{4}\sin w$$

$$= \frac{1}{\sqrt{2}}(\cos w - \sin w)$$

$$= \frac{1}{\sqrt{2}}\left[\sum_{n=0}^{\infty} \frac{(-1)^n w^{2n}}{2n!} - \sum_{n=0}^{\infty} \frac{(-1)^n w^{2n+1}}{(2n+1)!}\right]$$

$$= \frac{1}{\sqrt{2}}\left[1 - w - \frac{w^2}{2!} + \frac{w^3}{3!} + \frac{w^4}{4!} - \frac{w^5}{5!} - \frac{w^6}{6!} + \cdots\right]$$

$$= \frac{1}{\sqrt{2}}\left[1 - \left(z - \frac{\pi}{4}\right) - \frac{\left(z-\frac{\pi}{4}\right)^2}{2!} + \frac{\left(z-\frac{\pi}{4}\right)^3}{3!} + \frac{\left(z-\frac{\pi}{4}\right)^4}{4!} - \frac{\left(z-\frac{\pi}{4}\right)^5}{5!} - \frac{\left(z-\frac{\pi}{4}\right)^6}{6!} + \cdots\right]$$

Example 1.55: Find the Laurent series of $f(z)=\dfrac{e^z}{z(1-z)}$ about $z=1$. Find region of convergence.
Solution: Singularities of $f(z)$ are at $z = 0, 1$.

Thus, there will be two Laurent series of $f(z)$ about $z = 1$, one in region $0<|z-1|<1$ and other in region $|z-1|>1$.

In the region $0<|z-1|<1$,

$$f(z)=\frac{e^z}{z(1-z)}=-\frac{e\,.\,e^{(z-1)}}{(z-1)[1+(z-1)]}=-\frac{e\,.\,e^{(z-1)}}{(z-1)}[1+(z-1)]^{-1}$$

$$=-\frac{e}{(z-1)}\left[1+\frac{(z-1)}{1!}+\frac{(z-1)^2}{2!}+\frac{(z-1)^3}{3!}+\frac{(z-1)^4}{4!}+\frac{(z-1)^5}{5!}+\cdots\right].$$

$$\left[1-(z-1)+(z-1)^2-(z-1)^3+(z-1)^4-(z-1)^5+\cdots\right]$$

$$=-\frac{e}{(z-1)}\left[1+(-1+1)(z-1)+\left(1-1+\frac{1}{2!}\right)(z-1)^2+\left(-1+1-\frac{1}{2!}+\frac{1}{3!}\right)(z-1)^3\right.$$

$$\left.+\left(1-1+\frac{1}{2!}-\frac{1}{3!}+\frac{1}{4!}\right)(z-1)^4+\left(-1+1-\frac{1}{2!}+\frac{1}{3!}-\frac{1}{4!}+\frac{1}{5!}\right)(z-1)^5+\cdots\right]$$

$$=-\frac{e}{(z-1)}\left[1+\sum_{n=2}^{\infty}(-1)^n\left(\frac{1}{2!}-\frac{1}{3!}+\frac{1}{4!}-\cdots+\frac{(-1)^n}{n!}\right)(z-1)^n\right]$$

$$=-\frac{e}{(z-1)}-e\sum_{n=1}^{\infty}(-1)^{n+1}\left(\frac{1}{2!}-\frac{1}{3!}+\frac{1}{4!}-\cdots+\frac{(-1)^{n+1}}{(n+1)!}\right)(z-1)^n$$

and in the region $|z-1|>1$,

$$f(z)=\frac{e^z}{z(1-z)}=-\frac{e\,.\,e^{(z-1)}}{(z-1)[1+(z-1)]}=-\frac{e\,.\,e^{(z-1)}}{(z-1)^2}\left\{1+\frac{1}{(z-1)}\right\}^{-1}$$

$$=-\frac{e}{(z-1)^2}\left[1+\frac{(z-1)}{1!}+\frac{(z-1)^2}{2!}+\frac{(z-1)^3}{3!}+\frac{(z-1)^4}{4!}+\frac{(z-1)^5}{5!}+\cdots\right].$$

$$\left[1-\frac{1}{(z-1)}+\frac{1}{(z-1)^2}-\frac{1}{(z-1)^3}+\frac{1}{(z-1)^4}-\frac{1}{(z-1)^5}+\cdots\right]$$

$$=-\frac{e}{(z-1)^2}\left[\left\{\left(1-\frac{1}{1!}+\frac{1}{2!}-\frac{1}{3!}+\frac{1}{4!}-\cdots\right)+\left(\frac{1}{1!}-\frac{1}{2!}+\frac{1}{3!}-\frac{1}{4!}+\cdots\right)(z-1)\right.\right.$$

$$+\left(\frac{1}{2!}-\frac{1}{3!}+\frac{1}{4!}-\frac{1}{5!}+\cdots\right)(z-1)^2+\left(\frac{1}{3!}-\frac{1}{4!}+\frac{1}{5!}-\frac{1}{6!}+\cdots\right)(z-1)^3$$

$$+\left(\frac{1}{4!}-\frac{1}{5!}+\frac{1}{6!}-\frac{1}{7!}+\cdots\right)(z-1)^4+\left(\frac{1}{5!}-\frac{1}{6!}+\frac{1}{7!}-\frac{1}{8!}+\cdots\right)(z-1)^5$$

$$\left.+\left(\frac{1}{6!}-\frac{1}{7!}+\frac{1}{8!}-\cdots\right)(z-1)^6+\cdots\right\}$$

$$+\left\{\left(-1+1-\frac{1}{2!}+\frac{1}{3!}-\frac{1}{4!}+\cdots\right)\frac{1}{(z-1)}+\left(1-1+\frac{1}{2!}-\frac{1}{3!}+\frac{1}{4!}-\cdots\right)\frac{1}{(z-1)^2}\right.$$

$$+\left(-1+1-\frac{1}{2!}+\frac{1}{3!}-\frac{1}{4!}+\cdots\right)\frac{1}{(z-1)^3}+\left(1-1+\frac{1}{2!}-\frac{1}{3!}+\frac{1}{4!}-\cdots\right)\frac{1}{(z-1)^4}$$

$$\left.\left.+\left(-1+1-\frac{1}{2!}+\frac{1}{3!}-\frac{1}{4!}+\cdots\right)\frac{1}{(z-1)^5}+\cdots\right\}\right]$$

$$=-\frac{e}{(z-1)^2}\left[\left\{\frac{1}{e}+\left(1-\frac{1}{e}\right)(z-1)+\frac{1}{e}(z-1)^2+\left(\frac{1}{2!}-\frac{1}{e}\right)(z-1)^3-\left(\frac{1}{2!}-\frac{1}{3!}-\frac{1}{e}\right)(z-1)^4\right.\right.$$

$$\left.+\left(\frac{1}{2!}-\frac{1}{3!}+\frac{1}{4!}-\frac{1}{e}\right)(z-1)^5-\left(\frac{1}{2!}-\frac{1}{3!}+\frac{1}{4!}-\frac{1}{5!}-\frac{1}{e}\right)(z-1)^6+\cdots\right\}$$

$$\left.+\left\{\left(-\frac{1}{e}\right)\frac{1}{(z-1)}+\frac{1}{e}\frac{1}{(z-1)^2}+\left(-\frac{1}{e}\right)\frac{1}{(z-1)^3}+\frac{1}{e}\frac{1}{(z-1)^4}+\left(-\frac{1}{e}\right)\frac{1}{(z-1)^5}+\cdots\right\}\right]$$

$$=-1+\sum_{n=1}^{\infty}(-1)^n\left[e\left\{\frac{1}{2!}-\frac{1}{3!}+\frac{1}{4!}-\cdots+\frac{(-1)^{n+1}}{(n+1)}\right\}-1\right](z-1)^n+\frac{1-e}{(z-1)}-\sum_{n=2}^{\infty}\frac{(-1)^n}{(z-1)^n}$$

EXERCISE 1.3

1. Find all possible Taylor's and Laurent series expansions of the function $f(z)=\dfrac{1}{(1-z)}$ about $z=0$.
2. Expand the function $f(z)=\dfrac{1}{z}$ about the point $z=2$ in Taylor's series.
3. Find the Taylor series expansion of function $f(z)=\dfrac{1}{(z-1)(z-3)}$ about the point $z=4$. Find its region of convergence.
4. Find the first three terms of the Taylor series expansion of $f(z)=\dfrac{1}{z^2+4}$ about $z=-i$. Find the region of convergence.
5. Find the Laurent's expansion of $f(z)=\dfrac{1}{z(z-1)^2}$ about the point $z=1$.
6. Obtain the Taylor's or Laurent's series which represents the function $f(z)=\dfrac{1}{(1+z^2)(z+2)}$ when
   (i) $1<|z|<2$ (ii) $|z|>2$

7. Find the Taylor's or Laurent's series which represents the function $\frac{z^2-1}{z^2+5z+6}$ in the region
   (i) $|z|<2$ (ii) $2<|z|<3$
   (iii) $|z|>3$

8. Find all possible Laurent's series of $f(z)=\frac{7z^2+9z-18}{z^3-9z}$ about its singular points.

9. Obtain the Taylor series expansion of $f(z)=\frac{1}{z^2+(1+2i)z+2i}$ about $z=0$.

10. Find the Laurent's series for $f(z)=\frac{7z-2}{z^3-z^2-2z}$ in the region given by $0<|z+1|<1$.

11. Expand $\frac{1}{(1+z^2)(2+z^2)}$ in powers of z, when
    (i) $|z|<1$ (ii) $1<|z|<\sqrt{2}$
    (iii) $|z|>\sqrt{2}$

12. Find Laurent expansion of $f(z)=\frac{e^{2z}}{(z-1)^3}$ about $z=1$.

13. Obtain the Taylor series expansion of $f(z)=e^z$ about the point (i) $z=0$ (ii) $z=2$

14. Obtain the first three terms of the Laurent's series expansion of the function $f(z)=\frac{1}{(e^z-1)}$ about the point $z=0$ valid in the region $0<|z|<2\pi$.

ANSWERS 1.3

1. $\sum_{n=0}^{\infty} z^n; |z|<1, -\sum_{n=1}^{\infty}\frac{1}{z^n}; |z|>1$

2. $\sum_{n=0}^{\infty}\frac{(-1)^n(z-2)^n}{2^{n+1}}; |z-2|<2$

3. $\frac{1}{2}\sum_{n=0}^{\infty}(-1)^n\left(1-\frac{1}{3^{n+1}}\right)(z-4)^n;\quad |z-4|<1$

4. $\frac{1}{3}, \frac{2i}{9}(z+i), \frac{-7}{27}(z+i)^2$; region of convergence $|z+i|<1$

5. $-\frac{1}{(z-1)}+\frac{1}{(z-1)^2}+\sum_{n=0}^{\infty}(-1)^n(z-1)^n; 0<|z-1|<1$ and $\sum_{n=3}^{\infty}\frac{(-1)^{n-1}}{(z-1)^n}; |z-1|>1$

6. (i) $\frac{1}{5}\left[\sum_{n=0}^{\infty}\frac{(-1)^n z^n}{2^{n+1}}-\sum_{n=0}^{\infty}\left\{\frac{(-1)^n}{z^{2n+1}}-\frac{2(-1)^n}{z^{2n+2}}\right\}\right]$; $1<|z|<2$

   (ii) $\frac{1}{5}\sum_{n=1}^{\infty}\left[\frac{2^{2n}-(-1)^n}{z^{2n+1}}-\frac{2\{2^{2n}-(-1)^n\}}{z^{2n+2}}\right]$; $|z|>2$

7. (i) $-\frac{1}{6}+\sum_{n=1}^{\infty}(-1)^n\left(\frac{3}{2^{n+1}}-\frac{8}{3^{n+1}}\right)z^n; |z|<2$

(ii) $-\frac{5}{3}+\sum_{n=1}^{\infty}(-1)^{n+1}\frac{8}{3^{n+1}}z^n+\sum_{n=1}^{\infty}\frac{3.(-2)^{n-1}}{z^n}; 2<|z|<3$

(iii) $1+\sum_{n=1}^{\infty}\left[(-1)^n\left\{\frac{8(3)^{n-1}-3(2)^{n-1}}{z^n}\right\}\right]; |z|>3$

8. **About $z=0$** (i) $\frac{2}{z}+\sum_{n=0}^{\infty}\{(-1)^n-4\}\frac{z^n}{3^{n+1}}; 0<|z|<3$

(ii) $\frac{7}{z}+\sum_{n=2}^{\infty}\{(-1)^{n-1}+4\}\frac{3^{n-1}}{z^n}; |z|>3$

About $z=3$ (i) $\frac{4}{(z-3)}+\sum_{n=0}^{\infty}(-1)^n\left[\frac{2}{3^{n+1}}+\frac{1}{6^{n+1}}\right](z-3)^n;\ 0<|z-3|<3$

(ii) $\frac{6}{(z-3)}+2\sum_{n=2}^{\infty}\frac{(-1)^{n-1}3^{n-1}}{(z-3)^n}+\frac{1}{6}\sum_{n=0}^{\infty}\frac{(-1)^n(z-3)^n}{6^n};\ 3<|z-3|<6$

(iii) $\frac{7}{(z-3)}+\sum_{n=2}^{\infty}(-1)^{n-1}\frac{\left[2(3)^{n-1}+6^{n-1}\right]}{(z-3)^n}; |z-3|>6$

About $z=-3$ (i) $\frac{1}{(z+3)}-\frac{2}{3}\sum_{n=0}^{\infty}\left(\frac{1}{3^n}+\frac{1}{6^n}\right)(z+3)^n;\ 0<|z+3|<3$

(ii) $\frac{3}{(z+3)}+2\sum_{n=2}^{\infty}\frac{3^{n-1}}{(z+3)^n}-\frac{2}{3}\sum_{n=0}^{\infty}\frac{(z+3)^n}{6^n}; 3<|z+3|<6$

(iii) $\frac{7}{(z+3)}+\sum_{n=2}^{\infty}\frac{\left[2(3)^{n-1}+4(6)^{n-1}\right]}{(z+3)^n};\ |z+3|>6$

9. $\frac{1}{(1-2i)}\left[\sum_{n=0}^{\infty}\left\{\left(\frac{1}{2i}\right)^{n+1}-1\right\}(-1)^n z^n\right];\ |z|<1$

10. $\frac{-3}{z+1}-\sum_{n=0}^{\infty}\left(1+\frac{2}{3^{n+1}}\right)(z+1)^n;\ 0<|z+1|<1$

11. (i) $\sum_{n=0}^{\infty}(-1)^n\left\{1-\frac{1}{2^{n+1}}\right\}z^{2n},\quad |z|<1$

(ii) $\sum_{n=0}^{\infty}\frac{(-1)^n}{z^{2n+2}}-\sum_{n=0}^{\infty}(-1)^n\frac{z^{2n}}{2^{n+1}},\quad 1<|z|<\sqrt{2}$

(iii) $\sum_{n=1}^{\infty}\frac{(-1)^{n-1}(2^n-1)}{z^{2n+2}},\ |z|>\sqrt{2}$

12. $e^2\left[\dfrac{1}{(z-1)^3}+\dfrac{2}{(z-1)^2}+\dfrac{2}{(z-1)}+\sum_{n=0}^{\infty}\dfrac{2^{n+3}(z-1)^n}{(n+3)!}\right]$ $;|z-1|>0$

13. (i) $\sum_{n=0}^{\infty}\dfrac{z^n}{n!}$ (ii) $e^2\sum_{n=0}^{\infty}\dfrac{(z-2)^n}{n!}$

14. first three terms are $\dfrac{1}{z},-\dfrac{1}{2},\dfrac{z}{12}$.

1.12 ZEROS AND SINGULARITIES OF COMPLEX FUNCTIONS

1.12.1 Zeros of an Analytic Function

A point $z = z_0$ is called a zero of function $f(z)$ if $f(z)$ is analytic at z_0 and $f(z_0) = 0$.

In neighbourhood of zero z_0 of $f(z)$, by Taylor series

$$f(z) = a_0 + a_1(z - z_0) + a_2(z - z_0)^2 + a_3(z - z_0)^3 + \ldots\ldots\ldots\ldots$$

where
$$a_n = \frac{1}{n!}f^{(n)}(z_0)$$

If $a_0 = 0$ and $a_1 \neq 0$ then $z = z_0$ is called simple zero (or zero of order 1) of $f(z)$.

If $a_0 = a_1 = a_2 = \ldots\ldots = a_{n-1} = 0$ and $a_n \neq 0$ then $f(z)$ is said to have a zero of order n at $z = z_0$.

In the neighbourhood of z_0 which is zero of order n of $f(z)$

$$f(z) = a_n(z - z_0)^n + a_{n+1}(z - z_0)^{n+1} + \ldots\ldots ; a_n \neq 0$$

$\therefore$
$$f(z) = (z - z_0)^n g(z)$$

where
$$g(z) = a_n + a_{n+1}(z - z_0) + a_{n+2}(z - z_0)^2 + \ldots.. \neq 0 \quad \text{for} \quad z = z_0$$

1.12.2 Singularities of a Function

A point $z = z_0$ at which the function $f(z)$ is not defined or the function is not analytic is called singularity of $f(z)$. A rational function $f(z) = \dfrac{P(z)}{Q(z)}$ has singular point at $z = z_0$ if $Q(z_0) = 0$ and $P(z_0) \neq 0$ because in this case $f(z)$ is not defined at $z = z_0$.

There are following two types of singularities of a function.

(i) non-isolated singularity

(ii) isolated singularity

Non-isolated singularity

Singularity $z = z_0$ is called non-isolated singularity of $f(z)$ if every neighbourhood of $z = z_0$ contains singularity of $f(z)$ other than z_0 also.

Now, we shall prove that every neighbourhood of non-isolated singularity has infinite number of singularities of $f(z)$.

Theorem 1.21 If $z = z_0$ is non-isolated singularity of $f(z)$ then every neighbourhood of $z = z_0$ contains infinite number of singularities of $f(z)$.

Proof: If possible, let the neighbourhood $|z - z_0| < r$ of non-isolated singularity z_0 contains finite number of singularities $z_1, z_2, \ldots\ldots z_n$ of $f(z)$. Let $r = \min_{1 \le k \le n} |z_0 - z_k|$ then the neighbourhood $|z - z_0| < r$ of z_0 contains no singularity of $f(z)$. Thus z_0 is not non-isolated singularity of $f(z)$ which contradicts and thus, every neighbourhood of z_0 contains infinite no. of singularities of $f(z)$.

Isolated singularities

If there exists a neighbourhood of singularity $z = z_0$ of $f(z)$ which contains no singularity of $f(z)$ other than $z = z_0$ then $z = z_0$ is called isolated singularity of $f(z)$.

For example $f(z) = \cot \frac{\pi}{z}$ has singularities at $z = 0, \pm 1, \pm \frac{1}{2}, \pm \frac{1}{3}, \ldots\ldots$. Out of these singularities $z = 0$ is non-isolated and all other singularities are isolated singularities.

Isolated singularities are of three types.

Let $|z - z_0| < \delta$ be a neighbourhood of $z = z_0$ in which $f(z)$ has no singularity other than $z = z_0$. Laurent series expansion of $f(z)$ about $z = z_0$ in this neighbourhood may contain none, finite number or infinite number of terms of negative powers of $z - z_0$. In case of no term of negative power of $(z - z_0)$, $z = z_0$ is *removable isolated singularity*. In case of finite number of terms say n of negative powers of $(z - z_0)$, $z = z_0$ is a *pole of order n*. Pole of orders 1, 2 and 3 are also called simple pole, double pole or triple pole, respectively. In case of infinite no. of terms of negative powers of $(z - z_0)$, the isolated singularity $z = z_0$ is called *essential singularity*.

Remark 1.12:

(i) If $f(z)$ has a zero of nth order at $z = z_0$ then $\frac{1}{f(z)}$ has pole of order n at $z = z_0$ and vice versa.

(ii) If $f(z)$ has a zero of order n at $z = z_0$ then $(f(z))^k$ has zero of order kn at $z = z_0$.

(iii) If $f(z)$ has a zero of order n at $z = z_0$ then kth derivative of $f(z)$ i.e., $f^{(k)}(z)$ has zero of order $n - k$ at $z = z_0$ when $1 \le k \le n-1$.

(iv) If a function $f(z)$ is analytic for all z except possible a finite no. of poles then $f(z)$ is called a *meromorphic function*. For example, function $\frac{\sin z}{(z-1)(z+3)^2}$ has poles at $z = 1$ and $z = -3$ only and hence it is meromorphic function.

1.12.3 Method to Find Type of Isolated Singularity

If $z = z_1$ is isolated singularity of $f(z)$ then its type can be found by writing Laurent expansion of $f(z)$ about $z = z_1$. Also without writing the Laurent expansion, we can find the type of isolated singularity by the below mentioned results obtained from Laurent expansion of $f(z)$ about $z = z_1$.

(i) If $\lim_{z \to z_1} f(z)$ exists then z_1 is removable singularity.

(ii) If $\lim_{z \to z_1} (z - z_1)^k f(z) = \lambda$ where $\lambda \ne 0$ then z_1 is a pole of order k.

(iii) If $\lim_{z \to z_1} (z - z_1)^n f(z)$ is infinite for all non-negative integer n, then z_1 is essential singularity.

Example 1.56: What type of singularities have the following functions?

(i) $\dfrac{\tan z}{z}$ (ii) $e^{z^{-3}}$

Solution: (i) $f(z) = \dfrac{\tan z}{z}$ has singularities at $z = 0$ and the points where tan z is undefined, i.e., $z = (2n+1)\dfrac{\pi}{2}$; $n \in I$

$$\lim_{z\to 0} f(z) = \lim_{z\to 0}\frac{\tan z}{z} = \lim_{z\to 0}\frac{\sec^2 z}{1} \qquad (L' \text{ Hospital rule})$$

$$= 1$$

$\therefore f(z)$ has removable singularity at $z = 0$

$$\lim_{z\to n\pi+\frac{\pi}{2}}\left[z-\left(n\pi+\frac{\pi}{2}\right)\right] f(z) = \lim_{z\to n\pi+\frac{\pi}{2}}\frac{\sin z}{z}\cdot\frac{z-\left(n\pi+\frac{\pi}{2}\right)}{\cos z}$$

$$= \frac{\sin\left(n\pi+\frac{\pi}{2}\right)}{n\pi+\frac{\pi}{2}}\lim_{z\to n\pi+\frac{\pi}{2}}\frac{1}{-\sin z} \qquad (L' \text{ Hospital rule})$$

$$= \frac{(-1)^n}{n\pi+\frac{\pi}{2}}\,\frac{1}{-\sin\left(n\pi+\frac{\pi}{2}\right)} = -\frac{(-1)^n}{n\pi+\frac{\pi}{2}}\cdot\frac{1}{(-1)^n}$$

$$= \frac{-2}{(2n+1)\pi} \neq 0; \quad n \in I$$

$\therefore f(z)$ has simple poles at each $z = n\pi + \dfrac{\pi}{2}$; $n \in I$

(ii) $f(z) = e^{z^{-3}} = e^{\frac{1}{z^3}}$ has only singularity at $z = 0$

$$f(z) = 1 + \frac{1}{z^3} + \frac{1}{2!z^6} + \frac{1}{3!z^9} + \frac{1}{4!z^{12}} + \cdots$$

This expansion has infinite no. of terms of negative powers of z

$\therefore$ $f(z)$ has essential singularity at $z = 0$.

Example 1.57: Find the type of singularity of the function $f(z) = \dfrac{1-e^{2z}}{z^3}$ at $z = 0$.

Solution: $f(z) = \dfrac{1}{z^3}\left[1-\left(1+2z+\dfrac{(2z)^2}{2!}+\dfrac{(2z)^3}{3!}+\dfrac{(2z)^4}{4!}+\ldots\right)\right]$

$$= -\frac{2}{z^2} - \frac{2}{z} - \frac{4}{3} - \frac{2z}{3} - \cdots$$

$\therefore f(z)$ has double pole at $z = 0$.

Example 1.58: Find the location and type of singularity of the following functions

(i) $\dfrac{e^{-z}}{(z+2)^3}$ (ii) $\dfrac{(z+1)-\tan(z+1)}{(z+1)^3}$

Solution: (i) $f(z)=\dfrac{e^{-z}}{(z+2)^3}$ has singularity only at $z = -2$

$$\lim_{z\to -2}(z+2)^3 f(z)=\lim_{z\to -2} e^{-z}=e^2\neq 0$$

$\therefore f(z)$ has pole of order 3 at $z = -2$.

(ii) $f(z)=\dfrac{z+1-\tan(z+1)}{(z+1)^3}$ has singularities at $z = -1$ and $z=-1+(2n+1)\dfrac{\pi}{2};\ n\in I$

$$f(z)=\frac{1}{(z+1)^3}\left[(z+1)-\left((z+1)-\frac{(z+1)^3}{3}+\frac{2}{15}(z+1)^5-\ldots\right)\right]$$

$$=\frac{1}{3}-\frac{2}{15}(z+1)^2+\ldots.$$

$\therefore f(z)$ has removable singularity at $z = -1$

$$\lim_{z\to -1+(2n+1)\frac{\pi}{2}}\left[z+1-(2n+1)\frac{\pi}{2}\right]f(z)$$

$$=\lim_{z\to -1+(2n+1)\frac{\pi}{2}}\frac{\left[z+1-\left(n\pi+\frac{\pi}{2}\right)\right]}{(z+1)^2}-\frac{\sin(z+1)}{(z+1)^3}\cdot\frac{\left[z+1-\left(n\pi+\frac{\pi}{2}\right)\right]}{\cos(z+1)}$$

$$=0-\frac{\sin\left(n\pi+\frac{\pi}{2}\right)}{\left(n\pi+\frac{\pi}{2}\right)^3}\lim_{z\to -1+(2n+1)\frac{\pi}{2}}\frac{z+1-\left(n\pi+\frac{\pi}{2}\right)}{\cos(z+1)}$$

$$=-\frac{8\sin\left(n\pi+\frac{\pi}{2}\right)}{(2n+1)^3\pi^3}\lim_{z\to -1+(2n+1)\frac{\pi}{2}}\frac{1}{-\sin(z+1)}\qquad (L'\text{ Hospital rule})$$

$$=\frac{8\sin\left(n\pi+\frac{\pi}{2}\right)}{(2n+1)^3\pi^3}\cdot\frac{1}{\sin\left(n\pi+\frac{\pi}{2}\right)}=\frac{8}{(2n+1)^3\pi^3}\neq 0;\ n\in I$$

$\therefore f(z)$ has simple poles at each of $z=-1+n\pi+\dfrac{\pi}{2};\ n\in I$

Example 1.59: Find the singularities of $f(z)=\dfrac{z}{\left(z^2+4\right)^2}$ and indicate the character of the singularities.

Solution: $f(z)=\dfrac{z}{\left(z^2+4\right)^2}=\dfrac{z}{(z+2i)^2(z-2i)^2}$

$f(z)$ has singularities at $z=\pm 2i$

$$\lim_{z\to 2i}(z-2i)^2 f(z)=\lim_{z\to 2i}\frac{z}{(z+2i)^2}=\frac{2i}{-16}=-\frac{1}{8}i\neq 0$$

$\therefore f(z)$ has isolated singularity which is double pole at $z=2i$.

$$\lim_{z\to -2i}(z+2i)^2 f(z)=\lim_{z\to -2i}\frac{z}{(z-2i)^2}=\frac{-2i}{-16}=\frac{1}{8}i\neq 0$$

$\therefore f(z)$ has isolated singularity double pole at $z=-2i$.

Example 1.60: Find the nature and the location of singularities of $f(z)=\dfrac{1}{z\left(e^z-1\right)}$.

Solution: $f(z)$ has singularities at $z=0$ and $z=2n\pi i;\ n\in I,\ n\neq 0$

$$\lim_{z\to 0} z^2 f(z)=\lim_{z\to 0}\frac{z}{e^z-1}=\lim_{z\to 0}\frac{1}{e^z}\qquad (L'\ \text{Hospital rule})$$

$$=1\neq 0$$

$$\lim_{z\to 2n\pi i}(z-2n\pi i)f(z)=\lim_{z\to 2n\pi i}\frac{1}{z}\cdot\frac{z-2n\pi i}{e^z-1}$$

$$=\frac{1}{2n\pi i}\lim_{z\to 2n\pi i}\frac{z-2n\pi i}{e^z-1}$$

$$=\frac{1}{2n\pi i}\lim_{z\to 2n\pi i}\frac{1}{e^z}\qquad (L'\ \text{Hospital rule})$$

$$=\frac{-i}{2n\pi}(1)=\frac{-i}{2n\pi};\ n\in I,\ n\neq 0$$

$\therefore f(z)$ has isolated singularities at $z=0$, $z=2n\pi i$, $n\in I$, $n\neq 0$.
Pole of order two at $z=0$ and simple poles at $z=2n\pi i$; $n\in I$, $n\neq 0$.

Example 1.61: Find the nature and the location of the singularities of the following functions

(i) $f(z)=\tan\left(\dfrac{1}{z}\right)$ (ii) $f(z)=\operatorname{cosec}\dfrac{\pi}{z}$

Solution:

(i) $f(z)=\tan\left(\dfrac{1}{z}\right)$ has singularities at $z=0,\ \dfrac{1}{(2n+1)\frac{\pi}{2}};\ n\in I$

Now $\lim_{n\to\infty} \dfrac{1}{(2n+1)\frac{\pi}{2}} = 0$

Thus, every neighbourhood of $z = 0$ contains infinite no. of singularities among $z = \dfrac{1}{(2n+1)\frac{\pi}{2}}$; $n \in I$

$\therefore$ $z = 0$ is non-isolated singularity.

$$\lim_{z\to\frac{1}{(2n+1)\frac{\pi}{2}}} \left[z - \frac{1}{(2n+1)\frac{\pi}{2}}\right] f(z) = \lim_{z\to\frac{1}{(2n+1)\frac{\pi}{2}}} \frac{z - \frac{1}{(2n+1)\frac{\pi}{2}}}{\cot\frac{1}{z}}$$

$$= \lim_{z\to\frac{1}{(2n+1)\frac{\pi}{2}}} \frac{1}{\left(-\operatorname{cosec}^2\frac{1}{z}\right)\left(-\frac{1}{z^2}\right)} \quad (L' \text{ Hospital rule})$$

$$= \frac{1}{\left[(2n+1)\frac{\pi}{2}\right]^2} \neq 0;\ n \in I$$

$\therefore$ $f(z)$ has simple poles at each of $z = \dfrac{1}{(2n+1)\frac{\pi}{2}}$; $n \in I$

(ii) $f(z) = \operatorname{cosec}\dfrac{\pi}{z}$ has singularities at $z = 0$ and $z = \dfrac{1}{n}$; $n \in I$; $n \neq 0$

Now, $\lim_{n\to\infty} \dfrac{1}{n} = 0$

Thus, every neighbourhood of $z = 0$ contains infinite $\dfrac{1}{n}$ for $n \in I, n \neq 0$

$\therefore$ $z = 0$ is non-isolated singularity.

$$\lim_{z\to\frac{1}{n}} \left(z - \frac{1}{n}\right) f(z) = \lim_{z\to\frac{1}{n}} \frac{z - \frac{1}{n}}{\sin\frac{\pi}{z}}$$

$$= \lim_{z\to\frac{1}{n}} \frac{1}{-\frac{\pi}{z^2}\cos\frac{\pi}{z}} \quad (L' \text{ Hospital rule})$$

$$= \frac{(-1)^{n+1}}{n^2\pi} \neq 0;\ n \in I;\ n \neq 0$$

$\therefore$ $f(z)$ has isolated singularities simple poles at each of $z = \dfrac{1}{n}$; $n \in I$; $n \neq 0$

1.13 RESIDUE

If $z = z_0$ is isolated singularity of $f(z)$ then there exists neighbourhood $|z - z_0| < \delta$ of z_0 in which $f(z)$ has only singularity $z = z_0$. Thus, we can have Laurent series expansion of $f(z)$ about $z = z_0$ in this neighbourhood.

In this expansion, the coefficient a_{-1} of $(z - z_0)^{-1}$ i.e.,

$$a_{-1} = \frac{1}{2\pi i}\int_C f(z)dz \text{ where } C: \; |z - z_0| = r < \delta$$

is called residue of $f(z)$ at $z = z_0$ and in short it is written as Res $(z_0) = a_{-1}$

$$\therefore \qquad \text{Res } (z_0) = \frac{1}{2\pi i}\int_C f(z)dz \tag{1.40}$$

1.13.1 Residue at a Removable Singularity

Laurent series expansion of $f(z)$ about removable singularity does not contain any term of negative powers of $(z - z_0)$

$$\therefore \qquad \text{Res } (z_0) = a_{-1} = \text{Coefficient of } (z - z_0)^{-1} = 0$$

1.13.2 Residue at a Simple Pole

If $z = z_0$ is simple pole of $f(z)$, then its Laurent series expansion about $z = z_0$ is

$$f(z) = \sum_{n=0}^{\infty} a_n (z - z_0)^n + \frac{a_{-1}}{(z - z_0)}; \; a_{-1} \neq 0$$

$$\therefore \qquad \lim_{z \to z_0} (z - z_0) f(z) = a_{-1} = \text{Res}(z_0) \neq 0$$

If $f(z) = \dfrac{\phi(z)}{\psi(z)}$ and $f(z)$ has simple pole at $z = z_0$ such that $\phi(z_0) \neq 0$ then $\psi(z)$ has simple zero at $z = z_0$ and hence $\psi(z_0) = 0$ and $\psi'(z_0) \neq 0$.

Now,
$$\lim_{z \to z_0} (z - z_0) f(z) = \lim_{z \to z_0} \frac{(z - z_0)\phi(z)}{\psi(z)}$$

$$= \phi(z_0) \lim_{z \to z_0} \frac{1}{\psi'(z)} \qquad (L' \text{ Hospital rule})$$

$$= \frac{\phi(z_0)}{\psi'(z_0)}$$

1.13.3 Residue at Pole of Order *m*

If $f(z)$ has pole of order m at $z = z_0$ then in some neighbourhood of z_0, Laurent series expansion of $f(z)$ is

$$f(z) = \left[a_0 + a_1(z-z_0) + a_2(z-z_0)^2 + \cdots\right] + \left[\frac{a_{-1}}{z-z_0} + \frac{a_{-2}}{(z-z_0)^2} + \cdots\cdots + \frac{a_{-m}}{(z-z_0)^m}\right], \; a_{-m} \neq 0$$

$$\therefore \; (z-z_0)^m f(z) = \left[a_0(z-z_0)^m + a_1(z-z_0)^{m+1} + a_2(z-z_0)^{m+2} + \cdots\right]$$

$$+\left[a_{-1}(z-z_0)^{m-1} + a_{-2}(z-z_0)^{m-2} + \cdots\cdots + a_{-m}\right]$$

$$\therefore \; \frac{d^{m-1}}{dz^{m-1}}\left\{(z-z_0)^m f(z)\right\} = \left[a_0\frac{m!}{1!}(z-z_0) + a_1\frac{(m+1)!}{2!}(z-z_0)^2\right.$$

$$\left. + a_2\frac{(m+2)!}{3!}(z-z_0)^3 + \ldots\right] + a_{-1}(m-1)!$$

$$\therefore \; \lim_{z\to z_0} \frac{d^{m-1}}{dz^{m-1}}\left\{(z-z_0)^m f(z)\right\} = (m-1)!\; a_{-1}$$

$$\therefore \; \text{Res}\,(z_0) = a_{-1} = \frac{1}{(m-1)!}\lim_{z\to z_0}\frac{d^{m-1}}{dz^{m-1}}\left\{(z-z_0)^m f(z)\right\}$$

1.13.4 Residue at an Isolated Essential Singularity

When $z = z_0$ is essential singularity of $f(z)$ then expand $f(z)$ in Laurent series about $z = z_0$, the coefficient of $(z - z_0)^{-1}$ will be residue. In this case it is the only way to find residue at $z = z_0$.

Example 1.62: Determine the poles of the function $f(z) = \dfrac{z^2}{(z-1)(z-2)^2}$ and the residue at each pole.

Solution: $f(z)$ has simple pole at $z = 1$ and double pole at $z = 2$

$$\text{Res}\,(1) = \lim_{z\to 1}(z-1)f(z) = \lim_{z\to 1}\frac{z^2}{(z-2)^2} = 1$$

$$\text{Res}\,(2) = \frac{1}{1!}\left(\frac{d}{dz}(z-2)^2 f(z)\right)_{z=2} = \frac{d}{dz}\frac{z^2}{z-1}\bigg|_{z=2}$$

$$= \frac{(z-1)2z - z^2}{(z-1)^2}\bigg|_{z=2} = \frac{z^2-2z}{(z-1)^2}\bigg|_{z=2} = \frac{4-4}{1} = 0$$

Example 1.63: Determine poles and residues of $f(z)=\dfrac{1}{z^4+1}$ at each of its poles.

Solution: $z^4+1=0 \Rightarrow z^4=-1=i^2 \Rightarrow z^2=\pm i=\frac{1}{2}(1\pm i)^2$

$$\therefore \qquad z=\pm\frac{1}{\sqrt{2}}(1\pm i)$$

$\therefore$ Pole of $f(z)$ are simple poles at $z=z_1, z_2, z_3, z_4$

where $\quad z_1=\frac{1}{\sqrt{2}}(1+i),\ z_2=\frac{1}{\sqrt{2}}(1-i),\ z_3=-\frac{1}{\sqrt{2}}(1+i),\ z_4=-\frac{1}{\sqrt{2}}(1-i)$

$$\text{Res }(z_j)=\lim_{z\to z_j}\frac{(z-z_j)}{z^4+1}=\lim_{z\to z_j}\frac{1}{4z^3}\quad (L'\text{Hospital rule})$$

$$=\frac{1}{4z_j^3}=\frac{z_j}{4z_j^{\,4}}=-\frac{z_j}{4}\quad (\because z_j^{\,4}=-1)\ ;\ j=1,2,3,4$$

$$\therefore \qquad \text{Res}(z_1)=-\frac{1}{4\sqrt{2}}(1+i),\ \text{Res}(z_2)=-\frac{1}{4\sqrt{2}}(1-i)$$

$$\text{Res}(z_3)=\frac{1}{4\sqrt{2}}(1+i),\ \text{Res}(z_4)=\frac{1}{4\sqrt{2}}(1-i)$$

Example 1.64: Find the residues of $\dfrac{1}{z(z+2)^3}$ at $z=0$ and $z=-2$.

Solution: $f(z)=\dfrac{1}{z(z+2)^3}$ has simple pole at $z=0$ and pole of order 3 at $z=-2$

$$\text{Res }(0)=\lim_{z\to 0} z\,f(z)=\lim_{z\to 0}\frac{1}{(z+2)^3}=\frac{1}{8}$$

$$\text{Res }(-2)=\frac{1}{2!}\left(\frac{d^2}{dz^2}(z+2)^3 f(z)\right)_{z=-2}=\frac{1}{2}\left(\frac{d^2}{dz^2}\frac{1}{z}\right)_{z=-2}=\frac{1}{2}\frac{2}{z^3}\bigg|_{z=-2}=-\frac{1}{8}$$

Example 1.65: Compute the residues at all the singular points of $f(z)$ where $f(z)$ is given by

(i) $\dfrac{z^2}{z^2-2z+2}$ (ii) $\left(\dfrac{z+1}{z-1}\right)^3$ (iii) $\dfrac{z^2-2z}{(z+1)^2(z^2+4)}$ (iv) $\dfrac{1}{(z+1)^3}$

(v) $\dfrac{(z+3)^3}{(z-1)^4}$ (vi) $\dfrac{z^2}{z^n-1}$; $n\in N$ (vii) $\dfrac{1}{z^3+z^5}$

Solution: (i) $f(z)=\dfrac{z^2}{z^2-2z+2}$ has simple poles at $z=\dfrac{2\pm\sqrt{4-8}}{2}=1\pm i$

$$\text{Res }(1+i)=\lim_{z\to 1+i}(z-1-i)f(z)=\lim_{z\to 1+i}\frac{z^2}{z-1+i}=\frac{(1+i)^2}{2i}=1$$

$$\text{Res }(1-i) = \lim_{z\to 1-i}(z-1+i)f(z) = \lim_{z\to 1-i} \frac{z^2}{z-1-i} = \frac{(1-i)^2}{-2i} = 1$$

(ii) $f(z) = \left(\dfrac{z+1}{z-1}\right)^3$

$$f(z) = \frac{1}{(z-1)^3}(z-1+2)^3 = \frac{1}{(z-1)^3}\left[(z-1)^3 + 6(z-1)^2 + 12(z-1) + 8\right]$$

$$= 1 + \frac{6}{z-1} + \frac{12}{(z-1)^2} + \frac{8}{(z-1)^3}$$

$f(z)$ has pole of order 3 at $z = 1$.

$\therefore$ Res (1) = coeff of $\dfrac{1}{z-1} = 6$

(iii) $f(z) = \dfrac{z^2-2z}{(z+1)^2(z^2+4)}$ has double pole at $z = -1$, simple poles at $\pm 2i$

$$\text{Res }(-1) = \frac{d}{dz}(z+1)^2 f(z)\bigg|_{z=-1} = \frac{d}{dz}\frac{z^2-2z}{z^2+4}\bigg|_{z=-1} = \frac{(z^2+4)(2z-2)-(z^2-2z)2z}{(z^2+4)^2}\bigg|_{z=-1}$$

$$= \frac{5(-4)-(3)(-2)}{25} = -\frac{14}{25}$$

$$\text{Res }(2i) = \lim_{z\to 2i}(z-2i)f(z) = \lim_{z\to 2i}\frac{z^2-2z}{(z+1)^2(z+2i)} = \frac{-4-4i}{(1+2i)^2(4i)}$$

$$= \frac{-1-i}{(-3+4i)i} = \frac{-1-i}{-4-3i} = \frac{(1+i)(4-3i)}{16+9} = \frac{7+i}{25}$$

$$\text{Res }(-2i) = \lim_{z\to -2i}(z+2i)f(z) = \lim_{z\to -2i}\frac{z^2-2z}{(z+1)^2(z-2i)} = \frac{-4+4i}{(1-2i)^2(-4i)}$$

$$= \frac{1-i}{(-3-4i)i} = \frac{1-i}{4-3i} = \frac{(1-i)(4+3i)}{25} = \frac{7-i}{25}$$

(iv) $f(z) = \dfrac{1}{(z+1)^3}$ has triple pole at $z = -1$

$$\text{Res }(-1) = \frac{1}{2!}\lim_{z\to -1}\frac{d^2}{dz^2}(z+1)^3 f(z) = \frac{1}{2}\lim_{z\to -1}\frac{d^2}{dz^2}(1) = 0$$

Other Method

$f(z) = \dfrac{1}{(z+1)^3}$ is Laurent expansion

$\therefore$ Res (-1) = coeff of $\dfrac{1}{z+1} = 0$

(v) $f(z)=\dfrac{(z+3)^3}{(z-1)^4}$ has pole of order 4 at $z = 1$

$$f(z)=\frac{1}{(z-1)^4}[z-1+4]^3=\frac{1}{(z-1)^4}\left[(z-1)^3+12(z-1)^2+48(z-1)+64\right]$$

$$=\frac{1}{z-1}+\frac{12}{(z-1)^2}+\frac{48}{(z-1)^3}+\frac{64}{(z-1)^4}$$

$\therefore$ $\quad \text{Res}(1)=\text{coeff of } \dfrac{1}{z-1}=1$

(vi) $f(z)=\dfrac{z^2}{z^n-1}$,

Poles of $f(z)$ are simple poles at nth roots of unity, i.e., at $z_k=\rho^k$; $k = 0, 1, 2, \ldots, n-1$

where $\rho=\cos\dfrac{2\pi}{n}+i\sin\dfrac{2\pi}{n}=e^{i\frac{2\pi}{n}}$

$$\text{Res }(z_k)=\lim_{z\to z_k}\frac{(z-z_k)z^2}{z^n-1}=\lim_{z\to z_k}z^2\lim_{z\to z_k}\frac{z-z_k}{z^n-1}$$

$$=z_k^2\lim_{z\to z_k}\frac{1}{nz^{n-1}} \qquad (L' \text{ Hospital rule})$$

$$=\frac{1}{nz_k^{\,n-3}}=\frac{z_k^3}{n}\quad (\because\ z_k^n=1)$$

$\therefore$ $\quad \text{Res }(z_k)=\dfrac{1}{n}e^{\frac{6\pi ik}{n}}$; $k=0,1,2,\cdots\cdots,n-1$

(vii) $f(z)=\dfrac{1}{z^3+z^5}=\dfrac{1}{z^3(z^2+1)}$ has simple poles at $z=\pm i$ and triple pole at $z = 0$

$$\text{Res }(i)=\lim_{z\to i}\frac{z-i}{z^3(z-i)(z+i)}=\lim_{z\to i}\frac{1}{z^3(z+i)}=\frac{1}{-i(2i)}=\frac{1}{2}$$

$$\text{Res }(-i)=\lim_{z\to -i}(z+i)f(z)=\lim_{z\to -i}\frac{1}{z^3(z-i)}=\frac{1}{i(-2i)}=\frac{1}{2}$$

Now $\quad f(z)=\dfrac{1}{z^3}(1+z^2)^{-1}=\dfrac{1}{z^3}(1-z^2+z^4-z^6+\cdots\cdots\cdots)$

$$=\frac{1}{z^3}-\frac{1}{z}+z-z^3+\cdots\cdots\cdots$$

$\therefore$ $\quad \text{Res }(0)=\text{coeff of } \dfrac{1}{z}=-1$

Example 1.66: Determine the poles of the function $\dfrac{z}{\cos z}$ and the residue at each pole.

Solution: $f(z)=\dfrac{z}{\cos z}$ has simple poles at $z=(2n+1)\dfrac{\pi}{2}$; $n\in I$

$$\text{Res}\left(n\pi+\frac{\pi}{2}\right)=\lim_{z\to n\pi+\frac{\pi}{2}} z\cdot\frac{z-n\pi-\frac{\pi}{2}}{\cos z}$$

$$=\left(n\pi+\frac{\pi}{2}\right)\lim_{z\to n\pi+\frac{\pi}{2}}\frac{z-n\pi-\frac{\pi}{2}}{\cos z}$$

$$=\left(n\pi+\frac{\pi}{2}\right)\lim_{z\to n\pi+\frac{\pi}{2}}\frac{1}{-\sin z}\quad (L'\text{ Hospital rule})$$

$$=-\left(n\pi+\frac{\pi}{2}\right)\frac{1}{\sin\left(n\pi+\frac{\pi}{2}\right)}=(-1)^{n+1}(2n+1)\frac{\pi}{2};\quad n\in I$$

Example 1.67: Find the residue of $\dfrac{z^3}{(z-1)^4(z-2)(z-3)}$ at $z=1$.

Solution: $f(z)=\dfrac{z^3}{(z-1)^4(z-2)(z-3)}$ has pole of order 4 at $z=1$

$$(z-1)^4 f(z)=\frac{z^3}{(z-2)(z-3)}$$

$$=z+5+\frac{19}{z-3}+\frac{8}{(z-2)(z-3)}$$

Synthetic division

```
2) 1  0  0  0
      2  4  8
3) 1  2  4 |8
      3  15|
   1  5| 19
```

$$=z+5+\frac{19}{z-3}-\frac{8}{z-2}+\frac{8}{z-3}$$

(By suppression method)

$$=z+5+\frac{27}{z-3}-\frac{8}{z-2}$$

$$\text{Res}(1)=\frac{1}{3!}\lim_{z\to 1}\frac{d^3}{dz^3}(z-1)^4 f(z)=\frac{1}{6}\lim_{z\to 1}\frac{d^3}{dz^3}\left(z+5+\frac{27}{z-3}-\frac{8}{z-2}\right)$$

$$=\frac{1}{6}\lim_{z\to 1}\left[\frac{-27\times 6}{(z-3)^4}+\frac{8\times 6}{(z-2)^4}\right]=\frac{1}{6}\left[\frac{-27\times 6}{16}+48\right]$$

$$=-\frac{27}{16}+8=\frac{101}{16}$$

Other Method

$$f(z)=\frac{z^3}{(z-1)^4}\left[\frac{1}{z-3}-\frac{1}{z-2}\right]$$

$$=\frac{(1+z-1)^3}{(z-1)^4}\left[\frac{1}{-2+z-1}-\frac{1}{-1+z-1}\right]$$

$$=\frac{1+3(z-1)+3(z-1)^2+(z-1)^3}{(z-1)^4}\left[-\frac{1}{2}\left(1-\frac{z-1}{2}\right)^{-1}+(1-(z-1))^{-1}\right]$$

$$=\left[\frac{1}{(z-1)^4}+\frac{3}{(z-1)^3}+\frac{3}{(z-1)^2}+\frac{1}{z-1}\right]\left[-\frac{1}{2}\left(1+\frac{z-1}{2}+\frac{(z-1)^2}{4}+\frac{(z-1)^3}{8}+\cdots\cdots\right)\right.$$

$$\left.+\left(1+(z-1)+(z-1)^2+(z-1)^3+\cdots\cdots\right)\right]$$

$$\therefore \text{ Res }(1)=\text{coeff of }\frac{1}{z-1}=-\frac{1}{16}+1-\frac{3}{8}+3-\frac{3}{4}+3-\frac{1}{2}+1$$

$$=8-\frac{1+6+12+8}{16}=8-\frac{27}{16}$$

$$=\frac{101}{16}$$

Example 1.68: Compute the residues at all the singular points of

(i) $\dfrac{ze^{iz}}{z^2+a^2}$ (ii) $\dfrac{1-e^{2z}}{z^4}$ (iii) $\dfrac{e^{z^2}}{(z-i)^3}$ (iv) $\dfrac{1-\cos z}{z}$

(v) $\cot z$ (vi) $\sec z$

Solution: (i) $f(z)=\dfrac{ze^{iz}}{z^2+a^2}$

Singularities of $f(z)$ are simple poles at $z=\pm ia$

$$\text{Res}(ia)=\lim_{z\to ia}(z-ia)f(z)=\lim_{z\to ia}\frac{ze^{iz}}{z+ia}=\frac{iae^{-a}}{2ia}=\frac{1}{2}e^{-a}$$

$$\text{Res}(-ia)=\lim_{z\to -ia}(z+ia)f(z)=\lim_{z\to -ia}\frac{ze^{iz}}{z-ia}=\frac{1}{2}e^{a}$$

(ii) $f(z)=\dfrac{1-e^{2z}}{z^4}$ has pole of order 3 at origin

$$\because \quad f(z)=\frac{1}{z^4}\left[1-\left(1+2z+\frac{(2z)^2}{2!}+\frac{(2z)^3}{3!}+\frac{(2z)^4}{4!}\cdots\right)\right]$$

$$=-\frac{2}{z^3}-\frac{2}{z^2}-\frac{4}{3z}-\frac{2}{3}-\frac{2^5}{5!}z-\frac{2^6}{6!}z^2-\frac{2^7}{7!}z^3\cdots$$

$$\therefore \quad \text{Res }(0)=\text{coeff. of }\frac{1}{z}=-\frac{4}{3}$$

(iii) $f(z)=\dfrac{e^{z^2}}{(z-i)^3}$ has pole of order 3 at $z=i$

$$\text{Res }(i) = \frac{1}{2!}\lim_{z\to i}\frac{d^2}{dz^2}(z-i)^3 f(z)=\frac{1}{2}\lim_{z\to i}\frac{d^2}{dz^2}e^{z^2}$$

$$=\frac{1}{2}\lim_{z\to i}\frac{d}{dz}2ze^{z^2}=\frac{1}{2}\lim_{z\to i}\left(4z^2+2\right)e^{z^2}=-\frac{1}{e}$$

(iv) $f(z)=\dfrac{1-\cos z}{z}$

$$\lim_{z\to 0} f(z)=\lim_{z\to 0}\frac{1-\cos z}{z}=\lim_{z\to 0}\frac{\sin z}{1}=0 \qquad (L'\text{Hospital rule})$$

$\therefore f(z)$ has removable singularity at $z=0$

$\therefore$ Res $(0)=0$

(v) $f(z)=\cot z=\dfrac{\cos z}{\sin z}$ has simple poles at $z=n\pi$, $n\in I$

$$\text{Res }(n\pi)=\lim_{z\to n\pi}(z-n\pi)f(z)$$

$$=\lim_{z\to n\pi}\cos z\cdot\frac{z-n\pi}{\sin z}$$

$$=(-1)^n\lim_{z\to n\pi}\frac{z-n\pi}{\sin z}$$

$$=(-1)^n\lim_{z\to n\pi}\frac{1}{\cos z} \qquad (L'\text{Hospital rule})$$

$$=(-1)^n\cdot\frac{1}{(-1)^n}=1;\ n\in I$$

(vi) $f(z)=\sec z=\dfrac{1}{\cos z}$ has simple poles at $z=(2n+1)\dfrac{\pi}{2}$, $n\in I$

$$\text{Res}\left(n\pi+\frac{\pi}{2}\right)=\lim_{z\to n\pi+\frac{\pi}{2}}\frac{z-n\pi-\frac{\pi}{2}}{\cos z}$$

$$=\lim_{z\to n\pi+\frac{\pi}{2}}\frac{1}{-\sin z} \qquad (L'\text{Hospital rule})$$

$$=-\frac{1}{\sin\left(n\pi+\frac{\pi}{2}\right)}=(-1)^{n+1};\ n\in I$$

1.14 EVALUATION OF CONTOUR INTEGRALS USING RESIDUES

In this section, we discuss the application of residues in evaluating integrals of $f(z)$ over a simple closed curve. Following result is used to evaluate these integrals.

Theorem 1.22 Cauchy Residue Theorem

Let C be a simple closed curve and $f(z)$ be analytic on and inside C except at a finite no. of isolated singularities $z_1, z_2, \ldots\ldots, z_n$ lying inside C then

$$\oint_C f(z)\,dz = 2\pi i \sum_{k=1}^{n} \text{Res}(z_k)$$

$$= 2\pi i \text{ (sum of residues of } f(z) \text{ at its isolated singularities inside } C)$$

Proof: $z_1, z_2, \ldots\ldots, z_n$ are isolated singularities and hence there exist non-intersecting circles C_k with centre at z_k; $k = 1, 2, \ldots, n$ lying inside C such that each C_k contains only one singularity z_k inside it. By extension of Cauchy–Goursat theorem for multiply connected domains (Remark 1.9), we have

$$\oint_C f(z)\,dz = \sum_{k=1}^{n} \oint_{C_k} f(z)\,dz$$

By equation (1.40)

$$\oint_{C_k} f(z)\,dz = 2\pi i \ \text{Res}(z_k);\ k = 1, 2, \ldots, n$$

Thus,

$$\oint_C f(z)dz = 2\pi i \sum_{k=1}^{n} \text{Res}(z_k)$$

$$= 2\pi i \ [\text{sum of residues of } f(z) \text{ at its isolated singularities inside } C]$$

Example 1.69: Evaluate the following integral using residue theorem

$$\int_C \frac{z^2 dz}{(z-1)^2 (z+2)} \text{ where } C: |z| = 3$$

Solution: $f(z) = \dfrac{z^2}{(z-1)^2 (z+2)}$ has pole of order two at $z = 1$ and simple pole at $z = -2$. Both of these poles lie inside $|z| = 3$.

$$\text{Res}(1) = \lim_{z\to 1} \frac{d}{dz}(z-1)^2 f(z) = \lim_{z\to 1} \frac{d}{dz} \frac{z^2}{z+2}$$

$$= \lim_{z\to 1} \frac{d}{dz}\left(z - 2 + \frac{4}{z+2}\right) = \lim_{z\to 1}\left(1 - \frac{4}{(z+2)^2}\right) = 1 - \frac{4}{9} = \frac{5}{9}$$

$$\text{Res}(-2) = \lim_{z\to -2}(z+2) f(z) = \lim_{z\to -2} \frac{z^2}{(z-1)^2} = \frac{4}{9}$$

∴ By Cauchy residues theorem

$$\oint_C \frac{z^2 dz}{(z-1)^2 (z+2)} = 2\pi i \text{ (Sum of residues of } f(z) \text{ at isolated singularities inside } C)$$

$$= 2\pi i \left(\frac{5}{9} + \frac{4}{9}\right) = 2\pi i$$

Example 1.70: State residue theorem and use it to evaluate $\int_C \frac{dz}{z^8(z+4)}$ where C is the circle

(i) $|z| = 2$ (ii) $|z+2| = 3$

Solution: Statement of Cauchy residue theorem is already given in Theorem 1.22.

Now, $f(z) = \frac{1}{z^8(z+4)}$ has pole of order 8 at $z = 0$ and simple pole at $z = -4$

$$\text{Res}(0) = \frac{1}{7!}\lim_{z\to 0}\frac{d^7}{dz^7}z^8 f(z) = \frac{1}{7!}\lim_{z\to 0}\frac{d^7}{dz^7}\frac{1}{(z+4)}$$

$$= \frac{1}{7!}\lim_{z\to 0}\frac{(-1)^7 7!}{(z+4)^8} = -\frac{1}{4^8}$$

$$\text{Res}(-4) = \lim_{z\to -4}(z+4)f(z) = \lim_{z\to -4}\frac{1}{z^8} = \frac{1}{(-4)^8} = \frac{1}{4^8}$$

(i) $z = 0$ lies inside C and $z = -4$ lies outside C

$\therefore$ By Cauchy residues theorem

$$\int_C f(z)dz = 2\pi i\left(-\frac{1}{4^8}\right) = -\frac{2\pi i}{4^8}.$$

(ii) $|0+2| = 2 < 3$, $|-4+2| = 2 < 3$

$\therefore$ Both $z = 0$ and $z = -4$ lie inside C

$\therefore$ By Cauchy residue theorem

$$\int_C f(z)dz = 2\pi i \text{ (Sum of residues of } f(z) \text{ at isolated singularities inside } C)$$

$$= 2\pi i\left(-\frac{1}{4^8} + \frac{1}{4^8}\right) = 0$$

Example 1.71: Using residue theorem, evaluate $\int_C \frac{dz}{z^4+1}$ which C is the circle $x^2 + y^2 = 2x$.

Solution: $C: x^2 + y^2 = 2x$ or $(x-1)^2 + y^2 = 1$ is circle $|z-1| = 1$

Now, $f(z) = \frac{1}{z^4+1}$ has simples poles at zeros of $z^4 + 1 = 0$

i.e., $z^4 = -1 = i^2 \Rightarrow z^2 = \pm i = \frac{1}{2}(\pm 2i)$

$\therefore$ $z^2 = \frac{1}{2}(1\pm i)^2 \Rightarrow z = \pm\frac{1}{\sqrt{2}}(1\pm i)$

$\therefore$ $f(z)$ has simple poles at z_1, z_2, z_3, z_4

where $z_1 = \frac{1}{\sqrt{2}}(1+i)$, $z_2 = \frac{1}{\sqrt{2}}(1-i)$, $z_3 = -\frac{1}{\sqrt{2}}(1+i)$,

$z_4 = -\frac{1}{\sqrt{2}}(1-i)$

Now, $\left|\frac{1}{\sqrt{2}} \pm \frac{i}{\sqrt{2}} - 1\right| = \sqrt{\frac{\left(\sqrt{2}-1\right)^2 + 1}{2}} < 1$

and $\left|-\frac{1}{\sqrt{2}} \pm \frac{i}{\sqrt{2}} - 1\right| = \sqrt{\frac{\left(\sqrt{2}+1\right)^2 + 1}{2}} > 1$

$\therefore$ Only z_1 and z_2 lie inside C.

$$\text{Res}(z_1) = \lim_{z \to z_1} \frac{z - z_1}{z^4 + 1} = \lim_{z \to z_1} \frac{1}{4z^3} \qquad (\text{L}' \text{ Hospital rule})$$

$$= \frac{z_1}{4z_1^4} = -\frac{z_1}{4} \qquad (\because z_1^4 = -1)$$

Similarly, $\text{Res}(z_2) = -\frac{z_2}{4}$

$\therefore$ By Cauchy residue theorem

$$\int_C \frac{dz}{z^4 + 1} = 2\pi i \text{ (Sum of residues of } f(z) \text{ at isolated singularities inside } C)$$

$$= 2\pi i\left(-\frac{z_1}{4} - \frac{z_2}{4}\right) = -\frac{\pi}{2} i(z_1 + z_2) = -\frac{\pi}{2} i\sqrt{2} = -\frac{\pi i}{\sqrt{2}}.$$

Example 1.72: Evaluate by residue theorem $\int_C \frac{e^z dz}{(z+1)^2 (z-2)}$ where C is the circle $|z-1| = 3$.

Solution: $f(z) = \frac{e^z}{(z+1)^2 (z-2)}$ has simple pole at $z = 2$ and double pole at $z = -1$

$$|2-1| = 1 < 3, \ |-1-1| = 2 < 3$$

$\therefore$ Both $z = 2$ and $z = -1$ lie inside C

$$\text{Res (2)} = \lim_{z \to 2}(z-2) f(z) = \lim_{z \to 2} \frac{e^z}{(z+1)^2} = \frac{e^2}{9}$$

$$\text{Res (-1)} = \lim_{z \to -1} \frac{d}{dz}(z+1)^2 f(z) = \lim_{z \to -1} \frac{d}{dz} \frac{e^z}{z-2}$$

$$= \lim_{z \to -1} \frac{(z-2)e^z - e^z}{(z-2)^2} = \frac{(-1-2)e^{-1} - e^{-1}}{(-1-2)^2} = -\frac{4}{9e}$$

$\therefore$ By Cauchy residue theorem

$$\oint_C f(z) dz = 2\pi i \text{ (Sum of residues of } f(z) \text{ at isolated singularities inside } C)$$

$$= 2\pi i\left(\frac{e^2}{9} - \frac{4}{9e}\right) = \frac{2\pi i}{9}\left(e^2 - \frac{4}{e}\right)$$

Example 1.73: Evaluate the following integrals

(i) $\int_C \frac{dz}{z\sin z}$ where $C:|z|=1$ (ii) $\oint_C ze^{\frac{1}{z}}dz$ where $C:|z|=1$

(iii) $\oint_C \frac{dz}{\sinh 2z}$ where $C:|z|=2$ (iv) $\int_C \frac{\coth z}{z-i}dz$ where $C:|z|=2$

(v) $\oint_C e^{-\frac{1}{z}}\sin\frac{1}{z}dz$ where $C:|z|=1$ (vi) $\int \tan z\, dz$ where $C:|z|=2$

Solutions: (i) $f(z)=\dfrac{1}{z\sin z}$ has poles at $z=n\pi$; $n\in I$. Double pole at $z=0$ and simple poles at $z=n\pi$, $n\neq 0$, $n\in I$. Only $z=0$ lies inside $C:|z|=1$.

Now, $$f(z)=\frac{1}{z\sin z}=\frac{1}{z\left(z-\frac{z^3}{3!}+\frac{z^5}{5!}-\cdots\right)}=\frac{1}{z^2}\left[1-\left(\frac{z^2}{3!}-\frac{z^4}{5!}+\cdots\right)\right]^{-1}$$

$$=\frac{1}{z^2}\left(1+\frac{z^2}{3!}+\cdots\right)$$

$\therefore$ Res (0) = Coeff of $\frac{1}{z}=0$

$\therefore$ By Cauchy residue theorem

$$\oint_C f(z)dz = 2\pi i \text{ (Sum of residues of } f(z) \text{ at isolated singularities inside } C) = 0$$

(ii) $f(z)=ze^{\frac{1}{z}}$ has singularity at $z=0$ only which lies inside $C:|z|=1$.

$$f(z)=z\left(1+\frac{1}{z}+\frac{1}{2!z^2}+\frac{1}{3!z^3}+\cdots\right)$$

$\therefore$ Res (0) = Coeff of $\frac{1}{z}=\frac{1}{2}$

$\therefore$ By Cauchy residue theorem

$$\oint_C f(z)dz = 2\pi i\left(\frac{1}{2}\right)=\pi i$$

(iii) $f(z)=\dfrac{1}{\sinh 2z}$ has singularity at $z=0$ only which lies inside $C:|z|=2$.

$$\lim_{z\to 0} z f(z)=\lim_{z\to 0}\frac{z}{\sinh 2z}=\lim_{z\to 0}\frac{1}{2\cosh 2z} \quad (L' \text{ Hospital rule})$$

$$=\frac{1}{2}\neq 0$$

$\therefore f(z)$ has simple pole at $z = 0$

and Res $(0) = \dfrac{1}{2}$

$\therefore$ By Cauchy residue theorem

$$\oint_C f(z)\,dz = 2\pi i\left(\frac{1}{2}\right) = \pi i$$

(iv) $f(z) = \dfrac{\coth z}{z - i}$ has simple poles at $z = 0, i$.

Both $z = 0, i$ lie inside $C : |z| = 2$

$$\text{Res }(0) = \lim_{z\to 0}\frac{z\cosh z}{(z-1)\sinh z} = \frac{1}{-i}\lim_{z\to 0}\frac{z}{\sinh z}$$

$$= i\lim_{z\to 0}\frac{1}{\cosh z} = i \qquad (L'\text{ Hospital rule})$$

$$\text{Res }(i) = \lim_{z\to i}(z-i)f(z) = \lim_{z\to i}\coth z = \coth i$$

$\therefore$ By Cauchy residue theorem

$$\oint_C f(z)dz = 2\pi i\ (\text{Sum of residues of } f(z) \text{ at isolated singularities inside } C) = 2\pi i\left(i + \coth i\right)$$

(v) $f(z) = e^{-\frac{1}{z}}\sin\dfrac{1}{z}$ has singularity at $z = 0$ only which lies inside $C : |z| = 1$.

$$f(z) = \left(1 - \frac{1}{z} + \frac{1}{2!z^2} - \frac{1}{3!z^3} + \cdots\right)\left(\frac{1}{z} - \frac{1}{3!z^3} + \cdots\right)$$

$$\text{Res }(0) = \text{Coeff of } \frac{1}{z} = 1$$

$\therefore$ By Cauchy residue theorem

$$\oint_C f(z)\,dz = 2\pi i \qquad (\text{Sum of residues of } f(z) \text{ at isolated singularities inside } C)$$

$$= 2\pi i\ (1) = 2\pi i$$

(vi) $f(z) = \tan z$ has simple poles at $z = (2n + 1)\left(\dfrac{\pi}{2}\right)$; $n \in I$ of which only $z = \pm\dfrac{\pi}{2}$ lie inside $C : |z| = 2$

$$\text{Res}\left(\frac{\pi}{2}\right) = \lim_{z\to\frac{\pi}{2}}\frac{\left(z - \frac{\pi}{2}\right)\sin z}{\cos z} = (1)\lim_{z\to\frac{\pi}{2}}\frac{z - \frac{\pi}{2}}{\cos z}$$

$$= \lim_{z\to -\frac{\pi}{2}}\frac{1}{-\sin z} \qquad (L'\text{ Hospital rule})$$

$$= -1$$

$$\text{Res}\left(-\frac{\pi}{2}\right) = \lim_{z\to\frac{-\pi}{2}} \frac{\left(z+\frac{\pi}{2}\right)\sin z}{\cos z} = (-1)\lim_{z\to\frac{-\pi}{2}} \frac{z+\frac{\pi}{2}}{\cos z}$$

$$= -\lim_{z\to\frac{-\pi}{2}} \frac{1}{-\sin z} \quad (L' \text{ Hospital rule})$$

$$= -1$$

∴ By Cauchy residue theorem

$$\oint_C f(z)dz = 2\pi i \text{ (Sum of residues of } f(z) \text{ at isolated singularities inside } C)$$

$$= 2\pi i\,(-1-1) = -4\pi i$$

Example 1.74: Evaluate the integral $\int_C \frac{e^{2z}}{(z+1)^n}dz,\ C:\ |z|=2.$

Solution: $f(z)=\frac{e^{2z}}{(z+1)^n}$ has pole of order n at $z=-1$ which lies inside $C:\ |z|=2$

$$\text{Res}\,(-1) = \frac{1}{(n-1)!}\cdot\lim_{z\to-1}\frac{d^{n-1}}{dz^{n-1}}(z+1)^n f(z)$$

$$= \frac{1}{(n-1)!}\lim_{z\to-1}\frac{d^{n-1}}{dz^{n-1}}e^{2z} = \frac{1}{(n-1)!}\lim_{z\to-1} 2^{n-1}e^{2z}$$

$$= \frac{2^{n-1}e^{-2}}{(n-1)!}$$

∴ By Cauchy residue theorem

$$\oint_C f(z)\,dz = 2\pi i \text{ (Sum of residues of } f(z) \text{ at isolated singularities inside } C)$$

$$= 2\pi i\cdot\frac{2^{n-1}e^{-2}}{(n-1)!} = \frac{2^n\pi i}{(n-1)!e^2}$$

EXERCISE 1.4

1. Write the zeros of the following functions and their order

   (i) $z^2\sin z$ (ii) $\left(\frac{z+1}{z^2+1}\right)^3$

   (iii) $\frac{(z+1)(z-2)}{(z-3)(z+3)}$

2. Find the type of singularities of the functions at $z=0$

   (i) $f(z)=\frac{z-\sin z}{z^3}$

   (ii) $f(z)=\frac{\sin z}{z^r}, r\geq 2$ is a positive integer

3. Find the location and type of singularity of the following functions
   (i) $\dfrac{\sin(z-2)}{(z-2)}$
   (ii) $(z+1)^2 e^{1/(z+1)}$
   (iii) $\dfrac{2z}{(z-1)^3(z+2)}$

4. Show that the function
   (i) cosec z has a simple pole at $z = 0$.
   (ii) $\dfrac{1}{z^2-1}$ has simple poles at $z = 1$ and $z = -1$.
   (iii) $\dfrac{e^z}{z^3}$ has a pole of order 3 at $z = 0$
   (iv) $z\sin\dfrac{1}{z}$ has essential singularity at $z = 0$.

5. Classify the singular point $z = 0$ of the functions
   (i) $\dfrac{e^z}{z+\sin z}$ (ii) $\dfrac{e^z}{z-\sin z}$

6. Compute the residues at all the singular points of $f(z)$ where $f(z)$ is given by
   (i) $\dfrac{z}{(z+1)(z-2)}$
   (ii) $\dfrac{2z+1}{(z^2-z-2)}$
   (iii) $\dfrac{z+1}{z^2-2z}$
   (iv) $\dfrac{z^3}{(z-1)(z-2)(z-3)}$
   (v) $\dfrac{1+z+z^2}{(z-1)^2(z+2)}$
   (vi) $\dfrac{z^2}{(z-1)^2(z+2)}$
   (vii) $\dfrac{(z+1)}{(z^2-16)(z+2)}$
   (viii) $\dfrac{z^2}{(z^2+3z+2)^2}$
   (ix) $\dfrac{z^2}{(z^2+1)^2}$

7. Find the sum of the residues of the function $f(z) = \dfrac{\sin z}{z\cos z}$ at its poles inside the circle $|z| = 2$.

8. Compute the residues at all the singular points of
   (i) $z\sin\dfrac{1}{z}$ (ii) $z\cos\dfrac{1}{z}$ (iii) $\dfrac{1-e^{2z}}{z^3}$
   (iv) $\dfrac{ze^z}{(z-a)^3}$ (v) $\dfrac{\sin z}{z^2}$ (vi) $\dfrac{e^{z^2}}{z^3}$

9. Evaluate the following integrals using residue theorem
   (i) $\int_C \dfrac{1+z}{z(2-z)}dz$ where C is the circle $|z| = 1$
   (ii) $\int_C \dfrac{(z+3)}{(z+1)^2(z-2)}dz$ where $C : |z| = 3$
   (iii) $\oint_C \dfrac{2z-1}{z(z+1)(z-3)}dz$ where $C : |z| = 2$
   (iv) $\oint_C \dfrac{1}{z^3(z+4)}dz$ where $C : |z+2| = 3$
   (v) $\int_C \dfrac{4-3z}{z(z-1)(z-2)}dz$ where $C : |z| = \dfrac{3}{2}$

10. Evaluate the integral $\oint_C \dfrac{z-3}{z^2+2z+5}dz$ using residue theorem, where C is the circle
   (i) $|z| = 1$ (ii) $|z+1-i| = 2$
   (iii) $|z+1+i| = 2$

11. Evaluate the contour integral $I = \oint_C \frac{dz}{(e^z - 1)}, C: |z| = 1$

12. Evaluate $I = \oint_C \frac{z^2+4}{z^3+2z^2+2z} dz$ where C is
   (i) $|z| = 1$ (ii) $|z+1-i| = 1$
   (iii) $|z+1+i| = 1$ (iv) $|z-1| = 5$
   (v) rectangle with vertices at $2 + i$, $6 + i$, $2 + 4i$ and $6 + 4i$

13. Evaluate the integral
$$\int_C \frac{e^z}{(z+1)^n} dz, C: |z| = 2$$

14. Evaluate the following integral
   (i) $\int_C \frac{dz}{\cosh z}$ where C is $|z| = 2$
   (ii) $\int_C e^{1/z^2} dz$ where $C: |z| = 2$
   (iii) $\int_C \frac{\sin z}{z^6} dz$ where $C: |z| = 2$
   (iv) $\oint_C \sin\left(\frac{1}{z}\right) dz$ where $C: |z| = 1$
   (v) $\oint_C \frac{e^z}{(z+1)^2} dz$ where $C: |z-3| = 3$
   (vi) $\oint_C (z+1)\cot\left(\frac{z}{2}\right) dz$ where $C: |z| = 1$
   (vii) $\oint_C \frac{e^z dz}{\cos \pi z}$ where $C: |z| = 1$

15. Prove that $\int_C e^{\frac{1}{z}} \sin\frac{1}{z} dz = 2\pi i$ where C is the circle $|z| = 1$.

ANSWERS 1.4

1. (i) $z = n\pi$; $n \in I, n \neq 0$ and order = 1, $z = 0$ is a zero of order 3.
   (ii) $z = -1$ is a zero of order 3 (iii) Simple zeros at $z = -1$ and $z = 2$
2. (i) removable singularity at $z = 0$ (ii) pole of order $(r - 1)$ at $z = 0$
3. (i) removable singularity at $z = 2$ (ii) essential singularity at $z = -1$
   (iii) pole of order 3 at $z = 1$ and simple pole at $z = -2$
5. (i) simple pole at $z = 0$ (ii) $z = 0$ is a pole of order 3
6. (i) Res $(-1) = \frac{1}{3}$, Res $(2) = \frac{2}{3}$ (ii) Res $(-1) = \frac{1}{3}$, Res $(2) = \frac{5}{3}$
   (iii) Res $(0) = -\frac{1}{2}$, Res $(2) = \frac{3}{2}$ (iv) Res $(1) = \frac{1}{2}$, Res $(2) = -8$, Res $(3) = \frac{27}{2}$
   (v) Res $(1) = \frac{2}{3}$, Res $(-2) = \frac{1}{3}$ (vi) Res $(1) = \frac{5}{9}$, Res $(-2) = \frac{4}{9}$
   (vii) Res $(4) = \frac{5}{48}$, Res $(-4) = \frac{-3}{16}$, Res $(-2) = \frac{1}{12}$
   (viii) Res $(-1) = -4$, Res $(-2) = 4$ (ix) Res $(i) = \frac{-i}{4}$, Res $(-i) = \frac{i}{4}$
7. 0
8. (i) Res $(0) = 0$ (ii) Res $(0) = \frac{-1}{2}$ (iii) Res $(0) = -2$
   (iv) Res $(a) = e^a\left(\frac{a}{2}+1\right)$ (v) Res $(0) = 1$ (vi) Res $(0) = 1$

9. (i) πi (ii) 0 (iii) $-5\pi i/6$ (iv) 0 (v) $2\pi i$
10. (i) 0 (ii) $\pi(-2+i)$ (iii) $\pi(2+i)$
11. $2\pi i$ 12. (i) $4\pi i$ (ii) $-\pi(3+i)$ (iii) $\pi(3-i)$ (iv) $2\pi i$ (v) 0
13. $\dfrac{2\pi i}{e(n-1)!}$ 14. (i) 0 (ii) 0 (iii) $\dfrac{\pi i}{60}$ (iv) $2\pi i$ (v) 0 (vi) $4\pi i$ (vii) $-4i\sinh\dfrac{1}{2}$

1.15 APPLICATION OF CAUCHY RESIDUE THEOREM TO EVALUATE REAL INTEGRALS

Cauchy residue theorem is used to evaluate certain real integrals. These integrals are first transformed to associate contour integrals. The contour integrals are then evaluated using Cauchy residue theorem.

1.15.1 Integration Around the Unit Circle

Integrals of the type $\int_0^{2\pi} f(\cos\theta, \sin\theta)d\theta$ where f is a rational function of $\cos\theta$ and $\sin\theta$ can be obtained by setting $z = e^{i\theta}$ and thus

$$\cos\theta = \frac{z+z^{-1}}{2}, \sin\theta = \frac{z-z^{-1}}{2i}, dz = ie^{i\theta}d\theta = izd\theta$$

$$\therefore \qquad d\theta = \frac{dz}{iz}$$

As θ moves from 0 to 2π, z moves on unit circle $|z|=1$ once anticlockwise.

Thus, the integral takes the form $\int_C F(z)dz$

where $$F(z) = \frac{1}{iz}f\left(\frac{z+z^{-1}}{2}, \frac{z-z^{-1}}{2i}\right), \qquad C:|z|=1$$

Now, after finding isolated singularities of $F(z)$ inside C and residues at these singularities, we find value of integral using Cauchy residue theorem.

Example 1.75: Evaluate $\int_0^{2\pi}\frac{d\theta}{a+b\cos\theta}$; $a>|b|>0$ and using it prove that $\int_0^{\pi}\frac{d\theta}{17-8\cos\theta} = \frac{\pi}{15}$.

Solution: Let $I = \int_0^{2\pi}\frac{d\theta}{a+b\cos\theta}$; $a>|b|>0$

Put $z = e^{i\theta}$ $\quad\therefore\ dz = ie^{i\theta}d\theta = izd\theta$

$$\therefore \qquad d\theta = \frac{dz}{iz}$$

$$\cos\theta = \frac{e^{i\theta}+e^{-i\theta}}{2} = \frac{z+\frac{1}{z}}{2} = \frac{z^2+1}{2z}$$

As θ moves from 0 to 2π, z moves on unit circle $C : |z| = 1$ anticlockwise.

$$\therefore \quad I = \int_C \frac{dz}{iz\left[a + \frac{b(z^2+1)}{2z}\right]} = \frac{2}{i}\int_C \frac{dz}{bz^2 + 2az + b} \tag{1}$$

$$f(z) = \frac{1}{bz^2 + 2az + b} \text{ has simple poles at}$$

$$z_1 = \frac{-a + \sqrt{a^2 - b^2}}{b} \quad \text{and} \quad z_2 = \frac{-a - \sqrt{a^2 - b^2}}{b}$$

Now,
$$|z_2| = \frac{a + \sqrt{a^2 - b^2}}{|b|} > \frac{a}{|b|} > 1.$$

Also,
$$z_1 z_2 = \text{product of roots of quadratic equation } bz^2 + 2az + b = 0$$
$$= 1$$

$$\therefore \quad |z_1| = \frac{1}{|z_2|} < 1$$

$\therefore$ Only $z = z_1$ lies inside C

$$\text{Res }(z_1) = \lim_{z \to z_1} (z - z_1) f(z) = \lim_{z \to z_1} \frac{z - z_1}{bz^2 + 2az + b}$$

$$= \lim_{z \to z_1} \frac{1}{2bz + 2a} \qquad (L' \text{ Hospital rule})$$

$$= \frac{1}{2} \cdot \frac{1}{bz_1 + a} \qquad = \frac{1}{2\sqrt{a^2 - b^2}}$$

$\therefore$ By Cauchy residue theorem

$$\oint_C f(z)\,dz = 2\pi i \cdot \frac{1}{2\sqrt{a^2 - b^2}} = \frac{\pi i}{\sqrt{a^2 - b^2}}$$

$\therefore$ from (1)

$$I = \frac{2}{i} \frac{\pi i}{\sqrt{a^2 - b^2}} = \frac{2\pi}{\sqrt{a^2 - b^2}}$$

Take $a = 17, b = -8$ (condition $a > |b| > 0$ is satisfied)

$$\therefore \quad \int_0^{2\pi} \frac{d\theta}{17 - 8\cos\theta} = \frac{2\pi}{\sqrt{17^2 - 8^2}} = \frac{2\pi}{\sqrt{9(25)}} = \frac{2\pi}{15}$$

$$\Rightarrow \quad 2\int_0^{\pi} \frac{d\theta}{17 - 8\cos\theta} = \frac{2\pi}{15} \qquad (\because 17 - 8\cos(2\pi - \theta) = 17 - 8\cos\theta)$$

$$\therefore \quad \int_0^{\pi} \frac{d\theta}{17 - 8\cos\theta} = \frac{\pi}{15}$$

Example 1.76: Apply the calculus of residue to evaluate

$$\int_0^{\pi} \frac{a\,d\theta}{a^2+\sin^2\theta};\ a>0$$

Solution: Let $I = \int_0^{\pi} \frac{a\,d\theta}{a^2+\frac{1-\cos 2\theta}{2}} = \int_0^{\pi} \frac{2a\,d\theta}{2a^2+1-\cos 2\theta}$

Substitute $2\theta = \phi$ $\quad \therefore\ 2d\theta = d\phi$

As θ varies from 0 to π, ϕ varies from 0 to 2π

$$\therefore \qquad I = \int_0^{2\pi} \frac{a\,d\phi}{2a^2+1-\cos\phi};\ a>0$$

Substitute $e^{i\phi} = z$ $\qquad \therefore\ ie^{i\phi}d\phi = dz$

$$\therefore \qquad d\phi = \frac{dz}{iz}$$

$$\cos\phi = \frac{e^{i\phi}+e^{-i\phi}}{2} = \frac{z+\frac{1}{z}}{2} = \frac{z^2+1}{2z}$$

As ϕ moves from 0 to 2π, z moves on unit circle $C:|z|=1$ anticlockwise.

$$\therefore \qquad I = \oint_C \frac{adz}{iz\left(2a^2+1-\frac{z^2+1}{2z}\right)} = -\frac{2a}{i}\oint_C \frac{dz}{z^2-2\left(2a^2+1\right)z+1}$$

$$\therefore \qquad I = 2ai\oint_C \frac{dz}{z^2-2\left(2a^2+1\right)z+1}$$

Now, $\qquad f(z) = \frac{1}{z^2-2\left(2a^2+1\right)z+1}$ has simple poles at z_1 and z_2.

where

$$z_1 = \frac{2\left(2a^2+1\right)+\sqrt{4\left(2a^2+1\right)^2-4}}{2}$$

$$= 2a^2+1+\sqrt{\left(2a^2+1\right)^2-1}$$

$$= 2a^2+1+\sqrt{2a^2\left(2a^2+2\right)}$$

$$= 2a^2+1+2a\sqrt{a^2+1}$$

and $\qquad z_2 = 2a^2+1-2a\sqrt{a^2+1}$

As $a>0$, $|z_1| = 2a^2+1+2a\sqrt{a^2+1} > 1$

and $\qquad z_1 z_2 =$ product of roots of quadratic equation $z^2-2\left(2a^2+1\right)z+1=0$

$= 1$

$\therefore \quad |z_2| = \dfrac{1}{|z_1|} < 1$

$\therefore$ Only simple pole z_2 lies inside C.

$$\text{Res}\,(z_2) = \lim_{z\to z_2}(z-z_2)\,f(z) = \lim_{z\to z_2}\frac{z-z_2}{z^2-2(2a^2+1)z+1}$$

$$= \lim_{z\to z_2}\frac{1}{2z-2(2a^2+1)} \qquad (L' \text{ Hospital rule})$$

$$= \frac{1}{2\left[2a^2+1-2a\sqrt{a^2+1}\right]-2(2a^2+1)} = -\frac{1}{4a\sqrt{a^2+1}}$$

$\therefore$ By Cauchy residue theorem

$$I = 2ai\cdot 2\pi i\cdot\left(-\frac{1}{4a\sqrt{a^2+1}}\right) = \frac{\pi}{\sqrt{a^2+1}}$$

Example 1.77: Use calculus of residues to evaluate the integral $\displaystyle\int_0^{2\pi}\frac{d\theta}{5-4\sin\theta}$.

Solution: Let $I = \displaystyle\int_0^{2\pi}\frac{d\theta}{5-4\sin\theta}$

Substitute $z = e^{i\theta}$ $\quad \therefore dz = ie^{i\theta}d\theta = izd\theta$

$$\therefore \qquad d\theta = \frac{dz}{iz}$$

$$\sin\theta = \frac{e^{i\theta}-e^{-i\theta}}{2i} = \frac{z-\frac{1}{z}}{2i} = \frac{z^2-1}{2iz}$$

As θ moves from 0 to 2π, z moves on unit circle $C: |z| = 1$ anticlockwise.

$$\therefore \qquad I = \oint_C \frac{dz}{iz\left(5-4\dfrac{z^2-1}{2iz}\right)}$$

$$= \oint_C \frac{dz}{5iz-2z^2+2} = -\oint_C \frac{dz}{2z^2-5iz-2}$$

$$= -\oint_C \frac{dz}{2\left(z-\dfrac{i}{2}\right)(z-2i)} = -\frac{1}{2}\oint_C \frac{dz}{\left(z-\dfrac{i}{2}\right)(z-2i)}$$

$f(z) = \dfrac{1}{\left(z-\dfrac{i}{2}\right)(z-2i)}$ has simple poles at $z = \dfrac{i}{2}, 2i$ of which only $z = \dfrac{i}{2}$ lies inside C

$$\therefore \text{Res}\left(\frac{i}{2}\right) = \lim_{z\to\frac{i}{2}}\left(z-\frac{i}{2}\right)f(z) = \lim_{z\to\frac{i}{2}}\frac{1}{z-2i} = -\frac{2}{3i} = \frac{2i}{3}$$

$\therefore$ By Cauchy residue theorem

$$I = -\frac{1}{2}\cdot 2\pi i\left(\frac{2i}{3}\right) = \frac{2\pi}{3}$$

Example 1.78: Evaluate $\int_0^{2\pi}\frac{d\theta}{1-2a\cos\theta+a^2}$ where a is a complex constant and

(i) $|a|<1$ (ii) $|a|>1$

Solution: Let $I = \int_0^{2\pi}\frac{d\theta}{1-2a\cos\theta+a^2}$

Substitute $z = e^{i\theta}$

$\therefore \quad dz = ie^{i\theta}d\theta = izd\theta$

$\therefore \quad d\theta = \frac{dz}{iz}$

$$\cos\theta = \frac{e^{i\theta}+e^{-i\theta}}{2} = \frac{z+\frac{1}{z}}{2} = \frac{z^2+1}{2z}$$

As θ moves from 0 to 2π, z moves on unite circle $C: |z|=1$ anticlockwise.

$$I = \oint_C \frac{dz}{iz\left(1-2a\frac{z^2+1}{2z}+a^2\right)} = i\oint_C \frac{dz}{az^2-\left(1+a^2\right)z+a} \tag{1}$$

Let
$$f(z) = \frac{1}{az^2-\left(1+a^2\right)z+a}$$

$$= \frac{1}{(az-1)(z-a)} \text{ has simple poles at } z = \frac{1}{a}, a$$

$$\text{Res } (a) = \lim_{z\to a}(z-a)f(z) = \lim_{z\to a}\frac{1}{az-1} = \frac{1}{a^2-1}$$

$$\text{Res }\left(\frac{1}{a}\right) = \lim_{z\to\frac{1}{a}}\left(z-\frac{1}{a}\right)f(z) = \lim_{z\to\frac{1}{a}}\frac{1}{a(z-a)} = \frac{1}{a\left(\frac{1}{a}-a\right)} = \frac{1}{1-a^2}$$

(i) When $|a|<1$ then $z=a$ lies inside C and $z=\frac{1}{a}$ lies outside

$\therefore$ By Cauchy residue theorem

$$I = i\cdot 2\pi i\cdot\frac{1}{a^2-1} = \frac{2\pi}{1-a^2}$$

(ii) When $|a|>1$ then $\left|\frac{1}{a}\right|<1$, thus only $\frac{1}{a}$ lies inside C

$\therefore$ By Cauchy residue theorem

$$I = i\cdot 2\pi i\cdot\frac{1}{1-a^2} = \frac{2\pi}{a^2-1}$$

Example 1.79: Apply calculus of residues to evaluate $\int_0^{\pi} \cos^6 \theta \, d\theta$

Solution: Let $I = \int_0^{\pi} \cos^6 \theta \, d\theta = \int_0^{\pi} \left(\frac{1+\cos 2\theta}{2} \right)^3 d\theta$

Substitute $2\theta = \phi \quad \therefore \; d\theta = \frac{1}{2} d\phi$; when θ varies from 0 to π, ϕ varies from 0 to 2π

$$\therefore \qquad I = \int_0^{2\pi} \frac{(1+\cos\phi)^3}{8} \frac{d\phi}{2} = \frac{1}{16} \int_0^{2\pi} (1+\cos\phi)^3 \, d\phi$$

Substitute $\quad z = e^{i\phi}$

$$\therefore \qquad dz = ie^{i\phi} d\phi = iz d\phi$$

$$\therefore \qquad d\phi = \frac{dz}{iz}$$

$$\cos\phi = \frac{e^{i\phi} + e^{-i\phi}}{2} = \frac{z + \frac{1}{z}}{2} = \frac{z^2+1}{2z}$$

As ϕ moves from 0 to 2π, z moves on unit circle $C : |z| = 1$, anticlockwise.

$$\therefore \qquad I = \frac{1}{16} \oint_C \left(1 + \frac{z^2+1}{2z}\right)^3 \frac{dz}{iz} = \frac{1}{128i} \oint_C \frac{\left(z^2+2z+1\right)^3}{z^4} dz$$

$$= \frac{1}{128i} \oint_C \frac{(z+1)^6}{z^4} dz$$

Now, $f(z) = \frac{(z+1)^6}{z^4}$ has pole of order 4 at $z = 0$ and $z = 0$ lies inside C

$$\text{Res }(0) = \frac{1}{3!} \lim_{z\to 0} \frac{d^3}{dz^3} z^4 f(z) = \frac{1}{6} \lim_{z\to 0} \frac{d^3}{dz^3} (z+1)^6$$

$$= \frac{1}{6} \lim_{z\to 0} \; 6 \cdot 5 \cdot 4 \cdot (z+1)^3 = 20$$

$\therefore$ By Cauchy residue theorem

$$I = \frac{1}{128i} \cdot 2\pi i (20) = \frac{5\pi}{16}$$

Example 1.80: Apply calculus of residues to evaluate $\int_0^{2\pi} e^{\cos\theta} \cos(n\theta - \sin\theta) \, d\theta$ where n is any non-negative integer and hence show that $\int_0^{2\pi} e^{\cos\theta} \cos(\sin\theta) \, d\theta = 2\pi$.

Solution: Let $I = \int_0^{2\pi} e^{\cos\theta} e^{i(n\theta - \sin\theta)} \, d\theta = \int_0^{2\pi} e^{\cos\theta - i\sin\theta} e^{in\theta} \, d\theta$

$$= \int_0^{2\pi} e^{e^{-i\theta}} \left(e^{i\theta}\right)^n d\theta$$

Substitute $e^{-i\theta} = z \quad \therefore \; -ie^{-i\theta} d\theta = dz \Rightarrow d\theta = -\frac{dz}{iz}$

As θ varies from 0 to 2π, z moves on unite circle $C^* : |z| = 1$ clockwise. Taking anticlockwise direction positive, we have

$$I = -\oint_C \frac{e^z}{z^n}\left(-\frac{1}{iz}\right)dz = -i\oint_C \frac{e^z}{z^{n+1}}dz \text{ where } C : |z| = 1 \text{ anticlockwise}$$

$f(z) = \dfrac{e^z}{z^{n+1}}$ has only pole of order $n+1$ at $z = 0$ which lies inside C

$$\text{Res }(0) = \frac{1}{n!}\lim_{z\to 0}\frac{d^n}{dz^n} z^{n+1} f(z) = \frac{1}{n!}\lim_{z\to 0}\frac{d^n}{dz^n} e^z$$

$$= \frac{1}{n!}\lim_{z\to 0} e^z = \frac{1}{n!}$$

$\therefore$ By Cauchy residue theorem

$$I = -i\ 2\pi i\left(\frac{1}{n!}\right) = \frac{2\pi}{n!}$$

$$\int_0^{2\pi} e^{\cos\theta}\left[\cos(n\theta - \sin\theta) + i\sin(n\theta - \sin\theta)\right]d\theta = \frac{2\pi}{n!}$$

Equate real parts

$$\int_0^{2\pi} e^{\cos\theta}\cos(n\theta - \sin\theta)\,d\theta = \frac{2\pi}{n!}$$

Take $n = 0$

$$\int_0^{2\pi} e^{\cos\theta}\cos(\sin\theta)\,d\theta = 2\pi \qquad (\because \cos(-\sin\theta) = \cos(\sin\theta))$$

Example 1.81: Apply calculus of residues to show that

$$\int_0^{2\pi}\frac{\sin^2\theta}{a + b\cos\theta}d\theta = \frac{2\pi}{b^2}\left(a - \sqrt{a^2 - b^2}\right) \quad \text{where} \quad 0 < |b| < a$$

Solution: Let $I = \displaystyle\int_0^{2\pi}\frac{\sin^2\theta\ d\theta}{a + b\cos\theta} = \frac{1}{2}\int_0^{2\pi}\frac{1 - \cos 2\theta}{a + b\cos\theta}d\theta$

$$= \text{Real part of } \frac{1}{2}\int_0^{2\pi}\frac{1 - e^{2i\theta}}{a + b\cos\theta}d\theta \tag{1}$$

Let $\quad I_1 = \displaystyle\int_0^{2\pi}\frac{1 - e^{2i\theta}}{a + b\cos\theta}d\theta$

Substitute $\quad e^{i\theta} = z$

$\therefore \quad ie^{i\theta}d\theta = dz$

$$d\theta = \frac{dz}{iz}$$

$\therefore$

$$\cos\theta = \frac{e^{i\theta} + e^{-i\theta}}{2} = \frac{z + \frac{1}{z}}{2} = \frac{z^2 + 1}{2z}$$

As θ varies from 0 to 2π, z moves on unite circle $C:|z|=1$ anticlockwise.

$$\therefore \quad I_1 = \oint_C \frac{1-z^2}{a+b\dfrac{z^2+1}{2z}}\cdot\frac{dz}{iz} = \frac{2}{i}\oint_C \frac{(1-z^2)\,dz}{bz^2+2az+b}$$

$$f(z) = \frac{1-z^2}{bz^2+2az+b} \text{ has simple poles at } z_1 \text{ and } z_2$$

where $z_1 = \dfrac{-a+\sqrt{a^2-b^2}}{b},\ z_2 = \dfrac{-a-\sqrt{a^2-b^2}}{b}$

$$|z_2| = \frac{a+\sqrt{a^2-b^2}}{|b|} > 1 \quad \because a > |b| > 0$$

and $z_1 z_2$ = products of roots of equation ($bz^2 + 2az + b = 0$)

$= 1$

$$\therefore \quad |z_1| = \frac{1}{|z_2|} < 1$$

$\therefore$ Only z_1 lies inside C

$$\text{Res}\,(z_1) = \lim_{z\to z_1}(z-z_1)f(z) = \lim_{z\to z_1}\frac{(1-z^2)(z-z_1)}{bz^2+2az+b}$$

$$= (1-z_1^2)\lim_{z\to z_1}\frac{z-z_1}{bz^2+2az+b} = (1-z_1^2)\lim_{z\to z_1}\frac{1}{2bz+2a} \qquad (L'\text{ Hospital rule})$$

$$= \frac{(1-z_1^2)}{2(bz_1+a)} = \frac{1-\left(\dfrac{-a+\sqrt{a^2-b^2}}{b}\right)^2}{2\sqrt{a^2-b^2}} = \frac{b^2-\left(a^2+a^2-b^2-2a\sqrt{a^2-b^2}\right)}{2b^2\sqrt{a^2-b^2}}$$

$$= \frac{b^2-a^2+a\sqrt{a^2-b^2}}{b^2\sqrt{a^2-b^2}} = \frac{-\left(\sqrt{a^2-b^2}\right)^2+a\sqrt{a^2-b^2}}{b^2\sqrt{a^2-b^2}}$$

$$= -\frac{\sqrt{a^2-b^2}-a}{b^2}$$

$\therefore$ By Cauchy residue theorem

$$I_1 = \frac{2}{i}\,2\pi i\left[-\frac{\sqrt{a^2-b^2}-a}{b^2}\right] = \frac{4\pi\left(a-\sqrt{a^2-b^2}\right)}{b^2}$$

$\therefore$ from (1) I = real part of $\dfrac{1}{2}I_1 = \dfrac{2\pi}{b^2}\left[a-\sqrt{a^2-b^2}\right]$

1.15.2 Improper Real Integrals of the form $\int_{-\infty}^{\infty} f(x)\,dx$ or $\int_{0}^{\infty} f(x)\,dx$ where f(z) has no Real Singularity

The improper integral $\int_{-\infty}^{\infty} f(x)\,dx$ is defined as

$$I = \int_{-\infty}^{\infty} f(x)\,dx = \lim_{R\to\infty}\int_{-R}^{0} f(x)\,dx + \lim_{S\to\infty}\int_{0}^{S} f(x)\,dx$$

and $\lim_{R\to\infty}\int_{-R}^{R} f(x)dx$, if exists, is called Cauchy principal value of *I*. Sometimes, Cauchy principal value of *I* exists but *I* does not exist as for $f(x) = x$. But if *I* exists then Cauchy principal value must exist and Cauchy principal value must be equal to *I*. We shall be taking only those cases in which *I* exists and hence we shall be finding Cauchy principal value.

We shall be using Jordan lemma

$$\sin\theta \geq \frac{2\theta}{\pi} \text{ when } 0 \leq \theta \leq \frac{\pi}{2} \tag{1.41}$$

and following result

Result: If $f(R, \theta) \to 0$ as $R \to \infty$ uniformly over all values of θ, $0 \leq \theta \leq \pi$ then

$$\int_0^{\pi} f(R, \theta)d\theta \to 0 \;\&\; \int_0^{\pi/2} f(R, \theta)\,d\theta \to 0$$

Proof: $f(R, \theta) \to 0$ as $R \to \infty$ uniformly

$\therefore$ Corresponding to $\varepsilon > 0$ there exist R_1 such that

$$|f(R, \theta)| < \varepsilon \text{ when } R > R_1 \text{ for all values of } \theta \text{ in } [0, \pi]$$

$\therefore$
$$\left|\int_0^{\pi} f(R, \theta)d\theta\right| \leq \int_0^{\pi}|f(R, \theta)||d\theta| < \varepsilon\int_0^{\pi}|d\theta| = \pi\varepsilon$$

But ε is arbitrary

$\therefore$
$$\int_0^{\pi} f(R, \theta) \to 0 \quad \text{as } R \to \infty \text{ uniformly}$$

Similarly, $\int_0^{\pi/2} f(R, \theta)d\theta \to 0$ as $R \to \infty$ uniformly.

To evaluate $\int_{-\infty}^{\infty} f(x)\,dx$ or $\int_{0}^{\infty} f(x)\,dx$ we assume that $f(z)$ satisfies the following conditions

(i) $f(z)$ has no singularity on real axis
(ii) $f(z)$ is analytic in the upper half of z-plane except at a finite no. of isolated singularities $z_1, z_2, \ldots, z_n$ in this half plane.
(iii) $z f(z) \to 0$ uniformly as $R \to \infty$ through the values $0 \leq \theta \leq \pi$.

Let us consider $\int_C f(z)dz$ where C consists of $C_R : |z| = R$; Im $(z) > 0$ and real axis from $-R$ to R as shown in Figure 1.19.

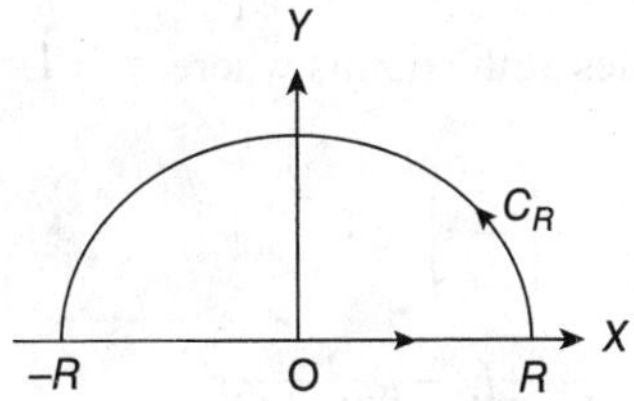

Figure 1.19

$$\therefore \int_C f(z)\,dz = \int_{C_R} f(z)\,dz + \int_{-R}^{R} f(x)\,dx \tag{1.42}$$

$z_1, z_2, \ldots, z_n$ are isolated singularities of $f(z)$ inside C

$\therefore$ By Cauchy residue theorem

$$\int_C f(z)\,dz = 2\pi i \sum_{k=1}^{n} \text{Res}\,(z_k) \tag{1.43}$$

On $C_R, z = Re^{i\theta}, 0 < \theta < \pi, dz = Rie^{i\theta} d\theta = iz d\theta$

Now, $$\left| \int_{C_R} f(z)\,dz \right| \le \int_0^{\pi} |f(z)\,iz|\,d\theta = \int_0^{\pi} |z\,f(z)|\,d\theta$$

$$\to 0 \text{ as } R \to \infty \tag{1.44}$$

$\therefore$ from equation (1.42) as $R \to \infty$

$$\int_{-\infty}^{\infty} f(x)\,dx = 2\pi i \sum_{k=1}^{n} \text{Res}\,(z_k) \qquad \text{(Using (1.43) and (1.44))}$$

Remark 1.13: To evaluate $\int_{-\infty}^{\infty} \cos ax\, f(x)\,dx$ or $\int_{-\infty}^{\infty} \sin ax\, f(x)\,dx$ we shall consider $\int_C e^{iaz} f(z)dz$ and shall proceed as above and in the last, real and imaginary parts as required will be equated. In the examples 1.82 to 1.86 we shall use the figure of this article.

Example 1.82: Using contour integration evaluate

(i) $\int_0^{\infty} \frac{dx}{x^6+1}$ (ii) $\int_0^{\infty} \frac{dx}{\left(x^2+a^2\right)^2}$

Solution: (i) Consider $\oint_C \frac{dz}{z^6+1}$

where C consists of semicircle $C_R : |z| = R;$ Im $(z) > 0$ and real axis from $-R$ to R where R is large

$$\therefore \oint_C \frac{dz}{z^6+1} = \int_{C_R} \frac{dz}{z^6+1} + \int_{-R}^{R} \frac{dx}{x^6+1} \tag{1}$$

Now, $f(z) = \dfrac{1}{z^6+1}$ has poles at the points where $z^6+1=0$ or $z^6=-1=i^6$

$\therefore \quad \left(\dfrac{z}{i}\right)^6 = 1 \Rightarrow \left(\dfrac{z}{i}\right)^3 = \pm 1 \Rightarrow \left(\dfrac{z}{\pm i}\right)^3 = 1$

$\therefore \quad \dfrac{z}{\pm i} = 1,\ \omega,\ \omega^2 \quad \Rightarrow \quad z = \pm i,\ \pm i\omega,\ \pm i\omega^2$

$\therefore$ Simple poles at $i, -i, i\omega, -i\omega, i\omega^2, -i\omega^2$

i.e.,
$$i,\ -i,\ i\omega = i\frac{-1+i\sqrt{3}}{2} = \frac{-\sqrt{3}}{2} - \frac{i}{2},\ -i\omega = \frac{\sqrt{3}}{2} + \frac{i}{2},$$
$$i\omega^2 = i\frac{-1-i\sqrt{3}}{2} = \frac{\sqrt{3}}{2} - \frac{i}{2},\ -i\omega^2 = -\frac{\sqrt{3}}{2} + \frac{i}{2}$$

Out of these poles, those poles whose imaginary part is positive lie within C.

$\therefore$ $i, -i\omega, -i\omega^2$ lie inside C

Let these are z_1, z_2, z_3 respectively.

$$\text{Res}\,(z_j) = \lim_{z \to z_j} \frac{(z - z_j)}{z^6+1} = \lim_{z \to z_j} \frac{1}{6z^5} \qquad (L' \text{ Hospital rule})$$
$$= \frac{1}{6z_j^5} = \frac{z_j}{6z_j^6} = -\frac{z_j}{6} \qquad \left(\because\ z_j^6 = -1\right)$$

$\therefore$ By Cauchy residue theorem

$$\oint_C \frac{dz}{z^6+1} = 2\pi i \text{ (Sum of residues at isolated singularities within } C)$$
$$= 2\pi i\left(-\frac{z_1}{6} - \frac{z_2}{6} - \frac{z_3}{6}\right) = -\frac{\pi i}{3}\left(z_1 + z_2 + z_3\right)$$
$$= -\frac{\pi i}{3}\left(i - i\omega - i\omega^2\right) = \frac{\pi}{3}\left(1 - \omega - \omega^2\right)$$
$$= \frac{2\pi}{3} \qquad (\because \omega + \omega^2 = -1) \tag{2}$$

On C_R, $z = Re^{i\theta} \quad \therefore dz = i\,Re^{i\theta}\,d\theta$

$$\therefore \quad \left|\int_{C_R} \frac{dz}{z^6+1}\right| = \left|\int_0^\pi \frac{i\,Re^{i\theta}\,d\theta}{R^6 e^{6i\theta}+1}\right| \le \int_0^\pi \frac{\left|i\,Re^{i\theta}\right|}{\left|R^6 e^{6i\theta}\right| - 1}\,|d\theta| \qquad \left(\because \frac{1}{|z_1 + z_2|} \le \frac{1}{|z_1| - |z_2|}\right)$$

$$\therefore \quad \left|\int_{C_R} \frac{dz}{z^6+1}\right| \le \frac{R}{R^6-1}\pi \to 0 \quad \text{as} \quad R \to \infty \tag{3}$$

$\therefore$ As $R \to \infty$, from (1), (2) and (3)

$$\int_{-\infty}^{\infty} \frac{dx}{x^6+1} = \frac{2\pi}{3}$$

$$\Rightarrow \qquad 2\int_0^\infty \frac{dx}{x^6+1} = \frac{2\pi}{3} \qquad \left(\because \frac{1}{x^2+1} \text{ is even function}\right)$$

$$\therefore \qquad \int_0^\infty \frac{dx}{x^6+1} = \frac{\pi}{3}.$$

(ii) Consider $\oint\limits_C \frac{dz}{\left(z^2+a^2\right)^2}; a > 0$

where C is the contour consisting of semicircle C_R; $|z| = R, \text{Im}(z) > 0$ and real axis from $-R$ to R where R is large

$$\therefore \qquad \oint\limits_C \frac{dz}{\left(z^2+a^2\right)^2} = \int\limits_{C_R} \frac{dz}{\left(z^2+a^2\right)^2} + \int\limits_{-R}^{R} \frac{dx}{\left(x^2+a^2\right)^2} \qquad (1)$$

Now, $f(z) = \dfrac{1}{\left(z^2+a^2\right)^2}$ has double poles at $z = \pm ia$

Out of which $z = ia$ lies inside C $\quad (\because a > 0)$

$$\text{Res}\,(ia) = \lim_{z\to ia} \frac{d}{dz}(z-ia)^2 f(z) = \lim_{z\to ia} \frac{d}{dz}\frac{1}{(z+ia)^2}$$

$$= \lim_{z\to ia} \frac{-2}{(z+ia)^3} = \frac{-2}{-8ia^3} = \frac{1}{4ia^3}$$

$\therefore$ By Cauchy residue theorem

$$\oint\limits_C \frac{dz}{\left(z^2+a^2\right)^2} = 2\pi i\left(\frac{1}{4ia^3}\right) = \frac{\pi}{2a^3} \qquad (2)$$

On C_R, $z = Re^{i\theta} \quad dz = i\,Re^{i\theta}\,d\theta$

$$\therefore \qquad \left|\int\limits_{C_R} \frac{dz}{\left(z^2+a^2\right)^2}\right| \le \int\limits_{C_R} \frac{|dz|}{\left|z^2+a^2\right|^2} \le \int\limits_{C_R} \frac{|dz|}{\left(\left|z^2\right|-a^2\right)^2} = \int\limits_0^\pi \frac{R\ d\theta}{\left(R^2-a^2\right)^2}$$

$$= \frac{\pi R}{(R^2-a^2)^2} \to 0 \text{ as } \quad R\to\infty \qquad (3)$$

$\therefore$ As $R \to \infty$, from (1), (2) and (3)

$$\int_{-\infty}^\infty \frac{dx}{\left(x^2+a^2\right)^2} = \frac{\pi}{2a^3}$$

$$\Rightarrow \qquad 2\int_0^\infty \frac{dx}{\left(x^2+a^2\right)^2} = \frac{\pi}{2a^3} \qquad \left(\because \frac{1}{\left(x^2+a^2\right)^2} \text{ is even function}\right)$$

$$\therefore \qquad \int_0^\infty \frac{dx}{\left(x^2+a^2\right)^2} = \frac{\pi}{4a^3}$$

Example 1.83: Apply calculus of residues to evaluate

$$\int_{-\infty}^{\infty}\frac{x^2dx}{\left(x^2+a^2\right)\left(x^2+b^2\right)};\ a,b>0 \text{ and hence find the value of } \int_{-\infty}^{\infty}\frac{x^2}{\left(x^2+1\right)\left(x^2+4\right)}dx$$

Solution: (i) Consider $\oint_C \frac{z^2dz}{\left(z^2+a^2\right)\left(z^2+b^2\right)};\ a,b>0$.

where C is the contour consisting of $C_R:|z|=R,\ \text{Im}(z)>0,$ and real axis from $-R$ to R where R is large

$$\therefore \qquad \oint_C \frac{z^2dz}{\left(z^2+a^2\right)\left(z^2+b^2\right)}=\int_{C_R}\frac{z^2dz}{\left(z^2+a^2\right)\left(z^2+b^2\right)}+\int_{-R}^{R}\frac{x^2dx}{\left(x^2+a^2\right)\left(x^2+b^2\right)} \qquad (1)$$

Now, $f(z)=\dfrac{z^2}{\left(z^2+a^2\right)\left(z^2+b^2\right)}$ has simple poles at $z=\pm ia,\pm ib$ of which $z=ia,\ ib$ are inside $C \quad (\because\ a,b>0)$

$$\text{Res}\,(ia)=\lim_{z\to ia}\frac{(z-ia)z^2}{(z^2+a^2)(z^2+b^2)}$$

$$=\frac{-a^2}{-a^2+b^2}\lim_{z\to ia}\frac{z-ia}{(z-ia)(z+ia)}$$

$$=\frac{-a^2}{b^2-a^2}\cdot\frac{1}{2ia}=\frac{ia}{2(b^2-a^2)}$$

Interchanging a and b $\qquad (\because\ f(z)$ does not change by interchange)

$$\text{Res}\,(ib)=\frac{ib}{2(a^2-b^2)}$$

$\therefore$ By Cauchy residue theorem

$$\oint_C \frac{z^2dz}{\left(z^2+a^2\right)\left(z^2+b^2\right)}=2\pi i\left(\frac{ia}{2(b^2-a^2)}+\frac{ib}{2(a^2-b^2)}\right)$$

$$=\frac{-\pi(a-b)}{b^2-a^2}=\frac{\pi(b-a)}{b^2-a^2}=\frac{\pi}{a+b} \qquad (2)$$

and $\left|\int_{C_R}\frac{z^2dz}{(z^2+a^2)(z^2+b^2)}\right|\le\int_{C_R}\frac{|z|^2|dz|}{\left(|z|^2-a^2\right)\left(|z|^2-b^2\right)}=\int_{C_R}\frac{R^2}{\left(R^2-a^2\right)\left(R^2-b^2\right)}|dz|$

$$=\frac{\pi R^3}{(R^2-a^2)(R^2-b^2)}\to 0 \text{ as } R\to\infty \qquad \left(\because \int_{C_R}|dz|=\pi R\right) \qquad (3)$$

$\therefore$ As $R\to\infty$, from (1), (2) and (3)

$$\int_{-\infty}^{\infty}\frac{x^2}{\left(x^2+a^2\right)\left(x^2+b^2\right)}dx=\frac{\pi}{a+b}$$

Take $a = 1 > 0, b = 2 > 0$

$$\int_{-\infty}^{\infty} \frac{x^2}{(x^2+1)(x^2+4)}dx = \frac{\pi}{3}$$

Example 1.84: Evaluate by the method of complex variables, the integral $\int_{-\infty}^{\infty} \frac{x^2}{(x^2+1)^3}dx$.

Solution: Consider $\oint_C \frac{z^2}{(z^2+1)^3}dz$

where C is the contour consisting of $C_R : |z| = R,\ \text{Im}(z) > 0$ and real axis from $-R$ to R where R is large

$$\therefore \quad \oint_C \frac{z^2}{(z^2+1)^3}dz = \int_{C_R} \frac{z^2}{(z^2+1)^3}dz + \int_{-R}^{R} \frac{x^2}{(x^2+1)^3}dx \quad (1)$$

Now, $f(z) = \frac{z^2}{(z^2+1)^3}$ has poles of order 3 at $z = \pm i$ out of which only $z = i$ lies inside C

$$\text{Res}\,(i) = \frac{1}{2!}\lim_{z\to i}\frac{d^2}{dz^2}(z-i)^3 f(z)$$

$$= \frac{1}{2}\lim_{z\to i}\frac{d^2}{dz^2}\frac{z^2}{(z+i)^3}$$

Now,
$$\frac{z^2}{(z+i)^3} = \frac{z^2+1-1}{(z+i)^3} = \frac{(z+i)(z-i)}{(z+i)^3} - \frac{1}{(z+i)^3}$$

$$= \frac{z-i}{(z+i)^2} - \frac{1}{(z+i)^3} = \frac{z+i-2i}{(z+i)^2} - \frac{1}{(z+i)^3}$$

$$= \frac{1}{z+i} - \frac{2i}{(z+i)^2} - \frac{1}{(z+i)^3}$$

$$\therefore \quad \text{Res}\,(i) = \frac{1}{2}\lim_{z\to i}\frac{d^2}{dz^2}\left(\frac{1}{z+i} - \frac{2i}{(z+i)^2} - \frac{1}{(z+i)^3}\right)$$

$$= \frac{1}{2}\lim_{z\to i}\left(\frac{2}{(z+i)^3} - \frac{12i}{(z+i)^4} - \frac{12}{(z+i)^5}\right)$$

$$= \frac{1}{2}\left(\frac{2}{-8i} - \frac{12i}{16} - \frac{12}{32i}\right) = \frac{1}{2}\left(\frac{i}{4} - \frac{3i}{4} + \frac{3i}{8}\right)$$

$$= -\frac{i}{16}$$

$\therefore$ By Cauchy residue theorem

$$\oint_C \frac{z^2}{(z^2+1)^3}dz = 2\pi i\left(-\frac{i}{16}\right) = \frac{\pi}{8} \quad (2)$$

$$\left|\int_{C_R} \frac{z^2}{(z^2+1)^3} dz\right| \le \int_{C_R} \frac{|z|^2}{(|z|^2-1)^3} |dz| = \frac{R^2}{(R^2-1)^3} \pi R \to 0 \quad \text{as} \quad R \to \infty \tag{3}$$

$\therefore$ As $R \to \infty$, from (1), (2) and (3)

$$\int_{-\infty}^{\infty} \frac{x^2}{(x^2+1)^3} dx = \frac{\pi}{8}$$

Example 1.85: Evaluate the following integrals

(i) $\int_{-\infty}^{\infty} \frac{\cos ax}{x^2+b^2} dx\,;\, a,\, b > 0$ (ii) $\int_{-\infty}^{\infty} \frac{\sin ax}{x^2+b^2} dx;\, a,\, b > 0$

Solution: (i) Consider $\oint_C \frac{e^{iaz}}{z^2+b^2} dz$

where C consists of semicircle $C_R : |z| = R;$ Im $(z) > 0$ and real axis from $-R$ to R for large R

$$\therefore \quad \oint_C \frac{e^{iaz}}{z^2+b^2} dz = \int_{C_R} \frac{e^{iaz}}{z^2+b^2} dz + \int_{-R}^{R} \frac{e^{iax}}{x^2+b^2} dx \tag{1}$$

Now, $f(z) = \frac{e^{iaz}}{z^2+b^2}$ has simple poles at $z = \pm\, ib$ of which $z = ib$ lies inside C $(\because\ b > 0)$

$$\text{Res } (ib) = \lim_{z \to ib} \frac{e^{iaz}}{(z+ib)} = \frac{e^{-ab}}{2\, ib}$$

$\therefore$ By Cauchy residue theorem

$$\oint_C \frac{e^{iaz}}{z^2+b^2} dz = 2\pi i \,\frac{e^{-ab}}{2\ ib} = \frac{\pi\, e^{-ab}}{b} \tag{2}$$

and

$$\left|\int_{C_R} \frac{e^{iaz}}{z^2+b^2} dz\right| \le \int_{C_R} \frac{\left|e^{iaR(\cos\theta + i\sin\theta)}\right|}{|z|^2-b^2} |dz| \qquad \left(\because \text{ On } C_R,\ z = Re^{i\theta}\right)$$

$$= \int_{C_R} \frac{e^{-a\,R\,\sin\theta}}{R^2-b^2} |dz| < \frac{1}{R^2-b^2} \pi R \to 0 \quad \text{as} \quad R \to \infty \tag{3}$$

$\left(\because\ a\,R\sin\,\theta > 0 \text{ for } 0 < \theta < \pi \text{ on } C_R\ \therefore\ e^{-a\,R\,\sin\theta} < 1\right)$

from (1), (2) and (3) as $R \to \infty$

$$\int_{-\infty}^{\infty} \frac{e^{iax}}{x^2+b^2} dx = \frac{\pi\, e^{-ab}}{b}$$

$$\Rightarrow \quad \int_{-\infty}^{\infty} \frac{\cos ax + i\sin ax}{x^2+b^2} dx = \frac{\pi\, e^{-ab}}{b}$$

Equate real and imaginary parts

$$\int_{-\infty}^{\infty} \frac{\cos ax}{x^2+b^2} dx = \frac{\pi\, e^{-ab}}{b}, \qquad \int_{-\infty}^{\infty} \frac{\sin ax}{x^2+b^2} dx = 0$$

Example 1.86: Apply residue theorem to evaluate

$$\int_0^{\infty} \frac{\cos ax}{x^2+1}\, dx\,;\, a>0$$

Solution: Taking $b=1$ in Example 1.85, we have

$$\int_{-\infty}^{\infty} \frac{\cos ax}{x^2+1}\, dx = \pi\, e^{-a}$$

$$\Rightarrow \qquad 2\int_0^{\infty} \frac{\cos ax}{x^2+1}\, dx = \pi\, e^{-a} \qquad \left(\because \frac{\cos ax}{x^2+1} \text{ is even function}\right)$$

$$\therefore \qquad \int_0^{\infty} \frac{\cos ax}{x^2+1}\, dx = \frac{\pi}{2}\, e^{-a}$$

1.15.3 Some Special Improper Real Integrals

In the above examples, there were finite number of poles within the contour C. If there are infinite number of poles within C for large R then this contour C cannot be taken and some other suitable contour will have to be taken. We give an example to explain it.

Example 1.87: Show that $\int\limits_{-\infty}^{\infty} \frac{e^{ax}}{e^x+1}\, dx = \frac{\pi}{\sin a\pi}; 0<a<1.$

Solution: Here $\frac{e^{az}}{e^z+1}$ has simple poles at $z = (2n+1)\pi i$ for all non-negative integers n inside the contour C of the above examples.

$\therefore$ We consider $\oint\limits_C \frac{e^{az}}{e^z+1}\, dz$ where C is the rectangle $ABCD$ with vertices $A(-R, 0)$, $B(R, 0)$, $C(R, 2\pi)$, $D(-R, 2\pi)$ where R is large

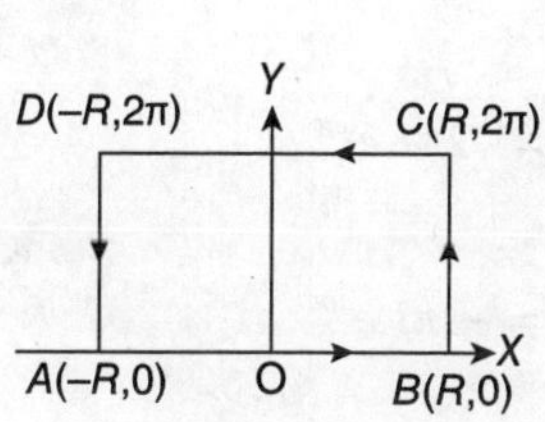

Figure 1.20

$$\therefore \qquad \oint\limits_C \frac{e^{az}}{e^z+1}\, dz = \int\limits_{-R}^{R} \frac{e^{ax}}{e^x+1}\, dx + \int_0^{2\pi} \frac{e^{a(R+iy)}}{e^{R+iy}+1}\, idy + \int\limits_{R}^{-R} \frac{e^{a(x+2i\pi)}}{e^{x+2i\pi}+1}\, dx + \int_{2\pi}^{0} \frac{e^{a(-R+iy)}}{e^{-R+iy}+1}\, idy \qquad (1)$$

Now, $f(z) = \frac{e^{az}}{e^z+1}$ has simple pole at $z = (2n+1)i\pi;\ n \in I$ of which only $z = i\pi$ lies inside C.

$$\text{Res}(i\pi) = \lim_{z\to i\pi} \frac{(z - i\pi)}{e^z + 1} e^{az}$$

$$= e^{ai\pi} \lim_{z\to i\pi} \frac{z - i\pi}{e^z + 1}$$

$$= e^{ai\pi} \lim_{z\to i\pi} \frac{1}{e^z} \qquad (L' \text{ Hospital rule})$$

$$= -e^{ai\pi}$$

∴ By Cauchy residue theorem

$$\oint_C \frac{e^{az}}{e^z + 1} dz = -2\pi i \; e^{ai\pi} \tag{2}$$

Now, $$\left| \int_0^{2\pi} \frac{e^{a(R+iy)}}{e^{R+iy} + 1} \, i dy \right| \le \int_0^{2\pi} \frac{\left|e^{a(R+iy)}\right|}{\left|e^{R+iy}\right| - 1} dy = \int_0^{2\pi} \frac{e^{aR}}{e^R - 1} dy$$

$$= \frac{2\pi e^{aR}}{e^R - 1}$$

and $$\lim_{R\to\infty} \frac{e^{aR}}{e^R - 1} = \lim_{R\to\infty} \frac{ae^{aR}}{e^R} = a \lim_{R\to\infty} e^{(a-1)R} \quad (L' \text{ Hospital rule})$$

$$= 0 \qquad (\because \; 0 < a < 1)$$

Thus, as $R \to \infty$, $$\int_0^{2\pi} \frac{e^{a(R+iy)}}{e^{R+iy} + 1} i \, dy \to 0 \tag{3}$$

and $$\left| \int_{2\pi}^{0} \frac{e^{a(-R+iy)}}{e^{-R+iy} + 1} \, i dy \right| \le \int_0^{2\pi} \frac{e^{-aR}}{1 - e^{-R}} \, dy = \frac{2\pi e^{-aR}}{1 - e^{-R}} \to 0 \text{ as } R \to \infty \quad (\because \; a > 0) \tag{4}$$

∴ As $R \to \infty$, from (1) to (4)

$$\int_{-\infty}^{\infty} \frac{e^{ax}}{e^x + 1} dx + \int_{\infty}^{-\infty} \frac{e^{a(x+2i\pi)}}{e^x + 1} dx = -2\pi i \; e^{ai\pi}$$

$$\Rightarrow \qquad \left(1 - e^{2i\pi a}\right) \int_{-\infty}^{\infty} \frac{e^{ax}}{e^x + 1} dx = -2\pi i \; e^{ai\pi}$$

$$\therefore \qquad \int_{-\infty}^{\infty} \frac{e^{ax}}{e^x + 1} dx = -\frac{2\pi i \; e^{ai\pi}}{1 - e^{2i\pi a}} = \frac{2\pi i}{e^{i\pi a} - e^{-i\pi a}} = \frac{\pi}{\sin \pi a}$$

1.15.4 Improper Integrals with Singularities on Real Axis

We assume that $f(z)$ has same conditions as in (1.15.2) except that $f(z)$ has singularities simple poles on real axis at $z = x_1, x_2 \ldots x_m$. Now considering the contour C consisting of C_R, l_1, l_2, ... l_{m+1}, C_1, $C_{2\ldots}$ C_m as shown in Figure1.21.

where C_R : $|z| = R$; Im(z) > 0 and C_k : $|z - x_k| = \varepsilon$, Im(z) > 0, where R is large and ε is small

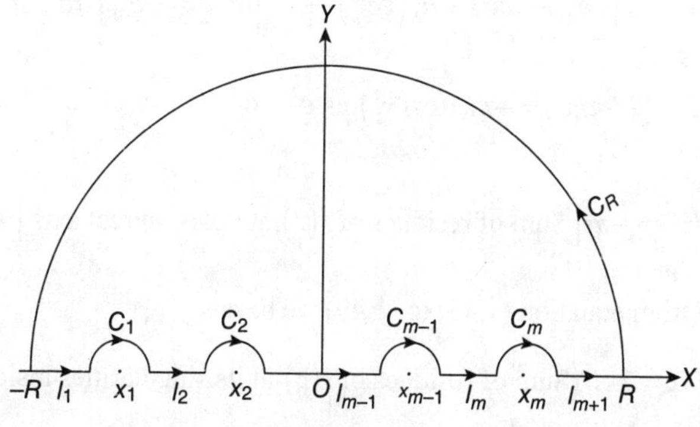

Figure 1.21

We shall have

$$\int_C f(z)\,dz = \int_{C_R} f(z)dz + \sum_{k=1}^{m} \int_{C_k(\text{clockwise})} f(z)dz + \sum_{k=1}^{m+1}\int_{l_k} f(x)dx \tag{1.45}$$

As in (1.15.2)

$$\int_C f(z)dz = 2\pi i\,\big(\text{Sum of residues of } f(z) \text{ at its isolated singularities inside } C\big) \tag{1.46}$$

$$\sum_{k=1}^{m+1}\int_{l_k} f(x)dx = \int_{-\infty}^{\infty} f(x)dx \text{ as } \varepsilon \to 0 \tag{1.47}$$

$$\int_{C_R} f(z)dz \;\to 0 \text{ as } R \to \infty \tag{1.48}$$

(Prove it as in (1.15.2))

Now, we find $\displaystyle\int_{C_k(\text{clockwise})} f(z)dz$

In the semicircle $C_k : |z - x_k| = \varepsilon, z - x_k = \varepsilon\, e^{i\theta}$ and θ varies from π to 0.
By Laurent's expansion, as x_k is simple pole of $f(z)$, we have

$$f(z) = \left\{a_0 + a_1(z - x_k) + a_2(z - x_k)^2 + ...\right\} + \frac{a_{-1}}{z - x_k}$$

Now, $z - x_k = \varepsilon e^{i\theta} \Rightarrow dz = i\varepsilon e^{i\theta} d\theta$ where θ vaies from π to 0.

$$\therefore \quad \int_{C_k} f(z)dz = \int_{\pi}^{0}\left\{a_0 + a_1(z - x_k) + a_2(z - x_k)^2 + ... + \frac{a_{-1}}{z - x_k}\right\} i\varepsilon e^{i\theta} d\theta$$

$$= \int_{\pi}^{0}\left[\left\{a_0 + a_1\varepsilon e^{i\theta} + a_2\left(\varepsilon e^{i\theta}\right)^2 + ...\right\} + \frac{a_{-1}}{\varepsilon e^{i\theta}}\right] i\varepsilon e^{i\theta} d\theta$$

$$= \varepsilon\int_{\pi}^{0}\left[a_0 + a_1\varepsilon e^{i\theta} + a_2\left(\varepsilon e^{i\theta}\right)^2 + ...\right]ie^{i\theta}d\theta + ia_{-1}\int_{\pi}^{0}d\theta \to ia_{-1}\left(\theta\right)_{\pi}^{0} \text{ as } \varepsilon \to 0$$

$$\therefore \qquad \int_{C_k} f(z)dz \to -\pi i a_{-1} = -\pi i \operatorname{Res}\left(x_k\right) \text{ as } \varepsilon \to 0$$

$$\therefore \qquad \sum_{k=1}^{m}\int_{C_k} f(z)dz \to -\pi i\left[\text{Sum of residues of } f(z) \text{ at poles on real axis}\right] \text{ as } \varepsilon \to 0 \qquad (1.49)$$

$\therefore$ As $R\to\infty$, $\varepsilon \to 0$ from equation (1.45) to (1.49) we have

$$\int_{-\infty}^{\infty} f(x)dx = 2\pi i\left(\text{Sum of residues of } f(z) \text{ at its singularities inside } C\right)$$

$$+\pi i\left(\text{Sum of residues of } f(z) \text{ at simple poles on real axis}\right)$$

Remark 1.14: In the examples 1.88 to 1.91 we shall be defining the contour C but the figure (as in the article) will not be drawn.

Example 1.88: Evaluate $\int_0^{\infty}\frac{\sin mx}{x}dx$; $m > 0$ and hence find the value of $\int_0^{\infty}\frac{\sin x}{x}dx$.

Solution: Consider $\oint_C \frac{e^{imz}}{z}dz$

where C consists of C_R: $|z| = R$, Im$(z) > 0$, real axis from $-R$ to $-r$, C_r: $|z| = r$, Im$(z) > 0$, real axis from r to R where R is large and r is small. Here, C_r is clockwise.

$$\therefore \qquad \oint_C \frac{e^{imz}}{z}dz = \int_{C_R}\frac{e^{imz}}{z}dz + \int_{-R}^{-r}\frac{e^{imx}}{x}dx + \int_{C_r}\frac{e^{imz}}{z}dz + \int_{r}^{R}\frac{e^{imx}}{x}dx \qquad (1)$$

Now, $\frac{e^{imz}}{z}$ is analytic on and inside C

$\therefore$ By Cauchy–Goursat theorem

$$\oint_C \frac{e^{imz}}{z}dz = 0 \qquad (2)$$

On C_R, $z = Re^{i\theta}$; $0<\theta<\pi$ $\therefore$ $dz = i\,Re^{i\theta}d\theta$

$$\left|\int_{C_R}\frac{e^{imz}}{z}dz\right| = \left|\int_0^{\pi}\frac{e^{imR(\cos\theta + i\sin\theta)}}{Re^{i\theta}}i\,Re^{i\theta}d\theta\right|$$

$$\le \int_0^{\pi}\frac{e^{-mR\sin\theta}}{R}\cdot Rd\theta = 2\int_0^{\pi/2}e^{-mR\sin\theta}d\theta \qquad \left(\because\ e^{-mR\sin(\pi-\theta)} = e^{-mR\sin\theta}\right)$$

$$\leq 2\int_0^{\frac{\pi}{2}} e^{\frac{-2mR\theta}{\pi}}\, d\theta$$

$$\left(\because \text{ By Jordan inequality when } 0 \leq \theta \leq \frac{\pi}{2} \text{ then } \sin\theta \geq \frac{2\theta}{\pi} \text{ and hence } -\sin\theta \leq \frac{-2\theta}{\pi}\right)$$

$$= -\frac{\pi}{mR} e^{-\frac{2mR}{\pi}\theta}\Bigg|_0^{\frac{\pi}{2}}$$

$$= \frac{\pi}{mR}\left(1 - e^{-mR}\right) \to 0 \text{ as } R \to \infty \qquad (\because m > 0) \tag{3}$$

On $C_r, z = re^{i\theta} \therefore\ dz = ire^{i\theta}\, d\theta$ and θ varies from π to 0. ($\because$ clockwise)

$$\therefore \qquad \int_{C_r} \frac{e^{imz}}{z} dz = \int_\pi^0 \frac{e^{im\, re^{i\theta}}}{re^{i\theta}}\, ire^{i\theta}\, d\theta = i\sum_{n=0}^{\infty}\int_\pi^0 \frac{\left(imre^{i\theta}\right)^n}{n!} d\theta$$

$$= i\int_\pi^0 d\theta + \sum_{n=1}^{\infty} i\frac{(im)^n r^n}{n!}\int_\pi^0 e^{in\theta}\, d\theta \to -\pi i + 0 \qquad \text{as } r \to 0 \tag{4}$$

$\therefore$ As $R \to \infty$, $r \to 0$ from (1) to (4) we have

$$\int_{-\infty}^{0} \frac{e^{imx}}{x} dx - \pi i + \int_0^\infty \frac{e^{imx}}{x} dx = 0$$

$$\Rightarrow \qquad \int_{-\infty}^{\infty} \frac{e^{imx}}{x} dx = \pi i$$

$$\therefore \qquad \int_{-\infty}^{\infty} \frac{\cos mx + i\sin mx}{x} dx = \pi i$$

Equate imaginary parts

$$\int_{-\infty}^{\infty} \frac{\sin mx}{x} dx = \pi$$

$$\Rightarrow \qquad 2\int_0^\infty \frac{\sin mx}{x} dx = \pi \qquad \left(\because \frac{\sin mx}{x} \text{ is even function}\right)$$

$$\therefore \qquad \int_0^\infty \frac{\sin mx}{x} dx = \frac{\pi}{2}; m > 0$$

Take $m = 1$

$$\int_0^\infty \frac{\sin x}{x} dx = \frac{\pi}{2}$$

Note: The above value is in Cauchy principal value sense. As the integral $\int_{-\infty}^{\infty} \frac{\sin mx}{x} dx;\ m > 0$ is convergent so the value of the integral will be same as the value in Cauchy principal value sense.

Example 1.89: Show that $\int_{0}^{\infty} \frac{\cos 2ax - \cos 2bx}{x^2} dx = \pi(b-a);\ a,b > 0.$

Solution: Consider $\oint_C \frac{e^{2iaz} - e^{2ibz}}{z^2} dz;\ a > 0,\ b > 0$

where C consists of $C_R : |z| = R, \text{Im}(z) > 0,$ real axis from $-R$ to $-r$, $C_r : |z| = R, \text{Im}(z) > 0,$ (clockwise) and real axis from r to R where R is large and r is small.

$$\therefore \oint_C \frac{e^{2iaz} - e^{2ibz}}{z^2} dz = \int_{C_R} \frac{e^{2iaz} - e^{2ibz}}{z^2} dz + \int_{-R}^{-r} \frac{e^{2iax} - e^{2ibx}}{x^2} dx + \int_{C_r} \frac{e^{2iaz} - e^{2ibz}}{z^2} dz + \int_{r}^{R} \frac{e^{2iax} - e^{2ibx}}{x^2} dx \quad (1)$$

Now, $\frac{e^{2iaz} - e^{2ibz}}{z^2}$ is analytic on and inside C

$\therefore$ By Cauchy–Goursat theorem

$$\oint_C \frac{e^{2iaz} - e^{2ibz}}{z^2} dz = 0 \quad (2)$$

On C_R, $z = Re^{i\theta}$ $\quad \therefore dz = i\,Re^{i\theta} d\theta; \quad 0 < \theta < \pi$

$$\therefore \left| \int_{C_R} \frac{e^{2iaz} - e^{2ibz}}{z^2} dz \right| = \left| \int_0^{\pi} \frac{e^{2iaR(\cos\theta + i\sin\theta)} - e^{2ibR(\cos\theta + i\sin\theta)}}{R^2 e^{2i\theta}} i\,Re^{i\theta} d\theta \right|$$

$$\leq \int_0^{\pi} \frac{e^{-2aR\sin\theta} + e^{-2bR\sin\theta}}{R^2} .Rd\theta$$

$$< \int_0^{\pi} \frac{1+1}{R} d\theta$$

$\left(\because\ aR\sin\theta > 0, bR\sin\theta > 0 \text{ for } a > 0, b > 0, 0 < \theta < \pi \therefore e^{-2aR\sin\theta} < 1, e^{-2bR\sin\theta} < 1\right)$

$$= \frac{2\pi}{R} \to 0 \text{ as } R \to \infty \quad (3)$$

On C_r, $z = re^{i\theta}$, $dz = ir\,e^{i\theta} d\theta$ and θ varies from π to 0 $(\because$ clockwise$)$

$$\therefore \int_{C_r} \frac{e^{2iaz} - e^{2ibz}}{z^2} dz = \int_{\pi}^{0} \frac{e^{2iare^{i\theta}} - e^{2ibre^{i\theta}}}{r^2 e^{2i\theta}} ir\,e^{i\theta} d\theta$$

$$= \int_{\pi}^{0} \frac{\sum_{n=0}^{\infty} \frac{(2iare^{i\theta})^n}{n!} - \sum_{n=0}^{\infty} \frac{(2ibre^{i\theta})^n}{n!}}{re^{i\theta}} id\theta$$

$$= \int_{\pi}^{0} \frac{\sum_{n=1}^{\infty} \frac{\left(2iare^{i\theta}\right)^n - \left(2ibre^{i\theta}\right)^n}{n!}}{re^{i\theta}}\, id\theta$$

$$= \sum_{n=1}^{\infty} \frac{(2ir)^n \left(a^n - b^n\right)}{r\,n!} \int_{\pi}^{0} \frac{e^{in\theta}}{e^{i\theta}}\, id\theta$$

$$= 2i(a-b)\int_{\pi}^{0} \frac{e^{i\theta}}{e^{i\theta}}\, id\theta + \sum_{n=2}^{\infty} \frac{(2i)^n r^{n-1}\left(a^n - b^n\right)}{n!} \int_{\pi}^{0} ie^{i(n-1)\theta}\, d\theta$$

$$= 2\pi(a-b) + \sum_{n=2}^{\infty} \frac{(2i)^n r^{n-1}\left(a^n - b^n\right)}{(n-1)n!}\left(1-(-1)^{n-1}\right)$$

$$\therefore \quad \int_{C_r} \frac{e^{2iaz} - e^{2ibz}}{z^2}\, dz \to 2\pi(a-b) \text{ as } r \to 0 \qquad (4)$$

$\therefore$ As $R \to \infty$, $r \to 0$ from (1) to (4)

$$\int_{-\infty}^{0} \frac{e^{2iax} - e^{2ibx}}{x^2}\, dx + 2\pi(a-b) + \int_{0}^{\infty} \frac{e^{2iax} - e^{2ibx}}{x^2}\, dx = 0$$

$$\therefore \quad \int_{-\infty}^{\infty} \frac{e^{2iax} - e^{2ibx}}{x^2}\, dx = 2\pi(b-a)$$

$$\Rightarrow \quad \int_{-\infty}^{\infty} \frac{(\cos 2ax - \cos 2bx) + i(\sin 2ax - \sin 2bx)}{x^2}\, dx = 2\pi(b-a)$$

Equate real parts

$$\int_{-\infty}^{\infty} \frac{\cos 2ax - \cos 2bx}{x^2}\, dx = 2\pi(b-a)$$

$$\therefore \quad \int_{0}^{\infty} \frac{\cos 2ax - \cos 2bx}{x^2}\, dx = \pi(b-a) \quad \left(\because \ \frac{\cos 2ax - \cos 2bx}{x^2} \text{ is an even function}\right)$$

Example 1.90: Evaluate the integral $I = \int_{0}^{\infty} \frac{\sin ax}{x\left(x^2 + b^2\right)}\, dx; \quad a, b > 0$

Solution: Consider $\oint_C \frac{e^{iaz}}{z\left(z^2 + b^2\right)}\, dz; \ a, b > 0$

where C consists of $C_R : |z| = R, \operatorname{Im}(z) > 0$ real axis from $-R$ to $-r$, $C_r : |z| = r, \operatorname{Im}(z) > 0$ (clockwise), real axis from r to R where R is large and r is small.

$$\therefore \oint_C \frac{e^{iaz}}{z\left(z^2 + b^2\right)}\, dz = \int_{C_R} \frac{e^{iaz}}{z\left(z^2 + b^2\right)}\, dz + \int_{-R}^{-r} \frac{e^{iax}}{x\left(x^2 + b^2\right)}\, dx + \int_{C_r} \frac{e^{iaz}}{z\left(z^2 + b^2\right)}\, dz + \int_{r}^{R} \frac{e^{iax}}{x\left(x^2 + b^2\right)}\, dx \quad (1)$$

Now, $f(z)=\dfrac{e^{iaz}}{z\left(z^2+b^2\right)}$ has simple poles at $z=0,\ \pm ib$ of which $z=ib$ lies inside C ($\because\ b>0$)

$$\text{Res }(ib)=\lim_{z\to ib}(z-ib)f(z)=\lim_{z\to ib}\frac{e^{iaz}}{z(z+ib)}=\frac{e^{-ab}}{ib(2ib)}=-\frac{e^{-ab}}{2b^2}$$

$\therefore$ By Cauchy residue theorem

$$\oint_C \frac{e^{iaz}}{z\left(z^2+b^2\right)}dz=2\pi i\left(-\frac{e^{-ab}}{2b^2}\right)=-\frac{\pi i e^{-ab}}{b^2} \tag{2}$$

On C_R, $z=R\mathrm{e}^{i\theta}$ $\quad\therefore dz=iR\mathrm{e}^{i\theta}\,d\theta,\ 0<\theta<\pi$

$$\therefore \quad \left|\oint_{C_R}\frac{e^{iaz}}{z\left(z^2+b^2\right)}dz\right|\le\int_{C_R}\left|\frac{e^{iaR(\cos\theta+i\sin\theta)}}{Re^{i\theta}\left(R^2e^{2i\theta}+b^2\right)}Rie^{i\theta}d\theta\right|$$

$$\le\int_0^{\pi}\frac{e^{-aR\sin\theta}}{R^2-b^2}d\theta<\int_0^{\pi}\frac{d\theta}{R^2-b^2}$$

$$\left(\because a>0, R>0, 0<\sin\theta<1 \text{ for } 0<\theta<\pi\ \therefore\ e^{-aR\sin\theta}<1\right)$$

$$=\frac{\pi}{R^2-b^2}\to 0 \text{ as } R\to\infty \tag{3}$$

Now, $f(z)=\dfrac{e^{iaz}}{z\left(z^2+b^2\right)}$ has simple pole at $z=0$

$\therefore$ By Laurent expansion about $z=0$

$$f(z)=\left(a_0+a_1z+a_2z^2+\cdots\right)+\frac{a_{-1}}{z}$$

On C_r, $z=re^{i\theta}$ $\quad dz=ire^{i\theta}d\theta$, θ varies from π to 0

$$\therefore \quad \int_{C_r}f(z)\,dz=\int_{\pi}^{0}\left[\left(a_0+a_1re^{i\theta}+a_2r^2e^{2i\theta}+\cdots\right)+\frac{a_{-1}}{re^{i\theta}}\right]ire^{i\theta}\,d\theta$$

$$=i\int_{\pi}^{0}\left(a_0+a_1re^{i\theta}+a_2r^2e^{2i\theta}+\cdots\right)re^{i\theta}\,d\theta+i\,a_{-1}(-\pi)\ \to 0-\pi i\,a_{-1} \quad \text{as } r\to 0$$

But $a_{-1}=\text{Res}(0)=\lim\limits_{z\to 0}\dfrac{e^{iaz}}{z^2+b^2}=\dfrac{1}{b^2}$

$\therefore$ As $r\to 0$

$$\int_{C_r}\frac{e^{iaz}}{z\left(z^2+b^2\right)}dz=-\frac{\pi i}{b^2} \tag{4}$$

$\therefore$ from (1) to (4) as $R\to\infty,\ r\to 0$

$$\int_{-\infty}^{0}\frac{e^{iax}}{x\left(x^2+b^2\right)}dx-\frac{\pi i}{b^2}+\int_0^{\infty}\frac{e^{iax}}{x\left(x^2+b^2\right)}dx=-\frac{\pi i e^{-ab}}{b^2}$$

$$\therefore \quad \int_{-\infty}^{\infty} \frac{e^{iax}}{x\left(x^2+b^2\right)} dx = \frac{\pi i}{b^2}\left(1-e^{-ab}\right)$$

$$\Rightarrow \quad \int_{-\infty}^{\infty} \frac{\cos ax + i\sin ax}{x\left(x^2+b^2\right)} dx = \frac{\pi i}{b^2}\left(1-e^{-ab}\right)$$

Equate imaginary parts

$$\int_{-\infty}^{\infty} \frac{\sin ax}{x\left(x^2+b^2\right)} dx = \frac{\pi}{b^2}\left(1-e^{-ab}\right)$$

$$\therefore \quad 2\int_{0}^{\infty} \frac{\sin ax}{x\left(x^2+b^2\right)} dx = \frac{\pi}{b^2}\left(1-e^{-ab}\right) \qquad \left(\because \frac{\sin ax}{x\left(x^2+b^2\right)} \text{ is even function}\right)$$

$$\int_{0}^{\infty} \frac{\sin ax}{x\left(x^2+b^2\right)} dx = \frac{\pi}{2b^2}\left(1-e^{-ab}\right)$$

Example 1.91: Evaluate the integral $\int_0^\infty \frac{\sin^2 x}{x^2} dx$.

Solution: Let $I = \int_0^\infty \frac{\sin^2 x}{x^2} dx = \int_0^\infty \frac{1-\cos 2x}{2x^2} dx$

Consider $\oint\limits_C \frac{1-e^{2iz}}{2z^2} dz$

where C consists of $C_R : |z| = R, \operatorname{Im}(z) > 0$, real axis from $-R$ to $-r$, $C_r : |z| = r, \operatorname{Im}(z) > 0$ clockwise, real axis from r to R

$$\therefore \quad \oint\limits_C \frac{1-e^{2iz}}{2z^2} dz = \int\limits_{C_R} \frac{1-e^{2iz}}{2z^2} dz + \int_{-R}^{-r} \frac{1-e^{2ix}}{2x^2} dx + \int\limits_{C_r} \frac{1-e^{2iz}}{2z^2} dz + \int_r^R \frac{1-e^{2ix}}{2x^2} dx \tag{1}$$

Now, $\frac{1-e^{2iz}}{2z^2}$ is analytic on and inside C

$\therefore$ By Cauchy–Goursat theorem

$$\oint\limits_C \frac{1-e^{2iz}}{2z^2} dz = 0 \tag{2}$$

On C_R, $z = Re^{i\theta}$ $\quad \therefore dz = i\,Re^{i\theta} d\theta;\ 0 < \theta < \pi$

$$\therefore \quad \left|\int\limits_{C_R} \frac{1-e^{2iz}}{2z^2} dz\right| \le \int\limits_{C_R} \left|\frac{1-e^{2iz}}{2z^2}\right| |dz| \le \int\limits_{C_R} \frac{1+\left|e^{2iz}\right|}{2R^2} |dz|$$

$$= \int_0^\pi \frac{1+\left|e^{2iR(\cos\theta + i\sin\theta)}\right|}{2R^2} R\,d\theta$$

$$= \int_0^{\pi} \frac{1+e^{-2R\sin\theta}}{2R} d\theta < \int_0^{\pi} \frac{1+1}{2R} d\theta \qquad \left(\because e^{-2R\sin\theta} < 1 \text{ as } 0 < \sin\theta < 1 \text{ for } 0 < \theta < \pi\right)$$

$$\to 0 \quad \text{as } R \to \infty \tag{3}$$

On C_r, $z = re^{i\theta}$ $\therefore$ $dz = ire^{i\theta}\, d\theta$, θ varies from π to 0 (clockwise)

$$\therefore \qquad \int_{C_r} \frac{1-e^{2iz}}{2z^2} dz = \int_{\pi}^{0} \frac{1-e^{2ire^{i\theta}}}{2r^2e^{2i\theta}} ire^{i\theta} d\theta = \int_{\pi}^{0} \frac{-\sum_{n=1}^{\infty} \frac{\left(2ie^{i\theta}\right)^n r^n}{n!}}{2re^{i\theta}} id\theta$$

$$= \int_{\pi}^{0} -i \sum_{n=1}^{\infty} \frac{\left(2ie^{i\theta}\right)^n r^{n-1}}{n!\, 2e^{i\theta}} d\theta$$

$$= \int_{\pi}^{0} \frac{-i}{2} \cdot \left[2i + \sum_{n=2}^{\infty} \frac{(2i)^n}{n!} \left(e^{i\theta}\right)^{n-1} r^{n-1}\right] d\theta$$

$$= \frac{2}{2}(-\pi) = -\pi \text{ as } r \to 0 \tag{4}$$

$\therefore$ As $R \to \infty$, $r \to 0$ from (1) to (4)

$$\int_{-\infty}^{0} \frac{1-e^{2ix}}{2x^2} dx - \pi + \int_{0}^{\infty} \frac{1-e^{2ix}}{2x^2} dx = 0$$

$$\therefore \qquad \int_{-\infty}^{\infty} \frac{1-e^{2ix}}{2x^2} dx = \pi$$

$$\Rightarrow \qquad \int_{-\infty}^{\infty} \frac{1-\cos 2x - i\sin 2x}{2x^2} dx = \pi$$

Equate real parts

$$\int_{-\infty}^{\infty} \frac{1-\cos 2x}{2x^2} = \pi$$

$$\therefore \qquad \int_{-\infty}^{\infty} \frac{\sin^2 x}{x^2} = \pi$$

$$\Rightarrow \qquad \int_0^{\infty} \frac{\sin^2 x}{x^2} = \frac{\pi}{2} \qquad \left(\because \frac{\sin^2 x}{x^2} \text{ is even function}\right)$$

EXERCISE 1.5

1. Using contour integration evaluate

   (i) $\int_0^{2\pi} \frac{d\theta}{(2+\sin\theta)}$

   (ii) $\int_0^{2\pi} \frac{\cos\theta}{(3+\sin\theta)} d\theta$

   (iii) $\int_0^{2\pi} \frac{\cos 3\theta}{5-4\cos\theta} d\theta$

   (iv) $\int_0^{\pi} \frac{1+2\cos\theta}{5+4\cos\theta} d\theta$

2. Using contour integration, prove that $\int_0^{2\pi} \frac{d\theta}{(1-2a\sin\theta+a^2)} = \frac{2\pi}{(1-a^2)}, 0<|a|<1$

3. Using contour integration, evaluate $\int_0^{\pi} \frac{d\theta}{(2+\cos\theta)^2}$.

4. Evaluate $\int_0^{2\pi} \frac{\cos n\theta}{(1+2a\cos\theta+a^2)} d\theta$ and $\int_0^{2\pi} \frac{\sin n\theta}{(1+2a\cos\theta+a^2)} d\theta$ given $a^2 < 1$ and n is a positive integer.

5. Using contour integration evaluate

   (i) $\int_0^{\infty} \frac{dx}{(x^4+1)}$

   (ii) $\int_0^{\infty} \frac{dx}{(x^4+16)}$

6. Apply residue theorem to evaluate $\int_0^{\infty} \frac{x^2}{(x^2+a^2)^2} dx, a>0.$

7. Evaluate the integral $\int_0^{\infty} \frac{x^2+2}{(x^2+1)(x^2+4)} dx.$

8. Show that $\int_0^{\infty} \frac{x^2}{(x^2+a^2)^3} dx = \frac{\pi}{16a^3}, a>0.$

9. Use contour integration to evaluate the real integral $\int_{-\infty}^{\infty} \frac{dx}{(x^2+a^2)^3}; a>0$ and hence find the value of $\int_0^{\infty} \frac{dx}{(x^2+1)^3}$.

10. Show that

    (i) $\int_{-\infty}^{\infty} \frac{x^2}{(x^2+1)} \cdot \frac{dx}{(x^2+2x+2)} = \frac{\pi}{8}$

    (ii) $\int_{-\infty}^{\infty} \frac{x^2}{(x^2+1)^2(x^2+2x+2)} dx = \frac{7\pi}{50}$

11. Evaluate $\int_{-\infty}^{\infty} \frac{\cos mx}{(x^2+a^2)(x^2+b^2)} dx;$ $a, b>0$ and $a \neq b; m>0.$

12. Using complex variable technique evaluate $\int_0^{\infty} \frac{\cos 3x}{(x^2+1)(x^2+4)} dx.$

13. Evaluate $\int_0^{\infty} \frac{\cos mx}{(x^2+a^2)^2} dx; m, a>0.$

14. Show that $\int_{-\infty}^{\infty} \frac{\sin x}{(x^2+4x+5)} dx = -\frac{\pi\sin 2}{e}.$

15. Evaluate the integral $\int_0^{\infty} \frac{x\sin x}{(x^2+a^2)^2} dx; a>0.$

16. By integrating e^{-z^2} round the rectangle whose vertices are $-R$, R, $R+ia$, $-R+ia$ evaluate the integral $\int_{-\infty}^{\infty} e^{-x^2} \cos 2ax \, dx.$

17. By integrating e^{-z^2} round the rectangle whose vertices are 0, R, $R + ia$, ia show that

(i) $\int_0^\infty e^{-x^2} \cos 2ax \, dx = \frac{e^{-a^2}}{2}\sqrt{\pi}$

(ii) $\int_0^\infty e^{-x^2} \sin 2ax \, dx = e^{-a^2} \int_0^a e^{y^2} dy$

18. Show that $\int_0^\infty \frac{\sin x}{x} dx = \frac{\pi}{2}$ and $\int_0^\infty \frac{\cos x}{x} dx = 0.$

19. Show that $\int_0^\infty \frac{\sin x}{x\left(x^2+a^2\right)} dx = \frac{\pi}{2a^2}(1-e^{-a})$ and $\int_0^\infty \frac{\cos x}{x\left(x^2+a^2\right)} dx = 0, a > 0.$

ANSWERS 1.5

1. (i) $2\pi/\sqrt{3}$ (ii) 0 (iii) $\pi/12$ (iv) 0

3. $\frac{2\pi}{3\sqrt{3}}$

4. (i) $\frac{2\pi(-1)^n a^n}{1-a^2}$ (ii) 0

5. (i) $\frac{\pi}{2\sqrt{2}}$ (ii) $\frac{\pi\sqrt{2}}{32}$

6. $\frac{\pi}{4a}$

7. $\frac{\pi}{3}$

9. $\frac{3\pi}{8a^5}, \frac{3\pi}{16}$

11. $\frac{\pi}{\left(a^2-b^2\right)}\left(\frac{e^{-bm}}{b}-\frac{e^{-am}}{a}\right)$

12. $\frac{\pi}{2}\left(\frac{e^{-3}}{3}-\frac{e^{-6}}{6}\right)$

13. $\frac{\pi(am+1)e^{-am}}{4a^3}$

15. $\frac{\pi e^{-a}}{4a}$

16. $\sqrt{\pi}e^{-a^2}$

1.16 CONFORMAL MAPPING

For every point (x, y) in the z-plane in domain of f, the relation $w = f(z)$ defines a corresponding point (u, v) in the w-plane. If a point $P(z_0)$ maps into the point P^* (w_0) then w_0 is known as image of z_0. If P moves along a curve C in z-plane then P^* will move along a corresponding curve C^* in the w-plane.

Now, let two curves C_1 and C_2 in z-plane intersect at $P(z_0)$ and the corresponding curves C_1^*, C_2^* in w-plane intersect at $P^*(w_0)$. If the angle of intersection of the curves C_1 and C_2 at P in z-plane is same in magnitude and sense as the angle of intersection of curves C_1^* and C_2^* at P^* in w-plane then the transformation is called conformal. Thus, if the sense of rotation as well as the magnitude of angle is preserved, the transformation is said to be **conformal**. If only the magnitude of the angle is preserved, then the transformation is **isogonal**.

The point z where $f'(z) = 0$ is called **critical point** of this transformation. The point z where $f(z)$ is defined and $f'(z) \neq 0$ is called **ordinary point**. The point z where $w = f(z) = z$ is called **fixed point or invariant point** of the transformation. Now, we prove sufficient conditions for a transformation to be conformal at a point.

Theorem 1.23 The mapping $w = f(z)$ is conformal at each point z where $f(z)$ is analytic and $f'(z) \neq 0$.

Proof: Let $w = f(z)$ be analytic at $P(z_0)$ and $f'(z_0) \neq 0$. Let C be a continuous curve in z-plane with parametric equation

$$z(t) = x(t) + iy(t);\ t_1 \leq t \leq t_2 \tag{1.50}$$

passing through $P(z_0)$. The image C^* of this curve in w-plane is $w(t) = f(z(t));\ t_1 \leq t \leq t_2$ passing through $P^*\ (w_0)$ the image of $P(z_0)$.

From equation (1.50), slope of tangent to C at $P = \dfrac{dy}{dx} = \dfrac{\dot{y}(t)}{\dot{x}(t)}$ at $t = t_0$ corresponding to point P where $\dot{y}(t) = \dfrac{dy}{dt}$, $\dot{x}(t) = \dfrac{dx}{dt}$

If θ_0 is inclination of this tangent at P, then

$$\tan\theta_0 = \frac{\dot{y}(t)}{\dot{x}(t)}$$

$$\therefore \quad \theta_0 = \tan^{-1}\frac{\dot{y}(t)}{\dot{x}(t)} = \arg \dot{z}(t) \text{ at } P$$

$$\therefore \quad \dot{z}(t_0) = r_0 e^{i\theta_0} \tag{1.51}$$

Similarly, $\dot{w}(t_0) = \rho_0 e^{i\phi_0}$ (1.52)

where ϕ_0 is inclination of tangent at P^* to C^*.

Now,
$$w(t) = f(z(t))$$
$$\therefore \quad \dot{w}(t) = f'(z(t))\dot{z}(t)$$
$$\therefore \quad \dot{w}(t_0) = f'(z(t_0))\dot{z}(t_0)$$

Using equations (1.51) and (1.52)

$$\rho_0 e^{i\phi_0} = R_0 e^{i\psi_0} r_0 e^{i\theta_0} \text{ where } f'(z(t_0)) = R_0 e^{i\psi_0} \text{ if } f'(z(t_0)) \neq 0$$

$$\therefore \quad \rho_0 e^{i\phi_0} = R_0 r_0 e^{i(\psi_0 + \theta_0)}$$

$$\therefore \quad \rho_0 = R_0 r_0,\ \phi_0 = \psi_0 + \theta_0$$

$$\therefore \quad \left.\begin{aligned} \psi_0 &= \phi_0 - \theta_0 \\ \text{and } \rho_0 &= R_0 r_0 \end{aligned}\right\}$$

Thus, the tangent to curve C at P is rotated by an angle

$$\psi_0 = Arg\left(f'(z_0)\right) \qquad \text{where} \quad z_0 = z(t_0)$$

Angle of rotation ψ_0 is independent of curve C and depends only on z_0.

Thus if C_1, C_2 be any two curves passing through $P(z_0)$ with their images C_1^*, C_2^* respectively meeting at $P^*(w_0)$ then angle of rotation of both C_1 and C_2 will be ψ_0.

If θ_1 and θ_2 are inclination of tangents to C_1 and C_2 at $P(z_0)$ respectively and ϕ_1 and ϕ_2 be inclination of tangents to C_1^* and C_2^* at $P^*(w_0)$ then

$$\psi_0 = \phi_1 - \theta_1 = \phi_2 - \theta_2$$

$$\therefore \quad \phi_2 - \phi_1 = \theta_2 - \theta_1$$

Thus, the transformation is conformal at P.

Remark 1.15: (i) The angle through which tangent at $P(z_0)$ to a curve rotates by transformation $w = f(z)$ is called angle of rotation.

$\therefore$ Angle of rotation at $P = \text{Arg}\,(f'(z_0))$

(ii) Element of arc r_0 through P is magnified by $R_0 = |f'(z_0)|$. It is called coefficient of magnification of transformation at P.

Theorem 1.24 A harmonic function $\phi(x, y)$ remains harmonic under the conformal mapping $w = f(z)$.

Proof: Since $\phi(x, y)$ is harmonic

$$\therefore \quad \frac{\partial^2\phi}{\partial x^2} + \frac{\partial^2\phi}{\partial y^2} = 0 \tag{1.53}$$

Also, $w = f(z) = u + iv$ is conformal, so $f(z)$ is analytic and hence u and v are harmonic.

$$\therefore \quad \frac{\partial^2 u}{\partial x^2} + \frac{\partial^2 u}{\partial y^2} = 0 \tag{1.54}$$

$$\frac{\partial^2 v}{\partial x^2} + \frac{\partial^2 v}{\partial y^2} = 0 \tag{1.55}$$

Also, $f(z)$ is analytic so u and v satisfy C–R equations

$$\frac{\partial u}{\partial x} = \frac{\partial v}{\partial y}, \quad \frac{\partial u}{\partial y} = -\frac{\partial v}{\partial x} \tag{1.56}$$

Now, $$\frac{\partial\phi}{\partial x} = \frac{\partial\phi}{\partial u}\frac{\partial u}{\partial x} + \frac{\partial\phi}{\partial v}\frac{\partial v}{\partial x}$$

$$\therefore \quad \frac{\partial^2\phi}{\partial x^2} = \frac{\partial\phi}{\partial u}\frac{\partial^2 u}{\partial x^2} + \frac{\partial}{\partial x}\left(\frac{\partial\phi}{\partial u}\right)\cdot\frac{\partial u}{\partial x} + \frac{\partial\phi}{\partial v}\frac{\partial^2 v}{\partial x^2} + \frac{\partial}{\partial x}\left(\frac{\partial\phi}{\partial v}\right)\frac{\partial v}{\partial x}$$

$$= \frac{\partial\phi}{\partial u}\frac{\partial^2 u}{\partial x^2} + \left(\frac{\partial^2\phi}{\partial u^2}\frac{\partial u}{\partial x} + \frac{\partial^2\phi}{\partial v\partial u}\frac{\partial v}{\partial x}\right)\frac{\partial u}{\partial x} + \frac{\partial\phi}{\partial v}\frac{\partial^2 v}{\partial x^2} + \left(\frac{\partial^2\phi}{\partial u\partial v}\frac{\partial u}{\partial x} + \frac{\partial^2\phi}{\partial v^2}\frac{\partial v}{\partial x}\right)\frac{\partial v}{\partial x}$$

$$= \frac{\partial \phi}{\partial u}\frac{\partial^2 u}{\partial x^2} + \frac{\partial \phi}{\partial v}\frac{\partial^2 v}{\partial x^2} + \frac{\partial^2 \phi}{\partial u^2}\left(\frac{\partial u}{\partial x}\right)^2 + \frac{\partial^2 \phi}{\partial v^2}\left(\frac{\partial v}{\partial x}\right)^2 + 2\frac{\partial^2 \phi}{\partial u \partial v}\frac{\partial u}{\partial x}\frac{\partial v}{\partial x} \tag{1.57}$$

$$\left(\because \quad \frac{\partial^2 \phi}{\partial u \partial v} = \frac{\partial^2 \phi}{\partial u \partial v}\right)$$

Similarly

$$\frac{\partial^2 \phi}{\partial y^2} = \frac{\partial \phi}{\partial u}\frac{\partial^2 u}{\partial y^2} + \frac{\partial \phi}{\partial v}\frac{\partial^2 v}{\partial y^2} + \frac{\partial^2 \phi}{\partial u^2}\left(\frac{\partial u}{\partial y}\right)^2 + \frac{\partial^2 \phi}{\partial v^2}\left(\frac{\partial v}{\partial y}\right)^2 + 2\frac{\partial^2 \phi}{\partial u \partial v}\frac{\partial u}{\partial y}\frac{\partial v}{\partial y} \tag{1.58}$$

Add equations (1.57) and (1.58)

$$\frac{\partial^2 \phi}{\partial x^2} + \frac{\partial^2 \phi}{\partial y^2} = \frac{\partial \phi}{\partial u}\left(\frac{\partial^2 u}{\partial x^2} + \frac{\partial^2 u}{\partial y^2}\right) + \frac{\partial \phi}{\partial v}\left(\frac{\partial^2 v}{\partial x^2} + \frac{\partial^2 v}{\partial y^2}\right) + \frac{\partial^2 \phi}{\partial u^2}\left\{\left(\frac{\partial u}{\partial x}\right)^2 + \left(\frac{\partial u}{\partial y}\right)^2\right\}$$

$$+ \frac{\partial^2 \phi}{\partial v^2}\left\{\left(\frac{\partial v}{\partial x}\right)^2 + \left(\frac{\partial v}{\partial y}\right)^2\right\} + 2\frac{\partial^2 \phi}{\partial u \partial v}\left(\frac{\partial u}{\partial x}\frac{\partial v}{\partial x} + \frac{\partial u}{\partial y}\frac{\partial v}{\partial y}\right)$$

Use equations (1.53) to (1.56)

$$0 = \frac{\partial^2 \phi}{\partial u^2}\left\{\left(\frac{\partial u}{\partial x}\right)^2 + \left(\frac{-\partial v}{\partial x}\right)^2\right\} + \frac{\partial^2 \phi}{\partial v^2}\left\{\left(\frac{\partial v}{\partial x}\right)^2 + \left(\frac{\partial u}{\partial x}\right)^2\right\} + 2\frac{\partial^2 \phi}{\partial u \partial v}\left\{\frac{\partial u}{\partial x}\frac{\partial v}{\partial x} + \left(\frac{-\partial v}{\partial x}\right)\left(\frac{\partial u}{\partial x}\right)\right\}$$

$$\therefore \quad \left(\frac{\partial^2 \phi}{\partial u^2} + \frac{\partial^2 \phi}{\partial v^2}\right)\left\{\left(\frac{\partial u}{\partial x}\right)^2 + \left(\frac{\partial v}{\partial x}\right)^2\right\} = 0 \tag{1.59}$$

Now, $f(z)$ is conformal

$$\therefore \quad f'(z) = \frac{\partial u}{\partial x} + i\frac{\partial v}{\partial x} \neq 0$$

$$\therefore \quad |f'(z)|^2 = \left(\frac{\partial u}{\partial x}\right)^2 + \left(\frac{\partial v}{\partial x}\right)^2 \neq 0$$

$\therefore$ from equation (1.59)

$$\frac{\partial^2 \phi}{\partial u^2} + \frac{\partial^2 \phi}{\partial v^2} = 0$$

Also all second order partial derivatives of ϕ w.r.t. u and v will be continuous. Hence, ϕ is harmonic in w-plane.

Example 1.92: Show that the mapping $w = e^z$ is conformal in the whole of the z-plane.

Solution: $w = f(z) = e^z$

$$\therefore \quad f'(z) = e^z \neq 0 \text{ for any } z$$

$\therefore f(z)$ is analytic and $f'(z) \neq 0$ for any z.

Hence, the given mapping is conformal in whole of z-plane.

Example 1.93: Show that the mapping $w = \sin z$ is conformal everywhere except at $z = (2n+1)\frac{\pi}{2}; n \in \text{I}$.

Solution: $w = f(z) = \sin z$

$\therefore\ f'(z) = \cos z$ is defined for all z

and $f'(z) = \cos z = 0$ for $z = (2n+1)\frac{\pi}{2}; n \in \text{I}$

Hence, the given mapping is conformal everywhere except at $z = (2n+1)\frac{\pi}{2}; n \in \text{I}$.

1.17 SOME STANDARD MAPPINGS

In this section, we discuss some of the important standard transformations.

1.17.1 Translation Mapping

The mapping $w = z + c;\ c = a + ib;\ a, b \in R$ is translation mapping.

We have $u + iv = x + iy + a + ib$

$$\therefore \qquad u = x + a,\ v = y + b$$

Hence point $P(x, y)$ in z-plane is mapped into the point $P^*(x + a, y + b)$ in the w-plane. Thus, if w-plane is superposed on the z-plane then the figure of w-plane is translated through vector $\vec{C} = a\hat{i} + b\hat{j}$.

If equation of curve in z-plane is given then setting $x = u - a$, $y = v - b$, we get the equation of the image in w-plane.

The regions in z-plane and w-plane have same shape, size and orientation and also mapping is conformal. If z_1 and z_2 be any two points in z-plane then corresponding points in w-plane are $w_1 = z_1 + c$, $w_2 = z_2 + c$ respectively. We have $|w_1 - w_2| = |z_1 - z_2|$. Thus, this mapping preserves the distance between the points.

1.17.2 Magnification/Contraction and Rotation

The mapping $w = cz;\ c = a + ib;\ a, b \in R,\ c \neq 0$ is magnification/contraction and rotation mapping.

We have $\quad u + iv = (a + ib)(x + iy)$

$$= ax - by + i(bx + ay)$$

$$\therefore \qquad u = ax - by,\ v = bx + ay$$

Also, $w = cz$

$$\Rightarrow \quad z = \frac{w}{c} = \frac{w\bar{c}}{c\bar{c}} = \frac{w\bar{c}}{|c|^2}$$

$$\therefore \quad x+iy=\frac{1}{a^2+b^2}(u+iv)(a-ib)=\frac{au+bv+i(av-bu)}{a^2+b^2}$$

$$\therefore \quad x=\frac{au+bv}{a^2+b^2},\ y=\frac{av-bu}{a^2+b^2}$$

Thus, on putting these values of x and y in the equation of the curve to be transformed we get the equation of the image.

If we write $c=|c|e^{i\alpha}$, $z=re^{i\theta}$, $w=Re^{i\phi}$ then we have

$$Re^{i\phi}=|c|re^{i(\theta+\alpha)} \qquad (\because w=cz)$$

$$\therefore \quad R=|c|r,\ \phi=\theta+\alpha$$

Thus, transformation $w = cz$ corresponding to rotation by an angle $\alpha = \text{Arg}(c)$ and magnification or contraction by $|c|$. If $|c| > 1$ then there will be magnification and if $|c| < 1$ then there will be contraction. If $|c| = 1$, then there is neither magnification nor contraction.

Shapes of figures in w-plane and z-plane are same and also the mapping is conformal.

1.17.3 Linear Transformation

The transformation $w = az + c$ where $a = a_1 + ia_2$, $c = c_1 + ic_2$ is called linear transformation. It is the combination of translation with rotation and magnification/contraction. This mapping is conformal. Shapes of figures in z-plane and w-plane are same.

Example 1.94: Find the images of the following regions or curves in the z-plane onto the w-plane under the given mappings

(i) The circle $|z-1|=2;\ w=2z$

(ii) The semicircular region $|z|<1,\ \text{Re}\,z>0;\ w=e^{i\pi/4}z$

(iii) The square with vertices at (1, 1), (3, 1), (3, 3), (1, 3);

(a) $w = z + (1 + 2i)$

(b) $w=\left(\sqrt{2}e^{-\pi i/4}\right)z+(1+2i)$

Solution: (i) $|z-1|=2,\ w=2z$

$$\therefore \quad \left|\frac{w}{2}-1\right|=2\Rightarrow|w-2|=4$$

$\therefore$ Image is circle $|w-2|=4$, i.e., circle with centre (2, 0) and radius 4.

(ii) $|z|<1,\ \text{Re}(z)>0, w=e^{\frac{i\pi}{4}}z\Rightarrow z=we^{\frac{-i\pi}{4}}$

$$\therefore \quad |z|<1\Rightarrow\left|we^{-\frac{i\pi}{4}}\right|<1\Rightarrow|w|<1$$

and $\text{Re}(z)>0\Rightarrow\text{Re}\left\{(u+iv)\left(\frac{1}{\sqrt{2}}-i\frac{1}{\sqrt{2}}\right)\right\}>0$

$$\Rightarrow \quad \frac{u}{\sqrt{2}}+\frac{v}{\sqrt{2}}>0 \Rightarrow u+v>0$$

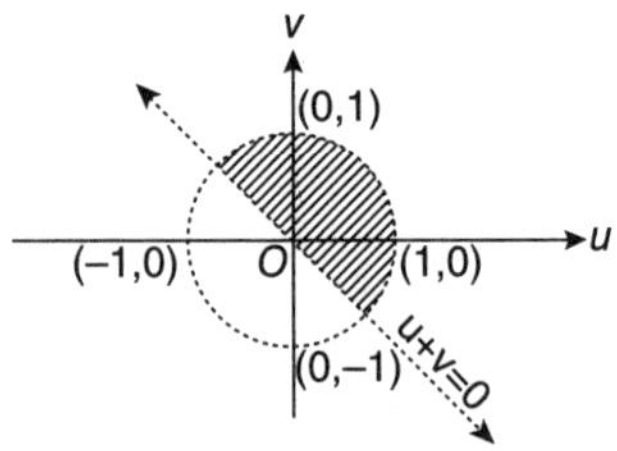

Figure 1.22

∴ Image is the interior of semicircle with centre (0, 0) and radius unity above the line $u + v = 0$.

(iii) (a) $w = z + (1 + 2i) = (x + 1) + i(y + 2)$

∴ The points (1, 1), (3, 1), (3, 3) and (1, 3) map to (2, 3), (4, 3), (4, 5) and (2, 5), respectively in the w-plane.

∴ Image is square with vertices (2, 3), (4, 3), (4, 5) and (2, 5).

(b) $$w = \sqrt{2}\, e^{-\pi i/4} z + (1+2i) = \sqrt{2}\left(\frac{1}{\sqrt{2}} - i\frac{1}{\sqrt{2}}\right)(x+iy) + (1+2i)$$

$$= (1-i)(x+iy) + (1+2i)$$

$$= (x+y+1) + i(-x+y+2)$$

∴ The points (1, 1), (3, 1), (3, 3) and (1, 3) map to (3, 2), (5, 0), (7, 2) and (5, 4), respectively in the w-plane.

∴ Image is square with vertices (3, 2), (5, 0), (7, 2) and (5, 4).

Example 1.95: Find the image of the triangle with vertices at i, $1 + i$, $1 - i$ in the z-plane under the transformation

(i) $w = 3z + 4 - 2i$

(ii) $w = e^{\frac{5\pi i}{3}} \cdot z - 2 + 4i$

Solution: (i) $w = 3z + 4 - 2i$

∴ w transforms i, $1 + i$, $1 - i$ to the points $3i + 4 - 2i = 4 + i$, $3(1 + i) + 4 - 2i = 7 + i$, $3(1 - i) + 4 - 2i = 7 - 5i$ respectively in the w-plane.

∴ Image is triangle with vertices $4 + i$, $7 + i$ and $7 - 5i$.

(ii) $$w = e^{\frac{5\pi i}{3}} z - 2 + 4i = \left(\cos\frac{5\pi}{3} + i\sin\frac{5\pi}{3}\right) z - 2 + 4i$$

$$= \left(\frac{1}{2} - i\frac{\sqrt{3}}{2}\right) z - 2 + 4i = \frac{1}{2}\left\{\left(1 - i\sqrt{3}\right) z - 4 + 8i\right\}$$

$\therefore$ w maps i to the point $\frac{1}{2}\left\{\left(1-i\sqrt{3}\right)i-4+8i\right\}=\frac{1}{2}\left\{\left(\sqrt{3}-4\right)+9i\right\}$

w maps $1+i$ to the point $\frac{1}{2}\left\{\left(1-i\sqrt{3}\right)(1+i)-4+8i\right\}=\frac{1}{2}\left\{\left(\sqrt{3}-3\right)+\left(9-\sqrt{3}\right)i\right\}$

w maps $1-i$ to the point $\frac{1}{2}\left\{\left(1-i\sqrt{3}\right)(1-i)-4+8i\right\}=\frac{1}{2}\left\{\left(-\sqrt{3}-3\right)+\left(7-\sqrt{3}\right)i\right\}$

$\therefore$ Image is triangle with vertices $\frac{1}{2}\left\{\left(\sqrt{3}-4\right)+9i\right\}$, $\frac{1}{2}\left\{\left(\sqrt{3}-3\right)+\left(9-\sqrt{3}\right)i\right\}$ and $\frac{1}{2}\left\{\left(-\sqrt{3}-3\right)+\left(7-\sqrt{3}\right)i\right\}$.

1.17.4 Inverse Transformation (Inversion and Reflection)

The mapping $w=\frac{1}{z}$ is called inverse transformation.

If $z=re^{i\theta}, w=\text{R}e^{i\phi}$ then $\text{R}e^{i\phi}=\frac{1}{r}e^{-i\theta}$

$\therefore$ $\text{R}=\frac{1}{r}$, $\phi=-\theta$

Thus, point (r, θ) in z-plane is mapped onto the point $\left(\frac{1}{r},-\theta\right)$ in w-plane. Thus, this mapping is combination of two mappings.

(i) inversion in unit circle which maps (r, θ) to $\left(\frac{1}{r}, \theta\right)$.

(ii) reflection in real axis which maps $\left(\frac{1}{r}, \theta\right)$ to $\left(\frac{1}{r},-\theta\right)$.

The points inside unit circle are mapped to points outside unit circle and vice versa. The points on unit circle are mapped to points on unit circle and hence unit circle maps onto unit circle.

Inverse transformation is conformal at all z except $z=0$ which is the critical point.

For invariant points, $w=\frac{1}{z}=z$.

$$\therefore \qquad z^2=1 \Rightarrow z=\pm 1$$

Thus, $z=\pm 1$ are invariant points.

Example 1.96: Show that under the mapping $w=\frac{1}{z}$ all circles and straight lines in the z-plane are transformed to circles and straight lines in the w-plane.

Solution: The equation

$$a\left(x^2+y^2\right)+bx+cy+d=0 \tag{1}$$

represents a circle if $a \neq 0$ and a straight line if $a=0$ in the z-plane.

Now, transformation is $w=\frac{1}{z}$.

or $$z=\frac{1}{w}\Rightarrow x+iy=\frac{1}{u+iv}=\frac{u-iv}{u^2+v^2}$$

$\therefore$ $$x=\frac{u}{u^2+v^2},\ y=\frac{-v}{u^2+v^2}$$

$\therefore$ Equation (1) in z-plane is transformed to equation

$$a\left[\left(\frac{u}{u^2+v^2}\right)^2+\left(\frac{-v}{u^2+v^2}\right)^2\right]+\frac{bu}{u^2+v^2}-\frac{cv}{u^2+v^2}+d=0$$

or $$\frac{a}{u^2+v^2}+\frac{bu}{u^2+v^2}-\frac{cv}{u^2+v^2}+d=0$$

or $$d\left(u^2+v^2\right)+bu-cv+a=0 \tag{2}$$

If $d=0$, then it is a straight line.

From (1), $d=0$ implies (1) passes through origin.

Thus if circle or straight line in z-plane passes through origin then it is transformed to a straight line in w-plane.

If $d\neq 0$, then (2) represents a circle.

Thus if circle or straight line in z-plane does not pass through origin then it is transformed to a circle in w-plane.

Thus, all circles and straight lines in z-plane are transformed to circles and straight lines in w-plane.

Remark 1.16:

(i) A circle in z-plane ($a\neq 0$) passing through origin ($d=0$) is transformed to a straight line in w-plane not passing through origin.

(ii) A circle in z-plane ($a\neq 0$) not passing through origin ($d\neq 0$) is transformed to a circle in w-plane not passing through origin.

(iii) A straight line in z-plane ($a=0$) passing through origin ($d=0$) is transformed to a straight line in w-plane passing through origin.

(iv) A straight line in z-plane ($a=0$) not passing through origin ($d\neq 0$) is transformed to a circle in w-plane passing through origin.

Example 1.97: Show that the mapping $w=\frac{4}{z}$ transforms the straight line $x=c,\ c\neq 0$ in the z-plane into a circle in the w-plane.

Solution: $w=\frac{4}{z}\Rightarrow z=\frac{4}{w}$

$\Rightarrow$ $$x+iy=\frac{4}{u+iv}=\frac{4(u-iv)}{u^2+v^2}$$

$\therefore$ $$x=\frac{4u}{u^2+v^2}$$

Thus, the straight line $x = c$ in z-plane is transformed to

$$\frac{4u}{u^2+v^2} = c$$

or $$c(u^2+v^2) - 4u = 0$$

which is a circle in w-plane $(c \neq 0)$.

Example 1.98: Find the image of the infinite strip

(i) $\frac{1}{4} < y < \frac{1}{2}$ (ii) $0 < y < \frac{1}{2}$ (iii) $1 < x < 2$

under the mapping $w = \frac{1}{z}$. Also, show the region graphically.

Solution: $w = \frac{1}{z} \Rightarrow z = \frac{1}{w}$

$$\therefore \quad x+iy = \frac{1}{u+iv} = \frac{u-iv}{u^2+v^2}$$

$$\therefore \quad x = \frac{u}{u^2+v^2},\ y = \frac{-v}{u^2+v^2}$$

(i) $\frac{1}{4} < y \Rightarrow \frac{1}{4} < \frac{-v}{u^2+v^2}$ or $u^2+v^2+4v < 0$

or $$u^2+(v+2)^2 < 2^2$$

or $$|w+2i| < 2$$

and $y < \frac{1}{2} \Rightarrow \frac{-v}{u^2+v^2} < \frac{1}{2}$ or $u^2+v^2+2v > 0$

or $$u^2+(v+1)^2 > 1$$

or $$|w+i| > 1$$

$\therefore$ Image of $\frac{1}{4} < y < \frac{1}{2}$ is interior of circle $|w+2i| = 2$ and exterior of circle $|w+i| = 1$. The region is shown graphically.

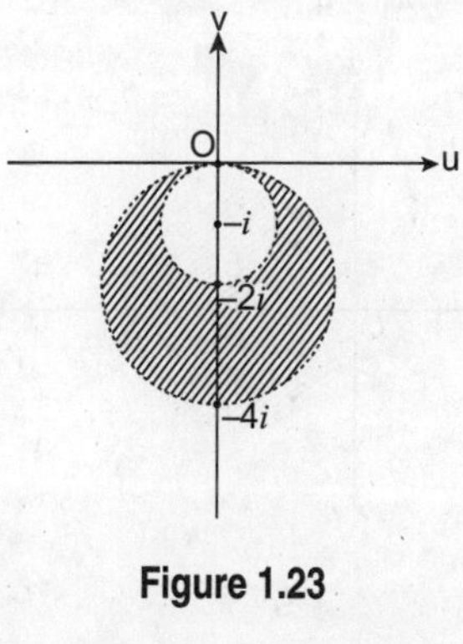

Figure 1.23

(ii) $0 < y \Rightarrow 0 < -\dfrac{v}{u^2+v^2} \Rightarrow v < 0$ and from part (i) $y < \dfrac{1}{2} \Rightarrow |w+i| > 1$

$\therefore$ Image of $0 < y < \dfrac{1}{2}$ is outside the circle $|w+i| = 1$ and below the real line.

The region is shown graphically.

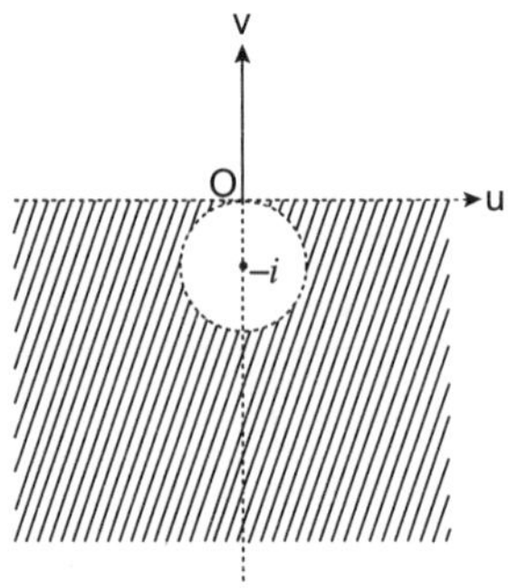

Figure 1.24

(iii) $1 < x \Rightarrow 1 < \dfrac{u}{u^2+v^2} \Rightarrow u^2+v^2-u < 0$

or $$\left(u-\frac{1}{2}\right)^2 + v^2 < \left(\frac{1}{2}\right)^2$$

or $$\left|w-\frac{1}{2}\right| < \frac{1}{2}$$

$x < 2 \Rightarrow \dfrac{u}{u^2+v^2} < 2 \Rightarrow u^2+v^2-\dfrac{u}{2} > 0$

or $$\left(u-\frac{1}{4}\right)^2 + v^2 > \left(\frac{1}{4}\right)^2$$

or $$\left|w-\frac{1}{4}\right| > \frac{1}{4}$$

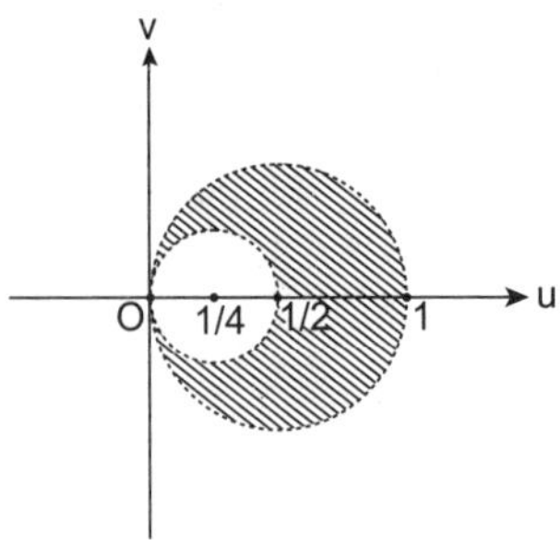

Figure 1.25

$\therefore$ Image is interior of circle $\left|w-\frac{1}{2}\right|=\frac{1}{2}$ and exterior of circle $\left|w-\frac{1}{4}\right|=\frac{1}{4}$. The region is shown graphically in above figure.

Example 1.99: Show that under the transformation $w=\frac{1}{z}$, the image of the hyperbola $x^2-y^2=1$ is the lemniscate $R^2=\cos 2\phi$ where $w=Re^{i\phi}$.

Solution: $$w=\frac{1}{z};\ w=Re^{i\phi},\ z=re^{i\theta}$$

$\therefore$ $$z=\frac{1}{w}$$

$\Rightarrow$ $$re^{i\theta}=\frac{1}{R}e^{-i\phi}$$

$\therefore$ $$r=\frac{1}{R},\ \theta=-\phi$$

$$x^2-y^2=1$$

$\Rightarrow$ $$(r\cos\theta)^2-(r\sin\theta)^2=1$$

$\Rightarrow$ $$r^2\left(\cos^2\theta-\sin^2\theta\right)=1$$

$\Rightarrow$ $$r^2\cos 2\theta=1$$

$\Rightarrow$ $$\frac{1}{R^2}\cos(-2\phi)=1\quad\left(\because\ r=\frac{1}{R},\ \theta=-\phi\right)$$

or $$R^2=\cos 2\phi$$

$\therefore$ Image of hyperbola $x^2-y^2=1$ is lemniscate $R^2=\cos 2\phi$.

Example 1.100: Find the image of the half-plane $y>\alpha$ under the mapping $w=\frac{1}{z}$, when
(i) $\alpha>0$ (ii) $\alpha<0$ (iii) $\alpha=0$

Solution: $w=\frac{1}{z}$

$\therefore$ $$z=\frac{1}{w}$$

$\Rightarrow$ $$x+iy=\frac{1}{u+iv}=\frac{u-iv}{u^2+v^2}$$

$\therefore$ $$x=\frac{u}{u^2+v^2},\ y=\frac{-v}{u^2+v^2}$$

Image of $y>\alpha$ is $\frac{-v}{u^2+v^2}>\alpha$

or $$\alpha\left(u^2+v^2\right)+v<0 \tag{1}$$

(i) $\alpha>0$

By (1), $u^2+v^2+\frac{v}{\alpha}<0$

$$\therefore \quad u^2+\left(v+\frac{1}{2\alpha}\right)^2<\left(\frac{1}{2\alpha}\right)^2$$

or
$$\left|w+\frac{1}{2\alpha}i\right|<\frac{1}{2\alpha}$$

$\therefore$ Image is interior of the circle $\left|w+\frac{1}{2\alpha}i\right|=\frac{1}{2\alpha}$

(ii) $\alpha<0$, from (1) $u^2+v^2+\frac{v}{\alpha}>0$

$$\therefore \quad u^2+\left(v+\frac{1}{2\alpha}\right)^2>\left(\frac{-1}{2\alpha}\right)^2$$

or
$$\left|w+\frac{1}{2\alpha}i\right|>\frac{-1}{2\alpha}$$

$\therefore$ Image is exterior of circle $\left|w+\frac{1}{2\alpha}i\right|=\frac{-1}{2\alpha}$

(iii) $\alpha=0$, from (1) $v<0$

$\therefore$ Image is open half plane below the real axis.

1.17.5 Square Transformation

The transformation $w=z^2$ is called square transformation.

We have $u+iv=(x+iy)^2=x^2-y^2+2ixy$

$$\therefore \quad u=x^2-y^2, v=2xy$$

Thus, any line parallel to x-axis say $y=c$ maps into $u=x^2-c^2, v=2cx$ where x is parameter.

Eliminating x

$$u=\frac{v^2}{4c^2}-c^2 \text{ or } v^2=4c^2\left(u+c^2\right)$$

which is right-handed parabola in w-plane with vertex $(-c^2, 0)$.

Any line parallel to y-axis say $x=b$ maps into $u=b^2-y^2,\ v=2by$ where y is parameter.

Eliminating y

$$u=b^2-\frac{v^2}{4b^2} \text{ or } v^2=-4b^2\left(u-b^2\right)$$

which is left-handed parabola in w-plane with vertex $(b^2, 0)$.

In polar co-ordinates, $z=re^{i\theta},\ w=Re^{i\phi}$

$$\therefore \quad w=z^2 \Rightarrow Re^{i\phi}=r^2e^{2i\theta}$$

$$\therefore \quad R=r^2,\ \phi=2\theta$$

Thus, any circle of radius r in z-plane is transformed to circle of radius r^2 in w-plane.

If $\theta = 0$ then $\phi = 0$. Thus, positive real axis in z-plane maps into +ve real axis in w-plane.

If $\theta = \frac{\pi}{2}$ then $\phi = \pi$. Thus, positive imaginary axis in z-plane maps into –ve real axis in w-plane.

The first quadrant in z-plane, i.e., $0 < \theta < \frac{\pi}{2}$ maps into $0 < \phi < \pi$, i.e., upper half of w-plane.

The angle in z-plane at origin maps into double the angle in w-plane. Hence, mapping is not conformal at origin.

Also, $\frac{dw}{dz} = 2z \neq 0 \text{ if } z \neq 0$ and w is analytic.

Thus, mapping is conformal at all z except $z = 0$ and $z = 0$ is only critical point.

Invariant points are given by $w = z^2 = z$.

i.e., $z = 0, 1$ are invariant points.

Example 1.101: Let the given mapping be $w = z^2$, show that

(i) The mapping is conformal everywhere except at $z = 0$.

(ii) Coefficient of magnification at $z = 1 + i$ is $2\sqrt{2}$.

(iii) Angle of rotation at $z = 1 + i$ is $\frac{\pi}{4}$.

(iv) The circle $|z-1| = 1$ maps to cardioid $\rho = 2(1+\cos\phi)$ where $w = \rho e^{i\phi}$ in the w-plane.

Solution: Mapping is $w = f(z) = z^2$

$$\therefore \quad f'(z) = 2z$$

$\therefore f(z)$ is analytic everywhere and $f'(z) = 0$ for $z = 0$.

(i) Thus, $w = f(z) = z^2$ is conformal everywhere except at $z = 0$.

(ii) $\{f'(z)\}_{z=1+i} = (2z)_{z=1+i} = 2(1+i)$

$$\therefore \quad |f'(1+i)| = 2\sqrt{2}$$

$\therefore$ Coefficient of magnification at $z = 1+i$ is $2\sqrt{2}$.

(iii) Angle of rotation at $(z = 1 + i)$

$$= \text{Arg}\{f'(1+i)\}$$

$$= \text{Arg}\{2(1+i)\} = \tan^{-1} 1 = \frac{\pi}{4}$$

(iv) The circle $|z-1| = 1$, i.e., $z-1 = e^{i\theta};\ 0 \le \theta < 2\pi$ maps to

$$w = z^2 = (1+e^{i\theta})^2 = \left\{e^{i\theta/2}\left(e^{-i\theta/2} + e^{i\theta/2}\right)\right\}^2$$

$$\therefore \quad \rho e^{i\phi} = e^{i\theta}\left(2\cos\frac{\theta}{2}\right)^2 = 4\cos^2\frac{\theta}{2}e^{i\theta}$$

$$\therefore \quad \rho = 4\cos^2\frac{\theta}{2} = 2(1+\cos\theta);\ \phi = \theta$$

$$\therefore \quad \rho = 2(1+\cos\phi)$$

$\therefore$ Circle $|z-1| = 1$ maps to cardioid $\rho = 2(1+\cos\phi)$.

Example 1.102: Determine the image of the region $\frac{1}{2} \le x \le 1$ and $\frac{1}{2} \le y \le 1$ in the w-plane under the mapping $w = z^2$. Also, show both the regions graphically.

Solution:

$$w = z^2$$

$$\therefore \quad u + iv = (x+iy)^2 = x^2 - y^2 + 2ixy$$

$$\therefore \quad u = x^2 - y^2,\ v = 2xy \tag{1}$$

The line $x = \frac{1}{2}$ is mapped to $u = \frac{1}{4} - y^2,\ v = y$ (from (1))

i.e.,
$$u = \frac{1}{4} - v^2 \text{ or } v^2 = -\left(u - \frac{1}{4}\right) \tag{2}$$

The line $x = 1$ is mapped to $u = 1 - y^2, v = 2y$ (from (1))

i.e.,
$$u = 1 - \frac{v^2}{4} \text{ or } v^2 = -4(u-1) \tag{3}$$

The line $y = \frac{1}{2}$ is mapped to $u = x^2 - \frac{1}{4},\ v = x$ (from (1))

i.e.,
$$u = v^2 - \frac{1}{4} \text{ or } v^2 = u + \frac{1}{4} \tag{4}$$

and the line $y = 1$ is mapped to $u = x^2 - 1,\ v = 2x$

i.e.,
$$u = \frac{v^2}{4} - 1 \text{ or } v^2 = 4(u+1) \tag{5}$$

$\therefore$ The region $\frac{1}{2} \le x \le 1$ is mapped to the region between the parabolas (2) and (3) including the parabolas and the region $\frac{1}{2} \le y \le 1$ is mapped to the region between the parabolas (4) and (5) including the parabolas. Also from (1), $v = 2xy$ and hence $v > 0$ for the region considered in z-plane. Therefore the rectangular area $\frac{1}{2} \le x \le 1$ and $\frac{1}{2} \le y \le 1$ is mapped to the shaded region in w-plane as shown graphically hereunder.

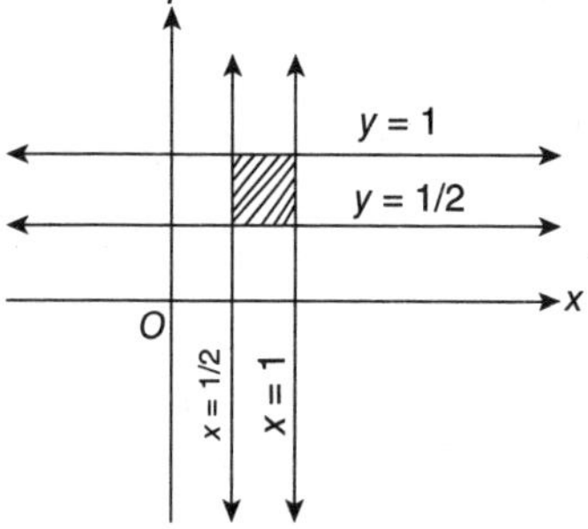

Figure 1.26

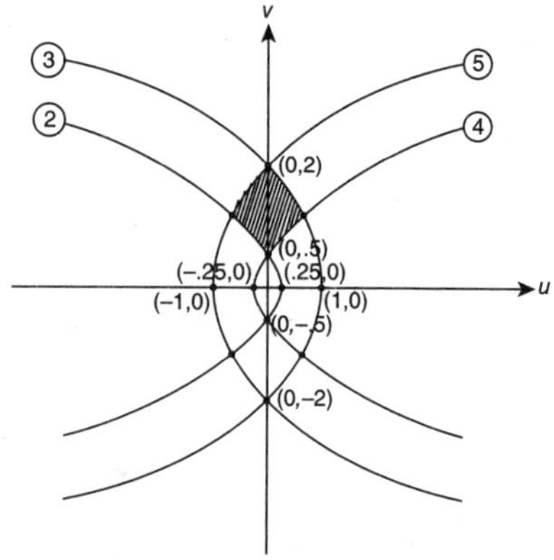

Figure 1.27

Example 1.103: Discuss the transformation $w = z + \frac{1}{z}$ and show that it maps the circle $|z| = a(a \neq 1)$ into an ellipse. Discuss the case when $a = 1$. Also, show that the radius vector $\arg(z) = \alpha$ ($\alpha > \pi/4$) is mapped to a branch of a hyperbola whose eccentricity is sec α.

Solution: For the transformation $w = f(z) = z + \frac{1}{z}$

$$f'(z) = 1 - \frac{1}{z^2} = 0 \text{ for } z = \pm 1$$

Transformation is not conformal at $z = \pm 1$.

$$w = z + \frac{1}{z}$$

$$\therefore \quad u + iv = r(\cos\theta + \sin\theta) + \frac{1}{r}(\cos\theta - i\sin\theta)$$

$$\therefore \quad u = \left(r + \frac{1}{r}\right)\cos\theta, \; v = \left(r - \frac{1}{r}\right)\sin\theta$$

$$\therefore \quad \frac{u^2}{\left(r + \frac{1}{r}\right)^2} + \frac{v^2}{\left(r - \frac{1}{r}\right)^2} = 1 \qquad (\because \cos^2\theta + \sin^2\theta = 1)$$

For circle $|z| = a(a \neq 1)$, we have $r = a$

$$\therefore \quad \frac{u^2}{\left(a + \frac{1}{a}\right)^2} + \frac{v^2}{\left(a - \frac{1}{a}\right)^2} = 1$$

which is an ellipse.

When $a = 1$, we have $r = 1$

$$\therefore \quad u = 2\cos\theta, \; v = 0$$

Thus, the image is line segment on real axis from $(-2, 0)$ to $(2, 0)$ ($\because -1 \leq \cos\theta \leq 1$) and length of line segment = 4.

For radius vector, $\arg z = \alpha \left(\alpha > \frac{\pi}{4}\right)$

$$z = r(\cos\alpha + i\sin\alpha)$$

$$\therefore \quad u = \left(r + \frac{1}{r}\right)\cos\alpha, \; v = \left(r - \frac{1}{r}\right)\sin\alpha$$

$$\therefore \quad \frac{u^2}{\cos^2\alpha} = r^2 + \frac{1}{r^2} + 2 \text{ and } \frac{v^2}{\sin^2\alpha} = r^2 + \frac{1}{r^2} - 2$$

$$\Rightarrow \quad \frac{u^2}{\cos^2\alpha} - \frac{v^2}{\sin^2\alpha} = 4$$

which is a branch of hyperbola whose eccentricity e is given by

$$e^2 - 1 = \frac{\sin^2\alpha}{\cos^2\alpha} = \tan^2\alpha$$

$$\therefore \quad e^2 = 1 + \tan^2\alpha = \sec^2\alpha \Rightarrow e = \sec\alpha > 1 \qquad \left(\because \alpha > \frac{\pi}{4}\right)$$

Example 1.104: Show that the image of the circle $|z| = 2$ under the transformation $w + 2i = z + \frac{1}{z}$ is an ellipse.

Solution: $w = u + iv, |z| = 2 \Rightarrow z = 2(\cos\theta + i\sin\theta)$

$$\therefore \quad w + 2i = z + \frac{1}{z} \Rightarrow u + iv + 2i = 2(\cos\theta + i\sin\theta) + \frac{1}{2}(\cos\theta - i\sin\theta)$$

Equate real and imaginary parts

$$u = \frac{5}{2}\cos\theta, \; v + 2 = \frac{3}{2}\sin\theta$$

$$\because \quad \cos^2\theta + \sin^2\theta = 1$$

$$\therefore \quad \frac{u^2}{\left(\frac{5}{2}\right)^2} + \frac{(v+2)^2}{\left(\frac{3}{2}\right)^2} = 1$$

which is an ellipse in w-plane.

Example 1.105: If $w = z + \frac{a^2}{z}$, prove that when z describes the circle $x^2 + y^2 = a^2$, w describes a line segment and find its length. Also prove that if z describes the circle $x^2 + y^2 = b^2$ where $b > a$, w describes an ellipse.

Solution: Circle $x^2 + y^2 = b^2$ is $|z| = b$

$$\therefore \quad z = b(\cos\theta + i\sin\theta)$$

$$\therefore \quad w = z + \frac{a^2}{z} \Rightarrow u + iv = b(\cos\theta + i\sin\theta) + \frac{a^2}{b}(\cos\theta - i\sin\theta)$$

$$\Rightarrow \quad u = \left(b + \frac{a^2}{b}\right)\cos\theta, \; v = \left(b - \frac{a^2}{b}\right)\sin\theta \tag{1}$$

For $b = a$, $u = 2a\cos\theta$, $v = 0$

As $-1 \le \cos\theta \le 1$, therefore w describes line segment on u-axis (real axis) from $(-2a, 0)$ to $(2a, 0)$ and length of line segment $= 4a$ $(a > 0)$.

$$\because \quad \cos^2\theta + \sin^2\theta = 1$$

$$\therefore \text{ From (1)}, \quad \frac{u^2}{\left(b + \frac{a^2}{b}\right)^2} + \frac{v^2}{\left(b - \frac{a^2}{b}\right)^2} = 1 \quad \text{or} \quad \frac{u^2}{(b^2 + a^2)^2} + \frac{v^2}{(b^2 - a^2)^2} = \frac{1}{b^2} (b > a) \tag{2}$$

$\therefore$ If z describes the circle $x^2 + y^2 = b^2, b > a$, then w describes ellipse (2).

1.17.6 Bilinear Transformation (Mobius Transformation or Fractional Transformation)

The transformation

$$w = \frac{az+b}{cz+d};\ ad-bc \neq 0$$

is called bilinear transformation or mobius transformation or fractional transformation.

Now, $$w = \frac{az+b}{cz+d} \Rightarrow cwz + dw = az + b$$

$\Rightarrow$ $$z = \frac{-dw+b}{cw-a}$$

which is also bilinear.

Now, $$w = \frac{az+b}{cz+d} \Rightarrow cwz + dw - az - b = 0$$

which is linear in both w and z and hence the name bilinear transformation.

Now, $$\frac{dw}{dz} = \frac{a(cz+d)-c(az+b)}{(cz+d)^2}$$

$$= \frac{ad-bc}{(cz+d)^2}$$

$\therefore\ \frac{dw}{dz} \neq 0\ (\text{as } ad-bc \neq 0)$ and w is analytic. Thus, the transformation is conformal for all z.

Bilinear transformation is clearly one-one transformation.

For invariant points, $w = \frac{az+b}{cz+d} = z$

$\therefore$ $$cz^2 + (d-a)z - b = 0$$

which is quadratic in z. Thus, its roots, at most two points will be invariant points. If roots coincide, i.e., when $(d-a)^2 + 4bc = 0$ then there is only one invariant point. In this case, bilinear transformation is called parabolic.

To find bilinear transformation, we require cross ratio of four points which we define hereunder.

1.17.7 Cross Ratio of Four Points

If t_1, t_2, t_3 and t_4 be any four numbers or complex points then $\frac{(t_1-t_2)(t_3-t_4)}{(t_1-t_4)(t_3-t_2)}$ is said to be their cross ratio and is denoted by (t_1, t_2, t_3, t_4).

Theorem 1.25 Bilinear transformation preserves cross ratio of four points.

Proof: Let the points z_1, z_2, z_3 and z_4 be the four points which map to w_1, w_2, w_3, w_4 of w-plane, respectively under the bilinear transformation $w = \frac{az+b}{cz+d}, ad-bc \neq 0.$

If these points are finite, then

$$w_j - w_k = \frac{az_j + b}{cz_j + d} - \frac{az_k + b}{cz_k + d} = \frac{(ad-bc)(z_j - z_k)}{(cz_j + d)(cz_k + d)};\ j, k = 1,2,3,4,\ j \neq k$$

$$\therefore \qquad (w_1, w_2, w_3, w_4) = \frac{(w_1 - w_2)\ (w_3 - w_4)}{(w_1 - w_4)\ (w_3 - w_2)}$$

$$= \frac{\dfrac{(ad-bc)(z_1 - z_2)}{(cz_1 + d)(cz_2 + d)} \cdot \dfrac{(ad-bc)(z_3 - z_4)}{(cz_3 + d)(cz_4 + d)}}{\dfrac{(ad-bc)(z_1 - z_4)}{(cz_1 + d)(cz_4 + d)} \cdot \dfrac{(ad-bc)(z_3 - z_2)}{(cz_3 + d)(cz_2 + d)}}$$

$$= \frac{(z_1 - z_2)(z_3 - z_4)}{(z_1 - z_4)(z_3 - z_2)} = (z_1, z_2, z_3, z_4)$$

Thus, the cross ratio of four points is invaried under bilinear transformation.

Remark 1.17:

(i) To find transformation which maps z_1, z_2, z_3 to w_1, w_2, w_3 respectively, we have

$$(w, w_1, w_2, w_3) = (z, z_1, z_2, z_3)$$

$$\therefore \qquad \frac{(w - w_1)(w_2 - w_3)}{(w - w_3)(w_2 - w_1)} = \frac{(z - z_1)(z_2 - z_3)}{(z - z_3)(z_2 - z_1)}$$

(ii) If one of z is infinite say $z_1 = \infty$ then

$$\frac{(w - w_1)(w_2 - w_3)}{(w - w_3)(w_2 - w_1)} = \frac{z_2 - z_3}{z - z_3} \qquad \left(\because \lim_{z_1 \to \infty} \frac{z - z_1}{z_2 - z_1} = 1\right)$$

(iii) If one of w is infinite say $w_1 = \infty$ then

$$\frac{w_2 - w_3}{w - w_3} = \frac{(z - z_1)(z_2 - z_3)}{(z - z_3)(z_2 - z_1)}$$

(iv) If one of z and one of w is infinite say $z_1 = \infty, w_3 = \infty$ then

$$\frac{w - w_1}{w_2 - w_1} = \frac{z_2 - z_3}{z - z_3}$$

(v) To find the bilinear transformation which maps a given region in z-plane to given region in w-plane, following points should be noted:

(a) If both the regions are not circles, interior of circles or exterior of circles, then take three points on boundary of region in z-plane and three points on boundary of region in w-plane and find the bilinear transformation and then check whether it suits the regions or not, otherwise multiply by -1.

(b) If both the regions are circles or interior of circles or exterior of circles then inverse points w.r.t. circles will transform to inverse points. Using this, find the transform.

Example 1.106: Find the bilinear transformation which maps the points $z = 0, 1, i$ in the z-plane onto the points $1 + i, -i, 2 - i$ in the w-plane respectively.

Solution: Let $z_1 = 0,\ z_2 = 1, z_3 = i$ and $w_1 = 1+i, w_2 = -i, w_3 = 2-i$

The bilinear transformation which maps z_1, z_2, z_3 to w_1, w_2, w_3 respectively is

$$\frac{(w-w_1)(w_2-w_3)}{(w-w_3)(w_2-w_1)} = \frac{(z-z_1)(z_2-z_3)}{(z-z_3)(z_2-z_1)}$$

$$\therefore \quad \frac{[w-(1+i)][-i-2+i]}{[w-(2-i)][-i-1-i]} = \frac{(z-0)(1-i)}{(z-i)(1-0)}$$

$$\Rightarrow \quad \frac{w-(1+i)}{w-(2-i)} = \frac{z}{z-i}\cdot\frac{(1-i)(1+2i)}{2} = \frac{z}{z-i}\cdot\frac{3+i}{2} = \frac{(3+i)z}{2z-2i}$$

Apply componendo and dividendo

$$\frac{2w-3}{1-2i} = \frac{(5+i)z-2i}{(1+i)z+2i}$$

$$\therefore \quad 2w-3 = \frac{(7-9i)z-2i-4}{(1+i)z+2i}$$

$$\therefore \quad 2w = \frac{(10-6i)z+4i-4}{(1+i)z+2i}$$

$$\Rightarrow \quad w = \frac{(5-3i)z+2i-2}{(1+i)z+2i}$$

which is the required bilinear transformation.

Example 1.107: Obtain the bilinear transformation which maps the points $z = 1, i, -1$ onto the points $w = i, 0, -i$ respectively and hence find

(i) the image of $|z| < 1$

(ii) the invariant points of this transformation

Solution: Let $z_1 = 1, z_2 = i, z_3 = -1$ and $w_1 = i, w_2 = 0, w_3 = -i$.

The transformation which maps z_1, z_2, z_3 to w_1, w_2, w_3 respectively is

$$\frac{(w-w_1)(w_2-w_3)}{(w-w_3)(w_2-w_1)} = \frac{(z-z_1)(z_2-z_3)}{(z-z_3)(z_2-z_1)}$$

$$\therefore \quad \frac{(w-i)(0+i)}{(w+i)(0-i)} = \frac{(z-1)(i+1)}{(z+1)(i-1)}$$

$$\Rightarrow \quad \frac{w-i}{w+i} = \frac{i(z-1)}{z+1}$$

Apply componendo and dividendo

$$\frac{2w}{2i}=\frac{(1+i)z+(1-i)}{(1-i)z+(1+i)}\Rightarrow w=\frac{(i-1)z+(i+1)}{(1-i)z+(1+i)}$$

$\therefore$ $$w=\frac{-z+i}{z+i}$$

which is the required transformation.

(i) $$\frac{w}{1}=\frac{-z+i}{z+i}$$

Apply componendo and dividendo

$$\frac{1-w}{1+w}=\frac{2z}{2i}\Rightarrow z=i\left(\frac{1-w}{1+w}\right)$$

$\therefore$ $$|z|<1\Rightarrow\left|\left(\frac{1-w}{1+w}\right)i\right|<1\Rightarrow|1-w|^2<|1+w|^2\quad(\because |i|=1)$$

$\Rightarrow$ distance of w from (1,0) < distance of w from (–1,0)

But right bisector of (1,0) and (–1,0) is $u=0$.

$\therefore$ Image of $|z|<1$ is $u>0$ which is open half plane to the right of imaginary axis.

(ii) Transformation is $w=\dfrac{-z+i}{z+i}$

for invariant points $w=z$

$\therefore$ $$\frac{-z+i}{z+i}=z\Rightarrow z^2+(1+i)z-i=0$$

$\therefore$ $$z=\frac{-(1+i)\pm\sqrt{(1+i)^2+4i}}{2}=\frac{-(1+i)\pm\sqrt{3}(1+i)}{2}$$

$\therefore$ Invariant points are $\left(\dfrac{\pm\sqrt{3}-1}{2}\right)(1+i)$

Example 1.108: Find the image of the half-plane $x + y > 0$ under the bilinear transformation $w=\dfrac{(z-1)}{(z+i)}$

Solution: The given transformation is

$$\frac{w}{1}=\frac{z-1}{z+i}$$

$\therefore$ $$wz+iw=z-1$$

$\therefore$ $$z=\frac{iw+1}{1-w}$$

$$\therefore \qquad x+iy=\frac{1+iw}{1-w}\cdot\frac{1-\bar{w}}{1-\bar{w}}=\frac{1-\bar{w}+iw-i\left|w\right|^2}{\left|1-w\right|^2}$$

$$=\frac{1-(u-iv)+i(u+iv)-i\left(u^2+v^2\right)}{\left|1-w\right|^2}$$

$$\therefore \qquad x=\frac{1-u-v}{\left|1-w\right|^2},\ y=\frac{v+u-\left(u^2+v^2\right)}{\left|1-w\right|^2}$$

$$\therefore \qquad x+y=\frac{1-u-v+v+u-\left(u^2+v^2\right)}{\left|1-w\right|^2}=\frac{1-\left(u^2+v^2\right)}{\left|1-w\right|^2}$$

$$\therefore \qquad x+y>0\Rightarrow 1-\left(u^2+v^2\right)>0$$

or $\qquad u^2+v^2<1\Rightarrow \left|w\right|<1$

$\therefore$ Image of half-plane $x+y>0$ is interior of circle $|w|=1$.

Example 1.109: Find the image of the annulus $1<|z|<2$ under the mapping $w=\dfrac{z}{z-1}$.

Solution: $w=\dfrac{z}{z-1}\Rightarrow zw-w=z$

$$\Rightarrow \qquad z=\frac{w}{w-1}$$

$\therefore\ 1<\left|z\right|<2$ is mapped to $1<\left|\dfrac{w}{w-1}\right|<2$

or $\qquad \left|w-1\right|<\left|w\right|<2\left|w-1\right| \qquad (1)$

Now, $\qquad \left|w-1\right|<\left|w\right|$

$$\Rightarrow \qquad \left|(u-1)+iv\right|^2<\left|u+iv\right|^2$$

$$\Rightarrow \qquad (u-1)^2+v^2<u^2+v^2$$

$$\Rightarrow \qquad -2u+1<0\Rightarrow u>\frac{1}{2} \qquad (2)$$

and $\qquad \left|w\right|<2\left|w-1\right|$

$$\Rightarrow \qquad \left|u+iv\right|^2<4.\left|(u-1)+iv\right|^2$$

$$\Rightarrow \qquad \left(u^2+v^2\right)<4\left[(u-1)^2+v^2\right]$$

$$\Rightarrow \qquad 3\left(u^2+v^2\right)-8u+4>0$$

$$\Rightarrow \qquad u^2+v^2-\frac{8}{3}u+\frac{4}{3}>0$$

$$\Rightarrow \qquad \left(u-\frac{4}{3}\right)^2+v^2>\left(\frac{2}{3}\right)^2$$

$$\Rightarrow \qquad \left|w-\frac{4}{3}\right|>\frac{2}{3} \tag{3}$$

$\therefore$ From (1), (2) and (3), the image of the annulus $1<|z|<2$ is part of exterior of circle $\left|w-\frac{4}{3}\right|=\frac{2}{3}$ in the region $u>\frac{1}{2}$.

Example 1.110: Find the fixed points of the bilinear transformation

(i) $w=\dfrac{3iz+1}{z+1}$ (ii) $w=\dfrac{3z-4}{z-1}$

Solution: For fixed points $w=z$

(i) For fixed points, $\dfrac{3iz+1}{z+1}=z$

$$\Rightarrow \qquad z^2+(1-3i)z-1=0$$

$$\Rightarrow \qquad z=\frac{-1+3i\pm\sqrt{(1-3i)^2+4}}{2}$$

$$\Rightarrow \qquad z=\frac{-1+3i\pm\sqrt{-4-6i}}{2}$$

$\therefore$ Fixed points are $\frac{1}{2}\left(-1+3i\pm\sqrt{-4-6i}\right)$

(ii) For fixed points $w=\dfrac{3z-4}{z-1}=z$

$$\therefore \qquad z^2-z=3z-4$$

$$\text{or} \qquad z^2-4z+4=0$$

$$\Rightarrow \qquad (z-2)^2=0$$

$$\Rightarrow \qquad z=2$$

$\therefore$ The fixed point is 2.

Example 1.111: Obtain the condition under which the mapping $w=\dfrac{az+b}{cz+d}$ maps a straight line of z-plane into a unit circle of w-plane.

Solution: $w=\dfrac{az+b}{cz+d}$

For unit circle $|w|=1$ in w-plane, $\left|\dfrac{az+b}{cz+d}\right|=1$

$$\Rightarrow \qquad \left|\frac{az+b}{cz+d}\right|^2 = 1 \Rightarrow \left(\frac{az+b}{cz+d}\right)\left(\frac{\bar{a}\,\bar{z}+\bar{b}}{\bar{c}\,\bar{z}+\bar{d}}\right) = 1$$

$$\therefore \qquad |a|^2|z|^2 + a\bar{b}z + \bar{a}b\bar{z} + |b|^2 = |c|^2|z|^2 + c\bar{d}z + \bar{c}d\bar{z} + |d|^2$$

or $$\left(|a|^2 - |c|^2\right)|z|^2 + \left(a\bar{b} - c\bar{d}\right)z + \left(\bar{a}b - \bar{c}d\right)\bar{z} + |b|^2 - |d|^2 = 0$$

For this to be straight line $|a|^2 - |c|^2 = 0$

$$\Rightarrow \qquad |a|^2 = |c|^2$$

$$\therefore \qquad |a| = |c|$$

which is the required condition.

Note: If a and c are real, then condition is $a = \pm c$.

Example 1.112: Find the transformation which maps the points $z = 1, -i, -1$ to points $w = i, 0, -i$ respectively. Show that this transformation maps the region outside the circle $|z| = 1$ into the half-space $R(w) > 0$.

Solution: Let $z_1 = 1,\ z_2 = -i,\ z_3 = -1$ and $w_1 = i, w_2 = 0, w_3 = -i$.

The transformation which maps z_1, z_2, z_3 to w_1, w_2, w_3 respectively is

$$\frac{(w-w_1)(w_2-w_3)}{(w-w_3)(w_2-w_1)} = \frac{(z-z_1)(z_2-z_3)}{(z-z_3)(z_2-z_1)}$$

$$\therefore \qquad \frac{(w-i)(0+i)}{(w+i)(0-i)} = \frac{(z-1)(-i+1)}{(z+1)(-i-1)}$$

or $$-\left(\frac{w-i}{w+i}\right) = -\left(\frac{z-1}{z+1}\right)\cdot\left(\frac{1-i}{1+i}\right)$$

or $$\frac{w-i}{w+i} = \frac{-i(z-1)}{(z+1)}$$

Apply componendo and dividendo

$$\frac{2w}{2i} = \frac{(1-i)z + (1+i)}{(1+i)z + (1-i)}$$

$$\Rightarrow \qquad w = \frac{(1+i)z - (1-i)}{(1+i)z + (1-i)} = \frac{z+i}{z-i}$$

which is the required transformation.

Region outside circle $|z| = 1$ is $|z| > 1$.

Now, $$\frac{w}{1} = \frac{z+i}{z-i}$$

Apply componendo and dividendo

$$\frac{2z}{2i} = \frac{w+1}{w-1} \Rightarrow z = i\left(\frac{w+1}{w-1}\right)$$

$\therefore$ $\quad |z| > 1$ is mapped to $\left| i\left(\frac{w+1}{w-1}\right) \right| > 1$

or $$|w+1| > |w-1|$$

or $$|(u+1)+iv|^2 > |(u-1)+iv|^2$$

or $$(u+1)^2 + v^2 > (u-1)^2 + v^2$$

or $$4u > 0 \Rightarrow u > 0$$

$\therefore$ Transformation maps the region outside the circle $|z| = 1$ into the open half plane to the right of imaginary axis.

Example 1.113: Show that mapping $w = \frac{i-z}{i+z}$ maps the real axis of the z-plane onto the circle $|w| = 1$, and the half-plane $y > 0$ onto the interior of the unit circle $|w| < 1$ in the w-plane.

Solution:

$$\frac{w}{1} = \frac{i-z}{i+z}$$

Apply componendo and dividendo

$$\frac{w+1}{w-1} = \frac{2i}{-2z}$$

$\therefore$ $$z = -i\left(\frac{w-1}{w+1}\right) = \frac{-i(w-1)(\overline{w}+1)}{|w+1|^2}$$

$\therefore$ $$x + iy = \frac{-i\left(|w|^2 + w - \overline{w} - 1\right)}{|w+1|^2} = \frac{-i\left(u^2 + v^2 + 2iv - 1\right)}{|w+1|^2}$$

Equate imaginary parts

$$y = \frac{-\left(u^2 + v^2\right) + 1}{|w+1|^2} \tag{1}$$

Real axis on z-plane is $y = 0$

It is mapped to $-\left(u^2 + v^2\right) + 1 = 0$

or $\quad u^2 + v^2 = 1$ or $|w| = 1$

$\therefore$ Real axis on z-plane is mapped to circle $|w| = 1$.

and from (1) $y > 0$ is mapped to $\frac{-\left(u^2 + v^2\right) + 1}{|w+1|^2} > 0$

i.e., $$u^2 + v^2 < 1 \text{ or } |w| < 1$$

which is interior of the unit circle $|w| = 1$.

Example 1.114: Given the bilinear transformation

$w = \dfrac{2z+3}{z-4}$, find the image of the circle $x^2 + y^2 - 4x = 0$ in the w-plane.

Solution: Circle $x^2 + y^2 - 4x = 0$ is $(x-2)^2 + y^2 = 2^2$ or $|z-2| = 2$.

Transformation is
$$w = \frac{2z+3}{z-4}$$

$\therefore$
$$wz - 4w = 2z + 3$$

$\Rightarrow$
$$(w-2)z = 4w + 3$$

$\Rightarrow$
$$z = \frac{4w+3}{w-2}$$

$\therefore$ $|z-2| = 2$ is mapped to $\left|\dfrac{4w+3}{w-2} - 2\right| = 2$

i.e.,
$$\left|\frac{2w+7}{w-2}\right| = 2$$

or
$$\left|w + \frac{7}{2}\right| = |w-2|$$

$\Rightarrow$
$$\left(u + \frac{7}{2}\right)^2 + v^2 = (u-2)^2 + v^2$$

$\Rightarrow$
$$11u + \frac{33}{4} = 0 \text{ or } 4u + 3 = 0$$

$\therefore$ Image of circle $x^2 + y^2 - 4x = 0$ in w-plane is line $4u + 3 = 0$.

Example 1.115: Find the bilinear transformation which maps the points $i, -i, 1$ of z-plane into $0, 1, \infty$ of the w-plane respectively.

Solution: Let $z_1 = i, z_2 = -i, z_3 = 1$ and $w_1 = 0, w_2 = 1, w_3 = \infty$.

The transformation which maps z_1, z_2, z_3 to w_1, w_2, w_3 respectively is
$$\frac{(w-w_1)(w_2-w_3)}{(w-w_3)(w_2-w_1)} = \frac{(z-z_1)(z_2-z_3)}{(z-z_3)(z_2-z_1)}$$

Take limit as $w_3 \to \infty$
$$\frac{w-w_1}{w_2-w_1} = \frac{(z-z_1)(z_2-z_3)}{(z-z_3)(z_2-z_1)}$$

$\therefore$
$$\frac{w}{1} = \frac{(z-i)(-i-1)}{(z-1)(-i-i)}$$

$\therefore$
$$w = \frac{1}{2}(1-i)\frac{(z-i)}{(z-1)}$$

Example 1.116: Determine the fractional transformation that maps

$$z_1 = 0, z_2 = 1, z_3 = \infty \text{ onto } w_1 = 1, w_2 = -i, w_3 = -1 \text{ respectively.}$$

Solution: The transformation which maps z_1, z_2, z_3 to w_1, w_2, w_3 respectively is

$$\frac{(w-w_1)(w_2-w_3)}{(w-w_3)(w_2-w_1)} = \frac{(z-z_1)(z_2-z_3)}{(z-z_3)(z_2-z_1)}$$

Take limit as $z_3 \to \infty$

$$\frac{(w-w_1)(w_2-w_3)}{(w-w_3)(w_2-w_1)} = \frac{z-z_1}{z_2-z_1}$$

$\therefore$ $$\frac{(w-1)(-i+1)}{(w+1)(-i-1)} = \frac{z-0}{1-0}$$

or $$\frac{w-1}{w+1} = \frac{z}{i}$$

$\Rightarrow$ $$wi - i = zw + z$$

$\Rightarrow$ $$w(i-z) = z + i \Rightarrow w = \frac{i+z}{i-z}$$

Example 1.117: Determine the mobius transformation having 1 and i as fixed (invariant) points and maps 0 to –1.

Solution: Let transformation be $w = \dfrac{az+b}{cz+d}$ where $ad - bc \neq 0$ (1)

1 and i are fixed points

$\therefore$ $$\frac{a+b}{c+d} = 1, \quad \frac{ai+b}{ci+d} = i$$

$\therefore$ $$a + b = c + d \quad (2)$$

$$ai + b = -c + di \quad (3)$$

Multiply (2) by i and subtract (3)

$$(i-1)\, b = (i+1)\, c$$

$\therefore$ $$c = \left(\frac{i-1}{i+1}\right) b = ib \quad (4)$$

(1) maps $z = 0$ to $w = -1$

$\therefore$ $$\frac{b}{d} = -1 \Rightarrow b = -d$$

$\therefore$ $$c = ib, \; d = -b$$

$\therefore$ from (2) $a + b = ib - b \Rightarrow a = (i-2)b$

$\therefore$ Transformation is

$$w = \frac{(i-2)bz + b}{ibz - b} = \frac{(i-2)z + 1}{iz - 1} = \frac{(1+2i)z - i}{z+i}$$

Example 1.118: Find a bilinear map which maps the upper half of the z-plane onto the right half of the w-plane.

Solution: Boundary of upper half of z-plane is Im $(z) = 0$, i.e., $y = 0$.

Boundary of right half of w-plane is Re $(w) = 0$, i.e., $u = 0$.

$$\therefore \qquad \text{Im}(z) = 0 \text{ is mapped to Re}(w) = 0$$

Suppose $z_1 = 0, z_2 = 1, z_3 = \infty$ are mapped to $w_1 = i, w_2 = -i$, and $w_3 = 2i$ respectively.

$$\therefore \quad \text{Mapping is } \frac{(w-w_1)(w_2-w_3)}{(w-w_3)(w_2-w_1)} = \frac{(z-z_1)(z_2-z_3)}{(z-z_3)(z_2-z_1)}$$

i.e.,
$$\frac{(w-i)(-i-2i)}{(w-2i)(-i-i)} = \frac{(z-0)}{(1-0)} \quad \text{(Take limit as } z_3 \to \infty)$$

or
$$\frac{w-i}{w-2i} = \frac{2}{3}z$$

or
$$3w - 3i = 2wz - 4iz$$

$$\therefore \qquad w = \frac{4iz-3i}{2z-3} = \frac{(4iz-3i)(2\bar{z}-3)}{|2z-3|^2}$$

$$\therefore \qquad u+iv = \frac{8i|z|^2 - 12i(x+iy) - 6i(x-iy) + 9i}{|2z-3|^2}$$

For $u > 0$, Real part of $(u + iv) > 0$

$$\therefore \qquad 12y - 6y > 0, \text{ i.e., } y > 0$$

$$\therefore \qquad w = \frac{4iz-3i}{2z-3} \text{ is required transformation.}$$

This is not unique.

EXERCISE 1.6

1. Find the images of the following regions in the z-plane onto the w-plane under the given mappings
   (i) The semicircular region $|z| < 1, \text{Im } z > 0; w = z + (2+i)$
   (ii) The half-plane $\text{Re } z > 0; w = (1-i)z + 2$
   (iii) The unit disk $|z| < 1; w = (1+i)z + 2i$
   (iv) The triangle with vertices at $(0,1),(1,-1),(1,1); w = z + (1-2i)$
   (v) The region $|z| \le 1; w = (1-i)z - 2i$
2. Show that the mapping $w = iz$ represents a rotation through an angle $\pi/2$.
3. Find and plot the image of the triangular region with vertices at (0,0), (1,0), (0,1) under the transformation $w = (1 - i)z + 3$.

4. Under the mapping $w = ze^{i\pi/4}$ find the region in w-plane corresponding to the triangular region on the z-plane bounded by the lines $x = 0$, $y = 0$ and $x + y = 1$.
5. Let a rectangular region $OABC$ with vertices at $O(0,0)$, $A(1,0)$, $B(1,2)$, $C(0,2)$ be defined in the z-plane. Find the image of the region in the w-plane, under the mapping
   (i) $w = z + 2 + i$ (ii) $w = 2z$
   (iii) $w = e^{i\pi/4} z$ (iv) $w = (1 - i) z - 2i$
6. Determine the region in the w-plane into which the rectangular region bounded by the lines $x = 0$, $y = 0$, $x = 1$, $y = 2$ in the z-plane is mapped under the transformation $w = (1 + i)z + 2 - i$.
7. Under the mapping $w = \dfrac{1}{z}$, find
   (i) The image of the circle $|z-2| = 3$
   (ii) The image of $|z-3i| = 3$
   (iii) The image of the disk $|z-1| \le 1$
   (iv) The image of the region $x + y > 1$
   The image of the region Re $z > 1$ and Im $z > 1$
8. Find the image of the region bounded by the lines $x - y < 2$ and $x + y > 2$ under the mapping $w = \dfrac{1}{z}$.
9. Find the critical points of the transformation $w = \dfrac{1}{z}$.
10. Under the mapping $w = \dfrac{(1+i)}{(z+i)}$, find the image of the region $|z-i| < 2$.
11. Under the mapping $w = \dfrac{i}{(z-i)}$, find the images of the following regions (i) Im $z < 0$ (ii) $|z| > 1$.
12. Under the mapping $w = f(z) = z^2$, find the image of the region bounded by the lines $x = 1$, $y = 1$ and $x + y = 1$.
13. Find the image of the curve $x^2 - y^2 = 4$ under the mapping $w = z^2$.
14. Determine the region of the w-plane into which the first quadrant of z-plane is mapped by the transformation $w = z^2$.
15. Determine the region onto which the sector $r < a, 0 \le \theta \le \dfrac{\pi}{4}$ is mapped by
   (i) $w = z^2$ (ii) $w = i\,z^2$ (iii) $w = \dfrac{i}{z^2}$, where $z = re^{i\theta}$ and $w = \text{R}e^{i\phi}$
16. Find bilinear transformation which maps the points $2, i, -2$ in the z-plane onto the points $1, i, -1$ in the w-plane.
17. Find the image of the closed half-disk $|z| \le 1, \text{Im}(z) \ge 0$ under the bilinear transformation $w = \dfrac{z}{(z+1)}$.
18. Find the fixed points of the bilinear transformation $w = \dfrac{z-1}{z+1}$ (ii) $w = \dfrac{z}{z-2}$
19. Show that both the transformations $w = \dfrac{z-i}{z+i}$ and $w = \dfrac{i-z}{i+z}$ transform the upper half-plane $\text{Im}(z) \ge 0$ into $|w| \le 1$.
20. Find the bilinear transformation that maps the points $z_1 = -i$, $z_2 = 0$, $z_3 = i$ into the points $w_1 = -1$, $w_2 = i$, $w_3 = 1$ respectively. Into what curves the y-axis is transformed to this transformation?
21. Show that the transformation $w = \dfrac{3-z}{z-2}$ transforms the circle with centre $\left(\dfrac{5}{2}, 0\right)$ and radius $\dfrac{1}{2}$ in the z-plane into the imaginary axis in the w-plane and the interior of the circle into the right half of the plane.

22. Find the bilinear transformation whose fixed points are –1 and 1.
23. Determine the bilinear transformation which maps $z_1 = 0, z_2 = 1, z_3 = \infty$ into $w_1 = i$, $w_2 = -1$, $w_3 = -i$ respectively.
24. Find the bilinear transformation which transforms the unit circle $|z| = 1$ into the real axis in such a way that the points $z = 1$, i and -1 are mapped into the points $w = 0$, 1 and ∞, respectively. Find the region into which the interior and exterior of the circle are mapped.
25. Determine the bilinear transformation which maps 0, 1, ∞ into i, -1, $-i$, respectively. Show that this transformation maps the interior of the unit circle in the z-plane into the half-plane Im $(w) > 0$.
26. Find the bilinear transformation which maps the points $z = 1, i, 2+i$ in the z-plane onto the points $w = i, 1, \infty$, in the w-plane.
27. (i) Determine the linear fractional transformation that sends the points $z = 0, -i, 2i$ into the points $w = 5i, \infty, \frac{-i}{3}$ respectively.
    (ii) Find the invariant points of this transformation.
    (iii) Find the image of $|z| < 1$ under this transformation.
28. Find a transformation $w = f(z)$ which maps
    (i) the real axis in the z-plane onto the real axis in the w-plane.
    (ii) the unit disk $|z| \le 1$ in the z-plane onto the half-plane $\text{Re}(w) \ge 0$ in the w-plane.
    (iii) the unit disk $|z| \le 1$ in the z-plane onto the unit disk $|w| \le 1$ in the w-plane.
29. Find a transformation $w = f(z)$ which maps the upper half-plane $\text{Im}(z) \ge 0$ onto the unit disk $|w| \le 1$.

ANSWERS 1.6

1. (i) Image is interior of semicircle with centre (2, 1) and radius unity above the line $v = 1$
   (ii) Image is open half-plane $u - v > 2$.
   (iii) Image is interior of circle with centre (0, 2) and radius $\sqrt{2}$.
   (iv) Image is triangle with vertices (1, –1), (2, –3) and (2, –1).
   (v) Image is inside and boundary of circle with centre (0, –2) and radius $\sqrt{2}$.
2. Image is triangular region with vertices $A(3, 0)$, $B(4, -1)$, $C(4, 1)$.
4. Image is triangular region bounded by the lines $v = -u$, $v = u$, $v = \frac{1}{\sqrt{2}}$
5. (i) Image is rectangle with vertices (2, 1), (3, 1), (3,3), (2, 3).
   (ii) Image is rectangle with vertices (0,0), (2, 0), (2,4), (0, 4).
   (iii) Image is rectangle with vertices $(0,0), \left(\frac{1}{\sqrt{2}}, \frac{1}{\sqrt{2}}\right), \left(\frac{-1}{\sqrt{2}}, \frac{3}{\sqrt{2}}\right), \left(\frac{-2}{\sqrt{2}}, \frac{2}{\sqrt{2}}\right)$
   (iv) Image is rectangle with vertices (0, –2), (1, –3), (3, –1), (2, 0).
6. Image is rectangular region bounded by the lines, $u + v = 1$, $u - v = 3$, $u + v = 3$, $u - v = -1$.

7. (i) Image is circle $\left|w+\frac{2}{5}\right|=\frac{3}{5}$.

(ii) Image is straight line $6v+1=0$.

(iii) Image is closed half-plane $u \geq \frac{1}{2}$.

(iv) $\left|w-\frac{1}{2}(1-i)\right|<\frac{1}{\sqrt{2}}$

(v) Image is the intersection of interiors of the circles $\left|w-\frac{1}{2}\right|=\frac{1}{2}$ and $\left|w+\frac{1}{2}i\right|=\frac{1}{2}$

8. Image is intersection of exterior of circle $\left|w-\frac{1}{4}(1+i)\right|=\frac{1}{2\sqrt{2}}$ and interior of circle $\left|w-\frac{1}{4}(1-i)\right|=\frac{1}{2\sqrt{2}}$.

9. $z=0$

10. Image is open half -plane $2u-2v-1>0$

11. (i) $\left|w+\frac{1}{2}\right|<\frac{1}{2}$ (ii) $u>-\frac{1}{2}$

12. The region is bounded by three parabolas $v^2=4(u+1), v^2=-4(u-1)$ $u^2=-2\left(v-\frac{1}{2}\right)$

13. $u=4$

14. Upper half of the w-plane

15. (i) sector $R<a^2$; $0 \leq \phi \leq \frac{\pi}{2}$ (ii) sector $R<a^2$; $\frac{\pi}{2} \leq \phi \leq \pi$

(iii) Region is first quadrant excluding inside and boundary of the circle $|w|=\frac{1}{a^2}$ in first quadrant.

16. $w=\frac{-3iz+2}{z-6i}$

17. $u \leq \frac{1}{2}$, $v \geq 0$

18. (i) $z=\pm i$ (ii) $z=0, 3$

20. $w=\frac{-i(z-1)}{(z+1)}$, $|w|=1$

22. $w=\frac{az+b}{bz+a}$ where a, b are any complex numbers such that $a^2 \neq b^2$.

23. $w=\frac{1}{i}\frac{(z-i)}{(z+i)}$

24. $w=\frac{i(1-z)}{(1+z)}$ which maps interior of unit circle to $v>0$ and exterior to $v<0$.

25. $w = -i\dfrac{(z-i)}{(z+i)}$

26. $w = \dfrac{(2+i)z-(1+2i)}{z-(2+i)}$

27. (i) $w = \dfrac{-3z+5i}{-iz+1}$ (ii) $i, -5i$ (iii) $v > 1$

28. (i) $w = \dfrac{(z-x_1)(x_2-x_3)}{(z-x_3)(x_2-x_1)}$ for any three different reals x_1, x_2, x_3

(ii) $w = \dfrac{i-z}{i+z}$ (not unique) (iii) $w = k\dfrac{z-\alpha}{\bar{\alpha}z-1}$; $|k| = 1,\ |\alpha| < 1$

29. $z = -\dfrac{1+i}{2}\ \dfrac{w-1}{w-i}$

Laplace Transform 2

2.1 INTRODUCTION

The study of Laplace transform is an essential part of mathematical background for engineers and scientists. Laplace transform provides easy and effective method to solve linear ordinary or partial differential equations under suitable initial and boundary conditions. Laplace transform changes the given differential equation under given conditions to an algebraic equation in terms of Laplace transform of dependent variable, then, solving this algebraic equation and finding the inverse Laplace transform, we get the required solution. The quality of this method is that the solution is found without finding the general solution. The Laplace transform method is most suitable to the physical systems in which the driving force (which is right-hand side of differential equation) is discontinuous or periodic or a large force acting for a short duration.

2.2 DEFINITION OF LAPLACE TRANSFORM AND INVERSE LAPLACE TRANSFORM

Let $f(t)$ be a given function defined for all $t > 0$. If $\int_0^{\infty} e^{-st} f(t)\, dt$ exists, then

$$F(s) = \int_0^{\infty} e^{-st} f(t)\, dt$$

is called the one-sided Laplace transform of the function $f(t)$ and is also denoted by $\pounds\{f(t)\}$ or $\bar{f}(s)$ for those values of s for which the integral exists. Thus

$$\pounds\{f(t)\} = F(s) = \bar{f}(s) = \int_0^{\infty} e^{-st} f(t)\, dt$$

If $F(s)$ is the Laplace transform of $f(t)$, i.e., $\pounds\{f(t)\} = F(s)$, then $f(t)$ is called inverse Laplace transform of $F(s)$ and we write

$$f(t) = \pounds^{-1}\{F(s)\}$$

The two-sided Laplace transform of $f(t)$ is defined by

$$\pounds\{f(t)\} = \int_{-\infty}^{\infty} e^{-st} f(t) dt$$

for all values of s real or complex for which the integral exists.

We shall be considering only one–sided Laplace transforms.

2.2.1 Piecewise Continuous Function

A function $f(t)$ is said to be piecewise continuous function on $[0, \infty)$ if for any interval $0 \le a \le t \le b$, there are at most a finite number of points $t_1, t_2, \ldots t_m$ in $[a, b]$ at which the function has finite jumps and hence discontinuous and is continuous in each subinterval (t_{k-1}, t_k).

2.2.2 Function of Exponential Order

A function $f(t)$ is called of exponential order k if there exist constants k and $M > 0$ such that

$$|f(t)| \le Me^{kt};\ t \ge 0$$

For example, for $t > 0$

$$\cosh t = \frac{e^t + e^{-t}}{2} < e^t, \quad t^n < e^t n!\ ; \quad n = 1, 2, 3, \ldots \text{ and } \sin t \le e^t\ (\because e^t \ge 1)$$

Therefore, these functions are all of exponential order. Thus, most of the functions are of exponential order.

However, for the function $f(t) = e^{t^2}$, we have $e^{t^2} > Me^{kt}$ for every large M and k for all $t > t_0$ where t_0 is sufficiently large number depending on M and k and thus this function is not of exponential order.

2.3 SUFFICIENT CONDITIONS FOR EXISTENCE OF LAPLACE TRANSFORM

Theorem 2.1 Let $f(t)$ be a function which is piecewise continuous on every finite interval in $t \ge 0$ and is of exponential order k for $t \ge 0$. Then, $£\{f(t)\}$ exists for $s > k$.

Proof: Since $f(t)$ is piecewise continuous, $e^{-st} f(t)$ is integrable over every finite interval on t-axis for $t \ge 0$.

We have
$$|£\{f(t)\}| = \left|\int_0^\infty e^{-st} f(t)dt\right| \le \int_0^\infty e^{-st}\,|f(t)|\,dt \tag{2.1}$$

Since $f(t)$ is exponential of order k, there exists $M > 0$ such that $|f(t)| \le Me^{kt};\ t \ge 0$.

$\therefore$ from (2.1)

$$|£\{f(t)\}| \le \int_0^\infty Me^{-(s-k)t}\,dt = \left\{-\frac{Me^{-(s-k)t}}{s-k}\right\}_0^\infty$$

$$= \frac{M}{(s-k)} \quad \text{for } s > k \qquad \left(\because \lim_{t\to\infty} e^{-(s-k)t} = 0 \text{ for } s > k\right)$$

Thus, $£\{f(t)\}$ exists for $s > k$.

Remark 2.1: It is important to note that the above theorem gives only the sufficient conditions for existence of Laplace transform of a function. The conditions are not necessary. For example, the function $f(t) = \dfrac{1}{\sqrt{t}}$ is not piecewise continuous in $[0, \infty)$ as $\lim\limits_{t\to 0+} f(t) = \infty$, however,

$$£(t^{-1/2}) = \int_0^\infty e^{-st}t^{-1/2}dt = \int_0^\infty e^{-x}\left(\frac{x}{s}\right)^{-1/2}\frac{1}{s}dx \qquad \text{(taking } x = st,\ s > 0)$$

$$= \frac{1}{\sqrt{s}}\int_0^\infty e^{-x}x^{\frac{1}{2}-1}dx = \frac{\Gamma\left(\frac{1}{2}\right)}{\sqrt{s}} = \sqrt{\frac{\pi}{s}}\ ; s > 0$$

$\therefore$ $£(t^{-1/2})$ exists.

2.4 PROPERTIES OF LAPLACE TRANSFORMS

(i) **If Laplace transform of a function exists, then it is unique.**

(ii) **Linearity Property of Laplace transforms**

Theorem 2.2: (Laplace transform operator is a linear operator) Let $f(t)$ and $g(t)$ be any two functions whose Laplace transforms exist for $s > k$ and a, b be any constants, then

$$£\{a\,f(t) + bg(t)\} = a£\{f(t)\} + b£\{g(t)\}$$

Proof: We have

$$£\{af(t) + bg(t)\} = \int_0^\infty e^{-st}\{af(t) + bg(t)\}dt$$

$$= a\int_0^\infty e^{-st}f(t)dt + b\int_0^\infty e^{-st}g(t)dt$$

$$= a£\{f(t)\} + b£\{g(t)\}\ ; s > k$$

(iii) **First Shifting Property or First Translation Property**

Theorem 2.3 If $£\{f(t)\} = F(s);\ s > k$, then

$$£\{e^{at}f(t)\} = F(s-a);\ s - a > k.$$

Proof: Since $F(s) = £\{f(t)\}$, therefore we have

$$F(s) = \int_0^\infty e^{-st}f(t)dt\ ; s > k$$

Therefore, $$F(s-a) = \int_0^\infty e^{-(s-a)t}f(t)\,dt = \int_0^\infty e^{-st}\{e^{at}f(t)\}dt\ ;\ \ s - a > k$$

$$= £\{e^{at}f(t)\}\ ; s - a > k$$

Thus, $$£\{e^{at}f(t)\} = F(s-a)\ ; s - a > k.$$

(iv) **Change of Scale Property**

Theorem 2.4 If $£\{f(t)\} = F(s)$, then $£\{f(at)\} = \dfrac{1}{a}F\left(\dfrac{s}{a}\right);\ a > 0$

Proof: We have $F(s) = £\{f(t)\} = \int_0^\infty e^{-st}f(t)dt$

Therefore, $$£\{f(at)\} = \int_0^\infty e^{-st}f(at)dt\ ; a > 0$$

$$= \int_0^{\infty} e^{-\frac{sx}{a}} f(x) \frac{1}{a} dx \qquad \left(\text{taking } t = \frac{x}{a}\right)$$

$$= \frac{1}{a} \int_0^{\infty} e^{-\frac{s}{a}t} f(t) dt = \frac{1}{a} F\left(\frac{s}{a}\right)$$

2.5 LAPLACE TRANSFORM OF ELEMENTARY FUNCTIONS

(i) $£(1) = \frac{1}{s}; s > 0$

We have $£(1) = \int_0^{\infty} e^{-st} dt = \left\{\frac{-e^{-st}}{s}\right\}_0^{\infty} = \frac{1}{s}; s > 0$

(ii) $£(e^{at}) = \frac{1}{s-a}; s > a$

Since $£(1) = \frac{1}{s}; s > 0$, therefore by using first shifting property

$$£(e^{at}) = \frac{1}{s-a}; s - a > 0, \text{ i.e., } s > a$$

(iii) $£(\sinh at) = \frac{a}{s^2 - a^2}; \; s > |a|$

We have $£(\sinh at) = £\left(\frac{e^{at} - e^{-at}}{2}\right)$

$$= \frac{1}{2} £(e^{at}) - \frac{1}{2} £(e^{-at}) \qquad \text{(by linearity property)}$$

$$= \frac{1}{2} \frac{1}{(s-a)} - \frac{1}{2} \cdot \frac{1}{(s+a)}; \; s > a \text{ and } s > -a$$

$$= \frac{1}{2} \left(\frac{1}{(s-a)} - \frac{1}{(s+a)}\right); \; s > |a|$$

$$= \frac{a}{(s^2 - a^2)}; \; s > |a|$$

(iv) $£(\cosh at) = \frac{s}{s^2 - a^2}; s > |a|$

As above, we have $£(\cosh at) = \frac{1}{2}\left(\frac{1}{(s-a)} + \frac{1}{(s+a)}\right) \qquad \left(\because \cosh at = \frac{e^{at} + e^{-at}}{2}\right)$

$$= \frac{s}{\left(s^2 - a^2\right)}; s > |a|$$

(v) $\pounds(\sin at) = \dfrac{a}{s^2+a^2}$, $\pounds(\cos at) = \dfrac{s}{s^2+a^2}$; $s > 0$

We have $\pounds(e^{iat}) = \dfrac{1}{s-ia}$; $s > 0$

$$= \frac{s+ia}{\left(s^2+a^2\right)}$$

But $e^{iat} = \cos at + i\sin at$. Therefore

$$\pounds(\cos at) + i\,\pounds(\sin at) = \frac{s+ia}{\left(s^2+a^2\right)};\ s>0$$

Equate real and imaginary parts

$$\pounds(\cos at) = \frac{s}{\left(s^2+a^2\right)} \text{ and } \pounds(\sin at) = \frac{a}{\left(s^2+a^2\right)};\ s>0$$

(vi) For negative a, $f(t) = e^{at}\cos wt$ and $f(t) = e^{at}\sin wt$ are called damped vibrations

$$\pounds(e^{at}\cos wt) = \frac{s-a}{(s-a)^2+w^2};\ s>a$$

$$\pounds(e^{at}\sin wt) = \frac{w}{(s-a)^2+w^2};\ s>a$$

We have $\pounds(\cos wt) = \dfrac{s}{s^2+w^2}$; and $\pounds(\sin wt) = \dfrac{w}{\left(s^2+w^2\right)}$; $s > 0$

Therefore, by first shifting property

$$\pounds(e^{at}\cos wt) = \frac{s-a}{(s-a)^2+w^2} \text{ and } \pounds(e^{at}\sin wt) = \frac{w}{(s-a)^2+w^2};\ s>a$$

(vii) $$\pounds(t^n) = \frac{\Gamma(n+1)}{s^{n+1}};\ s>0;\ n>-1$$

$$= \frac{n!}{s^{n+1}};\ s>0;\ \text{if } n \text{ is a non-negative integer}$$

We have $\pounds(t^n) = \int_0^\infty e^{-st}t^n\,dt = \int_0^\infty e^{-x}\left(\dfrac{x}{s}\right)^n \dfrac{1}{s}\,dx$; (taking $st = x$; $s > 0$)

$$= \frac{1}{s^{n+1}}\int_0^\infty e^{-x}x^{n+1-1}\,dx = \frac{\Gamma(n+1)}{s^{n+1}};\ s>0;\ n+1>0$$

For n to be non-negative integer $\Gamma(n+1) = n!$

Therefore, $\pounds(t^n) = \dfrac{n!}{s^{n+1}}$; $s > 0$; n is a non-negative integer

As a particular case, $\pounds(t^{-1/2}) = \dfrac{\Gamma\left(\frac{1}{2}\right)}{\sqrt{s}} = \sqrt{\dfrac{\pi}{s}}$ $\quad\left(\because \Gamma\left(\dfrac{1}{2}\right) = \sqrt{\pi}\right)$

Laplace transforms of some elementary functions obtained in 2.5 are given here in a table.

	f(t)	£{f(t)}		f(t)	£{f(t)}
1	1	$\frac{1}{s}, s > 0$	6	cos *at*	$\frac{s}{s^2+a^2}, s > 0$
2	e^{at}	$\frac{1}{s-a}, s > a$	7	$e^{at}\cos wt$	$\frac{s-a}{(s-a)^2+w^2}, s > a$
3	sinh *at*	$\frac{a}{s^2-a^2}, s > \lvert a\rvert$	8	$e^{at}\sin wt$	$\frac{w}{(s-a)^2+w^2}, s > a$
4	cosh *at*	$\frac{s}{s^2-a^2}, s > \lvert a\rvert$	9	t^n $(n > -1)$	$\frac{\Gamma(n+1)}{s^{n+1}}, s > 0$
5	sin *at*	$\frac{a}{s^2+a^2}, s > 0$	10	t^n (*n* is a non-negative integer)	$\frac{n!}{s^{n+1}}, s > 0$

2.6 LAPLACE TRANSFORMS OF DERIVATIVES AND INTEGRALS

Theorem 2.5 (Laplace transform of derivative)

Suppose $f(t)$ is continuous for all $t \geq 0$, $f(t)$ is of exponential order k and $f(t)$ has piecewise continuous derivative in every finite interval in $t \geq 0$. Then, $£\{f'(t)\}$ exists when $s > k$ and

$$£\{f'(t)\} = s£\{f(t)\} - f(0).$$

Proof: We first take the case when $f'(t)$ is continuous for all $t \geq 0$. Then

$$£(f'(t)) = \int_0^\infty e^{-st} f'(t)\,dt = e^{-st} f(t)\Big|_0^\infty - \int_0^\infty -se^{-st} f(t)\,dt \tag{2.2}$$

Now, $f(t)$ is of exponential order k, so there exists $M > 0$ such that $|f(t)| \leq Me^{kt}$

$$\therefore \qquad \left|e^{-st} f(t)\right| < Me^{-(s-k)t}$$

$$\therefore \qquad \lim_{t\to\infty} e^{-st} f(t) = 0 \qquad \text{if } s > k$$

Therefore, from (2.2)

$$£\{f'(t)\} = -f(0) + s£\{f(t)\}$$

Hence, $£\{f'(t)\}$ exists and $£\{f'(t)\} = s£\{f(t)\} - f(0)$.

If $f'(t)$ is piecewise continuous, then the proof is quite similar by breaking up the range of integration in $\int_0^\infty e^{-st} f'(t)\,dt$ into parts such that $f'(t)$ is continuous in each part.

Theorem 2.6 (Laplace transform of nth order derivative) ($n \in N$)

Suppose $f(t), f'(t), \ldots\ldots, f^{(n-1)}(t)$ are continuous for $t \geq 0$, each of exponential order k and $f^{(n)}(t)$ is piecewise continuous in every finite interval in $t \geq 0$. Then, $£\{f^{(n)}(t)\}$ exists when $s > k$ and $£\{f^{(n)}(t)\} = s^n £\{f(t)\} - s^{n-1} f(0) - s^{n-2} f'(0) - s^{n-3} f''(0) - \cdots - f^{(n-1)}(0)$.

Proof: The theorem will be proved by the principle of mathematical induction on n. For $n = 1$, the result is proved in Theorem 2.5. Let the theorem be true for $n = m$. Then

$$£\{f^{(m)}(t)\} = s^m £\{f(t)\} - s^{m-1} f(0) - s^{m-2} f'(0) - \cdots - s f^{(m-2)}(0) - f^{(m-1)}(0) \tag{2.3}$$

The application of Theorem 2.5 gives

$$\begin{aligned} £\{f^{(m+1)}(t)\} &= s£\{f^{(m)}(t)\} - f^{(m)}(0) \\ &= s^{m+1} £\{f(t)\} - s^m f(0) - s^{m-1} f'(0) - \cdots - s f^{(m-1)}(0) - f^{(m)}(0) \qquad \text{(from (2.3))} \end{aligned}$$

Hence, the theorem is true for $n = m + 1$.

$\therefore$ By the principle of mathematical induction, the theorem is true for all $n \in N$.

Theorem 2.7 (Laplace transform of integral function)

Let $£\{f(t)\} = F(s)$. If $f(t)$ is of exponential order k and piecewise continuous at every finite interval in $t \geq 0$, then

$$£\left(\int_0^t f(x)dx \right) = \frac{1}{s} F(s) \qquad (s > 0, s > k)$$

Proof: Let $$g(t) = \int_0^t f(x)dx \tag{2.4}$$

Since $f(t)$ is of exponential order k, there exists $M > 0$ such that $|f(t)| \leq Me^{kt}$ (2.5)

We can assume $k > 0$ because if (2.5) is satisfied for some negative k, it is also satisfied for positive k.

Now for $t > 0$, $|g(t)| = \left| \int_0^t f(x)dx \right| \leq \int_0^t |f(x)| dx \leq M \int_0^t e^{kx} dx$ (from (2.5))

$$= \frac{M}{k}(e^{kt} - 1) < \frac{M}{k} e^{kt}, \ (k > 0)$$

$\therefore$ $g(t)$ is also of exponential order k.

Since $f(t)$ is piecewise continuous at every finite interval in $t \geq 0$, therefore $g(t)$ is continuous for all $t \geq 0$ and $g'(t) = f(t)$ except for points at which $f(t)$ is discontinuous. Hence, $g'(t)$ is piecewise continuous in every finite interval in $t \geq 0$.

Now, $£\{f(t)\} = £\{g'(t)\} = s£\{g(t)\} - g(0)$; $s > k$ (by Theorem (2.5))

But $g(0) = 0$ from (2.4).

$$\therefore \qquad £\{f(t)\} = sL\{g(t)\} \text{ or } £\{g(t)\} = \frac{1}{s}£\{f(t)\};\ s > 0,\ s > k$$

$$\therefore \qquad £\left(\int_0^t f(x)dx\right) = \frac{F(s)}{s};\ s > 0,\ s > k \qquad (\because £\{f(t)\} = F(s))$$

Other Method $£\left(\int_0^t f(x)dx\right) = \int_0^\infty e^{-st}\left(\int_0^t f(x)dx\right)dt$ by definition of Laplace transform

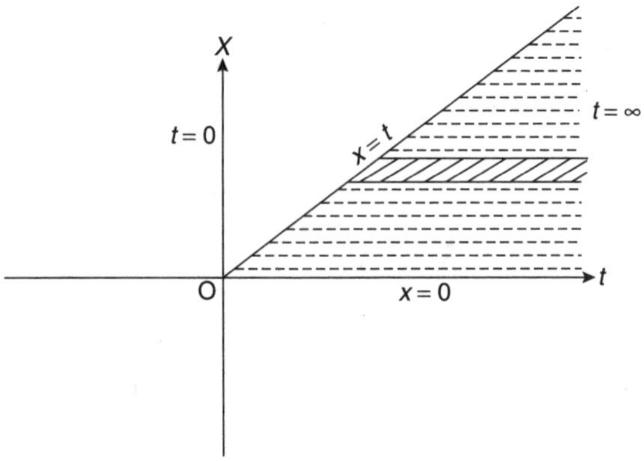

Figure 2.1

$$= \int_0^\infty \int_x^\infty e^{-st} f(x)\, dt\, dx \qquad \text{(by changing order of integration)}$$

$$= \int_0^\infty \left\{\frac{-e^{-st}}{s}\right\}_x^\infty f(x)\, dx = \frac{1}{s}\int_0^\infty e^{-sx} f(x)\, dx,\ s > 0$$

$$= \frac{1}{s} F(s);\ s > 0,\ s > k$$

2.7 DIFFERENTIATION AND INTEGRATION OF LAPLACE TRANSFORM

Theorem 2.8 (Differentiation of Laplace Transform)

Let $f(t)$ be piecewise continuous on every finite interval in $t \geq 0$ and be of exponential order. If $£\{f(t)\} = F(s)$, then

$$\frac{d}{ds}F(s) = -£\{tf(t)\} \qquad (2.6)$$

$$\frac{d^n}{ds^n}F(s) = (-1)^n £\{t^n f(t)\} \qquad (2.7)$$

Proof: We have $F(s)=\int_0^{\infty} e^{-st} f(t)\, dt$

$\therefore$ By Leibnitz's rule for differentiating under the integral sign

$$\frac{dF}{ds}=\int_0^{\infty}\left[\frac{\partial}{\partial s}e^{-st}f(t)\right]dt$$

$$=\int_0^{\infty} -te^{-st}\, f(t)\, dt=-\int_0^{\infty} e^{-st}\{t f(t)\}\, dt$$

$$=-\pounds\{t f(t)\}$$

Now we prove (2.7) for $n > 1$ by the principle of mathematical induction.

From (2.6), the result (2.7) is true for $n = 1$.

Let (2.7) be true for $n = m$. Then

$$\frac{d^m}{ds^m}F(s)=(-1)^m\, \pounds\{t^m f(t)\}$$

$$=(-1)^m \int_0^{\infty} e^{-st}\, t^m f(t)\, dt$$

$$\therefore \qquad \frac{d^{m+1}}{ds^{m+1}}F(s)=(-1)^m \int_0^{\infty}\left[\frac{\partial}{\partial s}e^{-st}t^m f(t)\right]dt$$

$$=(-1)^{m+1}\int_0^{\infty} e^{-st}t^{m+1}f(t)\, dt=(-1)^{m+1}\, \pounds\{t^{m+1} f(t)\}$$

$\therefore$ (2.7) is true for $n = m + 1$.

Hence, the result follows by mathematical induction.

Remark 2.2: From the above theorem, we observe that if $\pounds\{f(t)\} = F(s)$, then

$$\pounds\{t^n f(t)\}=(-1)^n\, \frac{d^n}{ds^n}F(s).$$

We can use it as a formula to calculate Laplace transform of a function $f(t)$ when it is multiplied by some positive integral power of t.

Theorem 2.9 (Integration of Laplace Transform)

Let $f(t)$ be piecewise continuous on $[0, \infty)$ and be of exponential order k and $\lim_{t\to 0^+}\left(\frac{f(t)}{t}\right)$ exists, then

$$\pounds\left\{\frac{f(t)}{t}\right\}=\int_s^{\infty}F(x)\, dx;\ s>k \qquad \text{where } \pounds\{f(t)\}=F(s)$$

Proof: We have $\quad F(s)=\int_0^{\infty} e^{-st} f(t)\, dt;\ s>k$

Integrate from s to ∞

$$\therefore \qquad \int_s^{\infty}F(x)\, dx=\int_s^{\infty}\int_0^{\infty} e^{-xt} f(t)\, dt\, dx$$

$$= \int_0^{\infty}\int_s^{\infty} e^{-xt} f(t)\,dx\,dt \qquad \text{(by changing order of integration)}$$

$$= \int_0^{\infty}\left\{\frac{e^{-xt}}{-t}\right\}_s^{\infty} f(t)\,dt = \int_0^{\infty} e^{-st}\frac{f(t)}{t}dt \tag{2.8}$$

Now, under the given conditions $\frac{f(t)}{t}$ satisfies conditions of existence of its Laplace transform.

Thus, right-hand side of (2.8) is $£\left\{\frac{f(t)}{t}\right\}$

$$\therefore \qquad £\left\{\frac{f(t)}{t}\right\} = \int_s^{\infty} F(x)\,dx$$

Now, we solve some examples.

Example 2.1: Find the Laplace transform of the function $f(t) = \sqrt{t^n}$; $n \geq 1$ and n is odd integer.

Solution: We have $£\left\{t^n\right\} = \frac{\Gamma(n+1)}{s^{n+1}}$ if $n > -1$

Therefore, $$£\left\{t^{n/2}\right\} = \frac{\Gamma\left(\frac{n}{2}+1\right)}{s^{\frac{n}{2}+1}} \tag{1}$$

But $$\Gamma\left(\frac{n}{2}+1\right) = \frac{n}{2}\left(\frac{n}{2}-1\right)\left(\frac{n}{2}-2\right)....\left(\frac{1}{2}\right)\Gamma\left(\frac{1}{2}\right)$$

$$= \frac{n(n-2)(n-4)......1}{2^{(n+1)/2}}\sqrt{\pi} \qquad \left(\because \Gamma\left(\frac{1}{2}\right) = \sqrt{\pi}\right)$$

$$= \frac{(n+1)!\sqrt{\pi}}{2^{(n+1)/2}\left[2.4.6....(n+1)\right]} = \frac{(n+1)!}{2^{n+1}\left(\frac{n+1}{2}\right)!}\sqrt{\pi}$$

$\therefore$ From (1)

$$£\left\{t^{n/2}\right\} = \frac{(n+1)!\sqrt{\pi}}{2^{n+1}\left(\frac{n+1}{2}\right)!\sqrt{s^{n+2}}}$$

Example 2.2: Find the Laplace transform of the functions

(i) $7e^{2t} + 5\cosh t + 7t^3 + 5\sin 3t + 2$ (ii) $\sinh\frac{t}{2}\sin\frac{\sqrt{3}}{2}t$ (iii) $\sin at \sin bt$

(iv) $\sinh^3 2t$ (v) $t^2 e^{-2t}$ (vi) $\left(\sqrt{t} - \frac{1}{\sqrt{t}}\right)^3$

Solution:

(i) Let $f(t) = 7e^{2t} + 5\cosh t + 7t^3 + 5\sin 3t + 2$

$\therefore$ $£\{f(t)\} = 7£(e^{2t}) + 5£(\cosh t) + 7£(t^3) + 5£(\sin 3t) + 2£(1)$

$$= \frac{7}{s-2} + \frac{5s}{s^2-1} + \frac{7(3!)}{s^4} + \frac{5(3)}{s^2+3^2} + \frac{2}{s}$$

$$= \frac{7}{s-2} + \frac{5s}{s^2-1} + \frac{42}{s^4} + \frac{15}{s^2+9} + \frac{2}{s}$$

(ii) $$£\left(\sinh\frac{t}{2}\sin\frac{\sqrt{3}}{2}t\right) = £\left(\frac{e^{\frac{t}{2}} - e^{\frac{-t}{2}}}{2}\sin\frac{\sqrt{3}}{2}t\right)$$

$$= \frac{1}{2}£\left(e^{\frac{t}{2}}\sin\frac{\sqrt{3}}{2}t\right) - \frac{1}{2}£\left(e^{\frac{-t}{2}}\sin\frac{\sqrt{3}}{2}t\right)$$

$$= \frac{1}{2}\frac{\frac{\sqrt{3}}{2}}{\left(s-\frac{1}{2}\right)^2 + \left(\frac{\sqrt{3}}{2}\right)^2} - \frac{1}{2}\frac{\frac{\sqrt{3}}{2}}{\left(s+\frac{1}{2}\right)^2 + \left(\frac{\sqrt{3}}{2}\right)^2} \quad \text{(by first shifting property)}$$

$$= \sqrt{3}\left[\frac{1}{(2s-1)^2+3} - \frac{1}{(2s+1)^2+3}\right]$$

(iii) Let $f(t) = \sin at \sin bt = \frac{1}{2}\left[\cos(a-b)t - \cos(a+b)t\right]$

$$\therefore \quad £\{f(t)\} = \frac{1}{2}£\{\cos(a-b)t\} - \frac{1}{2}£\{\cos(a+b)t\}$$

$$= \frac{1}{2}\cdot\frac{s}{s^2+(a-b)^2} - \frac{1}{2}\cdot\frac{s}{s^2+(a+b)^2}$$

$$= \frac{s}{2}\cdot\left[\frac{(a+b)^2-(a-b)^2}{\{s^2+(a-b)^2\}\{s^2+(a+b)^2\}}\right]$$

$$= \frac{2abs}{\{s^2+(a-b)^2\}\{s^2+(a+b)^2\}}$$

(iv) We have $\sinh 6t = 3\sinh 2t + 4\sinh^3 2t$

$$\therefore \quad \sinh^3 2t = \frac{1}{4}\sinh 6t - \frac{3}{4}\sinh 2t$$

$$\therefore \qquad £(\sinh^3 2t) = \frac{1}{4}\ £(\sinh 6t) - \frac{3}{4}\ £(\sinh 2t)$$

$$= \frac{1}{4}\left[\frac{6}{s^2 - 6^2} - 3\cdot\frac{2}{s^2 - 2^2}\right]$$

$$= \frac{6}{4}\cdot\frac{32}{\left(s^2-4\right)\left(s^2-36\right)} = \frac{48}{\left(s^2-4\right)\left(s^2-36\right)}$$

(v) We have $£\,(t^2) = \frac{2!}{s^3}$

$\therefore$ By first shifting property, $£(t^2e^{-2t}) = \frac{2!}{(s+2)^3} = \frac{2}{(s+2)^3}$

(vi) We have $\left(\sqrt{t} - \frac{1}{\sqrt{t}}\right)^3 = t^{3/2} - 3t^{1/2} + 3t^{-1/2} - t^{-3/2}$

Since Laplace transform of t^n exists only when $n > -1$ therefore $£\,(t^{-3/2})$ does not exist. Hence, $£\left(\sqrt{t} - \frac{1}{\sqrt{t}}\right)^3$ does not exist. However, if we use

$£(t^n) = \frac{\Gamma(n+1)}{s^{n+1}}$, $s > 0$ as a formula with the recurrence relation $\Gamma(n+1) = n\Gamma(n)$ for negative non-integral values n, then

$$£\left(\sqrt{t} - \frac{1}{\sqrt{t}}\right)^3 = £\,(t^{3/2}) - 3\,£\,(t^{1/2}) + 3\,£\,(t^{-1/2}) - £\,(t^{-3/2})$$

$$= \frac{\Gamma\left(\frac{5}{2}\right)}{s^{5/2}} - \frac{3\Gamma\left(\frac{3}{2}\right)}{s^{3/2}} + \frac{3\Gamma\left(\frac{1}{2}\right)}{s^{1/2}} - \frac{\Gamma\left(-\frac{1}{2}\right)}{s^{-1/2}}$$

$$= \frac{3}{2}\cdot\frac{1}{2}\sqrt{\pi}\cdot\frac{1}{s^{5/2}} - 3\cdot\frac{1}{2}\sqrt{\pi}\cdot\frac{1}{s^{3/2}} + 3\sqrt{\pi}\cdot\frac{1}{s^{1/2}} + \frac{2\sqrt{\pi}}{s^{-1/2}} \qquad \left(\begin{array}{l}\because \Gamma\frac{1}{2} = \frac{-1}{2}\Gamma\left(\frac{-1}{2}\right)\\ \therefore \Gamma\left(\frac{-1}{2}\right) = -2\sqrt{\pi}\end{array}\right)$$

$$= \frac{\sqrt{\pi}}{4s^{5/2}}\left[8s^3 + 12s^2 - 6s + 3\right]$$

Example 2.3: Find the Laplace transforms of

(i) $te^{-t}\sin 3t$

(ii) $t^n\, e^t \sin 4t$ ($n \in N$) and deduce the Laplace transforms of $t^2\, e^t \sin 4t$ and $t^3\, e^t \sin 4t$.

Solution:

(i) We have $\qquad £\,(t) = \frac{1}{s^2}$

$$\therefore \qquad £\{te^{-(1-3i)t}\} = \frac{1}{(s+1-3i)^2}\ ;\ s > -1 \qquad \text{(by first shifting property)}$$

$$\therefore \quad £\{te^{-t}(\cos 3t + i\sin 3t)\} = \frac{(s+1+3i)^2}{\left\{(s+1)^2+9\right\}^2}$$

$$\therefore \quad £(te^{-t}\cos 3t) + i£(te^{-t}\sin 3t) = \frac{(s+1)^2 - 9 + 6(s+1)i}{\left\{(s+1)^2+9\right\}^2}$$

Equate imaginary parts

$$£(te^{-t}\sin 3t) = \frac{6(s+1)}{\left\{(s+1)^2+9\right\}^2}; s > -1$$

Other Method We have $£(\sin 3t) = \dfrac{3}{s^2+3^2}; s > 0$

$$\therefore \quad £(t\sin 3t) = \frac{-d}{ds}\left(\frac{3}{s^2+3^2}\right) = \frac{6s}{(s^2+9)^2}; s > 0$$

$\therefore$ By first shifting property

$$£(e^{-t}t\sin 3t) = \frac{6(s+1)}{\left\{(s+1)^2+9\right\}^2}; s > -1$$

(ii) For $n \in N$, we have

$$£(t^n) = \frac{n!}{s^{n+1}}; s > 0$$

$$\therefore \quad £\{t^n e^{(1+4i)t}\} = \frac{n!}{(s-1-4i)^{n+1}}; s > 1$$

$$\therefore \quad £\{(t^n e^t(\cos 4t + i\sin 4t)\} = \frac{n!(s-1+4i)^{n+1}}{\left[(s-1)^2+16\right]^{n+1}}; s > 1$$

$\therefore £(t^n e^t \cos 4t) + i£(t^n e^t \sin 4t)$

$$= \frac{n!}{\left[(s-1)^2+16\right]^{n+1}}\left[(s-1)^{n+1} + {}^{n+1}C_1(s-1)^n(4i) + {}^{n+1}C_2(s-1)^{n-1}(4i)^2 + \cdots + {}^{n+1}C_{n+1}(4i)^{n+1}\right]$$

Equate imaginary parts

$$£(t^n e^t \sin 4t) = \frac{n!}{\left[(s-1)^2+16\right]^{n+1}}\left[{}^{n+1}C_1(s-1)^n(4) - {}^{n+1}C_3(s-1)^{n-2}4^3 + {}^{n+1}C_5(s-1)^{n-4}4^5\cdots\right]$$

Hence $£(t^n e^t \sin 4t) = \dfrac{n!}{\left[(s-1)^2+16\right]^{n+1}}\left[4.{}^{n+1}C_1(s-1)^n - 4^3.{}^{n+1}C_3(s-1)^{n-2} + 4^5.{}^{n+1}C_5(s-1)^{n-4}\cdots\right]$

The last term within the brackets will be $4^{n+1}i^n\,{}^{n+1}C_{n+1}$ if n is even and $4^n i^{n-1}\,{}^{n+1}C_n$ if n is odd; $s > 1$.

Take $n = 2$ and 3 respectively

$$£\,(t^2 e^t \sin 4t) = \frac{2}{\left[(s-1)^2+16\right]^3}\left[4\cdot {}^3C_1(s-1)^2 - 4^3\cdot {}^3C_3\right] = \frac{8\left[3(s-1)^2-16\right]}{\left[(s-1)^2+16\right]^3}$$

and $$£\,(t^3 e^t \sin 4t) = \frac{6}{\left[(s-1)^2+16\right]^4}\left[4\cdot {}^4C_1(s-1)^3 - 4^3\cdot {}^4C_3(s-1)\right] = \frac{96(s-1)\left[(s-1)^2-16\right]}{\left[(s-1)^2+16\right]^4}$$

Other Method We have $£\,(\sin 4t) = \dfrac{4}{\left(s^2+4^2\right)}$; $s > 0$

$$\therefore \qquad £\,(t^n \sin 4t) = (-1)^n \frac{d^n}{ds^n}\left(\frac{4}{s^2+16}\right)$$

$$= (-1)^n \frac{d^n}{ds^n}\left[\frac{4}{(s+4i)(s-4i)}\right]$$

$$= (-1)^n \cdot \frac{1}{2i}\frac{d^n}{ds^n}\left[\frac{1}{(s-4i)} - \frac{1}{(s+4i)}\right] \quad \text{(by suppression method)}$$

But we know that $\dfrac{d^n}{dx^n}(ax+b)^{-1} = \dfrac{(-1)^n n!a^n}{(ax+b)^{n+1}}$

$$\therefore \qquad £\,(t^n \sin 4t) = (-1)^n \cdot \frac{1}{2i}\left[\frac{(-1)^n n!}{(s-4i)^{n+1}} - \frac{(-1)^n n!}{(s+4i)^{n+1}}\right]$$

$$= \frac{n!}{2i}\left[\frac{(s+4i)^{n+1}}{(s^2+16)^{n+1}} - \frac{(s-4i)^{n+1}}{(s^2+16)^{n+1}}\right]$$

$$= \frac{n!}{2i(s^2+16)^{n+1}}\left[\left\{s^{n+1} + {}^{n+1}C_1 s^n(4i) + {}^{n+1}C_2 s^{n-1}(4i)^2 + \cdots + {}^{n+1}C_{n+1}(4i)^{n+1}\right\}\right.$$

$$\left. - \left\{s^{n+1} - {}^{n+1}C_1 s^n(4i) + {}^{n+1}C_2 s^{n-1}(4i)^2 \cdots + {}^{n+1}C_{n+1}(-1)^{n+1}(4i)^{n+1}\right\}\right]$$

$$= \frac{n!}{(s^2+16)^{n+1}}\left[4.{}^{n+1}C_1 s^n - 4^3.{}^{n+1}C_3 s^{n-2} + 4^5.{}^{n+1}C_5 s^{n-4}\cdots\right];\ s > 0$$

$\therefore$ By first shifting property

$$£\,(e^t t^n \sin 4t) = \frac{n!}{\left[(s-1)^2+16\right]^{n+1}}\left[4.{}^{n+1}C_1(s-1)^3 - 4^3.{}^{n+1}C_3(s-1)^{n-2} + 4^5.{}^{n+1}C_5(s-1)^{n-4}\cdots\right];$$

$s > 1.$

Last term is as above in the first method.

Now, the remaining results can be deduced as above in the first method.

Example 2.4: Evaluate $£\left(\frac{\sin at}{t}\right)$, $a > 0$. Does $£\left(\frac{\cos at}{t}\right)$ exist?

Solution: $$£\,(\sin at) = \frac{a}{s^2+a^2}\,;\ s > 0$$

$\therefore$ $$£\left(\frac{\sin at}{t}\right) = \int_s^\infty \frac{a}{\left(x^2+a^2\right)}\,dx = a\cdot\left\{\frac{1}{a}\tan^{-1}\frac{x}{a}\right\}_s^\infty$$

$$= \frac{\pi}{2} - \tan^{-1}\frac{s}{a} = \cot^{-1}\frac{s}{a};\ s > 0$$

and $$£\,(\cos at) = \frac{s}{s^2+a^2}\,;\ s > 0$$

If $L\left(\frac{\cos at}{t}\right)$ exists, then $L\left(\frac{\cos at}{t}\right) = \int_s^\infty \frac{x}{x^2+a^2}\,dx = \left\{\frac{1}{2}\log\left(x^2+a^2\right)\right\}_s^\infty$

But $\lim\limits_{x\to\infty}\frac{1}{2}\log\left(x^2+a^2\right) = \infty$

$\therefore$ $£\left(\frac{\cos at}{t}\right)$ does not exist.

Example 2.5: Find the Laplace transform of the following functions

(i) $\frac{e^{-t}\sin t}{t}$

(ii) $\frac{e^{at}-\cos bt}{t}$

Solution: (i) We have $£\,(\sin t) = \frac{1}{s^2+1}\,;\ s > 0$

$\therefore$ By first shifting property

$$£\left(e^{-t}\sin t\right) = \frac{1}{(s+1)^2+1};\ s > -1$$

Therefore, $£\left(\frac{e^{-t}\sin t}{t}\right) = \int_s^\infty \frac{1}{(x+1)^2+1}\,dx = \left[\tan^{-1}(x+1)\right]_s^\infty$

$$= \frac{\pi}{2} - \tan^{-1}(s+1) = \cot^{-1}(s+1);\ s > -1$$

(ii) $£\,(e^{at} - \cos bt) = £\,(e^{at}) - £\,(\cos bt) = \frac{1}{(s-a)} - \frac{s}{(s^2+b^2)};\ s > a,\ s > 0$

$$\therefore\ £\left(\frac{e^{at}-\cos bt}{t}\right) = \int_s^\infty\left(\frac{1}{x-a} - \frac{x}{x^2+b^2}\right)dx$$

$$= \left[\log\,(x-a) \;-\frac{1}{2}\log\,\left(x^2+b^2\right)\right]_s^\infty$$

$$= \left\{\log\frac{(x-a)}{\sqrt{x^2+b^2}}\right\}_s^\infty \;=\; \lim_{x\to\infty}\,\log\left\{\frac{\left(1-\frac{a}{x}\right)}{\sqrt{1+\left(\frac{b}{x}\right)^2}}\right\} - \log\frac{s-a}{\sqrt{s^2+b^2}}$$

$$= 0-\log\,\frac{s-a}{\sqrt{s^2+b^2}} = \log\,\frac{\sqrt{s^2+b^2}}{s-a};\; s>a,\, s>0$$

Example 2.6: Show that

(i) $£\left[\int_0^t \frac{\cos at-\cos bt}{t}\,dt\right] = \frac{1}{2s}\log\left(\frac{s^2+b^2}{s^2+a^2}\right)$

(ii) $£\left[\int_0^t \frac{e^t\sin t}{t}\,dt\right] = \frac{1}{s}\;\cot^{-1}(s-1)$

Solution: (i) $£\,(\cos at - \cos bt) = \dfrac{s}{\left(s^2+a^2\right)} - \dfrac{s}{\left(s^2+b^2\right)}$

$$\therefore \qquad £\left(\frac{\cos at-\cos bt}{t}\right) = \int_s^\infty\left(\frac{x}{x^2+a^2}-\frac{x}{x^2+b^2}\right)dx$$

$$= \left[\frac{1}{2}\log\,\left(x^2+a^2\right) - \frac{1}{2}\log\,\left(x^2+b^2\right)\right]_s^\infty$$

$$= \left[\frac{1}{2}\log\,\left(\frac{x^2+a^2}{x^2+b^2}\right)\right]_s^\infty = \frac{1}{2}\log\,\left(\frac{s^2+b^2}{s^2+a^2}\right)$$

$$\therefore \qquad £\left(\int_0^t\frac{\cos at-\cos bt}{t}\,dt\right) = \frac{1}{s}£\left(\frac{\cos at-\cos bt}{t}\right) = \frac{1}{2s}\log\left(\frac{s^2+b^2}{s^2+a^2}\right)$$

(by Laplace transform of integral function)

(ii) $\qquad £(\sin t) = \dfrac{1}{s^2+1}$

$$\therefore \qquad £\left(\frac{\sin t}{t}\right) = \int_s^\infty\frac{1}{x^2+1}\,dx = \{\tan^{-1}x\}_s^\infty = \frac{\pi}{2}-\tan^{-1}s = \cot^{-1}s$$

$$\therefore \qquad £\left(e^t\,\frac{\sin t}{t}\right) = \cot^{-1}(s-1) \qquad \text{(by first shifting property)}$$

$$\therefore \qquad £\left(\int_0^t\frac{e^t\sin t}{t}\,dt\right) = \frac{1}{s}£\left(\frac{e^t\sin t}{t}\right) = \frac{1}{s}\cot^{-1}(s-1)$$

Example 2.7: Find Laplace transform of

(i) $\dfrac{\cos 2t - \cos 3t}{t}$ (ii) $\displaystyle\int_0^t \frac{\sin u}{u}\,du$

Solution: (i) Take $a = 2$ and $b = 3$ in Example 2.6 (i)

$$\therefore \quad £\left(\frac{\cos 2t - \cos 3t}{t}\right) = \frac{1}{2}\log\left(\frac{s^2+9}{s^2+4}\right)$$

(ii) By Example 2.6 (ii)

$$£\left(\frac{\sin t}{t}\right) = \cot^{-1} s$$

$\therefore$ By Laplace transform of integral function

$$£\left(\int_0^t \frac{\sin u}{u}\,du\right) = \frac{1}{s}£\left(\frac{\sin t}{t}\right) = \frac{1}{s}\cot^{-1} s$$

Example 2.8: Find the Laplace transforms of the following functions

(i) $\displaystyle\int_0^{t/2} \frac{1-e^{-2x}}{x}\,dx$ (ii) $t\displaystyle\int_0^t e^{-u}\sin 2u\,du$

Solution: (i) Put $2x = u$ so that $2dx = du$

$\therefore$ The given integral becomes $\displaystyle\int_0^t \frac{1-e^{-u}}{u}\,du$

Now, $$£\,(1-e^{-t}) = £\,(1) - £(e^{-t}) = \frac{1}{s} - \frac{1}{s+1};\ s > 0$$

$$\therefore \quad £\left(\frac{1-e^{-t}}{t}\right) = \int_s^\infty \left(\frac{1}{x} - \frac{1}{x+1}\right)dx = \left[\log x - \log(x+1)\right]_s^\infty$$

$$= \left\{\log\left(\frac{x}{x+1}\right)\right\}_s^\infty = \log\left(\frac{s+1}{s}\right);\ s > 0$$

$$\therefore \quad £\left[\int_0^t \left(\frac{1-e^{-u}}{u}\right)du\right] = \frac{1}{s}£\left(\frac{1-e^{-t}}{t}\right) = \frac{1}{s}\log\left(\frac{s+1}{s}\right)$$

Hence $$£\left[\int_0^{t/2} \frac{1-e^{-2x}}{x}\,dx\right] = \frac{1}{s}\log\left(\frac{s+1}{s}\right);\ s > 0$$

(ii) $$£\,(\sin 2t) = \frac{2}{s^2+2^2};\ s > 0$$

$$\therefore \quad £\,(e^{-t}\sin 2t) = \frac{2}{(s+1)^2+2^2};\ s > -1 \quad \text{(by first shifting property)}$$

$$\therefore \quad £\left[\int_0^t e^{-u}\sin 2u\,du\right] = \frac{1}{s}\ £(e^{-t}\sin 2t);\ s>0$$

$$= \frac{1}{s}\cdot\frac{2}{\left[(s+1)^2+2^2\right]} = \frac{2}{s\left(s^2+2s+5\right)};\ s>0$$

$$\therefore \quad £\left[t\int_0^t e^{-u}\sin 2u\,du\right] = \frac{-d}{ds}\left(\frac{2}{s^3+2s^2+5s}\right) = \frac{2\left(3s^2+4s+5\right)}{\left(s^3+2s^2+5s\right)^2};\ s>0$$

Example 2.9: Find the Laplace transforms of $\frac{1-\cos t}{t^2}$

Solution: We have $\frac{1-\cos t}{t^2} = \frac{1}{t^2}\left[1-\left(1-\frac{t^2}{2!}+\frac{t^4}{4!}-\frac{t^6}{6!}+\frac{t^8}{8!}\cdots\right)\right]$

$$= \frac{1}{2!}-\frac{t^2}{4!}+\frac{t^4}{6!}-\frac{t^6}{8!}+\cdots$$

$$\therefore \quad £\left(\frac{1-\cos t}{t^2}\right) = \frac{1}{2!s}-\frac{2!}{4!s^3}+\frac{4!}{6!s^5}-\frac{6!}{8!s^7}+\cdots$$

$$= \frac{1}{1.2}\cdot\frac{1}{s}-\frac{1}{3.4}\cdot\frac{1}{s^3}+\frac{1}{5.6}\cdot\frac{1}{s^5}-\frac{1}{7.8}\cdot\frac{1}{s^7}+\cdots$$

$$= \left(1-\frac{1}{2}\right)\cdot\frac{1}{s}-\left(\frac{1}{3}-\frac{1}{4}\right)\cdot\frac{1}{s^3}+\left(\frac{1}{5}-\frac{1}{6}\right)\frac{1}{s^5}-\left(\frac{1}{7}-\frac{1}{8}\right)\frac{1}{s^7}+\cdots$$

$$= \left\{\frac{1}{s}-\frac{1}{3}\left(\frac{1}{s}\right)^3+\frac{1}{5}\left(\frac{1}{s}\right)^5-\frac{1}{7}\left(\frac{1}{s}\right)^7+\cdots\right\}$$

$$-\frac{s}{2}\left[\frac{1}{s^2}-\frac{1}{2}\left(\frac{1}{s^2}\right)^2+\frac{1}{3}\left(\frac{1}{s^2}\right)^3-\frac{1}{4}\left(\frac{1}{s^2}\right)^4+\cdots\right]$$

$$= \tan^{-1}\frac{1}{s}-\frac{s}{2}\log\left(1+\frac{1}{s^2}\right) = \cot^{-1}s-\frac{s}{2}\log\left(1+\frac{1}{s^2}\right);\ s>0$$

Second Method

$$£(1-\cos t) = \frac{1}{s}-\frac{s}{s^2+1};\ s>0$$

$$\therefore \quad £\left(\frac{1-\cos t}{t}\right) = \int_s^\infty\left(\frac{1}{x}-\frac{x}{x^2+1}\right)dx = \left\{\log x-\frac{1}{2}\log(x^2+1)\right\}_s^\infty$$

$$= \left\{\log\frac{x}{\sqrt{x^2+1}}\right\}_s^\infty = \log\left(\frac{\sqrt{s^2+1}}{s}\right)$$

$$\therefore \quad £\left(\frac{1-\cos t}{t^2}\right)=\int_s^\infty \log\left(\frac{\sqrt{x^2+1}}{x}\right)dx=\frac{1}{2}\int_s^\infty \log\left(\frac{x^2+1}{x^2}\right)dx$$

$$=\frac{1}{2}\int_s^\infty \left\{\log\left(x^2+1\right)-2\log x\right\}dx$$

$$=\frac{1}{2}\left[\left\{\log\left(x^2+1\right)-2\log x\right\}.x\right]_s^\infty-\frac{1}{2}\int_s^\infty\left(\frac{2x}{x^2+1}-\frac{2}{x}\right)x\,dx \quad \text{(integrating by parts)}$$

$$=\left[\frac{x}{2}\log\left(\frac{x^2+1}{x^2}\right)\right]_s^\infty+\int_s^\infty\frac{1}{x^2+1}dx$$

$$\lim_{x\to\infty}\frac{x}{2}\log\left(1+\frac{1}{x^2}\right)=\lim_{y\to 0+}\frac{1}{2y}\log\left(1+y^2\right)=\lim_{y\to 0+}\frac{2y/(1+y^2)}{2}=0 \quad (L'\text{ Hospital rule})$$

$$\therefore \quad £\left(\frac{1-\cos t}{t^2}\right)=-\frac{s}{2}\log\left(\frac{s^2+1}{s^2}\right)+(\tan^{-1}x)_s^\infty$$

$$=\frac{-s}{2}\log\left(\frac{s^2+1}{s^2}\right)+\frac{\pi}{2}-\tan^{-1}s$$

$$=\cot^{-1}s-\frac{s}{2}\log\left(1+\frac{1}{s^2}\right);\ s>0$$

Example 2.10: Find $£\left(\sin\sqrt{t}\right)$ and hence show that $£\left(\frac{\cos\sqrt{t}}{\sqrt{t}}\right)=\sqrt{\frac{\pi}{s}}\,e^{-\frac{1}{4s}}$.

Solution: We have $\sin\sqrt{t}=\sqrt{t}-\frac{t^{3/2}}{3!}+\frac{t^{5/2}}{5!}-\frac{t^{7/2}}{7!}+\cdots$

$$=\sum_{n=0}^\infty\frac{(-1)^n\,t^{n+\frac{1}{2}}}{(2n+1)!}$$

$$\therefore \quad £\left(\sin\sqrt{t}\right)=\sum_{n=0}^\infty\frac{(-1)^n}{(2n+1)!}L\left(t^{n+\frac{1}{2}}\right)=\sum_{n=0}^\infty\frac{(-1)^n}{(2n+1)!}\frac{\Gamma\left(n+\frac{3}{2}\right)}{s^{n+\frac{3}{2}}}$$

$$=\sum_{n=0}^\infty\frac{(-1)^n\left(n+\frac{1}{2}\right)\left(n-\frac{1}{2}\right)\ldots\left(\frac{1}{2}\right)\Gamma\left(\frac{1}{2}\right)}{(2n+1)!\ s^{n+\frac{3}{2}}}$$

$$=\sum_{n=0}^\infty\frac{(-1)^n\,(2n+1)(2n-1)\ldots1\sqrt{\pi}}{2^{n+1}\,(2n+1)!\qquad s^{n+3/2}}$$

$$= \sum_{n=0}^{\infty} \frac{(-1)^n \sqrt{\pi}}{2^{n+1}(2.4.6....2n)s^{n+3/2}}$$

$$= \frac{1}{2s}\sqrt{\frac{\pi}{s}} \sum_{n=0}^{\infty} \frac{(-1)^n}{n!}\left(\frac{1}{4s}\right)^n = \frac{\sqrt{\pi}}{2s^{3/2}} e^{-1/4s}$$

$\therefore$ $$£\left(\frac{d}{dt}\sin\sqrt{t}\right) = s\,£\left(\sin\sqrt{t}\right) - \left(\sin\sqrt{t}\right)_{t=0} \quad \text{(by Laplace transform of derivative)}$$

$\therefore$ $$£\left(\frac{1}{2}\frac{\cos\sqrt{t}}{\sqrt{t}}\right) = \frac{\sqrt{\pi}}{2s^{1/2}} e^{-1/4s} - 0$$

$\therefore$ $$£\left(\frac{\cos\sqrt{t}}{\sqrt{t}}\right) = \sqrt{\frac{\pi}{s}}\, e^{-1/4s}$$

Example 2.11: Find the Laplace transforms of $t^5e^t \sin 4t$ and $t^5e^t \cos 4t$.

Solution: We have $$£\,(t^5) = \frac{5!}{s^6} = \frac{120}{s^6};\ s > 0$$

$\therefore$ $$£\left\{t^5 e^{(1+4i)t}\right\} = \frac{120}{(s-1-4i)^6};\ s > 1 \quad \text{(by first shifting property)}$$

$\therefore$ $$£\left\{t^5e^t\left(\cos 4t + i\sin 4t\right)\right\} = \frac{120(s-1+4i)^6}{\left[(s-1)^2+16\right]^6}$$

$\therefore$ $$£\left\{\left(t^5e^t\cos 4t\right) + i\left(t^5e^t\sin 4t\right)\right\}$$

$$= \frac{120}{\left[(s-1)^2+16\right]^6}\Big[(s-1)^6 + {}^6C_1(s-1)^5(4i) + {}^6C_2(s-1)^4(4i)^2 + {}^6C_3(s-1)^3(4i)^3$$
$$+ {}^6C_4(s-1)^2(4i)^4 + {}^6C_5(s-1)(4i)^5 + (4i)^6\Big]$$

Equate real and imaginary parts

$$£\,(t^5\,e^t\cos 4t) = \frac{120}{\left[(s-1)^2+16\right]^6}\left[(s-1)^6 - 240(s-1)^4 + 3840(s-1)^2 - 4096\right]$$

and $$£(t^5\,e^t\sin 4t) = \frac{960(s-1)}{\left[(s-1)^2+16\right]^6}\left[3(s-1)^4 - 160(s-1)^2 + 768\right];\ s > 1$$

Example 2.12: Find the Laplace transforms of $erf(z)$ and $erf\,(\sqrt{z})$.

Solution: $erf(z)$ is defined as $\frac{2}{\sqrt{\pi}}\int_0^z e^{-u^2}\,du$ and is called error function and

$erf_c(z)$ is defined as $1 - erf(z)$ and is called complementary error function.

Now, $$\pounds\{erf(z)\} = \int_0^\infty e^{-sz} \frac{2}{\sqrt{\pi}} \int_0^z e^{-u^2}\,du\,dz$$

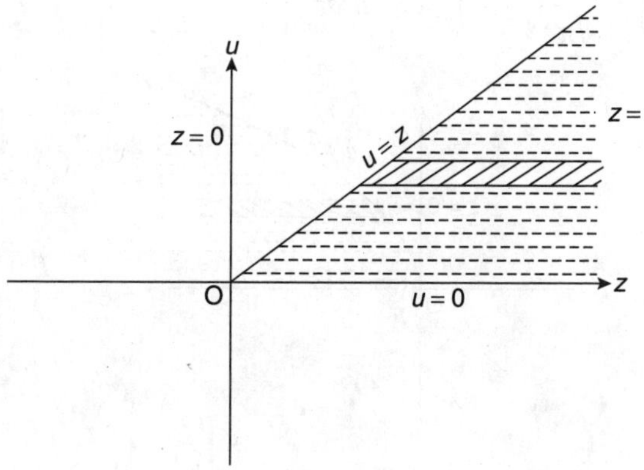

Figure 2.2

$\therefore$ $$\pounds\{erf(z)\} = \frac{2}{\sqrt{\pi}} \int_0^\infty \int_u^\infty e^{-sz} e^{-u^2}\,dz\,du \quad \text{(by changing order of integration)}$$

$$= \frac{2}{\sqrt{\pi}} \int_0^\infty e^{-u^2} \left\{ \frac{-e^{-sz}}{s} \right\}_u^\infty du,\ s > 0$$

$$= \frac{2}{s\sqrt{\pi}} \int_0^\infty e^{-u^2} \cdot e^{-su}\,du = \frac{2e^{\frac{1}{4}s^2}}{s\sqrt{\pi}} \int_0^\infty e^{-\left(u+\frac{1}{2}s\right)^2} du$$

$$= \frac{2e^{\frac{1}{4}s^2}}{s\sqrt{\pi}} \int_{s/2}^\infty e^{-t^2}\,dt \quad \text{Take } u + \frac{1}{2}s = t$$

$$= \frac{2e^{\frac{1}{4}s^2}}{s\sqrt{\pi}} \left[\int_0^\infty e^{-t^2}\,dt - \int_0^{\frac{s}{2}} e^{-t^2}\,dt \right]$$

$$= \frac{2e^{\frac{1}{4}s^2}}{s\sqrt{\pi}} \left[\frac{\sqrt{\pi}}{2} - \int_0^{\frac{s}{2}} e^{-t^2}\,dt \right] = \frac{e^{\frac{1}{4}s^2}}{s} \left[1 - \frac{2}{\sqrt{\pi}} \int_0^{\frac{s}{2}} e^{-t^2}\,dt \right]$$

$$= \frac{e^{\frac{1}{4}s^2}}{s} erf_c\left(\frac{s}{2}\right) \quad \text{(by definition of complementary error function)}$$

Hence, $$\pounds\{erf(z)\} = \frac{e^{\frac{1}{4}s^2}}{s} erf_c\left(\frac{s}{2}\right);\ s > 0$$

Now, $$erf\left(\sqrt{z}\right) = \frac{2}{\sqrt{\pi}}\int_0^{\sqrt{z}} e^{-u^2}\,du$$

$\therefore$ $$\pounds\left\{erf\left(\sqrt{z}\right)\right\} = \frac{2}{\sqrt{\pi}}\int_0^{\infty} e^{-sz}\int_0^{\sqrt{z}} e^{-u^2}\,du\,dz$$

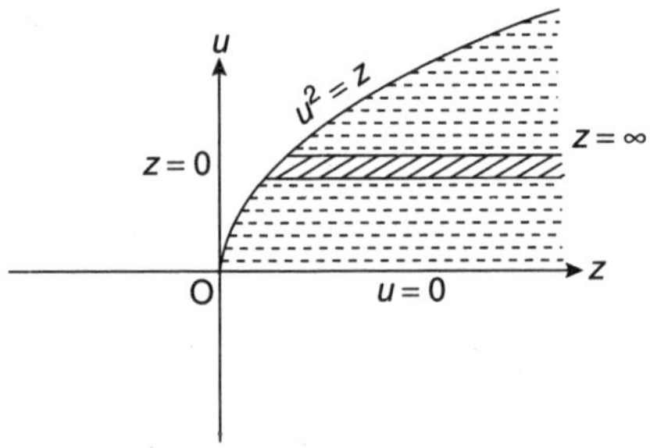

Figure 2.3

$$= \frac{2}{\sqrt{\pi}}\int_0^{\infty}\int_{u^2}^{\infty} e^{-sz}e^{-u^2}\,dz\,du \quad \text{(by changing order of integration)}$$

$$= \frac{2}{\sqrt{\pi}}\int_0^{\infty}\left\{-\frac{e^{-sz}}{s}\right\}_{u^2}^{\infty} e^{-u^2}\,du,\ s>0$$

$$= \frac{2}{s\sqrt{\pi}}\int_0^{\infty} e^{-(1+s)u^2}\,du = \frac{2}{s\sqrt{1+s}\sqrt{\pi}}\int_0^{\infty} e^{-t^2}\,dt \qquad \left(\text{taking } \sqrt{1+s}\ u = t\right)$$

$$= \frac{2}{s\sqrt{1+s}\sqrt{\pi}}\cdot\frac{\sqrt{\pi}}{2} = \frac{1}{s\sqrt{1+s}}$$

Hence, $$\pounds\left\{erf\left(\sqrt{z}\right)\right\} = \frac{1}{s\sqrt{1+s}};\ s>0.$$

Example 2.13: Find the Laplace transforms of Bessel functions $J_0(x)$ and $J_1(x)$. Also, deduce the Laplace transform of $J_0(ax)$.

Solution: We have $$J_0(x) = \sum_{r=0}^{\infty}\frac{(-1)^r}{(r!)^2}\left(\frac{x}{2}\right)^{2r} \qquad \text{(from definition of } J_n(x) \text{ for } n=0)$$

$\therefore$ $$\pounds\{J_0(x)\} = \sum_{r=0}^{\infty}\frac{(-1)^r}{(r!)^2\,2^{2r}}\,\pounds\left(x^{2r}\right)$$

$$= \sum_{r=0}^{\infty} \frac{(-1)^r (2r)!}{(r!)^2 2^{2r} s^{2r+1}}$$

$$= \frac{1}{s}\left[1 + \sum_{r=1}^{\infty} \frac{(-1)^r}{(r!)^2} \frac{(1\cdot 3\cdot 5\cdots\cdots(2r-1))(2\cdot 4\cdot 6\cdots\cdots 2r)}{2^{2r}}\left(\frac{1}{s^2}\right)^r\right]$$

$$= \frac{1}{s}\left[1 + \sum_{r=1}^{\infty} \frac{\left(-\frac{1}{2}\right)\left(-\frac{3}{2}\right)\left(-\frac{5}{2}\right)\cdots\left(-\frac{2r-1}{2}\right)}{r!}\left(\frac{1}{s^2}\right)^r\right]$$

$$\therefore \qquad £\{J_0(x)\} = \frac{1}{s}\left(1+\frac{1}{s^2}\right)^{-1/2} = \frac{1}{\sqrt{s^2+1}} \qquad (1)$$

Now, we know that $\qquad J_0'(x) = -J_1(x)$

$$\therefore \qquad J_1(x) = -J_0'(x)$$

$$\therefore \qquad £\{J_1(x)\} = -£\{J_0'(x)\}$$

$$= -[s\,£\{J_0(x)\} - J_0(0)] \text{ (by Laplace transform of derivative)}$$

$$= -s\cdot\frac{1}{\sqrt{s^2+1}} + 1 \text{ [from equation (1) and } J_0(0) = 1]$$

$$\therefore \qquad £\{J_1(x)\} = 1 - \frac{s}{\sqrt{s^2+1}}$$

Further, by change of scale property

$$£\{J_0(ax)\} = \frac{1}{a}F\left(\frac{s}{a}\right) \text{ where } F(s) = £\{J_0(x)\}$$

$$\therefore \qquad £\{J_0(ax)\} = \frac{1}{a}.\frac{1}{\sqrt{\left(\frac{s}{a}\right)^2+1}} \qquad \text{(from (1))}$$

Hence, $\qquad £\{J_0(ax)\} = \dfrac{1}{\sqrt{s^2+a^2}}$

2.8 EVALUATION OF REAL INTEGRALS USING LAPLACE TRANSFORM

Some real integrals can be obtained by finding Laplace transform of appropriate functions and giving suitable values to s as explained in the following examples.

Example 2.14: Evaluate

(i) $\int_0^\infty te^{-2t}\cos t\,dt$ (ii) $\int_0^\infty t^3e^{-t}\sin t\,dt$ (iii) $\int_0^\infty \frac{\sin^2 t}{t^2}dt$

(iv) $\int_0^\infty \frac{\cos at-\cos bt}{t}dt$ (v) $\int_0^\infty \frac{\sin mt}{t}dt$, $m>0$ and deduce the value of $\int_0^\infty \frac{\sin t}{t}dt$

Solution: (i) we have $£(t)=\frac{1}{s^2};\ s>0$

$\therefore \quad £\left(te^{it}\right)=\frac{1}{(s-i)^2};\ s>0$ (by first shifting property)

$\Rightarrow \quad £\{t\,(\cos t+i\sin t)\}=\frac{(s+i)^2}{(s^2+1)^2}$

$\therefore \quad £(t\cos t)+i\,£(t\sin t)=\frac{s^2-1+2is}{(s^2+1)^2}$

Equate real parts

$$£(t\cos t)=\frac{s^2-1}{(s^2+1)^2};\ s>0$$

$\therefore \quad \int_0^\infty e^{-st}t\cos t\,dt=\frac{s^2-1}{(s^2+1)^2};\ s>0$

Take $s=2$

$$\int_0^\infty te^{-2t}\cos t\,dt=\frac{3}{25}$$

(ii) $£(t^3)=\frac{3!}{s^4}=\frac{6}{s^4};\ s>0$

$\therefore \quad £(t^3e^{it})=\frac{6}{(s-i)^4}$

$\Rightarrow \quad £\{t^3(\cos t+i\sin t)\}=\frac{6(s+i)^4}{(s^2+1)^4}$

$\therefore \quad £(t^3\cos t)+i\,£(t^3\sin t)=\frac{6\left[s^4+4s^3i-6s^2-4si+1\right]}{\left(s^2+1\right)^4}$

Equate imaginary parts

$$£(t^3\sin t)=\frac{24\left(s^3-s\right)}{\left(s^2+1\right)^4}=\frac{24s\left(s^2-1\right)}{\left(s^2+1\right)^4};\ s>0$$

$\therefore \quad \int_0^\infty e^{-st}t^3\sin t\,dt=\frac{24s\left(s^2-1\right)}{(s^2+1)^4};\ s>0$

Take $s = 1$

$$\int_0^\infty t^3 e^{-t} \sin t = 0$$

(iii) $$\frac{\sin^2 t}{t^2} = \frac{1-\cos 2t}{2t^2} = \frac{1}{2t^2}\left[1-\left\{1-\frac{(2t)^2}{2!}+\frac{(2t)^4}{4!}-\frac{(2t)^6}{6!}+\cdots\right\}\right]$$

$$= 1-\frac{2^3}{4!}t^2+\frac{2^5}{6!}t^4-\frac{2^7}{8!}t^6+\cdots\cdots$$

$\therefore$ $$£\left(\frac{\sin^2 t}{t^2}\right) = \frac{1}{s}-\frac{2^3}{4!}.\frac{2!}{s^3}+\frac{2^5}{6!}.\frac{4!}{s^5}-\frac{2^7}{8!}.\frac{6!}{s^7}+\cdots$$

$$= \frac{1}{2}.\frac{2}{s}-\frac{1}{3.4}\left(\frac{2}{s}\right)^3+\frac{1}{5.6}\left(\frac{2}{s}\right)^5-\frac{1}{7.8}\left(\frac{2}{s}\right)^7+\cdots$$

$$= \left(1-\frac{1}{2}\right)\frac{2}{s}-\left(\frac{1}{3}-\frac{1}{4}\right)\left(\frac{2}{s}\right)^3+\left(\frac{1}{5}-\frac{1}{6}\right)\left(\frac{2}{s}\right)^5-\left(\frac{1}{7}-\frac{1}{8}\right)\left(\frac{2}{s}\right)^7+\cdots$$

$$= \left[\frac{2}{s}-\frac{1}{3}\left(\frac{2}{s}\right)^3+\frac{1}{5}\left(\frac{2}{s}\right)^5-\frac{1}{7}\left(\frac{2}{s}\right)^7+\cdots\right]$$

$$-\left[\frac{1}{s}-\frac{1}{4}\left(\frac{2}{s}\right)^3+\frac{1}{6}\left(\frac{2}{s}\right)^5-\frac{1}{8}\left(\frac{2}{s}\right)^7+\cdots\right]$$

$$= \tan^{-1}\frac{2}{s}-\frac{s}{4}\left[\frac{4}{s^2}-\frac{1}{2}\left(\frac{4}{s^2}\right)^2+\frac{1}{3}\left(\frac{4}{s^2}\right)^3-\frac{1}{4}\left(\frac{4}{s^2}\right)^4+\cdots\right]$$

$\therefore$ $$£\left(\frac{\sin^2 t}{t^2}\right) = \tan^{-1}\frac{2}{s}-\frac{s}{4}\log\left(1+\frac{4}{s^2}\right);\ s>0$$

$\therefore$ $$\int_0^\infty e^{-st}\frac{\sin^2 t}{t^2}dt = \tan^{-1}\frac{2}{s}-\frac{s}{4}\log\left(1+\frac{4}{s^2}\right);\ s>0$$

$\therefore$ $$\lim_{s\to 0^+}\int_0^\infty e^{-st}\frac{\sin^2 t}{t^2}dt = \lim_{s\to 0^+}\left(\tan^{-1}\frac{2}{s}-\frac{s}{4}\log\frac{s^2+4}{s^2}\right)$$

$\therefore$ $$\int_0^\infty \lim_{s\to 0^+} e^{-st}\frac{\sin^2 t}{t^2}dt = \frac{\pi}{2}-\frac{1}{4}\lim_{s\to 0^+}\frac{\log(s^2+4)-\log s^2}{\frac{1}{s}}$$

$$= \frac{\pi}{2}-\frac{1}{4}\lim_{s\to 0^+}\frac{\frac{2s}{s^2+4}-\frac{2s}{s^2}}{\frac{-1}{s^2}} \quad (L\text{'Hospital rule})$$

$$= \frac{\pi}{2}+\frac{1}{4}\lim_{s\to 0^+}\frac{-8s}{s^2+4} = \frac{\pi}{2}+0 = \frac{\pi}{2}$$

$\therefore$ $$\int_0^\infty \frac{\sin^2 t}{t^2}dt = \frac{\pi}{2}$$

(iv) By Example 2.6 (i)

$$£\left(\frac{\cos at - \cos bt}{t}\right) = \frac{1}{2}\log\left(\frac{s^2+b^2}{s^2+a^2}\right)$$

$$\therefore \quad \int_0^\infty e^{-st}\left(\frac{\cos at - \cos bt}{t}\right)dt = \frac{1}{2}\log\left(\frac{s^2+b^2}{s^2+a^2}\right); s > 0$$

$$\therefore \quad \lim_{s\to 0^+}\int_0^\infty e^{-st}\,\frac{\cos at - \cos bt}{t}\,dt = \lim_{s\to 0^+}\frac{1}{2}\log\left(\frac{s^2+b^2}{s^2+a^2}\right)$$

$$\therefore \quad \int_0^\infty \frac{\cos at - \cos bt}{t}\,dt = \frac{1}{2}\log\frac{b^2}{a^2} = \log\left|\frac{b}{a}\right|$$

(v) $$£(\sin mt) = \frac{m}{s^2+m^2}; s > 0, m > 0$$

$$\therefore \quad £\left(\frac{\sin mt}{t}\right) = \int_s^\infty \frac{m}{x^2+m^2}\,dx = \left\{\tan^{-1}\frac{x}{m}\right\}_s^\infty; m > 0$$

$$= \frac{\pi}{2} - \tan^{-1}\frac{s}{m} = \cot^{-1}\frac{s}{m}$$

$$\therefore \quad \int_0^\infty e^{-st}\,\frac{\sin mt}{t}\,dt = \cot^{-1}\frac{s}{m}; s > 0, m > 0$$

$$\therefore \quad \lim_{s\to 0^+}\int_0^\infty e^{-st}\,\frac{\sin mt}{t}\,dt = \lim_{s\to 0^+}\cot^{-1}\frac{s}{m}; m > 0$$

$$\therefore \quad \int_0^\infty \frac{\sin mt}{t}\,dt = \frac{\pi}{2}; m > 0$$

Take $m = 1$

$$\int_0^\infty \frac{\sin t}{t}\,dt = \frac{\pi}{2}$$

Example 2.15: Given $£\left(2\sqrt{\frac{t}{\pi}}\right) = \frac{1}{s^{3/2}}$, show that $£\left(\frac{1}{\sqrt{\pi t}}\right) = \frac{1}{\sqrt{s}}$.

Solution: Let $f(t) = 2\sqrt{\frac{t}{\pi}}$. Then

$$f'(t) = \frac{2}{\sqrt{\pi}}\cdot\frac{1}{2\sqrt{t}} = \frac{1}{\sqrt{\pi t}}$$

$\therefore$ By Laplace transform of derivative, we have

$$£\{f'(t)\} = s\,£\{f(t)\} - f(0)$$

$$\therefore \quad £\left\{\frac{1}{\sqrt{\pi t}}\right\} = s£\left(2\sqrt{\frac{t}{\pi}}\right) - 0 = s.\frac{1}{s^{3/2}} - 0 = \frac{1}{\sqrt{s}}$$

Example 2.16: Given that $£\left(\frac{\sin t}{t}\right) = \cot^{-1} s$, find $£\left(\frac{\sin at}{t}\right)$, $a > 0$.

Solution: By change of scale property

$$£\{f(at)\} = \frac{1}{a}F\left(\frac{s}{a}\right), a > 0 \text{ where } £\{f(t)\} = F(s)$$

$$\therefore \quad £\left(\frac{\sin at}{at}\right) = \frac{1}{a}\cot^{-1}\left(\frac{s}{a}\right)$$

$$\Rightarrow \quad £\left(\frac{\sin at}{t}\right) = \cot^{-1}\left(\frac{s}{a}\right), \; a > 0$$

Heaviside Function or Unit Step Function

Heaviside function $H(t)$, also called unit step function, $U(t)$ or $u_0(t)$ is defined by

$$H(t) = U(t) = u_0(t) = \begin{cases} 0, & \text{if } t < 0 \\ 1, & \text{if } t \geq 0 \end{cases}$$

If the jump discontinuity is at $t = a$, then we define $H(t–a) = U(t – a)$ also denoted by $u_a(t)$ as

$$H(t–a) = U(t-a) = u_a(t) = \begin{cases} 0, & \text{if } t < a \\ 1, & \text{if } t \geq a \end{cases} ; a \geq 0$$

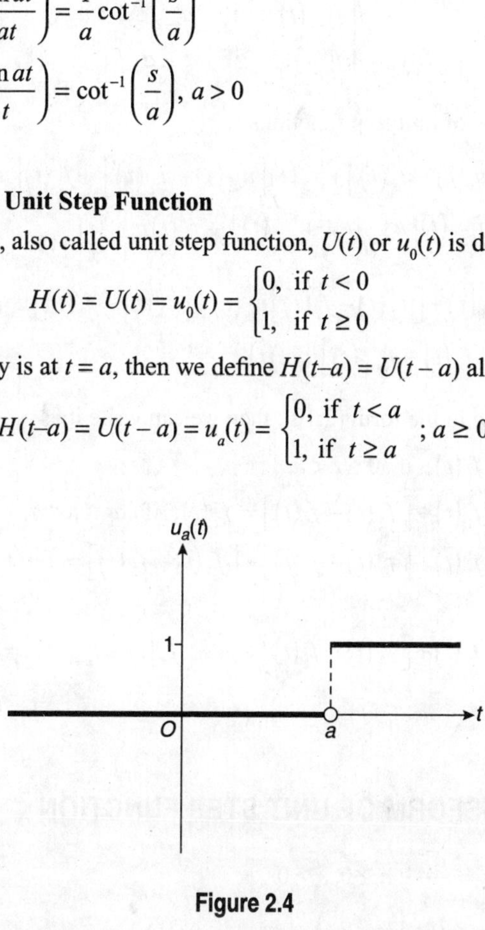

Figure 2.4

Unit step function $u_a(t)$

We now explain the writing of function with jumps in terms of unit step function and vice versa. Let $a < b$. Then, by the definition of unit step function

$$u_a(t) - u_b(t) = \begin{cases} 0, & \text{if } t < a \\ 1, & \text{if } a \leq t < b \\ 0, & \text{if } t \geq b \end{cases}$$

Therefore, if the given function is of the form

$$f(t)=\begin{cases} f_1(t), & \text{if} \quad 0\le t<a_1 \\ f_2(t), & \text{if} \quad a_1\le t<a_2 \\ f_3(t), & \text{if} \quad a_2\le t<a_3 \\ \vdots & \\ f_{n-1}(t), & \text{if} \quad a_{n-2}\le t<a_{n-1} \\ f_n(t), & \text{if} \quad t\ge a_{n-1} \end{cases} \tag{2.9}$$

We can express it in terms of unit step functions as

$$f(t)=f_1(t)\left[u_0(t)-u_{a_1}(t)\right]+f_2(t)\left[u_{a_1}(t)-u_{a_2}(t)\right]+f_3(t)\left[u_{a_2}(t)-u_{a_3}(t)\right]$$
$$+\cdots+f_{n-1}(t)\left[u_{a_{n-2}}(t)-u_{a_{n-1}}(t)\right]+f_n(t)u_{a_{n-1}}(t)$$

i.e.,

$$f(t)=f_1(t)u_0(t)+\left[f_2(t)-f_1(t)\right]u_{a_1}(t)+\left[f_3(t)-f_2(t)\right]u_{a_2}(t)$$
$$+\cdots+\left[f_n(t)-f_{n-1}(t)\right]u_{a_{n-1}}(t) \tag{2.10}$$

Conversely, if $f(t)$ is given in the form (2.10), then we can write it as

$$f(t)=\begin{cases} f_1(t), \text{ if } 0\le t<a_1 \\ f_1(t)+\left[f_2(t)-f_1(t)\right]=f_2(t), \text{ if } a_1\le t<a_2 \\ f_1(t)+\left[f_2(t)-f_1(t)\right]+\left[f_3(t)-f_2(t)\right]=f_3(t), \text{ if } a_2\le t<a_3 \\ \vdots \\ f_1(t)+\left[f_2(t)-f_1(t)\right]+\cdots+\left[f_n(t)-f_{n-1}(t)\right]=f_n(t), \text{ if } a_{n-1}\le t \end{cases}$$

which is (2.9).

2.9 LAPLACE TRANSFORM OF UNIT STEP FUNCTION

Theorem 2.10 $\pounds\{u_a(t)\}=\dfrac{e^{-as}}{s}; a\ge 0, s>0$

Proof: We have $u_a(t)=\begin{cases} 0; & t<a \\ 1; & t\ge a \end{cases}; \quad a\ge 0$

$$\therefore \qquad \pounds\{u_a(t)\}=\int_0^\infty e^{-st}u_a(t)\,dt=\int_0^a e^{-st}\cdot 0\,dt+\int_a^\infty e^{-st}\cdot 1\,dt$$

$$=\int_a^\infty e^{-st}dt=\left\{\frac{e^{-st}}{-s}\right\}_a^\infty=\frac{e^{-as}}{s}; a\ge 0, s>0$$

Theorem 2.11 (second shifting theorem)
If $\pounds\{f(t)\} = F(s)$; $s > k$ and $a \geq 0$ be any real number, then $\pounds\{f(t-a)\, u_a(t)\} = e^{-as} F(s)$.

Proof: We have by definition of $u_a(t)$, $f(t-a)\, u_a(t) = \begin{cases} 0 & ; t < a \\ f(t-a); & t \geq a \end{cases}$

$$\therefore \quad \pounds\{f(t-a)\, u_a(t)\} = \int_a^{\infty} e^{-st} f(t-a)\,dt$$

$$= \int_0^{\infty} e^{-s(a+x)} f(x)\,dx \quad \text{(taking } t = a + x\text{)}$$

$$= e^{-as} \int_0^{\infty} e^{-sx} f(x)\,dx = e^{-as} F(s)$$

Example 2.17: Find the Laplace transform of the following functions
(i) $t^2 u_3(t)$ (ii) $\sin 2t\, U(t-\pi)$ (iii) $e^{-3t}\, U(t-2)$ (iv) $e^{-t}[1-u_2(t)]$ (v) $t^4 u_2(t)$

Solution:

(i) $f(t) = t^2 u_3(t) = (t-3+3)^2\, u_3(t)$

$$= \left[(t-3)^2 + 6(t-3) + 9\right] u_3(t)$$

$$\therefore \quad \pounds\{f(t)\} = e^{-3s}\left[\frac{2}{s^3} + \frac{6}{s^2} + \frac{9}{s}\right]; s > 0$$

(ii) $f(t) = \sin 2t\, U(t-\pi) = -\sin(2\pi - 2t)\, U(t-\pi) = \sin\{2(t-\pi)\}\, U(t-\pi)$

$$\therefore \quad \pounds\{f(t)\} = e^{-\pi s} \frac{2}{s^2+4} = \frac{2e^{-\pi s}}{s^2+4}; s > 0$$

(iii) $f(t) = e^{-3t}\, U(t-2) = e^{-3(t-2)-6} \cdot U(t-2) = e^{-6} e^{-3(t-2)}\, U(t-2)$

$$\therefore \quad \pounds\{f(t)\} = e^{-6}\, \pounds\, \{e^{-3(t-2)}\, U(t-2)\} = e^{-6}\, e^{-2s} \frac{1}{s+3} = \frac{e^{-2(s+3)}}{s+3};\ s > -3$$

(iv) $f(t) = e^{-t}\left[1 - u_2(t)\right] = e^{-t} - e^{-2}\, e^{-(t-2)} u_2(t)$

$$\therefore \quad \pounds\{f(t)\} = \pounds\, \{e^{-t}\} - e^{-2}\, \pounds\{e^{-(t-2)} u_2(t)\} = \frac{1}{s+1} - e^{-2}.e^{-2s}.\frac{1}{s+1} = \frac{1-e^{-2(s+1)}}{s+1}; s > -1$$

(v) $f(t) = t^4 u_2(t) = (t-2+2)^4\, u_2(t)$

$$= [(t-2)^4 + 8\,(t-2)^3 + 24\,(t-2)^2 + 32\,(t-2) + 16]\, u_2(t)$$

$$\therefore \quad \pounds\{f(t)\} = e^{-2s}\left[\frac{4!}{s^5} + \frac{8(3!)}{s^4} + \frac{24(2!)}{s^3} + \frac{32}{s^2} + \frac{16}{s}\right]$$

$$= \frac{8e^{-2s}}{s}\left(2 + \frac{4}{s} + \frac{6}{s^2} + \frac{6}{s^3} + \frac{3}{s^4}\right); s > 0$$

Example 2.18: Find the Laplace transform of functions

(i) $f(t) = \begin{cases} 0; & 0 \le t < 2 \\ K; & t \ge 2 \end{cases}$ where K is a constant

(ii) $f(t) = \begin{cases} \cos\left(t - \dfrac{2\pi}{3}\right); & t \ge \dfrac{2\pi}{3} \\ 0 & ; \; t < \dfrac{2\pi}{3} \end{cases}$

(iii) $f(\text{t}) = \begin{cases} t; & 0 \le t \le 3 \\ 0; & t > 3 \end{cases}$

(iv) $f(t) = \begin{cases} \sin at & ; \; 0 < t < \dfrac{\pi}{a} \\ 0 & ; \; \dfrac{\pi}{a} < t \end{cases}$

(v) $f(t) = \begin{cases} \dfrac{2}{T}t & ; \; 0 < t < \dfrac{T}{2} \\ 2 - \dfrac{2t}{T} & ; \; \dfrac{T}{2} < t < T \\ 0 & ; \; t > T \end{cases}$

Solution:

(i) $f(t) = K\, u_2(t)$

$$\therefore\ \pounds\{f(t)\} = K\, \pounds\{u_2(t)\} = \frac{Ke^{-2s}}{s};\ s > 0 \quad \left(\because\ L\{u_a(t)\} = \frac{e^{-as}}{s}\right)$$

(ii) $f(t) = \cos\left(t - \dfrac{2\pi}{3}\right) u_{\frac{2\pi}{3}}(t)$

$$\therefore\ \pounds\{f(t)\} = e^{\frac{-2\pi}{3}s}\, \pounds(\cos t) = \frac{e^{\frac{-2\pi}{3}s} \cdot s}{(s^2+1)};\ s > 0 \quad \text{(by second shifting theorem)}$$

(iii) $f(t) = t[u_0(t) - u_3(t)] = t\, u_0(t) - [(t-3) + 3]\, u_3(t)$

$\therefore$ By second shifting theorem

$$\pounds\{f(t)\} = \frac{1}{s^2} - \left(\frac{1}{s^2} + \frac{3}{s}\right)e^{-3s} = \frac{1}{s^2}\left(1 - e^{-3s}\right) - \frac{3}{s}e^{-3s}$$

(iv) $f(t) = \sin at \left[u_0(t) - u_{\frac{\pi}{a}}(t) \right]$

$$= \sin at\, u_0(t) - \sin\left\{a\left(\frac{\pi}{a} - t\right)\right\} u_{\frac{\pi}{a}}(t)$$

$$= \sin at\, u_0(t) + \sin\left\{a\left(t - \frac{\pi}{a}\right)\right\} u_{\frac{\pi}{a}}(t)$$

$\therefore$ By second shifting theorem

$$\pounds\{f(t)\} = \frac{a}{s^2 + a^2} + e^{-\frac{\pi}{a}s} \frac{a}{s^2 + a^2} = \frac{a\left(1 + e^{\frac{-\pi}{a}s}\right)}{s^2 + a^2}$$

(v) $f(t) = \dfrac{2t}{T}\left[u_0(t) - u_{\frac{T}{2}}(t) \right] + \left(2 - \dfrac{2t}{T}\right)\left[u_{\frac{T}{2}}(t) - u_T(t) \right]$

$$\therefore \quad f(t) = \frac{2t}{T} u_0(t) - \frac{4}{T}\left(t - \frac{T}{2}\right) u_{\frac{T}{2}}(t) + \frac{2}{T}(t - T) u_T(t)$$

$\therefore$ By second shifting theorem

$$\pounds\{f(t)\} = \frac{2}{T}.\frac{1}{s^2} - \frac{4}{T} e^{-\frac{T}{2}s}.\frac{1}{s^2} + \frac{2}{T} e^{-Ts}.\frac{1}{s^2} = \frac{2}{Ts^2}\left(1 - e^{\frac{-Ts}{2}}\right)^2$$

$$\text{or } \pounds\{f(t)\} = \frac{2e^{\frac{-Ts}{2}}\left(e^{\frac{Ts}{4}} - e^{\frac{-Ts}{4}}\right)^2}{Ts^2} = \frac{8e^{\frac{-Ts}{2}} \sinh^2\left(\frac{Ts}{4}\right)}{Ts^2}$$

Remark 2.3: All the questions of above type can be solved directly by definition of Laplace transform. But it is always better to write the given function $f(t)$ in terms of unit step function to find $\pounds\{f(t)\}$.

We solve Example 2.18 (iv) and (v) below directly by definition

(iv) We have $\pounds\{f(t)\} = \displaystyle\int_0^{\infty} e^{-st} f(t)\,dt$

$$\therefore \qquad \pounds\{f(t)\} = \int_0^{\pi/a} e^{-st} \sin at\, dt + \int_{\pi/a}^{\infty} e^{-st}.0\, dt$$

$$= \left\{ \frac{e^{-st}}{\left(s^2 + a^2\right)} (-s \sin at - a \cos at) \right\}_0^{\frac{\pi}{a}}$$

$$= \frac{ae^{-s\pi/a} + a}{\left(s^2 + a^2\right)} = \frac{a\left(1 + e^{\frac{-\pi}{a}s}\right)}{\left(s^2 + a^2\right)}$$

(v) $$£\{f(t)\} = \int_0^{\frac{T}{2}} \frac{2}{T} t e^{-st}\, dt + \int_{\frac{T}{2}}^{T} \left(2 - \frac{2}{T}t\right) e^{-st}\, dt + \int_T^{\infty} 0 \cdot e^{-st}\, dt$$

$$= \frac{2}{T}\left\{ t\left(\frac{e^{-st}}{-s}\right) - 1\cdot\left(\frac{e^{-st}}{s^2}\right)\right\}_0^{\frac{T}{2}} + \left\{\left(2 - \frac{2}{T}t\right)\left(\frac{e^{-st}}{-s}\right) - \left(\frac{-2}{T}\right)\left(\frac{e^{-st}}{s^2}\right)\right\}_{\frac{T}{2}}^{T}$$

$$= \frac{-2}{T} e^{\frac{-sT}{2}} \left(\frac{T}{2s} + \frac{1}{s^2}\right) + \frac{2}{Ts^2} + \frac{2}{Ts^2} e^{-ST} + \frac{e^{\frac{-sT}{2}}}{s} - \frac{2}{Ts^2} e^{\frac{-sT}{2}}$$

$$= \frac{2}{Ts^2} - \frac{4}{Ts^2} e^{\frac{-sT}{2}} + \frac{2}{Ts^2} e^{-sT} = \frac{2}{Ts^2}\left(1 - e^{\frac{-Ts}{2}}\right)^2$$

$$\therefore \quad £\{f(t)\} = \frac{2e^{\frac{-Ts}{2}}\left(e^{\frac{Ts}{4}} - e^{\frac{-Ts}{4}}\right)^2}{Ts^2} = \frac{8e^{\frac{-Ts}{2}}}{Ts^2} \sinh^2\left(\frac{Ts}{4}\right)$$

Example 2.19: Find the Laplace transforms of the following functions

(i) $f(t) = \begin{cases} 2; & ; \quad 0 < t < \pi \\ 0 & ; \quad \pi < t < 2\pi \\ \sin t & ; \quad t > 2\pi \end{cases}$

(ii) $f(t) = \begin{cases} 2 + t^2 & ; \quad 0 < t < 2 \\ 6 & ; \quad 2 < t < 3 \\ 2t - 5 & ; \quad 3 < t < \infty \end{cases}$

(iii) $f(t) = |t - 1| + |t + 1|;\ t \geq 0$

(iv) $f(t) = \begin{cases} 0, & 0 \leq t \leq 3 \\ (t-3)^2, & t > 3 \end{cases}$

(v) $f(t) = \begin{cases} t^2 & \text{for} \quad 0 < t < 1 \\ 4t & \text{for} \quad t > 1 \end{cases}$

(vi) $f(t) = \begin{cases} t^2 & ; \quad 0 < t < 2 \\ t - 1 & ; \quad 2 < t < 3 \\ 7 & ; \quad t > 3 \end{cases}$

(vii) $f(t) = \begin{cases} t - 1 & ; \quad 1 < t < 2 \\ 3 - t & ; \quad 2 < t < 3 \\ 0 & ; \quad t > 3,\ 0 < t < 1 \end{cases}$

Solution:

(i) $f(t) = 2[u_0(t) - u_\pi(t)] + \sin t \cdot u_{2\pi}(t)$

$= 2\, u_0(t) - 2u_\pi(t) + \sin(t - 2\pi)\, u_{2\pi}(t)$

$\therefore\ \pounds\{f(t)\} = 2\pounds\{u_0(t)\} - 2\pounds\{u_\pi(t)\} + \pounds\{\sin(t - 2\pi)\, u_{2\pi}(t)\}$

$$= \frac{2}{s} - \frac{2e^{-\pi s}}{s} + e^{-2\pi s} \cdot \frac{1}{s^2+1} = \frac{2}{s}\left(1 - e^{-\pi s}\right) + \frac{e^{-2\pi s}}{s^2+1};\ s > 0$$

(ii) $f(t) = (2 + t^2)[u_0(t) - u_2(t)] + 6\,[u_2(t) - u_3(t)] + [2(t - 5)\, u_3(t)]$

$= (2 + t^2)\, u_0(t) + (4 - t^2)\, u_2(t) + (2t - 11)\, u_3(t)$

$= (2 + t^2)\, u_0(t) + [4 - (t - 2 + 2)^2]\, u_2(t) + [2(t - 3 + 3) - 11]\, u_3(t)$

$= (2 + t^2)\, u_0(t) - [(t - 2)^2 + 4(t - 2)]\, u_2(t) + [2(t - 3) - 5]\, u_3(t)$

$$\therefore\ \pounds\{f(t)\} = \left(\frac{2}{s} + \frac{2}{s^3}\right) - e^{-2s}\left(\frac{2}{s^3} + \frac{4}{s^2}\right) + e^{-3s}\left(\frac{2}{s^2} - \frac{5}{s}\right);\ s > 0$$

(iii) $f(t) = 1 - t + t + 1 = 2;\ 0 \le t \le 1$

$= t - 1 + t + 1 = 2t;\ t \ge 1$

$\therefore\quad f(t) = 2[u_0(t) - u_1(t)] + 2\,t\,u_1(t)$

$= 2u_0(t) + 2(t - 1)\, u_1(t)$

$$\therefore\ \pounds\{f(t)\} = \frac{2}{s} + \frac{2e^{-s}}{s^2} = \frac{2}{s}\left(1 + \frac{e^{-s}}{s}\right);\ s > 0$$

(iv) $f(t) = (t - 3)^2\, u_3(t)$

$$\therefore\ \pounds\{f(t)\} = e^{-3s} \cdot \frac{2}{s^3} = \frac{2e^{-3s}}{s^3};\ s > 0$$

(v) $f(t) = t^2\,[u_0(t) - u_1(t)] + 4t\, u_1(t)$

$= t^2\, u_0(t) + (4t - t^2)\, u_1(t)$

$= t^2\, u_0(t) + [4(t - 1) + 4 - (t - 1 + 1)^2]\, u_1(t)$

$= t^2\, u_0(t) + [3 + 2(t - 1) - (t - 1)^2]\, u_1(t)$

$$\therefore\ \pounds\{f(t)\} = \frac{2}{s^3} + e^{-s}\left[\frac{3}{s} + \frac{2}{s^2} - \frac{2}{s^3}\right] = \frac{1}{s}e^{-s}\left(3 + \frac{2}{s}\right) + \frac{2}{s^3}\left(1 - e^{-s}\right);\ s > 0$$

(vi) $f(t) = t^2[u_0(t) - u_2(t)] + (t - 1)\,[u_2(t) - u_3(t)] + 7u_3(t)$

$= t^2\, u_0(t) + (t - 1 - t^2)\, u_2(t) - (t - 8)\, u_3(t)$

$= t^2\, u_0(t) + [(t - 2) + 1 - (t - 2 + 2)^2]\, u_2(t) - [(t - 3) - 5]\, u_3(t)$

$= t^2\, u_0(t) - [(t - 2)^2 + 3\,(t - 2) + 3]\, u_2(t) - [(t - 3) - 5]\, u_3(t)$

$$\therefore\ \pounds\{f(t)\} = \frac{2}{s^3} - e^{-2s}\left[\frac{2}{s^3} + \frac{3}{s^2} + \frac{3}{s}\right] - e^{-3s}\left(\frac{1}{s^2} - \frac{5}{s}\right);\ s > 0$$

(vii) $f(t) = (t - 1)[u_1(t) - u_2(t)] + (3 - t)\,[u_2(t) - u_3(t)]$

$= (t - 1)\, u_1(t) + (4 - 2t)\, u_2(t) - (3 - t)\, u_3(t)$

$$= (t-1)\,u_1(t) - 2(t-2)\,u_2(t) + (t-3)\,u_3(t)$$

$$\therefore\ \mathcal{L}\{f(t)\} = e^{-s}\cdot\frac{1}{s^2} - 2e^{-2s}\cdot\frac{1}{s^2} + e^{-3s}\cdot\frac{1}{s^2}$$

$$= \frac{1}{s^2}e^{-s}(1-2e^{-s}+e^{-2s}) = \frac{1}{s^2}e^{-s}(1-e^{-s})^2$$

$$= \frac{1}{s^2}e^{-2s}(e^{s/2}-e^{-s/2})^2 = \frac{4\sinh^2\frac{s}{2}}{s^2e^{2s}}$$

Example 2.20: Express the functions defined by the following graphs (i), (ii) and (iii) in terms of unit step functions and find their Laplace transforms.

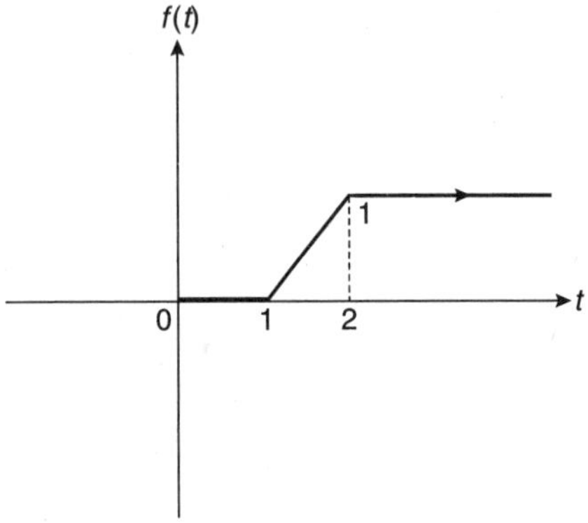

Figure 2.5 (i)

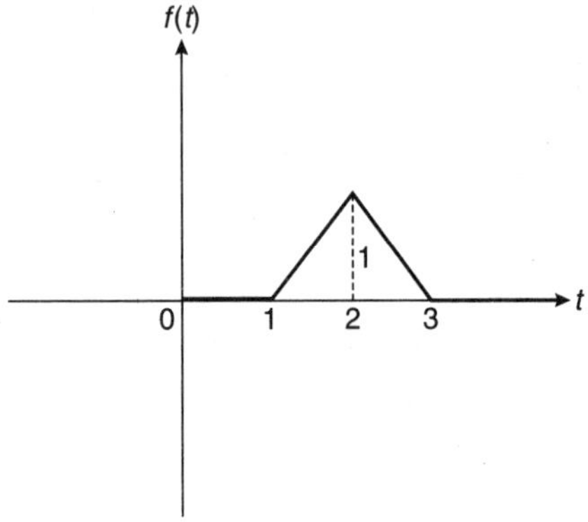

Figure 2.5 (ii)

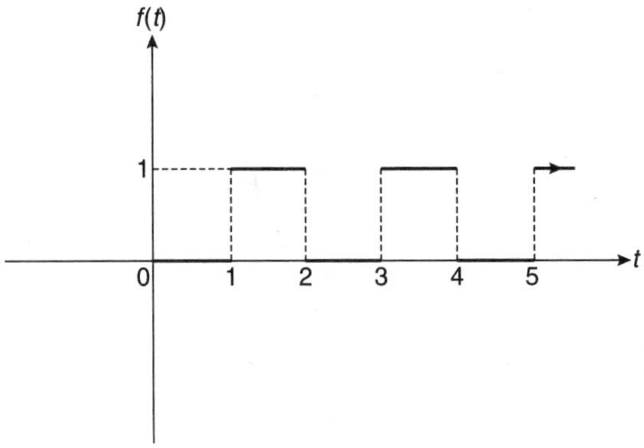

Figure 2.5 (iii)

Solution: (i) From graph (i)

$$f(t) = \begin{cases} 0 & ;\ 0 \le t \le 1 \\ t-1 & ;\ 1 \le t \le 2 \\ 1 & ;\ \quad t \ge 2 \end{cases}$$

$\therefore \qquad f(t) = (t-1)\,[u_1(t) - u_2(t)] + u_2(t) = (t-1)\,u_1(t) - (t-2)\,u_2(t)$

$\therefore$ By second shifting theorem

$$£\{f(t)] = \frac{1}{s^2}e^{-s} - \frac{1}{s^2}e^{-2s} = \frac{e^{-s}}{s^2}(1-e^{-s})\,;\, s>0$$

(ii) From graph (ii)

$$f(t) = \begin{cases} 0 & ;\ 0 \le t \le 1 \\ t-1 & ;\ 1 \le t \le 2 \\ 3-t & ;\ 2 \le t \le 3 \\ 0 & ;\ t \ge 3 \end{cases}$$

$\therefore \qquad f(t) = (t-1)\,[u_1(t) - u_2(t)] + (3-t)\,[u_2(t) - u_3(t)]$

$\therefore \qquad £\{f(t)\} = \dfrac{4\sinh^2 \dfrac{s}{2}}{s^2 e^{2s}}$ from Example (2.19 (vii))

(iii) From graph (iii)

$$\begin{aligned} f(t) &= 0;\ 0 \le t < 1 \\ &= 1;\ 1 < t < 2 \\ &= 0;\ 2 < t < 3 \\ &= 1;\ 3 < \text{t} < 4 \end{aligned}$$

and so on.

$\therefore \qquad f(t) = [u_1(t) - u_2(t)] + [u_3(t) - u_4(t)] + \cdots$

$\therefore$ By Laplace transform of unit function

$$£\{f(t)\} = \frac{e^{-s}}{s} - \frac{e^{-2s}}{s} + \frac{e^{-3s}}{s} + \frac{e^{-4s}}{s} + \cdots\cdots \qquad \text{(infinite G. P.)}$$

$$= \frac{\dfrac{e^{-s}}{s}}{1+e^{-s}} = \frac{1}{se^{s/2}(e^{s/2}+e^{-s/2})} = \frac{1}{2se^{s/2}\cosh\dfrac{s}{2}};\, s>0$$

2.10 LAPLACE TRANSFORM OF UNIT IMPULSE FUNCTION (DIRAC–DELTA FUNCTION)

It frequently occurs in many problems in electrical engineering, physics and mechanical engineering that a large force acts for a very short duration. To deal with such problems, we define unit impulse function or Dirac-delta function.

Define

$$\delta_\varepsilon(t) = \begin{cases} 0 \ ; \ t < 0 \\ \dfrac{1}{\varepsilon} \ ; \ 0 \le t < \varepsilon \\ 0 \ ; \ t \ge \varepsilon \end{cases}$$

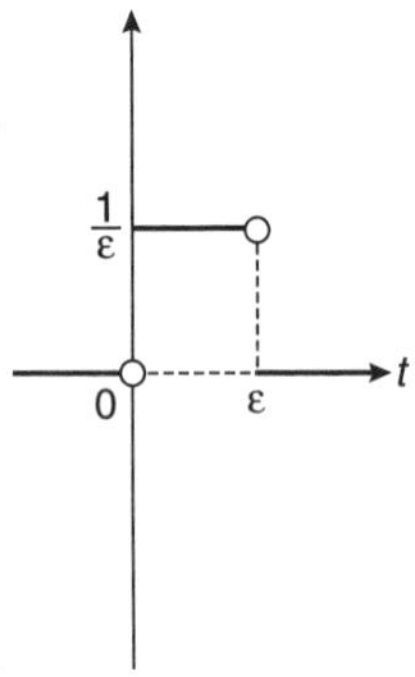

Figure 2.6

In terms of unit step function, we have

$$\delta_\varepsilon(t) = \frac{1}{\varepsilon}\{u_0(t) - u_\varepsilon(t)\}$$

The pulse has the height $\dfrac{1}{\varepsilon}$ and is of duration ε. As $\varepsilon \to 0^+$ the amplitude of pulse $\to \infty$. Thus, as $\varepsilon \to 0^+$ we define

$$\delta(t) = \lim_{\varepsilon \to 0^+} \delta_\varepsilon(t)$$

This delta function is called unit impulse function.

Its impulse $= \displaystyle\int_0^\varepsilon \frac{1}{\varepsilon} dt = 1$

It is clear from Figure 2.6 that as $\varepsilon \to 0^+$, the height of strip increases indefinitely and width decreases in such a way that its area is always unity.

The delta function can be made to act at any other point. The delta function $\delta(t - a)$ is defined by

$$\delta(t-a) = \lim_{\varepsilon \to 0^+} \delta_\varepsilon(t-a); \ a \ge 0$$

where

$$\delta_\varepsilon(t-a) = \begin{cases} 0 \ ; \ t < a \\ \dfrac{1}{\varepsilon} \ ; \ a \le t < a + \varepsilon \\ 0 \ ; \ t \ge a + \varepsilon \end{cases}$$

It acts at $t = a$.

Now, $$\delta_\varepsilon(t-a) = \frac{u_a(t)-u_{a+\varepsilon}(t)}{\varepsilon}$$

$\therefore$ $$\delta(t-a) = \lim_{\varepsilon\to 0^+} \delta_\varepsilon(t-a) = \lim_{\varepsilon\to 0^+} \frac{u_a(t)-u_{a+\varepsilon}(t)}{\varepsilon}$$

$$= \lim_{\varepsilon\to 0^+} \frac{u(t-a)-u(t-a-\varepsilon)}{\varepsilon} = u'(t-a)$$

Actually $\delta(t-a)$ is not a function in the ordinary sense but is the so-called 'generalized function' because we have

$$\delta(t-a) = \infty, \text{ if } t = a$$

$$= 0, \text{ otherwise}$$

and $$\int_0^\infty \delta(t-a)\,dt = 1$$

But an ordinary function which is everywhere zero except at a single point must have integral zero. Even then, in impulse problems it is convenient to operate on $\delta(t-a)$ as though it is an ordinary function.

Theorem 2.12 Filtering property of Dirac-delta function and its Laplace transform

Let $f(t)$ be continuous and integrable in $[0, \infty)$. Then, $\int_0^\infty f(t)\,\delta(t-a)dt = f(a)$ and $£\{\delta(t-a)\} = e^{-as}$

Proof: We have by definition of Dirac-delta function

$$\delta_\varepsilon(t-a) = \begin{cases} 0 \ ; \ t < a \\ \dfrac{1}{\varepsilon} \ ; \ a \le t < a+\varepsilon \\ 0 \ ; \ t > a+\varepsilon \end{cases}$$

and $$\delta(t-a) = \lim_{\varepsilon\to 0^+} \delta_\varepsilon(t-a)$$

Now, $$\int_0^\infty f(t)\delta_\varepsilon(t-a)dt = \int_a^{a+\varepsilon} \frac{1}{\varepsilon} f(t)dt$$

Using the mean value theorem of integral calculus we obtain

$$\int_0^\infty f(t)\,\delta_\varepsilon(t-a)\,dt = \int_a^{a+\varepsilon} \frac{1}{\varepsilon} f(t)\,dt = \frac{f(t_0)}{\varepsilon}\int_a^{a+\varepsilon} dt = f(t_0) \text{ for some } t_0,\ a \le t_0 \le a+\varepsilon$$

$$\therefore \quad \lim_{\varepsilon\to 0^+} \int_0^\infty f(t)\,\delta_\varepsilon(t-a)\,dt = \lim_{\varepsilon\to 0^+} f(t_0) = f(a)$$

$$\therefore \quad \int_0^\infty f(t)\,\delta(t-a)\,dt = f(a)$$

Taking $f(t) = e^{-st}$ we have

$$\int_0^\infty e^{-st}\delta(t-a)dt = e^{-as}$$

$$\therefore \qquad £\{\delta(t-a)\} = e^{-as}$$

Example 2.21: Evaluate (i) $£\left[\frac{1}{t}\{\delta(t-a)\}\right]$ (ii) $\int_0^\infty \sin 2t\ \delta\left(t-\frac{\pi}{4}\right)dt$

Solution: (i) We have $£\{\delta(t-a)\} = e^{-as}$

$$\therefore \qquad £\left\{\frac{1}{t}\delta(t-a)\right\} = \int_s^\infty e^{-ax}dx \quad \text{(by integration of Laplace transform)}$$

$$= \left\{-\frac{1}{a}e^{-ax}\right\}_s^\infty = \frac{e^{-as}}{a};\ a>0,\ s>0$$

(ii) Let $f(t) = \sin 2t$

$\therefore$ By filtering property, $\int_0^\infty \sin 2t\ \delta(t-a)dt = \sin 2a$

Take $a = \frac{\pi}{4}$, $\int_0^\infty \sin 2t\ \delta\left(t-\frac{\pi}{4}\right)dt = \sin\frac{\pi}{2} = 1$

2.11 LAPLACE TRANSFORM OF PERIODIC FUNCTIONS

Theorem 2.13 Let $f(t)$ be piecewise continuous for $t \ge 0$ and of exponential order k and periodic with period T. Then

$$£\{f(t)\} = \frac{1}{1-e^{-sT}}\int_0^T e^{-st} f(t)dt;\ s>0$$

Proof: We have $£\{f(t)\} = \int_0^\infty e^{-st} f(t)dt = \lim_{m\to\infty}\sum_{k=0}^m \int_{kT}^{(k+1)T} e^{-st} f(t)dt$

Put $\qquad t = kT + x$

$\therefore \qquad dt = dx$ and $f(t) = f(kT+x) = f(x)$ $\quad$ [$\because$ f is periodic with period T]

$$\therefore \qquad £\{f(t)\} = \lim_{m\to\infty}\sum_{k=0}^m \int_0^T e^{-s(kT+x)} f(x)dx$$

$$= \lim_{m\to\infty}\sum_{k=0}^m e^{-skT}\int_0^T e^{-sx} f(x)\,dx$$

$$= \int_0^T e^{-st} f(t)dt.\sum_{k=0}^\infty e^{-ksT} \qquad (2.11)$$

But $\sum_{k=0}^{\infty} e^{-ksT}$ is infinite G. P. with first term 1 and common ratio e^{-sT}, so

$$\sum_{k=0}^{\infty} e^{-ksT} = \frac{1}{\left(1-e^{-sT}\right)}\ ; s>0\ (\because e^{-sT} < 1 \text{ for } s>0)$$

$\therefore$ From (2.11)

$$£\{f(t)\} = \frac{1}{1-e^{-sT}} \int_0^T e^{-st} f(t)dt;\ s>0.$$

Example 2.22: If $f(t) = t^2;\ 0 < t < 2$ and $f(t+2) = f(t)$ for $t > 2$, find $£\{f(t)\}$.
Solution: Here, $f(t) = t^2;\ 0 < t < 2$ is periodic function with period 2.

$$\therefore \qquad £\{f(t)\} = \frac{1}{\left(1-e^{-2s}\right)} \int_0^2 e^{-st} f(t)dt$$

$$= \frac{1}{\left(1-e^{-2s}\right)} \int_0^2 t^2 e^{-st} dt$$

$$= \frac{1}{\left(1-e^{-2s}\right)} \left[t^2 \cdot \left(\frac{e^{-st}}{-s}\right) - 2t\left(\frac{e^{-st}}{s^2}\right) + 2.\left(\frac{-e^{-st}}{s^3}\right)\right]_0^2$$

$$= \frac{1}{\left(1-e^{-2s}\right)} \left[\frac{-4e^{-2s}}{s} - \frac{4e^{-2s}}{s^2} - \frac{2e^{-2s}}{s^3} + \frac{2}{s^3}\right]$$

$$= \frac{-2e^{-2s}}{s^3\left(1-e^{-2s}\right)} \left(2s^2+2s+1-e^{2s}\right)$$

$$= \frac{2}{s^3\left(1-e^{2s}\right)} \left(2s^2+2s+1-e^{2s}\right);\ s>0$$

Example 2.23: Find the Laplace transform of the rectified semiwave function defined by

$$f(t) = \begin{cases} \sin wt\ , & 0 < t \le \pi/w \\ 0\ , & \dfrac{\pi}{w} < t < \dfrac{2\pi}{w} \end{cases}$$

when $f(t)$ is periodic with period $\dfrac{2\pi}{w}$.

Solution: $f(t)$ is periodic with period $\dfrac{2\pi}{w}$.

$$\therefore \qquad £\{f(t)\} = \frac{1}{\left(1-e^{-2\pi s/w}\right)} \int_0^{\frac{2\pi}{w}} e^{-st} f(t)dt$$

$$= \frac{1}{\left(1-e^{-2\pi s/w}\right)}\int_0^{\frac{\pi}{w}} e^{-st}\sin wt\,dt$$

$$= \frac{1}{\left(1-e^{-2\pi s/w}\right)}\left[\frac{e^{-st}}{s^2+w^2}(-s\sin wt - w\cos wt)\right]_0^{\pi/w}$$

$$= \frac{1}{\left(1-e^{-2\pi s/w}\right)}\left[\frac{e^{-\pi s/w}}{s^2+w^2}w + \frac{w}{(s^2+w^2)}\right]$$

$$= \frac{w}{\left(1-e^{-2\pi s/w}\right)}\frac{(1+e^{-\pi s/w})}{(s^2+w^2)} = \frac{w}{(s^2+w^2)(1+e^{-\pi s/w})}\,;\ s>0$$

Example 2.24: Plot the 2π-periodic function $f(t)$ given by

$$f(t) = \begin{cases} t & ;\quad 0<t<\pi \\ \pi - t & ;\quad \pi<t<2\pi \end{cases}$$

and find its Laplace transform.

Solution: Graph of the given 2π-periodic function is shown hereunder.

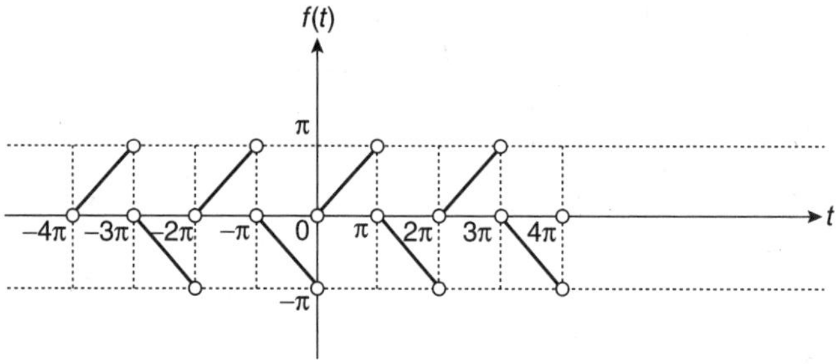

Figure 2.7

○ indicates that the point is not in the graph.

$$\pounds\{f(t)\} = \frac{1}{\left(1-e^{-2\pi s}\right)}\int_0^{2\pi} e^{-st} f(t)\,dt$$

$$= \frac{1}{\left(1-e^{-2\pi s}\right)}\left[\int_0^{\pi} te^{-st}dt + \int_{\pi}^{2\pi}(\pi - t)e^{-st}dt\right]$$

$$= \frac{1}{\left(1-e^{-2\pi s}\right)}\left[\int_0^{\pi} te^{-st}dt - \int_0^{\pi} xe^{-s(\pi+x)}dx\right] \quad \text{(taking } t=\pi+x \text{ in the second integral)}$$

$$= \frac{1}{\left(1-e^{-2\pi s}\right)}\left[\int_0^{\pi} te^{-st}dt - e^{-\pi s}\int_0^{\pi} te^{-st}dt\right]$$

$$= \frac{\left(1-e^{-\pi s}\right)}{\left(1-e^{-2\pi s}\right)}\int_0^{\pi} te^{-st}\,dt = \frac{1}{\left(1+e^{-\pi s}\right)}\left[t\cdot\left(\frac{e^{-st}}{-s}\right)-1.\left(\frac{e^{-st}}{s^2}\right)\right]_0^{\pi}$$

$$= \frac{1}{\left(1+e^{-\pi s}\right)}\left[-\frac{\pi}{s}e^{-\pi s}+\frac{1}{s^2}\left(1-e^{-\pi s}\right)\right];\ s>0$$

Example 2.25: Find the Laplace transform of the triangular wave function of period $2c$ given by

$$f(t)=\begin{cases} t & ;\quad 0<t<c \\ 2c-t & ;\quad c<t<2c \end{cases}$$

Solution: $f(t)$ is periodic function of period $2c$.

$$\therefore\quad £\{f(t)\} = \frac{1}{\left(1-e^{-2cs}\right)}\int_0^{2c} e^{-st} f(t)\,dt$$

$$= \frac{1}{\left(1-e^{-2cs}\right)}\left[\int_0^{c} te^{-st}\,dt + \int_c^{2c}(2c-t)e^{-st}\,dt\right]$$

$$= \frac{1}{\left(1-e^{-2cs}\right)}\left[\left\{t\cdot\left(\frac{e^{-st}}{-s}\right)-1\cdot\left(\frac{e^{-st}}{s^2}\right)\right\}_0^{c}+\left\{(2c-t)\left(\frac{e^{-st}}{-s}\right)-(-1)\left(\frac{e^{-st}}{s^2}\right)\right\}_c^{2c}\right]$$

$$= \frac{1}{\left(1-e^{-2cs}\right)}\left[\frac{-c}{s}e^{-cs}+\frac{1}{s^2}\left(1-e^{-cs}\right)+\frac{c}{s}e^{-cs}+\frac{1}{s^2}\left(e^{-2cs}-e^{-cs}\right)\right]$$

$$= \frac{\left(1-e^{-cs}\right)^2}{s^2\left(1-e^{-2cs}\right)} = \frac{1}{s^2}\left(\frac{1-e^{-cs}}{1+e^{-cs}}\right) = \frac{1}{s^2}\left(\frac{e^{\frac{cs}{2}}-e^{-\frac{cs}{2}}}{e^{\frac{cs}{2}}+e^{-\frac{cs}{2}}}\right) = \frac{1}{s^2}\tanh\left(\frac{cs}{2}\right);\ s>0$$

EXERCISE 2.1

1. Find the Laplace transform of the following functions
   (i) $1+2e^{-4t}+9\cosh 4t+7t^3+5\sin 2t$
   (ii) $\sin 2t\sin 3t$
   (iii) $\sin^3 2t$
   (iv) $\cos^2 3t$
   (v) $\cos(at+b)$
   (vi) $\cos\sqrt{t}$
2. Find the Laplace transforms of
   (i) $e^{-t}\left(\sqrt{t}+\frac{1}{\sqrt{t}}\right)$
   (ii) $e^{-3t}\cosh 4t\sin 3t$
   (iii) $(1+te^{-t})^3$
   (iv) $e^{-2t}(3\cos 4t-2\sin 5t)$
   (v) $2e^{t}\sin 4t\cos 2t$
3. Show that $£\{t\sin at\} = \frac{2as}{\left(s^2+a^2\right)^2}$ and
   $£\{t\cos at\} = \frac{s^2-a^2}{\left(s^2+a^2\right)^2}$
4. Find the Laplace transforms of
   (i) $te^{-t}\sin 2t$
   (ii) $t^2e^{t}\sin 4t$
   (iii) $t\sin 2t\cosh t$

(iv) t^3e^{-3t}
(v) $t\cos^3 t$
(vi) $te^{2t}(\cos t - \sin t)$

5. If $£\{t \sin at\} = \dfrac{2as}{(s^2+a^2)^2}$, evaluate
(i) $£\{at\cos at + \sin at\}$
(ii) $£\{2\cos at - at\sin at\}$

6. Find the Laplace transform of the following functions
(i) $\dfrac{e^{-at}-e^{-bt}}{t}$
(ii) $\dfrac{\sin^3 t}{t}$
(iii) $\dfrac{e^{-2t}\sin 2t\cosh t}{t}$
(iv) $\dfrac{\sin^2 t}{t^2}$

7. Find the Laplace transform of the following functions
(i) $\int_0^t e^{-t}t^4\,dt$
(ii) $\int_0^t t\cosh t\,dt$
(iii) $\int_0^t \cos^2 x\,dx$
(iv) $e^{-4t}\int_0^t t\sin 3t\,dt$

8. Evaluate $\int_0^\infty \left(\dfrac{\sin 2t+\sin 3t}{te^t}\right)dt$ using Laplace transform.

9. Using Laplace transform, evaluate
(i) $\int_0^\infty \dfrac{e^{-t}\sin^2 t}{t}\,dt$
(ii) $\int_0^\infty e^{-4t}\cosh^3 t\,dt$

10. Using Laplace transform show that

$$\int_{t=0}^{\infty}\int_{u=0}^{t}\frac{e^{-t}\sin u}{u}\,du\,dt=\frac{\pi}{4}$$

11. Find the Laplace transform of the following functions
(i) $f(t)=\begin{cases} e^t & \text{for } 0<t<1 \\ 0 & \text{for } t>1\end{cases}$
(ii) $f(t)=\begin{cases} \dfrac{t}{T} & \text{for } 0<t<T \\ 1 & \text{for } t>T\end{cases}$
(iii) $f(t)=\begin{cases} (t-1)^3, & t>1 \\ 0 & , t<1\end{cases}$
(iv) $f(t)=\begin{cases} \sin(t-\alpha) & , t>\alpha \\ 0 & , t<\alpha\end{cases}$

12. Write the given function in terms of Heaviside's unit step function and hence find $£\{f(t)\}$.
(i) $f(t)=\begin{cases} e^{-t} & , 0<t<3 \\ 0 & , t>3\end{cases}$
(ii) $f(t)=\begin{cases} 1 , & 0\le t<1 \\ t , & 1\le t<2 \\ t^2, & 2\le t<\infty\end{cases}$

13. Find the Laplace transforms of
(i) $\dfrac{\delta(t-\pi)}{t}$
(ii) $e^{-\pi t}\delta(t-a)$
(iii) $\sin 2t\ \delta\left(t-\dfrac{\pi}{4}\right)-t^2\ \delta(t-2)$
(iv) $t^2\ U(t-2)-\cosh t\ \delta(t-2)$
(v) $e^{-t}\sin t\ u_\pi(t)$
(vi) $t\ U(t-4)-t^3\ \delta(t-2)$

14. Evaluate $\int_0^\infty e^{-4t}\ \delta(t-3)\,dt$

15. For the periodic function $f(t)$ of period 4, defined by

$$f(t)=\begin{cases} 3t, & 0<t<2 \\ 6, & 2<t<4\end{cases},\ \text{find } £\{f(t)\}$$

16. Find the Laplace transform of the square wave function of period 'a' defined as

$$f(t)=\begin{cases}1 & \text{for } 0<t<a/2\\ -1 & \text{for } a/2<t<a\end{cases}$$

17. Find the Laplace transform of the periodic function

$$f(t)=\frac{kt}{T} \text{ for } 0<t<T,\ f(t+T)=f(t)$$

18. Find the Laplace transform of full rectified sine wave defined by the expression

$$f(t)=\begin{cases}\sin t & ;\ 0<t\le\pi\\ -\sin t & ;\ \pi<t<2\pi\end{cases}$$

$f(t)=f(t+\pi)$. Also, draw its graph.

19. Find the Laplace transform of the periodic function defined by the triangular wave

$$f(t)=\begin{cases}t/a & ,\ 0\le t\le a\\ (2a-t)/a & ,\ a\le t\le 2a\end{cases}$$

and $f(t+2a)=f(t)$.

ANSWERS 2.1

1. (i) $\frac{1}{s}+\frac{2}{s+4}+\frac{9s}{s^2-16}+\frac{42}{s^4}+\frac{10}{s^2+4};\ s>4$ (ii) $\frac{12s}{(s^2+1)(s^2+25)};\ s>0$

(iii) $\frac{48}{(s^2+4)(s^2+36)};\ s>0$ (iv) $\frac{(s^2+18)}{s(s^2+36)};\ s>0$

(v) $\frac{s\cos b-a\sin b}{s^2+a^2};\ s>0$ (vi) $\sum_{n=0}^{\infty}\frac{(-1)^n}{(2n)!}\frac{n!}{s^{n+1}};\ s>0$

2. (i) $\frac{\sqrt{\pi}}{2(s+1)^{3/2}}+\sqrt{\frac{\pi}{s+1}};\ s>-1$ (ii) $\frac{3(s^2+6s+34)}{(s^2-2s+10)(s^2+14s+58)};\ s>1$

(iii) $\frac{1}{s}+\frac{3}{(s+1)^2}+\frac{6}{(s+2)^3}+\frac{6}{(s+3)^4};\ s>0$ (iv) $\frac{3(s+2)}{\left(s^2+4s+20\right)}-\frac{10}{\left(s^2+4s+29\right)};\ s>-2$

(v) $\frac{6}{s^2-2s+37}+\frac{2}{s^2-2s+5};\ s>1$

4. (i) $\frac{4(s+1)}{(s^2+2s+5)^2};\ s>-1$ (ii) $\frac{8(3s^2-6s-13)}{\left(s^2-2s+17\right)^3};\ s>1$

(iii) $\frac{2(s-1)}{\left(s^2-2s+5\right)^2}+\frac{2(s+1)}{\left(s^2+2s+5\right)^2};\ s>1$ (iv) $\frac{6}{\left(s+3\right)^4};\ s>-3$

(v) $\frac{1}{4}\left[\frac{-s^2+9}{\left(s^2+9\right)^2}+\frac{s^2+3}{\left(s^2+1\right)^2}\right];\ s>0$ (vi) $\frac{s^2-6s+7}{\left(s^2-4s+5\right)^2};\ s>2$

5. (i) $\dfrac{2as^2}{\left(s^2+a^2\right)^2}$ (ii) $\dfrac{2s^3}{\left(s^2+a^2\right)^2}$

6. (i) $\log\dfrac{s+b}{s+a}$; $s>-a, s>-b$ (ii) $\dfrac{1}{4}\left(3\cot^{-1}s-\cot^{-1}\dfrac{s}{3}\right)$; $s>0$

(iii) $\dfrac{1}{2}\left\{\cot^{-1}\left(\dfrac{s+1}{2}\right)+\cot^{-1}\left(\dfrac{s+3}{2}\right)\right\}$; $s>-1$

(iv) $\dfrac{1}{4}\left\{-s\log\dfrac{s^2+4}{s^2}+4\cot^{-1}\dfrac{s}{2}\right\}$; $s>0$

7. (i) $\dfrac{24}{s(s+1)^5}$; $s>0$ (ii) $\dfrac{s^2+1}{s(s^2-1)^2}$; $s>1$

(iii) $\dfrac{s^2+2}{s^2(s^2+4)}$; $s>0$ (iv) $\dfrac{6}{(s^2+8s+25)^2}$; $s>0$

8. $3\pi/4$

9. (i) $\dfrac{1}{4}\log 5$ (ii) $\dfrac{12}{35}$

11. (i) $\dfrac{e^{1-s}-1}{1-s}$ (ii) $\dfrac{1-e^{-ST}}{Ts^2}$ (iii) $\dfrac{6e^{-S}}{s^4}$ (iv) $\dfrac{e^{-\alpha s}}{s^2+1}$

12. (i) $\dfrac{1-e^{-3(s+1)}}{s+1}$; $s>-1$ (ii) $\dfrac{1}{s}+\dfrac{2}{s}e^{-2s}+\dfrac{e^{-s}}{s^2}+\dfrac{3}{s^2}e^{-2s}+\dfrac{2}{s^3}e^{-2s}$; $s>0$

13. (i) $\dfrac{e^{-s\pi}}{\pi}$ (ii) $e^{-a(s+\pi)}$ (iii) $e^{\frac{-\pi s}{4}}-4e^{-2s}$

(iv) $e^{-2s}\left(\dfrac{2}{s^3}+\dfrac{4}{s^2}+\dfrac{4}{s}-\cosh 2\right)$; $s>0$ (v) $\dfrac{-e^{-\pi(s+1)}}{(s^2+2s+2)}$; $s>-1$

(vi) $e^{-4s}\left(\dfrac{1}{s^2}+\dfrac{4}{s}\right)-8e^{-2s}$; $s>0$

14. e^{-12}

15. $\dfrac{3}{s^2(1+e^{-2s})}-\dfrac{6e^{-4s}}{s(1-e^{-4s})}$

16. $\dfrac{1}{s}\tanh\left(\dfrac{as}{4}\right)$, $s>0$

17. $\frac{k}{s^2T} - \frac{ke^{-sT}}{s(1-e^{-sT})}; s > 0$

18. $\frac{1+e^{-\pi s}}{(1-e^{-\pi s})(s^2+1)}; s > 0$

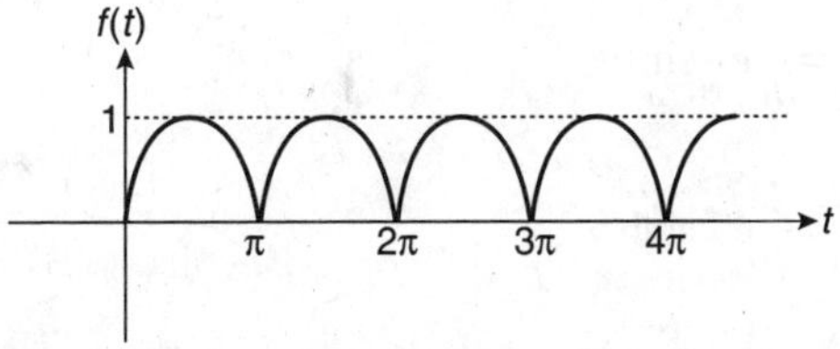

Figure 2.8

19. $\frac{1}{as^2}\tanh\left(\frac{as}{2}\right), s > 0$

2.12 INVERSE LAPLACE TRANSFORM

On the basis of Laplace transform of elementary functions and first shifting property, the following results of inverse Laplace transforms hold

(i) $£^{-1}\left(\frac{1}{s}\right) = 1$

(ii) $£^{-1}\left(\frac{1}{s^n}\right) = \frac{t^{n-1}}{(n-1)!}; n \in N$

(iii) $£^{-1}\left(\frac{1}{s^n}\right) = \frac{t^{n-1}}{\Gamma(n)}; n > 0$

(iv) $£^{-1}\left(\frac{1}{s^2+a^2}\right) = \frac{1}{a}\sin at$

(v) $£^{-1}\left(\frac{1}{s^2-a^2}\right) = \frac{1}{a}\sinh at$

(vi) $£^{-1}\left(\frac{s}{s^2+a^2}\right) = \cos at$

(vii) $£^{-1}\left(\frac{s}{s^2-a^2}\right) = \cosh at$

(viii) $£^{-1}\left(\frac{1}{s-a}\right) = e^{at}$

(ix) $\pounds^{-1}\left\{\dfrac{1}{(s-a)^n}\right\}=\dfrac{e^{at}t^{n-1}}{(n-1)!};\ n\in N$

(x) $\pounds^{-1}\left\{\dfrac{1}{(s-a)^n}\right\}=e^{at}\dfrac{t^{n-1}}{\Gamma(n)};\ n>0$

(xi) $\pounds^{-1}\left\{\dfrac{1}{(s-a)^2+b^2}\right\}=\dfrac{1}{b}e^{at}\sin bt$

(xii) $\pounds^{-1}\left\{\dfrac{1}{(s-a)^2-b^2}\right\}=\dfrac{1}{b}e^{at}\sinh bt$

(xiii) $\pounds^{-1}\left\{\dfrac{s-a}{(s-a)^2+b^2}\right\}=e^{at}\cos bt$

(xiv) $\pounds^{-1}\left\{\dfrac{s-a}{(s-a)^2-b^2}\right\}=e^{at}\cosh bt$

2.13 USE OF PARTIAL FRACTIONS TO FIND INVERSE LAPLACE TRANSFORM

After resolving the given function $F(s)$ into partial fractions and using above formulae, we can find inverse Laplace transform of $F(s)$, suppression method may be used to write partial fractions.

Example 2.26: Find the inverse Laplace transform of

(i) $\dfrac{3s+5\sqrt{2}}{s^2+8}$ (ii) $\dfrac{6}{2s-3}-\dfrac{3+4s}{9s^2-16}+\dfrac{8-6s}{16s^2+9}$

(iii) $\dfrac{s+4}{s(s-1)(s^2+4)}$ (iv) $\dfrac{s^2+2s-4}{(s^2+2s+5)(s^2+2s+2)}$ (v) $\dfrac{s}{s^4+4a^4}$

Solution:

(i) $$\pounds^{-1}\left\{\frac{3s+5\sqrt{2}}{s^2+8}\right\}=3\pounds^{-1}\left\{\frac{s}{s^2+\left(2\sqrt{2}\right)^2}\right\}+\frac{5}{2}\pounds^{-1}\left\{\frac{2\sqrt{2}}{s^2+\left(2\sqrt{2}\right)^2}\right\}$$

$$=3\cos 2\sqrt{2}t+\frac{5}{2}\sin 2\sqrt{2}t$$

(ii) $£^{-1}\left\{\frac{6}{2s-3}-\frac{3+4s}{9s^2-16}+\frac{8-6s}{16s^2+9}\right\}$

$$=£^{-1}\left\{\frac{3}{s-\frac{3}{2}}\right\}-\frac{1}{4}£^{-1}\left\{\frac{4/3}{s^2-\left(\frac{4}{3}\right)^2}\right\}-\frac{4}{9}£^{-1}\left\{\frac{s}{s^2-\left(\frac{4}{3}\right)^2}\right\}+\frac{2}{3}£^{-1}\left\{\frac{3/4}{s^2+\left(\frac{3}{4}\right)^2}\right\}$$

$$-\frac{3}{8}£^{-1}\left\{\frac{s}{s^2+\left(\frac{3}{4}\right)^2}\right\}$$

$$=3e^{\frac{3}{2}t}-\frac{1}{4}\sinh\frac{4t}{3}-\frac{4}{9}\cosh\frac{4t}{3}+\frac{2}{3}\sin\frac{3t}{4}-\frac{3}{8}\cos\frac{3t}{4}$$

(iii) By suppression method

Let $\frac{s+4}{s(s-1)\left(s^2+4\right)}\equiv\frac{(0+4)}{s(0-1)(0+4)}+\frac{1+4}{(s-1)1.(1+4)}+\frac{As+B}{s^2+4}$

$$\equiv\frac{-1}{s}+\frac{1}{s-1}+\frac{As+B}{s^2+4}$$

$\therefore$ $s+4\equiv-(s-1)\left(s^2+4\right)+s\left(s^2+4\right)+(As+B)\,s\,(s-1)$

Equate coefficients of s^3 and s^2

$$\left.\begin{aligned}0&=-1+1+A\\0&=1-A+B\end{aligned}\right\}\Rightarrow A=0,\ B=A-1=-1$$

$\therefore$ $\frac{s+4}{s(s-1)\left(s^2+4\right)}=-\frac{1}{s}+\frac{1}{s-1}-\frac{1}{s^2+4}=-\frac{1}{s}+\frac{1}{s-1}-\frac{1}{2}\frac{2}{s^2+2^2}$

$\therefore$ $£^{-1}\left\{\frac{s+4}{s(s-1)\left(s^2+4\right)}\right\}=-1+e^t-\frac{1}{2}\sin 2t$

(iv) $\frac{s^2+2s-4}{(s^2+2s+5)\left(s^2+2s+2\right)}=\frac{s^2+2s+5-9}{(s^2+2s+5)\left(s^2+2s+2\right)}$

$$=\frac{1}{s^2+2s+2}-3\left(\frac{1}{s^2+2s+2}-\frac{1}{s^2+2s+5}\right)$$

$$=\frac{3}{s^2+2s+5}-\frac{2}{s^2+2s+2}$$

$$=\frac{3}{2}\ \frac{2}{(s+1)^2+2^2}-2\cdot\frac{1}{(s+1)^2+1^2}$$

$\therefore$ $£^{-1}\left\{\frac{s^2+2s-4}{\left(s^2+2s+5\right)\left(s^2+2s+2\right)}\right\}=\frac{1}{2}e^{-t}\left(3\sin 2t-4\sin t\right)$

(v) $$\frac{s}{(s^4+4a^4)} = \frac{s}{(s^2+2a^2)^2-(2as)^2} = \frac{s}{(s^2-2as+2a^2)(s^2+2as+2a^2)}$$

$$= \frac{1}{4a}\left[\frac{1}{s^2-2as+2a^2} - \frac{1}{s^2+2as+2a^2}\right]$$

$$= \frac{1}{4a^2}\left[\frac{a}{(s-a)^2+a^2} - \frac{a}{(s+a)^2+a^2}\right]$$

$\therefore$ $$£^{-1}\left(\frac{s}{(s^4+4a^4)}\right) = \frac{1}{4a^2}\{e^{at}\sin at - e^{-at}\sin at\}$$

$$= \frac{1}{2a^2}\left(\frac{e^{at}-e^{-at}}{2}\right)\sin at = \frac{1}{2a^2}\sinh at \sin at$$

Example 2.27: Find the inverse Laplace transforms of

(i) $\dfrac{1}{(s^2-5s+6)}$ (ii) $\dfrac{(2s-3)}{(s^2+4s+13)}$ (iii) $\dfrac{1}{s(s+3)^2}$ (iv) $\dfrac{1}{s(s^2+9)}$

(v) $\dfrac{1}{s^2(s^2+4)}$ (vi) $\dfrac{1}{(s^3+4s^2)}$ (vii) $\dfrac{1}{(s^2-4s+8)}$ (viii) $\dfrac{4}{(s^2-s+2)}$

(ix) $\dfrac{3s-1}{(s-2)^2}$ (x) $\dfrac{6+s}{(s^2+6s+13)}$ (xi) $\dfrac{s^2}{(s^4-a^4)}$

Solution:

(i) $$\frac{1}{(s^2-5s+6)} = \frac{1}{(s-2)(s-3)} = \frac{1}{s-3} - \frac{1}{s-2}$$

$\therefore$ $$£^{-1}\left(\frac{1}{s^2-5s+6}\right) = £^{-1}\left\{\frac{1}{s-3}\right\} - £^{-1}\left\{\frac{1}{s-2}\right\} = e^{3t} - e^{2t}$$

(ii) $$£^{-1}\left\{\frac{2s-3}{s^2+4s+13}\right\} = £^{-1}\left\{\frac{2(s+2)-7}{(s+2)^2+3^2}\right\} = 2£^{-1}\left\{\frac{(s+2)}{(s+2)^2+3^2}\right\} - \frac{7}{3}£^{-1}\left\{\frac{3}{(s+2)^2+3^2}\right\}$$

$$= 2e^{-2t}\cos 3t - \frac{7}{3}e^{-2t}\sin 3t$$

$$= \frac{e^{-2t}}{3}(6\cos 3t - 7\sin 3t)$$

(iii) By suppression method

Let $$\frac{1}{s(s+3)^2} \equiv \frac{1}{9s} - \frac{1}{3(s+3)^2} + \frac{A}{s+3}$$

$\therefore$ $$1 \equiv \frac{1}{9}(s+3)^2 - \frac{1}{3}s + As(s+3)$$

Equate coefficient of s^2

$$0 = \frac{1}{9} + A \Rightarrow A = -\frac{1}{9}$$

$\therefore$ $$\frac{1}{s(s+3)^2} = \frac{1}{9s} - \frac{1}{9(s+3)} - \frac{1}{3(s+3)^2}$$

$\therefore$ $$£^{-1}\left\{\frac{1}{s(s+3)^2}\right\} = \frac{1}{9} - \frac{1}{9}e^{-3t} - \frac{1}{3}te^{-3t} = \frac{1}{9}\left(1 - e^{-3t} - 3te^{-3t}\right)$$

(iv) $$\frac{1}{s(s^2+9)} = \frac{s}{s^2(s^2+9)} = \frac{s}{9}\left[\frac{1}{s^2} - \frac{1}{s^2+9}\right] = \frac{1}{9}\left[\frac{1}{s} - \frac{s}{s^2+3^2}\right]$$

$\therefore$ $$£^{-1}\left\{\frac{1}{s(s^2+9)}\right\} = \frac{1}{9}(1 - \cos 3t)$$

(v) $$\frac{1}{s^2(s^2+4)} = \frac{1}{4}\left(\frac{1}{s^2} - \frac{1}{s^2+4}\right) = \frac{1}{4s^2} - \frac{1}{8}\left[\frac{2}{s^2+2^2}\right]$$

$\therefore$ $$£^{-1}\left\{\frac{1}{s^2(s^2+4)}\right\} = \frac{1}{4}t - \frac{1}{8}\sin 2t = \frac{1}{8}(2t - \sin 2t)$$

(vi) By suppression method

Let $$\frac{1}{s^3+4s^2} = \frac{1}{s^2(s+4)} \equiv \frac{1}{4s^2} + \frac{1}{16(s+4)} + \frac{A}{s}$$

$\therefore$ $$1 \equiv \frac{1}{4}(s+4) + \frac{1}{16}s^2 + As(s+4)$$

Equate coefficient of s^2

$$0 = \frac{1}{16} + A \Rightarrow A = -\frac{1}{16}$$

$\therefore$ $$\frac{1}{s^3+4s^2} = \frac{1}{4s^2} + \frac{1}{16(s+4)} - \frac{1}{16s}$$

$\therefore$ $$£^{-1}\left\{\frac{1}{s^3+4s^2}\right\} = \frac{1}{4}t + \frac{1}{16}e^{-4t} - \frac{1}{16} = \frac{1}{16}\left(4t + e^{-4t} - 1\right)$$

(vii) $$\frac{1}{s^2-4s+8}=\frac{1}{(s-2)^2+2^2}=\frac{1}{2}\left\{\frac{2}{(s-2)^2+2^2}\right\}$$

$\therefore$ $$£^{-1}\left(\frac{1}{s^2-4s+8}\right)=\frac{1}{2}e^{2t}\sin 2t$$

(viii) $$\frac{4}{(s^2-s+2)}=\frac{8}{\sqrt{7}}\frac{\frac{\sqrt{7}}{2}}{\left(s-\frac{1}{2}\right)^2+\left(\frac{\sqrt{7}}{2}\right)^2}$$

$\therefore$ $$£^{-1}\left(\frac{4}{s^2-s+2}\right)=\frac{8}{\sqrt{7}}e^{t/2}\sin\left(\frac{\sqrt{7}}{2}t\right)$$

(ix) $$\frac{3s-1}{(s-2)^2}=\frac{3(s-2)+5}{(s-2)^2}=\frac{3}{s-2}+\frac{5}{(s-2)^2}$$

$\therefore$ $$£^{-1}\left[\frac{3s-1}{(s-2)^2}\right]=(3+5t)e^{2t}$$

(x) $$\frac{6+s}{(s^2+6s+13)}=\frac{(s+3)+3}{(s+3)^2+2^2}=\frac{s+3}{(s+3)^2+2^2}+\frac{3}{2}\cdot\frac{2}{(s+3)^2+2^2}$$

$\therefore$ $$£^{-1}\left\{\frac{6+s}{s^2+6s+13}\right\}=e^{-3t}\left(\cos 2t+\frac{3}{2}\sin 2t\right)=\frac{1}{2}e^{-3t}(2\cos 2t+3\sin 2t)$$

(xi) $$\frac{s^2}{(s^4-a^4)}=\frac{s^2}{(s^2+a^2)(s^2-a^2)}=\frac{1}{2a}\left[\frac{a}{s^2-a^2}+\frac{a}{s^2+a^2}\right]$$

$\therefore$ $$£^{-1}\left(\frac{s^2}{s^4-a^4}\right)=\frac{1}{2a}(\sinh at+\sin at)$$

Example 2.28: Find the inverse Laplace transform of $\frac{1}{(s^2+1)(s^2+9)}$ and hence of the function $\frac{s}{(s^2+1)(s^2+9)}$.

Solution: $$\frac{1}{(s^2+1)(s^2+9)}=\frac{1}{8}\left(\frac{1}{s^2+1}-\frac{1}{3}\frac{3}{s^2+3^2}\right)$$ (by suppression method)

$$\therefore \qquad £^{-1}\left\{\frac{1}{(s^2+1)(s^2+9)}\right\}=\frac{1}{8}\left(\sin t-\frac{1}{3}\sin 3t\right)=\frac{1}{24}(3\sin t-\sin 3t)$$

$$=\frac{1}{6}\sin^3 t \qquad (\because \sin 3t = 3\sin t - 4\sin^3 t)$$

We know that

If £ $\{f(t)\} = F(s)$ then £ $\{f'(t)\} = sF(s)-f(0)$

Taking
$$f(t)=\frac{1}{6}\sin^3 t$$

$$f'(t)=\frac{1}{2}\sin^2 t\cos t \text{ and } f(0)=0$$

Also,
$$F(s)=\frac{1}{(s^2+1)(s^2+9)}$$

$$\therefore \qquad sF(s)=\frac{s}{(s^2+1)(s^2+9)}=£\left(\frac{1}{2}\sin^2 t\cos t\right)$$

$$\Rightarrow \qquad £^{-1}\left\{\frac{s}{(s^2+1)(s^2+9)}\right\}=\frac{1}{2}\sin^2 t\cos t=\frac{1}{4}\sin t\sin 2t$$

Example 2.29: Evaluate

(i) $£^{-1}\left\{\frac{s^2}{(s^2-a^2)(s^2-b^2)(s^2-c^2)}\right\}$ (ii) $£^{-1}\left\{\frac{s}{(s^2+a^2)^2}\right\}$ (iii) $£^{-1}\left\{\frac{1}{(s^2+a^2)^2}\right\}$

(iv) $£^{-1}\left\{\frac{1}{(s^2-9)^2}\right\}$ (v) $£^{-1}\left\{\frac{s}{(s^2-9)^2}\right\}$

Solution: (i) By suppression method

$$\frac{s^2}{(s^2-a^2)(s^2-b^2)(s^2-c^2)}=\frac{a^2}{(a^2-b^2)(a^2-c^2)(s^2-a^2)}+\frac{b^2}{(b^2-a^2)(b^2-c^2)(s^2-b^2)}$$
$$+\frac{c^2}{(c^2-a^2)(c^2-b^2)(s^2-c^2)}$$

$$\therefore \quad £^{-1}\left\{\frac{s^2}{(s^2-a^2)(s^2-b^2)(s^2-c^2)}\right\}=\frac{a\sinh at}{(a^2-b^2)(a^2-c^2)}+\frac{b\sinh bt}{(b^2-a^2)(b^2-c^2)}+\frac{c\sinh ct}{(c^2-a^2)(c^2-b^2)}$$

(ii)
$$\frac{s}{(s^2+a^2)^2}=\frac{s}{(s-ia)^2(s+ia)^2}=\frac{1}{4ia}\left[\frac{1}{(s-ia)^2}-\frac{1}{(s+ia)^2}\right]$$

$$\therefore \quad £^{-1}\left\{\frac{s}{\left(s^2+a^2\right)^2}\right\}=\frac{1}{4ia}\left[te^{iat}-te^{-iat}\right]=\frac{t}{2a}\left(\frac{e^{iat}-e^{-iat}}{2i}\right)=\frac{t}{2a}\sin at$$

(iii) $$\frac{1}{\left(s^2+a^2\right)^2}=\left\{\frac{1}{(s-ia)(s+ia)}\right\}^2=\left\{\frac{1}{2ia}\left(\frac{1}{s-ia}-\frac{1}{s+ia}\right)\right\}^2$$

$$=-\frac{1}{4a^2}\left\{\frac{1}{(s-ia)^2}+\frac{1}{(s+ia)^2}-\frac{2}{s^2+a^2}\right\}$$

$$\therefore \quad £^{-1}\left\{\frac{1}{\left(s^2+a^2\right)^2}\right\}=-\frac{1}{4a^2}\left\{te^{iat}+te^{-iat}-\frac{2}{a}\sin at\right\}$$

$$=\frac{1}{2a^3}\left\{\sin at-at\left(\frac{e^{iat}+e^{-iat}}{2}\right)\right\}$$

$$=\frac{1}{2a^3}\{\sin at-at\cos at\}$$

(iv) $$\frac{1}{\left(s^2-9\right)^2}=\left\{\frac{1}{(s-3)(s+3)}\right\}^2=\left\{\frac{1}{6}\left(\frac{1}{s-3}-\frac{1}{s+3}\right)\right\}^2$$

$$=\frac{1}{36}\left\{\frac{1}{(s-3)^2}+\frac{1}{(s+3)^2}-\frac{2}{s^2-9}\right\}$$

$$\therefore \quad £^{-1}\left\{\frac{1}{\left(s^2-9\right)^2}\right\}=\frac{1}{36}\left\{te^{3t}+te^{-3t}-\frac{2}{3}\sinh 3t\right\}$$

$$=\frac{1}{36}\left\{2t\left(\frac{e^{3t}+e^{-3t}}{2}\right)-\frac{2}{3}\sinh 3t\right\}$$

$$=\frac{1}{54}(3t\cosh 3t-\sinh 3t)$$

(v) $$\frac{s}{\left(s^2-9\right)^2}=\frac{s}{(s-3)^2(s+3)^2}=\frac{1}{12}\left\{\frac{1}{(s-3)^2}-\frac{1}{(s+3)^2}\right\}$$

$$\therefore \quad £^{-1}\left\{\frac{s}{\left(s^2-9\right)^2}\right\}=\frac{1}{12}\left\{te^{3t}-te^{-3t}\right\}=\frac{t}{6}\sinh 3t$$

Example 2.30: Show that

$$£^{-1}\left(\frac{1}{s}\sin\frac{1}{s}\right)=t-\frac{t^3}{(3!)^2}+\frac{t^5}{(5!)^2}-\frac{t^7}{(7!)^2}+\cdots\cdots\cdots\cdots$$

Solution: We have $\frac{1}{s}\sin\frac{1}{s}=\frac{1}{s}\left(\frac{1}{s}-\frac{1}{3!s^3}+\frac{1}{5!s^5}-\frac{1}{7!s^7}+\ldots\ldots\right)$

$$=\frac{1}{s^2}-\frac{1}{3!s^4}+\frac{1}{5!s^6}-\frac{1}{7!s^8}+\ldots..$$

$$\therefore \quad £^{-1}\left(\frac{1}{s}\sin\frac{1}{s}\right)=t-\frac{t^3}{(3!)^2}+\frac{t^5}{(5!)^2}-\frac{t^7}{(7!)^2}.+\ldots.\left(\because £^{-1}\left(\frac{1}{s^n}\right)=\frac{t^{n-1}}{(n-1)!}\right)$$

Example 2.31: Find the inverse Laplace transform of the following functions

(i) $\log\frac{s+a}{s+b}$ (ii) $\cot^{-1}\frac{s+3}{2}$ (iii) $\log\frac{s^2+9}{s(s+2)}$

(iv) $\frac{1}{s}\tan^{-1}\frac{a}{s}$; $s>0$ (v) $\frac{1}{s\sqrt{s+4}}$ (vi) $\frac{1}{s^3(s^2+1)}$

Solution:

(i) Let $£^{-1}\left\{\log\frac{s+a}{s+b}\right\}=f(t)$

$$\therefore \quad £\{f(t)\}=\log(s+a)-\log(s+b)$$

$$\therefore \quad £\{t\,f(t)\}=-\frac{d}{ds}\{\log(s+a)-\log(s+b)\} \quad \left(\because £\{t\,f(t)\}=-\frac{d}{ds}£\{f(t)\}\right)$$

$$=-\frac{1}{s+a}+\frac{1}{s+b}$$

$$\therefore \quad t\,f(t)=£^{-1}\left(\frac{1}{s+b}-\frac{1}{s+a}\right)=e^{-bt}-e^{-at}$$

$$\therefore \quad f(t)=\frac{e^{-bt}-e^{-at}}{t}$$

(ii) Let $£^{-1}\left(\cot^{-1}\frac{s+3}{2}\right)=f(t)$

$$\therefore \quad £\{f(t)\}=\cot^{-1}\frac{s+3}{2}$$

$$\therefore \quad £\{t\,f(t)\}=\frac{-d}{ds}\cot^{-1}\frac{s+3}{2}=\frac{1}{1+\left(\frac{s+3}{2}\right)^2}\cdot\frac{1}{2}=\frac{2}{(s+3)^2+2^2}$$

$\therefore \quad t\,f(t) = £^{-1}\left\{\dfrac{2}{(s+3)^2+2^2}\right\} = e^{-3t}\sin 2t$

$\therefore \quad f(t) = \dfrac{1}{t}e^{-3t}\sin 2t$

(iii) Let $£^{-1}\left\{\log\dfrac{s^2+9}{s(s+2)}\right\} = f(t)$

$\therefore \quad £\{f(t)\} = \log(s^2+9) - \log s - \log(s+2)$

$\therefore \quad £\{t\ f(t)\} = \dfrac{-d}{ds}£\{f(t)\} = -\dfrac{2s}{(s^2+9)} + \dfrac{1}{s} + \dfrac{1}{s+2}$

$\therefore \quad t\ f(t) = £^{-1}\left(\dfrac{1}{s} + \dfrac{1}{s+2} - 2\cdot\dfrac{s}{s^2+3^2}\right) = \left(1 + e^{-2t} - 2\cos 3t\right)$

$\Rightarrow \quad f(t) = \dfrac{1}{t}\left(1 + e^{-2t} - 2\cos 3t\right)$

(iv) Let $\quad £^{-1}\left(\tan^{-1}\dfrac{a}{s}\right) = f(t)$

$\therefore \quad £\{f(t)\} = \tan^{-1}\dfrac{a}{s}$

$\therefore \quad £\{t\ f(t)\} = \dfrac{-d}{ds}\tan^{-1}\dfrac{a}{s} = -\dfrac{d}{ds}\cot^{-1}\dfrac{s}{a}$

$$= \frac{1}{1+\dfrac{s^2}{a^2}}.\frac{1}{a} = \frac{a}{s^2+a^2}$$

$\therefore \quad t\,f(t) = £^{-1}\left(\dfrac{a}{s^2+a^2}\right) = \sin at$

$\therefore \quad f(t) = \dfrac{\sin at}{t}$

$\therefore \quad £^{-1}\left(\tan^{-1}\dfrac{a}{s}\right) = \dfrac{1}{t}\sin at$

$\therefore \quad £^{-1}\left(\dfrac{1}{s}\tan^{-1}\dfrac{a}{s}\right) = \int_0^t \dfrac{1}{u}\sin au\,du \qquad \left(\because\ £\left(\int_0^t f(u)\,du\right) = \dfrac{1}{s}£\{f(t)\}\right)$

(v) $\mathcal{L}^{-1}\left(\frac{1}{\sqrt{s+4}}\right)=e^{-4t}\mathcal{L}^{-1}\left(\frac{1}{\sqrt{s}}\right)=e^{-4t}\frac{t^{-1/2}}{\Gamma\left(\frac{1}{2}\right)}=\frac{e^{-4t}}{\sqrt{\pi t}}\quad\left(\because \mathcal{L}^{-1}\left(\frac{1}{s^n}\right)=\frac{t^{n-1}}{\Gamma n}\right).$

$$\therefore\quad \mathcal{L}^{-1}\left(\frac{1}{s\sqrt{s+4}}\right)=\int_0^t\frac{e^{-4t}}{\sqrt{\pi t}}dt$$

Put $4t = x^2$ so that $2\sqrt{t} = x$

$$\therefore\quad \frac{1}{\sqrt{t}}dt = dx$$

$$\therefore\quad \mathcal{L}^{-1}\left(\frac{1}{s\sqrt{s+4}}\right)=\int_0^{2\sqrt{t}}\frac{e^{-x^2}}{\sqrt{\pi}}dx=\frac{1}{2}.\frac{2}{\sqrt{\pi}}\int_0^{2\sqrt{t}}e^{-x^2}dx=\frac{1}{2}erf\left(2\sqrt{t}\right)$$

(vi) $\frac{1}{s^3\left(s^2+1\right)}=\frac{1}{s}\left\{\frac{1}{s^2\left(s^2+1\right)}\right\}=\frac{1}{s}\left\{\frac{1}{s^2}-\frac{1}{s^2+1}\right\}$

Now, $\mathcal{L}^{-1}\left\{\frac{1}{s^2}-\frac{1}{s^2+1}\right\}=t-\sin t$

$$\therefore\quad \mathcal{L}^{-1}\left\{\frac{1}{s}\left(\frac{1}{s^2}-\frac{1}{s^2+1}\right)\right\}=\int_0^t(x-\sin x)\,dx\quad\text{(by Laplace transform of integrals)}$$

$$=\left(\frac{x^2}{2}+\cos x\right)_0^t=\frac{t^2}{2}+\cos t-1$$

Example 2.32: Find the inverse Laplace transform of

(i) $\frac{s^2+2s-3}{s(s-3)(s+2)}$ (ii) $\log\frac{s-1}{s}$

Solution:

(i) By suppression method

$$\frac{s^2+2s-3}{s(s-3)(s+2)}=\frac{0+0-3}{s(0-3)(0+2)}+\frac{9+6-3}{3(s-3)(3+2)}+\frac{4-4-3}{(-2)(-2-3)(s+2)}$$

$$=\frac{1}{2s}+\frac{4}{5(s-3)}-\frac{3}{10(s+2)}$$

$$\therefore\quad \mathcal{L}^{-1}\left\{\frac{s^2+2s-3}{s(s-3)(s+2)}\right\}=\frac{1}{2}+\frac{4}{5}e^{3t}-\frac{3}{10}e^{-2t}$$

(ii) Taking $a = -1$ and $b = 0$ in Example 2.31 (i), we get

$$£^{-1}\left\{\log\frac{s-1}{s}\right\} = \frac{e^0 - e^t}{t} = \frac{1-e^t}{t}$$

Example 2.33: Find the inverse Laplace transform of the following functions

(i) $\dfrac{4e^{\frac{-\pi}{2}s}}{(s^2+16)}$ (ii) $\dfrac{(3s+1)}{s^2(s^2+4)}e^{-3s}$ (iii) $\dfrac{e^{-3s}}{(s+2)}$

(iv) $\dfrac{e^{4-3s}}{(s+4)^{5/2}}$ (v) $\dfrac{e^{-2s}}{s^2}$

Solution: (i) We have $L^{-1}\left(\dfrac{4}{s^2+16}\right) = \sin 4t$

$$£^{-1}\left\{e^{\frac{-\pi}{2}s}\frac{4}{s^2+16}\right\} = \sin\left\{4\left(t-\frac{\pi}{2}\right)\right\}u_{\pi/2}(t) \quad \text{(by second shifting theorem)}$$

$$= \sin 4t\, u_{\pi/2}(t)$$

$$\therefore \quad £^{-1}\left\{e^{\frac{-\pi}{2}s}\frac{4}{s^2+16}\right\} = \begin{cases} 0 & ; \quad 0 \le t < \pi/2 \\ \sin 4t & ; \quad t \ge \pi/2 \end{cases}$$

(ii)

$$\frac{3s+1}{s^2(s^2+4)} = \frac{(3s+1)}{4}\left(\frac{1}{s^2} - \frac{1}{s^2+4}\right) = \frac{1}{4}\left(\frac{3}{s} + \frac{1}{s^2} - 3.\frac{s}{s^2+2^2} - \frac{1}{2}.\frac{2}{s^2+2^2}\right)$$

$$\therefore \quad £^{-1}\left\{\frac{3s+1}{s^2(s^2+4)}\right\} = \frac{1}{4}\left(3 + t - 3\cos 2t - \frac{1}{2}\sin 2t\right)$$

$$\therefore \quad £^{-1}\left\{\frac{3s+1}{s^2(s^2+4)}e^{-3s}\right\} = \frac{1}{4}\left[3 + (t-3) - 3\cos\{2(t-3)\} - \frac{1}{2}\sin\{2(t-3)\}\right]u_3(t)$$

(by second shifting theorem)

$$= \frac{1}{4}\left[t - 3\cos(2t-6) - \frac{1}{2}\sin(2t-6)\right].u_3(t)$$

Hence, $£^{-1}\left\{\dfrac{3s+1}{s^2(s^2+4)}e^{-3s}\right\} = \begin{cases} 0 \; ; \quad 0 \le t < 3 \\ \dfrac{1}{4}\left[t - 3\cos(2t-6) - \dfrac{1}{2}\sin(2t-6)\right]; \; t \ge 3 \end{cases}$

(iii) We have $\mathcal{L}^{-1}\left(\frac{1}{s+2}\right) = e^{-2t}$

Hence, $\mathcal{L}^{-1}\left(\frac{1}{s+2}e^{-3s}\right) = e^{-2(t-3)} \cdot u_3(t) = \begin{cases} 0 & \text{if } 0 \le t < 3 \\ e^{-2(t-3)} & \text{if } t \ge 3 \end{cases}$

(iv) We have $\mathcal{L}^{-1}\left\{\frac{1}{(s+4)^{5/2}}\right\} = e^{-4t}\mathcal{L}^{-1}\left\{\frac{1}{s^{5/2}}\right\} = e^{-4t}\frac{t^{3/2}}{\Gamma 5/2} = \frac{e^{-4t}\, t^{3/2}}{\frac{3}{2}\cdot\frac{1}{2}\sqrt{\pi}} = \frac{4}{3\sqrt{\pi}}e^{-4t}\cdot t^{3/2}$

$$\therefore \quad \mathcal{L}^{-1}\left\{\frac{e^{4-3s}}{(s+4)^{5/2}}\right\} = e^4\mathcal{L}^{-1}\left\{\frac{e^{-3s}}{(s+4)^{5/2}}\right\} = e^4\cdot\frac{4}{3\sqrt{\pi}}e^{-4(t-3)}.(t-3)^{3/2}\, u_3(t)$$

$$= \frac{4}{3\sqrt{\pi}}e^{4(4-t)}(t-3)^{3/2}\cdot u_3(t) = \begin{cases} 0 & \text{if } 0 \le t < 3 \\ \frac{4}{3\sqrt{\pi}}e^{4(4-t)}(t-3)^{3/2} & \text{if } t \ge 3 \end{cases}$$

(v) $\mathcal{L}^{-1}\left(\frac{1}{s^2}\right) = t$

$$\therefore \quad \mathcal{L}^{-1}\left(\frac{e^{-2s}}{s^2}\right) = (t-2)u_2(t) = \begin{cases} 0 & ;\ 0 \le t < 2 \\ t-2 & ;\ t \ge 2 \end{cases}$$

2.14 CONVOLUTION THEOREM

We use convolution theorem to find inverse Laplace transform of product of two functions. Firstly, we shall define convolution of two functions.

Convolution

Let $f(t)$ and $g(t)$ be two functions defined in $[0, \infty)$. Then convolution of $f(t)$ and $g(t)$ denoted by $f(t)*g(t)$ or $(f*g)(t)$ is defined by $(f*g)(t) = \int_0^t f(u)g(t-u)\,du;\ t \ge 0$.

It can be easily proved that the convolution satisfies commutative, associative and distributive laws, i.e.,

(i) $f*g = g*f$ (ii) $f*(g*h) = (f*g)*h$

(iii) $f*(g+h) = f*g + f*h$

Theorem 2.14 (convolution theorem)

Let $f(t)$ and $g(t)$ be piecewise continuous functions on $[0, \infty)$ and be of exponential order, $F(s) = \mathcal{L}\{f(t)\}$ and $G(s) = \mathcal{L}\{g(t)\}$. Then, $\mathcal{L}^{-1}\{F(s)\cdot G(s)\} = (f*g)(t) = \int_0^t f(u).g(t-u)\,du$.

Proof: We have $\mathcal{L}\left(\int_0^t f(u)\cdot g(t-u)\,du\right) = \int_0^\infty e^{-st}\int_0^t f(u)\cdot g(t-u)\,du\,dt$

By changing the order of integration

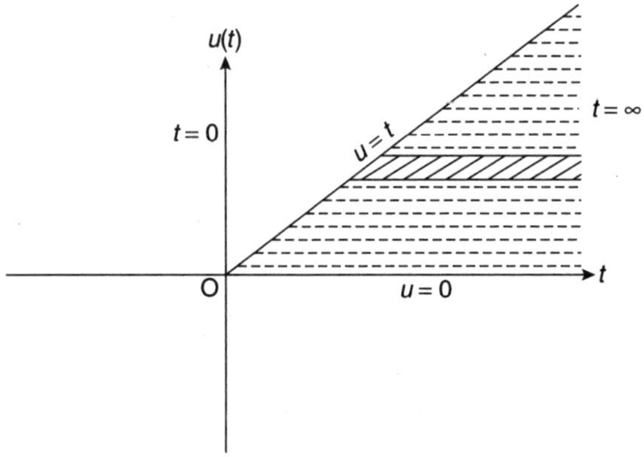

Figure 2.9

$$\pounds(f*g)(t)=\int_0^{\infty}\int_u^{\infty}e^{-st}f(u)\,g\,(t-u)\,dt\,du$$

$$=\int_0^{\infty}\left[\int_u^{\infty}e^{-st}g\,(t-u)\,dt\right]f(u)\,du$$

$$=\int_0^{\infty}\left[\int_0^{\infty}e^{-s(x+u)}\,g(x)\,dx\right]f(u)du \qquad \text{(taking } t-u=x\text{)}$$

$$=\int_0^{\infty}e^{-su}f(u)\,du.\int_0^{\infty}e^{-sx}g(x)\,dx$$

$$=\int_0^{\infty}e^{-st}f(t)\,dt.\int_0^{\infty}e^{-st}g(t)\,dt$$

$$=\pounds\{f(t)\}\cdot\pounds\{g(t)\}=F(s).G(s)$$

$$\therefore \quad \pounds^{-1}\{F(s)\cdot G(s)\}=(f*g)\,(t)=\int_0^{t}f(u)\,g(t-u)\,du$$

Example 2.34: Find the convolution sin at * cos at.

Solution: We have sin at * cos $at = \int_0^t \sin au\ \cos a(t-u)\,du$

$$=\frac{1}{2}\int_0^t 2\sin au\ \cos\,(at-au)\,du$$

$$= \frac{1}{2}\int_0^t [\sin at + \sin(2au - at)]\,du$$

$$= \frac{1}{2}\left[u \sin at - \frac{1}{2a}\cos\,(2au - at)\right]_0^t$$

$$= \frac{1}{2}\left[t \sin at - \frac{1}{2a}\cos at + \frac{1}{2a}\cos at\right]$$

$$= \frac{1}{2}\,t \sin at$$

Example 2.35: Use convolution theorem to find

(i) $£^{-1}\left(\frac{1}{s\sqrt{s+4}}\right)$ (ii) $£^{-1}\left\{\frac{s}{\left(s^2+a^2\right)^2}\right\}$ (iii) $£^{-1}\left\{\frac{1}{\left(s^2+a^2\right)^2}\right\}$

(iv) $£^{-1}\left\{\frac{s^2}{\left(s^2+a^2\right)\left(s^2+b^2\right)}\right\}$ (v) $£^{-1}\left\{\frac{1}{s^3\left(s^2+1\right)}\right\}$ (vi) $£^{-1}\left\{\frac{1}{(s-2)(s+3)}\right\}$

(vii) $£^{-1}\left\{\frac{s}{\left(s^2+1\right)\left(s^2+4\right)}\right\}$ (viii) $£^{-1}\left\{\frac{1}{s(s+1)(s+2)}\right\}$ (ix) $£^{-1}\left\{\frac{s^2}{\left(s^2+w^2\right)^2}\right\}$

Solution: (i) We have $£^{-1}\left(\frac{1}{s}\right) = 1,\ £^{-1}\left(\frac{1}{\sqrt{s+4}}\right) = e^{-4t}\cdot\frac{t^{-1/2}}{\Gamma\left(\frac{1}{2}\right)} = \frac{e^{-4t}}{\sqrt{\pi t}}$

$\therefore$ By convolution theorem

$$£^{-1}\left(\frac{1}{s\sqrt{s+4}}\right) = \frac{e^{-4t}}{\sqrt{\pi t}} * 1 = \int_0^t \frac{e^{-4u}}{\sqrt{\pi u}}\,du$$

Put $4u = x^2$ so that $2\sqrt{u} = x \Rightarrow \frac{1}{\sqrt{u}}du = dx.$

$$\therefore \qquad £^{-1}\left\{\frac{1}{s\sqrt{s+4}}\right\} = \frac{1}{\sqrt{\pi}}\int_0^{2\sqrt{t}} e^{-x^2}\,dx = \frac{1}{2}\cdot\frac{2}{\sqrt{\pi}}\int_0^{2\sqrt{t}} e^{-x^2}\,dx$$

$$= \frac{1}{2}\,erf\left(2\sqrt{t}\right)$$

(ii) We have $\pounds^{-1}\left(\dfrac{s}{s^2+a^2}\right)=\cos at,\ \pounds^{-1}\left(\dfrac{1}{s^2+a^2}\right)=\dfrac{1}{a}\ \sin at$

$\therefore$ By convolution theorem

$$\pounds^{-1}\left\{\frac{s}{(s^2+a^2)^2}\right\}=\left(\frac{1}{a}\sin at\right) * \cos at=\frac{1}{a}\ (\sin at * \cos at)$$

$$=\frac{1}{a}\cdot\frac{1}{2}t\sin at \qquad \text{(from Example 2.34)}$$

$$=\frac{1}{2a}t\sin at$$

(iii) We have $\pounds^{-1}\left(\dfrac{1}{s^2+a^2}\right)=\dfrac{1}{a}\ \sin at$

$\therefore$ By convolution theorem

$$\pounds^{-1}\left\{\frac{1}{(s^2+a^2)^2}\right\}=\left(\frac{1}{a}\sin at\right) * \left(\frac{1}{a}\sin at\right)=\frac{1}{a^2}\ (\sin at * \sin at)$$

$$=\frac{1}{a^2}\int_0^t \sin au\ \sin a(t-u)\ du$$

$$=\frac{1}{2a^2}\int_0^t 2\sin au\ \sin(at-au)\ du$$

$$=\frac{1}{2a^2}\int_0^t [\cos\ (2au-at)-\cos at]\ du$$

$$=\frac{1}{2a^2}\ \left[\frac{1}{2a}\sin\ (2au-at)-u\ \cos at\right]_0^t$$

$$=\frac{1}{2a^2}\ \left[\frac{1}{2a}\sin at+\frac{1}{2a}\sin at-t\cos at\right]$$

$$=\frac{1}{2a^3}\ [\sin at-at\cos at]$$

(iv) We have $\pounds^{-1}\left(\dfrac{s}{s^2+a^2}\right)=\cos at,\ \pounds^{-1}\left(\dfrac{s}{s^2+b^2}\right)=\cos bt$

$\therefore$ By convolution theorem

$$\pounds^{-1}\left\{\frac{s^2}{(s^2+a^2)(s^2+b^2)}\right\}=\cos at * \cos bt=\int_0^t \cos au\ \cos(bt-bu)\ du$$

$$=\frac{1}{2}\int_0^t 2\cos au\ \cos(bt-bu)\ du$$

$$= \frac{1}{2}\int_0^t [\cos\{(a-b)u+bt\}+\cos\{(a+b)u-bt\}]du$$

$$= \frac{1}{2}\left[\frac{1}{(a-b)}\sin\{(a-b)u+bt\}+\frac{1}{(a+b)}\sin\{(a+b)u-bt\}\right]_0^t$$

$$= \frac{1}{2}\left[\frac{1}{(a-b)}(\sin at-\sin bt)+\frac{1}{(a+b)}(\sin at+\sin bt)\right]$$

$$= \frac{1}{2}\left[\frac{2a}{a^2-b^2}\sin at-\frac{2b}{a^2-b^2}\sin bt\right]$$

$$= \frac{a\sin at-b\sin bt}{a^2-b^2}$$

(v) We have $£^{-1}\left(\frac{1}{s^3}\right)=\frac{1}{2}t^2,\ £^{-1}\left(\frac{1}{s^2+1}\right)=\sin t$

$\therefore$ By convolution theorem

$$£^{-1}\left\{\frac{1}{s^3(s^2+1)}\right\}=\frac{1}{2}\ t^2*\sin t=\int_0^t \frac{1}{2}\ u^2\sin(t-u)\ du=-\frac{1}{2}\int_0^t u^2\sin(u-t)\ du$$

$$=-\frac{1}{2}\left[u^2\cdot\{-\cos(u-t)\}-2u\cdot\{-\sin(u-t)\}+2\{\cos\ (u-t)\}\right]_0^t$$

$$=-\frac{1}{2}\left[-t^2+2-2\cos t\right]=\frac{t^2}{2}-(1-\cos t)=\frac{t^2}{2}-2\sin^2\frac{t}{2}$$

(vi) We have $£^{-1}\left(\frac{1}{s-2}\right)=e^{2t},\ £^{-1}\left(\frac{1}{s+3}\right)=e^{-3t}$

$\therefore$ By convolution theorem

$$£^{-1}\left\{\frac{1}{(s-2)(s+3)}\right\}=e^{2t}*e^{-3t}=\int_0^t e^{2u}e^{-3(t-u)}\ du$$

$$=e^{-3t}\int_0^t e^{5u}du=e^{-3t}\left(\frac{e^{5u}}{5}\right)_0^t=\frac{1}{5}e^{-3t}\left(e^{5t}-1\right)$$

$$=\frac{1}{5}\left(e^{2t}-e^{-3t}\right)$$

(vii) We have $£^{-1}\left(\frac{1}{s^2+1}\right)=\sin t,\ £^{-1}\left(\frac{s}{s^2+4}\right)=\cos 2t$

$\therefore$ By convolution theorem

$$£^{-1}\left\{\frac{s}{(s^2+1)(s^2+4)}\right\}=\cos 2t*\sin t=\int_0^t \cos 2u\ \sin(t-u)\ du$$

$$= \frac{1}{2}\int_0^t [\sin(u+t) - \sin(3u-t)]\ du$$

$$= \frac{1}{2}\left[-\cos(u+t) + \frac{1}{3}\cos(3u-t)\right]_0^t$$

$$= \frac{1}{2}\left[-\cos 2t + \cos t + \frac{1}{3}\cos 2t - \frac{1}{3}\cos t\right]$$

$$= \frac{1}{3}(\cos t - \cos 2t)$$

(viii) We have $£^{-1}\left(\dfrac{1}{s+1}\right) = e^{-t}$, $£^{-1}\left(\dfrac{1}{s(s+2)}\right) = £^{-1}\left(\dfrac{1}{(s+1)^2 - 1}\right)$

$$= e^{-t}\sinh t$$

∴ By convolution theorem

$$£^{-1}\left\{\frac{1}{s(s+1)(s+2)}\right\} = e^{-t} * e^{-t}\sinh t = \int_0^t e^{-u}\cdot e^{-(t-u)}\sinh(t-u)\ du$$

$$= e^{-t}\int_0^t \sinh(t-u)\ du = \left[-e^{-t}\cosh(t-u)\right]_0^t$$

$$= e^{-t}(\cosh t - 1)$$

(ix) We have $£^{-1}\left(\dfrac{s}{s^2+w^2}\right) = \cos wt$

∴ By convolution theorem

$$£^{-1}\left\{\frac{s^2}{(s^2+w^2)^2}\right\} = \cos wt * \cos wt = \int_0^t \cos wu \cos w(t-u)\ du$$

$$= \frac{1}{2}\int_0^t [\cos wt + \cos\ (2wu - wt)]du$$

$$= \frac{1}{2}\left[u\cos wt + \frac{1}{2w}\ \sin(2wu - wt)\right]_0^t$$

$$= \frac{1}{2}\left[t\cos wt + \frac{1}{2w}\ \sin wt + \frac{1}{2w}\ \sin wt\right]$$

$$= \frac{1}{2w}\ [tw\cos wt + \sin wt]$$

2.15 APPLICATIONS OF LAPLACE TRANSFORM TO SOLVE LINEAR DIFFERENTIAL EQUATIONS, SIMULTANEOUS LINEAR DIFFERENTIAL EQUATIONS AND INTEGRAL EQUATIONS

Laplace transform helps us in solving the initial value problems with constant and variable coefficients without finding complementary and particular integrals. This is also helpful in solving simultaneous differential equations and integral equations. Sometimes the following two theorems are also used to find the initial and final values of the function $f(t)$.

Theorem 2.15 Initial Value Theorem

Let $f(t)$ is continuous and has piecewise continuous derivative $f'(t)$ in every finite interval in $t \geq 0$ and both $f(t)$ and $f'(t)$ are of exponential order, then $\lim\limits_{t\to 0+} f(t) = \lim\limits_{s\to\infty} sF(s)$ provided the limits exist.

Proof: Let $f'(t)$ is of exponential order $k > 0$ then there exists $M > 0$, such that

$$f'(t) \leq Me^{kt}$$

By Theorem (2.5), Laplace transform of $f'(t)$ exists and $\pounds\left\{f'(t)\right\} = sF(s) - f(0)\ ;\ s > k$ where $F(s)$ is Laplace transform of $f(t)$.

We have

$$\left|\int_0^\infty e^{-st} f'(t)\, dt\right| \leq \int_0^\infty \left|e^{-st} f'(t)\right| dt$$

$$\leq M\int_0^\infty e^{-(s-k)t}\, dt$$

$$= -\left.\frac{M\ e^{-(s-k)t}}{s-k}\right|_0^\infty$$

$$= \frac{M}{s-k} \quad \text{as } e^{-(s-k)\,t} \to 0 \text{ as } t \to \infty \text{ when } s > k$$

$$\therefore \qquad \pounds\{f'(t)\} = [sF(s) - f(0)] \to 0 \quad \text{as } s \to \infty$$

$$\therefore \qquad \lim_{s\to\infty} sF(s) = \lim_{t\to 0+} f(t)$$

Theorem 2.16 Final Value Theorem

Let $f(t)$ is continuous and has piecewise continuous derivative $f'(t)$ in every finite interval in $t \geq 0$ and both $f(t)$ and $f'(t)$ are of exponential order, then

$$\lim_{t\to\infty} f(t) = \lim_{s\to 0+} sF(s) \text{ provided both the limits exist.}$$

Proof: By Theorem (2.5), Laplace transform of $f'(t)$ exists and $\pounds\{f'(t)\} = sF(s) - f(0); \quad s > k$

Then

$$\lim_{s\to 0+}[sF(s) - f(0)] = \lim_{s\to 0+} \pounds\{f'(t)\}$$

$$= \lim_{s\to 0+}\lim_{T\to\infty}\int_0^T e^{-st} f'(t)\, dt$$

$$= \lim_{s\to 0+}\lim_{T\to\infty}\left[e^{-st} f(t)\Big|_0^T + \int_0^T se^{-st} f(t)\, dt \right]$$

$$= \lim_{s\to 0+}\lim_{T\to\infty}\left[e^{-st} f(T) - f(0) + s\int_0^T e^{-st} f(t)\, dt \right]$$

$$= \lim_{s\to 0+}\lim_{T\to\infty} e^{-st} f(T) - f(0) + \lim_{s\to 0+} sF(s)$$

$$= \lim_{T\to\infty} f(T) - f(0) + 0$$

$$= \lim_{t\to\infty} f(t) - f(0)$$

$\therefore$
$$\lim_{t\to\infty} f(t) = \lim_{s\to 0+} sF(s)$$

Remark 2.4: Theorems (2.15) and (2.16) can be applied only when both the limits in the theorems exist and are finite.

Example 2.36: Using Laplace transform techniques, solve the following initial value problems

(i) $y' + 4y = t, \;\; y(0) = 1$

(ii) $y'' + ay' - 2a^2 y = 0, \;\; y(0) = 6, \;\; y'(0) = 0$

(iii) $\dfrac{d^2 y}{dx^2} - 3\dfrac{dy}{dx} + 2y = 4x + e^{3x}$ where $y(0) = 1, \;\; y'(0) = -1$

(iv) $y'' + 4y = 0, \;\; y(0) = 1, \;\; y'(0) = 6$

(v) $\dfrac{d^2 y}{dt^2} + 2\dfrac{dy}{dt} + 5y = e^{-t}\sin t, \quad y(0) = 1, \;\; y'(0) = -1$

(vi) $y'' + 4y' + 4y = 12t^2 e^{-2t}, \;\; y(0) = 2, \;\; y'(0) = 1$

(vii) $y'' - 3y' + 2y = 4e^{2t}, \;\; y(0) = -3, \;\; y'(0) = 5$

(viii) $\dfrac{d^2 y}{dt^2} + 2\dfrac{dy}{dt} + 5y = \dfrac{17}{2}\sin 2t; \; y = 2$ and $\dfrac{dy}{dt} = -4$ when $t = 0$

(ix) $y'' + 9y = 6\cos 3t, \;\; y(0) = 2, \;\; y'(0) = 0$

(x) $y'' + 9y = \sin 3t, \;\; y(0) = 0, \;\; y'(0) = 0$

Solution: (i) Initial value problem is

$$y' + 4y = t;\ y(0) = 1$$

Take Laplace transform of both sides

$$sY(s) - y(0) + 4Y(s) = \frac{1}{s^2} \text{ where } Y(s) = \pounds\{y(t)\}$$

$$\therefore \quad (s+4)\,Y(s) = \frac{1}{s^2} + 1 \qquad (\because y(0) = 1)$$

$$\therefore \quad Y(s) = \frac{1}{s^2(s+4)} + \frac{1}{(s+4)} = \frac{s^2+1}{s^2(s+4)}$$

By suppression method

$$\text{Let } \frac{s^2+1}{s^2(s+4)} \equiv \frac{17}{16(s+4)} + \frac{1}{4s^2} + \frac{A}{s}$$

$$\therefore \quad s^2 + 1 \equiv \frac{17}{16}s^2 + \frac{1}{4}(s+4) + As(s+4)$$

Equate coefficient of s^2

$$1 = \frac{17}{16} + A \Rightarrow A = -\frac{1}{16}$$

$$\therefore \quad Y(s) = \frac{17}{16(s+4)} + \frac{1}{4s^2} - \frac{1}{16s}$$

$$\therefore \quad y(t) = \pounds^{-1}\left\{\frac{17}{16(s+4)} + \frac{1}{4s^2} - \frac{1}{16s}\right\} = \frac{17}{16}e^{-4t} + \frac{t}{4} - \frac{1}{16} = \frac{1}{16}(17e^{-4t} + 4t - 1)$$

(ii) Initial value problem is

$$y'' + ay' - 2a^2y = 0,\ y(0) = 6,\ y'(0) = 0$$

Take Laplacc transform of both sides

$$s^2Y(s) - sy(0) - y'(0) + a\{sY(s) - y(0)\} - 2a^2Y(s) = 0 \text{ where } Y(s) = \pounds\{y(t)\}$$

$$\therefore \quad (s^2 + as - 2a^2)\,Y(s) = 6(s+a) \qquad (\because y(0) = 6 \text{ and } y'(0) = 0)$$

$$Y(s) = \frac{6(s+a)}{(s-a)(s+2a)} = \frac{4}{(s-a)} + \frac{2}{(s+2a)} \qquad \text{(by suppression method)}$$

$$\therefore \quad y(t) = \pounds^{-1}\left\{\frac{4}{(s-a)} + \frac{2}{(s+2a)}\right\} = 4e^{at} + 2e^{-2at}$$

(iii) Initial value problem is

$$\frac{d^2y}{dx^2}-3\frac{dy}{dx}+2y=4x+e^{3x},\ y(0)=1,\ \ y'(0)=-1$$

Take Laplace transform of both sides

$$s^2Y(s)-sy(0)-y'(0)-3\{sY(s)-y(0)\}+2Y(s)=\frac{4}{s^2}+\frac{1}{s-3}\text{ where }Y(s)=£\{y(x)\}$$

$$\therefore\ (s^2-3s+2)\ Y(s)=\frac{4}{s^2}+\frac{1}{(s-3)}+s-4 \qquad (\because y(0)=1,\ y'(0)=-1)$$

$$\therefore\ (s^2-3s+2)\ Y(s)=\frac{(s^2-3s+1)}{(s-3)}-\frac{4(s^2-1)}{s^2}=\frac{(s^2-3s+2)-1}{(s-3)}-\frac{4(s-1)(s+1)}{s^2}$$

or $$Y(s)=\frac{1}{s-3}-\frac{1}{(s-1)(s-2)(s-3)}-\frac{4(s+1)}{s^2(s-2)} \qquad (\because\ s^2-3s+2=(s-1)(s-2))$$

By suppression method

$$\frac{1}{(s-1)(s-2)(s-3)}=\frac{1}{(s-1)(1-2)(1-3)}+\frac{1}{(s-2)(2-1)(2-3)}+\frac{1}{(s-3)(3-1)(3-2)}$$

$$=\frac{1}{2(s-1)}-\frac{1}{(s-2)}+\frac{1}{2(s-3)}$$

and Let $$\frac{4(s+1)}{s^2(s-2)}\equiv\frac{4}{s^2(0-2)}+\frac{4\cdot(2+1)}{4(s-2)}+\frac{A}{s}=\frac{-2}{s^2}+\frac{3}{s-2}+\frac{A}{s}$$

$\therefore$ $$4(s+1)\equiv-2(s-2)+3s^2+As(s-2)$$

Equate coefficient of s^2

$$0=3+A$$

$\Rightarrow$ $$A=-3$$

$\therefore$ $$\frac{4(s+1)}{s^2(s-2)}=\frac{-2}{s^2}+\frac{3}{s-2}-\frac{3}{s}$$

$\therefore$ $$Y(s)=\frac{1}{s-3}-\frac{1}{2(s-1)}+\frac{1}{(s-2)}-\frac{1}{2(s-3)}+\frac{2}{s^2}-\frac{3}{(s-2)}+\frac{3}{s}$$

or $$Y(s)=\frac{3}{s}+\frac{2}{s^2}-\frac{1}{2(s-1)}-\frac{2}{(s-2)}+\frac{1}{2(s-3)}$$

$\therefore$ $$y(x)=£^{-1}\left\{\frac{3}{s}+\frac{2}{s^2}-\frac{1}{2(s-1)}-\frac{2}{(s-2)}+\frac{1}{2(s-3)}\right\}$$

$$=3+2x-\frac{1}{2}e^{x}-2e^{2x}+\frac{1}{2}e^{3x}$$

or $$y(x)=\frac{1}{2}(6+4x-e^{x}-4e^{2x}+e^{3x})$$

(iv) $y''+4y=0,\ y(0)=1,\ y'(0)=6$

Take Laplace transform of both sides

$$s^2Y(s)-sy(0)-y'(0)+4Y(s)=0 \quad \text{where } Y(s)=£\{y(t)\}$$

$$\therefore \quad (s^2+4)\,Y(s)=s+6 \qquad (\because y(0)=1, \text{ and } y'(0)=6)$$

$$\therefore \quad Y(s)=\frac{s+6}{s^2+4}=\frac{s}{s^2+2^2}+3\cdot\frac{2}{s^2+2^2}$$

$$\therefore \quad y(t)=£^{-1}\left(\frac{s}{s^2+2^2}+3\cdot\frac{2}{s^2+2^2}\right)=\cos 2t+3\sin 2t$$

(v) $y''+2y'+5y=e^{-t}\sin t,\ y(0)=1,\ y'(0)=-1$

Take Laplace transform of both sides

$$s^2Y(s)-sy(0)-y'(0)+2\{s.Y(s)-y(0)\}+5Y(s)=\frac{1}{(s+1)^2+1} \quad \text{where } Y(s)=£\{y(t)\}$$

$$(s^2+2s+5)\,Y(s)=\frac{1}{(s+1)^2+1}+s-1+2=\frac{1}{(s+1)^2+1}+(s+1) \qquad (\because y(0)=1,\ y'(0)=-1)$$

$$\therefore \quad Y(s)=\frac{1}{(s^2+2s+2)(s^2+2s+5)}+\frac{(s+1)}{(s+1)^2+2^2}$$

$$=\frac{(s^2+2s+5)-(s^2+2s+2)}{3(s^2+2s+2)(s^2+2s+5)}+\frac{(s+1)}{(s+1)^2+2^2}$$

$$=\frac{1}{3}\left[\frac{1}{(s+1)^2+1^2}-\frac{1}{2}\cdot\frac{2}{(s+1)^2+2^2}\right]+\frac{(s+1)}{(s+1)^2+2^2}$$

$$\therefore \quad y(t)=£^{-1}\{Y(s)\}=\frac{1}{3}(e^{-t}\sin t-\frac{1}{2}e^{-t}\sin 2t)+e^{-t}\cos 2t$$

$$=\frac{1}{6}e^{-t}(2\sin t-\sin 2t+6\cos 2t)$$

(vi) $y''+4y'+4y=12t^2e^{-2t};\ y(0)=2,\ y'(0)=1$

Take Laplace transform of both sides

$$s^2Y(s)-sy(0)-y'(0)+4\{sY(s)-y(0)\}+4Y(s)=\frac{24}{(s+2)^3} \quad \text{where } £\{y(t)\}=Y(s)$$

$$\therefore \quad (s^2+4s+4)\,Y(s)=\frac{24}{(s+2)^3}+2s+9=\frac{24}{(s+2)^3}+2(s+2)+5 \qquad (\because y(0)=2,\ y'(0)=1)$$

$$\therefore \qquad Y(s) = \frac{24}{(s+2)^5} + \frac{2}{(s+2)} + \frac{5}{(s+2)^2}$$

$$\therefore \qquad y(t) = £^{-1}\{Y(s)\} = e^{-2t}(t^4 + 2 + 5t) = (2 + 5t + t^4)e^{-2t}$$

(vii) $y'' - 3y' + 2y = 4e^{2t}$, $y(0) = -3$, $y'(0) = 5$

Take Laplace transform of both sides

$$s^2Y(s) - sy(0) - y'(0) - 3\{sY(s) - y(0)\} + 2Y(s) = \frac{4}{(s-2)} \text{ where } Y(s) = £\{y(t)\}$$

$$\therefore \quad (s^2 - 3s + 2)\ Y(s) = \frac{4}{(s-2)} - 3s + 5 + 9 = \frac{4}{(s-2)} - (3s - 14) \quad (\because y(0) = -3,\ y'(0) = 5)$$

$$\therefore \qquad Y(s) = \frac{4}{(s-1)(s-2)^2} - \frac{(3s-14)}{(s-1)(s-2)}$$

$$= 4\left[\frac{(s-1)-(s-2)}{(s-1)(s-2)^2}\right] - \frac{(3s-14)}{(s-1)(s-2)}$$

$$= \frac{4}{(s-2)^2} - \frac{4}{(s-1)(s-2)} - \frac{(3s-14)}{(s-1)(s-2)}$$

$$= \frac{4}{(s-2)^2} - \frac{(3s-10)}{(s-1)(s-2)}$$

$$= \frac{4}{(s-2)^2} - \frac{7}{(s-1)} + \frac{4}{(s-2)} \quad \text{(by suppression method)}$$

$$\therefore \qquad y(t) = £^{-1}\{Y(s)\}$$

$$= 4te^{2t} - 7e^t + 4e^{2t}$$

$$= e^t(4te^t + 4e^t - 7)$$

(viii) $\dfrac{d^2y}{dt^2} + 2\dfrac{dy}{dt} + 5y = \dfrac{17}{2}\sin 2t$; $y = 2$ and $\dfrac{dy}{dt} = -4$ when $t = 0$

Take Laplace transform of both sides

$$s^2Y(s) - sy(0) - y'(0) + 2\{sY(s) - y(0)\} + 5Y(s) = \frac{17}{(s^2+4)} \text{ where } Y(\mathrm{s}) = £\{y(t)\}$$

$$\therefore \qquad (s^2 + 2s + 5)\ Y(s) = \frac{17}{(s^2+4)} + 2s - 4 + 4 \qquad (\because y(0) = 2,\ y'(0) = -4)$$

or

$$Y(s) = \frac{17}{(s^2+4)(s^2+2s+5)} + \frac{2s}{(s^2+2s+5)}$$

Let $$\frac{17}{(s^2+2s+5)(s^2+4)} \equiv \frac{As+B}{(s^2+2s+5)} + \frac{Cs+D}{(s^2+4)}$$

$\therefore$ $$17 \equiv As(s^2+4) + B(s^2+4) + Cs(s^2+2s+5) + D(s^2+2s+5)$$

Equate coefficients of same powers of s

$$A + C = 0 \tag{1}$$

$$B + 2C + D = 0 \tag{2}$$

$$4A + 5C + 2D = 0 \tag{3}$$

$$4B + 5D = 17 \tag{4}$$

From equations (1) and (3)

$$A = -C,\ C + 2D = 0 \tag{5}$$

From equations (2) and (4)

$$B = \frac{17-5D}{4},\ \frac{17-5D}{4} + 2C + D = 0, \text{ i.e., } 8C - D = -17 \tag{6}$$

Solving equations (5) and (6)

$$A = 2, B = 3, C = -2, D = 1$$

$\therefore$ $$Y(s) = \frac{2s+3}{(s^2+2s+5)} + \frac{(-2s+1)}{(s^2+4)} + \frac{2s}{(s^2+2s+5)} = \frac{4s+3}{(s^2+2s+5)} + \frac{(-2s+1)}{(s^2+4)}$$

or $$Y(s) = \frac{4(s+1)-1}{\{(s+1)^2+2^2\}} - \frac{2s}{(s^2+4)} + \frac{1}{(s^2+4)}$$

$$Y(s) = \frac{4(s+1)}{(s+1)^2+2^2} - \frac{1}{2}.\frac{2}{(s+1)^2+2^2} - \frac{2s}{(s^2+4)} + \frac{1}{2}.\frac{2}{(s^2+4)}$$

$\therefore$ $$y(t) = £^{-1}\{Y(s)\} = 4e^{-t}\cos 2t - \frac{1}{2}e^{-t}\sin 2t - 2\cos 2t + \frac{1}{2}\sin 2t$$

$$= 2(2e^{-t}-1)\cos 2t + \frac{1}{2}(1-e^{-t})\sin 2t$$

(ix) $$y'' + 9y = 6\cos 3t,\ y(0) = 2,\ y'(0) = 0$$

Take Laplace transform of both sides

$$s^2Y(s) - sy(0) - y'(0) + 9Y(s) = \frac{6s}{(s^2+9)} \text{ where } Y(s) = £\{y(t)\}$$

or $$(s^2+9)Y(s) = \frac{6s}{(s^2+9)} + 2s \quad (\because y(0) = 2, y'(0) = 0)$$

$$\therefore \qquad Y(s)=\frac{6s}{\left(s^2+9\right)^2}+\frac{2s}{\left(s^2+9\right)}=\frac{6s}{(s-3i)^2(s+3i)^2}+\frac{2s}{\left(s^2+9\right)}$$

$$\text{or} \qquad Y(s)=\frac{1}{2i}\left\{\frac{1}{(s-3i)^2}-\frac{1}{(s+3i)^2}\right\}+2\frac{s}{\left(s^2+3^2\right)}$$

$$\therefore \qquad y(t)=£^{-1}\{Y(s)\}=\frac{1}{2i}\left\{te^{3it}-te^{-3it}\right\}+2\cos 3t$$

$$=t\left(\frac{e^{3it}-e^{-3it}}{2i}\right)+2\cos 3t=t\sin 3t+2\cos 3t$$

(x) $y''+9y=\sin 3t;\quad y(0)=0, y'(0)=0$

Take Laplace transform of both sides

$$s^2Y(s)-sy(0)-y'(0)+9Y(s)=\frac{3}{s^2+3^2} \quad \text{where } Y(s)=£\{y(t)\}$$

$$\therefore \qquad \left(s^2+9\right)Y(s)=\frac{3}{\left(s^2+9\right)} \qquad (\because y(0)=y'(0)=0)$$

$$\therefore \qquad Y(s)=\frac{3}{\left(s^2+9\right)^2}=\frac{3}{(s+3i)^2(s-3i)^2}$$

$$=3\left[\frac{1}{6i}\left(\frac{1}{s-3i}-\frac{1}{s+3i}\right)\right]^2$$

$$=\frac{-1}{12}\left[\frac{1}{(s-3i)^2}+\frac{1}{(s+3i)^2}-\frac{2}{s^2+3^2}\right]$$

$$\therefore \qquad y(t)=£^{-1}\{Y(s)\}=\frac{-1}{12}\left[te^{3it}+te^{-3it}-\frac{2}{3}\sin 3t\right]$$

$$=\frac{-1}{12}\left[2t\left(\frac{e^{3it}+e^{-3it}}{2}\right)-\frac{2}{3}\sin 3t\right]$$

$$=-\frac{1}{18}(3t\cos 3t-\sin 3t)=\frac{1}{18}(\sin 3t-3t\cos 3t)$$

Example 2.37: Find the general solution of the given problem $y''(t)+9y(t)=\cos 2t$ using the Laplace transform.

Solution: Given equation is $\qquad y''(t)+9y(t)=\cos 2t$

Take Laplace transform of both sides

$$s^2Y(s) - sy(0) - y'(0) + 9Y(s) = \frac{s}{s^2+4} \text{ where } Y(s) = \pounds\{y(t)\}$$

$\therefore$
$$(s^2+9)Y(s) = \frac{s}{s^2+4} + c_1 s + c_2 \text{ where we assume } c_1 = y(0) \text{ and } c_2 = y'(0)$$

or
$$Y(s) = \frac{s}{(s^2+9)(s^2+4)} + \frac{c_1 s}{(s^2+9)} + \frac{c_2}{(s^2+9)}$$

$$= \frac{s}{5}\left\{\frac{1}{s^2+4} - \frac{1}{s^2+9}\right\} + \frac{c_1 s}{(s^2+9)} + \frac{c_2}{(s^2+9)}$$

$\therefore$
$$y(t) = \pounds^{-1}\left\{\frac{1}{5}\cdot\frac{s}{s^2+2^2} - \frac{1}{5}\cdot\frac{s}{s^2+3^2} + c_1\frac{s}{s^2+3^2} + \frac{c_2}{3}\cdot\frac{3}{s^2+3^2}\right\}$$

$$= \frac{1}{5}\cos 2t - \frac{1}{5}\cos 3t + c_1 \cos 3t + \frac{c_2}{3}\sin 3t$$

$$= \left(c_1 - \frac{1}{5}\right)\cos 3t + \frac{c_2}{3}\sin 3t + \frac{1}{5}\cos 2t$$

$\therefore y(t) = C_1 \cos 3t + C_2 \sin 3t + \frac{1}{5}\cos 2t$ is the general solution of given equation where C_1 and C_2 are arbitrary constants.

Example 2.38: Solve the initial value problems

(i) $y'' + y' - 2y = \begin{cases} 3\sin t - \cos t & ;\ 0 < t < 2\pi \\ 3\sin 2t - \cos 2t\ ; & t > 2\pi \end{cases}$

with initial conditions $y(0) = 1, y'(0) = 0$

(ii) $y' + y = f(t);\ y(0) = 2$ where $f(t) = \begin{cases} 0 & ;\ 0 \le t < \pi/2 \\ \cos t\ ;\ t \ge \dfrac{\pi}{2} \end{cases}$

(iii) $y'' + 2y' + 5y = \delta(t-2),\ y(0) = 0,\ y'(0) = 0$

(iv) $y'' + 8y' + 17y = f(t),\ y(0) = 0,\ y'(0) = 0$

where $f(t)$ is periodic function with period 2π given by

$$f(t) = \begin{cases} 1\ ; & 0 < t < \pi \\ 0\ ; & \pi < t < 2\pi \end{cases}$$

(v) $\dfrac{d^2y}{dx^2} + 4y = E(x-2)$ where E is the unit step function and
$y(0) = 0$ and $y'(0) = 1$

Solution:

(i) In terms of unit step function

$$y''+y'-2y=(3\sin t-\cos t)\{u_0(t)-u_{2\pi}(t)\}+(3\sin 2t-\cos 2t)u_{2\pi}(t)$$

$$=(3\sin t-\cos t)u_0(t)+(-3\sin t+\cos t+3\sin 2t-\cos 2t)u_{2\pi}(t)$$

$$=(3\sin t-\cos t)u_0(t)+\{3\sin(2\pi-t)+\cos(2\pi-t)-3\sin(4\pi-2t)-\cos(4\pi-2t)\}u_{2\pi}(t)$$

$$=(3\sin t-\cos t)u_0(t)+\{-3\sin(t-2\pi)+\cos(t-2\pi)+3\sin 2(t-2\pi)-\cos 2(t-2\pi)\}u_{2\pi}(t)$$

Take Laplace transform of both sides

$$s^2Y(s)-sy(0)-y'(0)+sY(s)-y(0)-2Y(s)$$

$$=\frac{3}{s^2+1}-\frac{s}{s^2+1}+e^{-2\pi s}\left\{\frac{-3}{s^2+1}+\frac{s}{s^2+1}+\frac{6}{s^2+4}-\frac{s}{s^2+4}\right\} \text{ where } Y(s)=\pounds\,(y(t))$$

$$\therefore (s^2+s-2)Y(s)=s+1+\frac{3}{s^2+1}-\frac{s}{s^2+1}+e^{-2\pi s}\left(\frac{-3}{s^2+1}+\frac{s}{s^2+1}+\frac{6}{s^2+4}-\frac{s}{s^2+4}\right)$$

$$(\because y(0)=1,\ y'(0)=0)$$

$$\therefore \quad Y(s)=\frac{s+1}{s^2+s-2}+\frac{3-s}{(s^2+1)(s^2+s-2)}-\frac{(3-s)e^{-2\pi s}}{(s^2+1)(s^2+s-2)}+\frac{(6-s)e^{-2\pi s}}{(s^2+4)(s^2+s-2)}$$

$$=\frac{s+\frac{1}{2}}{\left(s+\frac{1}{2}\right)^2-\left(\frac{3}{2}\right)^2}+\frac{\frac{1}{2}}{\left(s+\frac{1}{2}\right)^2-\left(\frac{3}{2}\right)^2}-\left\{\frac{1}{s^2+1}-\frac{1}{s^2+s-2}\right\}$$

$$+\left\{\frac{1}{s^2+1}-\frac{1}{s^2+s-2}+\frac{1}{s^2+s-2}-\frac{1}{s^2+4}\right\}e^{-2\pi s}$$

$$=\frac{s+\frac{1}{2}}{\left(s+\frac{1}{2}\right)^2-\left(\frac{3}{2}\right)^2}+\frac{\frac{3}{2}}{\left(s+\frac{1}{2}\right)^2-\left(\frac{3}{2}\right)^2}-\frac{1}{s^2+1}+\left\{\frac{1}{s^2+1}-\frac{1}{2}\cdot\frac{2}{s^2+2^2}\right\}e^{-2\pi s}$$

$$\therefore \quad y(t)=\pounds^{-1}\{Y(s)\}=e^{\frac{-t}{2}}\left(\cosh\frac{3t}{2}+\sinh\frac{3t}{2}\right)-\sin t+\left(\sin(t-2\pi)-\frac{1}{2}\sin 2(t-2\pi)\right)u_{2\pi}(t)$$

$$\therefore \quad y(t)=e^{\frac{-t}{2}}e^{\frac{3t}{2}}-\sin t+\left\{-\sin(2\pi-t)+\frac{1}{2}\sin(4\pi-2t)\right\}u_{2\pi}(t)$$

$$= e^t - \sin t + \left(\sin t - \frac{1}{2}\sin 2t\right)u_{2\pi}(t)$$

$$\therefore \qquad y(t) = \begin{cases} e^t - \sin t & ; \ 0 \le t < 2\pi \\ e^t - \sin t + \sin t - \frac{1}{2}\sin 2t \ ; & t > 2\pi \end{cases}$$

$$\text{i.e.,} \qquad y(t) = \begin{cases} e^t - \sin t & ; \quad 0 \le t < 2\pi \\ e^t - \frac{1}{2}\sin 2t \ ; & t > 2\pi \end{cases}$$

(ii) In terms of unit step function $y' + y = \cos t . u_{\frac{\pi}{2}}(t) = \sin\left(\frac{\pi}{2} - t\right)u_{\frac{\pi}{2}}(t) = -\sin\left(t - \frac{\pi}{2}\right)u_{\frac{\pi}{2}}(t)$

$$\therefore \qquad y' + y = -\sin\left(t - \frac{\pi}{2}\right)u_{\frac{\pi}{2}}(t); \ y(0) = 2$$

Take Laplace transform of both sides

$$sY(s) - y(0) + Y(s) = -e^{-\frac{\pi}{2}s}\frac{1}{(s^2+1)} \qquad \text{where } Y(s) = \pounds\,(y(t))$$

$$\therefore \qquad (s+1)Y(s) = 2 - \frac{1}{s^2+1}e^{-\frac{\pi}{2}s} \qquad (\because \ y(0) = 2)$$

$$\therefore \qquad Y(s) = \frac{2}{s+1} - \frac{1}{(s+1)(s^2+1)}e^{-\frac{\pi}{2}s}$$

Let

$$\frac{1}{(s+1)(s^2+1)} \equiv \frac{1}{2(s+1)} + \frac{As+B}{(s^2+1)}$$

$$\therefore \qquad 1 \equiv \frac{1}{2}(s^2+1) + (As+B)(s+1)$$

Equatc coefficients of s^2 and s

$$0 = \frac{1}{2} + A$$

$$\therefore \qquad A = -\frac{1}{2}$$

$$0 = A + B$$

$$\therefore \qquad B = -A = \frac{1}{2}$$

$$\therefore \qquad Y(s) = \frac{2}{(s+1)} + \left\{-\frac{1}{2(s+1)} + \frac{1}{2}\frac{(s-1)}{(s^2+1)}\right\}e^{-\frac{\pi}{2}s}$$

$$= \frac{2}{(s+1)} + \left\{ -\frac{1}{2(s+1)} + \frac{1}{2}\frac{s}{s^2+1} - \frac{1}{2(s^2+1)} \right\} e^{-\frac{\pi}{2}s}$$

$\therefore \quad y(t) = £^{-1}\{Y(s)\} = 2e^{-t} + \left\{ -\frac{1}{2}e^{-\left(t-\frac{\pi}{2}\right)} + \frac{1}{2}\cos\left(t-\frac{\pi}{2}\right) - \frac{1}{2}\sin\left(t-\frac{\pi}{2}\right) \right\} u_{\frac{\pi}{2}}(t)$

$$= 2e^{-t} + \left\{ -\frac{1}{2}e^{\frac{\pi}{2}-t} + \frac{1}{2}\sin t + \frac{1}{2}\cos t \right\} u_{\frac{\pi}{2}}(t)$$

$$\therefore \quad y(t) = \begin{cases} 2e^{-t} & ; \ 0 \le t < \pi/2 \\ 2e^{-t} - \frac{1}{2}\left(e^{\frac{\pi}{2}-t} - \sin t - \cos t\right) & ; \ t \ge \frac{\pi}{2} \end{cases}$$

(iii) $y'' + 2y' + 5y = \delta(t-2)$; $y(0) = 0, y'(0) = 0$

Take Laplace transform of both sides

$$s^2Y(s) - sy(0) - y'(0) + 2\{sY(s) - y(0)\} + 5Y(s) = e^{-2s} \qquad \text{where } Y(s) = £\{y(t)\}$$

$\therefore \quad (s^2 + 2s + 5)\, Y(s) = e^{-2s} \qquad (\because y(0) = y'(0) = 0)$

$\therefore \quad Y(s) = \frac{e^{-2s}}{s^2 + 2s + 5} = \frac{1}{2}\cdot\frac{2}{(s+1)^2 + 2^2}e^{-2s}$

$\therefore \quad y(t) = £^{-1}\{Y(s)\} = \frac{1}{2}\{e^{-(t-2)}\sin 2(t-2)\}u_2(t)$

$$\therefore \quad y(t) = \begin{cases} 0 & ; \ 0 \le t < 2 \\ \frac{1}{2}e^{-(t-2)}\sin\{2(t-2)\} & ; \ t \ge 2 \end{cases}$$

(iv) $\quad f(t) = \begin{cases} 1 & ; \ 0 < t < \pi \\ 0 & ; \ \pi < t < 2\pi \end{cases}$

and $f(t)$ is periodic with period 2π

$\therefore \quad £(f(t)) = \frac{1}{(1-e^{-2\pi s})}\int_0^{2\pi} e^{-st} f(t)\,dt = \frac{1}{(1-e^{-2\pi s})}\int_0^{\pi} e^{-st}\,dt$

$$= \frac{1}{(1-e^{-2\pi s})}\left(\frac{e^{-st}}{-s}\right)_0^{\pi} = \frac{1-e^{-\pi s}}{s(1-e^{-2\pi s})} = \frac{1}{s(1+e^{-\pi s})}$$

Now, the given differential equation is

$$y'' + 8y' + 17y = f(t) \ ; y(0) = 0, y'(0) = 0$$

Take Laplace transform of both sides

$$s^2Y(s) - sy(0) - y'(0) + 8\{sY(s) - y(0)\} + 17Y(s) = \frac{1}{s(1+e^{-\pi s})} \qquad \text{where } \pounds\{y(t)\} = Y(s)$$

$$\therefore \quad (s^2 + 8s + 17)\, Y(s) = \frac{1}{s(1+e^{-\pi s})} \qquad (\because y(0) = y'(0) = 0)$$

$$\therefore \quad Y(s) = \frac{(1+e^{-\pi s})^{-1}}{s\{(s+4)^2+1\}} = \sum_{n=0}^{\infty} \frac{(-1)^n e^{-n\pi s}}{s\{(s+4)^2+1\}} \tag{1}$$

Now, $$\pounds^{-1}\left\{\frac{1}{(s+4)^2+1}\right\} = e^{-4t}\sin t$$

$$\therefore \quad \pounds^{-1}\left\{\frac{1}{s\{(s+4)^2+1\}}\right\} = \int_0^t e^{-4x}\sin x\, dx = \left\{\frac{e^{-4x}}{17}(-4\sin x - \cos x)\right\}_0^t$$

$$= \frac{1}{17}\{1 - e^{-4t}(4\sin t + \cos t)\}$$

$\therefore$ From equation (1)

$$y(t) = \pounds^{-1}\{Y(s)\} = \sum_{n=0}^{\infty}(-1)^n \frac{1}{17}\left[1 - e^{-4(t-n\pi)}\{4\sin(t-n\pi) + \cos(t-n\pi)\}\right]\cdot u_{n\pi}(t)$$

$$= \sum_{n=0}^{\infty}(-1)^n \frac{1}{17}\left[1 - e^{4n\pi}e^{-4t}\{4(-1)^n\sin t + (-1)^n\cos t\}\right].u_{n\pi}(t)$$

$$= \sum_{n=0}^{\infty}\frac{1}{17}\left[(-1)^n - e^{4n\pi}e^{-4t}\{4\sin t + \cos t\}\right].u_{n\pi}(t)$$

$$= \sum_{n=0}^{\infty}\frac{1}{17}\left[(-1)^n - e^{4n\pi}g(t)\right].u_{n\pi}(t) \text{ where } g(t) = e^{-4t}(4\sin t + \cos t)$$

$$\therefore \quad y(t) = \sum_{n=0}^{k}\frac{1}{17}\left[(-1)^n - e^{4n\pi}g(t)\right];\ k\pi < t < (k+1)\pi;\ k = 0, 1, 2, \ldots\ldots$$

$$= \frac{1}{17}\left[\frac{1}{2}\{1-(-1)^{k+1}\} - g(t).\frac{\{(e^{4\pi})^{k+1} - 1\}}{(e^{4\pi}-1)}\right];\ k\pi < t < (k+1)\pi \quad (\text{sum of } G.P)$$

$$\therefore \quad y(t) = \frac{1}{34}\left[(-1)^k + 1 - 2\frac{e^{4\pi(k+1)} - 1}{e^{4\pi} - 1}g(t)\right];\ k\pi < t < (k+1)\pi;\ k = 0, 1, 2, 3\ldots.$$

where $g(t) = e^{-4t}(4\sin t + \cos t)$

(v) $\dfrac{d^2y}{dx^2}+4y=E(x-2)$ where E is the unit step function and $y(0)=0$ and $y'(0)=1$

Take Laplace transform of both sides

$$s^2Y(s)-sy(0)-y'(0)+4Y(s)=\frac{1}{s}e^{-2s} \quad \text{where } Y(s)=\pounds\{y(x)\}$$

$\therefore$
$$(s^2+4)Y(s)=\frac{1}{s}e^{-2s}+1 \qquad (\because y(0)=0 \text{ and } y'(0)=1)$$

or
$$Y(s)=\frac{1}{s(s^2+4)}e^{-2s}+\frac{1}{(s^2+4)}$$

$$=\frac{(s^2+4)-s^2}{4s(s^2+4)}e^{-2s}+\frac{1}{(s^2+4)}$$

$$=\frac{1}{4}\left[\frac{1}{s}-\frac{s}{s^2+2^2}\right]e^{-2s}+\frac{1}{2}\frac{2}{(s^2+2^2)}$$

$\therefore$
$$y(x)=\pounds^{-1}\{Y(s)\}$$

$$=\frac{1}{4}\left[1-\cos\{2(x-2)\}\right]E(x-2)+\frac{1}{2}\sin 2x$$

$$=\frac{1}{2}\sin^2(x-2)E(x-2)+\frac{1}{2}\sin 2x$$

Example 2.39: Solve the following boundary value problem using the Laplace transform

$$(D^2+9)\,y=18t\;;\; y(0)=0,\; y\left(\frac{\pi}{2}\right)=1$$

Solution: Take Laplace transform of both sides of the given differential equation

$$\{s^2Y(s)-sy(0)-y'(0)\}+9Y(s)=\frac{18}{s^2} \quad \text{where } Y(s)=\pounds\{y(t)\}$$

Let $y'(0)=k$ (constant). Then

$$(s^2+9)Y(s)=\frac{18}{s^2}+k \qquad (\because x(0)=0)$$

$\therefore$
$$Y(s)=\frac{18}{s^2(s^2+9)}+\frac{k}{s^2+9}=2\left\{\frac{1}{s^2}-\frac{1}{s^2+9}\right\}+\frac{k}{s^2+9}$$

or
$$y(t)=\pounds^{-1}\{Y(s)\}=\pounds^{-1}\left\{\frac{2}{s^2}+\frac{k-2}{3}\cdot\frac{3}{s^2+3^2}\right\}$$

$\therefore$
$$y(t)=2t+\frac{(k-2)}{3}\sin 3t$$

Given $y\left(\frac{\pi}{2}\right)=1$

$\therefore$ $$1=2.\frac{\pi}{2}+\frac{(k-2)}{3}\sin\frac{3\pi}{2}=\pi-\frac{(k-2)}{3}$$

$\therefore$ $$3=3\pi-(k-2)$$

$\Rightarrow$ $$\frac{k-2}{3}=\pi-1$$

$\therefore$ $$y(t)=2t+(\pi-1)\sin 3t$$

Example 2.40: Using convolution theorem, solve the initial value problem $y''+16y=\cos 4t$, $y(0)=0, y'(0)=0$.

Solution: Given differential equation is $y''+16y=\cos 4t$

Take Laplace transform of both sides

$$s^2Y(s)-sy(0)-y'(0)+16Y(s)=\frac{s}{s^2+4^2} \quad \text{where} \quad Y(s)=£\{y(t)\}$$

$\therefore$ $$(s^2+16)Y(s)=\frac{s}{(s^2+16)} \qquad (\because\ y(0)=y'(0)=0)$$

$\therefore$ $$Y(s)=\frac{s}{(s^2+16)^2}$$

$\therefore$ $$y(t)=£^{-1}\{Y(s)\}=£^{-1}\left\{\frac{s}{(s^2+16)^2}\right\}$$

$$=\frac{1}{4}£^{-1}\left\{\frac{4}{(s^2+4^2)}\cdot\frac{s}{(s^2+4^2)}\right\}$$

Now, $$£(\sin 4t)=\frac{4}{s^2+4^2} \quad \text{and} \quad £(\cos 4t)=\frac{s}{s^2+4^2}$$

$\therefore$ By convolution theorem

$$y(t)=\sin 4t*\cos 4t$$

$$=\frac{1}{4}\int_0^t \sin 4u\cdot\cos 4(t-u)\,du$$

$$=\frac{1}{8}\int_0^t\{\sin 4t+\sin(8u-4t)\}\,du$$

$$= \frac{1}{8}\left[u\sin 4t - \frac{1}{8}\cos(8u-4t)\right]_0^t$$

$$= \frac{1}{8}\left[t\sin 4t - \frac{1}{8}\cos 4t + \frac{1}{8}\cos 4t\right] = \frac{t}{8}\sin 4t$$

Hence, the required solution is $y(t) = \frac{t}{8}\sin 4t.$

Now, we shall be taking examples of initial value problems with variable coefficients.

Example 2.41: Solve the differential equations

(i) $t\frac{d^2x}{dt^2} - (t+2)\frac{dx}{dt} + 3x = t-1$; $x(0) = 0,\ x(2) = 9$

(ii) $ty'' + 2y' + ty = \cos t$; $y(0) = 1$

(iii) $y'' + ty' - 2y = 6 - t,\ y(0) = 0,\ y'(0) = 1$

(iv) $ty'' + 2ty' + 2y = 2,\ y(0) = 1$

Solution: (i) $t\frac{d^2x}{dt^2} - (t+2)\frac{dx}{dt} + 3x = t-1$; $x(0) = 0,\ x(2) = 9$

Take Laplace transform of both sides

$$\frac{-d}{ds}\left\{s^2X(s) - sx(0) - x'(0)\right\} + \frac{d}{ds}\left\{sX(s) - x(0)\right\} - 2\left\{sX(s) - x(0)\right\} + 3X(s) = \frac{1}{s^2} - \frac{1}{s}$$

where $X(s) = £(x(t))$

$$\therefore \quad -s^2\frac{dX}{ds} - 2sX(s) + x(0) + s\frac{dX}{ds} + X(s) - 2sX(s) + 2x(0) + 3X(s) = \frac{1}{s^2} - \frac{1}{s}$$

$$\left(\because x'(0) \text{ and } x(0) \text{ are constants, therefore } \frac{d}{ds}x'(0) = \frac{d}{ds}x(0) = 0\right)$$

or $$\left(-s^2 + s\right)\frac{dX}{ds} - 4(s-1)X = \frac{1-s}{s^2} \qquad (\because x(0) = 0)$$

$$\therefore \quad \frac{dX}{ds} + \frac{4}{s}X = \frac{1}{s^3}$$

It is a linear differential equation.

$$I.F. = \int \frac{4}{s}ds = e^{4\log s} = e^{\log s^4} = s^4$$

$\therefore$ Solution is $s^4X(s) = \int \frac{1}{s^3}\cdot s^4\,ds + c$ where c is an arbitrary constant.

$$\therefore \quad X(s) = \frac{1}{s^4}\left(\frac{s^2}{2} + c\right) = \frac{1}{2s^2} + \frac{c}{s^4}$$

$\therefore \qquad x(t) = £^{-1}\{X(s)\} = \frac{t}{2} + \frac{c}{6}t^3$

Given $x(2) = 9$

$\therefore \qquad 9 = 1 + \frac{c}{6} \times 8$

$\Rightarrow \qquad c = 6$

Hence, $\qquad x(t) = \frac{t}{2} + t^3$

(ii) $ty'' + 2y' + ty = \cos t \;;\; y(0) = 1$

Take Laplace transform of both sides

$$\frac{-d}{ds}\{s^2Y(s) - sy(0) - y'(0)\} + 2\{sY(s) - y(0)\} - \frac{d}{ds}\{Y(s)\} = \frac{s}{s^2+1} \quad \text{where } Y(s) = £(y(t))$$

$$\therefore \qquad -s^2\frac{dY}{ds} - 2sY + y(0) + 2sY - 2y(0) - \frac{dY}{ds} = \frac{s}{s^2+1}$$

$$\therefore \qquad -(s^2+1)\frac{dY}{ds} = 1 + \frac{s}{s^2+1} \qquad (\because y(0) = 1)$$

$$\therefore \qquad -\frac{dY}{ds} = \frac{1}{(s^2+1)} + \frac{s}{(s^2+1)^2} = \frac{1}{s^2+1} + \frac{1}{2}\left\{-\frac{d}{ds}\frac{1}{(s^2+1)}\right\}$$

Take inverse Laplace transform of both sides

$$ty(t) = \sin t + \frac{1}{2}t\sin t$$

$$\therefore \qquad y(t) = \left(\frac{1}{t} + \frac{1}{2}\right)\sin t$$

(iii) $y'' + ty' - 2y = 6 - t \;;\; y(0) = 0, y'(0) = 1$

Take Laplace transform of both sides

$$s^2Y(s) - sy(0) - y'(0) - \frac{d}{ds}\{sY(s) - y(0)\} - 2Y(s) = \frac{6}{s} - \frac{1}{s^2}; \text{ where } Y(s) = £\{y(t)\}$$

$$\therefore \qquad s^2Y(s) - 1 - s\frac{dY}{ds} - Y(s) - 2Y(s) = \frac{6}{s} - \frac{1}{s^2} \qquad (\because y(0) = 0,\ y'(0) = 1)$$

$$\therefore \qquad s\frac{dY}{ds} + (3 - s^2)Y(s) = \frac{1}{s^2} - \frac{6}{s} - 1$$

$$\text{or} \qquad \frac{dY}{ds} + \left(\frac{3}{s} - s\right)Y(s) = \frac{1}{s^3}(-s^2 - 6s + 1)$$

It is a linear differential equation

$$I.F. = e^{\int\left(\frac{3}{s}-s\right)ds} = e^{\left(3\log s - \frac{s^2}{2}\right)} = s^3 \cdot e^{-s^2/2}$$

∴ Solution is

$$Y(s)\cdot s^3 e^{-s^2/2} = \int\left(-s^2-6s+1\right)e^{-s^2/2}ds + c \text{ where } c \text{ is an arbitrary constant.}$$

$$= \int s\left(-se^{-s^2/2}\right)ds + 6\int\left(-se^{-s^2/2}\right)ds + \int e^{-s^2/2}ds + c$$

$$= se^{-s^2/2} - \int e^{-s^2/2}ds + 6e^{-s^2/2} + \int e^{-s^2/2}ds + c \text{ (integrating by parts)}$$

$$= (s+6)\, e^{-s^2/2} + c$$

∴ $$Y(s) = \frac{1}{s^2} + \frac{6}{s^3} + \frac{c}{s^3}e^{s^2/2} \qquad (1)$$

But $$0 = \lim_{t\to 0+} y(t) = \lim_{s\to\infty} sY(s) \text{ (initial value problem)}$$

∴ From (1), $c = 0$

Hence, $$Y(s) = \frac{1}{s^2} + \frac{6}{s^3}$$

∴ $$y(t) = £^{-1}\{Y(s)\} = t + 3t^2$$

(iv) $ty'' + 2ty' + 2y = 2,\ \ y(0) = 1$

Take Laplace transform of both sides

$$-\frac{d}{ds}\{s^2Y(s) - sy(0) - y'(0)\} - 2\frac{d}{ds}\{sY(s) - y(0)\} + 2Y(s) = \frac{2}{s} \text{ where } Y(s) = £\{y(t)\}$$

∴ $$-s^2\frac{dY}{ds} - 2sY(s) + y(0) - 2s\frac{dY}{ds} - 2Y(s) + 2Y(s) = \frac{2}{s}$$

or $$\left(s^2+2s\right)\frac{dY}{ds} + 2sY(s) = 1 - \frac{2}{s} = \frac{s-2}{s} \qquad (\because y(0)=1)$$

∴ $$\frac{dY}{ds} + \frac{2}{s+2}Y = \frac{(s-2)}{s^2(s+2)}$$

It is a linear differential equation.

$$I.F. = e^{\int\frac{2}{s+2}ds} = e^{2\log(s+2)} = (s+2)^2$$

∴ Solution is

$$(s+2)^2Y(s) = \int\frac{(s-2)}{s^2(s+2)}\cdot(s+2)^2\, ds + c \text{ where } c \text{ is an arbitrary constant}$$

$$= \int \frac{s^2-4}{s^2}\,ds + c = \int\left(1-\frac{4}{s^2}\right)ds + c$$

$\therefore$
$$(s+2)^2 Y(s) = s + \frac{4}{s} + c$$

$\therefore$
$$Y(s) = \frac{s^2+4+cs}{s(s+2)^2} = \frac{\left(s^2+4s+4\right)-4s+cs}{s(s+2)^2}$$

or
$$Y(s) = \frac{1}{s} - \frac{4}{(s+2)^2} + \frac{c}{(s+2)^2}$$

$\therefore$
$$y(t) = \pounds^{-1}\{Y(s)\} = 1 - 4t\,e^{-2t} + ct\,e^{-2t} \qquad (1)$$

Now, $y(0) = 1$ is satisfied by (1)

$\therefore$ (1) is the required solution where c is an arbitrary constant.

Now, we solve some simultaneous linear differential equations.

Example 2.42: Solve the simultaneous equations

$$y_1' + 2y_2 = 2\sin 2t$$

$$2y_1 + y_2' = 2(1+\cos 2t)$$

where $y_1(0) = -2,\ y_2(0) = 1$

Solution: Let $Y_1(s) = \pounds\{y_1(t)\},\ Y_2(s) = \pounds\{y_2(t)\}$

We take Laplace transform of both sides of given differential equations

$$sY_1(s) - y_1(0) + 2Y_2(s) = \frac{4}{s^2+4}$$

$$2Y_1(s) + sY_2(s) - y_2(0) = 2\left(\frac{1}{s} + \frac{s}{s^2+4}\right)$$

Hence,
$$sY_1(s) + 2Y_2(s) = \frac{4}{s^2+4} - 2 \qquad (1)$$

$$2Y_1(s) + sY_2(s) = 1 + \frac{2}{s} + \frac{2s}{s^2+4} \qquad (2)$$

Solving these equations by Crammer rule

$$Y_1(s) = \frac{\begin{vmatrix} \frac{4}{s^2+4} - 2 & 2 \\ 1 + \frac{2}{s} + \frac{2s}{s^2+4} & s \end{vmatrix}}{\begin{vmatrix} s & 2 \\ 2 & s \end{vmatrix}} = \frac{\frac{4s}{s^2+4} - 2s - 2 - \frac{4}{s} - \frac{4s}{s^2+4}}{s^2-4}$$

$$\therefore \quad Y_1(s) = \frac{1}{s^2-4}\left(-2s-2-\frac{4}{s}\right) = -\frac{2s}{s^2-4} - \frac{2}{s^2-4} - \frac{4}{s(s^2-4)}$$

$$= -\frac{2s}{s^2-4} - \frac{2}{s^2-4} - \left[\frac{s}{s^2-4} - \frac{1}{s}\right]$$

$$= \frac{1}{s} - \frac{2}{s^2-4} - \frac{3s}{s^2-4}$$

and
$$Y_2(s) = \frac{\begin{vmatrix} s & \dfrac{4}{s^2+4}-2 \\ 2 & 1+\dfrac{2}{s}+\dfrac{2s}{s^2+4} \end{vmatrix}}{\begin{vmatrix} s & 2 \\ 2 & s \end{vmatrix}} = \frac{1}{(s^2-4)}\left(s+2+\frac{2s^2}{s^2+4}-\frac{8}{s^2+4}+4\right)$$

$$\therefore \quad Y_2(s) = \frac{1}{(s^2-4)}\left\{6+s+\frac{2(s^2+4-4)}{(s^2+4)} - \frac{8}{(s^2+4)}\right\}$$

$$= \frac{1}{(s^2-4)}\left\{6+s+2-\frac{16}{(s^2+4)}\right\} = \frac{8}{s^2-4} + \frac{s}{s^2-4} - 2\left(\frac{1}{s^2-4} - \frac{1}{s^2+4}\right)$$

$$\therefore \quad Y_2(s) = \frac{6}{s^2-4} + \frac{2}{s^2+4} + \frac{s}{s^2-4}$$

$$\therefore \quad y_1(t) = £^{-1}\{Y_1(s)\} = £^{-1}\left\{\frac{1}{s} - \frac{2}{s^2-2^2} - 3\cdot\frac{s}{s^2-2^2}\right\} = 1 - \sinh 2t - 3\cosh 2t$$

$$y_2(t) = £^{-1}\{Y_2(s)\} = £^{-1}\left\{3\cdot\frac{2}{s^2-2^2} + \frac{2}{s^2+2^2} + \frac{s}{s^2-2^2}\right\} = 3\sinh 2t + \sin 2t + \cosh 2t$$

Example 2.43: Solve by Laplace transform

$$(D-2)x - (D+1)y = 6\,e^{3t}$$

$$(2D-3)x + (D-3)y = 6\,e^{3t}$$

Given: $x = 3, y = 0$ when $t = 0$.

Solution: Take Laplace transform of both sides of given differential equations

$$sX(s) - x(0) - 2X(s) - \{sY(s) - y(0)\} - Y(s) = \frac{6}{s-3}$$

and
$$2\{sX(s) - x(0)\} - 3X(s) + \{sY(s) - y(0)\} - 3Y(s) = \frac{6}{s-3}$$

where
$$X(s) = £\{x(t)\} \text{ and } Y(s) = £\{y(t)\}$$

$$\therefore \qquad (s-2)X(s)-(s+1)Y(s)=\frac{6}{(s-3)}+3 \qquad (1)$$

and $$(2s-3)X(s)+(s-3)Y(s)=\frac{6}{(s-3)}+6 \qquad (2)$$

$$(\because\ x(0)=3 \text{ and } y(0)=0)$$

Now, multiplying equation (1) by $(2s-3)$ and (2) by $(s-2)$ and then subtracting we get

$$-[(2s-3)(s+1)+(s-3)(s-2)]Y(s)=\frac{6}{(s-3)}[(2s-3)-(s-2)]+3(2s-3)-6(s-2)$$

$$\Rightarrow \qquad -(3s^2-6s+3)Y(s)=\frac{6(s-1)}{(s-3)}+3=\frac{9s-15}{s-3}=\frac{3(3s-5)}{(s-3)}$$

or $$-(s-1)^2Y(s)=\frac{3s-5}{(s-3)}$$

$$\therefore \qquad Y(s)=\frac{-(3s-5)}{(s-1)^2(s-3)} \qquad (3)$$

By suppression method

$$\frac{-(3s-5)}{(s-1)^2(s-3)}\equiv\frac{-4}{4(s-3)}+\frac{2}{(-2)(s-1)^2}+\frac{A}{(s-1)}$$

$$\equiv-\frac{1}{(s-3)}-\frac{1}{(s-1)^2}+\frac{A}{(s-1)}$$

$$\therefore \qquad -(3s-5)=-(s-1)^2-(s-3)+A(s-1)(s-3)$$

Equating coefficient of s^2

$$0=-1+A$$

$$\Rightarrow \qquad A=1$$

$$\therefore \qquad Y(s)=-\frac{1}{(s-3)}-\frac{1}{(s-1)^2}+\frac{1}{(s-1)}$$

$$\therefore \qquad y(t)=\pounds^{-1}\{Y(s)\}=-e^{3t}-te^t+e^t$$

or $$y(t)=e^t(1-t-e^{2t})$$

Substituting the value of $Y(s)$ from equation (3) in equation (2), we get

$$(2s-3)X(s)=\frac{6}{(s-3)}+6+\frac{(3s-5)}{(s-1)^2}$$

$$\therefore \qquad X(s)=\frac{6}{(s-3)(2s-3)}+\frac{6}{(2s-3)}+\frac{(3s-5)}{(2s-3)(s-1)^2}$$

$$= \frac{2}{(s-3)} - \frac{4}{(2s-3)} + \frac{6}{(2s-3)} + \frac{2(2s-3)-(s-1)}{(2s-3)(s-1)^2}$$

$$= \frac{2}{(s-3)} + \frac{2}{(2s-3)} + \frac{2}{(s-1)^2} - \frac{1}{(2s-3)(s-1)}$$

$$= \frac{2}{(s-3)} + \frac{2}{(2s-3)} + \frac{2}{(s-1)^2} + \frac{1}{(s-1)} - \frac{2}{(2s-3)}$$

$$= \frac{2}{(s-3)} + \frac{2}{(s-1)^2} + \frac{1}{(s-1)}$$

$\therefore$ $$x(t) = £^{-1}\{X(s)\} = 2e^{3t} + 2te^{t} + e^{t}$$

$$= e^{t}(2e^{2t} + 2t + 1)$$

Hence, the solution is

$$x(t) = e^{t}(2e^{2t} + 2t + 1)$$

$$y(t) = e^{t}(1 - t - e^{2t})$$

Example 2.44: Solve the simultaneous equations

$$(D^2 - 3)x - 4y = 0$$

$$x + (D^2 + 1)y = 0$$

where $D \equiv \frac{d}{dt}$ and $t > 0$, given that $x = y = \frac{dy}{dt} = 0$ and $\frac{dx}{dt} = 2$ at $t = 0$

Solution: Take Laplace transform of both sides of given equations

$$s^2X(s) - sx(0) - x'(0) - 3X(s) - 4Y(s) = 0$$

and $$X(s) + s^2Y(s) - sy(0) - y'(0) + Y(s) = 0$$

where $$X(s) = £\{x(t)\},\ Y(s) = £\{y(t)\}$$

$\therefore$ $$(s^2 - 3)X(s) - 4Y(s) = 2 \quad (1)$$

and $$X(s) + (s^2 + 1)Y(s) = 0 \quad (2)$$

$$\left(\because\ x(0) = y(0) = y'(0) = 0 \text{ and } x'(0) = 2\right)$$

Multiply equation (2) by $(s^2 - 3)$ and then subtract from equation (1)

$$-\{4 + (s^2 + 1)(s^2 - 3)\}\ Y(s) = 2$$

i.e., $$-(4+s^4-2s^2-3)Y(s)=2$$

or $$Y(s)=\frac{-2}{\left(s^4-2s^2+1\right)}=-\frac{2}{\left(s^2-1\right)^2} \tag{3}$$

$\therefore$ $$Y(s)=-2\left\{\frac{1}{(s+1)(s-1)}\right\}^2=-2\left\{\frac{1}{2(s-1)}-\frac{1}{2(s+1)}\right\}^2$$

$\therefore$ $$Y(s)=-\frac{1}{2(s-1)^2}-\frac{1}{2(s+1)^2}+\frac{1}{\left(s^2-1\right)}$$

$$=-\frac{1}{2}\left[\frac{1}{(s-1)^2}+\frac{1}{(s+1)^2}-\frac{2}{\left(s^2-1\right)}\right]$$

$\therefore$ $$y(t)=£^{-1}\{Y(s)\}=-\frac{1}{2}(te^t+te^{-t}-2\sinh t)$$

$$=-\frac{1}{2}(2t\cosh t-2\sinh t)$$

$$=\sinh t-t\cosh t$$

From equations (2) and (3)

$$X(s)=\frac{2\left(s^2+1\right)}{\left(s^2-1\right)^2}=\frac{2\left(s^2+1\right)}{(s-1)^2(s+1)^2}=\frac{(s-1)^2+(s+1)^2}{(s-1)^2(s+1)^2}$$

$\therefore$ $$X(s)=\frac{1}{(s+1)^2}+\frac{1}{(s-1)^2}$$

$\therefore$ $$x(t)=£^{-1}\{X(s)\}=te^{-t}+te^t=2t\left(\frac{e^{-t}+e^t}{2}\right)$$

$\therefore$ $$x(t)=2t\cosh t$$

Hence, the solution is $x(t) = 2t \cosh t$ and $y(t) = \sinh t - t \cosh t$.

Example 2.45: Find the solution of the following problems

(i) $y(t)=t^2+\int_0^t y(u)\sin(t-u)\,du$

(ii) $F(t)+2\cdot\int_0^t F(t)\,dt=\cosh 2t$

(iii) $y'+5y+4\int_0^t y(u)\,du=f(t)\ ;y(0)=2$

where graph of $f(t)$ is

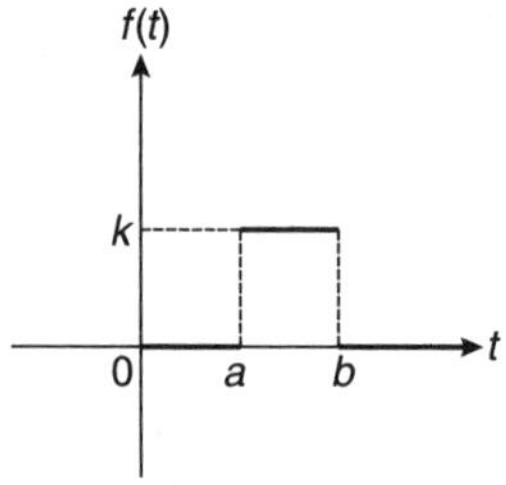

Figure 2.10

(iv) $f(t) = t + e^{-2t} + \int_0^t f(u)e^{2(t-u)}\,du$

Solution: (i) $y(t) = t^2 + \int_0^t y(u)\sin(t-u)\,du$

$$= t^2 + y(t) * \sin t$$

Take Laplace transform of both sides

$$Y(s) = \frac{2}{s^3} + Y(s)\frac{1}{(s^2+1)} \quad \text{where } Y(s) = £\{y(t)\}$$

$$\therefore \quad \left(1 - \frac{1}{s^2+1}\right)Y(s) = \frac{2}{s^3}$$

$$\Rightarrow \quad \frac{s^2}{(s^2+1)}Y(s) = \frac{2}{s^3}$$

$$\therefore \quad Y(s) = \frac{2(s^2+1)}{s^5} = \frac{2}{s^3} + \frac{1}{12}\frac{4!}{s^5}$$

$$\therefore \quad y(t) = £^{-1}\{Y(s)\} = t^2 + \frac{1}{12}t^4 = t^2\left(1 + \frac{1}{12}t^2\right)$$

(ii) $F(t) + 2\cdot\int_0^t F(t)\,dt = \cosh 2t$

Take Laplace transform of both sides

$$\overline{F}(s) + 2\frac{\overline{F}(s)}{s} = \frac{s}{s^2-4} \quad \text{where } £\{F(t)\} = \overline{F}(s)$$

$$\therefore \quad \frac{s+2}{s}\overline{F}(s) = \frac{s}{s^2-4}$$

$$\therefore \quad \overline{F}(s) = \frac{s^2}{(s+2)(s^2-4)} = \frac{s^2}{(s+2)^2(s-2)}$$

Let $\dfrac{s^2}{(s+2)^2(s-2)} \equiv \dfrac{4}{16(s-2)} + \dfrac{4}{(s+2)^2(-4)} + \dfrac{A}{s+2}$ (by suppression method)

$$\therefore \quad s^2 \equiv \frac{1}{4}(s+2)^2 - (s-2) + A(s+2)(s-2)$$

Equate coefficient of s^2

$$1 = \frac{1}{4} + A$$

$$\Rightarrow \quad A = \frac{3}{4}$$

$$\therefore \quad \overline{F}(s) = \frac{1}{4(s-2)} - \frac{1}{(s+2)^2} + \frac{3}{4(s+2)}$$

$$\therefore \quad F(t) = £^{-1}\left\{\overline{F}(s)\right\} = \frac{1}{4}e^{2t} - te^{-2t} + \frac{3}{4}e^{-2t}$$

$$\therefore \quad F(t) = \frac{1}{4}\left[e^{2t} + (3-4t)e^{-2t}\right]$$

(iii) From the graph

$$f(t) = k(u_a(t) - u_b(t))$$

$\therefore$ Equation is $y' + 5y + 4\int_0^t y(u)\,du = k\left(u_a(t) - u_b(t)\right)$

Take Laplace transform of both sides

$$sY(s) - y(0) + 5Y(s) + \frac{4}{s}Y(s) = \frac{k}{s}\left(e^{-as} - e^{-bs}\right) \qquad \text{where } Y(s) = £\{y(t)\}$$

$$\therefore \quad \frac{\left(s^2+5s+4\right)}{s}Y(s) = \frac{k}{s}\left(e^{-as} - e^{-bs}\right) + 2 \qquad (\because y(0) = 2)$$

$$\therefore \quad Y(s) = \frac{2s}{s^2+5s+4} + \frac{k\left(e^{-as} - e^{-bs}\right)}{\left(s^2+5s+4\right)}$$

$$= \frac{2s}{(s+1)(s+4)} + \frac{k}{(s+1)(s+4)}\left(e^{-as} - e^{-bs}\right)$$

$$= \frac{-2}{3(s+1)} + \frac{8}{3(s+4)} + k\left\{\frac{1}{3(s+1)} - \frac{1}{3(s+4)}\right\}\left(e^{-as} - e^{-bs}\right)$$

(by suppression method)

$$\therefore \quad y(t) = £^{-1}\{Y(s)\} = \frac{-2}{3}e^{-t} + \frac{8}{3}e^{-4t} + k\left\{\frac{1}{3}e^{-(t-a)} - \frac{1}{3}e^{-4(t-a)}\right\}u_a(t)$$

$$+k\left\{-\frac{1}{3}e^{-(t-b)} + \frac{1}{3}e^{-4(t-b)}\right\}u_b(t)$$

$$\therefore \quad y(t)=\begin{cases}\frac{2}{3}\left(4e^{-4t}-e^{-t}\right); & 0\le t<a\\ \frac{2}{3}\left(4e^{-4t}-e^{-t}\right)+\frac{k}{3}\left\{e^{-(t-a)}-e^{-4(t-a)}\right\}; & a\le t<b\\ \frac{2}{3}\left(4e^{-4t}-e^{-t}\right)+\frac{k}{3}\left\{e^{-(t-a)}-e^{-4(t-a)}-e^{-(t-b)}+e^{-4(t-b)}\right\}; & t\ge b\end{cases}$$

(iv) $$f(t)=t+e^{-2t}+\int_0^t f(u)\,e^{2(t-u)}du=t+e^{-2t}+f(t)*e^{2t}$$

Take Laplace transform of both sides

$$F(s)=\frac{1}{s^2}+\frac{1}{s+2}+F(s)\cdot\frac{1}{(s-2)} \qquad \text{where } F(s)=£\{f(t)\}$$

$$\therefore \quad \left(1-\frac{1}{s-2}\right)F(s)=\frac{1}{s^2}+\frac{1}{s+2}$$

$$\therefore \quad F(s)=\frac{(s-2)}{(s-3)s^2}+\frac{(s-2)}{(s+2)(s-3)}=\frac{1}{s}\left[\frac{2}{3s}+\frac{1}{3(s-3)}\right]+\frac{4}{5(s+2)}+\frac{1}{5(s-3)}$$

(by suppression method)

$$\therefore \quad F(s)=\frac{2}{3s^2}+\frac{1}{3}\left\{-\frac{1}{3s}+\frac{1}{3(s-3)}\right\}+\frac{4}{5(s+2)}+\frac{1}{5(s-3)}$$

$$=\frac{2}{3s^2}-\frac{1}{9s}+\frac{14}{45(s-3)}+\frac{4}{5(s+2)}$$

$$\therefore \quad f(t)=£^{-1}\{F(s)\}=\frac{2}{3}t-\frac{1}{9}+\frac{14}{45}e^{3t}+\frac{4}{5}e^{-2t}$$

$$=\frac{1}{45}\left(30t-5+14e^{3t}+36e^{-2t}\right)$$

EXERCISE 2.2

1. Find the inverse Laplace transform of the following functions

(i) $\dfrac{3(s^2-1)^2}{2s^5}+\dfrac{12}{4-3s}+\dfrac{4}{\sqrt{s}}$ (ii) $\dfrac{s^3}{(s^4-a^4)}$ (iii) $\dfrac{2s^2-1}{(s^2+1)(s^2+4)}$

(iv) $\dfrac{2s^3-6s+5}{s^3-6s^2+11s-6}$ (v) $\dfrac{s}{\left(s^2-a^2\right)^2}$ (vi) $\dfrac{5s+3}{(s-1)(s^2+2s+5)}$

(vii) $\dfrac{1+2s}{(s+2)^2(s-1)^2}$ (viii) $\dfrac{s+2}{(s+3)(s+1)^3}$ (ix) $\dfrac{1}{\left(s^2-a^2\right)^2}$

(x) $\dfrac{s}{s^4+s^2+1}$ (xi) $\dfrac{6s^3-21s^2+20s-7}{(s+1)(s-2)^3}$

2. Find the inverse Laplace transforms of

(i) $\log\left(\dfrac{s^2+a^2}{s^2+b^2}\right)$ (ii) $\tan^{-1}(s-1)$ (iii) $\dfrac{e^{-1/s}}{s}$ (iv) $\log\left(1+\dfrac{1}{s^2}\right)$

(v) $\dfrac{1}{(s^2+1)^3}$ (vi) $\left\{s\log\dfrac{s}{\sqrt{s^2+1}}+\cot^{-1}s\right\}$

3. Find the inverse Laplace transforms of

(i) $\dfrac{e^{-s}-3e^{-3s}}{s^2}$ (ii) $\dfrac{e^{-cs}}{s^2(s+a)}$; $c>0$ (iii) $\dfrac{e^{-s}}{\sqrt{s+1}}$

(iv) $\dfrac{e^{-3s}}{(s^2+8s+25)}$ (v) $\dfrac{se^{-s/2}+\pi e^{-s}}{(s^2+\pi^2)}$ (vi) $\left(\dfrac{se^{-as}}{s^2-w^2}\right)$; $a>0$

4. If $\pounds^{-1}\left\{\dfrac{s}{(s^2+1)^2}\right\}=\dfrac{1}{2}t\sin t$, find $\pounds^{-1}\left\{\dfrac{32s}{(16s^2+1)^2}\right\}$.

5. State convolution theorem and hence evaluate

(i) $\pounds^{-1}\left\{\dfrac{1}{(s-2)(s-3)}\right\}$

(ii) $\pounds^{-1}\left\{\dfrac{1}{s^2(s+1)^2}\right\}$

6. Use convolution theorem to evaluate the Laplace transform of $\dfrac{8s}{(s^2+16)(s^2+1)^2}$.

7. Using the convolution theorem, evaluate $\int_0^t \sin u.\cos(t-u)\,du$.

8. Use convolution theorem to solve the following integral equations

(i) $f(t)=t+6\int_0^t f(u).e^{(t-u)}du$

(ii) $f(t)=1+t+2\int_0^t \sin u.f(t-u)\,du$

(iii) $f(t)=t+e^{-2t}+\int_0^t f(u)\,e^{2(t-u)}du$

9. Solve the following boundary value problem using the Laplace transform

$$y''(t)+9y(t)=\cos 2t$$

$$y(0)=1,\ y(\pi/2)=-1$$

10. Using Laplace transform techniques, solve the following differential equations

(i) $y''' - 2y'' + 5y' = 0;\ y = 0,\ y' = 1$ at $t = 0$ and $y = 1$ at $t = \frac{\pi}{8}$

(ii) $(D^2 + n^2)x = a\sin(nt + \alpha);\ x = Dx = 0$ at $t = 0$.

(iii) $\frac{d^2y}{dx^2} + 4\frac{dy}{dx} + 8y = 1$ given that $y = 1$ and $\frac{dy}{dx} = 1$ at $x = 0$

(iv) $\frac{d^4x}{dt^2} - a^4x = 0$, where a is a contant and $x = 1,\ x' = x'' = x''' = 0$ at $t = 0$.

(v) $\frac{d^2x}{dt^2} + x = t\cos 2t$ given that $x(0) = x'(0) = 0$

(vi) $y''' - 3y'' + 3y' - y = t^2e^t$, given $y(0) = 1,\ y'(0) = 0,\ y''(0) = -2$

(vii) $y'' + 4y' + 3y = e^{-t}$, $y(0) = y'(0) = 1$

(viii) $y'' + 2y' - 3y = 3$, $y(0) = 4$, $y'(0) = -7$

(ix) $y'' - 5y' + 4y = e^{2t}$, $y(0) = \frac{19}{12},\ y'(0) = \frac{8}{3}$

(x) $y'' + 4y' + 13y = e^{-t}$, $y(0) = 0,\ y'(0) = 2$

(xi) $y'' + 3y' + 2y = t\delta(t-1)$, $y(0) = 0,\ y'(0) = 0$

11. A particle moves in a line so that its displacement x from a fixed point 0 at any time t, is given by $\frac{d^2x}{dt^2} + 4\frac{dx}{dt} + 5x + 80\sin 5t$

If initially particle is at rest at $x = 0$, find its displacement at any time t.

12. Solve $\frac{d^2x}{dt^2} + 4x = \phi(t)$ with $x(0) = x'(0) = 0$,

where $\phi(t) = 0$ when $0 < t < \pi$

$= \sin t$ when $\pi < t < 2\pi$

$= 0$ when $t > 2\pi$

13. Using convolution theorem, solve the initial value problem $y'' + 9y = \sin 3t,\ y(0) = 0,\ y'(0) = 0$

14. Solve the initial value problem $y' - 4y + 3\int_0^t y(u)\, du = t,\ y(0) = 1$

15. Solve $\frac{d^2y}{dt^2} + t\frac{dy}{dt} - y(t) = 0$ if $y(0) = 0,\ \left(\frac{dy}{dt}\right)_{t=0} = 1$.

16. Solve the problem $t\frac{d^2y}{dt^2} + \frac{dy}{dt} + 4ty = 0$ given that $y = 3$ and $\frac{dy}{dt} = 0$ when $t = 0$.

17. Using Laplace transform solve the following differential equation $y'' + 2ty' - y = t$ when $y(0) = 0$ and $y'(0) = 1$

18. Solve the following simultaneous equations by Laplace transform

(i) $\frac{dx}{dt} - y = e^t, \frac{dy}{dt} + x = \sin t;$ given $x(0) = 1, y(0) = 0.$

(ii) $\frac{dx}{dt} + y = \sin t, \frac{dy}{dt} + x = \cos t;$ given $x = 2$ and $y = 0$ when $t = 0$.

(iii) $\frac{dx}{dt} + \frac{dy}{dt} + x = e^{-t}, \frac{dx}{dt} + 2\frac{dy}{dt} + 2x + 2y = 0;$ given that $x(0) = -1, y(0) = 1.$

19. Solve the following simultaneous equations

$$\frac{dx}{dt} + x + 3\int_0^t y\,dt = \cos t + 3\sin t,\ 2\frac{dx}{dt} + 3\frac{dy}{dt} + 6y = 0$$

subject to the conditions $x = -3, y = 2$ at $t = 0$.

20. Solve the simultaneous equations $\frac{dx}{dt} - \frac{dy}{dt} - 2x + 2y = 1 - 2t,\ \frac{d^2x}{dt^2} + 2\frac{dy}{dt} + x = 0$ given $x = 0, y = 0, \frac{dx}{dt} = 0$ when $t = 0$

21. Solve the simultaneous equations $\frac{d^2x}{dt^2} + 5\frac{dy}{dt} - x = t,\ 2\frac{dx}{dt} - \frac{d^2y}{dt^2} + 4y = 2$ given that when $t = 0, x = 0, y = 0\ \frac{dx}{dt} = 0,\ \frac{dy}{dt} = 0$

ANSWERS 2.2

1. (i) $\frac{3}{2} - \frac{3}{2}t^2 + \frac{1}{16}t^4 - 4e^{\frac{4}{3}t} + \frac{4}{\sqrt{\pi t}}$ (ii) $\frac{1}{2}(\cosh at + \cos at)$

(iii) $-\sin t + \frac{3}{2}\sin 2t$ (iv) $\frac{1}{2}e^t - e^{2t} + \frac{5}{2}e^{3t}$ (v) $\frac{t}{2a}\sinh at$

(vi) $e^t - e^{-t}\cos 2t + \frac{3}{2}e^{-t}\sin 2t$ (vii) $\frac{t}{3}(e^t - e^{-2t})$

(viii) $\frac{1}{8}\left\{e^{-3t} + (2t^2 + 2t - 1)e^{-t}\right\}$ (ix) $\frac{at\cosh at - \sinh at}{2a^3}$

(x) $\frac{2}{\sqrt{3}}\sin\left(\frac{\sqrt{3}}{2}t\right)\sinh\frac{t}{2}$ (xi) $2e^{-t} + \left(4 + 3t - \frac{t^2}{2}\right)e^{2t}$

2. (i) $\frac{2}{t}(\cos bt - \cos at)$ (ii) $-\frac{1}{t}e^t\sin t$ (iii) $J_0\left(2\sqrt{t}\right)$

(iv) $\frac{2(1-\cos t)}{t}$ (v) $\frac{1}{8}\left[\left(3 - t^2\right)\sin t - 3t\cos t\right]$ (vi) $\frac{1-\cos t}{t^2}$

3. (i) $\begin{cases} 0 & ; \quad 0 \le t < 1 \\ t-1 & ; \quad 1 \le t < 3 \\ -2t+8 & ; \quad t \ge 3 \end{cases}$

(iii) $\frac{1}{a^2}\left\{a(t-c)-1+e^{-a(t-c)}\right\}U(t-c)$ (iv) $\frac{e^{-(t-1)}}{\sqrt{\pi(t-1)}}\cdot U(t-1)$

(v) $\frac{1}{3}e^{-4(t-3)}\sin(3t-9)U(t-3)$ (vi) $\sin \pi t\left[U\left(t-\frac{1}{2}\right)-U(t-1)\right]$

(vii) $\cosh\{w(t-a)\}\cdot U(t-a)$

4. $t\sin\frac{t}{4}$ 5. (i) $e^{3t}-e^{2t}$ (ii) $t+te^{-t}-2+2e^{-t}$

6. $\frac{1}{225}(60t\sin t+8\cos 4t-8\cos t)$ 7. $\frac{t}{2}\sin t$

8. (i) $\frac{1}{49}\left(6e^{7t}+7t-6\right)$ (ii) $2e^{t}-t-1$ (iii) $\frac{1}{45}\left(14e^{3t}-5+30t+36e^{-2t}\right)$

9. $y(t)=\frac{1}{5}(\cos 2t+4\cos 3t+4\sin 3t)$

10. (i) $y(t)=1+e^{t}(\sin 2t-\cos 2t)$ (ii) $x(t)=\frac{a}{2n^2}\{\cos\alpha\sin nt-nt\cos(\alpha+nt)\}$

(iii) $y(x)=\frac{1}{8}-\frac{e^{-2x}}{8}(\cos 2x-3\sin 2x)$ (iv) $x(t)=\frac{1}{2}(\cosh at+\cos at)$

(v) $x(t)=\frac{1}{9}(4\sin 2t-5\sin t-3t\cos 2t)$ (vi) $y(t)=e^{t}\left(1-t-\frac{t^2}{2}-\frac{t^5}{60}\right)$

(vii) $y(t)=\frac{7}{4}e^{-t}+\frac{1}{2}te^{-t}\frac{-3}{4}e^{-3t}$ (viii) $y(t)=-1+2e^{t}+3e^{-3t}$

(ix) $y(t)=\frac{-1}{2}e^{2t}+\frac{14}{9}e^{t}+\frac{19}{36}e^{4t}$ (x) $y(t)=\frac{1}{30}e^{-2t}(3e^{t}-3\cos 3t+19\sin 3t)$

(xi) $y(t)=e^{-(t-1)}U(t-1)-e^{-2(t-1)}U(t-1)$

11. $x(t)=-2(\cos 5t+\sin 5t)+2e^{-2t}(\cos t+7\sin t)$

12. $x(t)=\begin{cases} 0 & \text{for } \ 0<t<\pi \\ \frac{1}{6}\sin 2t+\frac{1}{3}\sin t & \text{for } \ \pi<t<2\pi \\ \frac{1}{3}\sin 2t & \text{for } \ t<2\pi. \end{cases}$

13. $y(t) = \frac{1}{18}(\sin 3t - 3t\cos 3t)$

14. $y(t) = \frac{1}{3} - e^t + \frac{5}{3}e^{3t}$ 15. $y = t$ 16. $y = 3\,J_0(2t)$ 17. $y = t$

18. (i) $x(t) = \frac{1}{2}(e^t + \cos t + 2\sin t - t\cos t),\ \ y(t) = \frac{1}{2}(t\sin t - e^t + \cos t - \sin t)$

(ii) $x(t) = e^{-t} + e^t,\ y(t) = \sin t + e^{-t} - e^t$

(iii) $x(t) = -e^{-t}(\cos t + \sin t),\ y(t) = e^{-t}(1 + \sin t)$

19. $x(t) = e^{-3t} + \sin t - 2\cos t,\ y(t) = 2e^{-3t} - \frac{2}{3}\sin t$

20. $x(t) = 2(1 - e^{-t} - te^{-t}),\ \ y(t) = 2 - t - 2(t+1)e^{-t}$

21. $x(t) = -t + 5\sin t - 2\sin 2t,\ y(t) = 1 - 2\cos t + \cos 2t$

2.16 APPLICATIONS OF LAPLACE TRANSFORM TO ENGINEERING PROBLEMS

Laplace transform is utilized as a tool for solving some of the engineering problems. We now discuss some of these applications.

2.16.1 Problems Related to Electrical Circuits

Consider the *LCR* circuit shown in the figure. It contains resistance *R* measured in ohms (Ω), inductance *L* measured in henrys, and capacitor *C* measured in farads. The voltage drop is equal to voltage of battery. If battery is not in the system, then total voltage drop will be zero. If $i(t)$ be current and $q(t)$ charge at any time *t*, then $i(t) = \frac{dq}{dt}$. Voltage drop through resistance *R* is $Ri = R\frac{dq}{dt}$, voltage drop in inductance *L* is $L\frac{di}{dt} = L\frac{d^2q}{dt^2}$, and voltage drop in capacitor of capacity *C* is $\frac{q}{C} = \frac{1}{C}\int_0^t i(t)dt$.

According to Kirchoff's laws, we have

(i) Total incoming current at a point is always equal to the total outgoing current;

(ii) Total voltage drop in a closed circuit or in a mesh is equal to total voltage of the electric source, i.e., battery etc.

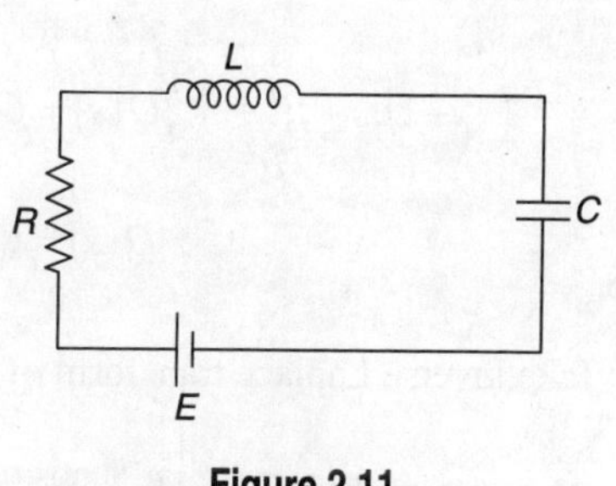

Figure 2.11

Example 2.46: A condenser of capacity C is charged to potential E and discharged at $t = 0$ through an inductance L and resistance R. Show that the charge q at any time t is given by

$$q = \frac{CE}{\eta} e^{-\mu t}\left[\mu \sin \eta t + \eta \cos \eta t\right] \text{ where } \mu = \frac{R}{2L} \text{ and } \eta^2 = \frac{1}{LC} - \frac{R^2}{4L^2}; L > \frac{CR^2}{4}$$

Solution:

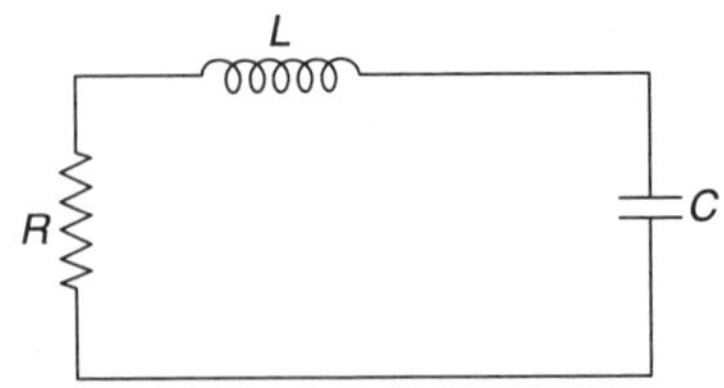

Figure 2.12

If $q(t)$ is charge at any time t then by Kirchoff's law, $q(t)$ satisfies the differential equation

$$L\frac{d^2q}{dt} + R\frac{dq}{dt} + \frac{q}{C} = 0$$

Take Laplace transform of both sides

$$L\left(s^2Q(s) - sq(0) - q'(0)\right) + R\left(sQ(s) - q(0)\right) + \frac{1}{C}Q(s) = 0$$

where $Q(s) = £(q(t))$

But it is given that $q(0) = EC$ and $q'(0) = i(0) = 0$

$$\therefore \quad \left(Ls^2 + Rs + \frac{1}{C}\right)Q(s) = CE\left(Ls + R\right)$$

$$\therefore \quad \left(s^2 + \frac{R}{L}s + \frac{1}{CL}\right)Q(s) = CE\left(s + \frac{R}{L}\right)$$

Let $\quad \frac{R}{L} = 2\mu, \ \frac{1}{CL} = \mu^2 + \eta^2 = \frac{R^2}{4L^2} + \eta^2$

$$\left(\text{as } L > \frac{CR^2}{4}, \text{ so } \frac{1}{CL} - \frac{R^2}{4L^2} > 0 \text{ and hence can be assumed } \eta^2\right)$$

$$\therefore \quad \left(s^2 + 2\mu s + \mu^2 + \eta^2\right)Q(s) = CE\left(s + 2\mu\right)$$

$$\therefore \quad Q(s) = CE\left[\frac{s+\mu}{(s+\mu)^2 + \eta^2} + \frac{\mu}{\eta}\frac{\eta}{(s+\mu)^2 + \eta^2}\right]$$

Take inverse Laplace transform of both sides

$$q(t) = CE\left[e^{-\mu t}\cos \eta(t) + \frac{\mu}{\eta}e^{-\mu t}\sin \eta t\right] = \frac{CE}{\eta}e^{-\mu t}\left[\eta \cos \eta t + \mu \sin \eta t\right]$$

$$\therefore \qquad q(t) = \frac{CE}{\eta} e^{-\mu t}\left[\mu \sin(\eta t) + \eta \cos(\eta t)\right]$$

where $\qquad \mu = \dfrac{R}{2L}, \eta^2 = \dfrac{1}{CL} - \mu^2 = \dfrac{1}{CL} - \dfrac{R^2}{4L^2}.$

Example 2.47: An impulsive voltage $E\delta(t)$ is applied to a circuit consisting of L, C, R in series with zero initial conditions. If i is the current at any subsequent time t, find the limit of i as $t \to 0$.

Solution:

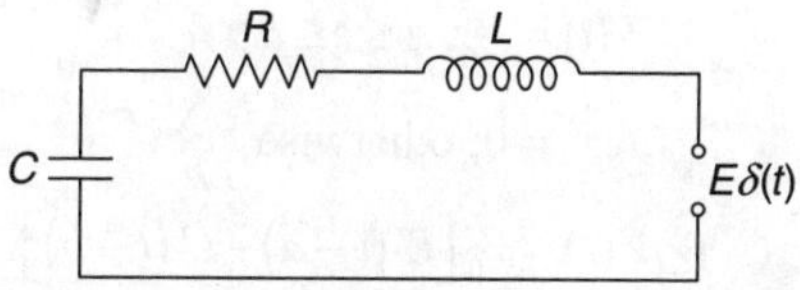

Figure 2.13

By Kirchoff's law $i(t)$ satisfy the differential equation

$$L\frac{di}{dt} + Ri + \frac{1}{C}\int_0^t i\,dt = E\delta(t)$$

Take Laplace transform of both sides

$$L(sI(s) - i(0)) + RI(s) + \frac{1}{Cs} I(s) = E$$

where $I(s) = \pounds(i(t))$

Initially $i(0) = 0$

$$\therefore \qquad \left(Ls + R + \frac{1}{Cs}\right) I(s) = E$$

$$\therefore \qquad I(s) = \frac{E}{Ls + R + \dfrac{1}{Cs}}$$

By initial value theorem

$$\lim_{t\to 0^+} i(t) = \lim_{s\to\infty} sI(s) = \lim_{s\to\infty} \frac{Es}{Ls + R + \dfrac{1}{Cs}}$$

$$= \lim_{s\to\infty} \frac{E}{L + \dfrac{R}{s} + \dfrac{1}{Cs^2}} = \frac{E}{L}.$$

Example 2.48: Find the current $i(t)$ in the circuit in figure given below, if a single square wave with voltage V_0 is applied. The circuit is assumed to be quiescent before the square wave is applied.

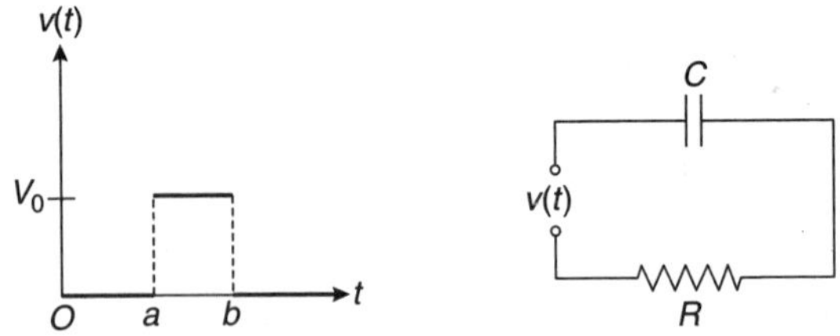

Figure 2.14 (a) and (b)

Solution:

$$V(t) = V_0;\ a \le t \le b$$

$$= 0;\ \text{otherwise}$$

$\therefore$
$$V(t) = V_0\left[U(t-a) - U(t-b)\right]$$

If $i(t)$ is current in the system at any time t then by Kirchoff's law

$$Ri + \frac{1}{C}\int_0^t i\,dt = V(t) = V_0\left[U(t-a) - U(t-b)\right]$$

Take Laplace transform of both sides

$$RI(s) + \frac{1}{Cs}I(s) = \frac{V_0}{s}\left[e^{-as} - e^{-bs}\right] \qquad \text{where } I(s) = \pounds(i(t))$$

$\therefore$
$$\left(s + \frac{1}{RC}\right)I(s) = \frac{V_0}{R}\left(e^{-as} - e^{-bs}\right)$$

$\therefore$
$$I(s) = \frac{V_0}{R}\left(\frac{e^{-as}}{s+\frac{1}{RC}} - \frac{e^{-bs}}{s+\frac{1}{RC}}\right)$$

Take inverse Laplace transform of both sides

$$i(t) = \frac{V_0}{R}\left(e^{-\frac{1}{RC}(t-a)}U(t-a) - e^{-\frac{1}{RC}(t-b)}U(t-b)\right)$$

Let $C_1 = \frac{V_0}{R}e^{\frac{a}{RC}},\ C_2 = \frac{V_0}{R}e^{\frac{b}{RC}}$

$\therefore$
$$i(t) = C_1 e^{-\frac{t}{RC}}U(t-a) - C_2 e^{-\frac{t}{RC}}U(t-b)$$

$\therefore$
$$i(t) = \begin{cases} 0 & ;\ 0 < t < a \\ C_1 e^{-\frac{t}{RC}} & ;\ a \le t < b \\ (C_1 - C_2)e^{-\frac{t}{RC}} & ;\ t \ge b \end{cases}$$

where $C_1 = \frac{V_0}{R} e^{\frac{a}{RC}}, \; C_2 = \frac{V_0}{R} e^{\frac{b}{RC}}$

Example 2.49: Given that $i = q = 0$ at $t = 0$, find charge q and current i in the below circuit for $t > 0$.

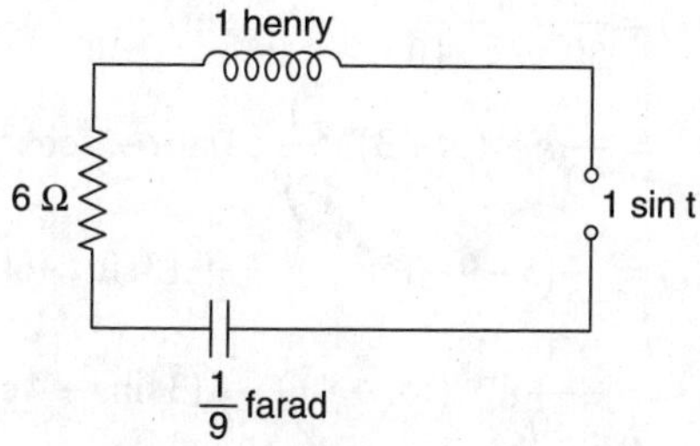

Figure 2.15

Solution: If L Henry inductance, R Ω resistance and C farad capacity capacitor is in the circuit having E e.m.f. source then by Kirchoff's law

$$L\frac{d^2q}{dt} + R\frac{dq}{dt} + \frac{q}{C} = E$$

It is given that $L = 1, R = 6, \; C = \frac{1}{9}, E = \sin t$

$$\therefore \qquad \frac{d^2q}{dt^2} + 6\frac{dq}{dt} + 9q = \sin t$$

Take Laplace transform of both sides

$$s^2Q(s) - sq(0) - q'(0) + 6\left(sQ(s) - q(0)\right) + 9Q(s) = \frac{1}{s^2+1}$$

where $Q(s) = \pounds\,(q(t))$

It is given that $q(0) = 0, q'(0) = i(0) = 0$

$$\therefore \quad \left(s^2 + 6s + 9\right)Q(s) = \frac{1}{s^2+1}$$

$$\therefore \qquad Q(s) = \frac{1}{\left(s^2+1\right)\left(s+3\right)^2}$$

$$\therefore \qquad Q(s) = \frac{1}{10(s+3)}\left[\frac{1}{s+3} - \frac{s-3}{s^2+1}\right] = \frac{1}{10(s+3)^2} - \frac{s+3-6}{10(s+3)\left(s^2+1\right)}$$

$$= \frac{1}{10(s+3)^2} - \frac{1}{10\left(s^2+1\right)} + \frac{3}{50}\left[\frac{1}{s+3} - \frac{s-3}{s^2+1}\right]$$

$$= \frac{1}{10(s+3)^2} - \frac{1}{10\left(s^2+1\right)} + \frac{3}{50}\cdot\frac{1}{s+3} - \frac{3}{50}\frac{s}{s^2+1} + \frac{9}{50}\frac{1}{s^2+1}$$

$$= \frac{3}{50(s+3)} + \frac{1}{10(s+3)^2} + \frac{2}{25(s^2+1)} - \frac{3}{50}\frac{s}{s^2+1}$$

Take inverse Laplace transform of both sides

$$q(t) = \frac{3}{50}e^{-3t} + \frac{t}{10}e^{-3t} + \frac{2}{25}\sin t - \frac{3}{50}\cos t$$

$$= \frac{1}{50}e^{-3t}(5t+3) + \frac{1}{50}(4\sin t - 3\cos t)$$

$$i(t) = q'(t) = \frac{1}{50}(5-9-15t)e^{-3t} + \frac{1}{50}(3\sin t + 4\cos t)$$

$$= -\frac{1}{50}e^{-3t}(15t+4) + \frac{1}{50}(3\sin t + 4\cos t)$$

Example 2.50: Inductance L, resistance R and capacitor of capacity C are connected in series in a circuit having e.m.f. $\delta(t)$. Find the charge at any time t given that initially the current and charge are zero.

Solution:

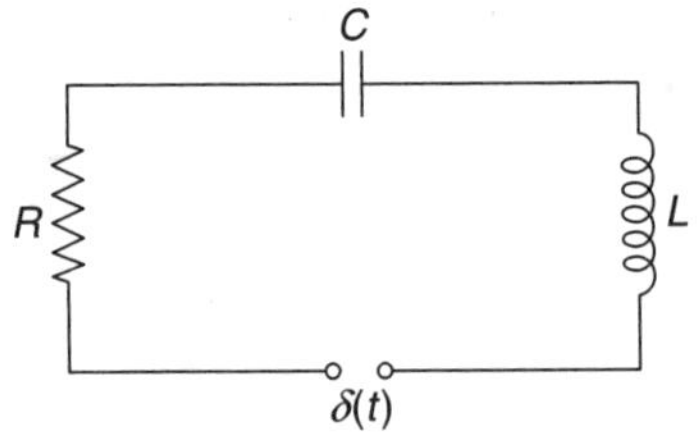

Figure 2.16

If $q(t)$ is charge at time t then by Kirchoff's law

$$L\frac{d^2q}{dt^2} + R\frac{dq}{dt} + \frac{q}{C} = \delta(t)$$

Take Laplace transform of both sides

$$L\left(s^2Q(s) - sq(0) - q'(0)\right) + R\left(sQ(s) - q(0)\right) + \frac{1}{C}Q(s) = 1 \qquad \text{where } Q(s) = \pounds(q(t))$$

It is given that $q(0) = 0,\ q'(0) = i(0) = 0$

$\therefore$
$$\left(Ls^2 + Rs + \frac{1}{C}\right)Q(s) = 1$$

or
$$\left(s^2 + \frac{R}{L}s + \frac{1}{CL}\right)Q(s) = \frac{1}{L}$$

Let
$$\frac{R}{2L} = \mu > 0 \text{ and } \lambda^2 = \frac{1}{LC} - \mu^2 = \frac{1}{LC} - \frac{R^2}{4L^2} = \frac{4L - CR^2}{4L^2C}$$

Here λ be zero or positive or purely imaginary according as $4L = CR^2,\ 4L > CR^2,\ 4L < CR^2$ respectively

$$\therefore \qquad \left(s^2 + 2\mu s + \mu^2 + \lambda^2\right)Q(s) = \frac{1}{L}$$

$$\therefore \qquad Q(s) = \frac{1}{L}\cdot\frac{1}{(s+\mu)^2 + \lambda^2}$$

If $\lambda = 0$, i.e., $4L = CR^2$ then

$$Q(s) = \frac{1}{L}\frac{1}{(s+\mu)^2}$$

Take inverse Laplace transform

$$q(t) = \frac{t}{L}e^{-\mu t}$$

$$\therefore \qquad q(t) = \frac{t}{L}e^{-\mu t}, \text{ when } 4L = CR^2 \text{ where } \mu = \frac{R}{2L}$$

If $\lambda > 0$,

i.e., $4L > CR^2$

then $Q(s) = \dfrac{1}{L\lambda}\cdot\dfrac{\lambda}{(s+\mu)^2 + \lambda^2}$

Take inverse Laplace transform

$$q(t) = \frac{1}{L\lambda}e^{-\mu t}\sin\lambda t$$

$$\therefore \qquad q(t) = \frac{1}{L\lambda}e^{-\mu t}\sin\lambda t, \qquad \text{when } 4L > CR^2 \text{ where } \mu = \frac{R}{2L},\ \lambda = \frac{\sqrt{4L - CR^2}}{2L\sqrt{C}}$$

If $\lambda < 0$, i.e., $4L < CR^2$

then replacing λ by $i\lambda$

$$q(t) = \frac{1}{Li\lambda}e^{-\mu t}\sin i\lambda t \text{ when } 4L < CR^2 \text{ where } \mu = \frac{R}{2L};\ \lambda = \frac{\sqrt{CR^2 - 4L}}{2L\sqrt{C}}$$

$$= \frac{1}{L\lambda}e^{-\mu t}\sinh\lambda t \qquad (\because \sin ix = i\sinh x)$$

$\therefore$ If $\mu = \dfrac{R}{2L}$

$$q(t) = \frac{t}{L}e^{-\mu t} \text{ if } 4L = CR^2$$

$$= \frac{1}{L\lambda}e^{-\mu t}\sin\lambda t \text{ if } 4L > CR^2 \text{ where } \lambda = \frac{\sqrt{4L - CR^2}}{2L\sqrt{C}}$$

$$= \frac{1}{L\lambda}e^{-\mu t}\sinh\lambda t; \text{ if } 4L < CR^2 \text{ where } \lambda = \frac{\sqrt{CR^2 - 4L}}{2L\sqrt{C}}$$

Example 2.51: If $i(0) = 0$ then find the current $i(t)$ at any time t in RL-network shown in the below figure.

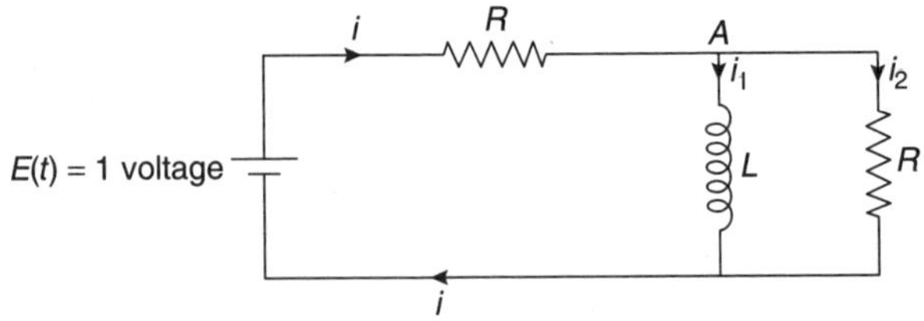

Figure 2.17

Solution: Since incoming current at A = outgoing current

$$\therefore \quad i = i_1 + i_2 \tag{1}$$

In the mesh containing both resistances and battery, by Kirchoff's law

$$Ri + Ri_2 = 1$$

Put value of i_2 from (1)

$$Ri + R(i - i_1) = 1$$

$$\therefore \quad 2Ri - Ri_1 = 1 \tag{2}$$

In the mesh containing R, L and battery, by Kirchoff's law

$$L\frac{di_1}{dt} + Ri = 1 \tag{3}$$

Take Laplace transform of both sides of (2) and (3)

$$2RI(s) - RI_1(s) = \frac{1}{s}$$

$$L\left(sI_1(s) - i_1(0)\right) + RI(s) = \frac{1}{s} \qquad \text{where } I(s) = \pounds(i(t)),\ I_1(s) = \pounds(i_1(t))$$

Since $i(0) = 0$

$\therefore \quad i_1(0) = 0$

$$\therefore \quad 2RI(s) - RI_1(s) = \frac{1}{s} \tag{4}$$

$$RI(s) + LsI_1(s) = \frac{1}{s} \tag{5}$$

By Crammer's rule

$$RI(s) = \frac{1}{2Ls + R}\begin{vmatrix} \frac{1}{s} & -R \\ \frac{1}{s} & Ls \end{vmatrix}$$

$$\therefore \qquad I(s) = \frac{Ls+R}{sR(2Ls+R)}$$

$$= \frac{1}{R}\left[\frac{1}{s} + \frac{-\frac{R}{2}+R}{-\frac{R}{2L}(2Ls+R)}\right] \qquad \text{(by suppression method)}$$

$$= \frac{1}{Rs} - \frac{L}{R(2Ls+R)}$$

$$= \frac{1}{Rs} - \frac{1}{2R}\frac{1}{s+\frac{R}{2L}}$$

Take inverse Laplace transform of both sides

$$i(t) = \frac{1}{R} - \frac{1}{2R}e^{-\frac{Rt}{2L}} = \frac{1}{2R}\left(2 - e^{-\frac{Rt}{2L}}\right)$$

2.16.2 Problem Related to Deflection of a Loaded Beam

Let y be deflection of a loaded beam at distance x from one end of the beam. Radius of curvature is given by

$$R = \frac{\left[1+\left(\frac{dy}{dx}\right)^2\right]^{\frac{3}{2}}}{\frac{d^2y}{dx^2}}$$

Since deflection is small, $\frac{dy}{dx}$ will be small, and hence $\left(\frac{dy}{dx}\right)^2$ will be negligible. Thus, we can take

$$\frac{d^2y}{dx^2} = \frac{1}{R}$$

If m be moment of the forces acting on the beam about the point at distance x from one end then

$$\frac{E}{R} = \frac{M}{I},$$

where I is moment of inertia about the neutral axis, i.e. the axis passing through the centre of gravity of the beam and E is young's modulus of the material of which beam is formed, which is assumed to be same throughout the beam.

$$\therefore \qquad EI\frac{d^2y}{dx^2} = M$$

$\frac{dM}{dx} = EI\frac{d^3y}{dx^3}$ is shearing stress at the point and $\frac{d^2M}{dx^2} = EI\frac{d^4y}{dx^4}$ is load per unit length on the beam at distance x from one end.

When a load w_0 act at a point at distance x from one end then this can be considered as the limiting case of uniform loading $\frac{w_0}{\varepsilon}$ per unit length over the portion of the beam between $x = a$ and $x = a + \varepsilon$ where ε is very small. Thus

$$w(x) = \frac{w_0}{\varepsilon};\ a < x < a + \varepsilon$$

$$= 0;\ \text{otherwise}$$

$$\therefore \qquad w(x) = w_0\,\delta(x - a)$$

It should be noted that boundary conditions will be as follows:

(i) If an end is clamped, built-in, or fixed end, then there cannot be deflection at this end, and at this end, tangent to deflection curve will be neutral axis. Thus, at this end $y = \frac{dy}{dx} = 0$.

(ii) If an end is hinged or simply supported end, then there cannot be deflection at this end, and also sum of moments M about this end will be zero. Thus, at this end $y = 0$, $\frac{d^2y}{dx^2} = 0$.

(iii) If an end is free and then sum of moments M and shearing stress $\frac{dM}{dx}$ will be zero at this end, and thus, at this end $\frac{d^2y}{dx^2} = 0$, $\frac{d^3y}{dx^3} = 0$.

Example 2.52: A weightless beam of length l is freely supported at its both ends and a concentrated load W acts at a point $x = a$ on it measured from one end. Find the resultant deflection and also the deflection under the load.

Solution: If y is deflection at distance x from A, then differential equation of deflection (load equation) is given by

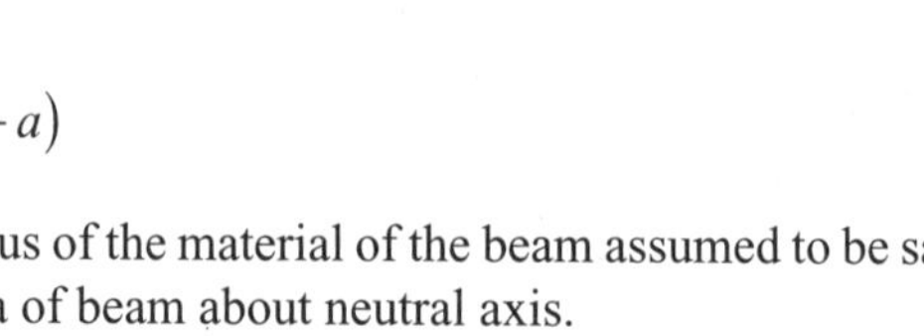

Figure 2.18

$$EI\frac{d^4y}{dx^4} = W\delta(x - a) \tag{1}$$

where E is Young's modulus of the material of the beam assumed to be same throughout the beam and I is moment of inertia of beam about neutral axis.

Take Laplace transform of both sides of (1)

$$EI\left(s^4Y(s) - s^3y(0) - s^2y'(0) - sy''(0) - y'''(0)\right) = We^{-as} \qquad \text{where } Y(s) = \pounds(y(x))$$

Now, end A is freely supported

$$\therefore \qquad y(0) = y''(0) = 0$$

Assuming $y'(0) = c_1, y'''(0) = c_2$

we have $$EI\left(s^4Y(s) - c_1s^2 - c_2\right) = We^{-as}$$

$$\therefore \quad s^4Y(s) = \frac{We^{-as}}{EI} + c_1s^2 + c_2$$

$$\therefore \quad Y(s) = \frac{We^{-as}}{EIs^4} + \frac{c_1}{s^2} + \frac{c_2}{s^4}$$

Take inverse Laplace transform of both sides

$$y(x) = \frac{W}{EI}\frac{(x-a)^3}{3!}U(x-a) + c_1x + c_2\frac{x^3}{3!}$$

$$\therefore \quad y(x) = \begin{cases} c_1x + c_2\dfrac{x^3}{6} & ;\ 0 \le x < a \\ \dfrac{W(x-a)^3}{6EI} + c_1x + c_2\dfrac{x^3}{6} & ;\ a \le x \le l \end{cases} \tag{2}$$

Now, end B is freely supported

$$\therefore \quad y(l) = y''(l) = 0$$

Now,

$$y(l) = \frac{W(l-a)^3}{6EI} + c_1l + c_2\frac{l^3}{6} = 0 \tag{3}$$

$$y''(x) = \frac{W(x-a)}{EI} + c_2x$$

$$\therefore \quad y''(l) = \frac{W(l-a)}{EI} + c_2l = 0$$

$$\therefore \quad c_2 = -\frac{W(l-a)}{EIl} \tag{4}$$

$\therefore$ from (3)

$$\frac{W(l-a)^3}{6EI} + c_1l - \frac{W(l-a)l^2}{6EI} = 0$$

$$\therefore \quad c_1 = -\frac{W(l-a)^3}{6EIl} + \frac{Wl(l-a)}{6EI} = W\frac{(l-a)(2al-a^2)}{6EIl} \tag{5}$$

Substituting from (4) and (5) in (2), deflection at distance x from A is

$$y(x) = \frac{-W(l-a)(a^2-2al)x}{6EIl} - \frac{W(l-a)x^3}{6EIl} = \frac{W(l-a)x}{6EIl}\left[2al - a^2 - x^2\right]; \quad 0 \le x < a$$

$$= \frac{W(x-a)^3}{6EI} + \frac{W(l-a)x}{6EIl}\left(2al - a^2 - x^2\right); \quad a \le x \le l$$

For deflection under the load, we have $x = a$.

$\therefore$ Deflection under the load is

$$y(a) = \frac{W(l-a)a}{6EIl}\left(2al - a^2 - a^2\right) = \frac{Wa^2(l-a)^2}{3EIl}$$

Example 2.53: A weightless beam of length L has its ends clamped at $x = 0$ and $x = L$. A concentrated load W acts vertically downward at the point $x = \frac{L}{3}$. Find the resulting deflection.

Solution: If y is deflection at distance x from A, then differential equation of deflection (load equation) is given by

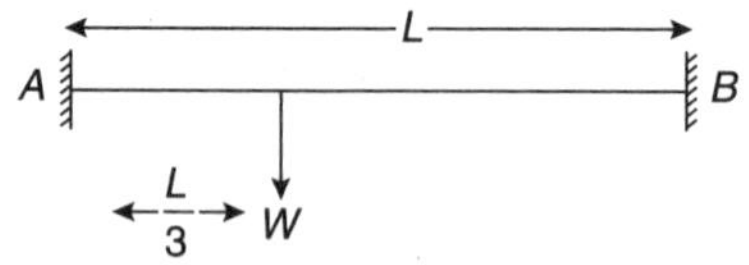

Figure 2.19

$$EI\frac{d^4y}{dx^4} = W\delta\left(x - \frac{L}{3}\right) \qquad (1)$$

where E is Young's modulus of the material of the beam assumed to be same throughout the beam and I is moment of inertia of beam about neutral axis.

Take Laplace transform of both sides of (1)

$$EI\left(s^4Y(s) - s^3y(0) - s^2y'(0) - sy''(0) - y'''(0)\right) = We^{-\frac{L}{3}s} \qquad \text{where } Y(s) = \pounds\,(y(x))$$

End A is clamped

$\therefore \quad y(0) = y'(0) = 0$

Let $y''(0) = c_1, y'''(0) = c_2$

$$\therefore \qquad EI\left(s^4Y(s) - c_1s - c_2\right) = We^{-\frac{L}{3}s}$$

$$\therefore \qquad Y(s) = \frac{W}{EI}\cdot\frac{e^{-\frac{L}{3}s}}{s^4} + \frac{c_1}{s^3} + \frac{c_2}{s^4}$$

Take inverse Laplace transform of both sides

$$y(x) = \frac{W}{6EI}\left(x - \frac{L}{3}\right)^3 U\left(x - \frac{L}{3}\right) + \frac{c_1x^2}{2} + \frac{c_2x^3}{6}$$

$$\therefore \qquad y(x) = \begin{cases} c_1\dfrac{x^2}{2} + c_2\dfrac{x^3}{6} & ;\ 0 \le x < \dfrac{L}{3} \\ \dfrac{W}{6EI}\left(x - \dfrac{L}{3}\right)^3 + c_1\dfrac{x^2}{2} + c_2\dfrac{x^3}{6} & ;\ \dfrac{L}{3} \le x \le L \end{cases} \qquad (2)$$

End B is clamped

$$\therefore \quad y(L) = y'(L) = 0$$

$$y(L) = \frac{4WL^3}{81EI} + \frac{c_1 L^2}{2} + c_2 \frac{L^3}{6} = 0 \tag{3}$$

$$y'(x) = \frac{W}{2EI}\left(x - \frac{L}{3}\right)^2 + c_1 x + \frac{c_2}{2} x^2; \ \frac{L}{3} < x \le L$$

$$\therefore \quad y'(L) = \frac{2WL^2}{9EI} + c_1 L + \frac{c_2}{2} L^2 = 0 \tag{4}$$

(3) and (4) can be written as

$$3c_1 + c_2 L + \frac{8}{27}\frac{WL}{EI} = 0$$

$$2c_1 + c_2 L + \frac{4}{9}\frac{WL}{EI} = 0$$

$$\therefore \quad \frac{c_1}{\frac{4}{9} - \frac{8}{27}} = \frac{c_2 L}{\frac{16}{27} - \frac{4}{3}} = \frac{WL/EI}{3-2}$$

$$\therefore \quad c_1 = \frac{4}{27}\frac{WL}{EI}, \ c_2 = -\frac{20}{27}\frac{W}{EI}$$

$\therefore$ from (2)

$$y(x) = \begin{cases} \dfrac{2}{27}\dfrac{WL}{EI}x^2 - \dfrac{10}{81}\dfrac{W}{EI}x^3 = \dfrac{2}{81}\dfrac{Wx^2}{EI}(3L - 5x) \ ; & 0 \le x < \dfrac{L}{3} \\ \dfrac{W}{6EI}\left(x - \dfrac{L}{3}\right)^3 + \dfrac{2}{81}\dfrac{Wx^2}{EI}(3L - 5x) \ ; & \dfrac{L}{3} \le x \le L \end{cases}$$

Example 2.54: A cantilever beam clamped at $x = 0$ and free at $x = l$ carries a uniform load w_0 per unit length. Show that deflection at any point is $y(x) = \frac{w_0 x^2}{EI}\left(x^2 - 4lx + 6l^2\right)$.

Solution: If y is deflection at distance x from A, then differential equation of deflection (load equation) is given by

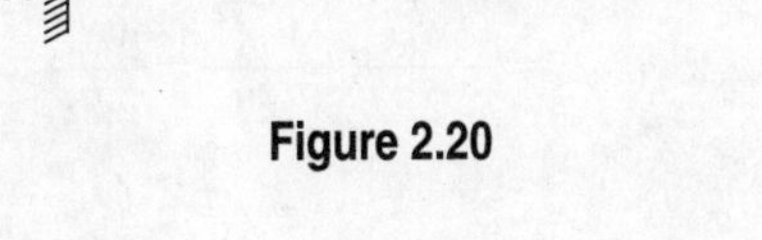

Figure 2.20

$$EI\frac{d^4 y}{dx^4} = w_0$$

$$\therefore \quad \frac{d^4 y}{dx^4} = \frac{w_0}{EI}$$

Take Laplace transform of both sides

$$s^4Y(s)-s^3y(0)-s^2y'(0)-sy''(0)-y'''(0)=\frac{w_0}{EIs}$$

where $Y(s)=£\,(y\,(x))$

Now, end A is clamped

$\therefore \quad y(0)=y'(0)=0$

Let $y''(0)=c_1, \qquad y'''(0)=c_2$

$$\therefore \qquad s^4Y(s)=\frac{w_0}{EIs}+c_1s+c_2$$

$$\therefore \qquad Y(s)=\frac{w_0}{EIs^5}+\frac{c_1}{s^3}+\frac{c_2}{s^4}$$

Take inverse Laplace transform of both sides

$$y(x)=\frac{w_0x^4}{4!EI}+\frac{c_1x^2}{2!}+\frac{c_2x^3}{3!} \tag{1}$$

End B is free end

$\therefore \quad y''(l)=y'''(l)=0.$

$$y'(x)=\frac{w_0x^3}{6EI}+c_1x+\frac{c_2x^2}{2}$$

$$y''(x)=\frac{w_0x^2}{2EI}+c_1+c_2x$$

$$y'''(x)=\frac{w_0x}{EI}+c_2$$

$$\therefore \qquad y''(l)=\frac{w_0l^2}{2EI}+c_1+c_2\,l=0 \tag{2}$$

$$y'''(l)=\frac{w_0l}{EI}+c_2=0$$

$$\therefore \qquad c_2=-\frac{w_0l}{EI}$$

$\therefore$ from (2),

$$c_1=-\frac{w_0l^2}{2EI}+\frac{w_0l^2}{EI}=\frac{w_0l^2}{2EI}$$

$\therefore$ from (1)

$$y(x)=\frac{w_0x^4}{24EI}+\frac{w_0l^2x^2}{4EI}-\frac{w_0lx^3}{6EI}$$

$$=\frac{w_0x^2}{24EI}\left(x^2-4lx+6l^2\right)$$

2.16.3 Problems Related to Mechanical Systems

If a particle of mass m moves along x-axis, then damping force due to medium (air, etc.) is proportional to its instantaneous velocity $\frac{dx}{dt}$ and act in the direction opposite to the direction of motion. Thus, damping force is $-a\frac{dx}{dt}$ where $a > 0$ is damping constant. If $f(t)$ represents all external forces acting on mass, then by Newton's second law of motion $m\frac{d^2x}{dt^2} = -k\frac{dx}{dt} + f(t)$

Mass-Spring System

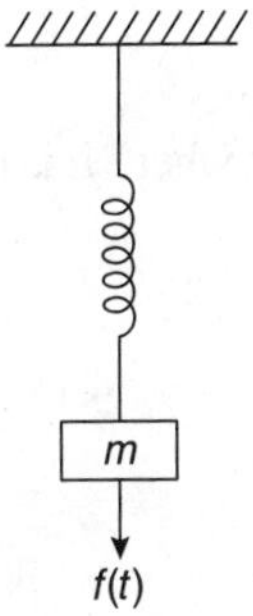

Figure 2.21

Let m be the mass suspended on a spring which is rigidly supported from one end. The rest position is denoted by $y = 0$. Downward displacement is taken as positive. Let $k > 0$ be spring constant, i.e., stiffness and $a > 0$ be damping constant. Then $a\frac{dx}{dt}$ is the damping force acting upward and kx is force due to stiffness of spring which opposes motion. If $f(t)$ is driving force then equation of motion of mass by Newton's second law of motion is

$$m\frac{d^2y}{dt^2} = -a\frac{dx}{dt} - kx + f(t)$$

Example 2.55: A mass m moves along x axis under the influence of a force which is proportional to its instantaneous speed and in a direction opposite to the direction of motion. Assuming that at $t = 0$, the particle is located at $x = a$ and moving to the right with speed V_0, find the position where the mass comes to rest.

Solution: Damping force is $-\mu\frac{dx}{dt}$ where μ is damping constant.

$\therefore$ By Newton's second law of motion, equation of motion of m is

$$m\frac{d^2x}{dt^2} = -\mu\frac{dx}{dt}$$

Take Laplace transform of both sides

$$m\left(s^2X(s)-sx(0)-x'(0)\right)=-\mu\left(sX(s)-x(0)\right) \qquad \text{where } X(s)=\pounds(x(t))$$

Initially $x(0)=a, x'(0)=V_0$

$$\therefore \qquad \left(ms^2+\mu s\right)X(s)=\left(ms+\mu\right)a+mV_0$$

$$\therefore \qquad X(s)=\frac{a}{s}+\frac{mV_0}{s\left(ms+\mu\right)}=\frac{a}{s}+\frac{mV_0}{\mu}\left(\frac{1}{s}-\frac{1}{s+(\mu/m)}\right) \qquad \text{(by suppression method)}$$

Take inverse Laplace transform of both sides

$$x(t)=a+\frac{mV_0}{\mu}\left(1-e^{\frac{-\mu}{m}t}\right) \tag{1}$$

which gives position of mass at any time t where μ is damping constant.

When the mass comes to rest then

$$\frac{dx}{dt}=\frac{mV_0}{\mu}\cdot\frac{\mu}{m}e^{-\frac{\mu}{m}t}=0$$

thus $\qquad t\to\infty$

and when $t\to\infty$, from (1)

$$x=a+\frac{mV_0}{\mu}$$

Thus, the mass comes to rest at distance $a+\frac{mV_0}{\mu}$ to the right from origin or at a distance $\frac{mV_0}{\mu}$ to the right from starting position where μ is damping constant.

Example 2.56: An electron of mass m is projected with velocity c into a uniform magnetic field of intensity k which is perpendicular to the direction of its motion. Find the position of electron at time t.

Solution: Let the electron of mass m is projected with velocity c into uniform field along x-axis from origin. If $P(x, y)$ is position of electron at time t, then components of velocity along x-axis and y-axis are $\frac{dx}{dt}$ and $\frac{dy}{dt}$ respectively. Now $-k$ (velocity) force is acting perpendicular to motion of electron due to magnetic field. Thus its components along x-axis and y-axis will be $-k\left(-\frac{dy}{dt}\right), -k\frac{dx}{dt}$ respectively.

$\therefore$ By Newton's second law of motion, equations of motion of electron are

$$m\frac{d^2x}{dt^2}=k\frac{dy}{dt}$$

$$m\frac{d^2y}{dt^2}=-k\frac{dx}{dt}$$

Take Laplace transform of both sides

$$m\left(s^2X(s)-sx(0)-x'(0)\right)=k\left(sY(s)-y(0)\right)$$

$$m\left(s^2Y(s)-sy(0)-y'(0)\right)=-k\left(sX(s)-x(0)\right)$$

where $Y(s) = £\,(y(t))$, $X(s) = £\,(x(t))$

Initially $x(0) = 0, y(0) = 0\; x'(0) = c, y'(0) = 0$

$$\therefore \qquad ms^2X(s)-ksY(s)=mc$$

$$ksX(s)+ms^2Y(s)=0$$

$$\therefore\; X(s)=\frac{cm^2s^2}{m^2s^4+k^2s^2}=\frac{cm^2}{m^2s^2+k^2}=\frac{c}{s^2+\left(\dfrac{k}{m}\right)^2}$$

$$Y(s)=\frac{-ckms}{m^2s^4+k^2s^2}=\frac{-ckm}{s\left(m^2s^2+k^2\right)}=\frac{-ckm}{k^2}\left(\frac{1}{s}-\frac{m^2s}{m^2s^2+k^2}\right) \quad \text{(by suppression method)}$$

$$=\frac{cm}{k}\left(\frac{s}{s^2+(k/m)^2}-\frac{1}{s}\right)$$

Taking inverse Laplace transform

$$x(t)=\frac{cm}{k}\sin\left(\frac{kt}{m}\right)$$

$$y(t)=\frac{cm}{k}\left(\cos\frac{kt}{m}-1\right)$$

which gives the position of electron at any time t.

Example 2.57: A pellet of mass m is fired into a viscous gas from a gun at time $t = 0$ with muzzle velocity V_0. Its initial position is origin and velocity is zero. Find its position at any time t.

Solution: If $a > 0$ is damping constant and pellet moves along x-axis, then by Newton's second law of motion, equation of motion of pellet is

$$m\frac{d^2x}{dt^2}=-a\frac{dx}{dt}+mV_0\delta(t)$$

where $\delta(t)$ is Dirac-delta function.

Take Laplace transform of both sides

$$m\left(s^2X(s)-sx(0)-x'(0)\right)=-a\left(sX(s)-x(0)\right)+mV_0 \qquad \text{where } X(s)=£(x(t))$$

Initially $x(0) = x'(0) = 0$

$$\therefore \qquad \left(ms^2+as\right)X(s)=mV_0$$

$$\therefore \qquad X(s) = \frac{mV_0}{s(ms+a)} = \frac{mV_0}{a}\left(\frac{1}{s} - \frac{1}{s+\frac{a}{m}}\right) \qquad \text{(by suppression method)}$$

Take inverse Laplace transform

$$x(t) = \frac{mV_0}{a}\left(1 - e^{\frac{-a}{m}t}\right)$$

where a is damping constant.

Example 2.58: Determine the response of the damped mass spring when there is a unit mass on the spring and driving force $r(t) = 10 \sin 2t$; $0 < t < \pi$ act on it. Assume that the spring constant and damping constant are 2 each and initially at $t = 0$ the mass on spring has displacement unity from equilibrium position and velocity at that time is 5 units in the opposite direction of displacement. Find the position at any time t.

Solution: Driving force $r(t)$ is

$$\begin{aligned} r(t) &= 10 \sin 2t \;;\; 0 < t < \pi \\ &= 10\sin 2t\left(U(t) - U(t-\pi)\right) \\ &= 10\sin 2t \;\; U(t) + 10\sin(2\pi - 2t)U(t-\pi) \\ &= 10\sin 2t \;\; U(t) - 10\sin 2(t-\pi)U(t-\pi) \end{aligned}$$

$\therefore$ By Newton's second Law of motion, equation of motion of unit mass is

$$\frac{d^2y}{dt^2} = -2\frac{dy}{dx} - 2y + 10\sin 2tU(t) - 10\sin\left(2(t-\pi)\right)U(t-\pi)$$

where y is downward displacement.

Take Laplace transform of both sides

$$s^2Y(s) - sy(0) - y'(0) = -2\left(sY(s) - y(0)\right) - 2Y(s) + \frac{20}{s^2+2^2} - \frac{20}{s^2+2^2}e^{-\pi s}$$

where $Y(s) = £\,(y(t))$

Now, $y(0) = 1$, $y'(0) = -5$

$$\therefore \qquad \left(s^2+2s+2\right)Y(s) = s + 2 - 5 + \frac{20}{s^2+4} - \frac{20e^{-\pi(s)}}{s^2+4}$$

$$\therefore \qquad Y(s) = \frac{s+1}{(s+1)^2+1} - \frac{4}{(s+1)^2+1} + \frac{20}{\left(s^2+4\right)\left(s^2+2s+2\right)}\left[1 - e^{-\pi s}\right] \qquad (1)$$

Let

$$\frac{20}{\left(s^2+4\right)\left(s^2+2s+2\right)} \equiv \frac{As+B}{s^2+4} + \frac{Cs+D}{s^2+2s+2}$$

$$\therefore \qquad 20 \equiv (As+B)\left(s^2+2s+2\right) + (Cs+D)\left(s^2+4\right)$$

Equate coefficients of s^3, s^2, s and constant

$$A + C = 0 \tag{2}$$

$$2A + B + D = 0 \tag{3}$$

$$2A + 2B + 4C = 0 \tag{4}$$

$$2B + 4D = 20 \tag{5}$$

from (2),

$$C = -A$$

$\therefore$ from (4),

$$B = A$$

$\therefore$ from (5),

$$D = \frac{20 - 2A}{4} = 5 - \frac{1}{2}A$$

$\therefore$ from (3),

$$2A + A + \frac{10 - A}{2} = 0$$

$\therefore$ $4A + 2A + 10 - A = 0$

$\therefore$ $A = -2$

$\therefore$ $A = -2, B = -2, C = 2, D = 6$

$\therefore$ $$\frac{20}{\left(s^2+4\right)\left(s^2+2s+2\right)} = \frac{-2\left(s+1\right)}{s^2+4} + \frac{2s+6}{s^2+2s+2}$$

$\therefore$ from (1)

$$Y(s) = \frac{s+1}{(s+1)^2+1} - \frac{4}{(s+1)^2+1} + \left[\frac{2(s+3)}{(s+1)^2+1} - \frac{2(s+1)}{s^2+4}\right](1-e^{-\pi s})$$

$$= \frac{s+1}{(s+1)^2+1} - \frac{4}{(s+1)^2+1} + \left[\frac{2(s+1)}{(s+1)^2+1} + \frac{4}{(s+1)^2+1} - \frac{2s}{s^2+4} - \frac{2}{s^2+4}\right](1-e^{-\pi s})$$

$$= \frac{3\left(s+1\right)}{(s+1)^2+1} - \frac{2s}{s^2+4} - \frac{2}{s^2+4} - \left[\frac{2(s+1)}{(s+1)^2+1} + \frac{4}{(s+1)^2+1} - \frac{2s}{s^2+4} - \frac{2}{s^2+4}\right]e^{-\pi s}$$

Take inverse Laplace transform of both sides

$$y(t) = 3e^{-t}\cos t - 2\cos 2t - \sin 2t - \Big[2e^{-(t-\pi)}\cos\left(t-\pi\right) + 4e^{-(t-\pi)}\sin\left(t-\pi\right)$$
$$-2\cos 2\left(t-\pi\right) - \sin 2\left(t-\pi\right)\Big]U\left(t-\pi\right)$$

$\therefore$ $$y(t) = 3e^{-t}\cos t - 2\cos 2t - \sin 2t;\ 0 \le t < \pi$$

$$= 3e^{-t}\cos t - 2\cos 2t - \sin 2t - 2e^{-(t-\pi)}\cos\left(\pi-t\right) + 4e^{-(t-\pi)}\sin\left(\pi-t\right)$$

$$+2\cos(2\pi-2t)-\sin(2\pi-2t);\ \ t\ge\pi$$

$$\therefore \qquad y(t)=3e^{-t}\cos t-2\cos 2t-\sin 2t;\quad 0\le t<\pi$$

$$=3e^{-t}\cos t+2e^{\pi-t}(\cos t+2\sin t);\quad t\ge\pi$$

which gives the position at any time t.

Example 2.59: Mechanical system in the given figure consist of two bodies each of mass unity on three springs each having spring constant k. If y_1, y_2 are displacement of bodies from their position of static equilibrium, then find y_1, y_2 at any time t assuming that $y_1(0) = y_2(0) = 1$, $y_1'(0)=\sqrt{3k}$, $y_2'(0)=-\sqrt{3k}$ and neglect the masses of the spring and damping.

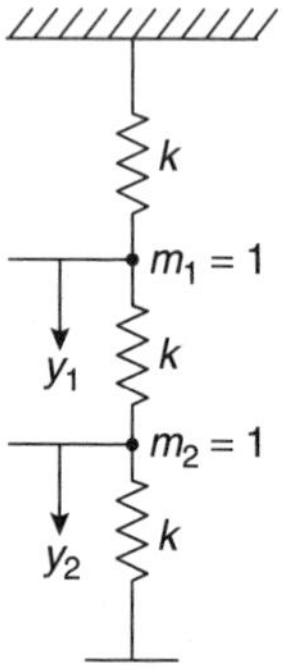

Figure 2.22

Solution: On the upper unit mass $-ky_1$ is the force of upper spring and $k(y_2 - y_1)$ is the force of middle spring.

$\therefore$ By Newton's second law of motion, its equation of motion is

$$\frac{d^2y_1}{dt^2}=-ky_1+k(y_2-y_1) \tag{1}$$

On the lower unit mass $-k(y_2 - y_1)$ is the force of middle spring and $-ky_2$ is the force of lower spring.

$\therefore$ By Newton's second law of motion, its equation of motion is

$$\frac{d^2y_2}{dt^2}=-k(y_2-y_1)-ky_2 \tag{2}$$

Take Laplace transform of both sides of (1) and (2)

$$s^2Y_1(s)-sy_1(0)-y_1'(0)=kY_2(s)-2kY_1(s)$$

$$s^2Y_2(s)-sy_2(0)-y_2'(0)=-2kY_2(s)+kY_1(s)$$

where $Y_1(s) = \pounds(y_1(t))$, $Y_2(s) = \pounds(y_2(t))$

Initially

$$y_1(0)=y_2(0)=1,\ y_1'(0)=\sqrt{3k},\ y_2'(0)=-\sqrt{3k}$$

$$\therefore \quad \left(s^2+2k\right)Y_1(s)-kY_2(s)=s+\sqrt{3k}$$

$$k\,Y_1(s)-\left(s^2+2k\right)Y_2(s)=-s+\sqrt{3k}$$

By Crammer's rule

$$Y_1(s)=\frac{1}{-\left(s^2+2k\right)^2+k^2}\begin{vmatrix} s+\sqrt{3k} & -k \\ -s+\sqrt{3k} & -\left(s^2+2k\right)\end{vmatrix}$$

$$=\frac{1}{\left(s^2+2k\right)^2-k^2}\left[\left(s+\sqrt{3k}\right)\left(s^2+2k\right)+k\left(s-\sqrt{3k}\right)\right]$$

$$=\frac{s^3+3ks+\sqrt{3k}\left(s^2+k\right)}{\left(s^2+2k\right)^2-k^2}=\frac{s\left(s^2+3k\right)+\sqrt{3k}\left(s^2+k\right)}{\left(s^2+3k\right)\left(s^2+k\right)}$$

$$=\frac{s}{s^2+k}+\frac{\sqrt{3k}}{s^2+3k}$$

By symmetry

$$Y_2(s)=\frac{s}{s^2+k}-\frac{\sqrt{3k}}{s^2+3k}$$

Take inverse Laplace transform

$$y_1(t)=\cos\left(\sqrt{k}\,t\right)+\sin\left(\sqrt{3k}\,t\right)$$

$$y_2(t)=\cos\left(\sqrt{k}\,t\right)-\sin\left(\sqrt{3k}\,t\right)$$

EXERCISE 2.3

1. A resistance R in series with inductance L is connected with e.m.f. $E(t)$. If the switch is connected at $t=0$ and disconnected at $t=a$, find the current i at any time t.
2. In the given circuit $L=1$ henry, $C=1$ farad, $V(t)=t$; $0<t<1$. Assuming that the current and charge on the capacitor vanish initially, find the current $i(t)$ at any time t.

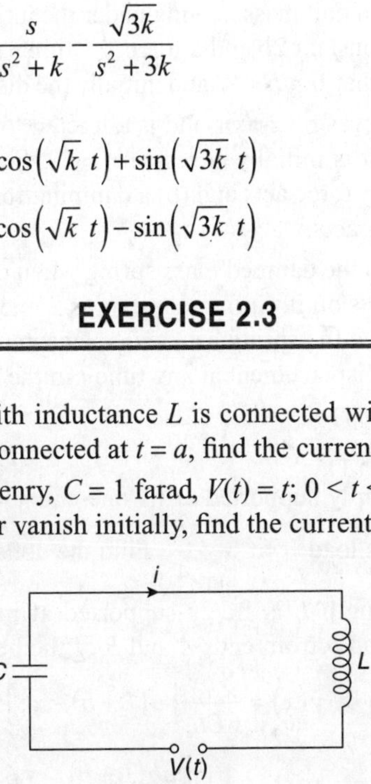

Figure 2.23

3. A voltage Ee^{-at} is applied at $t=0$ to a circuit of inductance L and resistance R in series.

   Show that the current at time t is $\dfrac{E}{R-aL}\left(e^{-at}-e^{\frac{-Rt}{L}}\right)$.

4. Constant voltage E is applied at $t = 0$ to a circuit containing inductance L, capacitance C and resistance R in series. Find the current i at time t if the initial current and charge are zero.
5. Given that the current i and charge q are zero at $t = 0$. Find current i at any time t in the following circuit

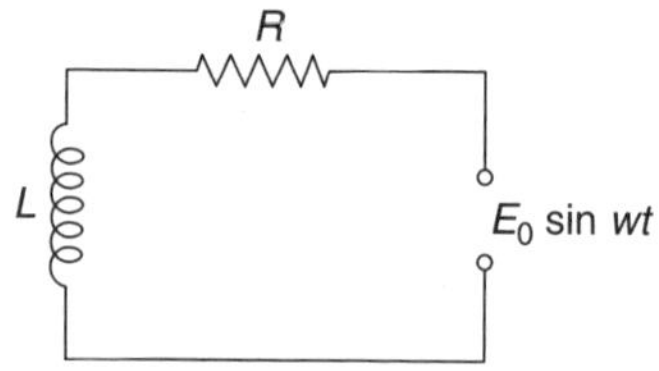

Figure 2.24

6. Currents i_1 and i_2 in two meshes of a given circuit are given bydifferential equations $\frac{di_1}{dt} - wi_2 = a\cos pt, \frac{di_2}{dt} + wi_1 = a\sin pt$. Find the currents i_1 and i_2 at any time t if $i_1 = i_2 = 0$ at $t = 0$.
7. Determine the position of a unit mass moving under the influence of driving force δt, damping force with damping constant $2b$ and a force λ^2 times the displacement in the opposite direction when it is given that $0 < b < \lambda$ and initially the displacement and velocity are zero.
8. A particle of mass 2 g moves on x-axis and is attracted towards origin O with a force numerically equal to $8x$. If it is initially at rest at $x = 10$, find its position at any subsequent time assuming (a) no other force acts and (b) a damping force numerically equal to 8 times the instantaneous velocity acts.
9. Determine the response to the damped mass spring when there is a unit mass on the spring and a driving force $r(t)$ acts on it. Spring constant is 2 and damping constant is 3. Assuming that at $t = 0$ the mass on the spring has no displacement from equilibrium position and velocity is zero, find the displacement at any time t in the cases
   (a) $r(t)$ is a square wave, $r(t) = 1;\ 1 \le t \le 2$
   (b) $r(t)$ is unit impulse at $t = 1$
10. A weightless beam is simply supported at its one end $x = 0$ and is clamped at the other end $x = l$ and is carrying a load W at $x = \frac{l}{4}$. Find the deflection of the beam at any point.
11. A weightless beam of length L is freely supported at its ends. A concentrated load W acts on it at distances a and b from ends A and B of the beam. Show that the deflection at distance x from A is given by $y(x) = \frac{Wbx}{6EIL}\left[a(L+b) - x^2\right];\ 0 \le x \le a$

$$= \frac{W}{6EI}\left[\frac{ab(L+b)}{L}x - \frac{b}{L}x^3 + (x-b)^3\right];\ a \le x \le L$$

12. A beam which is hinged at ends $x = 0$ and $x = l$ carries a uniform load w_0 per unit length. Find the deflection at any point.

13. A beam which is clamped at its ends $x = 0$ and $x = l$ carries a uniform load w_0 per unit length. Show that the deflection at any point is $y(x) = \dfrac{w_0 x^2 (l-x)^2}{24EI}$.

ANSWERS 2.3

1. $i(t) = \dfrac{E}{R}\left(1 - e^{\frac{-R}{L}t}\right);\ 0 \le t \le a;\ i(t) = \dfrac{E}{R} e^{\frac{-R}{L}t}\left(e^{\frac{Ra}{L}} - 1\right)\ ; t > a$

2. $i(t) = 1 - \cos t;\ 0 \le t \le 1;\ \ i(t) = \cos(t-1) - \cos t;\ t \ge 1$

4. $i(t) = \dfrac{E}{L} t\, e^{-\mu t}$ if $L = \dfrac{CR^2}{4}$; $i(t) = \dfrac{E}{\lambda L} e^{-\mu t} \sin \lambda t$ where $\lambda = \sqrt{\dfrac{1}{CL} - \dfrac{R^2}{4L^2}}$ if $L > \dfrac{CR^2}{4}$;

   $i(t) = \dfrac{E}{\lambda L} e^{-\mu t} \sinh \lambda t$ where $\lambda = \sqrt{\dfrac{R^2}{4C^2} - \dfrac{1}{CL}}$ if $L < \dfrac{CR^2}{4}$

   where μ in each case is $\dfrac{R}{2L}$.

5. $\dfrac{E_0}{L^2 w^2 + R^2}\left[wL\left(e^{-\frac{RT}{L}} - \cos wt\right) + R \sin wt\right]$

6. $i_1 = \dfrac{a}{p+w}(\sin wt + \sin pt), i_2 = \dfrac{a}{p+w}(\cos wt - \cos pt)$

7. $x(t) = e^{-bt}\left(\dfrac{1}{\sqrt{\lambda^2 - b^2}} \sin \sqrt{\lambda^2 - b^2}\ t\right)$

8. (a) $x(t) = 10 \cos 2t$ (b) $x(t) = 10(1+2t)e^{-2t}$

9. (a) $y(t) = 0; 0 \le t \le 1; y(t) = \dfrac{1}{2} - e^{-(t-1)} + \dfrac{1}{2} e^{-2(t-1)}; 1 \le t \le 2;$

   $y(t) = -e^{-(t-1)} + e^{-(t-2)} + \dfrac{1}{2} e^{-2(t-1)} - \dfrac{1}{2} e^{-2(t-2)}; t \ge 2$

   (b) $y(t) = 0;\ 0 \le t \le 1;\ y(t) = e^{-(t-1)} - e^{-2(t-1)};\ t \ge 1$

10. $y(x) = \dfrac{9Wx}{256EI}(l^2 - 3x^2); 0 \le x \le \dfrac{l}{4}; y(x) = \dfrac{W}{6EI}\left(x - \dfrac{l}{4}\right)^3 + \dfrac{9Wx}{256EI}(l^2 - 3x^2); \dfrac{l}{4} \le x \le l$

12. $y(x) = \dfrac{w_0 x}{24EI}(x^3 + l^3 - 2x^2 l)$

Fourier Series, Fourier Integrals and Fourier Transforms 3

3.1 INTRODUCTION

In Chapter 9, of Volume 1, we have derived the Fourier-Legendre series which expand a function $f(x)$ in $(-1,1)$ in terms of Legendre polynomials $P_n(x)$ and Fourier–Bessel series which expand a function $f(x)$ in $(0,R)$ in terms of Bessel functions of a given order $J_n(x)$. In these representations, we have used the orthogonality of Legendre polynomials and orthogonality of Bessel functions. Periodic phenomena occur quite frequently in motors, rotating machines, electrical and sound waves, motion of earth, heart beats, etc. Thus, it is an important practical problem to represent periodic functions in some simple series. Sine and cosine functions are periodic and form orthogonal system of functions and thus, Euler and Daniel Bernoulli worked and Fourier introduced the representation of periodic functions in sine and cosine term series which are called Fourier series. In this chapter, we discuss these series. Many functions including some discontinuous periodic functions can be expanded in a Fourier series and hence are, in certain sense, more universal than Taylor series expansions which cannot be represented for discontinuous functions. Fourier series solution method is a powerful tool in solving some ordinary and partial differential equations which we shall explain in the next chapter.

Fourier integrals and Fourier transforms extend the ideas and techniques of Fourier series to non-periodic functions defined for all x. They are helpful in solving ordinary and partial differential equations initial and boundary values problems.

3.1.1 Periodic Functions

A function $f(x)$ is called periodic if it is defined for all real x (except perhaps for some isolated x such as $\pm\frac{\pi}{2}, \pm\frac{3\pi}{2}, \ldots$ for $\tan x$) and there exists positive λ such that

$$f(x+\lambda) = f(x) \qquad \text{for all } x$$

This number λ is called a period of $f(x)$. If $f(x)$ is periodic and λ is its period, then

$$f(x+2\lambda) = f(x+\lambda+\lambda) = f(x+\lambda) = f(x)$$

and similarly $\quad f(x+3\lambda) = f(x)$

and $\quad f(x+n\lambda) = f(x);\ n \in N$

Hence, if λ is a period of $f(x)$, then $n\lambda$ for any natural number n is also a period of $f(x)$. The smallest positive λ, if exists, is called the fundamental period of $f(x)$. Fundamental period may or may not exist, for example $f(x) = 1$ for all x is a periodic function and every positive real number is its period but fundamental period does not exist.

If fundamental period of a periodic function exists, then it is unique. If $f(x)$ and $g(x)$ have fundamental period λ, then $h(x)=af(x)+bg(x)$ has fundamental period λ. Further, if $f(x)$ and $g(x)$ have fundamental periods λ_1 and λ_2 respectively then $h(x)=af(x)+bg(x)$ have fundamental period least common multiple of λ_1 and λ_2.

The graph of periodic functions is obtained by periodic extension of its graph in interval of any length.

For any natural number n, $\sin nx$ and $\cos nx$ have fundamental period $\frac{2\pi}{n}$ and $\tan nx$ has fundamental period $\frac{\pi}{n}$. But $a\sin nx + b\tan nx$ has fundamental period $\frac{2\pi}{n}$.

3.1.2 Trigonometric Series

Simple functions $1, \sin x, \cos x, \sin 2x, \cos 2x, \ldots, \sin nx, \cos nx, \ldots$ are periodic functions with period 2π.

The series $\frac{a_0}{2}+\sum_{n=1}^{\infty}(a_n \cos nx + b_n \sin nx)$, is called trigonometric series where a_0, a_1, a_2, … b_1, b_2, … are real constants and are called the coefficients of the series. Fundamental period of this series is 2π and hence if the series converges, its sum will be a function of fundamental period 2π. The set of functions $1, \sin x, \sin 2x, \ldots \cos x, \cos 2x, \ldots$ forming the series is called the trigonometrical system.

Similarly, the functions $1, \sin\frac{\pi x}{l}, \sin\frac{2\pi x}{l}, \sin\frac{3\pi x}{l}, \cdots, \cos\frac{\pi x}{l}, \cos\frac{2\pi x}{l}, \cos\frac{3\pi x}{l}$ are periodic functions with period, $2l$. The series

$$\frac{a_0}{2}+\sum_{n=1}^{\infty}\left(a_n \cos\frac{n\pi x}{l} + b_n \sin\frac{n\pi x}{l}\right)$$

is a trigonometric series. Fundamental period of this series is $2l$ and hence if the series converges, its sum will be a function of fundamental period $2l$.

Thus, trigonometric series can be used for representing any practically important periodic function *f*, simple or complicated, of any period $2l$. This series will then be called Fourier series of *f*.

3.1.3 Orthogonality of Trigonometric System

We have for $m, n \in N$

$$\int_c^{c+2l} \cos\frac{n\pi x}{l}\,dx = \frac{l}{n\pi}\left(\sin\frac{n\pi x}{l}\right)_c^{c+2l} = \frac{l}{n\pi}\left[\sin\frac{n\pi}{l}(c+2l)-\sin\frac{n\pi c}{l}\right]=0 \tag{3.1}$$

$$\int_c^{c+2l} \sin\frac{n\pi x}{l}\,dx = -\frac{l}{n\pi}\left(\cos\frac{n\pi x}{l}\right)_c^{c+2l} = -\frac{l}{n\pi}\left[\cos\frac{n\pi}{l}(c+2l)-\cos\frac{n\pi c}{l}\right]=0 \tag{3.2}$$

$$\begin{aligned}\int_c^{c+2l} \cos\frac{m\pi x}{l}\cos\frac{n\pi x}{l}\,dx &= \frac{1}{2}\int_c^{c+2l} 2\cos\frac{m\pi x}{l}\cos\frac{n\pi x}{l}\,dx \\ &= \frac{1}{2}\left[\int_c^{c+2l}\cos\frac{(m+n)\pi x}{l}\,dx + \int_c^{c+2l}\cos\frac{(m-n)\pi x}{l}\,dx\right]\end{aligned}$$

$$= \frac{1}{2}(0+0) = 0; \quad m \neq n \qquad \text{[from (3.1)]}$$

$$\int_c^{c+2l} \sin\frac{m\pi x}{l} \sin\frac{n\pi x}{l} dx = \frac{1}{2}\left[\int_c^{c+2l} \cos\frac{(m-n)\pi x}{l} dx - \int_c^{c+2l} \cos\frac{(m+n)\pi x}{l} dx\right]$$

$$= \frac{1}{2}(0-0) = 0; \quad m \neq n \qquad \text{[from (3.1)]}$$

$$\int_c^{c+2l} \sin\frac{m\pi x}{l} \cos\frac{n\pi x}{l} dx = \frac{1}{2}\left[\int_c^{c+2l} \sin\frac{(m+n)\pi x}{l} dx + \int_c^{c+2l} \sin\frac{(m-n)\pi x}{l} dx\right]$$

$$= \frac{1}{2}(0+0) = 0 \quad \text{for all } m, n \in N \qquad \text{[from (3.2)]}$$

Hence, the trigonometric system

$$1, \left\{\sin\frac{n\pi x}{l} : n \in N\right\}, \left\{\cos\frac{n\pi x}{l} : n \in N\right\}$$

is orthogonal with weight unity in any interval of length $2l$ and hence in $(-l, l)$ and $(0, 2l)$.

Taking $l = \pi$, the trigonometric system

$$1, \{\sin nx : n \in N\}, \{\cos nx : n \in N\}$$

is orthogonal with weight unity in any interval of length 2π and hence in $(-\pi, \pi)$ and $(0, 2\pi)$.

Also

$$\int_c^{c+2l} \cos^2\frac{n\pi x}{l} dx = \int_c^{c+2l} \frac{1}{2}\left(1+\cos\frac{2n\pi x}{l}\right) dx = \frac{1}{2}\left(x + \frac{l}{2n\pi}\sin\frac{2n\pi x}{l}\right)_c^{c+2l} = l \qquad (3.3)$$

$$\int_c^{c+2l} \sin^2\frac{n\pi x}{l} dx = \int_c^{c+2l} \frac{1}{2}\left(1-\cos\frac{2n\pi x}{l}\right) dx = \frac{1}{2}\left(x - \frac{l}{2n\pi}\sin\frac{2n\pi x}{l}\right)_c^{c+2l} = l \qquad (3.4)$$

Thus, integrals of $\cos^2\frac{n\pi x}{l}$ and $\sin^2\frac{n\pi x}{l}$ in any interval of length $2l$ is l.

3.1.4 Fourier Series

Fourier series arise from the practical task of representing a given periodic function $f(x)$ of period $2l$ in terms of cosine and sine functions. These series are trigonometric series whose coefficients are determined by Euler formulae discussed hereunder.

3.1.5 Euler Formulae for Fourier Coefficients

Suppose $f(x)$ is a periodic function with period $2l$ defined on $[c, c+2l]$ which can be expanded in uniformly convergent Fourier series

$$f(x) = \frac{a_0}{2} + \sum_{n=1}^{\infty}\left(a_n \cos\frac{n\pi x}{l} + b_n \sin\frac{n\pi x}{l}\right)$$

then
$$a_0 = \frac{1}{l}\int_c^{c+2l} f(x)\,dx$$
$$a_n = \frac{1}{l}\int_c^{c+2l} f(x)\cos\frac{n\pi x}{l}\,dx$$
$$b_n = \frac{1}{l}\int_c^{c+2l} f(x)\sin\frac{n\pi x}{l}\,dx$$

where a_0, a_n and b_n are called Fourier coefficients or Euler's coefficients.

Proof:
$$f(x) = \frac{a_0}{2} + \sum_{m=1}^{\infty}\left(a_m\cos\frac{m\pi x}{l} + b_m\sin\frac{m\pi x}{l}\right) \tag{3.5}$$

Integrate term by term within the limits c to $c+2l$ and use orthogonal property of trigonometric system:

$$\int_c^{c+2l} f(x)\,dx = \frac{a_0}{2}\int_c^{c+2l} dx + \sum_{m=1}^{\infty}\left[a_m\int_c^{c+2l}\cos\frac{m\pi x}{l}\,dx + b_m\int_c^{c+2l}\sin\frac{m\pi x}{l}\,dx\right]$$
$$= \frac{a_0}{2}(x)_c^{c+2l} + 0 + 0 = \frac{a_0}{2}(c+2l-c) = la_0$$

$\therefore$
$$a_0 = \frac{1}{l}\int_c^{c+2l} f(x)\,dx$$

Multiply (3.5) by $\cos\frac{n\pi x}{l}$ and integrate term by term within the limits c to $c+2l$ and use orthogonality property of trigonometric system and (3.3):

$$\int_c^{c+2l} f(x)\cos\frac{n\pi x}{l}\,dx = \frac{a_0}{2}\int_c^{c+2l}\cos\frac{n\pi x}{l}\,dx + \sum_{\substack{m=1\\m\neq n}}^{\infty} a_m\int_c^{c+2l}\cos\frac{m\pi x}{l}\cos\frac{n\pi x}{l}\,dx$$
$$+ a_n\int_c^{c+2l}\cos^2\frac{n\pi x}{l}\,dx + \sum_{m=1}^{\infty}\int_c^{c+2l}\sin\frac{m\pi x}{l}\cos\frac{n\pi x}{l}\,dx$$
$$= 0 + 0 + a_n l + 0 = la_n$$

$\therefore$
$$a_n = \frac{1}{l}\int_c^{c+2l} f(x)\cos\frac{n\pi x}{l}\,dx$$

Multiply (3.5) by $\sin\frac{n\pi x}{l}$ and integrate term by term within the limits c to $c+2l$ and use orthogonality property of trigonometric system and (3.4):

$$\int_c^{c+2l} f(x)\sin\frac{n\pi x}{l}\,dx = \frac{a_0}{2}\int_c^{c+2l}\sin\frac{n\pi x}{l}\,dx + \sum_{m=1}^{\infty} a_m\int_c^{c+2l}\cos\frac{m\pi x}{l}\sin\frac{n\pi x}{l}\,dx$$
$$+ \sum_{\substack{m=1\\m\neq n}}^{\infty} b_m\int_c^{c+2l}\sin\frac{m\pi x}{l}\sin\frac{n\pi x}{l}\,dx + b_n\int_c^{c+2l}\sin^2\frac{n\pi x}{l}\,dx$$
$$= 0 + 0 + 0 + b_n l$$

$\therefore$
$$b_n = \frac{1}{l}\int_c^{c+2l} f(x)\sin\frac{n\pi x}{l}\,dx$$

The expressions for Fourier coefficients

$$a_0 = \frac{1}{l}\int_c^{c+2l} f(x)\,dx$$

$$a_n = \frac{1}{l}\int_c^{c+2l} f(x)\cos\frac{n\pi x}{l}\,dx$$

$$b_n = \frac{1}{l}\int_c^{c+2l} f(x)\sin\frac{n\pi x}{l}\,dx$$

are called Euler formulae.

We can find a_n and b_n simultaneously by writing

$a_n + ib_n = \frac{1}{l}\int_c^{c+2l} f(x)e^{i\frac{n\pi x}{l}}\,dx$ and equating real and imaginary parts.

Remark 3.1: (i) If $f(x)$ is periodic with period 2π then Fourier series of $f(x)$ in $(0, 2\pi)$ is

$$f(x) = \frac{a_0}{2} + \sum_{n=1}^{\infty}(a_n \cos nx + b_n \sin nx)$$

and Euler formulae for coefficients are

$$a_0 = \frac{1}{\pi}\int_0^{2\pi} f(x)\,dx, \quad a_n = \frac{1}{\pi}\int_0^{2\pi} f(x)\cos nx\,dx, \quad b_n = \frac{1}{\pi}\int_0^{2\pi} f(x)\sin nx\,dx$$

and Fourier series of $f(x)$ in $(-\pi, \pi)$ is

$$f(x) = \frac{a_0}{2} + \sum_{n=1}^{\infty}(a_n \cos nx + b_n \sin nx)$$

and Euler formulae for coefficients are

$$a_0 = \frac{1}{\pi}\int_{-\pi}^{\pi} f(x)\,dx, \quad a_n = \frac{1}{\pi}\int_{-\pi}^{\pi} f(x)\cos nx\,dx, \quad b_n = \frac{1}{\pi}\int_{-\pi}^{\pi} f(x)\sin nx\,dx$$

(ii) If $f(x)$ is periodic with period $2l$ then Fourier series of $f(x)$ in $(0, 2l)$ is

$$f(x) = \frac{a_0}{2} + \sum_{n=1}^{\infty}\left(a_n \cos\frac{n\pi x}{l} + b_n \frac{n\pi x}{l}\right),$$

and Euler formulae for coefficients are

$$a_0 = \frac{1}{l}\int_0^{2l} f(x)\,dx, \; a_n = \frac{1}{l}\int_0^{2l} f(x)\cos\frac{n\pi x}{l}\,dx, \; b_n = \frac{1}{l}\int_0^{2l} f(x)\sin\frac{n\pi x}{l}\,dx,$$

and Fourier series of $f(x)$ in $(-l, l)$ is

$$f(x) = \frac{a_0}{2} + \sum_{n=1}^{\infty}\left(a_n \cos\frac{n\pi x}{l} + b_n \sin\frac{n\pi x}{l}\right)$$

and Euler formulae for coefficients are

$$a_0 = \frac{1}{l}\int_{-l}^{l} f(x)\,dx,\ a_n = \frac{1}{l}\int_{-l}^{l} f(x)\cos\frac{n\pi x}{l}\,dx,\ b_n = \frac{1}{l}\int_{-l}^{l} f(x)\sin\frac{n\pi x}{l}\,dx$$

We now state the Dirichlet's conditions of convergence of Fourier series; however, the proof is beyond the scope of this book.

3.1.6 Dirichlet's Conditions for Convergence of Fourier Series of *f*(*x*) in [*c*, *c* + 2*l*]

Following are the Dirichlet's conditions:

(i) $f(x)$ is periodic with period $2l$

(ii) $f(x)$ is piecewise continuous in $[c, c+2l]$

(iii) $f(x)$ has finite number of maxima or minima in $[c, c+2l]$

If a function $f(x)$ satisfies above mentioned Dirichlet's conditions, then the Fourier series of $f(x)$ is

$$\frac{a_0}{2} + \sum_{n=1}^{\infty}\left(a_n \cos\frac{n\pi x}{l} + b_n \sin\frac{n\pi x}{l}\right)$$

with Euler coefficients

$$a_0 = \frac{1}{l}\int_{c}^{c+2l} f(x)\,dx,\ a_n = \frac{1}{l}\int_{c}^{c+2l} f(x)\cos\frac{n\pi x}{l}\,dx,\ b_n = \frac{1}{l}\int_{c}^{c+2l} f(x)\sin\frac{n\pi x}{l}\,dx$$

is convergent. Its sum is $f(x)$, except at a point x_0 at which $f(x)$ is not continuous and sum of the series at discontinuity $x = x_0$ is $\frac{1}{2}\left[\lim_{x\to x_0^-} f(x) + \lim_{x\to x_0^+} f(x)\right]$ which we write as

$$\frac{1}{2}\left[f(x_0 - 0) + f(x_0 + 0)\right].$$

Thus, $f(x) \sim \frac{a_0}{2} + \sum_{n=1}^{\infty}\left(a_n \cos\frac{n\pi x}{l} + b_n \sin\frac{n\pi x}{l}\right)$ at the points where the series may or may not converge but $\sim$ can be replaced by equality sign $=$ at the points where the series converges.

For example, the function $f(x) = \frac{1}{3-x}$, $0 < x < 2\pi$ does not satisfy Dirichlet's conditions as $\lim_{x\to 3^-} f(x) = \infty$ and $\lim_{x\to 3^+} f(x) = -\infty$. Both these limits are infinite, so $f(x)$ is not piecewise continuous in $(0, 2\pi)$ as $3 \in (0, 2\pi)$. Thus, Fourier series expansion of $f(x)$ in $(0, 2\pi)$ does not exist.

3.1.7 Fourier Series of Even and Odd Functions

$f(x)$ is said to be an even function if $f(-x) = f(x)$ for all x in domain of f

and $f(x)$ is said to be an odd function if $f(-x) = -f(x)$ for all x in domain of f.

Now, Fourier series of $f(x)$ in $[-l,l]$ is

$$f(x)=\frac{a_0}{2}+\sum_{n=1}^{\infty}\left(a_n\cos\frac{n\pi x}{l}+b_n\sin\frac{n\pi x}{l}\right)$$

where $\quad a_0=\frac{1}{l}\int_{-l}^{l}f(x)\,dx,\ a_n=\frac{1}{l}\int_{-l}^{l}f(x)\cos\frac{n\pi x}{l}\,dx,\ b_n=\frac{1}{l}\int_{-l}^{l}f(x)\sin\frac{n\pi x}{l}\,dx$

If $f(x)$ is an even function in $[-l,l]$ then $f(x)\cos\frac{n\pi x}{\lambda}$ is even function and $f(x)\sin\frac{n\pi x}{l}$ is odd function in $[-l,l]$.

$$\therefore \qquad a_0=\frac{1}{l}\int_{-l}^{l}f(x)\,dx=\frac{2}{l}\int_{0}^{l}f(x)\,dx$$

$$a_n=\frac{1}{l}\int_{-l}^{l}f(x)\cos\frac{n\pi x}{l}\,dx=\frac{2}{l}\int_{0}^{l}f(x)\cos\frac{n\pi x}{l}\,dx$$

$$b_n=\frac{1}{l}\int_{-l}^{l}f(x)\sin\frac{n\pi x}{l}\,dx=0.$$

Thus, if $f(x)$ is an even function in $[-l,l]$ then its Fourier series is

$$f(x)=\frac{a_0}{2}+\sum_{n=1}^{\infty}a_n\cos\frac{n\pi x}{l},$$

where $\quad a_0=\frac{2}{l}\int_{0}^{l}f(x)\,dx,\ a_n=\frac{2}{l}\int_{0}^{l}f(x)\cos\frac{n\pi x}{l}\,dx.$

If $f(x)$ is an odd function in $[-l,l]$, then $f(x)\cos\frac{n\pi x}{l}$ is odd function and $f(x)\sin\frac{n\pi x}{l}$ is even function of x and hence $\frac{1}{l}\int_{-l}^{l}f(x)\,dx=0,\ \frac{1}{l}\int_{-l}^{l}f(x)\cos\frac{n\pi x}{l}\,dx=0$ and $\frac{1}{l}\int_{-l}^{l}f(x)\sin\frac{n\pi x}{l}\,dx=\frac{2}{l}\int_{0}^{l}f(x)\sin\frac{n\pi x}{l}\,dx.$

Thus, Fourier series of $f(x)$ is

$$f(x)=\sum_{n=1}^{\infty}b_n\sin\frac{n\pi x}{l}\ \text{ where } b_n=\frac{2}{l}\int_{0}^{l}f(x)\sin\frac{n\pi x}{l}\,dx$$

Example 3.1: Find the Fourier series of $f(x)=x,\ 0<x<2\pi$ and sketch the graph from $x=-4\pi$ to $x=4\pi$.

Solution: Taking $f(x)$ to be periodic function with period 2π, Fourier series expansion of $f(x)=x$ is

$$f(x)\sim\frac{a_0}{2}+\sum_{n=1}^{\infty}(a_n\cos nx+b_n\sin nx) \tag{1}$$

where $\quad a_0=\frac{1}{\pi}\int_{0}^{2\pi}x\,dx=\frac{1}{2\pi}\left(x^2\right)_0^{2\pi}=2\pi$

$$a_n + ib_n = \frac{1}{\pi}\int_0^{2\pi} xe^{inx}\,dx = \frac{1}{\pi}\left[-\frac{i}{n}xe^{inx} + \frac{1}{n^2}e^{inx}\right]_0^{2\pi}$$

$$= \frac{1}{\pi}\left[-\frac{2\pi i}{n}e^{i2\pi n} + \frac{1}{n^2}\left(e^{i2\pi n} - 1\right)\right]$$

$$= \frac{1}{\pi}\left[-\frac{2\pi i}{n}\right] = -\frac{2i}{n}$$

Equate real and imaginary parts

$$a_n = 0,\ b_n = -\frac{2}{n}$$

$\therefore$ Fourier-series expansion of $f(x)$ is

$$f(x) \sim \pi - 2\sum_{n=1}^{\infty}\frac{1}{n}\sin nx.$$

Graph of $f(x) = x,\ -4\pi < x < 4\pi$ is

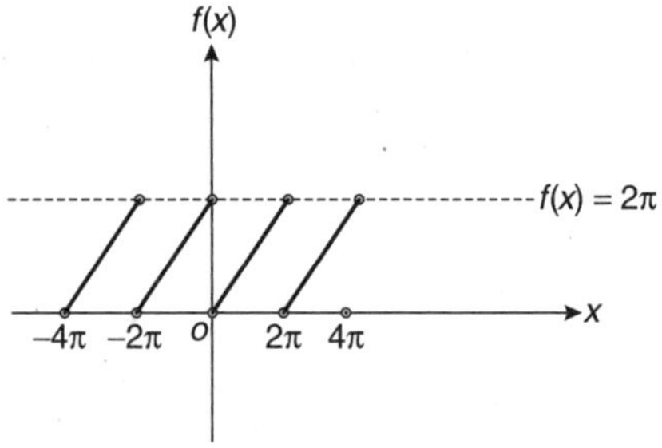

Figure 3.1

⊙ denotes that the point is not in the graph.

Example 3.2: Find a Fourier series to represent $x - x^2$ from $-\pi$ to π. Hence show that

$$\frac{1}{1^2} - \frac{1}{2^2} + \frac{1}{3^2} - \frac{1}{4^2} + \cdots = \frac{\pi^2}{12}$$

Solution: Let $f(x) = x - x^2,\ -\pi < x < \pi$

Considering $f(x)$ to be periodic with period 2π, Fourier-series expansion of $f(x)$ is

$$f(x) \sim \frac{a_0}{2} + \sum_{n=1}^{\infty}\left(a_n\cos nx + b_n\sin nx\right)$$

where $a_0 = \frac{1}{\pi}\int_{-\pi}^{\pi}\left(x - x^2\right)dx = -\frac{2}{\pi}\int_0^{\pi} x^2\,dx = -\frac{2}{\pi}\left(\frac{x^3}{3}\right)_0^{\pi} = -\frac{2\pi^2}{3}$

($\because$ x is odd and x^2 is even function)

$$a_n = \frac{1}{\pi}\int_{-\pi}^{\pi}\left(x - x^2\right)\cos nx\,dx = -\frac{2}{\pi}\int_0^{\pi} x^2 \cos nx\,dx \qquad (\because\ x\cos nx \text{ is odd and } x^2 \cos nx \text{ is even function})$$

$$= -\frac{2}{\pi}\left[x^2\left(\frac{1}{n}\sin nx\right) + \frac{2x}{n^2}\cos nx - \frac{2}{n^3}\sin nx\right]_0^{\pi}$$

$$= -\frac{4}{n^2}\cos n\pi = \frac{4(-1)^{n+1}}{n^2}$$

$$b_n = \frac{1}{\pi}\int_{-\pi}^{\pi}\left(x - x^2\right)\sin nx\,dx = \frac{2}{\pi}\int_0^{\pi} x\sin nx\,dx \qquad (\because\ x\sin nx \text{ is even and } x^2 \sin nx \text{ is odd function})$$

$$= \frac{2}{\pi}\left[x\left(-\frac{1}{n}\cos nx\right) - \left(-\frac{1}{n^2}\sin nx\right)\right]_0^{\pi} = -\frac{2}{n}(-1)^n$$

$$= \frac{2(-1)^{n+1}}{n}$$

$\therefore$ Fourier-series expansion of $f(x)$ is

$$f(x) \sim -\frac{\pi^2}{3} + 2\sum_{n=1}^{\infty}\frac{(-1)^{n+1}}{n^2}(2\cos nx + n\sin nx)$$

Taking $x = 0$

$$0 = -\frac{\pi^2}{3} + 4\sum_{n=1}^{\infty}\frac{(-1)^{n+1}}{n^2}$$

$$\therefore \qquad \frac{1}{1^2} - \frac{1}{2^2} + \frac{1}{3^2} - \frac{1}{4^2} + \cdots = \frac{\pi^2}{12}$$

Example 3.3: Obtain the Fourier series to represent $f(x) = \frac{1}{4}(\pi - x)^2,\ 0 < x < 2\pi$
Hence obtain the following relations

(i) $\frac{1}{1^2} + \frac{1}{2^2} + \frac{1}{3^2} + \cdots = \frac{\pi^2}{6}$ (ii) $\frac{1}{1^2} - \frac{1}{2^2} + \frac{1}{3^2} - \frac{1}{4^2} + \cdots = \frac{\pi^2}{12}$

(iii) $\frac{1}{1^2} + \frac{1}{3^2} + \frac{1}{5^2} + \cdots = \frac{\pi^2}{8}$

Solution: Considering $f(x)$ to be periodic function with period 2π, Fourier-series expansion of $f(x)$ is

$$f(x) \sim \frac{a_0}{2} + \sum_{n=1}^{\infty}\left(a_n \cos nx + b_n \sin nx\right)$$

where
$$a_0 = \frac{1}{\pi}\int_0^{2\pi}\frac{1}{4}(\pi - x)^2\,dx = -\frac{1}{12\pi}\left[(\pi - x)^3\right]_0^{2\pi} = \frac{1}{12\pi}\left[\pi^3 + \pi^3\right] = \frac{\pi^2}{6}$$

$$a_n + ib_n = \frac{1}{\pi}\int_0^{2\pi} \frac{1}{4}(\pi - x)^2 e^{inx}\,dx$$

$$= \frac{1}{4\pi}\left[(\pi - x)^2\left(-\frac{i}{n}e^{inx}\right) + 2(\pi - x)\left(-\frac{1}{n^2}e^{inx}\right) + 2\left(\frac{i}{n^3}e^{inx}\right)\right]_0^{2\pi}$$

$$= \frac{1}{4\pi}\left[-\frac{i\pi^2}{n} + \frac{i\pi^2}{n} + \frac{2\pi}{n^2} + \frac{2\pi}{n^2} + \frac{2i}{n^3} - \frac{2i}{n^3}\right] = \frac{1}{n^2}$$

Equate real and imaginary parts

$$a_n = \frac{1}{n^2}, \; b_n = 0$$

$\therefore$ Fourier-series expansion of $f(x)$ is

$$f(x) \sim \frac{\pi^2}{12} + \sum_{n=1}^{\infty} \frac{1}{n^2}\cos nx \qquad (1)$$

$$\therefore \quad f(0) = \frac{f(0+0) + f(0-0)}{2} = \frac{f(0+0) + f(2\pi - 0)}{2} = \frac{1}{2}\left[\frac{1}{4}\pi^2 + \frac{1}{4}\pi^2\right] = \frac{\pi^2}{12} + \sum_{n=1}^{\infty}\frac{1}{n^2}$$

$$\therefore \quad \sum_{n=1}^{\infty}\frac{1}{n^2} = \frac{\pi^2}{4} - \frac{\pi^2}{12} = \frac{\pi^2}{6}$$

$$\therefore \quad \frac{1}{1^2} + \frac{1}{2^2} + \frac{1}{3^2} + \cdots = \frac{\pi^2}{6} \qquad (2)$$

Taking $x = \pi$ in (1)

$$\frac{\pi^2}{12} + \sum_{n=1}^{\infty}\frac{(-1)^n}{n^2} = 0$$

$$\therefore \quad \frac{1}{1^2} - \frac{1}{2^2} + \frac{1}{3^2} - \frac{1}{4^2} + \cdots = \frac{\pi^2}{12} \qquad (3)$$

Add (2) and (3)

$$2\left(\frac{1}{1^2} + \frac{1}{3^2} + \frac{1}{5^2} + \cdots\right) = \frac{\pi^2}{6} + \frac{\pi^2}{12} = \frac{3\pi^2}{12}$$

$$\therefore \quad \frac{1}{1^2} + \frac{1}{3^2} + \frac{1}{5^2} + \cdots = \frac{\pi^2}{8}$$

Example 3.4: Obtain the Fourier series for $f(x) = e^{-x};\; 0 < x < 2\pi$ and deduce that $\dfrac{\pi}{2\sinh\pi} = \displaystyle\sum_{n=2}^{\infty}\frac{(-1)^n}{n^2+1}$.

Solution: Considering $f(x)$ to be periodic function with period 2π, Fourier-series expansion of $f(x)$ is

$$f(x) \sim \frac{a_0}{2} + \sum_{n=1}^{\infty}(a_n \cos nx + b_n \sin nx)$$

where $$a_0 = \frac{1}{\pi}\int_0^{2\pi} e^{-x}dx = -\frac{1}{\pi}\left(e^{-x}\right)_0^{2\pi} = \frac{1}{\pi}\left(1-e^{-2\pi}\right) = \frac{e^{-\pi}\left(e^{\pi}-e^{-\pi}\right)}{\pi} = \frac{2e^{-\pi}\sinh\pi}{\pi}$$

$$a_n + ib_n = \frac{1}{\pi}\int_0^{2\pi} e^{-x}e^{inx}dx = \frac{1}{\pi}\frac{1}{(-1+in)}\left(e^{-x}e^{inx}\right)_0^{2\pi}$$

$$= -\frac{1}{\pi}\frac{1+in}{1+n^2}\left[e^{-2\pi}-1\right] = \frac{(1+in)\left(1-e^{-2\pi}\right)}{\pi\left(1+n^2\right)}$$

$$= \frac{(1+in)e^{-\pi}\left(e^{\pi}-e^{-\pi}\right)}{\pi\left(1+n^2\right)} = \frac{2(1+in)e^{-\pi}\sinh\pi}{\pi\left(1+n^2\right)}$$

Equate real and imaginary parts

$$a_n = \frac{2e^{-\pi}\sinh\pi}{\pi\left(n^2+1\right)},\ b_n = \frac{2ne^{-\pi}\sinh\pi}{\pi\left(n^2+1\right)}$$

$\therefore$ Fourier-series expansion of $f(x)$ is

$$f(x) \sim \frac{1}{\pi}e^{-\pi}\sinh\pi\left[1+2\sum_{n=1}^{\infty}\frac{1}{n^2+1}(\cos nx + n\sin nx)\right]$$

Take $x = \pi$

$$e^{-\pi} = \frac{1}{\pi}e^{-\pi}\sinh\pi\left[1+2\sum_{n=1}^{\infty}\frac{(-1)^n}{n^2+1}\right]$$

$\therefore$ $$\frac{\pi}{2\sinh\pi} = \frac{1}{2}+\sum_{n=1}^{\infty}\frac{(-1)^n}{n^2+1} = \sum_{n=2}^{\infty}\frac{(-1)^n}{n^2+1}$$

Example 3.5: Expand $f(x) = x\sin x, 0 < x < 2\pi$ as a Fourier series and deduce that $\sum_{n=2}^{\infty}\frac{1}{n^2-1} = \frac{3}{4}$.

Solution: Considering $f(x)$ to be a periodic function with period 2π, Fourier-series expansion of $f(x)$ is

$$f(x) \sim \frac{a_0}{2}+\sum_{n=1}^{\infty}\left(a_n\cos nx + b_n\sin nx\right) \tag{1}$$

where $$a_0 = \frac{1}{\pi}\int_0^{2\pi} x\sin x\,dx = \frac{1}{\pi}\left[-x\cos x+\sin x\right]_0^{2\pi} = -2$$

$$a_n + ib_n = \frac{1}{\pi}\int_0^{2\pi} x\sin x\,e^{inx}dx = \frac{1}{\pi}\int_0^{2\pi} x\left(\frac{e^{ix}-e^{-ix}}{2i}\right)e^{inx}dx$$

$$= \frac{1}{2\pi i}\int_0^{2\pi} x\left(e^{i(n+1)x}-e^{i(n-1)x}\right)dx \tag{2}$$

$$= \frac{1}{2\pi i}\left[x\left(-\frac{i}{n+1}e^{i(n+1)x}\right)-\left(-\frac{1}{(n+1)^2}e^{i(n+1)x}\right)\right.$$

$$\left.-\left\{x\left(-\frac{i}{n-1}e^{i(n-1)x}\right)-\left(-\frac{1}{(n-1)^2}e^{i(n-1)x}\right)\right\}\right]_0^{2\pi} ;n \neq 1$$

$$= \frac{1}{2\pi i}\left[-\frac{2\pi i}{n+1}+\frac{1}{(n+1)^2}-\frac{1}{(n+1)^2}+\frac{2\pi i}{n-1}-\frac{1}{(n-1)^2}+\frac{1}{(n-1)^2}\right]; \; n \neq 1$$

$$= \left(\frac{1}{n-1}-\frac{1}{n+1}\right) = \frac{2}{n^2-1}; \; n \neq 1$$

Equate real and imaginary parts

$$a_n = \frac{2}{n^2-1}, \; b_n = 0; \; n \neq 1$$

From (2)

$$a_1 + ib_1 = \frac{1}{2\pi i}\int_0^{2\pi} x\left(e^{2ix}-1\right)dx$$

$$= \frac{1}{2\pi i}\left[x\left(-\frac{i}{2}e^{2ix}\right)+\frac{1}{4}e^{2ix}-\frac{x^2}{2}\right]_0^{2\pi}$$

$$= \frac{1}{2\pi i}\left[-\frac{2\pi i}{2}+\frac{1}{4}-\frac{1}{4}-2\pi^2\right]$$

$$= -\frac{1}{2}+\pi i$$

Equate real and imaginary parts

$$a_1 = -\frac{1}{2}, \; b_1 = \pi$$

$\therefore$ Fourier-series expansion of $f(x)$ is

$$f(x) \sim -1 - \frac{1}{2}\cos x + \pi \sin x + 2\sum_{n=2}^{\infty}\frac{1}{n^2-1}\cos nx \tag{3}$$

Other method to find a_n and b_n

$$a_n = \frac{1}{\pi}\int_0^{2\pi} x \sin x \cos nx \, dx$$

$$= \frac{1}{2\pi}\int_0^{2\pi} x\left[\sin(n+1)x - \sin(n-1)x\right]dx$$

$$= \frac{1}{2\pi}\left[x\left\{-\frac{1}{n+1}\cos(n+1)x + \frac{1}{n-1}\cos(n-1)x\right\}\right.$$

$$\left. - \left\{-\frac{1}{(n+1)^2}\sin(n+1)x + \frac{1}{(n-1)^2}\sin(n-1)x\right\}\right]_0^{2\pi} \quad ; n \neq 1$$

$$= \frac{1}{2\pi}\left[2\pi\left\{-\frac{1}{n+1} + \frac{1}{n-1}\right\}\right] = \frac{1}{n-1} - \frac{1}{n+1} = \frac{2}{n^2-1}; \; n \neq 1$$

$$a_1 = \frac{1}{\pi}\int_0^{2\pi} x \sin x \cos x \, dx = \frac{1}{2\pi}\int_0^{2\pi} x \sin 2x \, dx$$

$$= \frac{1}{2\pi}\left[x\left(-\frac{1}{2}\cos 2x\right) - \left(-\frac{1}{4}\sin 2x\right)\right]_0^{2\pi} = -\frac{1}{2}$$

$$b_n = \frac{1}{\pi}\int_0^{2\pi} x \sin x \sin nx \, dx$$

$$= \frac{1}{2\pi}\int_0^{2\pi} x\left[\cos(n-1)x - \cos(n+1)x\right]dx$$

$$= \frac{1}{2\pi}\left[x\left\{\frac{1}{n-1}\sin(n-1)x - \frac{1}{n+1}\sin(n+1)x\right\}\right.$$

$$\left. - \left\{\frac{-1}{(n-1)^2}\cos(n-1)x + \frac{1}{(n+1)^2}\cos(n+1)x\right\}\right]_0^{2\pi} \quad ; n \neq 1$$

$\therefore$ $$b_n = 0; \; n \neq 1$$

$$b_1 = \frac{1}{\pi}\int_0^{2\pi} x \sin^2 x \, dx = \frac{1}{2\pi}\int_0^{2\pi} x(1-\cos 2x)\,dx$$

$$= \frac{1}{2\pi}\left[\frac{x^2}{2} - \left\{x\left(\frac{1}{2}\sin 2x\right) - \left(-\frac{1}{4}\cos 2x\right)\right\}\right]_0^{2\pi} = \frac{1}{2\pi}\left[2\pi^2\right] = \pi$$

From (3)

$$f(0) = \frac{f(0+0)+f(0-0)}{2} = \frac{f(0+0)+f(2\pi-0)}{2}$$

$$= \frac{0+0}{2} = 0 = -1 - \frac{1}{2} + 2\sum_{n=2}^{\infty}\frac{1}{n^2-1}$$

$\therefore$ $$\sum_{n=2}^{\infty}\frac{1}{n^2-1} = \frac{3}{4}$$

Example 3.6: The function $f(x)$ is given by

$$f(x)=\begin{cases}-\pi; & -\pi<x<0\\ x\ ; & 0<x<\pi\end{cases}$$

$$f(x+2\pi)=f(x)$$

Draw its graph and find its Fourier series and hence show that $\frac{1}{1^2}+\frac{1}{3^2}+\frac{1}{5^2}+\cdots=\frac{\pi^2}{8}$.

Solution: Fourier-series expansion of $f(x)$ is

$$f(x)\sim\frac{a_0}{2}+\sum_{n=1}^{\infty}\left(a_n\cos nx+b_n\sin nx\right)$$

where

$$a_0=\frac{1}{\pi}\left[\int_{-\pi}^{0}-\pi\,dx+\int_0^{\pi}x\,dx\right]=\frac{1}{\pi}\left[\left(-\pi x\right)_{-\pi}^{0}+\left(\frac{x^2}{2}\right)_0^{\pi}\right]$$

$$=\frac{1}{\pi}\left[-\pi^2+\frac{\pi^2}{2}\right]=-\frac{\pi}{2}$$

$$a_n+ib_n=\frac{1}{\pi}\left[\int_{-\pi}^{0}-\pi e^{inx}dx+\int_0^{\pi}xe^{inx}dx\right]$$

$$=\frac{1}{\pi}\left[\frac{\pi i}{n}\left(e^{inx}\right)_{-\pi}^{0}+\left\{x\left(-\frac{i}{n}e^{inx}\right)-\left(-\frac{1}{n^2}e^{inx}\right)\right\}_0^{\pi}\right]$$

$$=\frac{1}{\pi}\left[\frac{\pi i}{n}\left(1-(-1)^n\right)-\frac{\pi i}{n}(-1)^n+\frac{1}{n^2}\left((-1)^n-1\right)\right]$$

$$=\frac{i}{n}\left[1-2(-1)^n\right]-\frac{1}{\pi n^2}\left[1-(-1)^n\right]$$

Equate real and imaginary parts

$$a_n=-\frac{1}{\pi n^2}\left[1-(-1)^n\right],\ b_n=\frac{1-2(-1)^n}{n}$$

$\therefore$

$$a_{2n}=0,\ a_{2n-1}=-\frac{2}{\pi(2n-1)^2};\ n=1,2,3,\ldots$$

$$b_{2n}=-\frac{1}{2n},\ b_{2n-1}=\frac{3}{2n-1};\ n=1,2,3,\ldots$$

$\therefore$ Fourier-series expansion of $f(x)$ is

$$f(x)\sim-\frac{\pi}{4}-\frac{2}{\pi}\sum_{n=1}^{\infty}\frac{\cos(2n-1)x}{(2n-1)^2}+\sum_{n=1}^{\infty}\left[\frac{3\sin(2n-1)x}{2n-1}-\frac{\sin 2nx}{2n}\right]$$

$$f(0)=\frac{f(0+0)+f(0-0)}{2}=\frac{0+(-\pi)}{2}=-\frac{\pi}{4}-\frac{2}{\pi}\sum_{n=1}^{\infty}\frac{1}{(2n-1)^2}$$

$\therefore$

$$\sum_{n=1}^{\infty}\frac{1}{(2n-1)^2}=\frac{\pi^2}{8}$$

$$\therefore \qquad \frac{1}{1^2}+\frac{1}{3^2}+\frac{1}{5^2}+\cdots=\frac{\pi^2}{8}$$

Graph of $f(x)$ is shown below

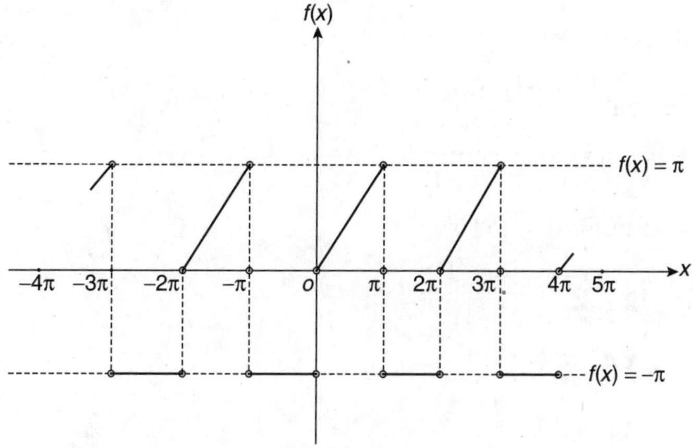

Figure 3.2

⊚ denotes that the point is not in the graph.

Example 3.7: Find the Fourier series to represent the function $f(x)$ given by

$$f(x)=\begin{cases} x & ; \ 0\le x\le \pi \\ 2\pi - x & ; \ \pi \le x \le 2\pi \end{cases} \quad \text{and deduce that } \frac{1}{1^2}+\frac{1}{3^2}+\frac{1}{5^2}+\cdots\cdots=\frac{\pi^2}{8}.$$

Solution: Fourier-series expansion (assuming $f(x)$ periodic of period 2π) of $f(x)$ is

$$f(x)=\frac{a_0}{2}+\sum_{n=1}^{\infty}\left(a_n\cos nx+b_n\sin nx\right)$$

where
$$a_0=\frac{1}{\pi}\left[\int_0^{\pi}x\,dx+\int_{\pi}^{2\pi}(2\pi-x)\,dx\right]=\frac{1}{\pi}\left[\left(\frac{x^2}{2}\right)_0^{\pi}+\left(2\pi x-\frac{x^2}{2}\right)_{\pi}^{2\pi}\right]$$

$$=\frac{1}{\pi}\left[\frac{\pi^2}{2}+4\pi^2-2\pi^2-2\pi^2+\frac{\pi^2}{2}\right]=\pi$$

$$a_n+ib_n=\frac{1}{\pi}\left[\int_0^{\pi}x\,e^{inx}dx+\int_{\pi}^{2\pi}(2\pi-x)e^{inx}dx\right]$$

$$=\frac{1}{\pi}\left[\left\{x\left(-\frac{i}{n}e^{inx}\right)-\left(-\frac{1}{n^2}e^{inx}\right)\right\}_0^{\pi}+\left\{(2\pi-x)\left(-\frac{i}{n}e^{inx}\right)\right.\right.$$

$$\left.\left.-(-1)\left(-\frac{1}{n^2}e^{inx}\right)\right\}_{\pi}^{2\pi}\right]$$

$$= \frac{1}{\pi}\left[-\frac{\pi}{n}i(-1)^n + \frac{1}{n^2}\left((-1)^n - 1\right) + \frac{i\pi}{n}(-1)^n - \frac{1}{n^2}\left(1-(-1)^n\right)\right]$$

$$= -\frac{2}{\pi n^2}\left[1-(-1)^n\right]$$

Equate real and imaginary parts

$$a_n = -\frac{2}{\pi n^2}\left[1-(-1)^n\right],\ b_n = 0; \qquad n = 1,2,3,\ldots$$

$$\therefore \qquad a_{2n} = 0,\quad a_{2n-1} = -\frac{4}{\pi(2n-1)^2}; \qquad n = 1,2,3,\ldots$$

$\therefore$ Fourier-series expansion of $f(x)$ is

$$f(x) = \frac{\pi}{2} - \frac{4}{\pi}\sum_{n=1}^{\infty}\frac{\cos(2n-1)x}{(2n-1)^2}$$

$$\therefore \qquad f(0) = 0 = \frac{\pi}{2} - \frac{4}{\pi}\sum_{n=1}^{\infty}\frac{1}{(2n-1)^2}$$

$$\therefore \qquad \sum_{n=1}^{\infty}\frac{1}{(2n-1)^2} = \frac{\pi^2}{8}$$

$$\frac{1}{1^2} + \frac{1}{3^2} + \frac{1}{5^2} + \cdots = \frac{\pi^2}{8}.$$

Example 3.8: If $f(x) = \begin{cases} 0 & , -\pi \le x \le 0 \\ \sin x & , 0 \le x \le \pi \end{cases}$

Prove that

$$f(x) = \frac{1}{\pi} + \frac{1}{2}\sin x - \frac{2}{\pi}\sum_{n=1}^{\infty}\frac{\cos 2n\,x}{4n^2 - 1}$$

Hence, show that

(i) $\frac{1}{1.3} + \frac{1}{3.5} + \frac{1}{5.7} + \cdots = \frac{1}{2}$ (ii) $\frac{1}{1.3} - \frac{1}{3.5} + \frac{1}{5.7} - \frac{1}{7.9} + \cdots = \frac{\pi - 2}{4}$

Solution: Considering $f(x)$ to be periodic with period 2π, Fourier-series expansion of $f(x)$ is

$$f(x) = \frac{a_0}{2} + \sum_{n=1}^{\infty}\left(a_n \cos nx + b_n \sin nx\right)$$

where
$$a_0 = \frac{1}{\pi}\int_0^{\pi}\sin x\,dx = \frac{1}{\pi}(-\cos x)_0^{\pi} = \frac{1}{\pi}(1+1) = \frac{2}{\pi}$$

$$a_n + i\,b_n = \frac{1}{\pi}\int_0^{\pi}\sin x\, e^{inx}dx = \frac{1}{\pi}\left[\frac{e^{inx}}{1-n^2}(i\,n\sin x - \cos x)\right]_0^{\pi},\ n \ne 1$$

$$= \frac{1}{\pi\left(1-n^2\right)}\left[1+(-1)^n\right];\quad n \ne 1$$

$$= \frac{-1}{\pi\left(n^2-1\right)}\left[1+(-1)^n\right];\quad n \ne 1$$

Equate real and imaginary parts

$$a_n = -\frac{1}{\pi\left(n^2-1\right)}\left[1+(-1)^n\right],\ b_n = 0;\ \ n = 2,3,\ldots$$

$$\therefore \qquad a_{2n} = -\frac{2}{\pi\left(4n^2-1\right)};\ n = 1,2,3,\ldots$$

$$a_{2n-1} = 0,\ b_n = 0; \quad n = 2,3,\ldots$$

$$a_1 = \frac{1}{\pi}\int_0^{\pi} \sin x \cos x\, dx = \frac{1}{2\pi}\int_0^{\pi} \sin 2x\, dx = -\frac{1}{4\pi}(\cos 2x)_0^{\pi} = 0$$

$$b_1 = \frac{1}{\pi}\int_0^{\pi} \sin^2 x\, dx = \frac{1}{2\pi}\int_0^{\pi}(1-\cos 2x)\, dx = \frac{1}{2\pi}\left[x - \frac{1}{2}\sin 2x\right]_0^{\pi} = \frac{1}{2}$$

$\therefore$ Fourier-series expansion of $f(x)$ is

$$f(x) = \frac{1}{\pi} + \frac{1}{2}\sin x - \frac{2}{\pi}\sum_{n=1}^{\infty}\frac{\cos 2nx}{4n^2-1} \tag{1}$$

Take $x = 0$

$$0 = \frac{1}{\pi} - \frac{2}{\pi}\sum_{n=1}^{\infty}\frac{1}{(2n-1)(2n+1)}$$

$$\therefore \qquad \frac{1}{1.3} + \frac{1}{3.5} + \frac{1}{5.7} + \cdots = \frac{1}{2}$$

In (1), take $x = \frac{\pi}{2}$

$$1 = \frac{1}{\pi} + \frac{1}{2} + \frac{2}{\pi}\sum_{n=1}^{\infty}\frac{(-1)^{n+1}}{(2n-1)(2n+1)}$$

$$\Rightarrow \qquad \frac{2}{\pi}\sum_{n=1}^{\infty}\frac{(-1)^{n+1}}{(2n-1)(2n+1)} = \frac{\pi-2}{2\pi}$$

$$\therefore \qquad \frac{1}{1.3} - \frac{1}{3.5} + \frac{1}{5.7} - \frac{1}{7.9} + \cdots = \frac{\pi-2}{4}$$

Example 3.9: Find the Fourier series of $f(x) = x$ in $-\pi < x < \pi,\ f(x+2\pi) = f(x)$.

Solution: $f(x)$ is odd function of x and hence Fourier series for $f(x)$ is

$$f(x) \sim \sum_{n=1}^{\infty} b_n \sin nx$$

where

$$b_n = \frac{2}{\pi}\int_0^{\pi} x \sin nx\, dx$$

$$= \frac{2}{\pi}\left[x\left(-\frac{1}{n}\cos nx\right) - \left(-\frac{1}{n^2}\sin nx\right)\right]_0^{\pi}$$

$$= \frac{2}{n}(-1)^{n+1}$$

$\therefore$ Fourier series for $f(x)$ is

$$f(x) \sim 2\sum_{n=1}^{\infty}\frac{(-1)^{n+1}}{n}\sin(nx)$$

Example 3.10: Find the Fourier-series representations for

(i) $f(x)=|\sin x|, -\pi < x < \pi$

(ii) $f(x)=|\cos x|, -\pi < x < \pi$

Also draw the graphs.

Solution: (i) Graph of $f(x)$ considering it periodic with period 2π is

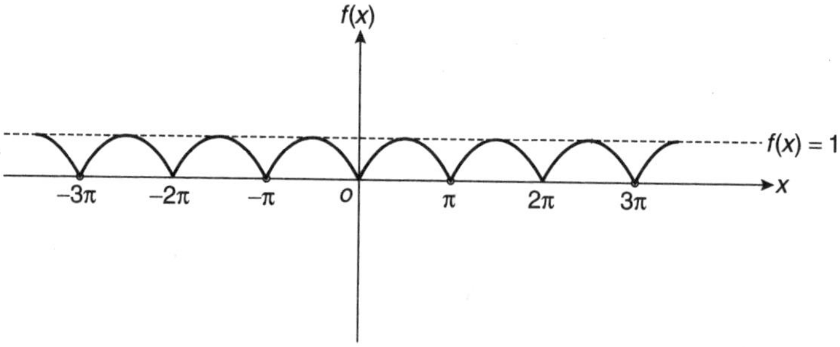

Figure 3.3

⊚ indicates point is not in graph.

$f(x)$ is even function of x. Its Fourier-series representation is

$$f(x) \sim \frac{a_0}{2} + \sum_{n=1}^{\infty} a_n \cos nx$$

where
$$a_0 = \frac{2}{\pi}\int_0^{\pi} \sin x \; dx = \left(-\frac{2}{\pi}\cos x\right)_0^{\pi} = \frac{4}{\pi}$$

$$a_n = \frac{2}{\pi}\int_0^{\pi} \sin x \cos nx \; dx = \frac{1}{\pi}\int_0^{\pi}\left[\sin(n+1)x - \sin(n-1)x\right]dx$$

$$= \frac{1}{\pi}\left[-\frac{1}{n+1}\cos(n+1)x + \frac{1}{n-1}\cos(n-1)x\right]_0^{\pi} \;;\; n \neq 1$$

$$= \frac{1}{\pi}\left[\frac{1}{n+1}\left\{(-1)^n+1\right\} - \frac{1}{n-1}\left\{(-1)^n+1\right\}\right]$$

$$= -\frac{2}{\pi\left(n^2-1\right)}\left[(-1)^n+1\right]$$

$\therefore$
$$a_{2n-1} = 0; \; n = 2,3,4,\ldots$$

$$a_{2n} = -\frac{4}{\pi\left(4n^2-1\right)}; \; n = 1,2,3,\ldots$$

$$a_1 = \frac{2}{\pi}\int_0^{\pi} \sin x \cos x \, dx = \frac{1}{\pi}\int_0^{\pi} \sin 2x \, dx = \frac{1}{2\pi}\left(-\cos 2x\right)_0^{\pi} = 0$$

$\therefore$ Fourier-series representation of $f(x)$ is

$$f(x) \sim \frac{2}{\pi} - \frac{4}{\pi}\sum_{n=1}^{\infty} \frac{\cos 2nx}{(2n-1)(2n+1)}$$

(ii) Consider $f(x)$ to be periodic with period 2π

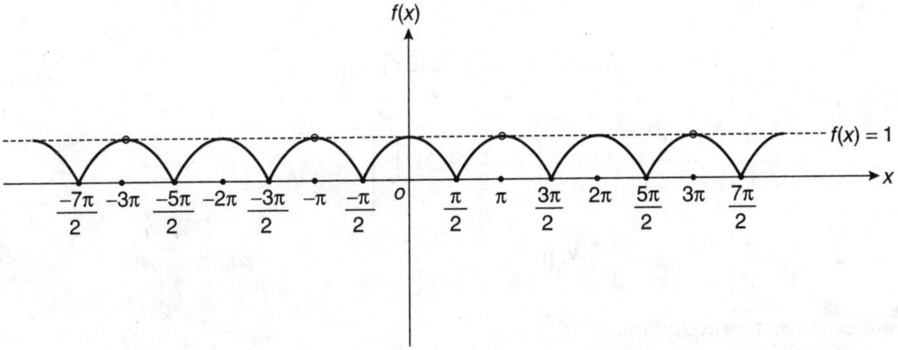

Figure 3.4

⊚ indicates that the point is not in the graph.

Here, $f(x)$ is an even function of x.

Fourier-series representation of $f(x)$ is

$$f(x) \sim \frac{a_0}{2} + \sum_{n=1}^{\infty} a_n \cos nx$$

where
$$a_0 = \frac{2}{\pi}\int_0^{\pi} |\cos x|\, dx = \frac{2}{\pi}\left[\int_0^{\frac{\pi}{2}} \cos x\, dx - \int_{\frac{\pi}{2}}^{\pi} \cos x\, dx\right] = \frac{2}{\pi}\left[(\sin x)_0^{\frac{\pi}{2}} - (\sin x)_{\frac{\pi}{2}}^{\pi}\right] = \frac{4}{\pi}$$

$$a_n = \frac{2}{n}\int_0^{\pi} |\cos x| \cos nx \; dx = \frac{1}{\pi}\left[\int_0^{\frac{\pi}{2}} 2\cos nx \, \cos x\, dx - \int_{\frac{\pi}{2}}^{\pi} 2\cos nx \, \cos x\, dx\right]$$

$$= \frac{1}{\pi}\left[\int_0^{\frac{\pi}{2}} \left[\cos(n+1)x + \cos(n-1)x\right] dx - \int_{\frac{\pi}{2}}^{\pi} \left[\cos(n+1)x + \cos(n-1)x\right] dx\right]$$

$$= \frac{1}{\pi}\left[\left\{\frac{1}{n+1}\sin(n+1)x + \frac{1}{n-1}\sin(n-1)x\right\}_0^{\frac{\pi}{2}}\right.$$

$$\left. - \left\{\frac{1}{n+1}\sin(n+1)x + \frac{1}{n-1}\sin(n-1)x\right\}_{\frac{\pi}{2}}^{\pi}\right]; \; n \neq 1$$

$$= \frac{2}{\pi}\left[\frac{1}{n+1}\sin(n+1)\frac{\pi}{2} + \frac{1}{n-1}\sin(n-1)\frac{\pi}{2}\right]; \;\; n \neq 1$$

$\therefore$
$$a_{2n-1} = 0; \; n = 2, 3, 4, \ldots$$

$$a_{2n} = \frac{2}{\pi}\left[\frac{1}{2n+1}(-1)^n - \frac{1}{2n-1}(-1)^n\right] = \frac{4(-1)^{n+1}}{\pi(2n-1)(2n+1)}; \;\; n = 1, 2, \ldots$$

$$a_1 = \frac{2}{\pi}\int_0^{\pi}|\cos x|\cos x\,dx = \frac{1}{\pi}\left[\int_0^{\frac{\pi}{2}} 2\cos^2 x\,dx - \int_{\frac{\pi}{2}}^{\pi} 2\cos^2 x\,dx\right]$$

$$= \frac{1}{\pi}\left[\int_0^{\frac{\pi}{2}}(1+\cos 2x)\,dx - \int_{\frac{\pi}{2}}^{\pi}(1+\cos 2x)\,dx\right]$$

$$= \frac{1}{\pi}\left[\left(x+\frac{1}{2}\sin 2x\right)_0^{\frac{\pi}{2}} - \left(x+\frac{1}{2}\sin 2x\right)_{\frac{\pi}{2}}^{\pi}\right]$$

$$= \frac{1}{\pi}\left[\frac{\pi}{2} - \left(\pi - \frac{\pi}{2}\right)\right] = 0$$

∴ Fourier-series representation of $f(x)$ is

$$f(x) \sim \frac{2}{\pi} + \frac{4}{\pi}\sum_{n=1}^{\infty}\frac{(-1)^{n+1}\cos 2nx}{(2n-1)(2n+1)}.$$

Example 3.11: Expand $f(x) = \cos ax$ as a Fourier series in $(-\pi, \pi)$ where a is fraction. What will happen to Fourier series if a is integer?

Solution: $f(x)$ is even function of x. Considering $f(x)$ to be periodic function of period 2π, Fourier-series expansion of $f(x)$ is

$$f(x) \sim \frac{a_0}{2} + \sum_{n=1}^{\infty} a_n \cos nx$$

where
$$a_0 = \frac{2}{\pi}\int_0^{\pi}\cos ax\,dx = \frac{2}{\pi a}(\sin ax)_0^{\pi} = \frac{2}{\pi a}\sin a\pi$$

$$a_n = \frac{2}{\pi}\int_0^{\pi}\cos ax\cos nx\,dx = \frac{1}{\pi}\int_0^{\pi}\left[\cos(n+a)x + \cos(n-a)x\right]dx$$

$$= \frac{1}{\pi}\left[\frac{1}{n+a}\sin(n+a)x + \frac{1}{n-a}\sin(n-a)x\right]_0^{\pi}$$

$$= \frac{1}{\pi}\left[\frac{(-1)^n}{n+a}\sin a\pi - \frac{(-1)^n}{n-a}\sin a\pi\right] = \frac{2a(-1)^{n+1}}{\pi\left(n^2-a^2\right)}\sin a\pi$$

∴ Fourier-series representation of $f(x)$ is

$$f(x) \sim \frac{1}{\pi a}\sin(a\pi) + \frac{2a}{\pi}\sin(a\pi)\sum_{n=1}^{\infty}\frac{(-1)^{n+1}}{n^2-a^2}\cos nx$$

$$= \frac{2a}{\pi}\sin(a\pi)\left[\frac{1}{2a^2} + \sum_{n=1}^{\infty}\frac{(-1)^{n+1}}{n^2-a^2}\cos nx\right]$$

When a is an integer, then Fourier-series expansion in $(-\pi, \pi)$ is

$f(x) = 1$ if $a = 0$

$= \cos ax$ if a is positive integer

$= \cos(-ax)$ if a is negative integer.

Example 3.12: Obtain Fourier expansion for $\sqrt{1-\cos x}$ in interval $-\pi < x < \pi$.

Solution: Here, $f(x) = \sqrt{1-\cos x} = \sqrt{2}\left|\sin\frac{x}{2}\right|$ is even function of x. Consider $f(x)$ to be periodic function with period 2π. Its Fourier-series expansion is

$$f(x) \sim \frac{a_0}{2} + \sum_{n=1}^{\infty} a_n \cos nx$$

where $$a_0 = \frac{2}{\pi}\int_0^{\pi} \sqrt{2}\left|\sin\frac{x}{2}\right| dx = \frac{2\sqrt{2}}{\pi}\int_0^{\pi} \sin\frac{x}{2}\, dx = \frac{4\sqrt{2}}{\pi}\left(-\cos\frac{x}{2}\right)_0^{\pi} = \frac{4\sqrt{2}}{\pi}$$

$$a_n = \frac{2}{\pi}\int_0^{\pi} \sqrt{2}\sin\frac{x}{2}\cos nx\, dx$$

$$= \frac{\sqrt{2}}{\pi}\int_0^{\pi}\left[\sin\left(n+\frac{1}{2}\right)x - \sin\left(n-\frac{1}{2}\right)x\right] dx$$

$$= \frac{\sqrt{2}}{\pi}\left[\frac{-2}{2n+1}\cos\frac{2n+1}{2}x + \frac{2}{2n-1}\cos\frac{2n-1}{2}x\right]_0^{\pi}$$

$$= \frac{2\sqrt{2}}{\pi}\left[\frac{1}{2n+1} - \frac{1}{2n-1}\right] = -\frac{4\sqrt{2}}{\pi\left[4n^2-1\right]}$$

$\therefore$ Fourier-series expansion is

$$f(x) \sim \frac{2\sqrt{2}}{\pi} - \frac{4\sqrt{2}}{\pi}\sum_{n=1}^{\infty}\frac{\cos nx}{4n^2-1}.$$

Example 3.13: Express $f(x) = |x|,\ -\pi < x < \pi$ as Fourier series.

Hence show that $\frac{1}{1^2} + \frac{1}{3^2} + \frac{1}{5^2} + \cdots = \frac{\pi^2}{8}$.

Solution: $f(x)$ is an even function of x. Considering $f(x)$ to be periodic function with period 2π, Fourier-series expansion of $f(x)$ is

$$f(x) \sim \frac{a_0}{2} + \sum_{n=1}^{\infty} a_n \cos nx$$

where $$a_0 = \frac{2}{\pi}\int_0^{\pi} x\, dx = \frac{2}{\pi}\left(\frac{x^2}{2}\right)_0^{\pi} = \pi$$

$$a_n = \frac{2}{\pi}\int_0^{\pi} x\cos nx\, dx = \frac{2}{\pi}\left[x\left(\frac{1}{n}\sin nx\right) - \left(-\frac{1}{n^2}\cos nx\right)\right]_0^{\pi}$$

$$= \frac{-2}{\pi n^2}\left[1-(-1)^n\right]$$

$$\therefore \qquad a_{2n} = 0, \quad a_{2n-1} = -\frac{4}{\pi(2n-1)^2}; \quad n = 1, 2, 3, \ldots$$

$\therefore$ Fourier-series expansion of $f(x)$ is

$$f(x) \sim \frac{\pi}{2} - \frac{4}{\pi}\sum_{n=1}^{\infty}\frac{\cos(2n-1)x}{(2n-1)^2}$$

Take $x = 0$

$$f(0) = 0 = \frac{\pi}{2} - \frac{4}{\pi}\sum_{n=1}^{\infty}\frac{1}{(2n-1)^2}$$

$$\therefore \qquad \frac{1}{1^2} + \frac{1}{3^2} + \frac{1}{5^2} + \cdots = \frac{\pi^2}{8}.$$

Example 3.14: Obtain the Fourier series for the function $f(x) = x^2, -\pi \le x \le \pi$. Sketch the graph of $f(x)$. Hence show that

(i) $\frac{1}{1^2} + \frac{1}{2^2} + \frac{1}{3^2} + \frac{1}{4^2} + \cdots = \sum_{n=1}^{\infty}\frac{1}{n^2} = \frac{\pi^2}{6}$

(ii) $\frac{1}{1^2} - \frac{1}{2^2} + \frac{1}{3^2} - \frac{1}{4^2} + \cdots = \frac{\pi^2}{12}$

(iii) $\frac{1}{1^2} + \frac{1}{3^2} + \frac{1}{5^2} + \cdots = \sum_{n=1}^{\infty}\frac{1}{(2n-1)^2} = \frac{\pi^2}{8}$

Solution: $f(x)$ is an even function. Considering $f(x)$ to be periodic function of period 2π, Fourier series of $f(x)$ is

$$f(x) = x^2 = \frac{a_0}{2} + \sum_{n=1}^{\infty} a_n \cos nx$$

where

$$a_0 = \frac{2}{\pi}\int_0^{\pi} x^2 dx = \frac{2}{\pi}\left(\frac{x^3}{3}\right)_0^{\pi} = \frac{2\pi^2}{3}$$

$$a_n = \frac{2}{\pi}\int_0^{\pi} x^2 \cos nx \, dx$$

$$= \frac{2}{\pi}\left[x^2\left(\frac{1}{n}\sin nx\right) - 2x\left(-\frac{1}{n^2}\cos nx\right) + 2\left(-\frac{1}{n^3}\sin nx\right)\right]_0^{\pi} = \frac{4}{n^2}(-1)^n$$

$\therefore$ Fourier series of $f(x)$ is

$$f(x) = \frac{\pi^2}{3} + 4\sum_{n=1}^{\infty}\frac{(-1)^n}{n^2}\cos nx \qquad (1)$$

Take $x = \pi$

$$f(\pi) = \pi^2 = \frac{\pi^2}{3} + 4\sum_{n=1}^{\infty}\frac{1}{n^2}$$

$$\therefore \qquad \frac{1}{1^2} + \frac{1}{2^2} + \frac{1}{3^2} + \cdots = \sum_{n=1}^{\infty}\frac{1}{n^2} = \frac{\pi^2}{6} \qquad (2)$$

Take $x = 0$ in (1)

$$f(0) = 0 = \frac{\pi^2}{3} + 4\sum_{n=1}^{\infty} \frac{(-1)^n}{n^2}$$

$$\therefore \quad \frac{1}{1^2} - \frac{1}{2^2} + \frac{1}{3^2} - \frac{1}{4^2} + \cdots = \frac{\pi^2}{12} \tag{3}$$

Add (2) and (3)

$$2\left(\frac{1}{1^2} + \frac{1}{3^2} + \frac{1}{5^2} \cdots\right) = \frac{\pi^2}{6} + \frac{\pi^2}{12} = \frac{3\pi^2}{12} = \frac{\pi^2}{4}$$

$$\Rightarrow \quad \frac{1}{1^2} + \frac{1}{3^2} + \frac{1}{5^2} \cdots = \frac{\pi^2}{8}$$

Graph of $f(x)$ is

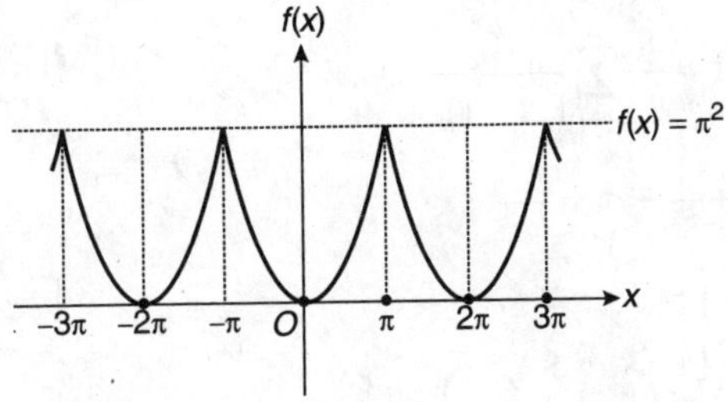

Figure 3.5

Example 3.15: Obtain Fourier series for $f(x)$ given by

$$f(x) = \begin{cases} 1 + \dfrac{2x}{\pi}, & -\pi \le x \le 0 \\ 1 - \dfrac{2x}{\pi}, & 0 \le x \le \pi \end{cases}$$

Hence, deduce that $\dfrac{1}{1^2} + \dfrac{1}{3^2} + \dfrac{1}{5^2} + \cdots = \dfrac{\pi^2}{8}$

Solution: $f(-x) = \begin{cases} 1 - \dfrac{2x}{\pi}, & -\pi \le -x \le 0 \text{ i.e., } 0 \le x \le \pi \\ 1 + \dfrac{2x}{\pi}, & 0 \le -x \le \pi \text{ i.e., } -\pi \le x \le 0 \end{cases}$

$\therefore f(x)$ is even function of x.

Considering $f(x)$ to be periodic function of period 2π, Fourier series for $f(x)$ is

$$f(x) = \frac{a_0}{2} + \sum_{n=1}^{\infty} a_n \cos nx$$

where $$a_0 = \frac{2}{\pi}\int_0^{\pi}\left(1 - \frac{2x}{\pi}\right)dx = \frac{2}{\pi}\left(x - \frac{x^2}{\pi}\right)_0^{\pi} = 0$$

$$a_n = \frac{2}{\pi}\int_0^{\pi}\left(1-\frac{2x}{\pi}\right)\cos nx \; dx$$

$$= \frac{2}{\pi}\left[\left(1-\frac{2x}{\pi}\right)\left(\frac{1}{n}\sin nx\right)-\left(-\frac{2}{\pi}\right)\left(-\frac{1}{n^2}\cos nx\right)\right]_0^{\pi}$$

$$= \frac{4}{\pi^2 n^2}\left[1-(-1)^n\right]$$

$\therefore$ $$a_{2n} = 0,\; a_{2n-1} = \frac{8}{\pi^2(2n-1)^2};\; n = 1,2,3,\ldots$$

$\therefore$ Fourier series for $f(x)$ is

$$f(x) = \frac{8}{\pi^2}\sum_{n=1}^{\infty}\frac{\cos(2n-1)x}{(2n-1)^2}$$

$\therefore$ $$f(0) = 1 = \frac{8}{\pi^2}\sum_{n=1}^{\infty}\frac{1}{(2n-1)^2}$$

$\therefore$ $$\frac{1}{1^2}+\frac{1}{3^2}+\frac{1}{5^2}+\cdots = \frac{\pi^2}{8}$$

Example 3.16: If $f(x) = \begin{cases} -k \;; & -\pi < x < 0 \\ k \;; & 0 < x < \pi \end{cases}$

and $f(x+2\pi) = f(x)$ for all x, obtain the Fourier series for $f(x)$. Deduce that

$$1-\frac{1}{3}+\frac{1}{5}-\frac{1}{7}+\cdots = \frac{\pi}{4}.$$

Solution: $f(-x) = \begin{cases} -k \;; & -\pi < -x < 0 \quad \text{i.e.,} \quad 0 < x < \pi \\ k \;; & 0 < -x < \pi \quad \text{i.e.,} \quad -\pi < x < 0 \end{cases}$

$\therefore$ $f(x)$ is odd function.

Fourier series for $f(x)$ is

$$f(x) \sim \sum_{n=1}^{\infty} b_n \sin nx$$

where $$b_n = \frac{2}{\pi}\int_0^{\pi} k\sin nx\, dx = \frac{2k}{\pi}\left(-\frac{1}{n}\cos nx\right)_0^{\pi} = \frac{2k}{n\pi}\left(1-(-1)^n\right)$$

$\therefore$ $$b_{2n} = 0;\; b_{2n-1} = \frac{4k}{\pi(2n-1)};\; n = 1,2,3,\ldots$$

$\therefore$ Fourier series for $f(x)$ is

$$f(x) \sim \frac{4k}{\pi}\sum_{n=1}^{\infty}\frac{\sin(2n-1)x}{2n-1}$$

$$f\left(\frac{\pi}{2}\right) = k = \frac{4k}{\pi}\sum_{n=1}^{\infty}\frac{(-1)^{n+1}}{2n-1}$$

$\therefore$ $$1-\frac{1}{3}+\frac{1}{5}-\frac{1}{7}+\cdots = \frac{\pi}{4}.$$

Example 3.17: Sketch the curve for the function

$$f(x)=\begin{cases}\pi+x; & -\pi\le x\le -\pi/2\\ \pi/2; & -\pi/2\le x\le \pi/2\\ \pi-x; & \pi/2\le x\le \pi\end{cases}$$

and find its Fourier expansion.

Solution: Considering $f(x)$ to be periodic function of period 2π, graph of $f(x)$ is

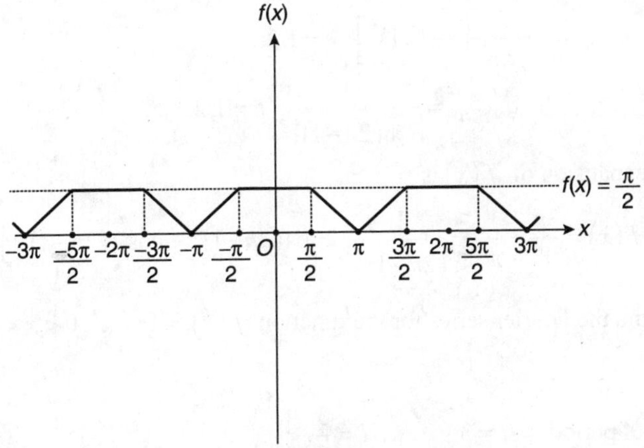

Figure 3.6

The graph is symmetrical about line $x=0$.

$\therefore\ f(x)$ is even function.

Fourier-series expansion of $f(x)$ is

$$f(x)=\frac{a_0}{2}+\sum_{n=1}^{\infty}a_n\cos nx$$

where

$$a_0=\frac{2}{\pi}\left[\int_0^{\pi/2}\frac{\pi}{2}dx+\int_{\frac{\pi}{2}}^{\pi}(\pi-x)\,dx\right]=\frac{2}{\pi}\left[\left(\frac{\pi x}{2}\right)_0^{\frac{\pi}{2}}+\left(\pi x-\frac{x^2}{2}\right)_{\frac{\pi}{2}}^{\pi}\right]$$

$$=\frac{2}{\pi}\left(\frac{\pi^2}{4}+\frac{\pi^2}{2}-\frac{\pi^2}{2}+\frac{\pi^2}{8}\right)=\frac{3\pi}{4}$$

$$a_n=\frac{2}{\pi}\left[\int_0^{\frac{\pi}{2}}\frac{\pi}{2}\cos nx\ dx+\int_{\frac{\pi}{2}}^{\pi}(\pi-x)\cos nx\,dx\right]$$

$$=\frac{2}{\pi}\left[\left\{\frac{\pi}{2n}\sin nx\right\}_0^{\frac{\pi}{2}}+\left\{\frac{\pi-x}{n}\sin nx-\frac{1}{n^2}\cos nx\right\}_{\frac{\pi}{2}}^{\pi}\right]$$

$$= \frac{2}{\pi}\left[\frac{\pi}{2n}\sin\frac{n\pi}{2} - \frac{\pi}{2n}\sin\frac{n\pi}{2} - \frac{1}{n^2}\left\{(-1)^n - \cos\frac{n\pi}{2}\right\}\right]$$

$$= -\frac{2}{\pi n^2}\left[(-1)^n - \cos\frac{n\pi}{2}\right]$$

$\therefore$ $$a_{2n-1} = \frac{2}{\pi(2n-1)^2};\ n = 1, 2, \ldots$$

$$a_{2n} = -\frac{2}{4\pi n^2}\left[1 - (-1)^n\right];\ n = 1, 2, \ldots$$

$\therefore$ $$a_{4n} = 0,\ a_{2(2n-1)} = -\frac{1}{\pi(2n-1)^2};\ n = 1, 2, 3, \ldots$$

$\therefore$ Fourier-series expansion of $f(x)$ is

$$f(x) = \frac{3\pi}{8} + \frac{1}{\pi}\sum_{n=1}^{\infty}\frac{1}{(2n-1)^2}\left[2\cos(2n-1)x - \cos 2(2n-1)x\right]$$

Example 3.18: Find the Fourier series for the function $f(x) = 2x - x^2, 0 < x < 3$ and deduce that $\sum_{n=1}^{\infty}\frac{1}{n^2} = \frac{\pi^2}{6}$.

Solution: Length of period $= 2l = 3$ $\quad\therefore\ l = \frac{3}{2}$

Fourier series for $f(x)$ is

$$f(x) \sim \frac{a_0}{2} + \sum_{n=1}^{\infty}\left(a_n\cos\frac{2n\pi x}{3} + b_n\sin\frac{2n\pi x}{3}\right)$$

where $$a_0 = \frac{2}{3}\int_0^3 (2x - x^2)\,dx = \frac{2}{3}\left(x^2 - \frac{x^3}{3}\right)_0^3 = 0$$

$$a_n + ib_n = \frac{2}{3}\int_0^3 (2x - x^2)e^{i\frac{2n\pi x}{3}}\,dx$$

$$= \frac{2}{3}\left[(2x - x^2)\left(-\frac{3i}{2n\pi}e^{i\frac{2n\pi x}{3}}\right) - (2 - 2x)\left(-\frac{9}{4n^2\pi^2}e^{i\frac{2n\pi x}{3}}\right) + (-2)\left(\frac{27i}{8n^3\pi^3}e^{i\frac{2n\pi x}{3}}\right)\right]_0^3$$

$$= \frac{2}{3}\left[\frac{9i}{2n\pi} - \frac{9}{n^2\pi^2} - \frac{9}{2n^2\pi^2}\right] = \frac{3i}{n\pi} - \frac{9}{n^2\pi^2}$$

Equate real and imaginary parts

$$a_n = -\frac{9}{n^2\pi^2},\ b_n = \frac{3}{n\pi};\ n = 1, 2, 3, \ldots$$

$\therefore$ Fourier series for $f(x)$ is

$$f(x) \sim \frac{3}{\pi}\sum_{n=1}^{\infty}\left(\frac{1}{n}\sin\frac{2n\pi x}{3} - \frac{3}{\pi n^2}\cos\frac{2n\pi x}{3}\right)$$

For $x = 3$

$$\frac{f(3-0)+f(3+0)}{2} = -\frac{9}{\pi^2}\sum_{n=1}^{\infty}\frac{1}{n^2}$$

$$\Rightarrow \qquad \frac{f(3-0)+f(0+0)}{2} = \frac{-9}{\pi^2}\sum_{n=1}^{\infty}\frac{1}{n^2}$$

$$\Rightarrow \qquad \sum_{n=1}^{\infty}\frac{1}{n^2} = -\frac{\pi^2}{9}\left(\frac{-3+0}{2}\right) = \frac{\pi^2}{6}$$

Example 3.19: (i) Express $f(x) = x^2$ in the Fourier series for $0 < x < 2$.

(ii) Find the Fourier series for $f(x) = x^2$ in $(0,4)$ and deduce that $\frac{\pi^2}{6} = 1 + \frac{1}{2^2} + \frac{1}{3^2} + \frac{1}{4^2} + \cdots$

Solution: (i) Fourier series for $f(x) = x^2$ in $(0,2)$ is

$$f(x) \sim \frac{a_0}{2} + \sum_{n=1}^{\infty}\left[a_n \cos(n\pi x) + b_n \sin(n\pi x)\right]$$

where $$a_0 = \int_0^2 x^2 dx = \frac{1}{3}\left(x^3\right)_0^2 = \frac{8}{3}$$

$$a_n + i b_n = \int_0^2 x^2 e^{in\pi x} dx$$

$$= \left[x^2\left(-\frac{i}{n\pi}e^{in\pi x}\right) - 2x\left(-\frac{1}{n^2\pi^2}e^{in\pi x}\right) + 2\left(\frac{i}{n^3\pi^3}e^{in\pi x}\right)\right]_0^2$$

$$= -\frac{4i}{n\pi} + \frac{4}{n^2\pi^2}$$

Equate real and imaginary parts

$$a_n = \frac{4}{n^2\pi^2},\; b_n = -\frac{4}{n\pi};\; n = 1,2,3,\ldots$$

$\therefore$ Fourier series for $f(x)$ is

$$f(x) \sim 4\left[\frac{1}{3} + \frac{1}{\pi}\sum_{n=1}^{\infty}\frac{1}{n}\left(\frac{1}{n\pi}\cos(n\pi x) - \sin(n\pi x)\right)\right]$$

(ii) Fourier series for $f(x) = x^2$ in $(0,4)$ is

$$f(x) \sim \frac{a_0}{2} + \sum_{n=1}^{\infty}\left[a_n \cos\frac{n\pi x}{2} + b_n \sin\frac{n\pi x}{2}\right].$$

where $$a_0 = \frac{1}{2}\int_0^4 x^2 dx = \frac{1}{6}\left(x^3\right)_0^4 = \frac{32}{3}$$

$$a_n + i\, b_n = \frac{1}{2}\int_0^4 x^2 e^{i\frac{n\pi x}{2}} dx$$

$$= \frac{1}{2}\left[x^2\left(\frac{-2i}{n\pi}e^{i\frac{n\pi x}{2}}\right) - 2x\left(-\frac{4}{n^2\pi^2}e^{i\frac{n\pi x}{2}}\right) + 2\left(\frac{8i}{n^3\pi^3}e^{i\frac{n\pi x}{2}}\right)\right]_0^4$$

$$= \frac{1}{2}\left[-\frac{32i}{n\pi} + \frac{32}{n^2\pi^2}\right] = -\frac{16i}{n\pi} + \frac{16}{n^2\pi^2}$$

Equate real and imaginary parts

$$a_n = \frac{16}{n^2\pi^2},\ b_n = -\frac{16}{n\pi};\ \ n = 1, 2, 3, \ldots$$

$\therefore$ Fourier series for $f(x)$ is

$$f(x) \sim 16\left[\frac{1}{3} + \frac{1}{\pi}\sum_{n=1}^{\infty}\frac{1}{n}\left(\frac{1}{n\pi}\cos\frac{n\pi x}{2} - \sin\frac{n\pi x}{2}\right)\right]$$

For $x = 4$

$$\frac{f(4-0) + f(4+0)}{2} = \frac{16}{3} + \frac{16}{\pi^2}\sum_{n=1}^{\infty}\frac{1}{n^2}$$

$$\Rightarrow \quad \frac{f(4-0) + f(0+0)}{2} = \frac{16}{3} + \frac{16}{\pi^2}\sum_{n=1}^{\infty}\frac{1}{n^2}$$

$$\therefore \quad \frac{16+0}{2} = \frac{16}{3} + \frac{16}{\pi^2}\sum_{n=1}^{\infty}\frac{1}{n^2}$$

$$\Rightarrow \quad \sum_{n=1}^{\infty}\frac{1}{n^2} = \pi^2\left(\frac{1}{2} - \frac{1}{3}\right) = \frac{\pi^2}{6}$$

Example 3.20: Find the Fourier series of

$$f(x) = \begin{cases} \pi x & ;\ 0 \le x < 1 \\ 0 & ;\ x = 1 \\ \pi(x-2) & ;\ 1 < x \le 2 \end{cases}$$

Hence deduce that $\frac{\pi}{4} = 1 - \frac{1}{3} + \frac{1}{5} - \frac{1}{7} + \cdots$

Solution: Fourier series of $f(x)$ is

$$f(x) \sim \frac{a_0}{2} + \sum_{n=1}^{\infty}\left[a_n\cos(n\pi x) + b_n\sin(n\pi x)\right]$$

where $$a_0 = \int_0^1 \pi x\,dx + \int_1^2 \pi(x-2)\,dx = \frac{\pi}{2}\left(x^2\right)_0^1 + \pi\left(\frac{x^2}{2} - 2x\right)_1^2 = \frac{\pi}{2} + \pi\left(-2 - \frac{1}{2} + 2\right) = 0$$

$$a_n + ib_n = \int_0^1 \pi x\,e^{in\pi x}\,dx + \int_1^2 \pi(x-2)e^{in\pi x}\,dx$$

$$= \pi\left[x\left(-\frac{i}{n\pi}e^{in\pi x}\right) - \left(-\frac{1}{n^2\pi^2}e^{in\pi x}\right)\right]_0^1 + \pi\left[(x-2)\left(-\frac{i}{n\pi}e^{in\pi x}\right) - \left(-\frac{1}{n^2\pi^2}e^{in\pi x}\right)\right]_1^2$$

$$= \pi\left[-\frac{i(-1)^n}{n\pi} - \frac{1}{n^2\pi^2}\left(1-(-1)^n\right) - \frac{i(-1)^n}{n\pi} + \frac{1}{n^2\pi^2}\left(1-(-1)^n\right)\right]$$

$$= -\frac{2i}{n}(-1)^n = \frac{2i}{n}(-1)^{n+1}$$

Equate real and imaginary parts

$$a_n = 0, \qquad b_n = \frac{2}{n}(-1)^{n+1};\ n = 1,2,\ldots$$

$\therefore$ Fourier series of $f(x)$ is

$$f(x) \sim 2\sum_{n=1}^{\infty} \frac{(-1)^{n+1}}{n}\sin(n\pi x)$$

$$f\left(\frac{1}{2}\right) = 2\sum_{n=1}^{\infty} \frac{(-1)^{n+1}}{n}\sin\frac{n\pi}{2}$$

Let $\qquad k_n = \dfrac{(-1)^{n+1}}{n}\sin\dfrac{n\pi}{2}$

$\therefore \qquad k_{2n} = 0,\ k_{2n-1} = \dfrac{(-1)^{n+1}}{2n-1};\ n = 1,2,3,\ldots$

$\therefore \qquad \dfrac{\pi}{2} = 2\displaystyle\sum_{n=1}^{\infty} \frac{(-1)^{n+1}}{2n-1} \Rightarrow 1 - \frac{1}{3} + \frac{1}{5} - \frac{1}{7} + \cdots = \frac{\pi}{4}$

Other method to find Fourier series

Let the substitution $x = y+1$ changes $f(x)$ to $g(y)$

$$g(y) = \begin{cases} \pi(y+1); & -1 \le y < 0 \\ \quad 0 \quad ; & y = 0 \\ \pi(y-1); & 0 < y \le 1 \end{cases}$$

$g(y)$ is odd function of y.

Its Fourier series is

$$g(y) \sim \sum_{n=1}^{\infty} b_n \sin(n\pi y)$$

where $\qquad b_n = 2\displaystyle\int_0^1 \pi(y-1)\sin n\pi y\,dy$

$$= 2\pi\left[(y-1)\left(-\frac{1}{n\pi}\cos n\pi y\right) - \left(-\frac{1}{n^2\pi^2}\sin n\pi y\right)\right]_0^1$$

$$= 2\pi\left[-\frac{1}{n\pi}\right] = -\frac{2}{n}$$

∴ Fourier series of $g(y)$ is

$$g(y) \sim -2\sum_{n=1}^{\infty}\frac{1}{n}\sin(n\pi y)$$

But $y = x-1$ changes $g(y)$ to $f(x)$

∴ $$f(x) \sim -2\sum_{n=1}^{\infty}\frac{1}{n}\sin\left[n\pi(x-1)\right]$$

or $$f(x) \sim 2\sum_{n=1}^{\infty}\frac{1}{n}\sin(n\pi - n\pi x)$$

or $$f(x) \sim 2\sum_{n=1}^{\infty}\frac{(-1)^{n+1}}{n}\sin(n\pi x)$$

Example 3.21: A sinusoidal voltage $E\sin\omega t$, where t is time, is passed through a half-wave rectifier that clips the negative portion of the wave. Find the Fourier series of the resulting periodic function

$$u(t) = \begin{cases} 0 & ; \ -L < t < 0 \\ E\sin\omega t & ; \ 0 < t < L \end{cases}$$

Take $p = 2L = \frac{2\pi}{\omega}$, $L = \frac{\pi}{\omega}$

Solution: Fourier series of $u(t)$ is

$$u(t) \sim \frac{a_0}{2} + \sum_{n=1}^{\infty}\left[a_n\cos\frac{n\pi t}{L} + b_n\sin\frac{n\pi t}{L}\right]$$

$$\sim \frac{a_0}{2} + \sum_{n=1}^{\infty}\left[a_n\cos n\omega t + b_n\sin n\omega t\right] \qquad \left(\because \frac{\pi}{L} = \omega\right)$$

where $$a_0 = \frac{1}{L}\int_0^L E\sin\omega t\,dt = \frac{E}{L}\left(-\frac{1}{\omega}\cos\omega t\right)_0^L = \frac{E}{\pi}(1-\cos\omega L)$$

$$= \frac{E}{\pi}(1-\cos\pi) = \frac{2E}{\pi} \qquad (\because L\omega = \pi)$$

$$a_n + ib_n = \frac{1}{L}\int_0^L E\sin\omega t\, e^{in\omega t}dt = \frac{E\omega}{\pi}\int_0^{\frac{\pi}{\omega}}\sin\omega t\, e^{in\omega t}dt$$

$$= \frac{E\omega}{\pi}\left[\frac{e^{in\omega t}}{\omega^2(1-n^2)}(in\omega\sin\omega t - \omega\cos\omega t)\right]_0^{\frac{\pi}{\omega}}; \quad n \neq 1$$

$$= \frac{E}{\pi(1-n^2)}\left[1+(-1)^n\right]; \quad n \neq 1$$

Equate real and imaginary parts

$$a_n = \frac{-E}{\pi(n^2-1)}\left[(-1)^n + 1\right], \ b_n = 0; \ n \neq 1$$

$$\therefore \qquad a_{2n} = -\frac{2E}{\pi\left(4n^2-1\right)}; \quad n=1,2,3,...$$

$$a_{2n-1} = 0; \quad n = 2,3,...$$

$$a_1 = \frac{1}{L}\int_0^L E\sin\omega t\cos\omega t\,dt$$

$$= \frac{E}{2L}\int_0^{\pi/\omega}\sin 2\omega t\,dt = \frac{-E}{4\omega L}\left(\cos 2\omega t\right)_0^{\pi/\omega} = 0$$

$$b_1 = \frac{1}{L}\int_0^L E\sin^2\omega t\,dt$$

$$= \frac{E}{2L}\int_0^{\pi/\omega}\left(1-\cos 2\omega t\right)dt = \frac{E}{2L}\left(t - \frac{1}{2\omega}\sin 2\omega t\right)_0^{\pi/\omega}$$

$$= \frac{E\pi}{2\omega L} = \frac{E}{2}$$

$\therefore$ Fourier series of $u(t)$ is

$$u(t) \sim \frac{E}{\pi} + \frac{E}{2}\sin\omega t - \frac{2E}{\pi}\sum_{n=1}^{\infty}\frac{\cos(2n\omega t)}{4n^2-1}$$

Example 3.22: Find the Fourier series for $f(x) = x^2$ in $-l < x < l$.
Solution: $f(x)$ is even function of x.
Fourier series for $f(x)$ is

$$f(x) \sim \frac{a_0}{2} + \sum_{n=1}^{\infty} a_n \cos\frac{n\pi x}{l}$$

where
$$a_0 = \frac{2}{l}\int_0^l x^2\,dx = \frac{2}{3l}\left(x^3\right)_0^l = \frac{2l^2}{3}$$

$$a_n = \frac{2}{l}\int_0^l x^2\cos\frac{n\pi x}{l}\,dx$$

$$= \frac{2}{l}\left[x^2\left(\frac{l}{n\pi}\sin\frac{n\pi x}{l}\right) - 2x\left(-\frac{l^2}{n^2\pi^2}\cos\frac{n\pi x}{l}\right)\right.$$

$$\left. +2\left(-\frac{l^3}{n^3\pi^3}\sin\frac{n\pi x}{l}\right)\right]_0^l = \frac{4l^2}{n^2\pi^2}(-1)^n$$

$\therefore$ Fourier series for $f(x)$ is

$$f(x) \sim \frac{l^2}{3} - \frac{4l^2}{\pi^2}\sum_{n=1}^{\infty}\frac{(-1)^{n+1}}{n^2}\cos\left(\frac{n\pi x}{l}\right)$$

Example 3.23: Sketch the graph for the function

$$f(x)=\begin{cases}0 & ; \quad -2\le x\le -1\\ 1+x & ; \quad -1\le x\le 0\\ 1-x & ; \quad 0\le x\le 1\\ 0 & ; \quad 1\le x\le 2\end{cases}$$

and obtain its Fourier series.

Solution: $f(x)$ is even function with period 4.

Its graph is shown below

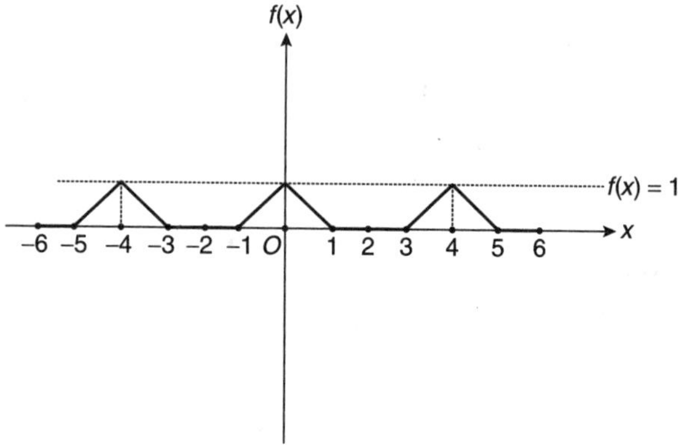

Figure 3.7

Fourier series of $f(x)$ is

$$f(x)=\frac{a_0}{2}+\sum_{n=1}^{\infty}a_n\cos\frac{n\pi x}{2}$$

where

$$a_0=\int_0^2 f(x)\,dx=\int_0^1(1-x)\,dx=\left(x-\frac{x^2}{2}\right)_0^1=\frac{1}{2}$$

$$a_n=\int_0^2 f(x)\cos\frac{n\pi x}{2}dx$$

$$=\int_0^1(1-x)\cos\frac{n\pi x}{2}dx$$

$$=\left[(1-x)\left(\frac{2}{n\pi}\sin\frac{n\pi x}{2}\right)+\left(-\frac{4}{n^2\pi^2}\cos\frac{n\pi x}{2}\right)\right]_0^1$$

$$=\frac{4}{n^2\pi^2}\left(1-\cos\frac{n\pi}{2}\right)$$

$\therefore$

$$a_{2n}=\frac{1}{n^2\pi^2}\left[1-(-1)^n\right],\quad a_{2n-1}=\frac{4}{\pi^2(2n-1)^2};\quad n=1,2,3,\ldots$$

$\therefore \qquad a_{4n} = 0,\ a_{2(2n-1)} = \dfrac{2}{\pi^2 (2n-1)^2},\ a_{2n-1} = \dfrac{4}{\pi^2 (2n-1)^2};\ n = 1,2,3,\ldots$

$\therefore$ Fourier series for $f(x)$ is

$$f(x) = \frac{1}{4} + \frac{2}{\pi^2} \sum_{n=1}^{\infty} \frac{1}{(2n-1)^2} \left[2\cos\frac{(2n-1)\pi x}{2} + \cos\{(2n-1)\pi x\} \right].$$

Example 3.24: Find the Fourier-series expansion of the periodic function

$$f(x) = \begin{cases} 2+x \ ; & -2 \le x \le 0 \\ 2-x \ ; & 0 < x \le 2 \end{cases}$$

and $f(x+4) = f(x)$. Sketch the graph and deduce that

$$1 + \frac{1}{3^2} + \frac{1}{5^2} + \frac{1}{7^2} + \cdots = \frac{\pi^2}{8}$$

Solution: $f(x)$ is an even function with period 4.
Fourier-series expansion of $f(x)$ is

$$f(x) = \frac{a_0}{2} + \sum_{n=1}^{\infty} a_n \cos\frac{n\pi x}{2}$$

where $\qquad a_0 = \int_0^2 (2-x)\,dx = \left(2x - \frac{x^2}{2} \right)_0^2 = 2$

$$a_n = \int_0^2 (2-x)\cos\frac{n\pi x}{2}\,dx$$

$$= \left[(2-x)\left(\frac{2}{n\pi}\sin\frac{n\pi x}{2} \right) + \left(-\frac{4}{n^2\pi^2}\cos\frac{n\pi x}{2} \right) \right]_0^2$$

$$= \frac{4}{n^2\pi^2}\left[1 - (-1)^n \right]$$

$\therefore \qquad a_{2n} = 0,\ a_{2n-1} = \dfrac{8}{\pi^2 (2n-1)^2};\quad n = 1,2,3,\ldots$

$\therefore$ Fourier-series expansion of $f(x)$ is

$$f(x) = 1 + \frac{8}{\pi^2} \sum_{n=1}^{\infty} \frac{1}{(2n-1)^2} \cos\frac{(2n-1)\pi x}{2}$$

$$f(0) = 2 = 1 + \frac{8}{\pi^2} \sum_{n=1}^{\infty} \frac{1}{(2n-1)^2}$$

$\therefore \qquad 1 + \dfrac{1}{3^2} + \dfrac{1}{5^2} + \dfrac{1}{7^2} + \cdots = \dfrac{\pi^2}{8}$

The graph is shown below

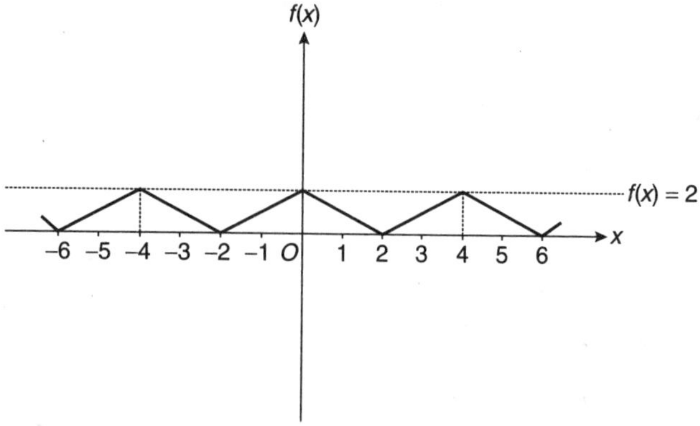

Figure 3.8

Example 3.25: Determine the Fourier series for the periodic triangle function $f(x)$ with period T defined for $0 < a \le \frac{T}{2}$ on $\left(-\frac{T}{2}, \frac{T}{2}\right)$ by

$$f(x) = \begin{cases} 1 - \frac{|x|}{a} & ; \quad |x| \le a \\ 0 & ; \quad a < |x| \le \frac{T}{2} \end{cases}$$

Solution: $$f(x) = \begin{cases} 1 - \frac{|x|}{a}; & -a \le x \le a \\ 0 & ; \quad -\frac{T}{2} \le x < -a, \quad a < x \le \frac{T}{2} \end{cases}$$

$f(x)$ is an even function.

Fourier series for $f(x)$ is

$$f(x) = \frac{a_0}{2} + \sum_{n=1}^{\infty} a_n \cos \frac{2n\pi x}{T}$$

where $$a_0 = \frac{4}{T}\int_0^a \left(1 - \frac{x}{a}\right)dx = \frac{4}{T}\left(x - \frac{x^2}{2a}\right)_0^a = \frac{2a}{T}$$

$$a_n = \frac{4}{T}\int_0^a \left(1 - \frac{x}{a}\right)\cos \frac{2n\pi x}{T}\,dx$$

$$= \frac{4}{T}\left[\left(1 - \frac{x}{a}\right)\left(\frac{T}{2n\pi}\sin \frac{2n\pi x}{T}\right) + \frac{1}{a}\left(\frac{-T^2}{4n^2\pi^2}\cos \frac{2n\pi x}{T}\right)\right]_0^a$$

$$=\frac{4}{T}\left[\frac{T^2}{4an^2\pi^2}\left(1-\cos\frac{2n\pi a}{T}\right)\right]=\frac{T}{a\pi^2n^2}\left(1-\cos\frac{2n\pi a}{T}\right)$$
$$=\frac{2T}{a\pi^2n^2}\sin^2\frac{n\pi a}{T}$$

$\therefore$ Fourier series for $f(x)$ is

$$f(x)=\frac{a}{T}+\frac{2T}{a\pi^2}\sum_{n=1}^{\infty}\frac{1}{n^2}\sin^2\frac{n\pi a}{T}\cos\frac{2n\pi x}{T}$$
$$=\frac{a\omega}{2\pi}+\frac{4}{a\pi\omega}\sum_{n=1}^{\infty}\frac{1}{n^2}\sin^2\frac{n\omega a}{2}\cos(n\omega x),$$

where $\omega=\dfrac{2\pi}{T}$.

Example 3.26: Obtain the Fourier series of the function *f* given by the following graph

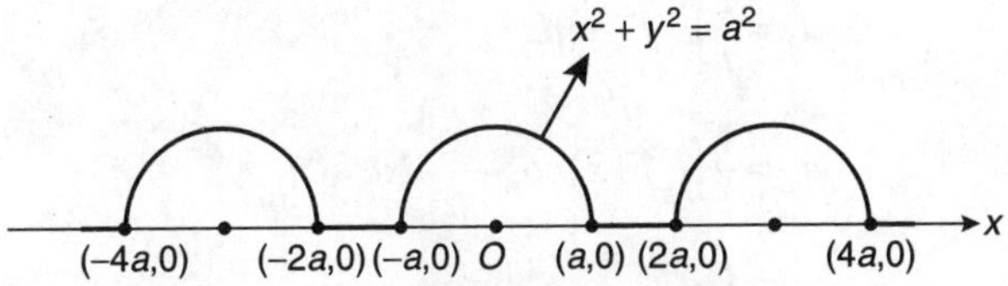

Figure 3.9

Solution: From the graph

$$f(x)=\begin{cases}0 & ;\ -3a/2\le x\le -a\\ \sqrt{a^2-x^2} & ;\ -a\le x\le a\\ 0 & ;\ a\le x\le 3a/2\end{cases}$$

and $f(x)$ is periodic with period $3a$.
Here, $f(x)$ is an even function.
Fourier series of $f(x)$ is

$$f(x)=\frac{a_0}{2}+\sum_{n=1}^{\infty}a_n\cos\frac{2n\pi x}{3a}$$

where $$a_0=\frac{4}{3a}\int_0^{3a/2}f(x)\,dx=\frac{4}{3a}\int_0^a\sqrt{a^2-x^2}\,dx$$
$$=\frac{4}{3a}\left[\frac{x\sqrt{a^2-x^2}}{2}+\frac{a^2}{2}\sin^{-1}\frac{x}{a}\right]_0^a=\frac{a\pi}{3}$$
$$a_n=\frac{4}{3a}\int_0^{3a/2}f(x)\cos\frac{2n\pi x}{3a}dx=\frac{4}{3a}\int_0^a\sqrt{a^2-x^2}\cos\frac{2n\pi x}{3a}dx$$

$\therefore$ Fourier series of $f(x)$ is

$$f(x)=\frac{a\pi}{6}+\sum_{n=1}^{\infty}a_n\cos\frac{2n\pi x}{3a}$$

where $$a_n=\frac{4}{3a}\int_0^a\sqrt{a^2-x^2}\cos\frac{2n\pi x}{3a}dx.$$

Example 3.27: If $f(x)$ is a periodic function in $[-l, l]$. Prove that at both the end points the Fourier series has the same value $\frac{a_0}{2}+\sum_{n=1}^{\infty}a_n(-1)^n$.

Solution: Fourier series of $f(x)$ in $[-l, l]$ is

$$f(x)=\frac{a_0}{2}+\sum_{n=1}^{\infty}a_n\cos\frac{n\pi x}{l}+\sum_{n=1}^{\infty}b_n\sin\frac{n\pi x}{l}$$

where $$a_0=\frac{1}{l}\int_{-l}^{l}f(x)dx$$

$$a_n=\frac{1}{l}\int_{-l}^{l}f(x)\cos\frac{n\pi x}{l}dx$$

$$b_n=\frac{1}{l}\int_{-l}^{l}f(x)\sin\frac{n\pi x}{l}dx$$

$$f(-l)=\frac{a_0}{2}+\sum_{n=1}^{\infty}a_n\cos\frac{n\pi(-l)}{l}+\sum_{n=1}^{\infty}b_n\sin\frac{n\pi(-l)}{l}$$

$$=\frac{a_0}{2}+\sum_{n=1}^{\infty}a_n\cos n\pi-\sum_{n=1}^{\infty}b_n\sin n\pi$$

$$=\frac{a_0}{2}+\sum_{n=1}^{\infty}a_n(-1)^n$$

$$f(l)=\frac{a_0}{2}+\sum_{n=1}^{\infty}a_n\cos\frac{n\pi l}{l}+\sum_{n=1}^{\infty}b_n\sin\frac{n\pi l}{l}$$

$$=\frac{a_0}{2}+\sum_{n=1}^{\infty}a_n(-1)^n$$

$\therefore$ At both ends Fourier series has the same value

$$\frac{a_0}{2}+\sum_{n=1}^{\infty}a_n(-1)^n$$

EXERCISE 3.1

1. Find the Fourier series of the function $f(x)=x+\pi$ if $-\pi<x<\pi$ and $f(x+2\pi)=f(x)$.
2. Obtain the Fourier series for the function $f(x)=2x+1,\ -\pi<x<\pi$. Hence deduce Fourier series for x and the line $y=mx+c$.
3. Obtain the Fourier series of $f(x)=\dfrac{\pi-x}{2}$ in the interval $(0,\ 2\pi)$. Deduce $\dfrac{\pi}{4}=1-\dfrac{1}{3}+\dfrac{1}{5}-\dfrac{1}{7}+\cdots$
4. Expand $f(x)=x^2,\ 0<x<2\pi$ in *a* Fourier series assuming that the function is of period 2π. Hence deduce that $\dfrac{\pi^2}{12}=\dfrac{1}{1^2}-\dfrac{1}{2^2}+\dfrac{1}{3^2}-\dfrac{1}{4^2}+\cdots$
5. Find the Fourier series for the function $f(x)=x+x^2,\quad -\pi<x<\pi.$

   Hence show that

   (i) $\dfrac{\pi^2}{6}=1+\dfrac{1}{2^2}+\dfrac{1}{3^2}+\dfrac{1}{4^2}+\cdots$

   (ii) $\dfrac{\pi^2}{12}=\dfrac{1}{1^2}-\dfrac{1}{2^2}+\dfrac{1}{3^2}-\dfrac{1}{4^2}+\cdots$
6. Find the Fourier series for $f(x)=x+\dfrac{x^2}{4},\ -\pi<x<\pi$
7. Find the Fourier series of $f(x)=\dfrac{3x^2-6x\pi+2\pi^2}{12}$ in the interval $(0,\ 2\pi)$. Hence deduce that $\dfrac{\pi^2}{6}=1+\dfrac{1}{2^2}+\dfrac{1}{3^2}+\cdots$
8. Find the Fourier series to represent e^{ax} in the interval $-\pi<x<\pi$ and hence derive series for $\dfrac{\pi}{\sinh\pi}$.
9. Show that in the range 0 to 2π the Fourier series expansion for e^x is

$$\frac{2e^{\pi}\sinh\pi}{\pi}\left[\frac{1}{2}+\sum_{n=1}^{\infty}\left(\frac{\cos nx}{n^2+1}\right)-\sum_{n=1}^{\infty}\left(\frac{n\sin nx}{n^2+1}\right)\right]$$

10. Expand in Fourier series the function f defined by $f(x)=\begin{cases}0, & -\pi<x<0\\ 1, & 0\le x<\pi\end{cases}$

    Deduce that sum of the Gregory series $1-\dfrac{1}{3}+\dfrac{1}{5}-\dfrac{1}{7}+\dfrac{1}{9}-\cdots$ is $\dfrac{\pi}{4}$.
11. Find the Fourier series of $f(x)=\begin{cases}-1, & 0<x<\pi\\ 2, & \pi<x<2\pi\end{cases}$
12. Find the Fourier series for the periodic function

$$f(x)=\begin{cases}0, & -\pi<x<0\\ x, & 0<x<\pi\end{cases}$$

$$f(x+2\pi)=f(x)$$

13. Find the Fourier series for $f(x)$ if

$$f(x)=\begin{cases}-\pi\ , & -\pi<x<0\\ x\ , & 0<x<\pi\\ -\pi/2\ , & x=0\end{cases}$$

Deduce that $\dfrac{1}{1^2}+\dfrac{1}{3^2}+\dfrac{1}{5^2}+\cdots=\dfrac{\pi^2}{8}$

14. If $f(x)=\begin{cases}x, & \dfrac{-\pi}{2}<x<\dfrac{\pi}{2}\\ 0, & \dfrac{\pi}{2}<x<\dfrac{3\pi}{2}\end{cases}$

    find the Fourier series of $f(x)$. Deduce that $\dfrac{\pi^2}{8}=\sum_{n=1}^{\infty}\dfrac{1}{(2n-1)^2}$.

15. Find the Fourier-series expansion of the periodic function of period 2π defined by

$$f(x)=\begin{cases} x \quad , & \dfrac{-\pi}{2}<x<\dfrac{\pi}{2} \\ \pi-x, & \dfrac{\pi}{2}<x<\dfrac{3\pi}{2} \end{cases}$$

16. Find the Fourier series of $f(x)=\begin{cases} 0 \ , & -\pi<x\le 0 \\ x^2, & 0\le x<\pi \end{cases}$

which is assumed to be periodic with period 2π.

17. Find the Fourier series of the function defined as

$f(x)=\begin{cases} x+\pi, & 0\le x<\pi \\ -x-\pi, & -\pi<x<0 \end{cases}$ and

$f(x+2\pi)=f(x)$.

18. Find the Fourier series of the function $f(x)=1-|x|, \quad -\pi<x<\pi$.

19. Find Fourier series of $f(x)=x^3$ in $(-\pi, \pi)$.

20. Expand $\cosh x$ in Fourier series in $-\pi<x<\pi$.

21. Find the Fourier series of $f(x)=e^{-|x|}$ in the interval $(-\pi, \pi)$.

22. Find the Fourier series for the function $f(x)=\sinh ax, \ -\pi<x<\pi$.

23. Expand the function $f(x)=\dfrac{\pi^2}{12}-\dfrac{x^2}{4}$ in Fourier series in the interval $(-\pi,\pi)$ and deduce that $\dfrac{\pi^2}{12}=\dfrac{1}{1^2}-\dfrac{1}{2^2}+\dfrac{1}{3^2}-\cdots$

24. Show that for $-\pi < x < \pi$,

$$\sin ax = \frac{2\sin a\pi}{\pi}\left(\frac{\sin x}{1^2-a^2}-\frac{2\sin 2x}{2^2-a^2}+\frac{3\sin 3x}{3^2-a^2}\cdots\right),$$

a is fraction.

25. Prove that in the interval $-\pi<x<\pi$,

$$x\cos x=-\frac{1}{2}\sin x+2\sum_{n=2}^{\infty}\frac{n(-1)^n}{(n^2-1)}\sin nx.$$

26. Find the Fourier series for the function

$$f(x)=\begin{cases} -1, & -\pi<x<-\pi/2 \\ 0, & -\pi/2<x<\pi/2 \\ 1, & \pi/2<x<\pi \end{cases}$$

27. Find the Fourier series of the function

$$f(x)=\begin{cases} \pi+2x, & -\pi<x<0 \\ \pi-2x, & 0\le x<\pi \end{cases}.$$

Hence deduce that $\dfrac{1}{1^2}+\dfrac{1}{3^3}+\dfrac{1}{5^2}+\cdots=\dfrac{\pi^2}{8}$.

28. Obtain Fourier series for $f(x)=\begin{cases} x \ , & -\pi<x<0 \\ -x, & 0<x<\pi \end{cases}$ and hence show that $\dfrac{1}{1^2}+\dfrac{1}{3^2}+\dfrac{1}{5^2}+\cdots=\dfrac{\pi^2}{8}$.

29. If $f(x)=\begin{cases} -1+x\ , & -\pi<x<0 \\ 1+x \ , & 0<x<\pi \end{cases}$ with period 2π, find the Fourier series for $f(x)$.

30. Find the Fourier series of

$$f(x)=\begin{cases} \cos x \ , & -\pi<x<0 \\ -\cos x, & 0<x<\pi \end{cases}$$

31. Determine the Fourier expansion for the function $f(x)$ given by

$$f(x)=\begin{cases} x+\dfrac{\pi}{2}, & -\pi<x<0 \\ \dfrac{\pi}{2}-x, & 0<x<\pi \end{cases}$$

32. Obtain the Fourier expansion for the function $f(x)$ with period 2π given by

$$f(x)=\begin{cases} -x^2, & -\pi<x<0 \\ x^2 \ , & 0<x<\pi \end{cases}$$

33. Find Fourier series for the function $f(x)=x-x^2, \ -1<x<1$.

34. Prove that $\frac{l}{2}-x=\frac{l}{\pi}\sum_{n=1}^{\infty}\frac{1}{n}\frac{\sin 2n\pi x}{l}$, $0<x<l$.

35. Obtain the Fourier series of $f(x)=\frac{\pi-x}{2}$ in $0<x<2$.

36. Expand $f(x)=e^{-x}$ as *a* Fourier series in the interval $(-l, l)$.

37. Find the Fourier series of $f(x)=4-x^2$ in (0, 2) and deduce that $\frac{\pi^2}{6}=\frac{1}{1^2}+\frac{1}{2^2}+\frac{1}{3^2}+\cdots$

38. Find the Fourier series of $f(x)=\pi x,\ 0<x\le 2$.

39. Find the Fourier series of
$$f(x)=\begin{cases}\pi x, & 0<x<1\\ 0, & 1<x<2\end{cases}$$

40. Find the Fourier series corresponding to the function $f(x)$ defined in (−2, 2) as follows
$$f(x)=\begin{cases}2, & -2\le x\le 0\\ x, & 0<x<2\end{cases}$$

41. Find the Fourier series for $f(x)$ defined in $(-1, 1)$ by
$$f(x)=\begin{cases}c_1, & -1<x<0\\ c_2, & 0<x<1\end{cases}$$

42. Obtain the Fourier series of
$$f(x)=\begin{cases}0, & -5<x<0\\ 3, & 0<x<5\end{cases}$$

43. Find the Fourier-series expansion of
$$f(x)=\begin{cases}1, & 0<x<1\\ 2, & 1<x<2\end{cases}$$

44. Find the Fourier series of
$$f(x)=\begin{cases}x, & -1<x<0\\ x+2, & 0<x<1\end{cases}$$

45. Find the Fourier series for the function
$$f(x)=\begin{cases}0, & -8<x<0\\ 4, & 0<x<4\\ 0, & 4<x<8\end{cases}$$

46. Find the Fourier series for the function
$$f(x)=\begin{cases}x, & 0<x<1\\ 1-x, & 1<x<2\end{cases}.$$
Deduce that $\frac{1}{1^2}+\frac{1}{3^2}+\frac{1}{5^2}+\cdots=\frac{\pi^2}{8}$.

47. Find the Fourier series for function
$$f(x)=\begin{cases}\pi x, & 0\le x\le 1\\ \pi(2-x), & 1\le x\le 2\end{cases}$$

48. Obtain Fourier-series expansion of $f(x)=x\cos\left(\frac{\pi x}{l}\right)$ in the interval $-l<x<l$.

49. Find Fourier-series expansion of $f(t)=1-t^2,\ -1\le t\le 1$.

50. Find the Fourier series of $f(x)=x|x|$ in the interval (−1, 1).

51. Obtain the Fourier-series expansion of $f(x)=|x|,\ -2\le x\le 2,\ f(x)=f(x+4)$.

52. Obtain the Fourier-series expansion of $f(x)=x^2-2,\ -2\le x\le 2$.

53. If $f(x)=x$ is defined in $-l<x<l$ with period $2l$, find the Fourier expansion of $f(x)$.

54. Find the Fourier series for
$$f(x)=\begin{cases}k(x-c), & -c<x<0\\ k(c+x), & 0<x<c\end{cases}$$

55. Find the Fourier-series expansion of the period function of period 1 given by
$$f(x)=\begin{cases}\frac{1}{2}+x, & -\frac{1}{2}<x\le 0\\ \frac{1}{2}-x, & 0<x<\frac{1}{2}\end{cases}$$

56. Find the Fourier series for periodic block function f with period $T>0$ and $0\le a\le T$ and defined by
$$f(x)=\begin{cases}1, & |x|\le a/2\le T/2\\ 0, & a/2<|x|\le T/2\end{cases}$$

ANSWERS 3.1

1. $f(x) \sim \pi + 2\sum_{n=1}^{\infty} \frac{(-1)^{n+1}}{n} \sin nx$

2. $f(x) \sim 1 + 4\sum_{n=1}^{\infty} \frac{(-1)^{n+1}}{n} \sin nx$

$x \sim 2\sum_{n=1}^{\infty} \frac{(-1)^{n+1}}{n} \sin nx$

$mx + c \sim c + 2m\sum_{n=1}^{\infty} \frac{(-1)^{n+1}}{n} \sin nx$

3. $f(x) \sim \sum_{n=1}^{\infty} \frac{1}{n} \sin nx$

4. $f(x) \sim \frac{4\pi^2}{3} + 4\sum_{n=1}^{\infty} \frac{1}{n^2} \cos nx - 4\pi \sum_{n=1}^{\infty} \frac{1}{n} \sin nx$

5. $f(x) \sim \frac{\pi^2}{3} + 4\sum_{n=1}^{\infty} \frac{(-1)^n \cos nx}{n^2} + 2\sum_{n=1}^{\infty} \frac{(-1)^{n+1} \sin nx}{n}$

6. $f(x) \sim \frac{\pi^2}{12} + \sum_{n=1}^{\infty} \frac{(-1)^n}{n^2} \cos nx + 2\sum_{n=1}^{\infty} \frac{(-1)^{n+1}}{n} \sin nx$

7. $f(x) \sim \sum_{n=1}^{\infty} \frac{1}{n^2} \cos nx$

8. $f(x) \sim \frac{\sinh a\pi}{\pi} \left[\frac{1}{a} + 2a\sum_{n=1}^{\infty} \frac{(-1)^n}{a^2 + n^2} \cos nx + 2\sum_{n=1}^{\infty} (-1)^{n+1} \frac{n}{a^2 + n^2} \sin nx \right]$

$\frac{\pi}{\sinh \pi} = 2\sum_{n=2}^{\infty} \frac{(-1)^n}{n^2 + 1} = 2\left[\frac{1}{2^2 + 1} - \frac{1}{3^2 + 1} + \frac{1}{4^2 + 1} - \cdots \right]$

10. $f(x) \sim \frac{1}{2} + \frac{2}{\pi} \sum_{n=1}^{\infty} \frac{\sin(2n-1)x}{(2n-1)}$

11. $f(x) \sim \frac{1}{2} - \frac{6}{\pi} \sum_{n=1}^{\infty} \frac{\sin(2n-1)x}{(2n-1)}$

12. $f(x) \sim \frac{\pi}{4} - \frac{2}{\pi} \sum_{n=1}^{\infty} \frac{\cos(2n-1)x}{(2n-1)^2} - \sum_{n=1}^{\infty} \frac{(-1)^n}{n} \sin nx$

13. $f(x) \sim -\frac{\pi}{4} - \frac{2}{\pi} \sum_{n=1}^{\infty} \frac{\cos(2n-1)x}{(2n-1)^2} + \sum_{n=1}^{\infty} \frac{1}{n} \left[1 - 2(-1)^n \right] \sin nx$

14. $f(x)\sim\frac{2}{\pi}\sum_{n=1}^{\infty}\frac{(-1)^{n+1}\sin(2n-1)x}{(2n-1)^2}+\sum_{n=1}^{\infty}\frac{(-1)^{n+1}\sin 2nx}{2n}$

15. $f(x)\sim\frac{4}{\pi}\sum_{n=1}^{\infty}(-1)^{n+1}\frac{\sin(2n-1)x}{(2n-1)^2}$

16. $f(x)\sim\frac{\pi^2}{6}+2\sum_{n=1}^{\infty}\frac{(-1)^n}{n^2}\cos nx-\frac{\pi}{2}\sum_{n=1}^{\infty}\frac{\sin 2nx}{n}+\sum_{n=1}^{\infty}\left[\frac{\pi}{2n-1}-\frac{4}{\pi(2n-1)^3}\right]\sin(2n-1)x$

17. $f(x)\sim\frac{\pi}{2}-\frac{4}{\pi}\sum_{n=1}^{\infty}\frac{\cos(2n-1)x}{(2n-1)^2}+4\sum_{n=1}^{\infty}\frac{\sin(2n-1)x}{(2n-1)}$

18. $f(x)\sim\frac{2-\pi}{2}+\frac{4}{\pi}\sum_{n=1}^{\infty}\frac{\cos(2n-1)x}{(2n-1)^2}$

19. $f(x)\sim 2\sum_{n=1}^{\infty}(-1)^n\frac{(6-n^2\pi^2)}{n^3}\sin nx$

20. $f(x)\sim\frac{\sinh\pi}{\pi}\left[1+2\sum_{n=1}^{\infty}\frac{(-1)^n}{(n^2+1)}\cos nx\right]$

21. $f(x)\sim\frac{4e^{-\pi/2}}{\pi}\left[\frac{1}{2}\sinh\frac{\pi}{2}+\sum_{n=1}^{\infty}\frac{\sinh\frac{\pi}{2}\cos 2nx}{4n^2+1}+\sum_{n=1}^{\infty}\frac{\cosh\frac{\pi}{2}\cos(2n-1)x}{(2n-1)^2+1}\right]$

22. $f(x)\sim\frac{2\sinh a\pi}{\pi}\sum_{n=1}^{\infty}\frac{(-1)^{n+1}n}{(n^2+a^2)}\sin nx$

23. $f(x)\sim\sum_{n=1}^{\infty}\frac{(-1)^{n+1}}{n^2}\cos nx$

26. $f(x)\sim\frac{2}{\pi}\sum_{n=1}^{\infty}\frac{1}{(2n-1)}\left[\sin(2n-1)x-\sin 2(2n-1)x\right]$

27. $f(x)\sim\frac{8}{\pi}\sum_{n=1}^{\infty}\frac{\cos(2n-1)x}{(2n-1)^2}$

28. $f(x)\sim\frac{-\pi}{2}+\frac{4}{\pi}\sum_{n=1}^{\infty}\frac{\cos(2n-1)x}{(2n-1)^2}$

29. $f(x)\sim\left(2+\frac{4}{\pi}\right)\sum_{n=1}^{\infty}\frac{\sin(2n-1)x}{(2n-1)}-\sum_{n=1}^{\infty}\frac{\sin 2nx}{n}$

30. $f(x) \sim \frac{-8}{\pi} \sum_{n=1}^{\infty} \frac{n \sin 2nx}{4n^2 - 1}$

31. $f(x) \sim \frac{4}{\pi} \sum_{n=1}^{\infty} \frac{\cos(2n-1)x}{(2n-1)^2}$

32. $f(x) \sim \frac{2}{\pi} \sum_{n=1}^{\infty} \left[\frac{\pi^2}{2n-1} - \frac{4}{(2n-1)^3} \right] \sin(2n-1)x - \pi \sum_{n=1}^{\infty} \frac{\sin 2nx}{n}$

33. $f(x) \sim -\frac{1}{3} + \frac{4}{\pi^2} \sum_{n=1}^{\infty} \frac{(-1)^{n+1}}{n^2} \cos n\pi x + \frac{2}{\pi} \sum_{n=1}^{\infty} \frac{(-1)^{n+1}}{n} \sin n\pi x$

35. $f(x) \sim \frac{\pi - 1}{2} + \frac{1}{\pi} \sum_{n=1}^{\infty} \frac{\sin n\pi x}{n\pi}$

36. $f(x) \sim \sinh l \left[\frac{1}{l} + 2l \sum_{n=1}^{\infty} \frac{(-1)^n}{l^2 + n^2\pi^2} \cos \frac{n\pi x}{l} + 2\pi \sum_{n=1}^{\infty} \frac{n(-1)^n}{l^2 + n^2\pi^2} \sin \frac{n\pi x}{l} \right]$

37. $f(x) \sim \frac{8}{3} - \frac{4}{\pi^2} \sum_{n=1}^{\infty} \frac{\cos n\pi x}{n^2} + \frac{4}{\pi} \sum_{n=1}^{\infty} \frac{\sin n\pi x}{n}$

38. $f(x) \sim \pi - 2 \sum_{n=1}^{\infty} \frac{\sin n\pi x}{n}$

39. $f(x) \sim \frac{\pi}{4} - \frac{2}{\pi} \sum_{n=1}^{\infty} \frac{1}{(2n-1)^2} \cos(2n-1)\pi x + \sum_{n=1}^{\infty} \frac{(-1)^{n+1}}{n} \sin n\pi x$

40. $f(x) \sim \frac{3}{2} - \frac{4}{\pi^2} \sum_{n=1}^{\infty} \frac{1}{(2n-1)^2} \cos(2n-1)\frac{\pi x}{2} - \frac{2}{\pi} \sum_{n=1}^{\infty} \frac{1}{n} \sin \frac{n\pi x}{2}$

41. $f(x) \sim \frac{c_1 + c_2}{2} + \frac{2(c_2 - c_1)}{\pi} \sum_{n=1}^{\infty} \frac{\sin(2n-1)\pi x}{(2n-1)}$

42. $f(x) \sim \frac{3}{2} + \frac{6}{\pi} \sum_{n=1}^{\infty} \frac{1}{(2n-1)} \sin \frac{(2n-1)\pi x}{5}$

43. $f(x) \sim \frac{3}{2} - \frac{2}{\pi} \sum_{n=1}^{\infty} \frac{\sin(2n-1)\pi x}{(2n-1)}$

44. $f(x) \sim 1 + \frac{2}{\pi} \sum_{n=1}^{\infty} \left[\frac{1 - 2(-1)^n}{n} \right] \sin n\pi x$

45. $f(x) \sim 1 + \frac{4}{\pi} \sum_{n=1}^{\infty} \frac{(-1)^{n+1}}{(2n-1)} \cos \frac{(2n-1)\pi x}{8} + \frac{4}{\pi} \sum_{n=1}^{\infty} \frac{1}{(2n-1)} \left[\sin \frac{(2n-1)\pi x}{8} + \sin \frac{(2n-1)\pi x}{4} \right]$

46. $f(x) \sim \frac{-4}{\pi^2}\sum_{n=1}^{\infty}\frac{1}{(2n-1)^2}\cos(2n-1)\pi x + \frac{2}{\pi}\sum_{n=1}^{\infty}\frac{\sin(2n-1)\pi x}{(2n-1)}$

47. $f(x) = \frac{\pi}{2} - \frac{4}{\pi}\sum_{n=1}^{\infty}\frac{1}{(2n-1)^2}\cos(2n-1)\pi x$

48. $f(x) \sim \frac{l}{\pi}\left[-\frac{1}{2}\sin\frac{\pi x}{l} + 4\sum_{n=2}^{\infty}\frac{n(-1)^n}{(n^2-1)}\sin\frac{n\pi x}{l}\right]$

49. $f(t) = \frac{2}{3} - \frac{4}{\pi^2}\sum_{n=1}^{\infty}\frac{(-1)^n}{n^2}\cos n\pi t$

50. $f(x) \sim -\frac{1}{\pi}\sum_{n=1}^{\infty}\left[\frac{1}{n}\sin(2n\pi x) + 2\left\{\frac{4}{(2n-1)^3\pi^2} - \frac{1}{(2n-1)}\right\}\sin(2n-1)\pi x\right]$

51. $f(x) = 1 - \frac{8}{\pi^2}\sum_{n=1}^{\infty}\frac{1}{(2n-1)^2}\cos\frac{(2n-1)\pi x}{2}$

52. $f(x) = -\frac{2}{3} + \frac{16}{\pi^2}\sum_{n=1}^{\infty}\frac{(-1)^n}{n^2}\cos\frac{n\pi x}{2}$

53. $f(x) \sim \frac{2l}{\pi}\sum_{n=1}^{\infty}\frac{(-1)^{n+1}}{n}\sin\frac{n\pi x}{l}$

54. $f(x) \sim \frac{2kc}{\pi}\sum_{n=1}^{\infty}\frac{1}{n}\left[1-2(-1)^n\right]\sin\frac{n\pi x}{c}$

55. $f(x) \sim \frac{1}{4} + \frac{2}{\pi^2}\sum_{n=1}^{\infty}\frac{1}{(2n-1)^2}\cos(2(2n-1)\pi x)$

56. $f(x) \sim \frac{a}{T} + \frac{2}{T\omega}\sum_{n=1}^{\infty}\frac{1}{n}\sin(na\omega)\cos(2n\omega x)$ where $\omega = \frac{\pi}{T}$

3.2 FOURIER HALF-RANGE SERIES

Suppose that a function $f(x)$ is defined on some finite interval [0, l]. It is possible to extend the definition of $f(x)$ to the interval [$-l$, 0] such that $f(x)$ is periodic function of period $2l$ and $f(x)$ is even function in [$-l$, l]. This can be done by defining $f(x) = f(-x);\ -l \le x \le 0$. Then, we can have Fourier-series expansion of this even extension of $f(x)$, i.e., extension of $f(x)$ to even function. This Fourier series will be cosine series which is called half-range cosine series of $f(x)$.

Thus, **half-range cosine series** of $f(x)$ in [0, l] is

$$f(x)=\frac{a_0}{2}+\sum_{n=1}^{\infty}a_n\cos\frac{n\pi x}{l}$$

where $$a_0=\frac{2}{l}\int_0^l f(x)\,dx,\; a_n=\frac{2}{l}\int_0^l f(x)\cos\frac{n\pi x}{l}\,dx$$

Similarly, we can extend the definition of $f(x)$ to interval $[-l, 0]$ such that $f(x)$ is periodic function of period $2l$ and $f(x)$ is odd function in $[-l, l]$. This can be done by defining $f(x)=-f(-x);\ -l\le x\le 0$. Then, we can have Fourier-series expansion of this odd extension of $f(x)$, i.e., extension of $f(x)$ to odd function. The Fourier series will be sine series which is half-range sine series of $f(x)$.
Thus, **half-range sine series** of $f(x)$ in [0, l] is $f(x)=\sum_{n=1}^{\infty}b_n\sin\frac{n\pi x}{l}$

where $$b_n=\frac{2}{l}\int_0^l f(x)\sin\frac{n\pi x}{l}\,dx$$

3.2.1 Convergence of Half-range Cosine Series

If $f(x)$ satisfies Dirichlet's conditions in [0, l] and $f(x)$ is continuous at $x_0\in(0,l)$, then half-range Fourier cosine series for $x=x_0$ converges to $f(x_0)$. If $f(x)$ is discontinuous at $x=x_0\in(0,l)$, then half-range Fourier cosine series for $x=x_0$ converges to $\frac{1}{2}\left[\lim_{x\to x_0^-}f(x)+\lim_{x\to x_0^+}f(x)\right]=\frac{1}{2}\left[f(x_0-0)+f(x_0+0)\right]$.

Now, $f(x)$ is even function with period $2l$

$\therefore$ $$f(0-0)=f(0+0)$$

and $$f(l+0)=f(-l-0)=f(2l-l-0)=f(l-0)$$

$\therefore$ for x = 0, Fourier cosine series converges to $f(0 + 0)$ and for $x = l$, Fourier cosine series converges to $f(l - 0)$.

3.2.2 Convergence of Half-range Sine Series

If $f(x)$ satisfies Dirichlet's conditions on [0, l] and $f(x)$ is continuous at $x_0\in(0,l)$, then half-range Fourier sine series for $x=x_0$ converges to $f(x_0)$. If $f(x)$ is discontinuous at $x=x_0\in(0,l)$, then half-range Fourier sine series for $x=x_0$ converges to $\frac{1}{2}\left[f(x_0-0)+f(x_0+0)\right]$.
Now, $f(x)$ is odd function with period $2l$

$\therefore$ $$f(0-0)=-f(0+0)$$

$$\therefore \qquad \frac{1}{2}\left[f(0-0)+f(0+0)\right]=0$$

$\therefore$ Half-range sine series for $x = 0$ has value 0

and
$$\begin{aligned} f(l+0) &= f(-2l+l+0) \\ &= f(-l+0) \\ &= -f(l-0) \end{aligned}$$

$$\therefore \qquad f(l+0)+f(l-0)=0$$

$\therefore$ Half-range sine series for $x = l$ has value 0.

Example 3.28: Prove that the function $f(x)=x$ can be expanded

(i) In a series of cosines in $0<x<\pi$ as

$$x=\frac{\pi}{2}-\frac{4}{\pi}\left[\frac{\cos x}{1^2}+\frac{\cos 3x}{3^2}+\frac{\cos 5x}{5^2}+\cdots\right]$$

hence deduce that $\frac{1}{1^2}+\frac{1}{3^2}+\frac{1}{5^2}+\cdots=\frac{\pi^2}{8}$

(ii) in a series of sines in $0<x<\pi$ as

$$x=2\left[\frac{\sin x}{1}-\frac{\sin 2x}{2}+\frac{\sin 3x}{3}-\frac{\sin 4x}{4}+\cdots\right]$$

hence deduce that $1-\frac{1}{3}+\frac{1}{5}-\frac{1}{7}+\cdots=\frac{\pi}{4}$

Solution:

(i) Fourier cosine series of $f(x)=x$ in $0<x<\pi$ is

$$f(x)=\frac{a_0}{2}+\sum_{n=1}^{\infty}a_n\cos nx$$

where
$$a_0=\frac{2}{\pi}\int_0^{\pi}x\,dx=\frac{1}{\pi}\left(x^2\right)_0^{\pi}=\pi$$

$$\begin{aligned} a_n &= \frac{2}{\pi}\int_0^{\pi}x\cos nx\,dx=\frac{2}{\pi}\left[x\left(\frac{1}{n}\sin nx\right)-\left(-\frac{1}{n^2}\cos nx\right)\right]_0^{\pi} \\ &= -\frac{2}{\pi n^2}\left[1-(-1)^n\right] \end{aligned}$$

$$\therefore \qquad a_{2n}=0,\ a_{2n-1}=-\frac{4}{\pi(2n-1)^2};\ n=1,2,3,\ldots$$

$\therefore$ Fourier cosine series of $f(x)$ is

$$f(x)=\frac{\pi}{2}-\frac{4}{\pi}\sum_{n=1}^{\infty}\frac{1}{(2n-1)^2}\cos[(2n-1)x];\quad 0<x<\pi$$

for $x=0$

$$f(0+0)=\frac{\pi}{2}-\frac{4}{\pi}\sum_{n=1}^{\infty}\frac{1}{(2n-1)^2}$$

$$\therefore\qquad 0=\frac{\pi}{2}-\frac{4}{\pi}\sum_{n=1}^{\infty}\frac{1}{(2n-1)^2}$$

$$\therefore\quad \frac{1}{1^2}+\frac{1}{3^2}+\frac{1}{5^2}+\cdots=\frac{\pi^2}{8}$$

(ii) Fourier sine series of $f(x)$ in $0<x<\pi$ is

$$f(x)=\sum_{n=1}^{\infty}b_n\sin nx$$

$$b_n=\frac{2}{\pi}\int_0^{\pi}x\sin nx\,dx=\frac{2}{\pi}\left[x\left(-\frac{1}{n}\cos nx\right)-\left(-\frac{1}{n^2}\sin nx\right)\right]_0^{\pi}$$

$$=\frac{2}{n}(-1)^{n+1}$$

$\therefore$ Fourier sine series of $f(x)$ is

$$f(x)=2\sum_{n=1}^{\infty}\frac{(-1)^{n+1}}{n}\sin(nx);\quad 0<x<\pi$$

$$f\left(\frac{\pi}{2}\right)=\frac{\pi}{2}=2\sum_{n=1}^{\infty}\frac{(-1)^{n+1}}{n}\sin\frac{n\pi}{2}$$

Let $\qquad k_n=\dfrac{(-1)^{n+1}}{n}\sin\dfrac{n\pi}{2}.$

$$\therefore\qquad k_{2n}=0,\ k_{2n-1}=\frac{1}{2n-1}\sin\left(n\pi-\frac{\pi}{2}\right)=\frac{(-1)^{n+1}}{2n-1};\ n=1,2,3,\ldots$$

$$\therefore\qquad f\left(\frac{\pi}{2}\right)=\frac{\pi}{2}=2\sum_{n=1}^{\infty}\frac{(-1)^{n+1}}{2n-1}$$

$$\therefore\quad 1-\frac{1}{3}+\frac{1}{5}-\frac{1}{7}+\cdots=\frac{\pi}{4}$$

Example 3.29: Find a series of cosines of multiples of x which will represent $x \sin x$ in the interval $(0,\pi)$ and show that $\frac{1}{1\cdot 3}-\frac{1}{3\cdot 5}+\frac{1}{5\cdot 7}-\frac{1}{7\cdot 9}+\cdots=\frac{\pi-2}{4}$.

Solution: Fourier cosine series of $x \sin x$ in $(0,\pi)$ is

$$x\sin x=\frac{a_0}{2}+\sum_{n=1}^{\infty}a_n\cos nx$$

where
$$a_0=\frac{2}{\pi}\int_0^{\pi}x\sin x\,dx=\frac{2}{\pi}\left[-x\cos x+\sin x\right]_0^{\pi}=2$$

$$a_n=\frac{2}{\pi}\int_0^{\pi}x\sin x\cos nx\,dx=\frac{1}{\pi}\int_0^{\pi}x\left[\sin(n+1)x-\sin(n-1)x\right]dx$$

$$=\frac{1}{\pi}\left[x\left\{-\frac{1}{n+1}\cos(n+1)x+\frac{1}{n-1}\cos(n-1)x\right\}\right.$$

$$\left.-\left\{-\frac{1}{(n+1)^2}\sin(n+1)x+\frac{1}{(n-1)^2}\sin(n-1)x\right\}\right]_0^{\pi}\ ;n\neq 1$$

$$=\frac{1}{n+1}(-1)^n-\frac{1}{n-1}(-1)^n=\frac{2}{n^2-1}(-1)^{n+1};\ n\neq 1$$

$$a_1=\frac{2}{\pi}\int_0^{\pi}x\sin x\cos x\,dx=\frac{1}{\pi}\int_0^{\pi}x\sin 2x\,dx$$

$$=\frac{1}{\pi}\left[x\left(-\frac{1}{2}\cos 2x\right)+\frac{1}{4}\sin 2x\right]_0^{\pi}=-\frac{1}{2}$$

∴ Fourier cosine series of $x \sin x$ in $(0,\pi)$ is

$$f(x)=x\sin x=1-\frac{1}{2}\cos x-2\sum_{n=2}^{\infty}\frac{(-1)^n}{n^2-1}\cos nx$$

∴
$$f\left(\frac{\pi}{2}\right)=\frac{\pi}{2}=1+2\sum_{n=2}^{\infty}\frac{(-1)^{n+1}}{n^2-1}\cos\frac{n\pi}{2}$$

Let
$$k_n=\frac{(-1)^{n+1}}{n^2-1}\cos\frac{n\pi}{2};\ n\geq 2$$

∴
$$k_{2n-1}=0;\ n=2,3,\ldots$$

$$k_{2n}=\frac{(-1)^{2n+1}(-1)^n}{4n^2-1}=\frac{(-1)^{n+1}}{4n^2-1};\ n=1,2,3,\ldots$$

∴
$$\sum_{n=1}^{\infty}\frac{(-1)^{n+1}}{(2n-1)(2n+1)}=\frac{\pi-2}{4}$$

∴
$$\frac{1}{1\cdot 3}-\frac{1}{3\cdot 5}+\frac{1}{5\cdot 7}-\frac{1}{7\cdot 9}+\cdots=\frac{\pi-2}{4}$$

Example 3.30: Prove that for $0 \le x \le \pi$

$$x(\pi - x) = \frac{\pi^2}{6} - \left[\frac{\cos 2x}{1^2} + \frac{\cos 4x}{2^2} + \frac{\cos 6x}{3^2} + \cdots\right]$$

Solution: Fourier cosine series of $x(\pi - x)$ in $0 \le x \le \pi$ is

$$x(\pi - x) = \frac{a_0}{2} + \sum_{n=1}^{\infty} a_n \cos nx$$

where
$$a_0 = \frac{2}{\pi}\int_0^{\pi}(\pi x - x^2)\,dx = \frac{2}{\pi}\left[\frac{\pi x^2}{2} - \frac{x^3}{3}\right]_0^{\pi} = \frac{\pi^2}{3}$$

$$a_n = \frac{2}{\pi}\int_0^{\pi}(\pi x - x^2)\cos nx\,dx$$

$$= \frac{2}{\pi}\left[(\pi x - x^2)\left(\frac{1}{n}\sin nx\right) - (\pi - 2x)\left(-\frac{1}{n^2}\cos nx\right) + (-2)\left(-\frac{1}{n^3}\sin nx\right)\right]_0^{\pi}$$

$$= \frac{2}{n^2}\left[(-1)^{n+1} - 1\right]$$

$\therefore$
$$a_{2n} = -\frac{1}{n^2}, \quad a_{2n-1} = 0; \quad n = 1, 2, 3, \ldots$$

$\therefore$ Fourier cosine series of $x(\pi - x)$ in $0 \le x \le \pi$ is

$$x(\pi - x) = \frac{\pi^2}{6} - \sum_{n=1}^{\infty}\frac{1}{n^2}\cos 2nx$$

$$= \frac{\pi^2}{6} - \left[\frac{\cos 2x}{1^2} + \frac{\cos 4x}{2^2} + \frac{\cos 6x}{3^2} + \cdots\right]; \quad 0 \le x \le \pi$$

Example 3.31: Determine half-range sine series for the function f defined by $f(t) = t^2 + t$, $0 \le t < \pi$.

Solution: Half-range sine series for $f(t) = t^2 + t$ in $0 \le t < \pi$ is

$$f(t) = t^2 + t = \sum_{n=1}^{\infty} b_n \sin nt$$

where
$$b_n = \frac{2}{\pi}\int_0^{\pi}(t^2 + t)\sin nt\,dt$$

$$= \frac{2}{\pi}\left[(t^2 + t)\left(-\frac{1}{n}\cos nt\right) - (2t + 1)\left(-\frac{1}{n^2}\sin nt\right) + 2\left(\frac{1}{n^3}\cos nt\right)\right]_0^{\pi}$$

$$= \frac{2}{\pi}\left[-\frac{\pi^2 + \pi}{n}(-1)^n - \frac{2}{n^3}\left(1 - (-1)^n\right)\right]$$

$$\therefore \qquad b_{2n} = -\frac{\pi+1}{n},\ b_{2n-1} = \frac{2(\pi+1)}{2n-1} - \frac{8}{\pi(2n-1)^3};\ n = 1,2,3,\ldots$$

$\therefore$ Half-range sine series for $f(t) = t^2 + t$ in $0 \le t < \pi$ is

$$f(t) = t^2 + t = \sum_{n=1}^{\infty}\left[-\frac{\pi+1}{n}\sin 2nt + \frac{2}{2n-1}\left\{\pi+1-\frac{4}{\pi(2n-1)^2}\right\}\sin(2n-1)t\right]$$

Example 3.32: If $f(x) = \begin{cases} x & ;\ 0 < x < \dfrac{\pi}{2} \\ \pi - x\ ; & \dfrac{\pi}{2} < x < \pi \end{cases}$

show that

(i) $f(x) = \dfrac{4}{\pi}\left[\sin x - \dfrac{\sin 3x}{3^2} + \dfrac{\sin 5x}{5^2} - \dfrac{\sin 7x}{7^2} + \cdots\right]$

(ii) $f(x) = \dfrac{\pi}{4} - \dfrac{2}{\pi}\left[\dfrac{\cos 2x}{1^2} + \dfrac{\cos 6x}{3^2} + \dfrac{\cos 10x}{5^2} + \cdots\right]$

Solution: (i) Half-range sine series of $f(x)$ in the given interval is

$$f(x) = \sum_{n=1}^{\infty} b_n \sin nx$$

where
$$b_n = \frac{2}{\pi}\left[\int_0^{\frac{\pi}{2}} x\sin nx\,dx + \int_{\frac{\pi}{2}}^{\pi}(\pi - x)\sin nx\,dx\right]$$

$$= \frac{2}{\pi}\left[\left\{x\left(-\frac{1}{n}\cos nx\right) + \frac{1}{n^2}\sin nx\right\}_0^{\frac{\pi}{2}} + \left\{(\pi - x)\left(-\frac{1}{n}\cos nx\right) - \frac{1}{n^2}\sin nx\right\}_{\frac{\pi}{2}}^{\pi}\right]$$

$$= \frac{2}{\pi}\left[-\frac{\pi}{2n}\cos\frac{n\pi}{2} + \frac{1}{n^2}\sin\frac{n\pi}{2} + \frac{\pi}{2n}\cos\frac{n\pi}{2} + \frac{1}{n^2}\sin\frac{n\pi}{2}\right]$$

$$= \frac{4}{\pi n^2}\sin\frac{n\pi}{2}$$

$$\therefore \qquad b_{2n} = 0,\ b_{2n-1} = \frac{4(-1)^{n+1}}{\pi(2n-1)^2};\ n = 1,2,\ldots$$

$\therefore$ Half-range sine series of $f(x)$ in the given interval is

$$f(x) = \frac{4}{\pi}\sum_{n=1}^{\infty}\frac{(-1)^{n+1}}{(2n-1)^2}\sin(2n-1)x$$

$$= \frac{4}{\pi}\left[\sin x - \frac{\sin 3x}{3^2} + \frac{\sin 5x}{5^2} - \frac{\sin 7x}{7^2} + \cdots\right]$$

(ii) Half-range cosine series of $f(x)$ in the given interval is

$$f(x) = \frac{a_0}{2} + \sum_{n=1}^{\infty} a_n \cos nx$$

where $$a_0 = \frac{2}{\pi}\left[\int_0^{\frac{\pi}{2}} x\,dx + \int_{\frac{\pi}{2}}^{\pi} (\pi - x)\,dx\right] = \frac{2}{\pi}\left[\left(\frac{x^2}{2}\right)_0^{\frac{\pi}{2}} + \left(\pi x - \frac{x^2}{2}\right)_{\frac{\pi}{2}}^{\pi}\right]$$

$$= \frac{2}{\pi}\left[\frac{\pi^2}{8} + \pi^2 - \frac{\pi^2}{2} - \frac{\pi^2}{2} + \frac{\pi^2}{8}\right] = \frac{\pi}{2}$$

$$a_n = \frac{2}{\pi}\left[\int_0^{\frac{\pi}{2}} x \cos nx\,dx + \int_{\frac{\pi}{2}}^{\pi} (\pi - x)\cos nx\,dx\right]$$

$$= \frac{2}{\pi}\left[\left\{x\left(\frac{1}{n}\sin nx\right) + \frac{1}{n^2}\cos nx\right\}_0^{\frac{\pi}{2}} + \left\{(\pi - x)\left(\frac{1}{n}\sin nx\right) - \frac{1}{n^2}\cos nx\right\}_{\frac{\pi}{2}}^{\pi}\right]$$

$$= \frac{2}{\pi}\left[\frac{\pi}{2n}\sin\frac{n\pi}{2} + \frac{1}{n^2}\cos\frac{n\pi}{2} - \frac{1}{n^2} - \frac{\pi}{2n}\sin\frac{n\pi}{2} - \frac{1}{n^2}(-1)^n + \frac{1}{n^2}\cos\frac{n\pi}{2}\right]$$

$$= \frac{2}{\pi}\left[\frac{2}{n^2}\cos\frac{n\pi}{2} - \frac{1}{n^2}\left(1 + (-1)^n\right)\right]$$

$\therefore$ $$a_{2n} = \frac{(-1)^n}{\pi n^2} - \frac{1}{\pi n^2}, \quad a_{2n-1} = 0; \quad n = 1, 2, 3, \ldots$$

$\therefore$ $$a_{4n} = 0,\ a_{2(2n-1)} = -\frac{2}{\pi(2n-1)^2}, \quad a_{2n-1} = 0; \ n = 1, 2, 3, \ldots$$

$\therefore$ Half-range cosine series of $f(x)$ in the given interval is

$$f(x) = \frac{\pi}{4} - \frac{2}{\pi}\sum_{n=1}^{\infty} \frac{1}{(2n-1)^2}\cos\left(2(2n-1)x\right)$$

$$= \frac{\pi}{4} - \frac{2}{\pi}\left[\frac{\cos 2x}{1^2} + \frac{\cos 6x}{3^2} + \frac{\cos 10x}{5^2} + \frac{\cos 14x}{7^2} + \cdots\right]$$

Example 3.33: Express $f(x)=x^2$ as a half-range cosine series for $0<x<2$.

Solution: Half-range cosine series for $f(x)=x^2$ in $0<x<2$ is

$$f(x)=x^2=\frac{a_0}{2}+\sum_{n=1}^{\infty}a_n\cos\frac{n\pi x}{2}$$

where
$$a_0=\int_0^2 x^2dx=\frac{1}{3}\left(x^3\right)_0^2=\frac{8}{3}$$

$$a_n=\int_0^2 x^2\cos\frac{n\pi x}{2}dx$$

$$=\left[x^2\left(\frac{2}{n\pi}\sin\frac{n\pi x}{2}\right)-2x\left(-\frac{4}{n^2\pi^2}\cos\frac{n\pi x}{2}\right)+2\left(-\frac{8}{n^3\pi^3}\frac{\sin n\pi x}{2}\right)\right]_0^2$$

$$=\frac{16}{n^2\pi^2}\cos n\pi=\frac{16}{n^2\pi^2}(-1)^n$$

$\therefore$ Half-range cosine series for $f(x)=x^2$ in $0<x<2$ is

$$f(x)=x^2=\frac{4}{3}-\frac{16}{\pi^2}\sum_{n=1}^{\infty}\frac{(-1)^{n+1}}{n^2}\cos\frac{n\pi x}{2}$$

Example 3.34: Obtain the half-range sine series for $f(x)=2-x$ for $0<x<2$ and hence deduce that $1-\frac{1}{3}+\frac{1}{5}-\frac{1}{7}+\cdots=\frac{\pi}{4}$.

Solution: Half-range sine series for $f(x)=2-x$ in $0<x<2$ is

$$f(x)=2-x=\sum_{n=1}^{\infty}b_n\sin\frac{n\pi x}{2}$$

where
$$b_n=\int_0^2(2-x)\sin\frac{n\pi x}{2}dx$$

$$=\left[(2-x)\left(-\frac{2}{n\pi}\cos\frac{n\pi x}{2}\right)+\left(-\frac{4}{n^2\pi^2}\sin\frac{n\pi x}{2}\right)\right]_0^2$$

$$=\frac{4}{n\pi}$$

$\therefore$ Half-range sine series for $f(x)=2-x$ in $0<x<2$ is

$$f(x)=2-x=\frac{4}{\pi}\sum_{n=1}^{\infty}\frac{1}{n}\sin\frac{n\pi x}{2}$$

Take $x=1$,
$$1=\frac{4}{\pi}\sum_{n=1}^{\infty}\frac{1}{n}\sin\frac{n\pi}{2}$$

$\Rightarrow$
$$1-\frac{1}{3}+\frac{1}{5}-\frac{1}{7}+\cdots=\frac{\pi}{4}$$

Example 3.35: If $f(x)=1-\frac{x}{l},\ 0<x<l$ find (i) Fourier cosine series and (ii) Fourier sine series of $f(x)$. Graph the corresponding periodic continuations of $f(x)$.

Solution: Fourier cosine series of $f(x)$ in $0<x<l$ is

$$f(x)=1-\frac{x}{l}=\frac{a_0}{2}+\sum_{n=1}^{\infty}a_n\cos\frac{n\pi x}{l}$$

where
$$a_0=\frac{2}{l}\int_0^l\left(1-\frac{x}{l}\right)dx=\frac{2}{l}\left(x-\frac{x^2}{2l}\right)_0^l=1$$

$$a_n=\frac{2}{l}\int_0^l\left(1-\frac{x}{l}\right)\cos\frac{n\pi x}{l}\,dx$$

$$=\frac{2}{l}\left[\left(1-\frac{x}{l}\right)\left(\frac{l}{n\pi}\sin\frac{n\pi x}{l}\right)+\frac{1}{l}\left(-\frac{l^2}{n^2\pi^2}\cos\frac{n\pi x}{l}\right)\right]_0^l$$

$$=-\frac{2}{n^2\pi^2}\left[(-1)^n-1\right]=\frac{2}{n^2\pi^2}\left[1-(-1)^n\right]$$

$$\therefore\quad a_{2n}=0,\ a_{2n-1}=\frac{4}{\pi^2(2n-1)^2};\ n=1,2,3,\ldots$$

$\therefore$ Fourier cosine series of $f(x)$ in $0<x<l$ is

$$f(x)=1-\frac{x}{l}=\frac{1}{2}+\frac{4}{\pi^2}\sum_{n=1}^{\infty}\frac{1}{(2n-1)^2}\cos\frac{(2n-1)\pi x}{l}$$

Graph of periodic continuation of $f(x)$ is

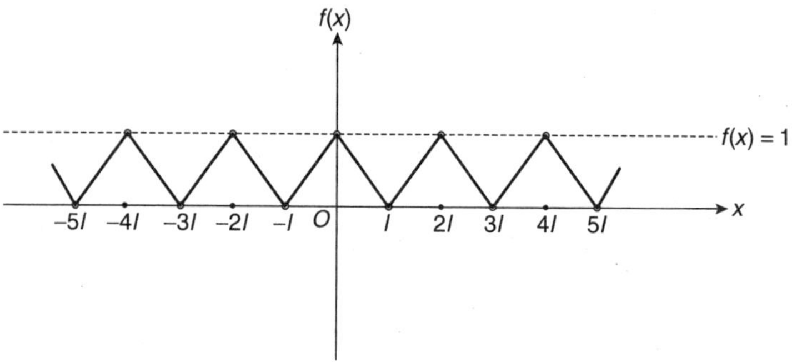

Figure 3.10

◎ indicates that the point is not in the graph.

(ii) Fourier sine series of $f(x)$ in $(0,l)$ is

$$f(x)=1-\frac{x}{l}=\sum_{n=1}^{\infty} b_n \sin\frac{n\pi x}{l}$$

where $$b_n=\frac{2}{l}\int_0^l\left(1-\frac{x}{l}\right)\sin\frac{n\pi x}{l}\,dx$$

$$=\frac{2}{l}\left[\left(1-\frac{x}{l}\right)\left(-\frac{l}{n\pi}\cos\frac{n\pi x}{l}\right)+\frac{1}{l}\left(-\frac{l^2}{n^2\pi^2}\sin\frac{n\pi x}{l}\right)\right]_0^l=\frac{2}{n\pi}$$

$\therefore$ Fourier sine series of $f(x)$ is

$$f(x)=1-\frac{x}{l}=\frac{2}{\pi}\sum_{n=1}^{\infty}\frac{1}{n}\sin\frac{n\pi x}{l};\quad 0<x<l$$

Graph of periodic continuation of $f(x)$ is

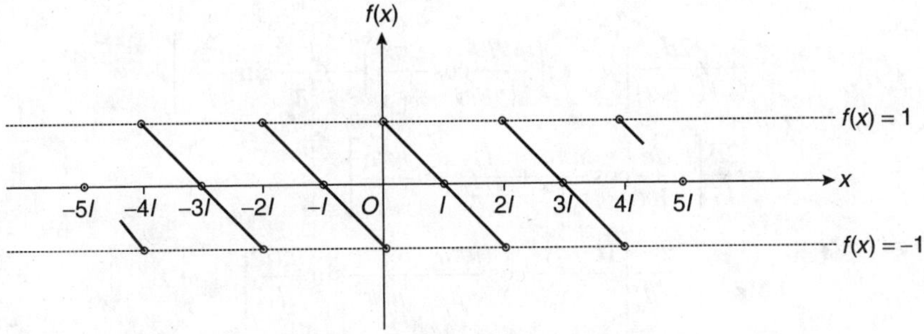

Figure 3.11

⊙ indicates that the point is not in the graph.

Example 3.36: Find the half period sine series for $f(x)$ given in the range $[0,l]$ by the curve *OPQ* in the following diagram

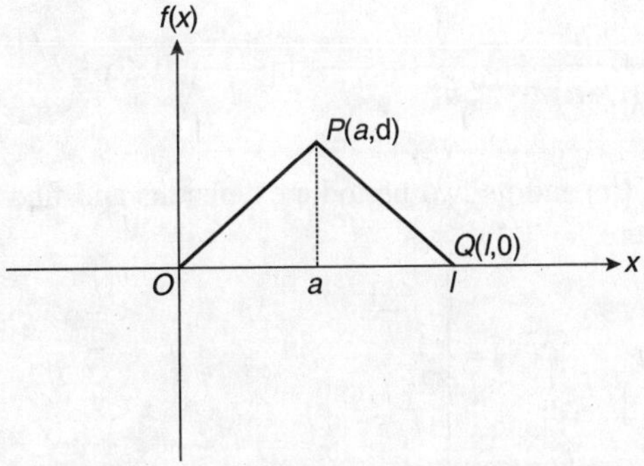

Figure 3.12

Solution: From the graph

$$f(x)=\begin{cases}\dfrac{d}{a}x & ;\ 0\le x\le a\\[2mm] \dfrac{d}{l-a}(l-x); & a\le x\le l\end{cases}$$

Half period sine series for $f(x)$ in range $[0,l]$ is

$$f(x)=\sum_{n=1}^{\infty}b_n\sin\frac{n\pi x}{l}$$

where
$$b_n=\frac{2}{l}\left[\int_0^a\frac{d}{a}x\sin\frac{n\pi x}{l}dx+\int_a^l\frac{d}{l-a}(l-x)\sin\frac{n\pi x}{l}dx\right]$$

$$=\frac{2d}{la}\left[x\left(-\frac{l}{n\pi}\cos\frac{n\pi x}{l}\right)+\frac{l^2}{n^2\pi^2}\sin\frac{n\pi x}{l}\right]_0^a$$

$$+\frac{2d}{l(l-a)}\left[(l-x)\left(-\frac{l}{n\pi}\cos\frac{n\pi x}{l}\right)-\frac{l^2}{n^2\pi^2}\sin\frac{n\pi x}{l}\right]_a^l$$

$$=\frac{2d}{la}\left[-\frac{la}{n\pi}\cos\frac{n\pi a}{l}+\frac{l^2}{n^2\pi^2}\sin\frac{n\pi a}{l}\right]$$

$$+\frac{2d}{l(l-a)}\left[\frac{(l-a)l}{n\pi}\cos\frac{n\pi a}{l}+\frac{l^2}{n^2\pi^2}\sin\frac{n\pi a}{l}\right]$$

$$=2d\left(\frac{l}{an^2\pi^2}+\frac{l}{(l-a)n^2\pi^2}\right)\sin\frac{n\pi a}{l}$$

$$=\frac{2dl^2}{a(l-a)n^2\pi^2}\sin\frac{n\pi a}{l}$$

$\therefore$ Fourier sine series of $f(x)$ is

$$f(x)=\frac{2dl^2}{a(l-a)\pi^2}\sum_{n=1}^{\infty}\frac{1}{n^2}\sin\frac{n\pi a}{l}\sin\frac{n\pi x}{l};\ 0\le x\le l$$

Example 3.37: Sketch $f(x)$ and its two periodic extensions and find the Fourier cosine series for the function $f(x)$ where

$$f(x)=\begin{cases}\dfrac{2kx}{l} & ;\ 0<x<\dfrac{l}{2}\\[2mm] \dfrac{2k}{l}(l-x) & ;\ \dfrac{l}{2}<x<l\end{cases}$$

Also find the Fourier sine series for $f(x)$.

Solution: Below is the graph of $f(x)$ and its two periodic extensions

(i) Graph of $f(x)$

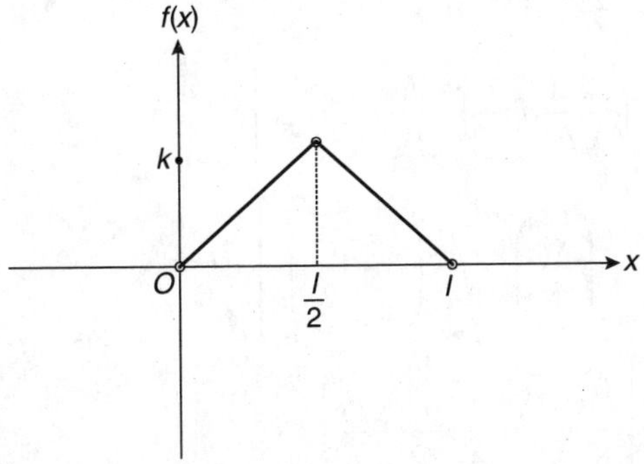

Figure 3.13

⊚ indicates the point is not in the graph.

(ii) Graph of even extension

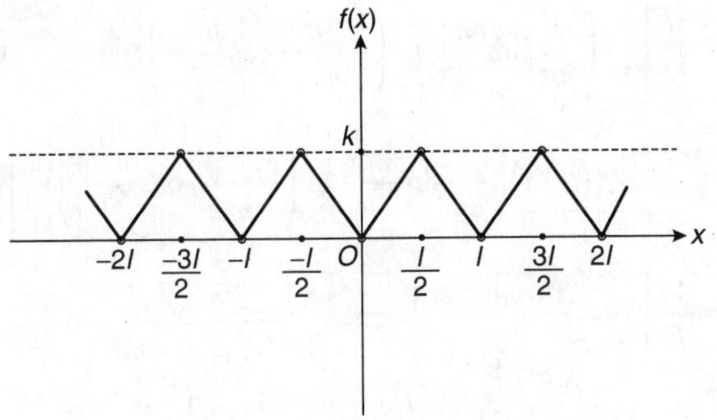

Figure 3.14

⊚ indicates the point is not in the graph.

(iii) Graph of odd extension

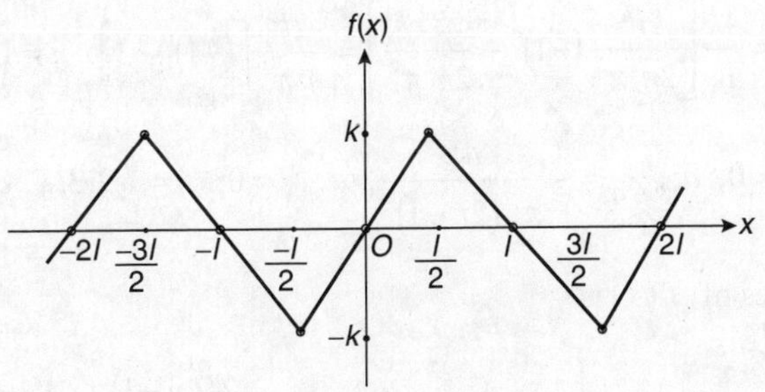

Figure 3.15

⊚ indicates the point is not in the graph.

Fourier cosine series is the Fourier series for even extension and it is

$$f(x) \sim \frac{a_0}{2} + \sum_{n=1}^{\infty} a_n \cos\frac{n\pi x}{l}$$

where
$$a_0 = \frac{2}{l}\left[\int_0^{\frac{l}{2}} \frac{2kx}{l}\,dx + \int_{\frac{l}{2}}^{l} \frac{2k}{l}(l-x)\,dx\right]$$

$$= \frac{2}{l}\left[\frac{2k}{l}\left(\frac{x^2}{2}\right)_0^{\frac{l}{2}} + \frac{2k}{l}\left(lx - \frac{x^2}{2}\right)_{\frac{l}{2}}^{l}\right]$$

$$= \frac{4k}{l^2}\left[\frac{l^2}{8} + l^2 - \frac{l^2}{2} - \frac{l^2}{2} + \frac{l^2}{8}\right] = k$$

$$a_n = \frac{2}{l}\cdot\frac{2k}{l}\left[\int_0^{\frac{l}{2}} x\cos\frac{n\pi x}{l}\,dx + \int_{\frac{l}{2}}^{l} (l-x)\cos\frac{n\pi x}{l}\,dx\right]$$

$$= \frac{4k}{l^2}\left[\left\{x\left(\frac{l}{n\pi}\sin\frac{n\pi x}{l}\right) - \left(-\frac{l^2}{n^2\pi^2}\cos\frac{n\pi x}{l}\right)\right\}_0^{\frac{l}{2}}\right.$$
$$\left. + \left\{(l-x)\left(\frac{l}{n\pi}\sin\frac{n\pi x}{l}\right) + \left(-\frac{l^2}{n^2\pi^2}\cos\frac{n\pi x}{l}\right)\right\}_{\frac{l}{2}}^{l}\right]$$

$$= \frac{4k}{l^2}\left[\frac{l^2}{2n\pi}\sin\frac{n\pi}{2} + \frac{l^2}{n^2\pi^2}\cos\frac{n\pi}{2} - \frac{l^2}{n^2\pi^2} - \frac{l^2}{2n\pi}\sin\frac{n\pi}{2}\right.$$
$$\left. - \frac{l^2}{n^2\pi^2}(-1)^n + \frac{l^2}{n^2\pi^2}\cos\frac{n\pi}{2}\right]$$

$$= \frac{4k}{l^2}\left[\frac{2l^2}{n^2\pi^2}\cos\frac{n\pi}{2} - \frac{l^2}{n^2\pi^2}\left\{1+(-1)^n\right\}\right]$$

∴
$$a_{2n} = \frac{4k}{l^2}\left[\frac{l^2}{2n^2\pi^2}(-1)^n - \frac{l^2}{2n^2\pi^2}\right] = \frac{2k}{n^2\pi^2}\left[(-1)^n - 1\right];\ n = 1, 2, \ldots$$

∴
$$a_{4n} = 0,\ a_{2(2n-1)} = -\frac{4k}{\pi^2(2n-1)^2},\ a_{2n-1} = 0;\ n = 1, 2, 3, \ldots$$

∴ Fourier cosine series of $f(x)$ is

$$f(x) \sim k\left[\frac{1}{2} - \frac{4}{\pi^2}\sum_{n=1}^{\infty}\frac{1}{(2n-1)^2}\cos\frac{2(2n-1)\pi x}{l}\right]$$

Fourier sine series is the Fourier series for odd extension and it is

$$f(x) \sim \sum_{n=1}^{\infty} b_n \sin\frac{n\pi x}{l}$$

where $$b_n = \frac{2}{l}\cdot\frac{2k}{l}\left[\int_0^{\frac{l}{2}} x\sin\frac{n\pi x}{l}\,dx + \int_{\frac{l}{2}}^{l}(l-x)\sin\frac{n\pi x}{l}\,dx\right]$$

$$= \frac{4k}{l^2}\left[\left\{x\left(-\frac{l}{n\pi}\cos\frac{n\pi x}{l}\right)+\frac{l^2}{n^2\pi^2}\sin\frac{n\pi x}{l}\right\}_0^{\frac{l}{2}}\right.$$

$$\left.+\left\{(l-x)\left(-\frac{l}{n\pi}\cos\frac{n\pi x}{l}\right)+\left(-\frac{l^2}{n^2\pi^2}\sin\frac{n\pi x}{l}\right)\right\}_{\frac{l}{2}}^{l}\right]$$

$$= \frac{4k}{l^2}\left[-\frac{l^2}{2n\pi}\cos\frac{n\pi}{2}+\frac{l^2}{n^2\pi^2}\sin\frac{n\pi}{2}+\frac{l^2}{2n\pi}\cos\frac{n\pi}{2}+\frac{l^2}{n^2\pi^2}\sin\frac{n\pi}{2}\right]$$

$$= \frac{8k}{n^2\pi^2}\sin\frac{n\pi}{2}$$

$\therefore$ $$b_{2n} = 0,\ b_{2n-1} = \frac{8k}{\pi^2(2n-1)^2}(-1)^{n+1};\quad n = 1,2,3,\ldots$$

$\therefore$ Fourier sine series of $f(x)$ is

$$f(x) \sim \frac{8k}{\pi^2}\sum_{n=1}^{\infty}\frac{(-1)^{n+1}}{(2n-1)^2}\sin\frac{(2n-1)\pi x}{l}$$

Example 3.38: Express $f(x) = \begin{cases} \frac{1}{4} - x\ ; & 0 < x < 1/2 \\ x - \frac{3}{4}\ ; & 1/2 < x < 1 \end{cases}$

as a Fourier series of sine terms.

Solution: Half-range sine series for $f(x)$ is

$$f(x) = \sum_{n=1}^{\infty} b_n \sin n\pi x,\quad 0 < x < 1/2;\ \ 1/2 < x < 1$$

where $$b_n = 2\int_0^1 f(x)\sin n\pi x\,dx$$

$$= 2\left[\int_0^{\frac{1}{2}}\left(\frac{1}{4}-x\right)\sin n\pi x\,dx + \int_{\frac{1}{2}}^{1}\left(x-\frac{3}{4}\right)\sin n\pi x\,dx\right]$$

$$= 2\left[\left\{\left(\frac{1}{4}-x\right)\left(-\frac{1}{n\pi}\cos n\pi x\right)+\left(-\frac{1}{n^2\pi^2}\sin n\pi x\right)\right\}_0^{\frac{1}{2}}\right.$$

$$\left.+\left\{\left(x-\frac{3}{4}\right)\left(-\frac{1}{n\pi}\cos n\pi x\right)+\frac{1}{n^2\pi^2}\sin n\pi x\right\}_{\frac{1}{2}}^{1}\right]$$

$$= \frac{1}{2n\pi}\left(\cos\frac{n\pi}{2}+1\right)-\frac{2}{n^2\pi^2}\sin\frac{n\pi}{2}-\frac{1}{2n\pi}(-1)^n-\frac{1}{2n\pi}\cos\frac{n\pi}{2}-\frac{2}{n^2\pi^2}\sin\frac{n\pi}{2}$$

$$= -\frac{4}{n^2\pi^2}\sin\frac{n\pi}{2}+\frac{1}{2n\pi}\left\{1-(-1)^n\right\}$$

$$\therefore \qquad b_{2n}=0,\ b_{2n-1}=\frac{4(-1)^n}{\pi^2(2n-1)^2}+\frac{1}{\pi(2n-1)};\quad n=1,2,3,\ldots$$

$\therefore$ Fourier sine series of $f(x)$ is

$$f(x)=\frac{1}{\pi}\sum_{n=1}^{\infty}\frac{1}{(2n-1)}\left[1+\frac{4(-1)^n}{\pi(2n-1)}\right]\sin(2n-1)\pi x,\ 0<x<1/2;\ 1/2<x<1$$

EXERCISE 3.2

1. Find a Fourier sine serics for $f(x)=k$ in $0<x<\pi$.
2. Find the half-range cosine series for $f(x)=\sin x$ in $0\le x\le\pi$ and hence show that $\sum_{n=1}^{\infty}\frac{1}{4n^2-1}=\frac{1}{2}$.
3. Find the Fourier (i) cosine series and (ii) sine series given by the function $f(x)=\pi-x$ in $(0,\pi)$.
4. Expand $\pi x-x^2$ in a half-range sine series in the interval $(0,\pi)$ (upto the first three terms). Deduce that $\frac{1}{1^3}-\frac{1}{3^3}+\frac{1}{5^3}-\cdots=\frac{\pi^3}{32}$.
5. Find the half-range cosine series for the function $f(x)=x^2$ in $0\le x\le\pi$ and hence find the sum of the series $1-\frac{1}{2^2}+\frac{1}{3^2}-\frac{1}{4^2}+\cdots$
6. Expand $x\sin x$ as sine series in $0<x<\pi$.
7. Find a cosine series for $f(x)=e^x$ in $0<x<\pi$.
8. Expand $f(x)=\cos x;\ 0<x<\pi$ in half-range sine series.
9. Represent the following function by a Fourier sine series: $$f(t)=\begin{cases} t, & 0<t\le\pi/2\\ \pi/2, & \pi/2<t\le\pi\end{cases}$$
10. If a function $f(x)$ is defined in $(0,\pi)$ as $$f(x)=\begin{cases}\pi/3, & 0<x<\pi/3\\ 0, & \pi/3<x<2\pi/3\\ -\pi/3, & 2\pi/3<x<\pi\end{cases}$$ then show that its cosine series is $$f(x)=\frac{2}{\sqrt{3}}\left[\cos x-\frac{1}{5}\cos 5x+\frac{1}{7}\cos 7x-\cdots\right]$$

and its sine series is

$$f(x)=\sin 2x+\frac{1}{2}\sin 4x+\frac{1}{4}\sin 8x+\frac{1}{5}\sin 10x+\cdots$$

11. Obtain the half-range sine series for the function $f(x)=x^2$ in the interval $0<x<3$.
12. Obtain the half-range sine series for e^x in $0<x<1$.
13. Find a Fourier sine series for $f(x)=ax+b$ in $0<x<l$.
14. Express $f(x)=x$ as a half-range cosine series in $0<x<2$.
15. Show that in the interval $(0,1)$, $\cos\pi x=\frac{8}{\pi}\sum_{n=1}^{\infty}\frac{n}{4n^2-1}\sin 2n\pi x$
16. Develop $\sin\left(\frac{\pi x}{l}\right)$ in half-range cosine series in the range $0<x<l$.
17. Find the half-range sine series for the function $f(t)=t-t^2$ in the interval $0<t<1$.
18. Find the half-range sine series for $f(x)=x$ in $0<x<2$.
19. Find the Fourier cosine series and sine series of the function $f(x)=1,\ 0\le x\le 2$.
20. Obtain the half-range sine series for $f(x)=lx-x^2$ in $(0,l)$ and hence show that $\frac{1}{1^3}-\frac{1}{3^3}+\frac{1}{5^3}-\frac{1}{7^3}+\cdots=\frac{\pi^3}{32}$.
21. Show that the series $\frac{l}{\pi}\sum_{n=1}^{\infty}\frac{1}{n}\sin\frac{2n\pi x}{l}$ represents $\frac{l}{2}-x$, when $0<x<l$.
22. Find the half-range cosine series of $f(x)=\begin{cases}1, & 0\le x\le 1\\ x, & 1\le x\le 2\end{cases}$
23. Find the Fourier cosine series of the function $f(x)=\begin{cases}x^2, & 0\le x\le 2\\ 4, & 2\le x\le 4\end{cases}$
24. Obtain the half-range cosine series for $f(x)$ defined by $f(x)=\begin{cases}k, & 0<x<a\\ 0, & a<x<l\end{cases}$
25. Expand $f(x)=\begin{cases}x, & 0<x<1\\ 2-x, & 1<x<2\end{cases}$ as a Fourier sine series. Hence deduce that $\frac{\pi^2}{8}=\frac{1}{1^2}+\frac{1}{3^2}+\frac{1}{5^2}+\cdots$
26. Obtain a half-range cosine series for $f(x)=\begin{cases}kx, & 0\le x\le l/2\\ k(l-x), & l/2\le x\le l\end{cases}$. Deduce the sum of the series $\frac{1}{1^2}+\frac{1}{3^2}+\frac{1}{5^2}+\cdots$ Also, prove that

$$f(x)=\frac{4kl}{\pi^2}\sum_{n=0}^{\infty}\frac{(-1)^n}{(2n+1)^2}\sin\frac{(2n+1)\pi x}{l}.$$

ANSWERS 3.2

1. $f(x)=\frac{4k}{\pi}\sum_{n=1}^{\infty}\frac{1}{(2n-1)}\sin(2n-1)x$
2. $f(x)=\frac{2}{\pi}-\frac{4}{\pi}\sum_{n=1}^{\infty}\frac{\cos 2n\,x}{(4n^2-1)}$

3. (i) $\pi - x = \frac{\pi}{2} + \frac{4}{\pi}\sum_{n=1}^{\infty}\frac{\cos(2n-1)x}{(2n-1)^2}$ (ii) $\pi - x = 2\sum_{n=1}^{\infty}\frac{\sin nx}{n}$

4. $f(x) = \frac{8}{\pi}\left[\frac{\sin x}{1^3} + \frac{\sin 3x}{3^3} + \frac{\sin 5x}{5^3} + \cdots\right]$

5. $x^2 = \frac{\pi^2}{3} - 4\sum_{n=1}^{\infty}\frac{(-1)^{n+1}}{n^2}\cos nx$

6. $x\sin x = \frac{\pi}{2}\sin x - \frac{16}{\pi}\sum_{n=1}^{\infty}\frac{n\sin 2x}{(4n^2-1)}$

7. $e^x = \frac{e^{\pi}-1}{\pi} + \frac{2}{\pi}\sum_{n=1}^{\infty}\frac{\left(e^{\pi}(-1)^n - 1\right)}{n^2+1}\cos nx$

8. $f(x) = \frac{8}{\pi}\sum_{n=1}^{\infty}\frac{n}{4n^2-1}\sin 2nx$

9. $f(t) = \sum_{n=1}^{\infty}\left[\frac{-\sin 2nt}{2n} + \left\{\frac{2(-1)^{n+1}}{\pi(2n-1)^2} + \frac{1}{(2n-1)}\right\}\sin(2n-1)t\right]$

11. $f(x) = \sum_{n=1}^{\infty}\left[\frac{18}{\pi^3(2n-1)^3}\left\{\pi^2(2n-1)^2 - 4\right\}\sin\frac{(2n-1)\pi x}{3} - \frac{9}{n\pi}\sin\frac{2n\pi x}{3}\right]$

12. $e^x = 2\pi\sum_{n=1}^{\infty}\frac{n\left\{1+e(-1)^{n+1}\right\}}{(1+n^2\pi^2)}\sin n\pi x$

13. $ax + b = \frac{1}{\pi}\sum_{n=1}^{\infty}\left[\frac{2(al+2b)}{(2n-1)}\sin\frac{(2n-1)\pi x}{l} - \frac{al}{n}\sin\frac{2n\pi x}{l}\right]$

14. $x = 1 - \frac{8}{\pi^2}\sum_{n=1}^{\infty}\frac{1}{(2n-1)^2}\cos\frac{(2n-1)\pi x}{2}$

16. $\sin\left(\frac{\pi x}{l}\right) = \frac{2}{\pi} - \frac{4}{\pi}\sum_{n=1}^{\infty}\frac{1}{(4n^2-1)}\cos\left(\frac{2n\pi x}{l}\right)$

17. $f(t) = \frac{8}{\pi^3}\sum_{n=1}^{\infty}\frac{1}{(2n-1)^3}\sin(2n-1)\pi t$

18. $f(x) = \frac{4}{\pi}\sum_{n=1}^{\infty}\frac{(-1)^{n+1}}{n}\sin\left(\frac{n\pi x}{2}\right)$

19. $f(x)=1;\ f(x)=\frac{4}{\pi}\sum_{n=1}^{\infty}\frac{1}{(2n-1)}\sin\frac{(2n-1)\pi x}{2}$

20. $lx-x^2=\frac{8l^3}{\pi^3}\sum_{n=1}^{\infty}\frac{1}{(2n-1)^3}\sin\frac{(2n-1)\pi x}{l}$

22. $f(x)=\frac{5}{4}+\frac{2}{\pi^2}\sum_{n=1}^{\infty}\frac{1}{(2n-1)^2}\left[\cos(2n-1)\pi x-2\cos\frac{(2n-1)\pi x}{2}\right]$

23. $f(x)=\frac{8}{3}+\frac{8}{\pi^3}\sum_{n=1}^{\infty}(-1)^n\left[\frac{8}{(2n-1)^3}\cos\frac{(2n-1)\pi x}{4}+\frac{\pi}{n^2}\cos\frac{n\pi x}{2}\right]$

24. $f(x)=\frac{ka}{l}+\frac{2k}{\pi}\sum_{n=1}^{\infty}\frac{1}{n}\sin\frac{n\pi a}{l}\cos\frac{n\pi x}{l}$

25. $f(x)=\frac{8}{\pi^2}\sum_{n=1}^{\infty}\frac{(-1)^{n+1}}{(2n-1)^2}\sin\frac{(2n-1)\pi x}{2}$

26. $f(x)=\frac{kl}{4}-\frac{2kl}{\pi^2}\sum_{n=1}^{\infty}\frac{1}{(2n-1)^2}\cos\left\{\frac{2(2n-1)\pi x}{l}\right\}$

3.3 OTHERS FORMULAE

3.3.1 Parseval's Formulae

(a) If $f(x)$ is continuous and differentiable in $(-l,l)$, then

$$\int_{-l}^{l}[f(x)]^2dx=l\left[\frac{1}{2}a_0^2+\sum_{n=1}^{\infty}(a_n^2+b_n^2)\right]$$

where $a_0=\frac{1}{l}\int_{-l}^{l}f(x)\,dx,\ a_n=\frac{1}{l}\int_{-l}^{l}f(x)\cos\frac{n\pi x}{l}dx,\ b_n=\frac{1}{l}\int_{-l}^{l}f(x)\sin\frac{n\pi x}{l}dx.$

(b) If $f(x)$ is continuous and differentiable in $(0,l)$, then

(i) $\int_0^l[f(x)]^2dx=\frac{l}{2}\left[\frac{a_0^2}{2}+\sum_{n=1}^{\infty}a_n^2\right]$

(ii) $\int_0^l[f(x)]^2dx=\frac{l}{2}\sum_{n=1}^{\infty}b_n^2$

where $a_0=\frac{2}{l}\int_0^l f(x)dx,\ a_n=\frac{2}{l}\int_0^l f(x)\cos\frac{n\pi x}{l}dx,\ b_n=\frac{2}{l}\int_0^l f(x)\sin\frac{n\pi x}{l}dx$

Proof: (a) Fourier series of $f(x)$ in $(-l,l)$ is

$$f(x)=\frac{a_0}{2}+\sum_{n=1}^{\infty}\left[a_n\cos\frac{n\pi x}{l}+b_n\sin\frac{n\pi x}{l}\right] \tag{3.6}$$

where

$$a_0=\frac{1}{l}\int_{-l}^{l}f(x)\,dx \tag{3.7}$$

$$a_n=\frac{1}{l}\int_{-l}^{l}f(x)\cos\frac{n\pi x}{l}\,dx \tag{3.8}$$

$$b_n=\frac{1}{l}\int_{-l}^{l}f(x)\sin\frac{n\pi x}{l}\,dx \tag{3.9}$$

Multiply (3.6) by $f(x)$ and integrate term by term within the limits $-l$ to l [term by term integration is possible as series on R.H.S. of (3.6) is uniformly convergent because of continuity and differentiability of $f(x)$ in $(-l,l)$]

$$\therefore\quad \int_{-l}^{l}[f(x)]^2\,dx=\frac{a_0}{2}\int_{-l}^{l}f(x)\,dx+\sum_{n=1}^{\infty}\left[a_n\int_{-l}^{l}f(x)\cos\frac{n\pi x}{l}\,dx+b_n\int_{-l}^{l}f(x)\sin\frac{n\pi x}{l}\,dx\right]$$

Using (3.7) to (3.9)

$$\int_{-l}^{l}[f(x)]^2\,dx=l\left[\frac{a_0^2}{2}+\sum_{n=1}^{\infty}\left(a_n^2+b_n^2\right)\right]$$

(b) (i) Half-range Fourier cosine series of $f(x)$ in $(0,l)$ is

$$f(x)=\frac{a_0}{2}+\sum_{n=1}^{\infty}a_n\cos\frac{n\pi x}{l} \tag{3.10}$$

where

$$a_0=\frac{2}{l}\int_{0}^{l}f(x)\,dx \tag{3.11}$$

$$a_n=\frac{2}{l}\int_{0}^{l}f(x)\cos\frac{n\pi x}{l}\,dx \tag{3.12}$$

Multiply (3.10) by $f(x)$ and integrate within the limits 0 to l

$$\int_{0}^{l}[f(x)]^2\,dx=\frac{a_0}{2}\int_{0}^{l}f(x)\,dx+\sum_{n=1}^{\infty}a_n\int_{0}^{l}f(x)\cos\frac{n\pi x}{l}\,dx$$

Using (3.11) and (3.12)

$$\int_{0}^{l}[f(x)]^2\,dx=\frac{l}{2}\left[\frac{a_0^2}{2}+\sum_{n=1}^{\infty}a_n^2\right]$$

(ii) Half-range Fourier sine series of $f(x)$ in $(0,l)$ is

$$f(x)=\sum_{n=1}^{\infty} b_n \sin\frac{n\pi x}{l} \tag{3.13}$$

where
$$b_n=\frac{2}{l}\int_0^l f(x)\sin\frac{n\pi x}{l}\,dx \tag{3.14}$$

Multiply (3.13) by $f(x)$ and integrate within the limits 0 to l

$$\int_0^l [f(x)]^2 dx=\sum_{n=1}^{\infty} b_n\int_0^l f(x)\sin\frac{n\pi x}{l}\,dx=\frac{l}{2}\sum_{n=1}^{\infty} b_n^2 \qquad \text{(from 3.14)}$$

3.3.2 Root Mean Square (R.M.S.) Value

Root mean square value of a periodic function is frequently used in mechanical vibrations and electric circuit theory. Root mean square value is also called the effective value of the function and it is the square root of mean value of square of function and

Root mean square value of $f(x)$ in $(a,b)=\sqrt{\dfrac{1}{b-a}\displaystyle\int_a^b [f(x)]^2\,dx}$

3.3.3 Complex Form of Fourier Series

Fourier series of $f(x)$ in $(-l,l)$ is

$$f(x)=\frac{a_0}{2}+\sum_{n=1}^{\infty}\left(a_n\cos\frac{n\pi x}{l}+b_n\sin\frac{n\pi x}{l}\right) \tag{3.15}$$

where
$$a_0=\frac{1}{l}\int_{-l}^{l} f(x)\,dx \tag{3.16}$$

$$a_n=\frac{1}{l}\int_{-l}^{l} f(x)\cos\frac{n\pi x}{l}\,dx \tag{3.17}$$

$$b_n=\frac{1}{l}\int_{-l}^{l} f(x)\sin\frac{n\pi x}{l}\,dx \tag{3.18}$$

Now, (3.15) can be written as

$$f(x)=\frac{a_0}{2}+\sum_{n=1}^{\infty}\left[a_n\frac{e^{i\frac{n\pi x}{l}}+e^{-i\frac{n\pi x}{l}}}{2}+b_n\frac{e^{i\frac{n\pi x}{l}}-e^{-i\frac{n\pi x}{l}}}{2i}\right]$$

$$=\frac{a_0}{2}+\sum_{n=1}^{\infty}\left[\frac{1}{2}(a_n-i\,b_n)e^{i\frac{n\pi x}{l}}+\frac{1}{2}(a_n+i\,b_n)e^{-i\frac{n\pi x}{l}}\right]$$

Defining $c_0 = \frac{a_0}{2}$, $c_n = \frac{1}{2}(a_n - ib_n)$, $c_{-n} = \frac{1}{2}(a_n + ib_n)$, we write $f(x)$ as

$$f(x) = c_0 + \sum_{n=1}^{\infty} c_n e^{i\frac{n\pi x}{l}} + c_{-n} e^{-i\frac{n\pi x}{l}} = \sum_{n=-\infty}^{\infty} c_n e^{i\frac{n\pi x}{l}} \tag{3.19}$$

where $$c_0 = \frac{a_0}{2} = \frac{1}{2l}\int_{-l}^{l} f(x)\,dx$$

$$c_n = \frac{1}{2}(a_n - i\,b_n) = \frac{1}{2}\left[\frac{1}{l}\int_{-l}^{l} f(x)\cos\frac{n\pi x}{l}dx - \frac{i}{l}\int_{-l}^{l} f(x)\sin\frac{n\pi x}{l}dx\right]$$

$$= \frac{1}{2l}\int_{-l}^{l} f(x)\left(\cos\frac{n\pi x}{l} - i\sin\frac{n\pi x}{l}\right)dx = \frac{1}{2l}\int_{-l}^{l} f(x)e^{-i\frac{n\pi x}{l}}dx$$

$$c_{-n} = \frac{1}{2}(a_n + ib_n) = \frac{1}{2}\left[\frac{1}{l}\int_{-l}^{l} f(x)\cos\frac{n\pi x}{l}dx + \frac{i}{l}\int_{-l}^{l} f(x)\sin\frac{n\pi x}{l}dx\right]$$

$$= \frac{1}{2l}\int_{-l}^{l} f(x)\left(\cos\frac{n\pi x}{l} + i\sin\frac{n\pi x}{l}\right)dx = \frac{1}{2l}\int_{-l}^{l} f(x)e^{\frac{in\pi x}{l}}dx$$

Hence, all the coefficients can be expressed as

$$c_n = \frac{1}{2l}\int_{-l}^{l} f(x)\; e^{-i\frac{n\pi x}{l}}dx;\quad n = 0, \pm 1, \pm 2, \ldots$$

Thus, complex form of Fourier series of $f(x)$ in $(-l, l)$ is

$$f(x) = \sum_{n=-\infty}^{\infty} c_n e^{i\frac{n\pi x}{l}}$$

where $$c_n = \frac{1}{2l}\int_{-l}^{l} f(x)e^{-i\frac{n\pi x}{l}}dx;\quad n = 0, \pm 1, \pm 2, \pm 3, \ldots$$

Remark 3.2: If $f(x)$ satisfies Dirichlet's conditions in $(-l, l)$ then complex form of Fourier series converges to $f(x)$ at a point of continuity and to $\frac{1}{2}\left[f(x+0) + f(x-0)\right]$ at a point of discontinuity.

Example 3.39: Obtain the Fourier series for $y = x^2$ in $-\pi < x < \pi$. Using the two values of y, show that $\frac{1}{1^4} + \frac{1}{2^4} + \frac{1}{3^4} + \frac{1}{4^4} + \cdots = \frac{\pi^4}{90}$.

Solution: $y = x^2$, $-\pi < x < \pi$ is an even function.

$\therefore$ Fourier series for y is

$$y = \frac{a_0}{2} + \sum_{n=1}^{\infty} a_n \cos nx;\quad -\pi < x < \pi$$

where $$a_0 = \frac{2}{\pi}\int_0^{\pi} x^2 dx = \frac{2}{\pi}\left(\frac{x^3}{3}\right)_0^{\pi} = \frac{2\pi^2}{3}.$$

and $$a_n = \frac{2}{\pi}\int_0^{\pi} x^2 \cos nx\, dx = \frac{2}{\pi}\left[x^2\left(\frac{1}{n}\sin nx\right) - 2x\left(-\frac{1}{n^2}\cos nx\right) + 2\left(-\frac{1}{n^3}\sin nx\right)\right]_0^{\pi}$$

$$= \frac{2}{\pi}\left[\frac{2\pi}{n^2}(-1)^n\right] = \frac{4(-1)^n}{n^2}$$

$\therefore$ Fourier series for y is

$$y = \frac{\pi^2}{3} + 4\sum_{n=1}^{\infty}\frac{(-1)^n}{n^2}\cos nx; \quad -\pi < x < \pi$$

Let $\bar{y}$ be root mean square value of y.

Then, $$\bar{y}^2 = \frac{1}{2\pi}\int_{-\pi}^{\pi} y^2 dx = \frac{1}{\pi}\int_0^{\pi} y^2 dx = \frac{1}{2}\left[\frac{a_0^2}{2} + \sum_{n=1}^{\infty} a_n^2\right] \quad \text{(by Parseval's formula)}$$

$$= \frac{1}{2}\left[\frac{2\pi^4}{9} + \sum_{n=1}^{\infty}\frac{16}{n^4}\right] \tag{1}$$

Also $$y = x^2, \quad -\pi < x < \pi$$

then, $$\bar{y}^2 = \frac{1}{2\pi}\int_{-\pi}^{\pi} x^4 dx = \frac{1}{\pi}\int_0^{\pi} x^4 dx = \frac{1}{\pi}\left(\frac{x^5}{5}\right)_0^{\pi} = \frac{\pi^4}{5} \tag{2}$$

Equating two values of $\bar{y}^2$ from (1) and (2)

$$\frac{\pi^4}{5} = \frac{1}{2}\left[\frac{2\pi^4}{9} + \sum_{n=1}^{\infty}\frac{16}{n^4}\right]$$

$$\therefore \quad 16\sum_{n=1}^{\infty}\frac{1}{n^4} = \frac{2\pi^4}{5} - 2\frac{\pi^4}{9} = \frac{8\pi^4}{45}$$

$$\therefore \quad \frac{1}{1^4} + \frac{1}{2^4} + \frac{1}{3^4} + \frac{1}{4^4} + \cdots = \frac{\pi^4}{90}$$

Example 3.40: Prove that for $0 < x < \pi$

(a) $x(\pi - x) = \frac{\pi^2}{6} - \left[\frac{\cos 2x}{1^2} + \frac{\cos 4x}{2^2} + \frac{\cos 6x}{3^2} + \cdots\right]$

(b) $x(\pi - x) = \frac{8}{\pi}\left[\frac{\sin x}{1^3} + \frac{\sin 3x}{3^3} + \frac{\sin 5x}{5^3} + \cdots\right]$

Deduce form (a) and (b) respectively that

$$\sum_{n=1}^{\infty}\frac{1}{n^4}=\frac{\pi^4}{90} \text{ and } \sum_{n=1}^{\infty}\frac{1}{n^6}=\frac{\pi^6}{945}$$

Solution: (a) Let $f(x)=x(\pi-x),\quad 0<x<\pi$

By half-range cosine series of $f(x)$, we have

$$f(x)=x(\pi-x)=\frac{a_0}{2}+\sum_{n=1}^{\infty}a_n\cos nx,\ 0<x<\pi$$

where $$a_0=\frac{2}{\pi}\int_0^{\pi}\left(\pi x-x^2\right)dx=\frac{2}{\pi}\left(\frac{\pi x^2}{2}-\frac{x^3}{3}\right)_0^{\pi}=\frac{2}{\pi}\left(\frac{\pi^3}{2}-\frac{\pi^3}{3}\right)=\frac{\pi^2}{3}$$

and $$a_n=\frac{2}{\pi}\int_0^{\pi}\left(\pi x-x^2\right)\cos nx\,dx$$

$$=\frac{2}{\pi}\left[\left(\pi x-x^2\right)\left(\frac{1}{n}\sin nx\right)-(\pi-2x)\left(-\frac{1}{n^2}\cos nx\right)+(-2)\left(-\frac{1}{n^3}\sin nx\right)\right]_0^{\pi}$$

$$=\frac{2}{\pi}\left[-\frac{\pi}{n^2}(-1)^n-\frac{\pi}{n^2}\right]=\frac{2}{n^2}\left[(-1)^{n+1}-1\right]$$

$\therefore$ $$a_{2n}=-\frac{4}{(2n)^2}=-\frac{1}{n^2},\ a_{2n-1}=0;\quad n=1,2,3,...$$

$\therefore$ Half-range cosine series of $f(x)$ in $(0,\pi)$ is

$$x(\pi-x)=\frac{\pi^2}{6}+\sum_{n=1}^{\infty}\left(-\frac{1}{n^2}\right)\cos 2nx$$

$$=\frac{\pi^2}{6}-\left[\frac{\cos 2x}{1^2}+\frac{\cos 4x}{2^2}+\frac{\cos 6x}{3^2}+\cdots\right]$$

By Parseval's formula

$$\frac{2}{\pi}\int_0^{\pi}\left[f(x)\right]^2dx=\frac{a_0^2}{2}+\sum_{n=1}^{\infty}a_n^2=\frac{a_0^2}{2}+\sum_{n=1}^{\infty}a_{2n}^2$$

$\therefore$ $$\frac{2}{\pi}\int_0^{\pi}\left(\pi x-x^2\right)^2dx=\frac{\pi^4}{18}+\sum_{n=1}^{\infty}\frac{1}{n^4}$$

$\therefore$ $$\frac{2}{\pi}\int_0^{\pi}\left(\pi^2x^2-2\pi x^3+x^4\right)dx=\frac{\pi^4}{18}+\sum_{n=1}^{\infty}\frac{1}{n^4}$$

$\therefore$ $$\sum_{n=1}^{\infty}\frac{1}{n^4}=-\frac{\pi^4}{18}+\frac{2}{\pi}\left(\frac{\pi^2x^3}{3}-\frac{\pi x^4}{2}+\frac{x^5}{5}\right)_0^{\pi}$$

$$= -\frac{\pi^4}{18} + \frac{2}{\pi}\left(\frac{\pi^5}{3} - \frac{\pi^5}{2} + \frac{\pi^5}{5}\right) = -\frac{\pi^4}{18} + \frac{2\pi^4}{30}(10-15+6)$$

$$= \pi^4\left(\frac{-5+6}{90}\right) = \frac{\pi^4}{90}$$

(b) $f(x) = x(\pi - x);\ \ 0 < x < \pi$

By half-range sine series of $f(x)$, we have

$$f(x) = x(\pi - x) = \sum_{n=1}^{\infty} b_n \sin nx,\ 0 < x < \pi$$

where
$$b_n = \frac{2}{\pi}\int_0^{\pi}(\pi x - x^2)\sin nx\, dx$$

$$= \frac{2}{\pi}\left[(\pi x - x^2)\left(\frac{-1}{n}\cos nx\right) - (\pi - 2x)\left(-\frac{1}{n^2}\sin nx\right) + (-2)\left(\frac{1}{n^3}\cos nx\right)\right]_0^{\pi}$$

$$= \frac{2}{\pi}\cdot\frac{2}{n^3}\left[1 - (-1)^n\right]$$

$$\therefore \qquad b_{2n} = 0,\ b_{2n-1} = \frac{8}{\pi(2n-1)^3};\ n = 1, 2, 3, \ldots$$

$$\therefore \qquad x(\pi - x) = \frac{8}{\pi}\sum_{n=1}^{\infty}\frac{\sin(2n-1)x}{(2n-1)^3}$$

$$= \frac{8}{\pi}\left[\frac{\sin x}{1^3} + \frac{\sin 3x}{3^3} + \frac{\sin 5x}{5^3} + \cdots\right]$$

By Parseval's formula

$$\frac{2}{\pi}\int_0^{\pi} x^2(\pi - x)^2\, dx = \sum_{n=1}^{\infty} b_n^2 = \sum_{n=1}^{\infty} b_{2n-1}^2$$

$$\therefore \qquad \frac{64}{\pi^2}\sum_{n=1}^{\infty}\frac{1}{(2n-1)^6} = \frac{2}{\pi}\int_0^{\pi}\left(x^4 - 2\pi x^3 + \pi^2 x^2\right)dx$$

$$= \frac{2}{\pi}\left[\frac{x^5}{5} - 2\pi\frac{x^4}{4} + \frac{\pi^2 x^3}{3}\right]_0^{\pi}$$

$$= \frac{2}{\pi}\left[\frac{\pi^5}{5} - \frac{\pi^5}{2} + \frac{\pi^5}{3}\right]$$

$$= 2\pi^4\left[\frac{6-15+10}{30}\right] = \frac{\pi^4}{15}$$

$\therefore$
$$\frac{1}{1^6}+\frac{1}{3^6}+\frac{1}{5^6}+\cdots = \frac{\pi^6}{960}$$

Add $\frac{1}{2^6}+\frac{1}{4^6}+\frac{1}{6^6}+\cdots$ to both sides

$$\sum_{n=1}^{\infty}\frac{1}{n^6} = \frac{\pi^6}{960}+\left(\frac{1}{2^6}+\frac{1}{4^6}+\frac{1}{6^6}+\cdots\right)$$

$$= \frac{\pi^6}{960}+\frac{1}{64}\left(\frac{1}{1^6}+\frac{1}{2^6}+\frac{1}{3^6}+\cdots\right)$$

$$= \frac{\pi^6}{960}+\frac{1}{64}\sum_{n=1}^{\infty}\frac{1}{n^6}$$

$\Rightarrow$
$$\frac{63}{64}\sum_{n=1}^{6}\frac{1}{n^6} = \frac{\pi^6}{960}$$

$\therefore$
$$\sum_{n=1}^{6}\frac{1}{n^6} = \frac{64\pi^6}{63(960)} = \frac{\pi^6}{945}$$

Example 3.41: Find the complex form of Fourier series of $f(x) = e^{-x}$ in $-1 \le x < 1$, $f(x+2) = f(x)$.

Solution: Complex form of Fourier series of $f(x)$ is

$$f(x) = \sum_{n=-\infty}^{\infty} c_n e^{in\pi x}; \quad -1 \le x < 1$$

where
$$c_n = \frac{1}{2}\int_{-1}^{1} e^{-x}e^{-in\pi x}\,dx = -\frac{1}{2(1+i\,n\pi)}\left(e^{-x}e^{-in\pi x}\right)_{-1}^{1}$$

$$= \left(\frac{1-i\,n\pi}{1+n^2\pi^2}\right)\left[\frac{e(-1)^n - e^{-1}(-1)^n}{2}\right]$$

$$= \frac{(-1)^n(1-i\,n\pi)}{1+n^2\pi^2}\sinh 1$$

$\therefore$ Complex form of Fourier series of $f(x)$ is

$$f(x) = \sinh 1\sum_{n=-\infty}^{\infty}\frac{(-1)^n(1-i\,n\pi)}{1+n^2\pi^2}e^{in\pi x}; \ -1 \le x < 1$$

Example 3.42: Obtain the complex form of the Fourier series of the functions

(i) $f(x)=\begin{cases}0, & -\pi<x<0\\ 1, & 0\le x\le \pi\end{cases}$ (ii) $f(x)=\begin{cases}1, & -2\le x<1\\ 0, & 1\le x<2\end{cases}$

Solution: (i) Complex form of Fourier series of $f(x)$ is

$$f(x)\sim\sum_{n=-\infty}^{\infty}c_n e^{inx}$$

where
$$c_n=\frac{1}{2\pi}\int_{-\pi}^{\pi}f(x)e^{-inx}dx=\frac{1}{2\pi}\int_0^{\pi}e^{-inx}dx$$

$$=\frac{1}{-2i\pi n}\left(e^{-inx}\right)_0^{\pi};\ n\neq 0$$

$$=\frac{1}{i\pi n}\left(\frac{1-(-1)^n}{2}\right)$$

$\therefore$
$$c_n=\begin{cases}0 & ;\ n \text{ is even}\\ -\dfrac{i}{\pi n}; & n \text{ is odd}\end{cases}$$

$$c_0=\frac{1}{2\pi}\int_0^{\pi}dx=\frac{1}{2}$$

$\therefore$ Complex form of Fourier series of $f(x)$ is

$$f(x)\sim\frac{1}{2}-\frac{i}{\pi}\left[\left(\frac{e^{ix}}{1}+\frac{e^{-ix}}{-1}\right)+\left(\frac{e^{3ix}}{3}+\frac{e^{-3ix}}{-3}\right)+\left(\frac{e^{5ix}}{5}+\frac{e^{-5ix}}{-5}\right)+\cdots\right]$$

(ii) Complex form of Fourier series of $f(x)$ is

$$f(x)\sim\sum_{n=-\infty}^{\infty}c_n\, e^{\frac{in\pi x}{2}}$$

where
$$c_n=\frac{1}{4}\int_{-2}^{2}f(x)e^{-\frac{in\pi x}{2}}dx$$

$$=\frac{1}{4}\int_{-2}^{1}e^{-\frac{in\pi x}{2}}dx=\frac{i}{2n\pi}\left(e^{-i\frac{n\pi x}{2}}\right)_{-2}^{1};\ n\neq 0$$

$$=\frac{i}{2n\pi}\left[e^{-\frac{in\pi}{2}}-e^{in\pi}\right]$$

$$= \frac{e^{i\frac{n\pi}{4}}}{n\pi} \frac{\left[e^{i\frac{3n\pi}{4}} - e^{-i\frac{3n\pi}{4}}\right]}{2i}$$

$$= \frac{e^{i\frac{n\pi}{4}}}{n\pi} \sin\frac{3n\pi}{4}; \quad n \neq 0$$

$$c_0 = \frac{1}{4}\int_{-2}^{2} f(x)\,dx = \frac{1}{4}\int_{-2}^{1} dx = \frac{1}{4}(x)_{-2}^{1} = \frac{3}{4}$$

$\therefore$ Complex form of Fourier series of $f(x)$ is

$$f(x) \sim \frac{3}{4} + \sum_{\substack{n=-\infty \\ n\neq 0}}^{\infty} \frac{e^{i\frac{n\pi}{4}}}{n\pi} \sin\frac{3n\pi}{4} e^{i\frac{n\pi x}{2}}$$

Example 3.43: Find complex form of Fourier series of $f(x) = e^{ax}$ in the interval $(-\pi, \pi)$ where 'a' is a nonzero real constant.

Hence deduce that $\dfrac{\pi}{a\sinh a\pi} = \displaystyle\sum_{n=-\infty}^{\infty} \frac{(-1)^n}{a^2+n^2}$.

Solution: Complex form of Fourier series of $f(x)$ is

$$f(x) \sim \sum_{n=-\infty}^{\infty} c_n e^{inx}$$

where
$$c_n = \frac{1}{2\pi}\int_{-\pi}^{\pi} e^{ax} e^{-inx}\,dx = \frac{1}{2\pi(a-in)}\left[e^{ax} e^{-inx}\right]_{-\pi}^{\pi}$$

$$= \frac{a+in}{2\pi(a^2+n^2)}\left[e^{a\pi}(-1)^n - e^{-a\pi}(-1)^n\right]; \ a \neq 0$$

$$= \frac{(-1)^n (a+in)}{\pi(a^2+n^2)} \sinh(a\pi) \qquad (\because a \neq 0, \ \therefore a^2+n^2 \neq 0)$$

$\therefore$ Complex form of Fourier series of $f(x)$ is

$$f(x) \sim \frac{1}{\pi}\sinh(a\pi) \sum_{n=-\infty}^{\infty} \frac{(-1)^n (a+in)}{a^2+n^2} e^{inx}$$

Taking $x = 0$

$$f(0) = \frac{1}{\pi}\sinh(a\pi) \sum_{n=-\infty}^{\infty} \frac{(-1)^n (a+in)}{a^2+n^2}$$

But $$f(0)=e^0=1$$

$\therefore$ $$1=\frac{1}{\pi}\sinh(a\pi)\sum_{n=-\infty}^{\infty}\frac{(-1)^n(a+in)}{a^2+n^2}$$

or $$\frac{\pi}{\sinh a\pi}=\sum_{n=-\infty}^{\infty}\frac{(-1)^n(a+in)}{a^2+n^2}$$

Equate real parts

$$\frac{\pi}{\sinh a\pi}=\sum_{n=-\infty}^{\infty}\frac{(-1)^n a}{a^2+n^2}$$

$\therefore$ $$\frac{\pi}{a\sinh a\pi}=\sum_{n=-\infty}^{\infty}\frac{(-1)^n}{a^2+n^2}.$$

EXERCISE 3.3

1. By using the sine series for $f(x)=1$ in $0<x<\pi$ show that $\frac{\pi^2}{8}=1+\frac{1}{3^2}+\frac{1}{5^2}+\frac{1}{7^2}+\cdots$
2. From the Fourier sine series for $f(x)=x(\pi-x)$ in $0<x<\pi$. Prove that $1+\frac{1}{3^6}+\frac{1}{5^6}+\frac{1}{7^6}+\ldots=\frac{\pi^6}{960}.$
3. If $f(x)=\begin{cases}\pi x & , 0<x<1\\ \pi(2-x), & 1<x<2\end{cases}$ using half-range cosine series, show that $\frac{1}{1^4}+\frac{1}{3^4}+\frac{1}{5^4}+\cdots=\frac{\pi^4}{96}.$
4. Given the series $x=2\sum_{n=1}^{\infty}\frac{(-1)^{n+1}}{n}\sin nx,$ show that $\sum_{n=1}^{\infty}\frac{1}{n^2}=\frac{\pi^2}{6}.$
5. Find the complex form of Fourier series of $f(x)=2x$ in $0<x<2\pi$.
6. Find the complex form of Fourier series of $f(x)=\sin ax$ in the interval $(-\pi,\pi)$ where a is not an integer.
7. Find the complex form of Fourier series of $f(x)=e^{2x}$, $0<x<2$.
8. Find the complex form of Fourier series of $f(x)=\cosh ax$ in $-\pi<x<\pi$ where a is not an integer.
9. Find the complex form of Fourier series of $f(x)=\begin{cases}-1, & -1<x<0\\ 1, & 0<x<1\end{cases}$
10. Obtain the complex form of Fourier series of $f(x)=e^x$ for $-\pi<x<\pi$.
11. Find the complex form of Fourier series of $f(x)=\begin{cases}x^2, & 0\le x<1\\ 1, & 1<x<2\end{cases}$
12. Find the complex form of Fourier series of $f(x)=\sinh ax$ in the interval $(-l,l)$.
13. Find the complex form of the Fourier series of $f(x)=\cos ax$ in $-\pi<x<\pi$ where a is not an integer.
14. Find the complex form of Fourier series of $f(x)=\cosh 2x+\sinh 2x$ in the interval $(-5,5)$.
15. Find complex form of Fourier series of $f(x)=\begin{cases}0 & , -\pi/2<x<0\\ \cos x, & 0<x<\pi/2\end{cases}$

ANSWERS 3.3

5. $f(x)=2\pi+2i\sum_{\substack{n=-\infty\\n\neq 0}}^{\infty}\frac{1}{n}e^{inx}$

6. $f(x)=\frac{i\sin a\pi}{\pi}\sum_{n=-\infty}^{\infty}\frac{(-1)^n n}{n^2-a^2}e^{inx}$

7. $f(x)=\frac{(e^4-1)}{2}\sum_{n=-\infty}^{\infty}\frac{2+in\pi}{4+n^2\pi^2}e^{in\pi x}$

8. $f(x)=\frac{a\sinh a\pi}{\pi}\sum_{n=-\infty}^{\infty}\frac{(-1)^n}{a^2+n^2}e^{inx}$

9. $f(x)\sim\frac{2}{i\pi}\left[\left(\frac{e^{i\pi x}}{1}+\frac{e^{-i\pi x}}{-1}\right)+\left(\frac{e^{3i\pi x}}{3}+\frac{e^{-3i\pi x}}{-3}\right)+\left(\frac{e^{5i\pi x}}{5}+\frac{e^{-5i\pi x}}{-5}\right)+\cdots\right]$

10. $f(x)=\frac{\sinh\pi}{\pi}\sum_{n=-\infty}^{\infty}\frac{(-1)^n(1+in)}{(1+n^2)}e^{inx}$

11. $f(x)\sim\frac{2}{3}+\frac{1}{2}\sum_{\substack{n=-\infty\\n\neq 0}}^{\infty}\left[\frac{i}{n\pi}+\frac{2(-1)^n}{n^2\pi^2}-\frac{2i}{n^3\pi^3}\{(-1)^n-1\}\right]e^{in\pi x}$

12. $f(x)=i\pi\sinh al\sum_{n=-\infty}^{\infty}\frac{(-1)^n n}{a^2l^2+n^2\pi^2}e^{\frac{in\pi x}{l}}$

13. $f(x)=\frac{a\sin a\pi}{\pi}\sum_{n=-\infty}^{\infty}\frac{(-1)^n}{a^2-n^2}e^{inx}$

14. $f(x)=\sinh 10\sum_{n=-\infty}^{\infty}\frac{(-1)^n(10+in\pi)}{(100+n^2\pi^2)}e^{in\pi x/5}$

15. $f(x)\sim\frac{1}{\pi}\sum_{n=-\infty}^{\infty}\frac{1}{(1-4n^2)}\{(-1)^n+2in\}e^{i2nx}$

3.4 HARMONIC ANALYSIS

If $f(x)$ is a period function with period $2l$ defined in $(0, 2l)$ then it can be expanded in Fourier series

$$f(x)=\frac{a_0}{2}+\sum_{n=1}^{\infty}a_n\cos\frac{n\pi x}{l}+\sum_{n=1}^{\infty}b_n\sin\frac{n\pi x}{l} \tag{3.20}$$

where

$$a_0=\frac{1}{l}\int_0^{2l}f(x)\,dx \tag{3.21}$$

$$a_n = \frac{1}{l}\int_0^{2l} f(x)\cos\frac{n\pi x}{l}\,dx \tag{3.22}$$

$$b_n = \frac{1}{l}\int_0^{2l} f(x)\sin\frac{n\pi x}{l}\,dx \tag{3.23}$$

Further, if $f(x)$ is given in $(0, l)$, we can have its even and odd extensions both of which are periodic functions of period $2l$. Even extension can be expanded in half-range Fourier cosine series

$$f(x) = \frac{a_0}{2} + \sum_{n=1}^{\infty} a_n \cos\frac{n\pi x}{l} \tag{3.24}$$

where

$$a_0 = \frac{2}{l}\int_0^{l} f(x)\,dx \tag{3.25}$$

$$a_n = \frac{2}{l}\int_0^{l} f(x)\cos\frac{n\pi x}{l}\,dx \tag{3.26}$$

and odd extension can be expanded in half-range Fourier sine series

$$f(x) = \sum_{n=1}^{\infty} b_n \sin\frac{n\pi x}{l} \tag{3.27}$$

where

$$b_n = \frac{2}{l}\int_0^{l} f(x)\sin\frac{n\pi x}{l}\,dx \tag{3.28}$$

If $f(x)$ is given an explicit function of x then Fourier coefficients in (3.21) to (3.23) or (3.25) to (3.26) or (3.28) can be evaluated by integrals as mentioned in these equations. In practice, however, the function is often given not by a formula but by graph or by table of corresponding values. In such cases, integrals cannot be evaluated and instead, the following alternate forms of these integrals are employed. As, for any function $F(x)$, $\frac{1}{b-a}\int_a^b F(x)\,dx$ is mean value of $y = F(x)$ over range (a, b), so equations (3.21) to (3.23) give

$$a_0 = 2\frac{1}{2l}\int_0^{2l} f(x)\,dx = 2\ [\text{mean value of } f(x) \text{ in } (0, 2l)]$$

$$a_n = 2\frac{1}{2l}\int_0^{2l} f(x)\cos\frac{n\pi x}{l}\,dx = 2\ \left[\text{mean value of } f(x)\cos\frac{n\pi x}{l} \text{ in } (0, 2l)\right]$$

$$b_n = 2\frac{1}{2l}\int_0^{2l} f(x)\sin\frac{n\pi x}{l}\,dx = 2\ \left[\text{mean value of } f(x)\sin\frac{n\pi x}{l} \text{ in } (0, 2l)\right]$$

and equations (3.25) to (3.26) give

$$a_0 = 2\frac{1}{l}\int_0^{l} f(x)\,dx = 2\ [\text{mean value of } f(x) \text{ in } (0, l)]$$

$$a_n = 2\frac{1}{l}\int_0^{l} f(x)\cos\frac{n\pi x}{l}\,dx = 2\ \left[\text{mean value of } f(x)\cos\frac{n\pi x}{l} \text{ in } (0, l)\right]$$

and equation (3.28) gives

$$b_n = 2\frac{1}{l}\int_0^{l} f(x)\sin\frac{n\pi x}{l}\,dx = 2\ \left[\text{mean value of } f(x)\sin\frac{n\pi x}{l} \text{ in } (0, l)\right]$$

Thus, in each case, Fourier coefficient is twice the mean value of function to be integrated in that interval. Now, the mean value in discrete data can be obtained by dividing the sum by number of values.

In (3.20), the term $a_1 \cos\frac{\pi x}{l} + b_1 \sin\frac{\pi x}{l}$ is called the fundamental or first harmonic. The term $a_2 \cos\frac{2\pi x}{l} + b_2 \sin\frac{2\pi x}{l}$ is called the second harmonic, the term $a_3 \cos\frac{3\pi x}{l} + b_3 \sin\frac{3\pi x}{l}$ is called third harmonic and so on.

Amplitude of first harmonic $= \sqrt{a_1^2 + b_1^2}$

Amplitude of second harmonic $= \sqrt{a_2^2 + b_2^2}$

Amplitude of third harmonic $= \sqrt{a_3^2 + b_3^2}$

and so on.

Remark 3.3:

(i) If data for function $f(x)$ at equidistant points is

x	x_0	x_1	x_2	$\cdots$	x_{k-1}	x_k
$f(x)$	y_0	y_1	y_2	$\cdots$	y_{k-1}	y_k

such that $y_k = y_0$ then $x_k = x_0 + 2l$ where $2l$ is period of $f(x)$. For finding mean of $f(x)$, $f(x)\cos\frac{n\pi x}{l}, f(x)\sin\frac{n\pi x}{l}$ in the period $2l$ the entry $f(x_k) = y_k$ should not be taken as we take $f(x_0) = y_0$. The result will be same if we obtain the integrals by trapezoidal rule.

(ii) If data for function $f(x)$ at equidistant points is

x	0	x_1	x_2	$\cdots$	x_{k-1}	x_k
$f(x)$	y_0	y_1	y_2	$\cdots$	y_{k-1}	y_k

and we are to find half-range sine-series, then for odd extension of $f(x)$, we must have $y_0 = 0$, otherwise odd extension cannot be possible. Further, if $y_k = 0$ then $x_k = l$ $(\because f(-l) = f(l)$ and $f(-l) = -f(l) \Rightarrow f(l) = 0)$ and the entry y_k should not be taken but if $y_k \neq 0$ then $x_k \neq l$ and $y_{k+1} = 0$ at $x_{k+1} = l$ and hence entry y_k should be taken.

(iii) If data for function $f(x)$ at equidistant points is

x	0	x_1	x_2	$\cdots$	x_{k-1}	x_k
$f(x)$	y_0	y_1	y_2	$\cdots$	y_{k-1}	y_k

and we are to find half-range cosine series. Then for even extension of $f(x)$, y_0 may or may not be zero. Further, for even extension $f(-l) = f(l)$ and hence $f(l)$ may or may not be zero. Hence y_k may or may not be zero, therefore, we have $x_k = l$. Thus, for even extension by trapezoidal rule

$$a_0 = \frac{2}{l}\int_0^l f(x)\,dx = \frac{2}{l}\left[\int_0^{x_1} f(x)\,dx + \int_{x_1}^{x_2} f(x)\,dx + \cdots + \int_{x_{k-1}}^{x_k} f(x)\,dx\right]$$

$$= \frac{2}{l}\left[(x_1 - 0)\frac{y_0 + y_1}{2} + (x_2 - x_1)\frac{y_1 + y_2}{2} + \cdots + (x_k - x_{k-1})\frac{y_{k-1} + y_k}{2}\right]$$

$$= \frac{2}{l} \cdot \frac{l}{k} \left[\frac{y_0 + y_1}{2} + \frac{y_1 + y_2}{2} + \cdots + \frac{y_{k-1} + y_k}{2} \right]$$

$$= \frac{2}{k} \left[\frac{y_0 + y_k}{2} + y_1 + y_2 + \cdots y_{k-1} \right]$$

Similarly,

$$a_n = \frac{2}{k} \left[\frac{y_0 + y_k \cos \frac{n\pi x_k}{l}}{2} + y_1 \cos \frac{n\pi x_1}{l} + y_2 \cos \frac{n\pi x_2}{l} + \cdots y_{k-1} \cos \frac{n\pi x_{k-1}}{l} \right] \quad (\because x_0 = 0)$$

Example 3.44: Given that x is discrete function of θ, tabulated as follows:

θ	0°	30°	60°	90°	120°	150°	180°	210°	240°	270°	300°	330°
X	70	886	1293	1400	1307	814	–70	–886	–1293	–1400	–1307	–814

Express x as a Fourier series in θ up to third harmonics.

Solution:

$\theta°$	x	$\cos\theta$	$\sin\theta$	$\cos 2\theta$	$\sin 2\theta$	$\cos 3\theta$	$\sin 3\theta$
0	70	1	0	1	0	1	0
30	886	$\sqrt{3}/2$	0.5	0.5	$\sqrt{3}/2$	0	1
60	1293	0.5	$\sqrt{3}/2$	–0.5	$\sqrt{3}/2$	–1	0
90	1400	0	1	–1	0	0	–1
120	1307	–0.5	$\sqrt{3}/2$	–0.5	$-\sqrt{3}/2$	1	0
150	814	$-\sqrt{3}/2$	0.5	0.5	$-\sqrt{3}/2$	0	1
180	–70	–1	0	1	0	–1	0
210	–886	$-\sqrt{3}/2$	–0.5	0.5	$\sqrt{3}/2$	0	–1
240	–1293	–0.5	$-\sqrt{3}/2$	–0.5	$\sqrt{3}/2$	1	0
270	–1400	0	–1	–1	0	0	1
300	–1307	0.5	$-\sqrt{3}/2$	–0.5	$-\sqrt{3}/2$	–1	0
330	–814	$\sqrt{3}/2$	–0.5	0.5	$-\sqrt{3}/2$	0	–1
	0						

$$a_0 = \frac{1}{6}\sum x = 0$$

$$a_1 = \frac{1}{6}\sum x\cos\theta = \frac{1}{6}\left[2\left\{70 + \frac{\sqrt{3}}{2}(886-814) + 0.5(1293-1307)\right\}\right]$$

$$= \frac{1}{3}\left[70 + \sqrt{3}(36) - 7\right] = 42$$

$$b_1 = \frac{1}{6}\sum x\sin\theta = \frac{1}{6}\left[2\left\{1400 + 0.5(886+814) + \frac{\sqrt{3}}{2}(1293+1307)\right\}\right]$$

$$= \frac{1}{3}\left[1400 + 850 + \sqrt{3}(1300)\right] = 1501$$

$$a_2 = \frac{1}{6}\sum x\cos 2\theta = 0$$

$$b_2 = \frac{1}{6}\sum x\sin 2\theta = 0$$

$$a_3 = \frac{1}{6}\sum x\cos 3\theta = \frac{1}{6}\left[2(70-1293+1307)\right] = 28$$

$$b_3 = \frac{1}{6}\sum x\sin 3\theta = \frac{1}{6}\left[2(886-1400+814)\right] = 100$$

$$x = \frac{a_0}{2} + a_1\cos\theta + b_1\sin\theta + a_2\cos 2\theta + b_2\sin 2\theta + a_3\cos 3\theta + b_3\sin 3\theta + \cdots$$

∴ up to three harmonics

$$x \cong 42\cos\theta + 1501\sin\theta + 28\cos 3\theta + 100\sin 3\theta$$

Example 3.45: Find first three harmonics for the function $f(\theta)$ given by the following table

$\theta°$	0	60	120	180	240	300	360
$f(\theta)$	0.8	0.6	0.4	0.7	0.9	1.1	0.8

Solution:

$\theta°$	$f(\theta)$	$\cos\theta$	$\sin\theta$	$\cos 2\theta$	$\sin 2\theta$	$\cos 3\theta$	$\sin 3\theta$
0	0.8	1	0	1	0	1	0
60	0.6	0.5	$\sqrt{3}/2$	–0.5	$\sqrt{3}/2$	–1	0
120	0.4	–0.5	$\sqrt{3}/2$	–0.5	$-\sqrt{3}/2$	1	0
180	0.7	–1	0	1	0	–1	0
240	0.9	–0.5	$-\sqrt{3}/2$	–0.5	$\sqrt{3}/2$	1	0
300	1.1	0.5	$-\sqrt{3}/2$	–0.5	$-\sqrt{3}/2$	–1	0

$$a_1 = \frac{2}{6}\sum f(\theta)\cos\theta = \frac{1}{3}\left[0.8-0.7+0.5\left(0.6-0.4-0.9+1.1\right)\right] = 0.1$$

$$b_1 = \frac{2}{6}\sum f(\theta)\sin\theta = \frac{1}{3}\left[\frac{\sqrt{3}}{2}(0.6+0.4-0.9-1.1)\right] = -0.3$$

$$a_2 = \frac{2}{6}\sum f(\theta)\cos 2\theta = \frac{1}{3}\left[0.8+0.7-0.5\left(0.6+0.4+0.9+1.1\right)\right] = 0$$

$$b_2 = \frac{2}{6}\sum f(\theta)\sin 2\theta = \frac{1}{3}\cdot\frac{\sqrt{3}}{2}(0.6-0.4+0.9-1.1) = 0$$

$$a_3 = \frac{2}{6}\sum f(\theta)\cos 3\theta = \frac{1}{3}[0.8-0.6+0.4-0.7+0.9-1.1] = -0.1$$

$$b_3 = \frac{2}{6}\sum f(\theta)\sin 3\theta = 0$$

$\therefore$ first three harmonics are

first harmonic $= a_1 \cos\theta + b_1 \sin\theta = 0.1 \cos\theta - 0.3 \sin\theta$

second harmonic $= a_2 \cos2\theta + b_2 \sin2\theta = 0$

third harmonic $= a_3 \cos3\theta + b_3 \sin3\theta = -0.1 \cos3\theta$

Example 3.46: Following values of y give the displacement of a certain machine part for the rotation x of the fly wheel

x	0	$\frac{\pi}{3}$	$\frac{2\pi}{3}$	π	$\frac{4\pi}{3}$	$\frac{5\pi}{3}$	2π
y	1.98	2.15	2.77	0.22	–0.31	1.43	1.98

Express y in Fourier series up to the third harmonics.

Solution:

x	y	cos x	sin x	cos $2x$	sin $2x$	cos $3x$	sin $3x$
0	1.98	1	0	1	0	1	0
$\frac{\pi}{3}$	2.15	0.5	$\sqrt{3}/2$	–0.5	$\sqrt{3}/2$	–1	0
$\frac{2\pi}{3}$	2.77	–0.5	$\sqrt{3}/2$	–0.5	$-\sqrt{3}/2$	1	0
π	0.22	–1	0	1	0	–1	0
$\frac{4\pi}{3}$	–0.31	–0.5	$-\sqrt{3}/2$	–0.5	$\sqrt{3}/2$	1	0
$\frac{5\pi}{3}$	1.43	0.5	$-\sqrt{3}/2$	–0.5	$-\sqrt{3}/2$	–1	0
	8.24						

$$\frac{1}{2}a_0 = \frac{1}{6}\sum y = \frac{1}{6}(8.24) = 1.37$$

$$a_1 = \frac{2}{6}\sum y\cos x = \frac{1}{3}\left[1.98 - 0.22 + 0.5(2.15 - 2.77 + 0.31 + 1.43)\right] = 0.77$$

$$b_1 = \frac{2}{6}\sum y\sin x = \frac{1}{3}\cdot\frac{\sqrt{3}}{2}[2.15 + 2.77 + 0.31 - 1.43] = \frac{\sqrt{3}}{6}(3.8) = 1.10$$

$$a_2 = \frac{2}{6}\sum y\cos 2x = \frac{1}{3}\left[1.98 + 0.22 - 0.5(2.15 + 2.77 - 0.31 + 1.43)\right] = -0.27$$

$$b_2 = \frac{2}{6}\sum y\sin 2x = \frac{1}{3}\frac{\sqrt{3}}{2}[2.15 - 2.77 - 0.31 - 1.43] = \frac{\sqrt{3}}{6}(-2.36) = -0.68$$

$$a_3 = \frac{2}{6}\sum y\cos 3x = \frac{1}{3}[1.98 - 2.15 + 2.77 - 0.22 - 0.31 - 1.43] = 0.21$$

$$b_3 = \frac{2}{6}\sum y\sin 3x = 0$$

y in Fourier series up to the third harmonics is

$$y = \frac{a_0}{2} + a_1\cos x + b_1\sin x + a_2\cos 2x + b_2\sin 2x + a_3\cos 3x + b_3\sin 3x$$

$$= 1.37 + (0.77)\cos x + (1.10)\sin x - (0.27)\cos 2x - (0.68)\sin 2x + (0.21)\cos 3x$$

Example 3.47: The turning moment T units of the crank shaft of a steam engine is given for a series of values of the crank angle θ in degrees

θ	0°	30°	60°	90°	120°	150°
T	0	5224	8097	7850	5499	2626

Find the first four terms in a series of sines to represent T. Also calculate T when $\theta = 75°$.

Solution:

$\theta°$	T	$\sin\theta$	$\sin 2\theta$	$\sin 3\theta$	$\sin 4\theta$
0	0	0	0	0	0
30	5224	0.5	$\sqrt{3}/2$	1	$\sqrt{3}/2$
60	8097	$\sqrt{3}/2$	$\sqrt{3}/2$	0	$-\sqrt{3}/2$
90	7850	1	0	−1	0
120	5499	$\sqrt{3}/2$	$-\sqrt{3}/2$	0	$\sqrt{3}/2$
150	2626	0.5	$-\sqrt{3}/2$	1	$-\sqrt{3}/2$

$$b_1 = \frac{2}{6}\sum T\sin\theta = \frac{1}{3}\left[\frac{\sqrt{3}}{2}(8097+5499)+7850+0.5(5224+2626)\right]$$

$$= \frac{1}{3}\left[\sqrt{3}(6798)+7850+3925\right] = 7850$$

$$b_2 = \frac{2}{6}\sum T\sin 2\theta = \frac{1}{3}\frac{\sqrt{3}}{2}[5224+8097-5499-2626] = \frac{\sqrt{3}}{6}(5196) = 1500$$

$$b_3 = \frac{2}{6}\sum T\sin 3\theta = \frac{1}{3}[5224-7850+2626] = 0$$

$$b_4 = \frac{2}{6}\sum T\sin 4\theta = \frac{1}{3}\frac{\sqrt{3}}{2}[5224-8097+5499-2626] = 0$$

∴ up to four terms

$$T = 7850\sin\theta + 1500\sin 2\theta + 0\sin 3\theta + 0\sin 4\theta + \cdots$$

$$T(75°) = 7850\sin 75° + 1500\sin 150° = \frac{7850(\sqrt{3}+1)}{2\sqrt{2}} + \frac{1500}{2} = 8332$$

Example 3.48: Obtain the constant term and the coefficients of the first two sine and first two cosine terms in the Fourier expansion of $y(x)$ tabulated below

x	0	1	2	3	4	5
y	9	18	24	28	26	20

Find also the amplitude of the first harmonic.

Solution:

x	y	$\frac{\pi x}{3}=\theta$	cos θ	sin θ	cos 2θ	sin 2θ
0	9	0	1	0	1	0
1	18	$\frac{\pi}{3}$	0.5	$\sqrt{3}/2$	–0.5	$\sqrt{3}/2$
2	24	$\frac{2\pi}{3}$	–0.5	$\sqrt{3}/2$	–0.5	$-\sqrt{3}/2$
3	28	π	–1	0	1	0
4	26	$\frac{4\pi}{3}$	–0.5	$-\sqrt{3}/2$	–0.5	$\sqrt{3}/2$
5	20	$\frac{5\pi}{3}$	0.5	$-\sqrt{3}/2$	–0.5	$-\sqrt{3}/2$
	125					

$$\frac{1}{2}a_0 = \frac{2}{12}\sum y = \frac{1}{6}(125) = 20.83$$

$$a_1 = \frac{2}{6}\sum y\cos\frac{\pi x}{3} = \frac{1}{3}\left[9-28+0.5(18-24-26+20)\right] = -8.33$$

$$b_1 = \frac{2}{6}\sum y\sin\frac{\pi x}{3} = \frac{1}{3}\frac{\sqrt{3}}{2}[18+24-26-20] = -1.15$$

$$a_2 = \frac{2}{6}\sum y\cos\frac{2\pi x}{3} = \frac{1}{3}\left[9+28-0.5(18+24+26+20)\right] = -2.33$$

$$b_2 = \frac{2}{6}\sum y\sin\frac{2\pi x}{3} = \frac{1}{3}\frac{\sqrt{3}}{2}[18-24+26-20] = 0$$

$\therefore$ Constant terms $= \frac{a_0}{2} = 20.83$

Coeff. of $\sin\frac{\pi x}{3} = b_1 = -1.15$, Coeff. of $\sin\frac{2\pi x}{3} = b_2 = 0$

Coeff. of $\cos\frac{\pi x}{3} = a_1 = -8.33$, Coeff. of $\cos\frac{2\pi x}{3} = -2.33$

Amplitude of first harmonic $= \sqrt{a_1^2 + b_1^2} = \sqrt{(-8.33)^2 + (-1.15)^2} = 8.41$

Example 3.49: A function $f(x)$ is defined as

$$f(x) = \begin{cases} x & ; \quad 0 \le x \le 1 \\ 2 - x ; & 1 \le x \le 2 \end{cases}$$

Using 12 ordinates, show that an approximate Fourier series for $f(x)$ is given by

$$f(x) = 0.5 - 0.415\cos\pi x - 0.056\cos 3\pi x - 0.030\cos 5\pi x$$

Solution: We have $\quad f(x+2) = 2 - (x+2);\ 1 \le x+2 \le 2$

But $\quad f(x+2) = f(x)$

$\therefore \quad f(x) = -x;\ -1 \le x \le 0$

$\therefore f(x)$ is an even function of x.

x	$f(x)$	$\cos \pi x$	$\cos 2\pi x$	$\cos 3\pi x$	$\cos 4\pi x$	$\cos 5\pi x$
0	0	1	1	1	1	1
1/6	1/6	$\sqrt{3}/2$	0.5	0	–0.5	$-\sqrt{3}/2$
1/3	1/3	0.5	–0.5	–1	–0.5	0.5
1/2	1/2	0	–1	0	1	0
2/3	2/3	–0.5	–0.5	1	–0.5	–0.5
5/6	5/6	$-\sqrt{3}/2$	0.5	0	–0.5	$\sqrt{3}/2$
1	1	–1	1	–1	1	–1
7/6	5/6	$-\sqrt{3}/2$	0.5	0	–0.5	$\sqrt{3}/2$
4/3	2/3	–0.5	–0.5	1	–0.5	–0.5
3/2	1/2	0	–1	0	1	0
5/3	1/3	0.5	–0.5	–1	–0.5	0.5
11/6	1/6	$\sqrt{3}/2$	0.5	0	–0.5	$-\sqrt{3}/2$

$$\frac{1}{2}a_0 = \frac{1}{2}\cdot\frac{2}{12}\sum f(x)$$
$$= \frac{1}{12}\left(\frac{1}{6}+\frac{2}{6}+\frac{3}{6}+\frac{4}{6}+\frac{5}{6}+\frac{6}{6}+\frac{5}{6}+\frac{4}{6}+\frac{3}{6}+\frac{2}{6}+\frac{1}{6}\right)$$
$$= \frac{1}{12}\left(\frac{36}{6}\right) = \frac{1}{2} = 0.5$$

$$a_1 = \frac{2}{12}\sum f(x)\cos\pi x$$
$$= \frac{1}{6}\left[\frac{\sqrt{3}}{2}\left(\frac{1}{6}-\frac{5}{6}-\frac{5}{6}+\frac{1}{6}\right)+0.5\left(\frac{1}{3}-\frac{2}{3}-\frac{2}{3}+\frac{1}{3}\right)-1\right]$$
$$= \frac{1}{6}\left[\frac{-2\sqrt{3}}{3}-\frac{1}{3}-1\right] = -0.4147$$

$$a_2 = \frac{2}{12}\sum f(x)\cos 2\pi x$$
$$= \frac{1}{6}\left[-\frac{1}{2}+1-\frac{1}{2}+0.5\left(\frac{1}{6}-\frac{1}{3}-\frac{2}{3}+\frac{5}{6}+\frac{5}{6}-\frac{2}{3}-\frac{1}{3}+\frac{1}{6}\right)\right]$$
$$= 0$$

$$a_3 = \frac{2}{12}\sum f(x)\cos 3\pi x = \frac{1}{6}\left[-\frac{1}{3}+\frac{2}{3}-1+\frac{2}{3}-\frac{1}{3}\right] = -\frac{1}{18} = -0.0556$$

$$a_4 = \frac{2}{12}\sum f(x)\cos 4\pi x$$
$$= \frac{1}{6}\left[\frac{1}{2}+1+\frac{1}{2}-0.5\left(\frac{1}{6}+\frac{1}{3}+\frac{2}{3}+\frac{5}{6}+\frac{5}{6}+\frac{2}{3}+\frac{1}{3}+\frac{1}{6}\right)\right]$$
$$= 0$$

$$a_5 = \frac{2}{12}\sum f(x)\cos 5\pi x$$
$$= \frac{1}{6}\left[\frac{\sqrt{3}}{2}\left(-\frac{1}{6}+\frac{5}{6}+\frac{5}{6}-\frac{1}{6}\right)+0.5\left(\frac{1}{3}-\frac{2}{3}-\frac{2}{3}+\frac{1}{3}\right)-1\right]$$
$$= \frac{1}{6}\left[\frac{2\sqrt{3}}{3}-\frac{1}{3}-1\right]$$
$$= -0.0298$$

∴ Approximate Fourier series is

$$f(x) = 0.5 - 0.415\cos\pi x - 0.056\cos 3\pi x - 0.030\cos 5\pi x$$

Example 3.50: Obtain the first three coefficients in the Fourier cosine series for y, where y is given in the following table:

x	0	1	2	3	4	5
y	4	8	15	7	6	2

Solution: Here, $l = 5$

$\therefore$ The Fourier cosine series for y is given by

$$y = \frac{a_0}{2} + a_1 \cos\frac{\pi x}{5} + a_2 \cos\frac{2\pi x}{5} + \cdots$$

x	y	$\frac{\pi x}{5}$	$\cos\frac{\pi x}{5}$	$\cos\frac{2\pi x}{5}$
0	4	0	1	1
1	8	$\frac{\pi}{5}$	0.8090	0.3090
2	15	$\frac{2\pi}{5}$	0.3090	–0.8090
3	7	$\frac{3\pi}{5}$	–0.3090	–0.8090
4	6	$\frac{4\pi}{5}$	–0.8090	0.3090
5	2	π	–1	1

$$\therefore \quad \frac{a_0}{2} = \frac{1}{2}\cdot\frac{2}{5}\left[\frac{4+2}{2} + (8+15+7+6)\right] = \frac{1}{5}(39) = 7.8$$

$$a_1 = \frac{2}{5}\left[\frac{4+(-2)}{2} + \{(8-6)0.8090 + (15-7)0.3090\}\right] = \frac{2}{5}(5.0900) = 2.036$$

$$a_2 = \frac{2}{5}\left[\frac{4+2}{2} + \{(8+6)0.3090 - (15+7)0.8090\}\right] = \frac{2}{5}(-10.4720) = -4.1888$$

$\therefore$ The first three coefficients in the Fourier cosine series are

$$\frac{a_0}{2} = 7.8,\ a_1 = 2.036 \text{ and } a_2 = -4.1888.$$

Example 3.51: Obtain the Fourier sine series for $f(x)$ containing three non-zero terms where $f(x)$ is given in the following table:

x	0	1	2	3	4	5
$f(x)$	0	10	15	8	5	3

Solution: Here, $l = 6$

$\therefore$ The Fourier sine series for $f(x)$ is given by

$$y = b_1 \sin\frac{\pi x}{6} + b_2 \sin\frac{2\pi x}{6} + b_3 \sin\frac{3\pi x}{6} + \cdots$$

x	f(x)	$\frac{\pi x}{6} = \theta$	sin θ	sin 2θ	sin 3θ
0	0	0	0	0	0
1	10	$\frac{\pi}{6}$	$1/2$	$\sqrt{3}/2$	1
2	15	$\frac{\pi}{3}$	$\sqrt{3}/2$	$\sqrt{3}/2$	0
3	8	$\frac{\pi}{2}$	1	0	-1
4	5	$\frac{2\pi}{3}$	$\sqrt{3}/2$	$-\sqrt{3}/2$	0
5	3	$\frac{5\pi}{6}$	$1/2$	$-\sqrt{3}/2$	1

$\therefore$
$$b_1 = \frac{2}{6}\sum f(x)\sin\theta = \frac{1}{3}\left[\left(5+8+\frac{3}{2}\right)+(15+5)\frac{\sqrt{3}}{2}\right]$$

$$= \frac{1}{3}\left[14.5+10\sqrt{3}\right] = 10.607$$

$$b_2 = \frac{2}{6}\sum f(x)\sin 2\theta = \frac{1}{3}\left[\frac{\sqrt{3}}{2}(10+15-5-3)\right] = 4.907$$

$$b_3 = \frac{2}{6}\sum f(x)\sin 3\theta = \frac{1}{3}[10-8+3] = \frac{5}{3} = 1.667$$

$\therefore$ Fourier sine series for $f(x)$ is

$$f(x) = 10.607\sin\frac{\pi x}{6} + 4.907\sin\frac{2\pi x}{6} + 1.667\sin\frac{3\pi x}{6} + \cdots$$

EXERCISE 3.4

1. Find the Fourier series as far as the second harmonic to represent the function given by the following table:

x	0	30°	60°	90°	120°	150°	180°	210°	240°	270°	300°	330°
$f(x)$	2.34	3.01	3.69	4.15	3.69	2.20	0.83	0.51	0.88	1.09	1.19	1.64

2. Given that y is discrete function of x, tabulated as follows:

$x°$	0	30	60	90	120	150	180	210	240	270	300	330	360
y	298	356	373	337	254	155	80	51	60	93	147	221	298

Express y as a Fourier series in x up to third harmonics.

3. The displacement y of a part of a mechanism is tabulated with corresponding angular movement x° of the crank. Express y as a Fourier series neglecting the harmonics above the third:

$x°$	0	30	60	90	120	150	180	210	240	270	300	330
y	1.80	1.10	0.30	0.16	0.50	1.30	2.16	1.25	1.30	1.52	1.76	2.00

4. Analyse the current i (amp) given by the table below into its constituent harmonics up to third.

$\theta°$	0	30	60	90	120	150	180	210	240	270	300	330
i	0	24	33.5	27.5	18.2	13	0	−24	−33.5	−27.5	−18.2	−13

5. A machine computes its cycle of operations every time as a certain pulley completes a revolution. The displacement $f(x)$ of a point on a certain portion of the machine is given in the table given below for 12 positions of the pulley, x being the angle in degree turned through by the pulley. Find a Fourier series to represent $f(x)$ for all values of x.

x	30°	60°	90°	120°	150°	180°	210°	240°	270°	300°	330°	360°
$f(x)$	7.976	8.026	7.204	5.676	3.674	1.764	0.552	0.262	0.904	2.492	4.736	6.824

6. In a machine the displacement y of a given point for a certain angle θ is observed as follows:

$\theta°$	0°	30°	60°	90°	120°	150°	180°	210°	240°	270°	300°	330°
y	7.9	8	7.2	5.6	3.6	1.7	0.5	0.2	0.9	2.5	4.7	6.8

Compute the coefficient of $\sin 2\theta$ in the Fourier series representing the above variation.

7. The following values of y give the displacement in inches of a certain machine part for the rotation x of the flywheel. Expand y in the form of a Fourier series:

x	0	$\frac{\pi}{6}$	$\frac{2\pi}{6}$	$\frac{3\pi}{6}$	$\frac{4\pi}{6}$	$\frac{5\pi}{6}$
y	0	9.2	14.4	17.8	17.3	11.7

8. The following table gives the variation of a periodic current over a period:

t (secs)	0	$\frac{T}{6}$	$\frac{T}{3}$	$\frac{T}{2}$	$\frac{2T}{3}$	$\frac{5T}{6}$	T
A(amp)	1.98	1.30	1.05	1.30	–0.88	–0.25	1.98

Show by harmonic analysis that there is a direct current part of 0.75 amp. in the variable current, and obtain the amplitude of the first harmonic.

9. Obtain the first three coefficients in the Fourier cosine series for y, where y is given by the following table:

x	0	1	2	3	4	5
y	4	10	15	8	5	3

ANSWERS 3.4

1. $f(x) = 2.10 + 0.56\cos x + 1.53\sin x - 0.52\cos 2x - 0.09\sin 2x \ldots$

2. $y = 202 + 107\cos x + 121\sin x - 39\cos 2x + 9\sin 2x + 2\cos 3x - \sin 3x$

3. $y = 1.26 + 0.04\cos x - 0.62\sin x + 0.53\cos 2x - 0.23\sin 2x - 0.1\cos 3x + 0.08\sin 3x$

4. $i = 5.7\cos\theta + 30.3\sin\theta - 5.1\cos 3\theta + 3.2\sin 3\theta$

5. $f(x) = 4.174 + 2.450\cos x + 3.160\sin x + 0.120\cos 2x$
$+ 0.034\sin 2x + 0.080\cos 3x + 0.010\sin x + \cdots$

6. –0.072

7. $y = 11.73 - 7.73\cos 2x - 1.56\sin 2x - 2.83\cos 4x + 0.115\sin 4x$

8. 1.07 amp.

9. 8.3, 2.6832 and –4.1888.

3.5 FOURIER INTEGRALS AND FOURIER TRANSFORMS

Fourier series are powerful tools in treating various problems involving periodic functions. Since, of course, many practical problems involve non-periodic functions, so we extend the method of Fourier series to Fourier integrals for such functions.

3.5.1 Fourier Series to Fourier Integrals

Theorem 3.1 Fourier Integral Theorem

If $f(x)$ is piecewise continuous in every finite interval and left-handed derivative, right-handed derivative at every point and $\int_{-\infty}^{\infty}|f(x)|\,dx$ exist then $f(x)$ can be represented by Fourier integral

$$f(x)=\int_0^{\infty}\left[A(\omega)\cos(\omega x)+B(\omega)\sin(\omega x)\right]d\omega,$$

where $$A(\omega)=\frac{1}{\pi}\int_{-\infty}^{\infty}f(t)\cos(\omega t)\,dt,\ B(\omega)=\frac{1}{\pi}\int_{-\infty}^{\infty}f(t)\sin(\omega t)\,dt.$$

Fourier integral converges to $f(x_0)$ if $f(x)$ is continuous at $x=x_0$ and converges to $\frac{1}{2}\left[f(x_0-0)+f(x_0+0)\right]$ if $f(x)$ is discontinuous at $x=x_0$.

Proof: Let $f_l(x)$ be a periodic function of period $2l$ defined in $(-l,l)$. Its Fourier series is

$$f_l(x)=\frac{a_0}{2}+\sum_{n=1}^{\infty}\left[a_n\cos(\omega_n x)+b_n\sin(\omega_n x)\right],\quad \text{where } \omega_n=\frac{n\pi}{l} \tag{3.29}$$

where $$a_0=\frac{1}{l}\int_{-l}^{l}f_l(t)\,dt,\ a_n=\frac{1}{l}\int_{-l}^{l}f_l(t)\cos(\omega_n t)\,dt,\ b_n=\frac{1}{l}\int_{-l}^{l}f_l(t)\sin(\omega_n t)\,dt.$$

Substituting values of a_0,a_n,b_n in (3.29)

$$f_l(x)=\frac{1}{2l}\int_{-l}^{l}f_l(t)\,dt+\frac{1}{l}\sum_{n=1}^{\infty}\left[\cos(\omega_n x)\int_{-l}^{l}f_l(t)\cos(\omega_n t)\,dt+\sin(\omega_n x)\int_{-l}^{l}f_l(t)\sin(\omega_n t)\,dt\right]$$

Let $\Delta\omega=\omega_{n+1}-\omega_n=\dfrac{(n+1)\pi}{l}-\dfrac{n\pi}{l}=\dfrac{\pi}{l}$ and hence $\dfrac{1}{l}=\dfrac{\Delta\omega}{\pi}$

then Fourier series of $f_l(x)$ is

$$f_l(x)=\frac{\Delta\omega}{2\pi}\int_{-l}^{l}f_l(t)\,dt$$
$$+\frac{1}{\pi}\sum_{n=1}^{\infty}\left[\cos(\omega_n x)\int_{-l}^{l}f_l(t)\cos(\omega_n t)\,dt+\sin(\omega_n x)\int_{-l}^{l}f_l(t)\sin(\omega_n t)\,dt\right]\Delta\omega \tag{3.30}$$

Take limit as $l \to \infty$ and hence $\Delta\omega \to 0$ in such a way that $\omega_n \to \omega$, the infinite series will become integral from 0 to ∞ and first term on R.H.S. of (3.30) will approach zero and $f_l(x) \to f(x)$

$$\therefore \qquad f(x) = \frac{1}{\pi}\int_0^\infty \left[\cos(\omega x)\int_{-\infty}^{\infty} f(t)\cos(\omega t)\,dt + \sin(\omega x)\int_{-\infty}^{\infty} f(t)\sin(\omega t)\,dt\right]d\omega$$

Let $$A(\omega) = \frac{1}{\pi}\int_{-\infty}^{\infty} f(t)\cos(\omega t)\,dt$$

$$B(\omega) = \frac{1}{\pi}\int_{-\infty}^{\infty} f(t)\sin(\omega t)\,dt$$

$$\therefore \qquad f(x) = \int_0^\infty \left[A(\omega)\cos(\omega x) + B(\omega)\sin(\omega x)\right]d\omega \tag{3.31}$$

Thus, Fourier integral of $f(x)$ is

$$f(x) = \int_0^\infty \left[A(\omega)\cos(\omega x) + B(\omega)\sin(\omega x)\right]d\omega$$

where $$A(\omega) = \frac{1}{\pi}\int_{-\infty}^{\infty} f(t)\cos(\omega t)\,dt,\ B(\omega) = \frac{1}{\pi}\int_{-\infty}^{\infty} f(t)\sin(\omega t)\,dt.$$

The convergence of Fourier integral follows from convergence of Fourier series of $f_l(x)$. Putting values of $A(\omega)$ and $B(\omega)$ in (3.31), we can write

$$f(x) = \frac{1}{\pi}\int_0^\infty \int_{-\infty}^{\infty} f(t)\cos(\omega x - \omega t)\,dt\,d\omega.$$

3.5.2 Fourier Cosine and Fourier Sine Integrals

Fourier integral of $f(x)$ is

$$f(x) = \int_0^\infty \left[A(\omega)\cos(\omega x) + B(\omega)\sin(\omega x)\right]d\omega$$

where $$A(\omega) = \frac{1}{\pi}\int_{-\infty}^{\infty} f(t)\cos(\omega t)\,dt,\ B(\omega) = \frac{1}{\pi}\int_{-\infty}^{\infty} f(t)\sin(\omega t)\,dt.$$

Case I: If $f(x)$ is an even function then $f(t)\cos(\omega t)$ is an even function and $f(t)\sin(\omega t)$ is an odd function

$$\therefore \qquad A(\omega) = \frac{1}{\pi}\int_{-\infty}^{\infty} f(t)\cos(\omega t)\,dt = \frac{2}{\pi}\int_0^\infty f(t)\cos(\omega t)\,dt \text{ and } B(\omega) = 0.$$

$\therefore$ Fourier integral of $f(x)$ is

$$f(x) = \frac{2}{\pi}\int_0^\infty \int_0^\infty f(t)\cos(\omega t)\cos(\omega x)\,dt\,d\omega;\ f \text{ is even.}$$

This Fourier integral of $f(x)$ is called Fourier cosine integral of $f(x)$.

Case II: If $f(x)$ is an odd function of x then $f(t)\cos(\omega t)$ is an odd function of t and $f(t)\sin(\omega t)$ is an even function of t.

$$\therefore \quad A(\omega)=0 \text{ and } B(\omega)=\frac{1}{\pi}\int_{-\infty}^{\infty} f(t)\sin(\omega t)\,dt=\frac{2}{\pi}\int_0^{\infty} f(t)\sin(\omega t)\,dt.$$

$\therefore$ Fourier integral of $f(x)$ is

$$f(x)=\frac{2}{\pi}\int_0^{\infty}\int_0^{\infty} f(t)\sin(\omega t)\sin(\omega x)\,dt\,d\omega;\ f \text{ is odd.}$$

This Fourier integral of $f(x)$ is called Fourier sine integral of $f(x)$.

3.5.3 Fourier Cosine and Sine Transforms

Case I: If $f(x)$ is an even function, the Fourier integral is the Fourier cosine integral

$$f(x)=\frac{2}{\pi}\int_0^{\infty}\int_0^{\infty} f(t)\cos(\omega t)\cos(\omega x)\,dt\,d\omega$$

We define

$$\mathcal{F}_c(f(x))=\int_0^{\infty} f(x)\cos(\omega x)\,dx=F_c(\omega) \text{ (say) } (c \text{ suggests cosine}) \tag{3.32}$$

then

$$f(x)=\frac{2}{\pi}\int_0^{\infty} F_c(\omega)\cos(\omega x)\,d\omega \tag{3.33}$$

$F_c(\omega)$ is called Fourier cosine transform of $f(x)$.

Formula (3.33) gives us back $f(x)$ from $F_c(\omega)$ and therefore we write

$$f(x)=\mathcal{F}_c^{-1}(F_c(\omega))=\frac{2}{\pi}\int_0^{\infty} F_c(\omega)\cos(\omega x)\,d\omega$$

and $f(x)$ is called inverse Fourier cosine transform of $F_c(\omega)$

Case II: If $f(x)$ is an odd function, the Fourier integral is the Fourier sine integral

$$f(x)=\frac{2}{\pi}\int_0^{\infty}\int_0^{\infty} f(t)\sin(\omega t)\sin(\omega x)\,dt\,d\omega$$

We define

$$\mathcal{F}_s(f(x))=\int_0^{\infty} f(x)\sin(\omega x)\,dx=F_s(\omega) \text{ (say)}$$

then

$$f(x)=\frac{2}{\pi}\int_0^{\infty} F_s(\omega)\sin(\omega x)\,d\omega.$$

$F_s(\omega)$ is called Fourier sine transform of $f(x)$ and $f(x)$ is inverse Fourier sine transform of $F_s(\omega)$

$$f(x)=\mathcal{F}_s^{-1}(F_s(\omega))$$

Remark 3.4:

(i) We can also take $F_c(\omega)=\sqrt{\frac{2}{\pi}}\int_0^{\infty} f(x)\cos(\omega x)\,dx$

then $$f(x)=\mathcal{F}_c^{-1}\left(F_c(\omega)\right)=\sqrt{\frac{2}{\pi}}\int_0^{\infty} F_c(\omega)\cos\omega x\,d\omega$$

(ii) We can take $$F_s(\omega)=\sqrt{\frac{2}{\pi}}\int_0^{\infty} f(x)\sin(\omega x)\,dx$$

then $$f(x)=\mathcal{F}_s^{-1}\left(F_s(\omega)\right)=\sqrt{\frac{2}{\pi}}\int_0^{\infty} F_s(\omega)\sin(\omega x)\,d\omega$$

3.5.4 Complex Form of the Fourier Integral

Real Fourier integral of $f(x)$ is

$$f(x)=\frac{1}{\pi}\int_0^{\infty}\int_{-\infty}^{\infty} f(t)\cos(\omega x-\omega t)\,dt\,d\omega$$

$$=\frac{1}{\pi}\int_{-\infty}^{\infty} f(t)\left[\int_0^{\infty}\cos(\omega x-\omega t)\,d\omega\right]dt \tag{3.34}$$

Now, $\cos(\omega x-\omega t)$ is an even function of ω

and $\sin(\omega x-\omega t)$ is an odd function of ω

$\therefore$ $$\frac{1}{2}\int_{-\infty}^{\infty}\cos(\omega x-\omega t)\,d\omega=\int_0^{\infty}\cos(\omega x-\omega t)\,d\omega \tag{3.35}$$

and $$\int_{-\infty}^{\infty}\sin(\omega x-\omega t)\,d\omega=0 \tag{3.36}$$

Multiply (3.36) by $\frac{i}{2}$ and add (3.35)

$$\int_0^{\infty}\cos(\omega x-\omega t)\,d\omega=\frac{1}{2}\int_{-\infty}^{\infty}\left[\cos(\omega x-\omega t)+i\sin(\omega x-\omega t)\right]d\omega$$

$$=\frac{1}{2}\int_{-\infty}^{\infty} e^{i(\omega x-\omega t)}\,d\omega$$

$\therefore$ from (3.34)

$$f(x)=\frac{1}{2\pi}\int_{-\infty}^{\infty} f(t)\left[\int_{-\infty}^{\infty} e^{i(\omega x-\omega t)}\,d\omega\right]dt$$

$$=\frac{1}{2\pi}\int_{-\infty}^{\infty}\int_{-\infty}^{\infty} f(t)e^{i(\omega x-\omega t)}\,dt\,d\omega$$

which is called complex form of Fourier integral.

3.5.5 Fourier Transform and Its Inverse

Complex form of Fourier integral of $f(x)$ is

$$f(x) = \frac{1}{2\pi}\int_{-\infty}^{\infty}\int_{-\infty}^{\infty} f(t)e^{i(\omega x-\omega t)}dt\,d\omega$$

$$= \frac{1}{2\pi}\int_{-\infty}^{\infty} e^{i\omega x}\int_{-\infty}^{\infty} f(t)e^{-i\omega t}dt\,d\omega.$$

We define $$\mathfrak{F}(f(x)) = \int_{-\infty}^{\infty} f(x)e^{-i\omega x}dx = F(\omega) \text{ (say)}$$

then $$f(x) = \frac{1}{2\pi}\int_{-\infty}^{\infty} e^{i\omega x}\,\mathfrak{F}(f(x))d\omega$$

$\mathfrak{F}(f(x))$ is called Fourier transform of $f(x)$ and

$f(x) = \mathfrak{F}^{-1}(F(\omega)) = \frac{1}{2\pi}\int_{-\infty}^{\infty} e^{i\omega x}F(\omega)d\omega$ is called inverse Fourier transform of $F(\omega)$.

Remark 3.5: We can take following pairs of Fourier transforms of $f(x)$ and inverse Fourier transforms.

	Fourier transform	Inverse Fourier transform
(i)	$F(\omega) = \int_{-\infty}^{\infty} f(x)e^{-i\omega x}dx$	$\frac{1}{2\pi}\int_{-\infty}^{\infty} F(\omega)e^{i\omega x}d\omega$
(ii)	$F(\omega) = \frac{1}{\sqrt{2\pi}}\int_{-\infty}^{\infty} f(x)e^{-i\omega x}dx$	$\frac{1}{\sqrt{2\pi}}\int_{-\infty}^{\infty} F(\omega)e^{i\omega x}d\omega$
(iii)	$F(\omega) = \int_{-\infty}^{\infty} f(x)e^{i\omega x}dx$	$\frac{1}{2\pi}\int_{-\infty}^{\infty} F(\omega)e^{-i\omega x}d\omega$
(iv)	$F(\omega) = \frac{1}{\sqrt{2\pi}}\int_{-\infty}^{\infty} f(x)e^{i\omega x}dx$	$\frac{1}{\sqrt{2\pi}}\int_{-\infty}^{\infty} F(\omega)e^{-i\omega x}d\omega$

Out of these pairs, we shall be taking pair (i).

Remark 3.6: Function $f(x)$ is called **self-reciprocal** if

$$\mathfrak{F}(f(x)) = \sqrt{2\pi}\,f(\omega)$$

and if (ii) pair is taken then $f(x)$ is **self-reciprocal** if

$$\mathfrak{F}(f(x)) = f(\omega)$$

3.5.6 Spectrum

Let $F(\omega)$ be Fourier transform of $f(x)$, then representation of $f(x)$ as

$$f(x)=\frac{1}{2\pi}\int_{-\infty}^{\infty}F(\omega)e^{i\omega x}\,d\omega$$

is called spectral representation of $f(x)$ and $F(\omega)$ is called spectral density of $f(x)$. Here, ω is called the frequency of transform.

The graph of $|F(\omega)|$ for values of ω is called amplitude spectrum of $f(x)$ and $|F(\omega)|^2$ is called energy of spectrum.

3.6 PROPERTIES OF FOURIER TRANSFORMS

(i) **Linearity property**

For any functions $f(x)$ and $g(x)$ and any constants a and b

(a) $\mathcal{F}(af+bg)=a\mathcal{F}f(x)+b\mathcal{F}g(x)$

(b) $\mathcal{F}_c(af+bg)=a\mathcal{F}_c(x)+b\mathcal{F}_cg(x)$, $\quad$ f and g are even functions

(c) $\mathcal{F}_s(af+bg)=a\mathcal{F}_sf(x)+b\mathcal{F}_sg(x)$; $\quad$ f and g are odd functions.

Proof:

(a) $$\mathcal{F}(af+bg)=\int_{-\infty}^{\infty}\left[af(x)+bg(x)\right]e^{-i\omega x}dx$$

$$=a\int_{-\infty}^{\infty}f(x)e^{-i\omega x}dx+b\int_{-\infty}^{\infty}g(x)e^{-i\omega x}dx$$

$$=a\mathcal{F}f(x)+b\mathcal{F}g(x)$$

(b) $$\mathcal{F}_c(af+bg)=\int_0^{\infty}\left[af(x)+bg(x)\right]\cos(\omega x)\,dx$$

$$=a\int_0^{\infty}f(x)\cos(\omega x)\,dx+b\int_0^{\infty}g(x)\cos(\omega x)\,dx$$

$$=a\mathcal{F}_cf(x)+b\mathcal{F}_cg(x)$$

(c) $$\mathcal{F}_s(af+bg)=\int_0^{\infty}\left[af(x)+bg(x)\right]\sin(\omega x)\,dx$$

$$=a\int_0^{\infty}f(x)\sin(\omega x)\,dx+b\int_0^{\infty}g(x)\sin(\omega x)\,dx$$

$$=a\mathcal{F}_s(f(x))+b\mathcal{F}_s(g(x))$$

(ii) **Shifting property**

(a) Shifting property on *x* axis

If $\mathcal{F}(f(x)) = F(\omega)$ and a is any real number, then

$$\mathcal{F}(f(x-a)) = e^{-i\omega a}F(\omega) \text{ and hence } \mathcal{F}^{-1}\left(e^{-i\omega a}F(\omega)\right) = f(x-a).$$

(b) Frequency shifting

If $\mathcal{F}(f(x)) = F(\omega)$ and a is any real number, then

$$\mathcal{F}\left(e^{iax}f(x)\right) = F(\omega - a) \text{ and hence } \mathcal{F}^{-1}\left(F(\omega-a)\right) = e^{iax}f(x).$$

Proof:

(a) $$\mathcal{F}(f(x-a)) = \int_{-\infty}^{\infty} f(x-a)e^{-i\omega x}dx = e^{-i\omega a}\int_{-\infty}^{\infty} f(x-a)e^{-i\omega(x-a)}dx$$

$$= e^{-i\omega a}\int_{-\infty}^{\infty} f(t)e^{-i\omega t}dt \qquad \text{(taking } x-a=t\text{)}$$

$$= e^{-i\omega a}F(\omega)$$

(b) $$\mathcal{F}\left(e^{iax}f(x)\right) = \int_{-\infty}^{\infty} e^{iax}f(x)e^{-i\omega x}dx$$

$$= \int_{-\infty}^{\infty} f(x)e^{-i(\omega-a)x}dx = F(\omega-a)$$

(iii) **Change of scale property for any $a > 0$**

(a) If $\mathcal{F}f(x) = F(\omega)$, then $\mathcal{F}f(ax) = \frac{1}{a}F\left(\frac{\omega}{a}\right)$.

(b) If $\mathcal{F}_c f(x) = F_c(\omega)$, then $\mathcal{F}_c f(ax) = \frac{1}{a}F_c\left(\frac{\omega}{a}\right)$.

(c) If $\mathcal{F}_s f(x) = F_s(\omega)$, then $\mathcal{F}_s f(ax) = \frac{1}{a}F_s\left(\frac{\omega}{a}\right)$.

Proof:

(a) $$\mathcal{F}f(ax) = \int_{-\infty}^{\infty} f(ax)e^{-i\omega x}dx = \int_{-\infty}^{\infty} f(t)e^{-\frac{i\omega t}{a}}\frac{1}{a}dt \qquad \text{(taking } ax=t\text{)}$$

$$= \frac{1}{a}F\left(\frac{\omega}{a}\right)$$

(b) $$\mathcal{F}_c f(ax) = \int_{0}^{\infty} f(ax)\cos(\omega x)dx = \int_{0}^{\infty} f(t)\cos\left(\frac{\omega t}{a}\right)\frac{1}{a}dt \qquad \text{(taking } ax=t\text{)}$$

$$= \frac{1}{a}F_c\left(\frac{\omega}{a}\right)$$

(c) $\mathcal{F}_s f(ax) = \int_0^{\infty} f(ax)\sin(\omega x)\,dx = \int_0^{\infty} f(t)\sin\left(\frac{\omega t}{a}\right)\frac{1}{a}\,dt$ (taking $ax = t$)

$$= \frac{1}{a} F_s\left(\frac{\omega}{a}\right)$$

(iv) **Modulus property (or modulation theorem)**

If $\mathcal{F} f(x) = F(\omega)$, and a is any real number then

(a) $\mathcal{F}\left(f(x)\cos(ax)\right) = \frac{1}{2}\left[F(\omega+a) + F(\omega-a)\right]$

(b) $\mathcal{F}\left(f(x)\sin(ax)\right) = \frac{i}{2}\left[F(\omega+a) - F(\omega-a)\right]$

(c) $\mathcal{F}^{-1}\left(F(\omega)\cos(a\omega)\right) = \frac{1}{2}\left[f(x+a) + f(x-a)\right]$

(d) $\mathcal{F}^{-1}\left(F(\omega)\sin(a\omega)\right) = \frac{-i}{2}\left[f(x+a) - f(x-a)\right]$

Proof:

(a) $\mathcal{F}\left(f(x)\cos(ax)\right) = \int_{-\infty}^{\infty} f(x)\cos(ax)e^{-i\omega x}\,dx$

$$= \int_{-\infty}^{\infty} f(x)\cdot\frac{e^{iax}+e^{-iax}}{2}e^{-i\omega x}\,dx$$

$$= \frac{1}{2}\int_{-\infty}^{\infty} f(x)e^{-i(\omega-a)x}\,dx + \frac{1}{2}\int_{-\infty}^{\infty} f(x)e^{-i(\omega+a)x}\,dx$$

$$= \frac{1}{2}\left[F(\omega-a) + F(\omega+a)\right]$$

$$= \frac{1}{2}\left[F(\omega+a) + F(\omega-a)\right].$$

(b) $\mathcal{F}\left(f(x)\sin(ax)\right) = \int_{-\infty}^{\infty} f(x)\sin(ax)e^{-i\omega x}\,dx$

$$= \int_{-\infty}^{\infty} f(x)\left(\frac{e^{iax}-e^{-iax}}{2i}\right)e^{-i\omega x}\,dx$$

$$= \frac{i}{2}\int_{-\infty}^{\infty} f(x)\left[e^{-iax} - e^{iax}\right]e^{-i\omega x}\,dx$$

$$= \frac{i}{2}\left[\int_{-\infty}^{\infty} f(x)e^{-i(\omega+a)x}\,dx - \int_{-\infty}^{\infty} f(x)e^{-i(\omega-a)x}\,dx\right]$$

$$= \frac{i}{2}\left[F(\omega+a) - F(\omega-a)\right].$$

(c) $\mathcal{F}^{-1}\left(F(\omega)\cos(\omega a)\right)=\frac{1}{2\pi}\int_{-\infty}^{\infty}F(\omega)\cos(\omega a)e^{i\omega x}d\omega=\frac{1}{2\pi}\int_{-\infty}^{\infty}F(\omega)\frac{e^{i\omega a}+e^{-i\omega a}}{2}e^{i\omega x}d\omega$

$$=\frac{1}{2}\left[\frac{1}{2\pi}\int_{-\infty}^{\infty}F(\omega)e^{i(x+a)\omega}d\omega+\frac{1}{2\pi}\int_{-\infty}^{\infty}F(\omega)e^{i(x-a)\omega}d\omega\right]$$

$$=\frac{1}{2}\left[f(x+a)+f(x-a)\right].$$

(d) $\mathcal{F}^{-1}\left(F(\omega)\sin(\omega a)\right)=\frac{1}{2\pi}\int_{-\infty}^{\infty}F(\omega)\sin(\omega a)e^{i\omega x}d\omega=\frac{1}{2\pi}\int_{-\infty}^{\infty}F(\omega)\frac{e^{i\omega a}-e^{-i\omega a}}{2i}e^{i\omega x}d\omega$

$$=\frac{1}{2i}\left[\frac{1}{2\pi}\int_{-\infty}^{\infty}F(\omega)e^{i(x+a)\omega}d\omega-\frac{1}{2\pi}\int_{-\infty}^{\infty}F(\omega)e^{i(x-a)\omega}d\omega\right]$$

$$=\frac{-i}{2}\left[f(x+a)-f(x-a)\right].$$

(v) **Modulation property of sine and cosine transforms**

If $\mathcal{F}_c f(x)=F_c(\omega)$, $\mathcal{F}_s f(x)=F_s(\omega)$ and a is any real number then

(a) $\mathcal{F}_s\left(f(x)\cos(ax)\right)=\frac{1}{2}\left[F_s(\omega+a)+F_s(\omega-a)\right]$

(b) $\mathcal{F}_s\left(f(x)\sin(ax)\right)=\frac{1}{2}\left[F_c(\omega-a)-F_c(\omega+a)\right]$

(c) $F_c\left(f(x)\cos(ax)\right)=\frac{1}{2}\left[F_c(\omega+a)+F_c(\omega-a)\right]$

(d) $\mathcal{F}_c\left(f(x)\sin(ax)\right)=\frac{1}{2}\left[F_s(\omega+a)-F_s(\omega-a)\right]$

(e) $\mathcal{F}_s^{-1}\left(F_s(\omega)\cos(a\omega)\right)=\frac{1}{2}\left[f(x+a)+f(x-a)\right]$

(f) $\mathcal{F}_c^{-1}\left(F_c(\omega)\cos(a\omega)\right)=\frac{1}{2}\left[f(x+a)+f(x-a)\right]$

Proof:

(a) $\mathcal{F}_s\left(f(x)\cos(ax)\right)=\int_0^{\infty}f(x)\cos(ax)\sin(\omega x)\,dx=\frac{1}{2}\int_0^{\infty}f(x)\left[\sin((\omega+a)x)+\sin((\omega-a)x)\right]dx$

$$=\frac{1}{2}\int_0^{\infty}f(x)\sin((\omega+a)x)\,dx+\frac{1}{2}\int_0^{\infty}f(x)\sin((\omega-a)x)\,dx$$

$$=\frac{1}{2}\left[F_s(\omega+a)+F_s(\omega-a)\right].$$

(b) $$\mathcal{F}_s\left(f(x)\sin(ax)\right)=\int_0^\infty f(x)\sin(ax)\sin(\omega x)\,dx=\frac{1}{2}\int_0^\infty f(x)\left[\cos((\omega-a)x)-\cos((\omega+a)x)\right]dx$$
$$=\frac{1}{2}\left[\int_0^\infty f(x)\cos((\omega-a)x)\,dx-\int_0^\infty f(x)\cos((\omega+a)x)\,dx\right]$$
$$=\frac{1}{2}\left[F_c(\omega-a)-F_c(\omega+a)\right].$$

(c) $$\mathcal{F}_c\left(f(x)\cos(ax)\right)=\int_0^\infty f(x)\cos(ax)\cos(\omega x)\,dx=\frac{1}{2}\int_0^\infty f(x)\left[\cos((\omega+a)x)+\cos((\omega-a)x)\right]dx$$
$$=\frac{1}{2}\left[\int_0^\infty f(x)\cos((\omega+a)x)\,dx+\int_0^\infty f(x)\cos((\omega-a)x)\,dx\right]$$
$$=\frac{1}{2}\left[F_c(\omega+a)+F_c(\omega-a)\right].$$

(d) $$\mathcal{F}_c\left(f(x)\sin(ax)\right)=\int_0^\infty f(x)\sin(ax)\cos(\omega x)\,dx=\frac{1}{2}\int_0^\infty f(x)\left[\sin((\omega+a)x)-\sin((\omega-a)x)\right]dx$$
$$=\frac{1}{2}\left[\int_0^\infty f(x)\sin((\omega+a)x)\,dx-\int_0^\infty f(x)\sin((\omega-a)x)\,dx\right]$$
$$=\frac{1}{2}\left[F_s(\omega+a)-F_s(\omega-a)\right].$$

(e) $$\mathcal{F}_s^{-1}\left(F_s(\omega)\cos(a\omega)\right)=\frac{2}{\pi}\int_0^\infty F_s(\omega)\cos(a\omega)\sin(\omega x)\,d\omega$$
$$=\frac{1}{2}\left[\frac{2}{\pi}\int_0^\infty F_s(\omega)\left[\sin((x+a)\omega)\,d\omega+\sin((x-a)\omega)\,d\omega\right]\right]$$
$$=\frac{1}{2}\left[\frac{2}{\pi}\int_0^\infty F_s(\omega)\sin((x+a)\omega)\,d\omega+\frac{2}{\pi}\int_0^\infty F_s(\omega)\sin((x-a)\omega)\,d\omega\right]$$
$$=\frac{1}{2}\left[f(x+a)+f(x-a)\right].$$

(f) $$\mathcal{F}_c^{-1}\left(F_c(\omega)\cos(a\omega)\right)=\frac{2}{\pi}\int_0^\infty F_c(\omega)\cos(a\omega)\cos(\omega x)\,d\omega$$
$$=\frac{1}{2}\left[\frac{2}{\pi}\int_0^\infty F_c(\omega)\left[\cos((x+a)\omega)+\cos((x-a)\omega)\right]d\omega\right]$$
$$=\frac{1}{2}\left[\frac{2}{\pi}\int_0^\infty F_c(\omega)\cos((x+a)\omega)\,d\omega+\frac{2}{\pi}\int_0^\infty F_c(\omega)\cos((x-a)\omega)\,d\omega\right]$$
$$=\frac{1}{2}\left[f(x+a)+f(x-a)\right]$$

(vi) **Conjugate property**

If $\mathcal{F}\, f(x) = F(\omega)$, then

(a) $\mathcal{F}\, \overline{f(-x)} = \overline{F(\omega)}$,

(b) $\mathcal{F}\, f(-x) = F(-\omega)$ and

(c) $\mathcal{F}\, \overline{\left(f(x)\right)} = \overline{F(-\omega)}$.

Proof:

(a)
$$\mathcal{F}\, \overline{f(-x)} = \int_{-\infty}^{\infty} \overline{f(-x)} e^{-i\omega x} dx$$
$$= \int_{\infty}^{-\infty} \overline{f(t)} e^{i\omega t} (-dt) \qquad \text{(taking } x = -t\text{)}$$
$$= \int_{-\infty}^{\infty} \overline{f(t)}\, e^{i\omega t} (dt)$$
$$= \int_{-\infty}^{\infty} \overline{f(t) e^{-i\omega t}} dt$$
$$= \overline{\left(\int_{-\infty}^{\infty} f(t) e^{-i\omega t} dt \right)} = \overline{F(\omega)}$$

(b)
$$\mathcal{F}\, f(-x) = \int_{-\infty}^{\infty} f(-x) e^{-i\omega x} dx$$
$$= \int_{\infty}^{-\infty} f(t) e^{i\omega t} (-dt) \qquad \text{(taking } x = -t\text{)}$$
$$= \int_{-\infty}^{\infty} f(t) e^{-i(-\omega)t} dt$$
$$= F(-\omega)$$

(c)
$$\mathcal{F}\left(\overline{f(x)}\right) = \int_{-\infty}^{\infty} \overline{f(x)} e^{-i\omega x} dx$$
$$= \int_{-\infty}^{\infty} \overline{f(x) e^{i\omega x}} dx$$
$$= \overline{\int_{-\infty}^{\infty} f(x) e^{-i(-\omega)x} dx}$$
$$= \overline{F(-\omega)}$$

(vii) **Differentiations of transforms w.r.t. frequency**

If $\mathcal{F}(f(x)) = F(\omega)$, $\mathcal{F}_s(f(x)) = F_s(\omega)$ and $\mathcal{F}_c(f(x)) = F_c(\omega)$ exist then

(a) $\mathcal{F}(x^n f(x)) = i^n \dfrac{d^n}{d\omega^n} F(\omega); \ n \in N$

(b) $\mathcal{F}_s(x^n f(x)) = (-1)^{\frac{n}{2}} \dfrac{d^n}{d\omega^n} F_s(\omega)$ if n is even

$= (-1)^{\frac{n+1}{2}} \dfrac{d^n}{d\omega^n} F_c(\omega)$ if n is odd

(c) $\mathcal{F}_c(x^n f(x)) = (-1)^{\frac{n}{2}} \dfrac{d^n}{d\omega^n} F_c(\omega)$ if n is even

$= (-1)^{\frac{n-1}{2}} \dfrac{d^n}{d\omega^n} F_s(\omega)$ if n is odd

Proof:

(a)
$$F(\omega) = \mathcal{F} f(x) = \int_{-\infty}^{\infty} f(x) e^{-i\omega x} dx$$

$\therefore$
$$\frac{d^n}{d\omega^n} F(\omega) = \int_{-\infty}^{\infty} f(x)(-ix)^n e^{-i\omega x} dx$$

$$= (-i)^n \int_{-\infty}^{\infty} f(x) x^n e^{-i\omega x} dx = \frac{1}{i^n} \mathcal{F}(x^n f(x))$$

$\therefore$
$$\mathcal{F}(x^n f(x)) = i^n \frac{d^n}{d\omega^n} F(\omega); \ n \in N$$

(b)
$$F_s(\omega) = \mathcal{F}_s f(x) = \int_0^{\infty} f(x) \sin(\omega x) dx$$

$$F_c(\omega) = \mathcal{F}_c(f(x)) = \int_0^{\infty} f(x) \cos(\omega x) dx$$

$\therefore$ if n is even natural number

then
$$\frac{d^n}{d\omega^n} F_s(\omega) = \int_0^{\infty} f(x)(-1)^{\frac{n}{2}} x^n \sin(\omega x) dx$$

$$= (-1)^{\frac{n}{2}} \int_0^{\infty} f(x) x^n \sin(\omega x) dx$$

$$= (-1)^{\frac{n}{2}} \mathcal{F}_s(x^n f(x))$$

$\therefore$
$$\mathcal{F}_s(x^n f(x)) = (-1)^{\frac{n}{2}} \frac{d^n}{d\omega^n} F_s(\omega); \ n = 2, 4, 6, \ldots$$

and if n is odd natural number

then
$$\frac{d^n}{d\omega^n} F_c(\omega) = \int_0^{\infty} f(x)(-1)^{\frac{n+1}{2}} x^n \sin(\omega x) dx$$

$$= (-1)^{\frac{n+1}{2}} \mathcal{F}_s\left(x^n f(x)\right)$$

$$\therefore \qquad \mathcal{F}_s\left(x^n f(x)\right) = (-1)^{\frac{n+1}{2}} \frac{d^n}{d\omega^n} F_c(\omega); \quad n = 1,3,5,\ldots$$

(c) It can be proved as proof of part (b).

(viii) **Transform of derivatives**

(a) Let $f(x)$ be continuous and $f^{(k)}(x)$ be piecewise continuous on every finite interval $(-l, l)$ and $\int_{-\infty}^{\infty} \left|f^{(k-1)}(x)dx\right|$ converges for $k = 1,2,\ldots,n$ and $f^{(k-1)}(x) \to 0$ as $x \to \pm\infty$ for $k = 1,2,\ldots n$

then $$\mathcal{F}\left(f^{(n)}(x)\right) = (i\omega)^n \mathcal{F}\left(f(x)\right).$$

(b) Let $f(x)$ be continuous on $[0,\infty)$, $f(x) \to 0$ as $x \to \infty$ and $f'(x)$ is piecewise continuous on every finite interval $[0, l]$ then $\mathcal{F}_c f'(x) = \omega \mathcal{F}_s\left(f(x)\right) - f(0)$ and $\mathcal{F}_s f'(x) = -\omega \mathcal{F}_c\left(f(x)\right)$.

(c) If $f(x)$ and $f'(x)$ are continuous in $(0,\infty)$, $f(x) \to 0$, $f'(x) \to 0$ as $x \to \infty$ and $f''(x)$ is piecewise continuous on every subinterval $[0, l]$ then

$$\mathcal{F}_c\left(f''(x)\right) = -\omega^2 \mathcal{F}_c\left(f(x)\right) - f'(0)$$

and $$\mathcal{F}_s\left(f''(x)\right) = -\omega^2 \mathcal{F}_s\left(f(x)\right) + \omega f(0)$$

Proof: (a) We shall prove the result by mathematical induction

$$\mathcal{F}\left(f'(x)\right) = \int_{-\infty}^{\infty} f'(x) e^{-i\omega x} dx$$

Integrating by parts

$$\mathcal{F}\left(f'(x)\right) = \left(e^{-i\omega x} f(x)\right)_{-\infty}^{\infty} - \int_{-\infty}^{\infty} (-i\omega) e^{-i\omega x} f(x) dx$$

$$= i\omega \int_{-\infty}^{\infty} f(x) e^{-i\omega x} dx = (i\omega) \mathcal{F}\left(f(x)\right)$$

$\therefore$ Result holds for $n = 1$

Let $$\mathcal{F}\left(f^{(k)}(x)\right) = (i\omega)^k \mathcal{F}\left(f(x)\right)$$

then $$\mathcal{F}\left(f^{(k+1)}(x)\right) = \int_{-\infty}^{\infty} e^{-i\omega x} f^{(k+1)}(x) dx$$

Integrate by parts

$$\mathcal{F}\left(f^{k+1}(x)\right) = \left(f^{(k)}(x) e^{-i\omega x}\right)_{-\infty}^{\infty} - \int_{-\infty}^{\infty} (-i\omega) e^{-i\omega x} f^{(k)}(x) dx$$

$$= (i\omega) \int_{-\infty}^{\infty} e^{-i\omega x} f^{(k)}(x) dx = (i\omega) \mathcal{F}\left(f^{(k)}(x)\right)$$

$$= (i\omega)(i\omega)^k \mathcal{F} f(x) = (i\omega)^{k+1} \mathcal{F} f(x)$$

Hence the result holds by the principle of mathematical induction.

$$\therefore \qquad \mathscr{F}\left(f^{(n)}(x)\right) = (i\omega)^n \, \mathscr{F} f(x); \quad n \in N$$

(b) $\mathscr{F}_c\left(f'(x)\right) = \int_0^\infty f'(x)\cos(\omega x)\,dx$

Integrate by parts

$$\mathscr{F}_c\left(f'(x)\right) = \left(f(x)\cos(\omega x)\right)_0^\infty - \int_0^\infty (-\omega)\,\sin(\omega x)\,f(x)\,dx$$

$$= -f(0) + \omega\int_0^\infty f(x)\sin(\omega x)\,dx$$

$$= \omega \mathscr{F}_s\left(f(x)\right) - f(0)$$

and
$$\mathscr{F}_s f'(x) = \int_0^\infty f'(x)\sin(\omega x)\,dx$$

Integrate by parts

$$\mathscr{F}_s\left(f'(x)\right) = \left(f(x)\sin(\omega x)\right)_0^\infty - \int_0^\infty \omega\cos(\omega x) f(x)\,dx$$

$$= -\omega\,\mathscr{F}_c\left(f(x)\right)$$

(c) $\mathscr{F}_c\left(f''(x)\right) = \int_0^\infty f''(x)\cos(\omega x)\,dx$

Integrate by parts

$$\mathscr{F}_c\left(f''(x)\right) = \left(f'(x)\cos(\omega x)\right)_0^\infty - \int_0^\infty (-\omega)\sin(\omega x)\,f'(x)\,dx$$

$$= -f'(0) + \omega\,\mathscr{F}_s\left(f'(x)\right)$$

$$= -f'(0) + \omega\left[-\omega\,\mathscr{F}_c\left(f(x)\right)\right] \qquad \text{(from part (b))}$$

$$= -\omega^2 \mathscr{F}_c\left(f(x)\right) - f'(0)$$

and
$$\mathscr{F}_s\left(f''(x)\right) = \int_0^\infty f''(x)\sin(\omega x)\,dx$$

$$= \left(f'(x)\sin(\omega x)\right)_0^\infty - \int_0^\infty \omega\cos(\omega x)\,f'(x)\,dx \qquad \text{(integration by parts)}$$

$$= -\omega \mathscr{F}_c\left(f'(x)\right) = -\omega\left[\omega\,\mathscr{F}_s\left(f(x)\right) - f(0)\right] \qquad \text{(from part (b))}$$

$$= -\omega^2 \mathscr{F}_s\left(f(x)\right) + \omega f(0)$$

(ix) **Fourier Transform of Integral Function**

Let $f(x)$ be piecewise continuous on every interval $(-l, l)$ and $\int_{-\infty}^\infty |f(x)|\,dx$ converges, then

$$\mathscr{F}\left[\int_{-\infty}^x f(t)\,dt\right] = \frac{1}{i\omega}\mathscr{F}\left(f(x)\right), \text{ provided } \mathscr{F}\left(f(x)\right) = F(\omega) = 0 \text{ for } \omega = 0$$

Proof:

Let $\qquad \phi(x)=\int_{-\infty}^{x} f(t)\,dt$

$\therefore \qquad \lim\limits_{x\to\infty}\phi(x)=\int_{-\infty}^{\infty} f(t)\,dt=\left(\mathcal{F}(f(x))\right)_{\omega=0}=0$ and $\phi'(x)=f(x)$

$\therefore \qquad \mathcal{F}(f(x))=\mathcal{F}(\phi'(x))=(i\omega)\mathcal{F}(\phi(x))$

$\therefore \qquad \mathcal{F}(\phi(x))=\dfrac{1}{i\omega}\mathcal{F}(f(x))$

$\therefore \qquad \mathcal{F}\left[\int_{-\infty}^{x} f(t)\,dt\right]=\dfrac{1}{i\omega}\mathcal{F}(f(x))$, provided $\mathcal{F}(f(x))=F(\omega)=0$ for $\omega=0$

3.7 CONVOLUTION THEOREM AND PARSEVAL'S IDENTITIES

3.7.1 Convolution

Let $f(x)$ and $g(x)$ be piecewise continuous on every interval $(-l, l)$ and $\int_{-\infty}^{\infty}|f(x)|\,dx$, $\int_{-\infty}^{\infty}|g(x)|\,dx$ converge then convolution of f and g denoted by $(f*g)(x)$ is defined as

$$(f*g)(x)=\int_{-\infty}^{\infty} f(\tau)g(x-\tau)\,d\tau$$

we have $(f*g)(x)=(g*f)(x)$

3.7.2 Convolution Theorem (or Faltung Theorem) for Fourier Transforms

Theorem 3.2 Let $f(x)$ and $g(x)$ be piecewise continuous on every interval (–l, l) and let $\int_{-\infty}^{\infty}|f(x)|\,dx$, $\int_{-\infty}^{\infty}|g(x)|\,dx$ converge.

Let $\mathcal{F}(f(x))=F(\omega)$, $\mathcal{F}(g(x))=G(\omega)$ then

(a) $\mathcal{F}\left[(f*g)(x)\right]=F(\omega)G(\omega)$ (convolution w.r.t. *x*)

(b) $\mathcal{F}(f(x)g(x))=\dfrac{1}{2\pi}(F*G)(\omega)$ (convolution w.r.t. frequency)

Proof: (a)
$$\begin{aligned}\mathcal{F}\left[(f*g)(x)\right]&=\int_{-\infty}^{\infty}(f*g)(x)e^{-i\omega x}\,dx\\&=\int_{-\infty}^{\infty}\int_{-\infty}^{\infty} f(\tau)g(x-\tau)\,d\tau\, e^{-i\omega x}\,dx\\&=\int_{-\infty}^{\infty} f(\tau)e^{-i\omega\tau}\int_{-\infty}^{\infty} g(x-\tau)e^{-i\omega(x-\tau)}\,dx\,d\tau\\&=\int_{-\infty}^{\infty} f(\tau)e^{-i\omega\tau}\int_{-\infty}^{\infty} g(u)e^{-i\omega u}\,du\,d\tau \qquad \text{(taking } x-\tau=u\text{)}\\&=\int_{-\infty}^{\infty} f(\tau)e^{-i\omega\tau}\,d\tau\int_{-\infty}^{\infty} g(u)e^{-i\omega u}\,du\\&=F(\omega)G(\omega)\end{aligned}$$

(b) $$\mathcal{F}^{-1}(F*G)(\omega)=\frac{1}{2\pi}\int_{-\infty}^{\infty}(F*G)(\omega)e^{i\omega x}d\omega$$

$$=\frac{1}{2\pi}\int_{-\infty}^{\infty}\int_{-\infty}^{\infty}F(\tau)G(\omega-\tau)d\tau\, e^{i\omega x}d\omega$$

$$=\frac{1}{2\pi}\int_{-\infty}^{\infty}\int_{-\infty}^{\infty}F(\tau)G(\omega-\tau)e^{i\omega x}d\omega\, d\tau$$

$$=\frac{1}{2\pi}\int_{-\infty}^{\infty}F(\tau)e^{i\tau x}\int_{-\infty}^{\infty}G(\omega-\tau)e^{i(\omega-\tau)x}d\omega\, d\tau$$

$$=\frac{1}{2\pi}\int_{-\infty}^{\infty}F(\tau)e^{i\tau x}\int_{-\infty}^{\infty}G(u)e^{iux}du\, d\tau \qquad \text{(taking } \omega-\tau=u\text{)}$$

$$=\frac{1}{2\pi}\int_{-\infty}^{\infty}F(\tau)e^{i\tau x}d\tau\int_{-\infty}^{\infty}G(u)e^{iux}du$$

$$=\frac{1}{2\pi}\int_{-\infty}^{\infty}F(\omega)e^{i\omega x}d\omega\int_{-\infty}^{\infty}G(\omega)e^{i\omega x}d\omega$$

$$=\frac{1}{2\pi}\left[2\pi f(x).2\pi g(x)\right]=2\pi f(x)g(x)$$

$\therefore$ $$\mathcal{F}^{-1}\frac{1}{2\pi}(F*G)(\omega)=f(x)g(x)$$

$\therefore$ $$\mathcal{F}(f(x)g(x))=\frac{1}{2\pi}(F*G)(\omega)$$

3.7.3 Parseval's Identities (Energy Theorem)

Theorem 3.3 If $f(x)$ and $g(x)$ are piecewise continuous in every finite interval (–*l*, *l*) and $\int_{-\infty}^{\infty}|f(x)|dx$ and $\int_{-\infty}^{\infty}|g(x)|dx$ converge and $\mathcal{F}f(x)=F(\omega)$, $\mathcal{F}g(x)=G(\omega)$, $\mathcal{F}_s f(x)=F_s(\omega)$, $\mathcal{F}_s g(x)=G_s(\omega)$, $\mathcal{F}_c f(x)=F_c(\omega)$ and $\mathcal{F}_c g(x)=G_c(\omega)$ then

(i) $\frac{1}{2\pi}\int_{-\infty}^{\infty}F(\omega)\overline{G}(\omega)d\omega=\int_{-\infty}^{\infty}f(x)\overline{g}(x)dx$

(ii) $\frac{1}{2\pi}\int_{-\infty}^{\infty}|F(\omega)|^2 d\omega=\int_{-\infty}^{\infty}|f(x)|^2 dx$

(iii) $\frac{2}{\pi}\int_0^{\infty}F_s(\omega)G_s(\omega)d\omega=\int_0^{\infty}f(x)g(x)dx$

(iv) $\frac{2}{\pi}\int_0^{\infty}F_c(\omega)G_c(\omega)d\omega=\int_0^{\infty}f(x)g(x)dx$

(v) $\frac{2}{\pi}\int_0^{\infty}|F_s(\omega)|^2 d\omega=\frac{2}{\pi}\int_0^{\infty}|F_c(\omega)|^2 d\omega=\int_0^{\infty}|f(x)|^2 dx$

Identities (ii) and (iv) are called Parseval's identities or energy theorems.

Proof: (i) $\mathcal{F}(g(x)) = G(\omega)$

$$\therefore \qquad g(x) = \mathcal{F}^{-1} G(\omega) = \frac{1}{2\pi}\int_{-\infty}^{\infty} G(\omega) e^{i\omega x} d\omega$$

Take conjugate of both sides

$$\bar{g}(x) = \frac{1}{2\pi}\int_{-\infty}^{\infty} \overline{G(\omega)} e^{-i\omega x} d\omega$$

$$\therefore \qquad \int_{-\infty}^{\infty} f(x)\bar{g}(x)\,dx = \frac{1}{2\pi}\int_{-\infty}^{\infty}\int_{-\infty}^{\infty} f(x)\overline{G(\omega)}\, e^{-i\omega x}\, d\omega\, dx$$

$$= \frac{1}{2\pi}\int_{-\infty}^{\infty}\int_{-\infty}^{\infty} f(x)\overline{G(\omega)} e^{-i\omega x}\, dx\, d\omega$$

$$= \frac{1}{2\pi}\int_{-\infty}^{\infty} \overline{G(\omega)}\left[\int_{-\infty}^{\infty} f(x) e^{-i\omega x} dx\right] d\omega$$

$$= \frac{1}{2\pi}\int_{-\infty}^{\infty} \overline{G(\omega)} F(\omega)\, d\omega = \frac{1}{2\pi}\int_{-\infty}^{\infty} F(\omega)\bar{G}(\omega)\, d\omega$$

(ii) In part (i), taking $g(x) = f(x)$

$$\int_{-\infty}^{\infty} f(x)\overline{f(x)}\, dx = \frac{1}{2\pi}\int_{-\infty}^{\infty} F(\omega)\overline{F(\omega)}\, d\omega$$

$$\therefore \qquad \int_{-\infty}^{\infty} |f(x)|^2\, dx = \frac{1}{2\pi}\int_{-\infty}^{\infty} |F(\omega)|^2 d\omega$$

(iii) $\qquad \mathcal{F}_s\, g(x) = G_s(\omega)$

$$\therefore \qquad g(x) = \mathcal{F}_s^{-1} G_s(\omega) = \frac{2}{\pi}\int_{0}^{\infty} G_s(\omega)\sin(\omega x)\, d\omega$$

$$\therefore \qquad \int_{0}^{\infty} f(x) g(x)\, dx = \frac{2}{\pi}\int_{0}^{\infty}\int_{0}^{\infty} f(x) G_s(\omega)\sin(\omega x)\, d\omega\, dx$$

$$= \frac{2}{\pi}\int_{0}^{\infty}\int_{0}^{\infty} f(x) G_s(\omega)\sin(\omega x)\, dx\, d\omega$$

$$= \frac{2}{\pi}\int_{0}^{\infty} G_s(\omega)\left[\int_{0}^{\infty} f(x)\sin(\omega x)\, dx\right] d\omega$$

$$= \frac{2}{\pi}\int_{0}^{\infty} G_s(\omega) F_s(\omega)\, d\omega$$

$$\therefore \qquad \frac{2}{\pi}\int_{0}^{\infty} F_s(\omega) G_s(\omega)\, d\omega = \int_{0}^{\infty} f(x) g(x)\, dx$$

(iv) $$\mathcal{F}_c\, g(x) = G_c(\omega)$$

$\therefore$ $$g(x) = \mathcal{F}_c^{-1} G_c(\omega) = \frac{2}{\pi}\int_0^\infty G_c(\omega)\cos(\omega x)\,d\omega$$

$\therefore$ $$\int_0^\infty f(x)g(x)\,dx = \frac{2}{\pi}\int_0^\infty\int_0^\infty f(x)G_c(\omega)\cos(\omega x)\,d\omega\,dx$$

$$= \frac{2}{\pi}\int_0^\infty\int_0^\infty f(x)G_c(\omega)\cos(\omega x)\,dx\,d\omega$$

$$= \frac{2}{\pi}\int_0^\infty G_c(\omega)\left[\int_0^\infty f(x)\cos(\omega x)dx\right]d\omega$$

$$= \frac{2}{\pi}\int_0^\infty G_c(\omega)F_c(\omega)\,d\omega$$

$\therefore$ $$\frac{2}{\pi}\int_0^\infty F_c(\omega)G_c(\omega)\,d\omega = \int_0^\infty f(x)g(x)\,dx$$

(v) In parts (iii) and (iv), take $g(x) = \overline{f(x)}$ and use

$$\mathcal{F}_s\left[\overline{f(x)}\right] = \overline{F_s(\omega)} \text{ and } \mathcal{F}_c\left[\overline{f(x)}\right] = \overline{F_c(\omega)} \text{ (clear from definitions)}$$

$\therefore$ $$\frac{2}{\pi}\int_0^\infty \left|F_s(\omega)\right|^2 d\omega = \frac{2}{\pi}\int_0^\infty \left|F_c(\omega)\right|^2 d\omega = \int_0^\infty \left|f(x)\right|^2 dx$$

3.7.4 Relation between Fourier and Laplace Transforms

If $\pounds\, g(x)$ exists and $\pounds\, g(x) = G(s)$, then

$$\mathcal{F}\left(e^{-ax}g(x)H(x)\right) = G(a+i\omega), \quad a>0; \text{ where } H(x) \text{ is unit step function.}$$

Proof: $$\mathcal{F}\left(e^{-ax}g(x)H(x)\right) = \int_0^\infty e^{-ax}g(x)e^{-i\omega x}dx;\ a>0$$

$$= \int_0^\infty g(x)e^{-(a+i\omega)x}dx$$

$$= \int_0^\infty g(x)e^{-sx}dx \text{ where } a+i\omega = s$$

$$= \pounds\left(g(x)\right) = G(s) = G(a+i\omega)$$

Example 3.52: Using Fourier integral, show that

$$\int_0^\infty \frac{1-\cos(\pi\lambda)}{\lambda}\sin(\lambda x)\,d\lambda = \begin{cases} \dfrac{\pi}{2}; & 0<x<\pi \\ 0\ ; & x>\pi \end{cases}$$

Solution: Consider the function

$$f(x) = \begin{cases} \dfrac{\pi}{2}; & 0 < x < \pi \\ 0; & x > \pi \end{cases}$$

Its Fourier sine integral is

$$\begin{aligned} f(x) &= \frac{2}{\pi}\int_0^\infty \int_0^\infty f(t)\sin(\lambda x)\sin(\lambda t)\,dt\,d\lambda \\ &= \frac{2}{\pi}\int_0^\infty \sin(\lambda x)\int_0^\pi \frac{\pi}{2}\sin(\lambda t)\,dt\,d\lambda \\ &= \int_0^\infty \sin(\lambda x)\left(-\frac{1}{\lambda}\cos(\lambda t)\right)_0^\pi d\lambda \\ &= \int_0^\infty \frac{1-\cos(\pi\lambda)}{\lambda}\sin(\lambda x)\,d\lambda \end{aligned}$$

$$\therefore \quad \int_0^\infty \frac{1-\cos(\pi\lambda)}{\lambda}\sin(\lambda x)\,d\lambda = \begin{cases} \dfrac{\pi}{2}; & 0 < x < \pi \\ 0; & x > \pi \end{cases}$$

Example 3.53: Use Fourier integral to prove that

$$\int_0^\infty \frac{\sin(\lambda x)\sin(\pi\lambda)}{1-\lambda^2}\,d\lambda = \begin{cases} \dfrac{\pi}{2}\sin x; & 0 < x < \pi \\ 0 \quad; & x > \pi \end{cases}$$

Solution: Consider the function

$$f(x) = \begin{cases} \dfrac{\pi}{2}\sin x; & 0 < x < \pi \\ 0 \quad; & x > \pi \end{cases}$$

Its Fourier sine integral is

$$\begin{aligned} f(x) &= \frac{2}{\pi}\int_0^\infty \int_0^\infty f(t)\sin(\lambda x)\sin(\lambda t)\,dt\,d\lambda \\ &= \frac{2}{\pi}\int_0^\infty \sin(\lambda x)\int_0^\pi \frac{\pi}{2}\sin t\sin(\lambda t)\,dt\,d\lambda \\ &= \int_0^\infty \sin(\lambda x)\int_0^\pi \frac{1}{2}\left[(\cos(\lambda-1)t) - (\cos(\lambda+1)t)\right]dt\,d\lambda \\ &= \int_0^\infty \frac{1}{2}\sin(\lambda x)\left[\frac{1}{(\lambda-1)}(\sin(\lambda-1)t) - \frac{1}{\lambda+1}(\sin(\lambda+1)t)\right]_0^\pi d\lambda \\ &= \int_0^\infty \frac{1}{2}\sin(\lambda x)\left[\frac{1}{(\lambda-1)}\sin(\lambda\pi-\pi) - \frac{1}{\lambda+1}\sin(\pi+\lambda\pi)\right]d\lambda \end{aligned}$$

$$= \int_0^\infty \frac{1}{2}\sin(\lambda x)\left[\frac{1}{(1-\lambda)}\sin(\pi-\lambda\pi)+\frac{1}{1+\lambda}\sin(\lambda\pi)\right]d\lambda$$

$$= \int_0^\infty \frac{1}{2}\sin(\lambda x)\left[\frac{1}{(1-\lambda)}\sin(\lambda\pi)+\frac{1}{1+\lambda}\sin(\lambda\pi)\right]d\lambda$$

$$= \int_0^\infty \frac{1}{2}\sin(\lambda x)\left[\left(\frac{1}{1-\lambda}+\frac{1}{1+\lambda}\right)\sin(\pi\lambda)\right]d\lambda$$

$$= \int_0^\infty \frac{\sin(\pi\lambda)\sin(\lambda x)}{1-\lambda^2}d\lambda$$

$$\therefore \quad \int_0^\infty \frac{\sin(\lambda x)\sin(\pi\lambda)}{1-\lambda^2}d\lambda = \begin{cases} \frac{\pi}{2}\sin x; & 0 < x < \pi \\ 0 \quad ; & x > \pi \end{cases}$$

Example 3.54: For $a > 0, t > 0$, show that

(i) $\int_0^\infty \frac{\cos(\omega t)}{\omega^2 + a^2}d\omega = \frac{\pi}{2a}e^{-at}$

(ii) $\int_0^\infty \frac{\omega\sin(\omega t)}{\omega^2 + a^2}d\omega = \frac{\pi}{2}e^{-at}$

Solution:

(i) Consider the function

$$f(t) = \frac{\pi}{2a}e^{-at}; \quad t > 0$$

Its Fourier cosine integral is

$$f(t) = \frac{2}{\pi}\int_0^\infty\int_0^\infty f(x)\cos(\omega t)\cos(\omega x)\, dx\, d\omega$$

$$= \frac{2}{\pi}\int_0^\infty \cos(\omega t)\int_0^\infty \frac{\pi}{2a}e^{-ax}\cos(\omega x)\, dx\, d\omega$$

$$= \frac{1}{a}\int_0^\infty \cos(\omega t)\left[\frac{e^{-ax}}{\omega^2+a^2}(-a\cos(\omega x)+\omega\sin(\omega x))\right]_0^\infty d\omega$$

$$= \frac{1}{a}\int_0^\infty \cos(\omega t)\frac{a}{\omega^2+a^2}d\omega \qquad (\because a > 0)$$

$$\therefore \quad \int_0^\infty \frac{\cos(\omega t)}{\omega^2+a^2}d\omega = f(t) = \frac{\pi}{2a}e^{-at}; \quad t > 0$$

(ii) Fourier sine integral of

$$f(t) = \frac{\pi}{2a}e^{-at}, t > 0 \text{ is}$$

$$f(t) = \frac{2}{\pi}\int_0^\infty \int_0^\infty f(x)\sin(\omega t)\sin(\omega x)\, dx\, d\omega$$

$$= \frac{2}{\pi}\int_0^\infty \sin(\omega t)\int_0^\infty \frac{\pi}{2a} e^{-ax} \sin(\omega x)\, dx\, d\omega$$

$$= \frac{1}{a}\int_0^\infty \sin(\omega t)\left[\frac{e^{-ax}}{\omega^2 + a^2}(-a\sin(\omega x) - \omega\cos(\omega x))\right]_0^\infty d\omega$$

$$= \frac{1}{a}\int_0^\infty \frac{\omega \sin(\omega t)}{\omega^2 + a^2} d\omega \qquad (\because a > 0)$$

$$\therefore \qquad \int_0^\infty \frac{\omega\sin(\omega t)}{\omega^2 + a^2} d\omega = a\, f(t) = \frac{\pi}{2} e^{-at};\ t > 0$$

Example 3.55: If $f(x) = \begin{cases} \sin x\ ; & 0 < x < \pi \\ 0\ ; & x < 0 \text{ and } x > \pi \end{cases}$ then prove that

$$f(x) = \frac{1}{\pi}\int_0^\infty \frac{\cos(\lambda x) + \cos(\lambda(\pi - x))}{1 - \lambda^2} d\lambda$$

and hence find the value of the integral $\int_0^\infty \frac{\cos\frac{\pi s}{2}}{1 - s^2} ds$.

Solution: Fourier integral of

$$f(x) = \begin{cases} \sin x\ ; & 0 < x < \pi \\ 0\ ; & x < 0 \text{ and } x > \pi \end{cases} \quad \text{is}$$

$$f(x) = \frac{1}{\pi}\int_0^\infty \int_{-\infty}^\infty f(t)\cos(\lambda t - \lambda x)\, dt\, d\lambda$$

$$= \frac{1}{\pi}\int_0^\infty \int_0^\pi \sin t\cos(\lambda t - \lambda x)\, dt\, d\lambda$$

$$= \text{Re}\ \frac{1}{\pi}\int_0^\infty \int_0^\pi e^{i(\lambda t - \lambda x)} \sin t\, dt\, d\lambda \qquad \text{(Re means 'real part of ')}$$

$$= \text{Re}\ \frac{1}{\pi}\int_0^\infty e^{-i\lambda x}\int_0^\pi e^{i\lambda t} \sin t\, dt\, d\lambda$$

$$= \text{Re}\ \frac{1}{\pi}\int_0^\infty e^{-i\lambda x}\left[\frac{e^{i\lambda t}}{1 - \lambda^2}(i\lambda\sin t - \cos t)\right]_0^\pi d\lambda$$

$$= \text{Re}\ \frac{1}{\pi}\int_0^\infty \frac{e^{-i\lambda x}}{1 - \lambda^2}\left[e^{i\lambda\pi} + 1\right] d\lambda$$

$$= \text{Re}\ \frac{1}{\pi}\int_0^\infty \frac{e^{i\lambda(\pi - x)} + e^{-i\lambda x}}{1 - \lambda^2} d\lambda$$

$$= \frac{1}{\pi}\int_0^\infty \frac{\cos(\lambda(\pi - x)) + \cos(\lambda x)}{1 - \lambda^2} d\lambda$$

$$\therefore \int_0^\infty \frac{\cos(\lambda x)+\cos(\lambda(\pi-x))}{1-\lambda^2}d\lambda = \pi f(x) = \begin{cases} \pi\sin x; & 0<x<\pi \\ 0 \quad ; & x<0 \text{ and } x>\pi \end{cases}$$

Take $x=\dfrac{\pi}{2}$ and replace λ by s

$$\int_0^\infty \frac{2\cos\frac{\pi s}{2}}{1-s^2}ds = \pi$$

$$\Rightarrow \qquad \int_0^\infty \frac{\cos\frac{\pi s}{2}}{1-s^2}ds = \frac{\pi}{2}$$

Example 3.56: Find the Fourier transforms of the functions

(i) $f(t)=e^{-a|t|};\ a>0$ (ii) $f(t)=te^{-a|t|};\ a>0$

(iii) $e^{-t}U(t)$ where $U(t)$ is unit step function

Solution: (i) $f(t)=e^{-a|t|};\ a>0$

$$\therefore \qquad f(t)=\begin{cases} e^{-at}; & t\geq 0 \\ e^{at}\ ; & t<0 \end{cases}$$

$$\therefore \qquad \mathcal{F}(f(t)) = \int_{-\infty}^{\infty} f(t)e^{-i\omega t}dt = \int_{-\infty}^{0} e^{at}e^{-i\omega t}dt + \int_0^\infty e^{-at}e^{-i\omega t}dt$$

$$= \frac{1}{a-i\omega}\left(e^{at}e^{-i\omega t}\right)_{-\infty}^{0} - \frac{1}{a+i\omega}\left(e^{-at}e^{-i\omega t}\right)_0^\infty$$

$$= \frac{1}{a-i\omega}+\frac{1}{a+i\omega} = \frac{2a}{\omega^2+a^2} \qquad (\because\ a>0)$$

(ii) $f(t)=te^{-a|t|};\ a>0$

$$\therefore \qquad f(t)=\begin{cases} t\,e^{at}\ ; & t<0 \\ t\,e^{-at}; & t\geq 0 \end{cases}$$

$$\therefore \qquad \mathcal{F}(f(t)) = \int_{-\infty}^{\infty} f(t)e^{-i\omega t}dt = \int_{-\infty}^{0} te^{at}e^{-i\omega t}dt + \int_0^\infty te^{-at}e^{-i\omega t}dt$$

$$= \left[t\left(\frac{1}{(a-i\omega)}e^{(a-i\omega)t}\right) - \left(\frac{1}{(a-i\omega)^2}e^{(a-i\omega)t}\right)\right]_{-\infty}^{0}$$

$$+\left[\left(\frac{t}{-(a+i\omega)}e^{-(a+i\omega)t} - \frac{1}{(a+i\omega)^2}e^{-(a+i\omega)t}\right)\right]_0^\infty$$

Now, $\lim\limits_{t\to-\infty} t\,e^{at} = \lim\limits_{t\to-\infty}\dfrac{t}{e^{-at}} = \lim\limits_{t\to-\infty}\dfrac{1}{-a\,e^{-at}}$ (L' Hospital rule)

$$= 0 \qquad (\because\ a>0)$$

and $$\lim_{t\to\infty} t\,e^{-at} = \lim_{t\to\infty} \frac{t}{e^{at}} = \lim_{t\to\infty} \frac{1}{a\,e^{at}} \qquad (L'\text{ Hospital rule})$$

$$= 0 \qquad (\because\ a > 0)$$

$\therefore$ $$\mathcal{F}\left(f(t)\right) = -\frac{1}{(a-i\omega)^2} + \frac{1}{(a+i\omega)^2} = \frac{-(a+i\omega)^2 + (a-i\omega)^2}{\left(\omega^2+a^2\right)^2}$$

$$= -\frac{4ai\omega}{\left(\omega^2+a^2\right)^2}$$

Other Method

from part (i)

$$\mathcal{F}\left(e^{-a|t|}\right) = \frac{2a}{\omega^2+a^2}$$

$\therefore$ $$\mathcal{F}\left(t\,e^{-a|t|}\right) = i\,\frac{d}{d\omega}\,\frac{2a}{\omega^2+a^2} = -\frac{4ai\omega}{\left(\omega^2+a^2\right)^2}$$

(iii) $$f(t) = e^{-t}U(t) = \begin{cases} 0 & ;\ t<0 \\ e^{-t}; & t \ge 0 \end{cases}$$

$\therefore$ $$\mathcal{F}\left(f(t)\right) = \int_{-\infty}^{\infty} f(t) e^{-i\omega t}\,dt = \int_0^{\infty} e^{-t} e^{-i\omega t}\,dt;$$

$$= -\frac{1}{(1+i\omega)}\left[e^{-(t+i\omega t)}\right]_0^{\infty} = \frac{1}{1+i\omega} = \frac{1-i\omega}{\omega^2+1}$$

Example 3.57: Find the Fourier transform of the function $f(x) = e^{-ax^2};\ a>0$ and hence find the Fourier transform of $e^{-x^2/2}$.

Solution: $$f(x) = e^{-ax^2};\ a>0$$

$\therefore$ $$\mathcal{F}\left(f(x)\right) = \int_{-\infty}^{\infty} e^{-ax^2} e^{-i\omega x}\,dx = \int_{-\infty}^{\infty} e^{-\left(ax^2+i\omega x\right)}dx$$

$$= \int_{-\infty}^{\infty} e^{-\left(\sqrt{a}\,x+\frac{i\omega}{2\sqrt{a}}\right)^2} \cdot e^{-\omega^2/4a}\,dx$$

$$= \frac{1}{\sqrt{a}} e^{-\omega^2/4a} \int_{-\infty}^{\infty} e^{-y^2}\,dy = \sqrt{\frac{\pi}{a}}\; e^{-\omega^2/4a} \qquad \left(\text{taking } \sqrt{a}x + \frac{i\omega}{2\sqrt{a}} = y\right)$$

Take $a = \dfrac{1}{2}$

$$\mathcal{F}\left(e^{-x^2/2}\right) = \sqrt{2\pi}\; e^{-\omega^2/2}$$

Example 3.58: Obtain the Fourier transform of the function $f(x)$ given by

$$f(x)=\begin{cases}1-x^2 & ; |x|\le 1\\ 0 & ; \text{otherwise}\end{cases}$$

and hence evaluate $\int_0^{\infty}\frac{x\cos x-\sin x}{x^3}\cos\frac{x}{2}\,dx.$

Solution:

$$\mathcal{F}(f(x))=\int_{-\infty}^{\infty}f(x)e^{-i\omega x}dx=\int_{-1}^{1}(1-x^2)e^{-i\omega x}dx$$

$$=\int_{-1}^{1}(1-x^2)(\cos(\omega x)-i\sin(\omega x))\,dx$$

Now, $(1-x^2)\cos(\omega x)$ is an even function of x

and $(1-x^2)\sin(\omega x)$ is an odd function of x

$\therefore$

$$\mathcal{F}(f(x))=2\int_0^1(1-x^2)\cos(\omega x)\,dx$$

$$=2\left[(1-x^2)\left(\frac{1}{\omega}\sin(\omega x)\right)+2x\left(-\frac{1}{\omega^2}\cos(\omega x)\right)-2\left(-\frac{1}{\omega^3}\sin(\omega x)\right)\right]_0^1$$

$$=4\left(-\frac{1}{\omega^2}\cos\omega+\frac{1}{\omega^3}\sin\omega\right)$$

$$=\frac{4}{\omega^3}(\sin\omega-\omega\cos\omega)=F(\omega)$$

By inversion formula

$$f(x)=\frac{1}{2\pi}\int_{-\infty}^{\infty}\frac{4}{\omega^3}(\sin\omega-\omega\cos\omega)e^{i\omega x}d\omega$$

$$=\frac{1}{2\pi}\int_{-\infty}^{\infty}\frac{4}{\omega^3}(\sin\omega-\omega\cos\omega)(\cos(\omega x)+i\sin(\omega x))\,dx$$

Now, $\frac{1}{\omega^3}(\sin\omega-\omega\cos\omega)\cos(\omega x)$ is an even function of ω

and $\frac{1}{\omega^3}(\sin\omega-\omega\cos\omega)\sin(\omega x)$ is an odd function of ω

$\therefore$

$$f(x)=\frac{4}{\pi}\int_0^{\infty}\frac{\cos(\omega x)}{\omega^3}(\sin\omega-\omega\cos\omega)\,d\omega$$

$\therefore$

$$\frac{4}{\pi}\int_0^{\infty}\frac{\cos(\omega x)}{\omega^3}(\sin\omega-\omega\cos\omega)d\omega=\begin{cases}1-x^2; & |x|\le 1\\ 0 & ; \text{otherwise}\end{cases}$$

Take $x=\frac{1}{2}$

$$\int_0^{\infty}\frac{1}{\omega^3}(\sin\omega-\omega\cos\omega)\cos\frac{\omega}{2}d\omega=\frac{\pi}{4}\left[1-\frac{1}{4}\right]=\frac{3\pi}{16}$$

$\therefore$

$$\int_0^{\infty}\frac{x\cos x-\sin x}{x^3}\cos\frac{x}{2}dx=-\frac{3\pi}{16}$$

Example 3.59: Find the Fourier transform of

$$f(x)=\begin{cases}1-|x|; & |x|<1\\ 0 \quad ; & |x|>1\end{cases}$$

and hence evaluate $\int_0^\infty \frac{\sin^2 x}{x^2}dx.$

Solution: $\mathcal{F}(f(x))=\int_{-\infty}^{\infty} f(x)e^{-i\omega x}dx$

$$=\int_{-1}^{1}(1-|x|)e^{-i\omega x}dx=\int_{-1}^{1}(1-|x|)(\cos(\omega x)-i\sin(\omega x))dx$$

Now, $(1-|x|)\cos(\omega x)$ is an even function of x

and $(1-|x|)\sin(\omega x)$ is an odd function of x

$$\therefore \quad \mathcal{F}(f(x))=2\int_0^1(1-x)\cos(\omega x)\,dx=2\left[(1-x)\left(\frac{1}{\omega}\sin(\omega x)\right)-\frac{1}{\omega^2}\cos(\omega x)\right]_0^1$$

$$=\frac{2}{\omega^2}(1-\cos\omega)=F(\omega)$$

By inversion formula

$$\frac{1}{2\pi}\int_{-\infty}^{\infty}\frac{2(1-\cos\omega)}{\omega^2}e^{i\omega x}d\omega=f(x)=\begin{cases}1-|x|; & |x|<1\\ 0 \quad ; & |x|>1\end{cases}$$

or

$$\int_{-\infty}^{\infty}\frac{1-\cos\omega}{\omega^2}(\cos\omega x+i\sin\omega x)d\omega=\begin{cases}\pi(1-|x|); & |x|<1\\ 0 \quad ; & |x|>1\end{cases}$$

Now, $\frac{1-\cos\omega}{\omega^2}\cos(\omega x)$ is an even function of ω and

$\frac{1-\cos\omega}{\omega^2}\sin(\omega x)$ is an odd function of ω

$$\therefore \quad 2\int_0^\infty\frac{1-\cos\omega}{\omega^2}\cos(\omega x)\,d\omega=\begin{cases}\pi(1-|x|); & |x|<1\\ 0 \quad ; & |x|>1\end{cases}$$

Take $x=0$

$$\int_0^\infty\frac{1-\cos\omega}{\omega^2}d\omega=\frac{\pi}{2}$$

$$\therefore \quad \int_0^\infty\frac{2\sin^2\frac{\omega}{2}}{\omega^2}d\omega=\frac{\pi}{2}$$

Take $\omega = 2x$

$$\int_0^{\infty} \frac{\sin^2 x}{x^2} dx = \frac{\pi}{2}$$

Example 3.60: Find the Fourier transform of $f(x) = \begin{cases} 1; & |x| < a \\ 0; & |x| > a \end{cases}$ where $a > 0$

and evaluate $\int_{-\infty}^{\infty} \frac{\sin(as)\cos(sx)}{s} ds$ and $\int_0^{\infty} \frac{\sin x}{x} dx.$

Solution: $\mathcal{F}(f(x)) = \int_{-\infty}^{\infty} f(x) e^{-isx} dx$

$$= \int_{-a}^{a} e^{-isx} dx = \int_{-a}^{a} (\cos(sx) - i\sin(sx)) dx$$

$$= 2\int_0^{a} \cos(sx) dx = \frac{2}{s}(\sin(sx))_0^a = \frac{2\sin(sa)}{s} = F(s)$$

By inversion formula

$$\frac{1}{2\pi}\int_{-\infty}^{\infty} \frac{2\sin(sa)}{s} e^{isx} ds = f(x) = \begin{cases} 1; & |x| < a \\ 0; & |x| > a \end{cases}$$

$$\therefore \quad \int_{-\infty}^{\infty} \frac{\sin(sa)}{s}(\cos(sx) + i\sin(sx)) ds = \begin{cases} \pi; & |x| < a \\ 0; & |x| > a \end{cases}$$

Now $\frac{\sin(sa)}{s}\sin(sx)$ is an odd function of s

$$\therefore \quad \int_{-\infty}^{\infty} \frac{\sin(sa)\cos(sx)}{s} ds = \begin{cases} \pi; & |x| < a \\ 0; & |x| > a \end{cases}$$

Take $x = 0$ and $a = 1$,

$$\int_{-\infty}^{\infty} \frac{\sin s}{s} ds = \pi$$

$$\Rightarrow \quad 2\int_0^{\infty} \frac{\sin s}{s} ds = \pi$$

$$\Rightarrow \quad \int_0^{\infty} \frac{\sin x}{x} dx = \frac{\pi}{2}$$

Example 3.61: Find the complex Fourier transform of dirac delta function $\delta(t-a)$. Deduce the Fourier transform of unity in terms of generalized function dirac delta.
Solution: By definition

$$\delta(t-a) = \lim_{\varepsilon \to 0+} \delta_{\varepsilon}(t-a)$$

where $$\delta_\varepsilon(t-a)=\begin{cases}\dfrac{1}{\varepsilon}; & a<t<a+\varepsilon\\ 0\ ; & \text{otherwise}\end{cases}$$

$\therefore$ $$\mathcal{F}\,\delta(t-a)=\lim_{\varepsilon\to 0+}\int_{-\infty}^{\infty}e^{-i\omega t}\,\delta_\varepsilon(t-a)\,dt$$

$$=\lim_{\varepsilon\to 0+}\int_a^{a+\varepsilon}\frac{1}{\varepsilon}e^{-i\omega t}\,dt$$

$$=\lim_{\varepsilon\to 0+}\left\{-\frac{1}{i\omega\varepsilon}\left(e^{-i\omega t}\right)_a^{a+\varepsilon}\right\}=-\frac{e^{-i\omega a}}{i\omega}\lim_{\varepsilon\to 0+}\frac{e^{-i\omega\varepsilon}-1}{\varepsilon}$$

$$=\frac{e^{-i\omega a}}{i\omega}\lim_{\varepsilon\to 0+}\frac{1-e^{-i\omega\varepsilon}}{\varepsilon}$$

$$=\frac{e^{-i\omega a}}{i\omega}\lim_{\varepsilon\to 0+}i\omega e^{-i\omega\varepsilon}\qquad \text{(L' Hospital rule)}$$

$$=e^{-i\omega a}$$

Note that if $a=0$, then $\mathcal{F}(\delta(t))=1$.

By inversion formula

$$\mathcal{F}^{-1}(1)=\frac{1}{2\pi}\int_{-\infty}^{\infty}e^{i\omega t}\,d\omega=\delta(t)$$

$\Rightarrow$ $$\int_{-\infty}^{\infty}e^{i\omega t}\,d\omega=2\pi\delta(t)$$

Interchange t and ω

$$\int_{-\infty}^{\infty}e^{i\omega t}\,dt=2\pi\delta(\omega)\tag{1}$$

Change ω to $-\omega$

$$\int_{-\infty}^{\infty}e^{-i\omega t}\,dt=2\pi\delta(-\omega)\tag{2}$$

From (1), $$\delta(\omega)=\frac{1}{2\pi}\int_{-\infty}^{\infty}e^{i\omega t}\,dt$$

$\therefore$ $$\delta(-\omega)=\frac{1}{2\pi}\int_{-\infty}^{\infty}e^{-i\omega t}\,dt=-\frac{1}{2\pi}\int_{\infty}^{-\infty}e^{i\omega x}\,dx\qquad (\text{taking } t=-x)$$

$$=\frac{1}{2\pi}\int_{-\infty}^{\infty}e^{i\omega t}\,dt=\delta(\omega)$$

$\therefore$ $\delta(\omega)$ is an even function of ω

$\therefore$ from (2), $$\int_{-\infty}^{\infty}e^{-i\omega t}\,dt=2\pi\delta(\omega)$$

$\therefore$ $$\mathcal{F}(1)=2\pi\delta(\omega)$$

Example 3.62: Find the Fourier transform of signum function sign(t) and Heaviside function $H(t)$.

Solution: Consider the function

$$f(t)=\begin{cases}-e^{at}; & t<0\\ e^{-at}\ ; & t>0\end{cases} \quad \text{where } a>0$$

$$\mathscr{F}(f(t))=\int_{-\infty}^{\infty} f(t)e^{-i\omega t}dt$$

$$=\int_{-\infty}^{0} -e^{at}e^{-i\omega t}dt+\int_{0}^{\infty} e^{-at}e^{-i\omega t}dt$$

$$=-\frac{1}{a-i\omega}\left(e^{at}e^{-i\omega t}\right)_{-\infty}^{0}-\frac{1}{a+i\omega}\left(e^{-at}e^{-i\omega t}\right)_{0}^{\infty}$$

$$=-\frac{1}{a-i\omega}+\frac{1}{a+i\omega}=-\frac{2i\omega}{\omega^2+a^2} \qquad (\because\ a>0)$$

when $a\to 0+,\ f(t)\to \text{sign}(t)$

$$\therefore \qquad \mathscr{F}(\text{sign}(t))=\lim_{a\to 0+}\frac{-2i\omega}{\omega^2+a^2}=-\frac{2i}{\omega}$$

Now, Heaviside function $H(t)$ is defined as

$$H(t)=\begin{cases}0 & ; \ t<0\\ 1 & ; \ t\geq 0\end{cases}$$

$$\therefore \qquad H(t)=\begin{cases}\frac{1}{2}[1+\text{sign}(t)], & t\neq 0\\ 1 & , t=0\end{cases}$$

$$\therefore \qquad \mathscr{F}(H(t))=\frac{1}{2}[\mathscr{F}(1)+\mathscr{F}\text{sign}(t)]$$

$$=\frac{1}{2}\left[2\pi\ \delta(\omega)-\frac{2i}{\omega}\right]$$

$$=\pi\ \delta(\omega)-\frac{i}{\omega}$$

It should be noted that Fourier transforms of Unity and Heaviside functions are only in terms of generalized function dirac delta which in actual practice is not a function. Thus, actually in terms of definitions of functions, Fourier transforms of Unity and Heaviside functions do not exist. The reason is $\lim\limits_{t\to-\infty}\cos\omega t,\ \lim\limits_{t\to\infty}\cos\omega t,\ \lim\limits_{t\to-\infty}\sin\omega t,\ \lim\limits_{t\to\infty}\sin\omega t$ do not exist.

Example 3.63: Find the Fourier transform of xe^{-ax^2}; $a>0$.

Solution:

$$\mathscr{F}\left(e^{-ax^2}\right)=\int_{-\infty}^{\infty} e^{-ax^2}e^{-i\omega x}dx$$

$$=\int_{-\infty}^{\infty} e^{-\left(\sqrt{a}x+\frac{i\omega}{2\sqrt{a}}\right)^2} e^{\frac{-\omega^2}{4a}}dx$$

$$= \int_{-\infty}^{\infty} e^{-y^2} e^{-\omega^2/4a} \frac{1}{\sqrt{a}} dy \qquad \left(y = \sqrt{a}x + \frac{i\omega}{2\sqrt{a}} \right)$$

$$= \sqrt{\frac{\pi}{a}}\, e^{-\omega^2/4a}$$

Now, if $\qquad f(x) = e^{-ax^2}$

then $\qquad f'(x) = -2ax\, e^{-ax^2}$

$\therefore \qquad \mathcal{F}\left(f'(x)\right) = \mathcal{F}\left(-2ax\, e^{-ax^2}\right) = (i\omega)\mathcal{F}\left(f(x)\right) = (i\omega)\sqrt{\frac{\pi}{a}}\, e^{-\omega^2/4a}$

$\therefore \qquad \mathcal{F}\left(x e^{-ax^2}\right) = \frac{-i\omega}{2a}\sqrt{\frac{\pi}{a}}\, e^{-\omega^2/4a}$

Example 3.64: Find the Fourier transform of $f(x) = \begin{cases} 1; & |x| < a \\ 0; & |x| > a \end{cases}$

and hence find $\mathcal{F}\left[f(x)\left(1 + \cos\frac{\pi x}{a}\right)\right]$.

Solution: $\qquad \mathcal{F}\left(f(x)\right) = \int_{-\infty}^{\infty} f(x) e^{-i\omega x} dx = \int_{-a}^{a} e^{-i\omega x} dx$

$$= \int_{-a}^{a} \left(\cos(\omega x) - i\sin(\omega x)\right) dx$$

$\cos(\omega x)$ is an even function of x

and $\sin(\omega x)$ is an odd function of x

$\therefore \qquad \mathcal{F} f(x) = 2\int_0^a \cos(\omega x)\, dx = \frac{2}{\omega}\left(\sin(\omega x)\right)_0^a = \frac{2}{\omega}\sin(\omega a)$

$\therefore$ By linearity and modulation properties

$$\mathcal{F}\left[f(x)\left(1 + \cos\frac{\pi x}{a}\right)\right] = \mathcal{F} f(x) + \mathcal{F}\left(f(x)\cos\frac{\pi x}{a}\right)$$

$$= \frac{2\sin(\omega a)}{\omega} + \frac{\sin\left(\omega + \frac{\pi}{a}\right)a}{\omega + \frac{\pi}{a}} + \frac{\sin\left(\omega - \frac{\pi}{a}\right)a}{\omega - \frac{\pi}{a}}$$

$$= \frac{2\sin(\omega a)}{\omega} - \frac{a\sin(\omega a)}{\pi + \omega a} + \frac{a\sin(\omega a)}{\pi - a\omega}$$

$$= \sin(\omega a)\left[\frac{2\left(\pi^2 - \omega^2 a^2\right) - a\omega(\pi - a\omega) + a\omega(\pi + \omega a)}{\omega(\pi + \omega a)(\pi - \omega a)}\right]$$

$$= \frac{2\pi^2 \sin(\omega a)}{\omega\left(\pi^2 - \omega^2 a^2\right)}$$

Example 3.65: Find the Fourier transform of $f(x)=e^{-ax}U(x)$, $a>0$ where $U(x)$ is unit step function. Hence, prove that $\mathcal{F}\left(x^n e^{-ax}U(x)\right)=\dfrac{n!}{(a+is)^{n+1}}$.

Solution:

$$\mathcal{F}(f(x))=\int_{-\infty}^{\infty} e^{-ax}U(x)e^{-isx}\,dx$$

$$=\int_0^{\infty} e^{-ax}e^{-isx}\,dx$$

$$=-\frac{1}{a+is}\left(e^{-ax}e^{-isx}\right)_0^{\infty}$$

$$=\frac{1}{a+is}$$

Differentiate n times w.r.t. 's'

$$\int_0^{\infty} e^{-ax}(-ix)^n e^{-isx}\,dx=\frac{(-1)^n\, n!\, i^n}{(a+is)^{n+1}}$$

$\therefore$

$$\int_0^{\infty} e^{-ax}x^n e^{-isx}\,dx=\frac{n!}{(a+is)^{n+1}}$$

$\Rightarrow$

$$\mathcal{F}\left(e^{-ax}x^n\,\mathrm{U}(x)\right)=\frac{n!}{(a+is)^{n+1}}$$

Example 3.66: Show that the Fourier transform of $e^{-x^2/2}$ is self-reciprocal.

Solution: We take $\mathcal{F}(f(t))=\dfrac{1}{\sqrt{2\pi}}\displaystyle\int_{-\infty}^{\infty} f(t)e^{-i\omega t}\,dt=F(\omega)$

then its reciprocal is $\dfrac{1}{\sqrt{2\pi}}\displaystyle\int_{-\infty}^{\infty} F(\omega)e^{i\omega t}\,d\omega$

Now,

$$f(x)=e^{-x^2/2}$$

$\therefore$

$$\mathcal{F}(f(x))=\frac{1}{\sqrt{2\pi}}\int_{-\infty}^{\infty} e^{-x^2/2}e^{-i\omega x}\,dx$$

$$=\frac{1}{\sqrt{2\pi}}\int_{-\infty}^{\infty} e^{-\left(\frac{x}{\sqrt{2}}+\frac{i\omega}{\sqrt{2}}\right)^2} e^{-\omega^2/2}\,dx$$

Put

$$\frac{x}{\sqrt{2}}+\frac{i\omega}{\sqrt{2}}=y$$

$\therefore$

$$\mathcal{F}f(x)=\frac{1}{\sqrt{2\pi}}\int_{-\infty}^{\infty} e^{-y^2}e^{-\omega^2/2}\sqrt{2}\,dy=\frac{1}{\sqrt{\pi}}\sqrt{\pi}\,e^{-\omega^2/2}$$

$$=e^{-\omega^2/2}=f(\omega)$$

Thus, Fourier transform of $e^{-x^2/2}$ is self-reciprocal, i.e., only variable name is changed.

Example 3.67: Find the Fourier transform of $f(t)=\dfrac{1}{5+it}$.

Solution: We have

$$\mathcal{F}\left(H(t)e^{-5t}\right)=\frac{1}{5+i\omega}$$

∵ By inversion formula

$$\frac{1}{2\pi}\int_{-\infty}^{\infty}\frac{e^{i\omega t}}{5+i\omega}\,d\omega=H(t)e^{-5t}$$

Interchange t and ω

$$\frac{1}{2\pi}\int_{-\infty}^{\infty}\frac{e^{i\omega t}}{5+it}\,dt=H(\omega)e^{-5\omega}$$

Change ω to $-\omega$

$$\int_{-\infty}^{\infty}\frac{e^{-i\omega t}}{5+it}\,dt=2\pi H(-\omega)e^{5\omega}$$

∴
$$\mathcal{F}\left(\frac{1}{5+it}\right)=\begin{cases}2\pi e^{5\omega}; & -\omega\geq 0\\ 0\ ; & -\omega<0\end{cases}$$

i.e.,
$$\mathcal{F}\left(\frac{1}{5+it}\right)=\begin{cases}2\pi e^{5\omega}; & \omega\leq 0\\ 0\ ; & \omega>0\end{cases}$$

Example 3.68: Find the inverse Fourier transforms of functions

(i) $\dfrac{e^{4i\omega}}{3+i\omega}$ (ii) $\dfrac{\sqrt{\pi}\,\omega e^{-\omega^2/8}}{4\sqrt{2}\,i}$

(iii) $\dfrac{1}{30+11i\omega-\omega^2}$ (iv) $\dfrac{1}{\left(4+\omega^2\right)\left(9+\omega^2\right)}$

Solution: (i) We have $\mathcal{F}^{-1}\left(\dfrac{1}{3+i\omega}\right)=H(t)e^{-3t}$

By shifting property

$$\mathcal{F}^{-1}\left(\frac{e^{4i\omega}}{3+i\omega}\right)=\mathcal{F}^{-1}\frac{e^{-(-4i)\omega}}{3+i\omega}=H(t-4)e^{-3(t-4)}$$

∴
$$\mathcal{F}^{-1}\left(\frac{e^{4i\omega}}{3+i\omega}\right)=\begin{cases}e^{-3(t-4)}\ ; & t\geq 4\\ 0\ ; & t<4\end{cases}$$

(ii) for $a>0$

$$\mathcal{F}\left(e^{-at^2}\right)=\int_{-\infty}^{\infty}e^{-at^2}e^{-i\omega t}\,dt=\int_{-\infty}^{\infty}e^{-\left(\sqrt{a}\,t+\frac{i\omega}{2\sqrt{a}}\right)^2}e^{-\omega^2/4a}\,dt$$

$$= \frac{1}{\sqrt{a}} e^{-\omega^2/4a} \int_{-\infty}^{\infty} e^{-y^2} dy \qquad \left(\text{taking } \sqrt{a}\, t + \frac{i\omega}{2\sqrt{a}} = y \right)$$

$$= \sqrt{\frac{\pi}{a}} e^{-\omega^2/4a}$$

Take $a = 2$

$$\mathcal{F}\left(e^{-2t^2}\right) = \sqrt{\frac{\pi}{2}} e^{-\omega^2/8}$$

$$\therefore \qquad \mathcal{F}\left(t\, e^{-2t^2}\right) = i \frac{d}{d\omega} \mathcal{F}\left(e^{-2t^2}\right) = i\sqrt{\frac{\pi}{2}}\left(-\frac{\omega}{4}\right) e^{-\omega^2/8}$$

$$\therefore \qquad \frac{\sqrt{\pi}\, \omega e^{-\omega^2/8}}{4\sqrt{2}\, i} = \mathcal{F}\left(t\, e^{-2t^2}\right)$$

$$\Rightarrow \qquad \mathcal{F}^{-1}\left(\frac{\sqrt{\pi}\, \omega e^{-\omega^2/8}}{4\sqrt{2}\, i}\right) = \left(t\, e^{-2t^2}\right)$$

(iii) $\mathcal{F}^{-1} \dfrac{1}{30 + 11i\omega - \omega^2} = \mathcal{F}^{-1} \dfrac{1}{(5+i\omega)(6+i\omega)}$

$$= \mathcal{F}^{-1}\left(\frac{1}{5+i\omega} - \frac{1}{6+i\omega}\right)$$

$$= \mathcal{F}^{-1}\left(\frac{1}{5+i\omega}\right) - \mathcal{F}^{-1}\left(\frac{1}{6+i\omega}\right)$$

$$= e^{-5t} H(t) - e^{-6t} H(t)$$

$$= \begin{cases} e^{-5t} - e^{-6t}; & t \ge 0 \\ 0 \qquad ; & t < 0 \end{cases}$$

(iv) $\mathcal{F}^{-1} \dfrac{1}{(4+\omega^2)(9+\omega^2)} = \mathcal{F}^{-1}\left[\dfrac{1}{5}\left(\dfrac{1}{4+\omega^2} - \dfrac{1}{9+\omega^2}\right)\right]$ (by suppression method)

$$= \frac{1}{5} \mathcal{F}^{-1}\left[\frac{1}{(2+i\omega)(2-i\omega)} - \frac{1}{(3+i\omega)(3-i\omega)}\right]$$

$$= \frac{1}{5} \mathcal{F}^{-1}\left[\frac{1}{4}\left\{\frac{1}{2+i\omega} + \frac{1}{2-i\omega}\right\} - \frac{1}{6}\left\{\frac{1}{3+i\omega} + \frac{1}{3-i\omega}\right\}\right]$$

(by suppression method)

Now, $\qquad \mathcal{F} e^{-at} H(t) = \dfrac{1}{a+i\omega}; \quad a > 0$

$$\therefore \qquad \mathcal{F} e^{at} H(-t) = \frac{1}{a - i\omega}; \; a > 0$$

$$\left(\because \text{ if } \mathcal{F}\{f(t)\} = F(\omega) \text{ then } \mathcal{F}\{f(-t)\} = F(-\omega)\right)$$

$$\therefore \qquad \mathcal{F}^{-1}\left\{\frac{1}{a + i\omega}\right\} = e^{-at} H(t); \;\; a > 0$$

and
$$\mathcal{F}^{-1}\left\{\frac{1}{a - i\omega}\right\} = e^{at} H(-t); \;\; a > 0$$

$$\therefore \qquad \mathcal{F}^{-1} \frac{1}{(4+\omega^2)(9+\omega^2)} = \frac{1}{5}\left[\frac{1}{4}\left\{e^{-2t} H(t) + e^{2t} H(-t)\right\} - \frac{1}{6}\left\{e^{-3t} H(t) + e^{3t} H(-t)\right\}\right]$$

$$\therefore \qquad \mathcal{F}^{-1} \frac{1}{(4+\omega^2)(9+\omega^2)} = \begin{cases} \frac{1}{5}\left(\frac{1}{4}e^{-2t} - \frac{1}{6}e^{-3t}\right) & ; \text{ if } t > 0 \\ \frac{1}{5}\left(\frac{1}{4}e^{2t} - \frac{1}{6}e^{3t}\right) & ; \text{ if } t < 0 \\ \frac{1}{5}\left(\frac{1}{2}\cosh 2t - \frac{1}{3}\cosh 3t\right) & ; \text{ if } t = 0 \end{cases}$$

or
$$\mathcal{F}^{-1} \frac{1}{(4+\omega^2)(9+\omega^2)} = \begin{cases} \frac{1}{60}\left(3e^{-2t} - 2e^{-3t}\right) & ; \; t > 0 \\ \frac{1}{60}\left(3e^{2t} - 2e^{3t}\right) & ; \; t < 0 \\ \frac{1}{30}\left(3\cosh 2t - 2\cosh 3t\right); & t = 0 \end{cases}$$

Example 3.69: Do Fourier sine and cosine transforms of exp (x) exist? Explain.

Solution: If Fourier sine transform of e^x exists, then

$$\mathcal{F}_s\left(e^x\right) = \int_0^\infty e^x \sin(\omega x)\, dx$$

$$= \left[\frac{e^x}{1+\omega^2}\left(\sin(\omega x) - \omega\cos(\omega x)\right)\right]_0^\infty$$

But $\lim\limits_{x\to\infty} e^x\left(\sin(\omega x) - \omega\cos(\omega x)\right)$ does not exist.

$\therefore \mathcal{F}_s\left(e^x\right)$ does not exist.

If Fourier cosine transform of e^x exists, then

$$\mathcal{F}_c\left(e^x\right) = \int_0^\infty e^x \cos(\omega x)\, dx$$

$$= \left[\frac{e^x}{1+\omega^2}\left(\cos(\omega x) + \omega\sin(\omega x)\right)\right]_0^\infty$$

But $\lim_{x\to\infty} e^{x}\left(\cos(\omega x)+\omega\sin(\omega x)\right)$ does not exist.

$\therefore$ Fourier cosine transform of e^{x} also does not exist.

Example 3.70: Find the Fourier sine transforms of

(i) $\dfrac{1}{x}$ (ii) $\dfrac{1}{x}e^{-ax};\ a>0$

Solution: (i) $\mathcal{F}_s\left(\dfrac{1}{x}\right)=\int_0^\infty \dfrac{1}{x}\sin(\omega x)\,dx$

Put $\omega x = y$

$$\therefore \quad \mathcal{F}_s\left(\frac{1}{x}\right)=\int_0^\infty \frac{\sin y}{y}\,dy=\frac{\pi}{2}$$

(ii)

$$\mathcal{F}_s\left(\frac{1}{x}e^{-ax}\right)=\int_0^\infty \frac{1}{x}e^{-ax}\sin(\omega x)\,dx \tag{1}$$

$$\therefore \quad \frac{d}{d\omega}\mathcal{F}_s\left(\frac{1}{x}e^{-ax}\right)=\int_0^\infty e^{-ax}\cos(\omega x)\,dx$$

$$=\left[\frac{e^{-ax}}{\omega^2+a^2}(-a\cos(\omega x)+\omega\sin(\omega x))\right]_0^\infty$$

$$=\frac{a}{\omega^2+a^2} \qquad (\because\ a>0)$$

$$\therefore \quad \mathcal{F}_s\left(\frac{1}{x}e^{-ax}\right)=\int\frac{a}{\omega^2+a^2}\,d\omega+c=\tan^{-1}\frac{\omega}{a}+c$$

For $\omega=0,\ \mathcal{F}_s\left(\dfrac{1}{x}e^{-ax}\right)=0$ (from (1))

$$\therefore \quad 0=0+c$$

$$\Rightarrow \quad c=0$$

$$\therefore \quad \mathcal{F}_s\left(\frac{1}{x}e^{-ax}\right)=\tan^{-1}\frac{\omega}{a}.$$

Example 3.71: Find the Fourier cosine transform of e^{-x^2}.

Solution:

$$\mathcal{F}_c\left(e^{-x^2}\right)=\int_0^\infty e^{-x^2}\cos(\omega x)\,dx$$

$$\therefore \quad \frac{d}{d\omega}\mathcal{F}_c\left(e^{-x^2}\right)=\int_0^\infty -x\,e^{-x^2}\sin(\omega x)\,dx=\frac{1}{2}\int_0^\infty \sin(\omega x)\left(-2x\,e^{-x^2}\right)dx$$

$$=\frac{1}{2}\left(\sin(\omega x)e^{-x^2}\right)_0^\infty-\frac{1}{2}\int_0^\infty \omega\cos(\omega x)e^{-x^2}\,dx \qquad \text{(integrating by parts)}$$

$$= -\frac{\omega}{2}\mathcal{F}_c\left(e^{-x^2}\right)$$

or $$\frac{d}{d\omega}\mathcal{F}_c(e^{-x^2}) + \frac{\omega}{2}\mathcal{F}_c(e^{-x^2}) = 0$$

It is Leibnitz's linear differential equation.

$$I.F = e^{\int \frac{\omega}{2} d\omega} = e^{\frac{\omega^2}{4}}$$

$\therefore$ Solution is $$e^{\frac{\omega^2}{4}}\mathcal{F}_c\left(e^{-x^2}\right) = K$$

$\therefore$ $$\mathcal{F}_c\left(e^{-x^2}\right) = K e^{-\frac{\omega^2}{4}}$$

For $\omega = 0$, $\mathcal{F}_c\left(e^{-x^2}\right) = \int_0^\infty e^{-x^2}\,dx = \frac{\sqrt{\pi}}{2} = K$

$\therefore$ $$\mathcal{F}_c\left(e^{-x^2}\right) = \frac{\sqrt{\pi}}{2}e^{-\frac{\omega^2}{4}}$$

Example 3.72: Find the Fourier cosine transform of $e^{-a^2x^2}$ and hence evaluate Fourier sine transform of $xe^{-a^2x^2}$.

Solution: $$\mathcal{F}_c\left(e^{-a^2x^2}\right) = \int_0^\infty e^{-a^2x^2}\cos(\omega x)\,dx \tag{1}$$

$$= \text{Re}\int_0^\infty e^{-a^2x^2}e^{i\omega x}\,dx$$

$$= \text{Re}\int_0^\infty e^{-\left(|a|x - \frac{i\omega}{2|a|}\right)^2} e^{-\frac{\omega^2}{4a^2}}\,dx$$

$$= e^{-\frac{\omega^2}{4a^2}}\int_0^\infty e^{-y^2}\frac{1}{|a|}\,dy \qquad \left(\text{taking } |a|x - \frac{i\omega}{2|a|} = y\right)$$

$$= \frac{\sqrt{\pi}}{2|a|}e^{-\frac{\omega^2}{4a^2}}$$

from (1), $$\frac{d}{d\omega}\mathcal{F}_c\left(e^{-a^2x^2}\right) = \int_0^\infty -xe^{-a^2x^2}\sin(\omega x)\,dx = -\mathcal{F}_s\left(xe^{-a^2x^2}\right)$$

$\therefore$ $$\mathcal{F}_s\left(xe^{-a^2x^2}\right) = -\frac{d}{d\omega}\left(\frac{\sqrt{\pi}}{2|a|}e^{-\frac{\omega^2}{4a^2}}\right)$$

$$= \frac{\sqrt{\pi}}{2|a|}\cdot\frac{\omega}{2a^2}e^{-\frac{\omega^2}{4a^2}}$$

$$= \frac{\sqrt{\pi}}{4|a|^3}\omega e^{-\frac{\omega^2}{4a^2}}.$$

Example 3.73: Find the Fourier cosine transform of $f(x)=\dfrac{1}{a^2+x^2}$. Hence derive Fourier sine transform of $\phi(x)=\dfrac{x}{a^2+x^2}$.

Solution:

$$\mathcal{F}_c\left(\frac{1}{a^2+x^2}\right)=\int_0^\infty \frac{\cos(\omega x)}{a^2+x^2}dx \tag{1}$$

$\therefore$

$$\frac{d}{d\omega}\mathcal{F}_c\left(\frac{1}{a^2+x^2}\right)=\int_0^\infty \frac{-x\sin(\omega x)}{a^2+x^2}dx=-\int_0^\infty \frac{x^2\sin(\omega x)}{x\left(a^2+x^2\right)}dx$$

$$=-\int_0^\infty \frac{\left(a^2+x^2-a^2\right)\sin(\omega x)}{x\left(a^2+x^2\right)}dx$$

$$=-\int_0^\infty \frac{\sin(\omega x)}{x}dx+a^2\int_0^\infty \frac{\sin(\omega x)}{x\left(a^2+x^2\right)}dx$$

$$=-\frac{\pi}{2}+a^2\int_0^\infty \frac{\sin(\omega x)}{x\left(a^2+x^2\right)}dx \tag{2}$$

$\therefore$

$$\frac{d^2}{d\omega^2}\mathcal{F}_c\left(\frac{1}{a^2+x^2}\right)=a^2\int_0^\infty \frac{\cos(\omega x)}{a^2+x^2}dx=a^2\mathcal{F}_c\left(\frac{1}{a^2+x^2}\right)$$

It is an ordinary differential equation.
Its solution is

$$\mathcal{F}_c\left(\frac{1}{a^2+x^2}\right)=c_1e^{-a\omega}+c_2e^{a\omega} \tag{3}$$

from (1),

when $\omega=0$,

$$\mathcal{F}_c\left(\frac{1}{a^2+x^2}\right)=\int_0^\infty \frac{dx}{a^2+x^2}=\left(\frac{1}{a}\tan^{-1}\frac{x}{a}\right)_0^\infty=\frac{\pi}{2a}$$

$\therefore$ from (3),

$$c_1+c_2=\frac{\pi}{2a} \tag{4}$$

from (1),

$$\lim_{a\to\infty}\mathcal{F}_c\left(\frac{1}{a^2+x^2}\right)=0$$

$\therefore$ from (3),

$$c_2=0$$

$\therefore$ from (4),

$$c_1=\frac{\pi}{2a}$$

Hence from (3),

$$\mathcal{F}_c\left(\frac{1}{a^2+x^2}\right)=\frac{\pi}{2a}e^{-a\omega}$$

$\therefore$

$$\int_0^\infty \frac{\cos(\omega x)}{a^2+x^2}dx=\frac{\pi}{2a}e^{-a\omega}$$

Differentiate w.r.t. 'ω'

$$\int_0^\infty \frac{-x\sin(\omega x)}{a^2+x^2}\,dx = -\frac{\pi}{2}e^{-a\omega}$$

$$\Rightarrow \quad \int_0^\infty \frac{x}{a^2+x^2}\sin(\omega x)\,dx = \frac{\pi}{2}e^{-a\omega}$$

$$\Rightarrow \quad \mathcal{F}_s(\phi(x)) = \mathcal{F}_s\left(\frac{x}{a^2+x^2}\right) = \frac{\pi}{2}e^{-a\omega}$$

Example 3.74: Find the Fourier sine and Fourier cosine transforms of

$$f(t) = \begin{cases} t; & 0 \le t \le l \\ 0; & t > l \end{cases}$$

Solution: We have

$$\mathcal{F}_c(f(t)) + i\mathcal{F}_s f(t) = \int_0^\infty f(t)(\cos(\omega t) + i\sin(\omega t))\,dt$$

$$= \int_0^l te^{i\omega t}\,dt$$

$$= \left(\frac{-it}{\omega}e^{i\omega t} + \frac{1}{\omega^2}e^{i\omega t}\right)_0^l$$

$$= -\frac{il}{\omega}e^{i\omega l} + \frac{1}{\omega^2}\left(e^{i\omega l} - 1\right)$$

$$= -\frac{il}{\omega}[\cos(\omega l) + i\sin(\omega l)] + \frac{1}{\omega^2}[\cos(\omega l) + i\sin(\omega l) - 1]$$

Equate real and imaginary parts

$$\mathcal{F}_c(f(t)) = \frac{l}{\omega}\sin(\omega l) + \frac{1}{\omega^2}\cos(\omega l) - \frac{1}{\omega^2} = \frac{1}{\omega^2}[l\omega\sin(\omega l) + \cos(\omega l) - 1]$$

$$\mathcal{F}_s(f(t)) = -\frac{l}{\omega}\cos(\omega l) + \frac{1}{\omega^2}\sin(\omega l) = \frac{1}{\omega^2}[\sin(\omega l) - l\omega\cos(\omega l)]$$

Example 3.75: Find the Fourier sine transform of $f(t) = \frac{1}{t}e^{-at};\ a > 0$. Deduce that $\mathcal{F}\left(\frac{1}{t}\right) = -i\,\pi$

Solution: We have

$$\mathcal{F}_s\left(e^{-at}\right) = \int_0^\infty e^{-at}\sin(\omega t)\,dt = \left[\frac{e^{-at}}{\omega^2+a^2}(-a\sin(\omega t) - \omega\cos(\omega t))\right]_0^\infty$$

$$= \frac{\omega}{\omega^2+a^2};\ a > 0$$

Integrate both sides w.r.t. 'a' within the limits a to ∞

$$\int_a^\infty \int_0^\infty e^{-at}\sin(\omega t)\,dt\,da = \int_a^\infty \frac{\omega}{\omega^2+a^2}\,da$$

$\Rightarrow$ $$\int_0^\infty \int_a^\infty e^{-at}\sin(\omega t)\,da\,dt = \left(\tan^{-1}\frac{a}{\omega}\right)_a^\infty \quad \text{(changing the order of integration)}$$

$\therefore$ $$\int_0^\infty \left(-\frac{1}{t}e^{-at}\right)_a^\infty \sin(\omega t)\,dt = \frac{\pi}{2} - \tan^{-1}\frac{a}{\omega}$$

$\therefore$ $$\int_0^\infty \frac{1}{t}e^{-at}\sin(\omega t)\,dt = \frac{\pi}{2} - \cot^{-1}\frac{\omega}{a} = \tan^{-1}\frac{\omega}{a}$$

$\therefore$ $$\mathcal{F}_s\left(\frac{1}{t}e^{-at}\right) = \tan^{-1}\frac{\omega}{a}$$

$\therefore$ $$\int_0^\infty \frac{e^{-at}}{t}\sin(\omega t)\,dt = \tan^{-1}\frac{\omega}{a};\ a>0$$

$\therefore$ $$\lim_{a\to 0+}\int_0^\infty \frac{e^{-at}}{t}\sin(\omega t)\,dt = \lim_{a\to 0+}\tan^{-1}\frac{\omega}{a}$$

$\therefore$ $$\int_0^\infty \frac{1}{t}\sin(\omega t)\,dt = \frac{\pi}{2} \qquad (1)$$

$$\mathcal{F}\left(\frac{1}{t}\right) = \int_{-\infty}^\infty \frac{1}{t}e^{-i\omega t}\,dt$$

$$= \int_{-\infty}^\infty \frac{1}{t}\left[\cos(\omega t) - i\sin(\omega t)\right]dt$$

$\frac{1}{t}\cos(\omega t)$ is an odd function of t and

$\frac{1}{t}\sin(\omega t)$ is an even function of t

$\therefore$ $$\mathcal{F}\left(\frac{1}{t}\right) = 2\int_0^\infty \frac{-i}{t}\sin(\omega t)\,dt = -2i\frac{\pi}{2} \qquad \text{from (1)}$$

$$= -i\pi.$$

Example 3.76: Find the inverse Fourier sine transforms of the functions

(i) $\frac{\omega}{\omega^2+1}$ (ii) $\frac{1}{\omega}e^{-a\omega};\ a>0$

Solution: (i) Let $$\mathcal{F}_s^{-1}\frac{\omega}{\omega^2+1} = f(x)$$

$\therefore$ $$\mathcal{F}_s(f(x)) = \frac{\omega}{\omega^2+1}$$

$\Rightarrow$ $$\mathcal{F}_s(f(x)) + \omega^2\mathcal{F}_s(f(x)) = \omega \qquad (1)$$

Now, $\mathcal{F}_s\left(f''(x)\right) = -\omega^2 \mathcal{F}_s f(x) + \omega f(0);$

when $\lim\limits_{x\to\infty} f(x) = \lim\limits_{x\to\infty} f'(x) = 0$

$\therefore \quad \mathcal{F}_s\left(f''(x)\right) + \omega^2 \mathcal{F}_s\left(f(x)\right) = \omega f(0)$ (2)

From (1) and (2), we are to find $f(x)$ such that

$$f''(x) = f(x); \ f(0) = 1, \ \lim_{x\to\infty} f(x) = \lim_{x\to\infty} f'(x) = 0$$

Now, solution of $f''(x) = f(x)$

is $f(x) = c_1 e^{-x} + c_2 e^{x}$ (3)

$f(0) = c_1 + c_2 = 1$ (4)

From (3), $\lim\limits_{x\to\infty} f(x) = 0$ is satisfied only when $c_2 = 0$

$\therefore$ from (4), $c_1 = 1$

$\therefore$ from (3), $f(x) = e^{-x}$ which satisfies $\lim\limits_{x\to\infty} f(x) = \lim\limits_{x\to\infty} f'(x) = 0$

(ii) $\mathcal{F}_s^{-1}\left(\frac{1}{\omega} e^{-a\omega}\right) = \frac{2}{\pi}\int_0^\infty \frac{1}{\omega} e^{-a\omega} \sin(\omega x)\, d\omega$ (1)

$$\therefore \quad \frac{d}{dx}\mathcal{F}_s^{-1}\left(\frac{1}{\omega} e^{-a\omega}\right) = \frac{2}{\pi}\int_0^\infty e^{-a\omega} \cos(\omega x)\, d\omega$$

$$= \frac{2}{\pi}\left[\frac{e^{-a\omega}}{x^2 + a^2}\left(-a\cos(\omega x) + \omega \sin(\omega x)\right)\right]_0^\infty$$

$$= \frac{2a}{\pi\left(x^2 + a^2\right)} \qquad (\because a > 0)$$

Integrate both sides w.r.t. x

$$\mathcal{F}_s^{-1}\left(\frac{1}{\omega} e^{-a\omega}\right) = \frac{2}{\pi}\tan^{-1}\frac{x}{a} + c$$

From (1), for $x = 0$, $\mathcal{F}_s^{-1}\left(\frac{1}{\omega} e^{-a\omega}\right) = 0 = 0 + c$

$\therefore \quad c = 0$

$$\therefore \quad \mathcal{F}_s^{-1}\left(\frac{1}{\omega} e^{-a\omega}\right) = \frac{2}{\pi}\tan^{-1}\frac{x}{a}.$$

Example 3.77: Solve the integral equation

$$\int_0^\infty f(x)\cos(sx)\,dx = \begin{cases} 1-s; & 0 \le s \le 1 \\ 0 \quad ; & s > 1 \end{cases}$$

and hence show that $\int_0^\infty \frac{\sin^2 t}{t^2}\,dt = \frac{\pi}{2}$.

Solution:
$$\int_0^\infty f(x)\cos sx\,dx = \begin{cases} 1-s; & 0 \le s \le 1 \\ 0 \quad ; & s > 1 \end{cases}$$

$\Rightarrow$
$$\mathfrak{F}_c f(x) = \begin{cases} 1-s; & 0 \le s \le 1 \\ 0 \quad ; & s > 1 \end{cases}$$

$\therefore$ by inversion formula

$$f(x) = \frac{2}{\pi}\int_0^\infty \left(\mathfrak{F}_c f(x)\right)\cos(sx)\,ds$$

$$= \frac{2}{\pi}\int_0^1 (1-s)\cos(sx)\,ds = \frac{2}{\pi}\left[(1-s)\left(\frac{1}{x}\sin(sx)\right) - \frac{1}{x^2}\cos(sx)\right]_0^1$$

$$= \frac{2}{\pi x^2}(1-\cos x);\ x > 0$$

$\therefore$
$$\mathfrak{F}_c f(x) = \int_0^\infty \frac{2(1-\cos x)}{\pi x^2}\cos(sx)\,dx = \begin{cases} 1-s\ ; & 0 \le s \le 1 \\ 0 \quad ; & s > 1 \end{cases}$$

$\therefore$
$$\lim_{s\to 0+} \mathfrak{F}_c\left(f(x)\right) = \int_0^\infty \frac{2(1-\cos x)}{\pi x^2}\,dx = 1$$

$\therefore$
$$\int_0^\infty \frac{1-\cos x}{x^2}\,dx = \frac{\pi}{2}$$

$\therefore$
$$\int_0^\infty \frac{2\sin^2 \frac{x}{2}}{x^2}\,dx = \frac{\pi}{2}$$

Put $\frac{x}{2} = y$

$$\int_0^\infty \frac{2\sin^2 y}{4y^2}\cdot 2\,dy = \frac{\pi}{2}$$

or
$$\int_0^\infty \frac{\sin^2 y}{y^2}\,dy = \frac{\pi}{2}$$

or
$$\int_0^\infty \frac{\sin^2 t}{t^2}\,dt = \frac{\pi}{2}$$

Example 3.78: Using convolution theorem, find $f(t)$ when

(i) $\mathcal{F}(f(t)) = \dfrac{1}{(1+i\omega)^2}$, $f(t)=0;\ t<0.$

(ii) $\mathcal{F}(f(t)) = \dfrac{1}{\omega(\omega^2-1)}$; $f(t)=0;\ t<0.$

Solution: (i) $f(t) = \mathcal{F}^{-1}\dfrac{1}{(1+i\omega)^2} = \mathcal{F}^{-1}\left(\dfrac{1}{1+i\omega}\cdot\dfrac{1}{1+i\omega}\right)$

and $\quad \mathcal{F}^{-1}\dfrac{1}{1+i\omega} = e^{-t}H(t)$

$\therefore$ by convolution theorem

$$\mathcal{F}^{-1}\frac{1}{(1+i\omega)^2} = e^{-t}H(t) * e^{-t}H(t)$$

$$= \int_{-\infty}^{\infty} e^{-\tau}H(\tau)e^{-(t-\tau)}H(t-\tau)\,d\tau$$

$$= e^{-t}\int_{-\infty}^{\infty} H(\tau)H(t-\tau)\,d\tau$$

Now, $$H(\tau)H(t-\tau) = \begin{cases} 1\ ; & \tau\ge 0,\ t-\tau\ge 0,\ \text{i.e., } t\ge\tau\ge 0 \\ 0\ ; & t<\tau<0 \end{cases}$$

$\therefore$ $$\mathcal{F}^{-1}\frac{1}{(1+i\omega)^2} = e^{-t}\int_0^t d\tau = te^{-t};\ t\ge 0$$

$\therefore$ $$f(t) = \begin{cases} te^{-t}\ ; & t\ge 0 \\ 0\ ; & t<0 \end{cases}$$

(ii) We have $\mathcal{F}^{-1}\left(\dfrac{1}{\omega}\right) = iH(t)$, $\mathcal{F}^{-1}\dfrac{1}{\omega^2-1} = \mathcal{F}^{-1}\dfrac{-1}{1-\omega^2} = -\sin(t)H(t)$

$\therefore$ by convolution theorem

$$\mathcal{F}^{-1}\frac{1}{\omega(\omega^2-1)} = -\sin(t)H(t) * iH(t)$$

$$= \int_{-\infty}^{\infty} -\sin\tau\, H(\tau)\,i\,H(t-\tau)\,d\tau$$

$$= -i\int_{-\infty}^{\infty} \sin\tau\, H(\tau)\, H(t-\tau)\,d\tau$$

Now, $$H(\tau)H(t-\tau) = \begin{cases} 1, & t\ge\tau\ge 0 \\ 0, & t<\tau<0 \end{cases}$$

$\therefore$ $$\mathcal{F}^{-1}\frac{1}{\omega(\omega^2-1)} = -\int_0^t i\sin\tau\, d\tau = i(\cos\tau)_0^t = -i(1-\cos t),\ t\ge 0$$

$\therefore$ $$f(t) = \begin{cases} -i(1-\cos t)\ ; & t\ge 0 \\ 0\ ; & t<0 \end{cases}$$

Example 3.79: Find the Fourier transform of $f(x)=\begin{cases}1-|x| & ; |x|<1 \\ 0 & ; |x|>1\end{cases}$ and hence find the value of $\int_0^\infty \frac{\sin^4 t}{t^4}\,dt.$

Solution:
$$\mathcal{F}(f(x))=\int_{-\infty}^{\infty} f(x)e^{-i\omega x}dx=\int_{-1}^{1}(1-|x|)e^{-i\omega x}dx$$
$$=\int_{-1}^{1}(1-|x|)(\cos(\omega x)-i\sin(\omega x))\,dx$$

Now, $(1-|x|)\cos(\omega x)$ is even function of x and

$(1-|x|)\sin(\omega x)$ is odd function of x

$$\therefore \qquad \mathcal{F}(f(x))=2\int_0^1(1-x)\cos(\omega x)\,dx$$
$$=2\left[(1-x)\left(\frac{1}{\omega}\sin(\omega x)\right)-\frac{1}{\omega^2}\cos(\omega x)\right]_0^1=\frac{2}{\omega^2}(1-\cos\omega)$$

By Parseval's identity
$$\frac{1}{2\pi}\int_{-\infty}^{\infty}\left|\mathcal{F}(f(x))\right|^2 d\omega=\int_{-\infty}^{\infty}|f(x)|^2\,dx$$

$$\therefore \quad \frac{1}{2\pi}\int_{-\infty}^{\infty}\frac{4(1-\cos\omega)^2}{\omega^4}d\omega=\int_{-1}^{1}(1-|x|)^2\,dx$$

or
$$\frac{4}{\pi}\int_0^\infty\frac{\left(2\sin^2\frac{\omega}{2}\right)^2}{\omega^4}d\omega=2\int_0^1(1-x)^2\,dx$$
$$\left(\because\ (1-|x|)^2 \text{ and } \frac{(1-\cos\omega)^2}{\omega^4} \text{ are even functions of } x \text{ and } \omega \text{ resp.}\right)$$

or
$$\frac{16}{\pi}\int_0^\infty\frac{\sin^4\frac{\omega}{2}}{\omega^4}d\omega=\frac{-2}{3}\left[(1-x)^3\right]_0^1$$

Put $\frac{\omega}{2}=t$

$$\frac{16}{\pi}\int_0^\infty\frac{\sin^4 t}{16t^4}.2\,dt=\frac{2}{3}$$

$$\Rightarrow \qquad \int_0^\infty\frac{\sin^4 t}{t^4}dt=\frac{\pi}{3}$$

Example 3.80: Using Parseval's identity, prove that

(i) $\int_0^\infty \left(\frac{\sin t}{t}\right)^2 dt = \frac{\pi}{2}$ (ii) $\int_0^\infty \frac{\sin(at)}{t(a^2+t^2)} dt = \frac{\pi}{2}\left(\frac{1-e^{-a^2}}{a^2}\right)$.

Solution:

(i) Let $f(x) = \begin{cases} 1; & -1 < x < 1 \\ 0; & \text{otherwise} \end{cases}$

$$\mathcal{F}(f(x)) = \int_{-1}^{1} e^{-i\omega x} dx = \int_{-1}^{1} (\cos(\omega x) - i\sin(\omega x))\, dx$$

$$= 2\int_0^1 \cos(\omega x)\, dx = \frac{2}{\omega}(\sin(\omega x))_0^1 = \frac{2}{\omega}\sin\omega$$

By Parseval's identity

$$\frac{1}{2\pi}\int_{-\infty}^{\infty} |\mathcal{F}(f(x))|^2 d\omega = \int_{-\infty}^{\infty} |f(x)|^2 dx$$

$\therefore$ $$\frac{1}{2\pi}\int_{-\infty}^{\infty} \frac{4\sin^2\omega}{\omega^2} d\omega = \int_{-1}^{1} dx = 2$$

$\Rightarrow$ $$\int_{-\infty}^{\infty} \frac{\sin^2 t}{t^2} dt = \pi$$

or $$2\int_0^\infty \left(\frac{\sin t}{t}\right)^2 dt = \pi$$

$\therefore$ $$\int_0^\infty \left(\frac{\sin t}{t}\right)^2 dt = \frac{\pi}{2}$$

(ii) Let $f(x) = e^{-ax};\ x > 0,\ g(x) = \begin{cases} 1\ ; & 0 < x < a \\ 0\ ; & x > a \end{cases}$ where $a > 0$

$$\mathcal{F}_c(f(x)) = \int_0^\infty e^{-ax}\cos(\omega x)\, dx = \left[\frac{e^{-ax}}{\omega^2 + a^2}(-a\cos(\omega x) + \omega\sin(\omega x))\right]_0^\infty$$

$$= \frac{a}{\omega^2 + a^2} = F_c(\omega)$$

$$\mathcal{F}_c(g(x)) = \int_0^a \cos(\omega x)\, dx = \frac{1}{\omega}(\sin(\omega x))_0^a = \frac{1}{\omega}\sin(\omega a) = G_c(\omega)$$

We have the identity

$$\frac{2}{\pi}\int_0^\infty F_c(\omega) G_c(\omega)\, d\omega = \int_0^\infty f(x) g(x)\, dx$$

$\therefore$ $$\frac{2}{\pi}\int_0^\infty \frac{a\sin(\omega a)}{\omega(a^2+\omega^2)} d\omega = \int_0^a e^{-ax} dx = -\frac{1}{a}\left(e^{-ax}\right)_0^a$$

$\therefore$ $$\int_0^\infty \frac{\sin(at)}{t(a^2+t^2)} dt = -\frac{\pi}{2a^2}\left(e^{-a^2} - 1\right) = \frac{\pi}{2}\left(\frac{1-e^{-a^2}}{a^2}\right).$$

Example 3.81: Evaluate $\int_0^\infty \frac{x^2}{(a^2+x^2)(b^2+x^2)}dx$ and hence find $\int_0^\infty \left(\frac{x}{x^2+1}\right)^2 dx$.

Solution: Let $f(x)=e^{-|a|x};\ x>0,\ g(x)=e^{-|b|x};\ x>0$

$$\mathcal{F}_s(f(x))=\int_0^\infty e^{-|a|x}\sin(\omega x)\,dx=\left[\frac{e^{-|a|x}}{\omega^2+a^2}\left(-|a|\sin(\omega x)-\omega\cos(\omega x)\right)\right]_0^\infty$$

$$=\frac{\omega}{\omega^2+a^2}=F_s(\omega)$$

Replacing a by b

$$\mathcal{F}_s(g(x))=\frac{\omega}{\omega^2+b^2}=G_s(\omega)$$

By Parseval's identity

$$\frac{2}{\pi}\int_0^\infty F_s(\omega)G_s(\omega)\,d\omega=\int_0^\infty f(x)g(x)\,dx$$

$$\therefore\quad \frac{2}{\pi}\int_0^\infty \frac{\omega^2}{(\omega^2+a^2)(\omega^2+b^2)}d\omega=\int_0^\infty e^{-(|a|+|b|)x}dx$$

$$\therefore\quad \int_0^\infty \frac{\omega^2}{(\omega^2+a^2)(\omega^2+b^2)}d\omega=\frac{-\pi}{2}\frac{1}{(|a|+|b|)}\left[e^{-(|a|+|b|)x}\right]_0^\infty$$

$$\therefore\quad \int_0^\infty \frac{x^2}{(x^2+a^2)(x^2+b^2)}dx=\frac{\pi}{2(|a|+|b|)}$$

Taking $a=b=1$

$$\int_0^\infty \left(\frac{x}{x^2+1}\right)^2 dx=\frac{\pi}{4}.$$

EXERCISE 3.5

1. Find the Fourier integral representation of the function $f(x)=\begin{cases}1, & |x|\le 1\\ 0, & |x|>1\end{cases}$ and hence evaluate $\int_0^\infty \frac{\sin\lambda\cos(\lambda x)}{\lambda}d\lambda.$ Also deduce the value of $\int_0^\infty \frac{\sin\lambda}{\lambda}d\lambda.$

2. Find the Fourier integral representation of the function $f(x)=\begin{cases}1-x^2, & |x|\le 1\\ 0, & |x|>1\end{cases}.$

3. Express $f(x)=\begin{cases}1, & 0\le x\le \pi\\ 0, & x>\pi\end{cases}$ as a Fourier sine integral and hence evaluate $\int_0^\infty \frac{1-\cos(\pi\lambda)}{\lambda}\sin(x\lambda)\,d\lambda.$

4. Determine the Fourier sine integral of the function $f(x)=\begin{cases} x \ , & 0<x<1 \\ 2-x, & 1<x<2. \\ 0 \ , & x>2 \end{cases}$

5. Using Fourier integral representation show that

$$\int_0^{\infty}\frac{\cos(x\alpha)+\alpha\sin(x\alpha)}{1+\alpha^2}\,d\alpha=\begin{cases} 0, & x<0 \\ \dfrac{\pi}{2}, & x=0. \\ \pi e^{-x}, & x>0 \end{cases}$$

6. Find (a) Fourier cosine integral and (b) Fourier sine integral of $f(x)=\begin{cases} \sin x, & 0\le x\le\pi \\ 0 \ , & x>\pi \end{cases}$ and show that

$$\int_0^{\infty}\frac{\sin(\omega x)\sin(\pi\omega)}{1-\omega^2}\,d\omega=\frac{\pi}{2}\sin x;\ 0\le x\le\pi.$$

7. Find the Fourier integral representation of the function

$$f(x)=\begin{cases} e^{ax} \ , & x\le 0 \\ e^{-ax} \ , & x\ge 0 \end{cases} \text{ for } a>0.$$

Hence show that

$$\int_0^{\infty}\frac{\cos(\omega x)}{\omega^2+a^2}\,d\omega=\frac{\pi}{2a}e^{-ax},\ x\ge 0.$$

8. Find the Fourier cosine integral of $f(x)=\begin{cases} \cos x, & |x|<\pi/2 \\ 0 \ , & |x|>\pi/2 \end{cases}$.

9. Find the Fourier cosine integral of $f(x)=e^{-x}\cos x;\ x\ge 0$

10. Find the Fourier integral representation of the function $f(x)=\begin{cases} 0, & x<0 \\ 1/2, & x=0. \\ e^{-x}, & x>0 \end{cases}$

11. Find the Fourier transform of the following functions defined on $(-\infty,\infty)$

(i) $f(x)=\begin{cases} 1, & 0<x<a \\ 0, & \text{otherwise} \end{cases}$.

(ii) $f(x)=\begin{cases} 1, & |x|\le a \\ 0, & |x|>a \end{cases}$.

(iii) $f(x)=\begin{cases} x, & 0<x<a \\ 0, & \text{otherwise} \end{cases}$.

(iv) $f(x)=\begin{cases} x, & |x|\le a \\ 0, & |x|>a \end{cases}$.

(v) $f(x)=\begin{cases} a, & -l<x<l \\ 0, & \text{otherwise} \end{cases}$.

(vi) $f(x)=\begin{cases} 0, & x<0 \\ e^{-\alpha x}, & x\ge 0 \end{cases}$.
where $\alpha>0$

(vii) $f(x)=\begin{cases} e^{-ax}, & x>0 \\ -e^{ax}, & x<0 \end{cases}$.
where $a>0$

(viii) $f(x)=\begin{cases} xe^{-x}, & x>0 \\ 0, & x<0 \end{cases}$.

(ix) $f(x)=\begin{cases} \dfrac{1}{2a}, & |x|\le a \\ 0, & |x|>a \end{cases}$.

(x) $f(x)=\begin{cases} a, & -l<x<0 \\ 0, & \text{otherwise} \end{cases}$.
where $a>0$

(xi) $f(x)=\begin{cases} a, & -l<x<0 \\ b, & 0<x<l \\ 0, & \text{otherwise} \end{cases}$.
where $a>0, b>0$

12. Find the Fourier transform of $f(x)=\begin{cases} \sin x, & 0<x<\pi \\ 0, & \text{otherwise} \end{cases}$.

Hence deduce that $\int_0^\infty \frac{\cos(\pi\omega/2)}{(1-\omega^2)} d\omega = \frac{\pi}{2}$.

13. Let $f(t) = \begin{cases} 1-\frac{t}{a}, & 0<t<a \\ 1+\frac{t}{a}, & -a<t<0. \\ 0, & \text{otherwise} \end{cases}$

Find $\mathcal{F}\{f(t)\}$.

14. Find the Fourier transform of $e^{-9t}U_0(t)$ where $U_0(t)$ is the unit step function.
15. Find the amplitude spectrum of the function

$f(t) = \begin{cases} 5, & -2 \le t \le 2 \\ 0, & \text{otherwise} \end{cases}$.

16. Find the Fourier transform of $f(x) = \begin{cases} a-|x|, & |x| \le a \\ 0, & |x| > a \end{cases}$.

Hence show that $\int_0^\infty \frac{\sin^2 x}{x^2} dx = \frac{\pi}{2}$.

17. Prove that $\mathcal{F}[\cos(ax)U(x)] = \frac{is}{(a^2-s^2)}$ where $U(x)$ is the unit step function.
18. Using modulation theorem, find the Fourier transform of $f(t)\cos bt$, where f is defined

by $f(t) = \begin{cases} 1, & |t| < a \\ 0, & |t| > a \end{cases}$.

19. Find the Fourier sine transform of $f(x)$ defined by $f(x) = \begin{cases} \sin x, & 0<x<a \\ 0, & x>a \end{cases}$.
20. Find the Fourier sine transform of e^{-ax}; $(a>0, x>0)$ and show that $\int_0^\infty \frac{x \sin mx}{1+x^2} dx = \frac{\pi}{2} e^{-m}$; $(m>0)$.
21. Find the Fourier sine transform of $e^{-|x|}$. Hence evaluate $\int_0^\infty \frac{x \sin mx}{1+x^2} dx \ (m>0)$.
22. Find the function $f(x)$ if

$$\int_0^\infty f(x)\sin(sx)\,dx = \begin{cases} 1, & 0 \le s < 1 \\ 2, & 1 \le s < 2. \\ 0, & s \ge 2 \end{cases}$$

23. Find the Fourier cosine transform of $f(x) = \begin{cases} \cos x, & 0<x<a \\ 0, & x>a \end{cases}$.
24. Find the Fourier cosine transform of $f(x) = e^{-2x} + 4e^{-3x}$.
25. Find $f(x)$ whose Fourier cosine transform is $\frac{\sin as}{s}$.
26. Find $f(x)$ if its Fourier cosine transform is $\frac{1}{(1+s^2)}$.
27. Solve the integral equation $\int_0^\infty f(x)\cos(sx)\,dx = e^{-s}\ (s \ge 0)$.
28. Find the Fourier sine and cosine transforms of the following functions

(i) $f(x) = \begin{cases} 1, & 0 \le x \le l \\ 0, & x > l \end{cases}$

(ii) $f(x) = \begin{cases} x, & 0<x<1 \\ 2-x, & 1<x<2 \\ 0, & x>2 \end{cases}$

(iii) $f(x) = 2e^{-5x} + 5e^{-2x}$

(iv) $f(x) = xe^{-ax}, a>0$

29. Find (a) Fourier cosine and (b) sine transform of $f(x) = e^{-ax}$ for $x>0$; $a>0$. Deduce that Laplace integrals $\int_0^\infty \frac{\cos \alpha x}{a^2+\alpha^2} d\alpha = \frac{\pi}{2a} e^{-ax}$ and $\int_0^\infty \frac{\alpha \sin \alpha x}{a^2+\alpha^2} d\alpha = -\frac{\pi}{2} e^{-ax}$.

30. Find Fourier sine and cosine transforms of x^{n-1}. Deduce that $\dfrac{1}{\sqrt{x}}$ is self-reciprocal with respect to both the transforms.

31. Using convolution theorem, find the inverse Fourier transform of $\dfrac{1}{\left(12+7is-s^2\right)}$.

32. Using convolution theorem, find the inverse Fourier transform of $\dfrac{1}{\left(6+5is-s^2\right)}$.

33. Using Parseval's identity for Fourier cosine and sine transforms of e^{-ax} evaluate

(i) $\displaystyle\int_0^\infty \frac{dx}{\left(a^2+x^2\right)^2}$ and

(ii) $\displaystyle\int_0^\infty \frac{x^2\,dx}{\left(a^2+x^2\right)^2}$ where $a>0$.

34. Using Parseval's identity, prove that $\displaystyle\int_0^\infty \frac{dt}{\left(a^2+t^2\right)\left(b^2+t^2\right)}=\frac{\pi}{2ab\left(a+b\right)}$.

35. Find the Fourier transform of $f\left(x\right)=\begin{cases}1, & |x|<a\\ 0, & |x|>a>0\end{cases}$. Hence prove that $\displaystyle\int_0^\infty \left(\frac{\sin t}{t}\right)^2 dt=\frac{\pi}{2}$.

36. Find the Fourier transform of $f\left(x\right)=\begin{cases}1-|x|, & |x|<1\\ 0, & |x|>1\end{cases}$. Hence find the value of $\displaystyle\int_0^\infty \frac{\sin^4 x}{x^4}\,dx$.

ANSWERS 3.5

1. $f\left(x\right)=\dfrac{2}{\pi}\displaystyle\int_0^\infty \frac{\cos\left(\lambda x\right)\sin\lambda}{\lambda}\,d\lambda;\quad \begin{cases}\pi/2, & |x|<1\\ 0, & |x|>1;\\ \pi/4, & |x|=1\end{cases}\quad \frac{\pi}{2}$

2. $f\left(x\right)=\dfrac{4}{\pi}\displaystyle\int_0^\infty \frac{\sin\lambda-\lambda\cos\lambda}{\lambda^3}\cos\lambda\; d\lambda$

3. $f\left(x\right)=\dfrac{2}{\pi}\displaystyle\int_0^\infty \frac{1-\cos\left(\pi\lambda\right)}{\lambda}\sin\left(x\lambda\right)d\lambda;\quad \begin{cases}\pi/2, & 0\le x<\pi\\ 0, & x>\pi\\ \pi/4, & x=\pi\end{cases}$

4. $f\left(x\right)=\dfrac{4}{\pi}\displaystyle\int_0^\infty \frac{1}{\lambda^2}\left(1-\cos\lambda\right)\sin\lambda\sin\left(\lambda x\right)d\lambda$

6. (a) $f\left(x\right)=\dfrac{2}{\pi}\displaystyle\int_0^\infty \frac{1+\cos\left(\lambda\pi\right)}{1-\lambda^2}\cos\left(\lambda x\right)d\lambda$

(b) $f\left(x\right)=\dfrac{2}{\pi}\displaystyle\int_0^\infty \frac{\sin\left(\lambda\pi\right)}{1-\alpha^2}\sin\left(\lambda x\right)d\lambda$

7. $f\left(x\right)=\dfrac{2a}{\pi}\displaystyle\int_0^\infty \frac{\cos\left(\lambda x\right)}{\lambda^2+a^2}\,d\lambda$

8. $f(x)=\frac{2}{\pi}\int_0^\infty \frac{\cos(\lambda x)\cos\left(\frac{\pi\lambda}{2}\right)}{(1-\lambda^2)}d\lambda$

9. $f(x)=\frac{2}{\pi}\int_0^\infty \frac{(\lambda^2+2)}{(\lambda^4+4)}\cos(\lambda x)\,d\lambda$

10. $f(x)=\frac{1}{\pi}\int_0^\infty \frac{\cos(\lambda x)+\lambda\sin(\lambda x)}{1+\lambda^2}d\lambda$

11. (i) $\frac{2}{\omega}e^{-i\omega a/2}\sin\frac{\omega a}{2}$ (ii) $\frac{2}{\omega}\sin(\omega a)$ (iii) $\frac{i}{\omega^2}\left(a\omega e^{-i\omega a}-2e^{-\frac{\omega a}{2}}\sin\frac{\omega a}{2}\right)$

(iv) $\frac{2i}{\omega^2}(a\omega\cos(\omega a)-\sin(\omega a))$ (v) $\frac{2a}{\omega}\sin(\omega l)$ (vi) $\frac{1}{\alpha+i\omega}$

(vii) $\frac{-2i\omega}{a^2+\omega^2}$ (viii) $\frac{1}{(1+i\omega)^2}$ (ix) $\frac{\sin(a\omega)}{a\omega}$ (x) $\frac{2a}{\omega}e^{\frac{i\omega l}{2}}\sin\frac{\omega l}{2}$

(xi) $\frac{2}{\omega}\sin\frac{\omega l}{2}\left(ae^{\frac{i\omega l}{2}}+be^{-\frac{i\omega l}{2}}\right)$

12. $\frac{1+e^{-i\pi\omega}}{1-\omega^2}$

13. $\frac{2}{a\omega^2}(1-\cos(\omega a))$

14. $\frac{1}{(9+i\omega)}$

15.

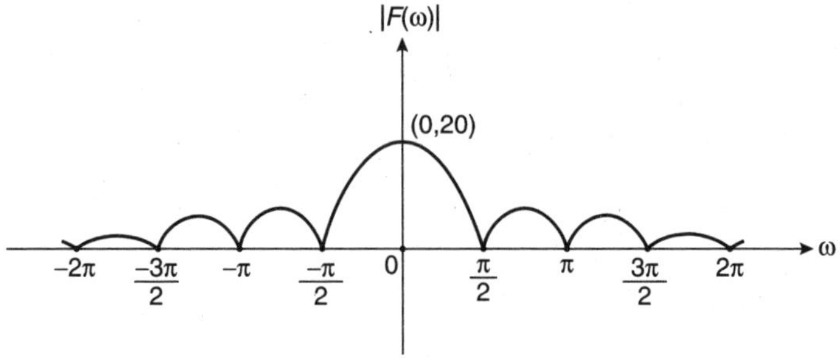

Figure 3.16

16. $\frac{2}{\omega^2}(1-\cos(\omega a))$

18. $\frac{\sin(\omega+b)a}{(\omega-b)}+\frac{\sin(\omega-b)a}{(\omega-b)}$

19. $\frac{1}{2}\left[\frac{\sin(\omega-1)a}{\omega-1}-\frac{\sin(\omega+1)a}{\omega+1}\right]$

20. $\frac{\omega}{\left(a^2+\omega^2\right)}$

21. $\frac{\omega}{\left(1+\omega^2\right)};\ \frac{\pi}{2}e^{-m}$

22. $\frac{2}{\pi x}(1+\cos x-2\cos 2x)$

23. $\frac{1}{2}\left[\frac{\sin((1+\omega)a)}{(1+\omega)}+\frac{\sin((1-\omega)a)}{(1-\omega)}\right]$

24. $2\left(\frac{1}{\omega^2+4}+\frac{6}{\omega^2+9}\right)$

25. $f(x)=\begin{cases}1, & x<a\\ 0, & x>a\end{cases}$

26. $e^{-x}, x\neq 0$

27. $\frac{2}{\pi\left(1+x^2\right)}$

28. (i) $\mathcal{F}_s(f(x))=\frac{1-\cos(\omega l)}{\omega}$, $\mathcal{F}_c(f(x))=\frac{\sin(\omega l)}{\omega}$

(ii) $\mathcal{F}_s(f(x))=\frac{2\sin\omega}{\omega^2}(1-\cos\omega)$, $\mathcal{F}_c(f(x))=\frac{2\cos\omega}{\omega^2}(1-\cos\omega)$

(iii) $\mathcal{F}_s(f(x))=\omega\left(\frac{2}{\omega^2+25}+\frac{5}{\omega^2+4}\right)$, $\mathcal{F}_c(f(x))=10\left(\frac{1}{\omega^2+25}+\frac{1}{\omega^2+4}\right)$

(iv) $\mathcal{F}_s(f(x))=\frac{2a\omega}{\left(a^2+\omega^2\right)^2}$, $\mathcal{F}_c(f(x))=\frac{a^2-\omega^2}{\left(a^2+\omega^2\right)^2}$

29. (a) $\mathcal{F}_c(f(x))=\frac{a}{\left(a^2+\omega^2\right)}$, (b) $\mathcal{F}_s(f(x))=\frac{\omega}{\left(a^2+\omega^2\right)}$

30. $\mathcal{F}_s\left(x^{n-1}\right)=\sqrt{\frac{2}{\pi}}\frac{\Gamma(n)}{\omega^n}\sin\frac{n\pi}{2}$, $\mathcal{F}_c\left(x^{n-1}\right)=\sqrt{\frac{2}{\pi}}\frac{\Gamma(n)}{\omega^n}\cos\frac{n\pi}{2}$

31. $\begin{cases} e^{-3t}\left(1-e^{-t}\right); & t \geq 0 \\ 0 & ; t < 0 \end{cases}$

32. $\begin{cases} e^{-2t}\left(1-e^{-t}\right), & t \geq 0 \\ 0 & , t = 0 \end{cases}$

33. (i) $\dfrac{\pi}{4a^3}$ (ii) $\dfrac{\pi}{4a}$

35. $\dfrac{2\sin a\omega}{\omega}$

36. $\dfrac{2(1-\cos\omega)}{\omega^2}; \dfrac{\pi}{3}$

3.8 APPLICATIONS OF FOURIER TRANSFORMS

Ordinary differential equations for the range of the variable $-\infty$ to ∞ can be solved by Fourier transform technique. If boundary conditions are not given then homogenous solution is found by writing auxiliary equation and particular integral is found by assuming boundary conditions and using Fourier transform.

Fourier transforms are also applied to solve boundary value problems. In one dimensional boundary value problems, the partial differential equation can easily be transformed into an ordinary differential equation by applying a suitable transform. The required solution is then obtained by solving this differential equation and taking the inverse transform. In a problem for finding $u(x,t)$ from partial differential equation, if $x \geq 0$ then Fourier sine or Fourier cosine transform is to be used. If $\left[\dfrac{\partial}{\partial x}u(x,t)\right]_{x=0}$ is given then Fourier cosine transform is used as Fourier cosine transform of $\dfrac{\partial^2 u}{\partial x^2}$ contains $\left[\dfrac{\partial}{\partial x}u(x,t)\right]_{x=0}$ and if $u(x,t)_{x=0}$ is given then we use Fourier sine transform.

Example 3.82: Find the solution of the following differential equation by using Fourier transform $\dfrac{dy}{dt} - 5y = u_0(t)e^{-5t}; \quad -\infty < t < \infty$ where $u_0(t)$ is unit step function.

Solution: *A.E.* is $m - 5 = 0$

$$\therefore \qquad m = 5$$

$$\therefore \qquad C.F. = c_1 e^{5t}$$

To find particular integral (P.I.), we take boundary conditions $\lim\limits_{t\to\pm\infty} y(t) = 0$.

Take Fourier transform of both sides of the given differential equation

$$(i\omega-5)Y(\omega)=\frac{1}{5+i\omega} \quad \text{where } Y(\omega)=\mathcal{F}\,y(t)$$

$$\therefore \qquad Y(\omega)=-\frac{1}{(5+i\omega)(5-i\omega)}$$

$$=-\frac{1}{10}\left(\frac{1}{5-i\omega}+\frac{1}{5+i\omega}\right) \quad \text{(by suppression method)}$$

$$\therefore \qquad y(t)=-\frac{1}{10}\left(\mathcal{F}^{-1}\frac{1}{5-i\omega}+\mathcal{F}^{-1}\frac{1}{5+i\omega}\right)$$

This $y(t)$ is particular integral

$$\therefore \qquad P.I.=-\frac{1}{10}\left(\mathcal{F}^{-1}\frac{1}{5-i\omega}+\mathcal{F}^{-1}\frac{1}{5+i\omega}\right) \tag{1}$$

Now, $$\mathcal{F}\,e^{-5t}u_0(t)=\frac{1}{5+i\omega}$$

$$\therefore \qquad \mathcal{F}\,e^{5t}u_0(-t)=\frac{1}{5-i\omega}$$

$$\left(\because \text{ if } \mathcal{F}(f(t))=F(\omega), \text{ then } \mathcal{F}\,f(-t)=F(-\omega)\right)$$

$$\therefore \qquad \mathcal{F}^{-1}\frac{1}{5-i\omega}=e^{5t}u_0(-t) \text{ and } \mathcal{F}^{-1}\frac{1}{5+i\omega}=e^{-5t}u_0(t)$$

$\therefore$ from (1)

$$P.I.=-\frac{1}{10}\left(e^{5t}u_0(-t)+e^{-5t}u_0(t)\right)=\begin{cases}-\dfrac{1}{10}e^{-5t}; & t>0\\[2mm] -\dfrac{1}{10}e^{5t}\ ; & t<0\\[2mm] -\dfrac{1}{5}\ ; & t=0\end{cases}$$

$\therefore$ general solution is

$$y(t)=\begin{cases}c_1e^{5t}-\dfrac{1}{10}e^{-5t} & ;\ t>0\\[2mm] c_1e^{5t}-\dfrac{1}{10}e^{5t}=c_2e^{5t}; & t<0\\[2mm] c_1-\dfrac{1}{5}=c & ;\ t=0\end{cases}$$

$\therefore$ general solution is

$$y(t)=\begin{cases} Ae^{5t}-\dfrac{1}{10}e^{-5t}; & t>0 \\ Ae^{5t} & ; t\le 0 \end{cases}$$

where A is an arbitrary constant.

Example 3.83: Solve the differential equation

$$\frac{d^2y}{dt^2}+3\frac{dy}{dt}+2y=e^{-|t|}$$

Solution: A.E. is

$$m^2+3m+2=0$$

$$\therefore \qquad m=-1,-2$$

$$\therefore \qquad C.F.=c_1e^{-t}+c_2e^{-2t}$$

For finding P.I., we assume boundary conditions

$$\lim_{t\to\pm\infty} y(t)=\lim_{t\to\pm\infty} y'(t)=0$$

Take Fourier transform of both sides of the given differential equation

$$\left[(i\omega)^2+3i\omega+2\right]Y(\omega)=\int_{-\infty}^{\infty} e^{-|t|}e^{-i\omega t}dt; \quad \left(Y(\omega)=\mathcal{F}\, y(t)\right)$$

$$=\int_{-\infty}^{0} e^{t}e^{-i\omega t}dt+\int_{0}^{\infty} e^{-t}e^{-i\omega t}dt$$

$$=\left[\frac{e^{(1-i\omega)t}}{1-i\omega}\right]_{-\infty}^{0}+\left[\frac{e^{-(1+i\omega)t}}{-(1+i\omega)}\right]_{0}^{\infty}$$

$$=\frac{1}{1-i\omega}+\frac{1}{1+i\omega}$$

$$\therefore \qquad Y(\omega)=\frac{1}{(1+i\omega)(2+i\omega)}\left[\frac{1}{1-i\omega}+\frac{1}{1+i\omega}\right]$$

$$=\frac{1}{(1+i\omega)(2+i\omega)(1-i\omega)}+\frac{1}{1+i\omega}\left[\frac{1}{1+i\omega}-\frac{1}{2+i\omega}\right]$$

$$=\frac{1}{2(1+i\omega)}+\frac{1}{6(1-i\omega)}-\frac{1}{3(2+i\omega)}+\frac{1}{(1+i\omega)^2}-\frac{1}{(1+i\omega)(2+i\omega)}$$

(by suppression method)

$$=\frac{1}{2(1+i\omega)}+\frac{1}{6(1-i\omega)}-\frac{1}{3(2+i\omega)}+\frac{1}{(1+i\omega)^2}-\frac{1}{1+i\omega}+\frac{1}{2+i\omega}$$

$$=\frac{1}{(1+i\omega)^2}-\frac{1}{2(1+i\omega)}+\frac{1}{6(1-i\omega)}+\frac{2}{3(2+i\omega)}$$

$$\therefore \qquad y(t) = \mathcal{F}^{-1}\frac{1}{(1+i\omega)^2} - \frac{1}{2}\mathcal{F}^{-1}\frac{1}{1+i\omega} + \frac{1}{6}\mathcal{F}^{-1}\frac{1}{1-i\omega} + \frac{2}{3}\mathcal{F}^{-1}\frac{1}{2+i\omega} \qquad (1)$$

Now, $\qquad \mathcal{F}e^{-at}u_0(t) = \dfrac{1}{a+i\omega};\ a>0$

$\therefore \qquad \mathcal{F}e^{at}u_0(-t) = \dfrac{1}{a-i\omega};\ a>0 \qquad (\because \text{ if } \mathcal{F}f(t) = F(\omega), \text{ then } \mathcal{F}f(-t) = F(-\omega))$

$$\therefore \qquad \mathcal{F}^{-1}\frac{1}{1+i\omega} = e^{-t}u_0(t),\ \mathcal{F}^{-1}\frac{1}{1-i\omega} = e^{t}u_0(-t),\ \mathcal{F}^{-1}\frac{1}{2+i\omega} = e^{-2t}u_0(t)$$

Also, $\qquad \mathcal{F}e^{-t}u_0(t) = \dfrac{1}{1+i\omega}$

$$\Rightarrow \qquad \mathcal{F}t\,e^{-t}u_0(t) = i\frac{d}{d\omega}\frac{1}{1+i\omega} = \frac{1}{(1+i\omega)^2}$$

$$\therefore \qquad \mathcal{F}^{-1}\frac{1}{(1+i\omega)^2} = t\,e^{-t}u_0(t)$$

From (1), $y(t)$ is P.I.

$$\therefore \qquad P.I. = t\,e^{-t}u_0(t) - \frac{1}{2}e^{-t}u_0(t) + \frac{1}{6}e^{t}u_0(-t) + \frac{2}{3}e^{-2t}u_0(t)$$

$$\therefore \qquad P.I. = \begin{cases} \dfrac{1}{6}e^{t} & ;\ t<0 \\ \left(t-\dfrac{1}{2}\right)e^{-t} + \dfrac{2}{3}e^{-2t} & ;\ t>0 \\ \dfrac{1}{3} & ;\ t=0 \end{cases}$$

$\therefore$ general solution is

$$y(t) = C.F. + P.I. = \begin{cases} C_1e^{-t} + C_2e^{-2t} + \dfrac{1}{6}e^{t}; & t<0 \\ (C_1+t)e^{-t} + C_2e^{-2t}; & t\ge 0 \end{cases}$$

where C_1 and C_2 are arbitrary constants.

Example 3.84: The temperature distribution $u(x,t)$ in a thin, homogeneous infinite bar can be modelled by the initial boundary value problem

$\dfrac{\partial u}{\partial t} = c^2\dfrac{\partial^2 u}{\partial x^2},\ -\infty < x < \infty,\ t>0,\ u(x,0) = f(x),\ u(x,t)$ is finite as $x \to \pm\infty$. Find $u(x,t)$, $t>0$.

Solution: $$\frac{\partial u}{\partial t} = c^2 \frac{\partial^2 u}{\partial x^2}$$

Take Fourier transforms of both sides w.r.t. x

$$\frac{d}{dt}U(\omega,t) = c^2 (i\omega)^2 U(\omega,t) \quad \text{where } \mathcal{F} u(x,t) = U(\omega,t)$$

$$= -c^2\omega^2 U(\omega,t)$$

Solution of this first order ordinary differential equation is

$$U(\omega,t) = Ke^{-c^2\omega^2 t} \tag{1}$$

Now, $$u(x,0) = f(x)$$

Take Fourier transform

$$U(\omega,0) = F(\omega) \quad \text{where } \mathcal{F} f(x) = F(\omega)$$

$\therefore$ from (1),

$$U(\omega,0) = K = F(\omega)$$

Hence, $$U(\omega,t) = F(\omega)e^{-c^2\omega^2 t}$$

Take inverse Fourier transform

$$u(x,t) = \frac{1}{2\pi}\int_{-\infty}^{\infty} F(\omega)e^{-c^2\omega^2 t}e^{i\omega x}d\omega \tag{2}$$

But $$F(\omega) = \mathcal{F} f(x) = \int_{-\infty}^{\infty} f(\xi)e^{-i\omega\xi}d\xi$$

$\therefore$ $$u(x,t) = \frac{1}{2\pi}\int_{-\infty}^{\infty} e^{-c^2\omega^2 t}e^{i\omega x}\int_{-\infty}^{\infty} f(\xi)e^{-i\omega\xi}d\xi d\omega$$

$$= \frac{1}{2\pi}\int_{-\infty}^{\infty}\int_{-\infty}^{\infty} f(\xi)e^{-c^2\omega^2 t}e^{-i\omega(\xi-x)}d\xi d\omega$$

$$= \frac{1}{2\pi}\int_{-\infty}^{\infty}\int_{-\infty}^{\infty} f(\xi)e^{-c^2\omega^2 t}e^{-i\omega(\xi-x)}d\omega\, d\xi$$ (by changing order of integration)

$$= \frac{1}{2\pi}\int_{-\infty}^{\infty}\int_{-\infty}^{\infty} f(\xi)e^{-c^2\omega^2 t}\left[\cos\omega(\xi-x) - i\sin\omega(\xi-x)\right]d\omega\, d\xi$$

But $e^{-c^2\omega^2 t}\cos\omega(\xi-x)$ is even function of ω and $e^{-c^2\omega^2 t}\sin\omega(\xi-x)$ is odd function of ω

$\therefore$ $$u(x,t) = \frac{1}{2\pi}\int_{-\infty}^{\infty} 2\int_{0}^{\infty} f(\xi)e^{-c^2\omega^2 t}\cos\omega(\xi-x)\,d\omega d\xi$$

$$= \frac{1}{\pi}\int_{0}^{\infty}\int_{-\infty}^{\infty} f(\xi)e^{-c^2\omega^2 t}\cos\omega(\xi-x)\,d\xi d\omega$$ (by changing order of integration)

Example 3.85: Find the solution in the above example if $f(x)=\begin{cases} V, & -l<x<l \\ 0, & \text{otherwise} \end{cases}$ where V is a constant.

Solution: Here, $\quad F(\omega)=\int_{-l}^{l} Ve^{-i\omega x}dx=\int_{-l}^{l} V(\cos(\omega x)-i\sin(\omega x))dx$

$\cos(\omega x)$ is an even function of x and $\sin(\omega x)$ is an odd function of x

$$\therefore \quad F(\omega)=2V\int_0^l \cos(\omega x)dx=\frac{2V}{\omega}(\sin(\omega x))_0^l=\frac{2V}{\omega}\sin(\omega l)$$

$\therefore$ from equation (2) in the above example

$$u(x,t)=\frac{1}{2\pi}\int_{-\infty}^{\infty}\frac{2V}{\omega}\sin(\omega l)e^{-c^2\omega^2 t}e^{i\omega x}d\omega$$

$$=\frac{1}{2\pi}\int_{-\infty}^{\infty}\frac{2V}{\omega}\sin(\omega l)e^{-c^2\omega^2 t}(\cos(\omega x)+i\sin(\omega x))d\omega$$

Now, $\frac{1}{\omega}\sin(\omega l)e^{-c^2\omega^2 t}\cos(\omega x)$ is an even function of ω

and $\frac{1}{\omega}\sin(\omega l)e^{-c^2\omega^2 t}\sin(\omega x)$ is an odd function of ω

$$\therefore \quad u(x,t)=\frac{2V}{\pi}\int_0^{\infty}\frac{1}{\omega}\sin(\omega l)e^{-c^2\omega^2 t}\cos(\omega x)d\omega$$

$$=\frac{V}{\pi}\int_0^{\infty}\frac{1}{\omega}e^{-c^2\omega^2 t}\left[\sin\omega(l+x)+\sin\omega(l-x)\right]d\omega$$

Put $\omega=\frac{1}{c\sqrt{t}}y$

$$u(x,t)=\frac{V}{\pi}\int_0^{\infty}e^{-y^2}\left[\frac{1}{y}\sin\frac{(l+x)y}{c\sqrt{t}}+\frac{1}{y}\sin\frac{(l-x)y}{c\sqrt{t}}\right]dy$$

$$=\frac{V}{\pi}\left[\frac{\pi}{2}erf\frac{(l+x)}{2c\sqrt{t}}+\frac{\pi}{2}erf\frac{(l-x)}{2c\sqrt{t}}\right]$$

$$=\frac{V}{2}\left[erf\frac{(l+x)}{2c\sqrt{t}}+erf\frac{(l-x)}{2c\sqrt{t}}\right]$$

Example 3.86: Solve $\frac{\partial^2 u}{\partial t^2}=\alpha^2\frac{\partial^2 u}{\partial x^2}$, $-\infty<x<\infty$, $t\ge 0$ with conditions $u(x,0)=f(x)$, $\frac{\partial u}{\partial t}(x,0)=g(x)$ and assuming u, $\frac{\partial u}{\partial x}\to 0$ as $x\to\pm\infty$.

Solution:
$$\frac{\partial^2 u}{\partial t^2}=\alpha^2\frac{\partial^2 u}{\partial x^2}$$

Take Fourier transform of both sides w.r.t. x

$$\frac{d^2}{dt^2}U(\omega,t)=-\alpha^2\omega^2U(\omega,t) \quad \text{where } \mathcal{F}u(x,t)=U(\omega,t)$$

Its solution is

$$U(\omega,t)=c_1\cos(\alpha\omega t)+c_2\sin(\alpha\omega t) \tag{1}$$

Now,
$$u(x,0)=f(x)$$

Take Fourier transform

$$U(\omega,0)=F(\omega) \quad \text{where } \mathcal{F}(f(x))=F(\omega)$$

$\therefore$ from (1),

$$U(\omega,0)=c_1=F(\omega)$$

$\therefore$
$$U(\omega,t)=F(\omega)\cos(\alpha\omega t)+c_2\sin(\alpha\omega t) \tag{2}$$

Now,
$$\frac{\partial}{\partial t}u(x,0)=g(x)$$

Take Fourier transform

$$\frac{d}{dt}U(\omega,0)=G(\omega) \quad \text{where } \mathcal{F}g(x)=G(\omega) \tag{3}$$

From (2)

$$\frac{d}{dt}U(\omega,t)=-\alpha\omega F(\omega)\sin(\alpha\omega t)+c_2\alpha\omega\cos(\alpha\omega t)$$

$\therefore$
$$\frac{d}{dt}U(\omega,0)=c_2\alpha\omega=G(\omega) \qquad \text{[from (3)]}$$

$\therefore$
$$c_2=\frac{1}{\alpha\omega}G(\omega)$$

$\therefore$ from (2)

$$U(\omega,t)=F(\omega)\cos(\alpha\omega t)+\frac{1}{\alpha\omega}G(\omega)\sin(\alpha\omega t)$$

$\therefore$
$$u(x,t)=\mathcal{F}^{-1}\left[F(\omega)\cos(\alpha\omega t)\right]+\frac{i}{\alpha}\mathcal{F}^{-1}\frac{G(\omega)}{i\omega}\sin(\alpha\omega t) \tag{4}$$

Now,
$$\mathcal{F}^{-1}\frac{G(\omega)}{i\omega}=\int_{-\infty}^{x}g(x)\,dx$$

By modulation theorem

$$\mathcal{F}^{-1}\left[F(\omega)\cos(\alpha\omega t)\right]=\frac{1}{2}\left[f(x+\alpha t)+f(x-\alpha t)\right] \quad (5)$$

$$\mathcal{F}^{-1}\left[\frac{G(\omega)}{i\omega}\sin(\alpha\omega t)\right]=-\frac{i}{2}\left[\int_{-\infty}^{x+\alpha t} g(x)\,dx-\int_{-\infty}^{x-\alpha t} g(x)\,dx\right]$$

$$=-\frac{i}{2}\left[\int_{-\infty}^{x-\alpha t} g(x)\,dx+\int_{x-\alpha t}^{x+\alpha t} g(x)\,dx-\int_{-\infty}^{x-\alpha t} g(x)\,dx\right]$$

$$=-\frac{i}{2}\int_{x-\alpha t}^{x+\alpha t} g(x)\,dx \quad (6)$$

Substituting from (5) and (6) in (4)

$$u(x,t)=\frac{1}{2}\left[f(x+\alpha t)+f(x-\alpha t)\right]+\frac{1}{2\alpha}\int_{x-\alpha t}^{x+\alpha t} g(y)\,dy$$

Example 3.87: Use Fourier sine transform to solve the equation $\frac{\partial v}{\partial t}=K\frac{\partial^2 v}{\partial x^2}$; $x>0,\ t>0$ subject to the conditions $v=v_0$ when $x=0,\ t>0$ and $v=0$ when $t=0,\ x>0$.

Solution: $$\frac{\partial v}{\partial t}=K\frac{\partial^2 v}{\partial x^2}$$

Take Fourier sine transform w.r.t. x of both sides

$$\frac{d}{dt}V_s(\omega,t)=K\left[-\omega^2 V_s(\omega,t)+\omega v(0,t)\right], \quad \text{where } \mathcal{F}_s\, v(x,t)=V_s(\omega,t)$$

But $$v(0,t)=v_0$$

$\therefore$ $$\frac{d}{dt}V_s(\omega,t)+K\omega^2 V_s(\omega,t)=K\omega v_0$$

It is Leibnitz linear differential equation

$$I.F.=e^{K\omega^2 t}$$

$\therefore$ solution is

$$e^{K\omega^2 t}V_s(\omega,t)=\int K\omega v_0\, e^{K\omega^2 t}\,dt+c$$

$$=\frac{v_0}{\omega}e^{K\omega^2 t}+c$$

$\therefore$ $$V_s(\omega,t)=\frac{v_0}{\omega}+c\,e^{-K\omega^2 t} \quad (1)$$

Now, $$v(x,0)=0$$

$\therefore$ $$V_s(\omega,0)=0$$

$\therefore$ from (1),

$$V_s(\omega,0)=\frac{v_0}{\omega}+c=0$$

$\Rightarrow$ $$c=-\frac{v_0}{\omega}$$

Hence, $$V_s(\omega,t)=\frac{v_0}{\omega}\left[1-e^{-K\omega^2 t}\right]$$

Take inverse Fourier sine transform

$$v(x,t)=\frac{2}{\pi}\int_0^\infty \frac{v_0}{\omega}\left(1-e^{-K\omega^2 t}\right)\sin(\omega x)\, d\omega$$

$$=\frac{2v_0}{\pi}\int_0^\infty \frac{\sin(\omega x)}{\omega}d\omega-\frac{2v_0}{\pi}\int_0^\infty e^{-K\omega^2 t}\frac{\sin(\omega x)}{\omega}d\omega$$

$$=\frac{2v_0}{\pi}\frac{\pi}{2}-\frac{2v_0}{\pi}\int_0^\infty e^{-y^2}\frac{1}{y}\sin\frac{xy}{\sqrt{Kt}}dy \qquad \left(\omega=\frac{y}{\sqrt{Kt}}\right)$$

$$=v_0-\frac{2v_0}{\pi}\cdot\frac{\pi}{2}erf\left(\frac{x}{2\sqrt{Kt}}\right)$$ (we prove this result in remark after this example)

$$=v_0\left(1-erf\frac{x}{2\sqrt{Kt}}\right)$$

$$=v_0\, erf_C\frac{x}{2\sqrt{Kt}}$$

Remark: From Q. No. 17 (i) Exercise 1.5, we have

$$\int_0^\infty e^{-x^2}\cos 2ax\, dx=\frac{\sqrt{\pi}}{2}e^{-a^2}$$

Changing x by y and a by x

$$\int_0^\infty e^{-y^2}\cos 2xy\, dy=\frac{\sqrt{\pi}}{2}e^{-x^2}$$

Integrate both sides w.r.t. x from 0 to $\frac{c}{2\sqrt{t}}$

$$\int_0^{\frac{c}{2\sqrt{t}}}\int_0^\infty e^{-y^2}\cos 2xy\, dy\, dx=\frac{\sqrt{\pi}}{2}\int_0^{\frac{c}{2\sqrt{t}}}e^{-x^2}dx$$

$\Rightarrow$ $$\int_0^\infty\int_0^{\frac{c}{2\sqrt{t}}}e^{-y^2}\cos 2xy\, dx\, dy=\frac{\sqrt{\pi}}{2}\frac{\sqrt{\pi}}{2}erf\left(\frac{c}{2\sqrt{t}}\right)$$ (by changing order of integration)

$\therefore$ $$\int_0^\infty \frac{e^{-y^2}}{2y}\left(\sin 2xy\right)_0^{\frac{c}{2\sqrt{t}}}dy=\frac{\pi}{4}erf\left(\frac{c}{2\sqrt{t}}\right)$$

$\therefore$ $$\int_0^\infty \frac{e^{-y^2}}{y}\left(\sin\frac{cy}{\sqrt{t}}\right)dy=\frac{\pi}{2}erf\left(\frac{c}{2\sqrt{t}}\right)$$

Example 3.88: Solve the equation $\frac{\partial u}{\partial t}=\frac{\partial^2 u}{\partial x^2};\quad 0<x<\infty,\ t>0$ subject to the conditions

(i) $u(0,t)=0\ ,\ t>0$ (ii) $u(x,0)=\begin{cases}1\ ;\ 0<x<1\\0\ ;\ x>1\end{cases}$

(iii) $u(x,t)$ is bounded

Solution: $$\frac{\partial u}{\partial t}=\frac{\partial^2 u}{\partial x^2};\quad 0<x<\infty,\ t>0$$

Take Fourier sine transform of both sides w.r.t. x

$$\frac{d}{dt}U_s(\omega,t)=-\omega^2U_s(\omega,t)+\omega u(0,t)\quad \text{where } \mathcal{F}_s\, u(x,t)=U_s(\omega,t)$$

But $$u(0,t)=0 \text{ for } t>0$$

$\therefore$ $$\frac{d}{dt}U_s(\omega,t)=-\omega^2U_s(\omega,t)$$

Solution of this first order ordinary differential equation is

$$U_s(\omega,t)=K\,e^{-\omega^2 t} \tag{1}$$

Now, $$u(x,0)=\begin{cases}1\ ;\ 0<x<1\\0\ ;\ x>1\end{cases}$$

$\therefore$ $$U_s(\omega,0)=\int_0^\infty u(x,0)\sin(\omega x)\,dx=\int_0^1\sin(\omega x)\,dx=-\frac{1}{\omega}\left(\cos(\omega x)\right)_0^1=\frac{1-\cos\omega}{\omega}$$

$\therefore$ from (1),

$$U_s(\omega,0)=K=\frac{1-\cos\omega}{\omega}$$

$\therefore$ $$U_s(\omega,t)=\frac{1-\cos\omega}{\omega}e^{-\omega^2 t}$$

Take inverse Fourier sine transform

$$u(x,t)=\frac{2}{\pi}\int_0^\infty\frac{1-\cos\omega}{\omega}e^{-\omega^2 t}\sin(\omega x)\,d\omega$$

$$=\frac{1}{\pi}\int_0^\infty\frac{e^{-\omega^2 t}}{\omega}\left[2\sin(\omega x)-2\sin(\omega x)\cos\omega\right]d\omega$$

$$=\frac{1}{\pi}\int_0^\infty\frac{e^{-\omega^2 t}}{\omega}\left[2\sin(\omega x)-\sin\omega(x+1)-\sin\omega(x-1)\right]d\omega$$

Put $\omega=\frac{y}{\sqrt{t}}$

$$u(x,t)=\frac{1}{\pi}\int_0^\infty e^{-y^2}\left[2\frac{\sin\frac{xy}{\sqrt{t}}}{y}-\frac{\sin\frac{(x+1)y}{\sqrt{t}}}{y}-\frac{\sin\frac{(x-1)y}{\sqrt{t}}}{y}\right]dy$$

$$=\frac{1}{\pi}\frac{\pi}{2}\left[2\,erf\frac{x}{2\sqrt{t}}-erf\frac{x+1}{2\sqrt{t}}-erf\frac{x-1}{2\sqrt{t}}\right]$$ (from remark of the previous example)

$$=erf\frac{x}{2\sqrt{t}}-\frac{1}{2}erf\frac{x+1}{2\sqrt{t}}-\frac{1}{2}erf\frac{x-1}{2\sqrt{t}}$$

Example 3.89: Solve the equation $\frac{\partial u}{\partial t}=2\frac{\partial^2 u}{\partial x^2}$; $0<x<\infty$, $t>0$ subject to the conditions

(i) $u(0,t)=0$, $t>0$ (ii) $u(x,0)=e^{-x}$, $x>0$

(iii) u and $\frac{\partial u}{\partial x}$ both tend to zero as $x\to\infty$

Solution:
$$\frac{\partial u}{\partial t}=2\frac{\partial^2 u}{\partial x^2};\quad 0<x<\infty,\ t>0$$

Take Fourier sine transform of both sides w.r.t. x

$$\frac{d}{dt}U_s(\omega,t)=2\left[-\omega^2U_s(\omega,t)+\omega u(0,t)\right]\quad \text{where } \mathcal{F}_s\,u(x,t)=U_s(\omega,t)$$

But $u(0,t)=0$ for $t>0$

$\therefore$
$$\frac{d}{dt}U_s(\omega,t)=-2\omega^2U_s(\omega,t)$$

Solution of this first order ordinary differential equation is

$$U_s(\omega,t)=Ke^{-2\omega^2t} \tag{1}$$

Now, $u(x,0)=e^{-x}$, $x>0$

$\therefore$
$$U_s(\omega,0)=\int_0^\infty u(x,0)\sin(\omega x)\,dx=\int_0^\infty e^{-x}\sin(\omega x)\,dx$$

$$=\left[\frac{e^{-x}}{(\omega^2+1)}(-\sin(\omega x)-\omega\cos(\omega x))\right]_0^\infty=\frac{\omega}{\omega^2+1}$$

from (1),

$$U_s(\omega,0)=K=\frac{\omega}{\omega^2+1}$$

$\therefore$
$$U_s(\omega,t)=\frac{\omega}{(\omega^2+1)}e^{-2\omega^2t}$$

Take inverse Fourier sine transform

$$u(x,t)=\frac{2}{\pi}\int_0^\infty\frac{\omega}{1+\omega^2}e^{-2\omega^2t}\sin(\omega x)\,d\omega$$

Example 3.90: Solve the Laplace's equation in the semi-infinite strip shown in the figure provided:

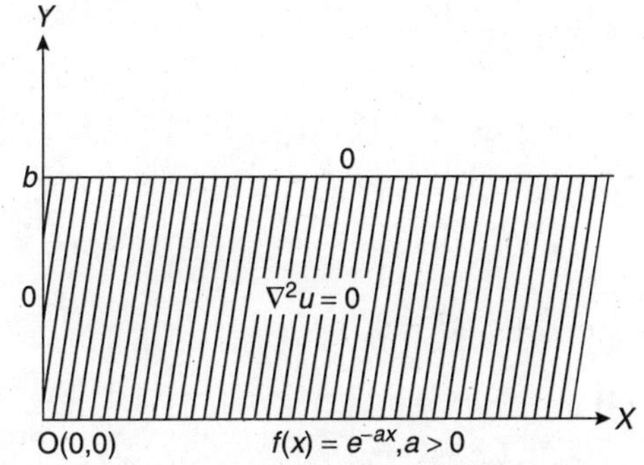

Figure 3.17

Solution: Laplace's equation in the semi-infinite strip is

$$\frac{\partial^2 u}{\partial x^2} + \frac{\partial^2 u}{\partial y^2} = 0; \quad 0 < x < \infty, 0 < y < b \tag{1}$$

subject to the conditions

(i) $u(0, y) = 0, \; 0 < y < b$ (ii) $u(x, b) = 0; \; 0 < x < \infty$

(iii) $u(x, 0) = f(x) = e^{-ax}; \; 0 < x < \infty$

Take Fourier sine transform w.r.t. x of both sides of (1)

$$-\omega^2 U_s(\omega, y) + \omega u(0, y) + \frac{d^2}{dy^2} U_s(\omega, y) = 0 \quad \text{where } \mathcal{F}_s u(x, y) = U_s(\omega, y)$$

But $u(0, y) = 0, \quad 0 < y < b$

$$\therefore \quad \frac{d^2}{dy^2} U_s(\omega, y) = \omega^2 U_s(\omega, y)$$

Its solution is

$$U_s(\omega, y) = c_1 e^{-\omega y} + c_2 e^{\omega y} \tag{2}$$

Now, $\quad u(x, 0) = f(x) = e^{-ax}; \; 0 < x < \infty$

$$\therefore \quad U_s(\omega, 0) = \int_0^\infty u(x, 0) \sin(\omega x)\, dx$$

$$= \int_0^\infty e^{-ax} \sin(\omega x)\, dx = \left[\frac{e^{-ax}}{\omega^2 + a^2} (-a \sin(\omega x) - \omega \cos(\omega x)) \right]_0^\infty$$

$$= \frac{\omega}{\omega^2 + a^2} \quad (\because a > 0)$$

∴ From (2)

$$U_s(\omega,0) = c_1 + c_2 = \frac{\omega}{\omega^2 + a^2} \tag{3}$$

Also, $u(x,b) = 0$

$\Rightarrow$ $U_s(\omega,b) = 0$

∴ from (2),

$$U_s(\omega,b) = c_1 e^{-\omega b} + c_2 e^{\omega b} = 0 \tag{4}$$

Solving equations (3) and (4)

$$c_1 = \frac{\omega e^{\omega b}}{\left(a^2+\omega^2\right)\left(e^{\omega b} - e^{-\omega b}\right)} = \frac{\omega e^{\omega b}}{2\left(a^2+\omega^2\right)\sinh(\omega b)}$$

$$c_2 = \frac{-\omega e^{-\omega b}}{\left(a^2+\omega^2\right)\left(e^{\omega b} - e^{-\omega b}\right)} = \frac{-\omega e^{-\omega b}}{2\left(a^2+\omega^2\right)\sinh(\omega b)}$$

∴ from (2)

$$U_s(\omega, y) = \frac{\omega\left[e^{\omega(b-y)} - e^{-\omega(b-y)}\right]}{2\left(a^2+\omega^2\right)\sinh(\omega b)} = \frac{\omega \sinh\left(\omega(b-y)\right)}{\left(a^2+\omega^2\right)\sinh(\omega b)}$$

$$= \frac{\omega\left[\sinh(\omega b)\cosh(\omega y) - \cosh(\omega b)\sinh(\omega y)\right]}{\left(a^2+\omega^2\right)\sinh(\omega b)}$$

$$= \frac{\omega}{(a^2+\omega^2)}\left[\cosh(\omega y) - \coth(\omega b)\sinh(\omega y)\right]$$

Take inverse Fourier sine transform

$$u(x,y) = \frac{2}{\pi}\int_0^\infty \frac{\omega}{(a^2+\omega^2)}\left[\cosh(\omega y) - \coth(\omega b)\sinh(\omega y)\right]\sin(\omega x)\, d\omega$$

Example 3.91: The steady state temperature distribution $u(x,y)$ in a thin homogenous semi-infinite plate is governed by the boundary value problem

$$\frac{\partial^2 u}{\partial x^2} + \frac{\partial^2 u}{\partial y^2} = 0; \quad 0 < x < l,\ 0 < y < \infty$$

subject to the conditions

(i) $u(0,y) = e^{-5y}$, $y > 0$ (ii) $u(l,y) = 0$, $y > 0$ (iii) $\frac{\partial u}{\partial y}(x,0) = 0;\ 0 < x < l$

Find the temperature distribution $u(x,y)$, $0 < x < l$, $y > 0$

Solution: $\dfrac{\partial^2 u}{\partial x^2}+\dfrac{\partial^2 u}{\partial y^2}=0;\ 0<x<l,\ y>0$

Take Fourier cosine transform w.r.t. y of both sides

$$\frac{d^2}{dx^2}U_c(x,\omega)+\left[-\omega^2 U_c(x,\omega)-\frac{\partial}{\partial y}u(x,0)\right]=0$$

where $\mathcal{F}_c\, u(x,y)=U_c(x,\omega)$

But $\dfrac{\partial}{\partial y}u(x,0)=0;\ 0<x<l$

$\therefore$ $\dfrac{d^2}{dx^2}U_c(x,\omega)=\omega^2 U_c(x,\omega)$

Its solution is

$$U_c(x,\omega)=c_1e^{-\omega x}+c_2e^{\omega x} \tag{1}$$

But $u(l,y)=0$

$\therefore$ $U_c(l,\omega)=0$

$\therefore$ from (1),

$$U_c(l,\omega)=c_1e^{-\omega l}+c_2e^{\omega l}=0 \tag{2}$$

Also, $u(0,y)=e^{-5y};\ y>0$

$\therefore$ $U_c(0,\omega)=\int_0^\infty u(0,y)\cos(\omega y)\,dy$

$$=\int_0^\infty e^{-5y}\cos(\omega y)\,dy=\left[\frac{e^{-5y}}{\omega^2+25}\left(-5\cos(\omega y)+\omega\sin(\omega y)\right)\right]_0^\infty$$

$$=\frac{5}{\omega^2+25}$$

$\therefore$ from (1),

$$U_c(0,\omega)=c_1+c_2=\frac{5}{\omega^2+25} \tag{3}$$

Solving (2) and (3)

$$c_1=\frac{-5e^{\omega l}}{(\omega^2+25)(e^{-\omega l}-e^{\omega l})}=\frac{5e^{\omega l}}{2(\omega^2+25)\sinh(\omega l)}$$

$$c_2=\frac{5e^{-\omega l}}{(\omega^2+25)(e^{-\omega l}-e^{\omega l})}=-\frac{5e^{-\omega l}}{2(\omega^2+25)\sinh(\omega l)}$$

Substituting in (1)

$$U_c(x,\omega)=\frac{5\left[e^{\omega(l-x)}-e^{-\omega(l-x)}\right]}{2(\omega^2+25)\sinh(\omega l)}=\frac{5\sinh(\omega l-\omega x)}{(\omega^2+25)\sinh(\omega l)}$$

$$=\frac{5\left[\sinh(\omega l)\cosh(\omega x)-\cosh(\omega l)\sinh(\omega x)\right]}{(\omega^2+25)\sinh(\omega l)}$$

$$=\frac{5}{(\omega^2+25)}\left[\cosh(\omega x)-\coth(\omega l)\sinh(\omega x)\right]$$

Take inverse Fourier cosine transform

$$u(x,y)=\frac{10}{\pi}\int_0^{\infty}\frac{1}{(\omega^2+25)}\left[\cosh(\omega x)-\coth(\omega l)\sinh(\omega x)\right]\cos(\omega y)\,d\omega$$

EXERCISE 3.6

1. If the initial temperature of an infinite bar is given by $\theta(x)=\begin{cases}\theta_0 \text{ for } |x|<a\\ 0 \text{ for } |x|>a\end{cases}$ determine the temperature at any point x and at any instant t.
2. Solve two dimensional Laplace equation $\frac{\partial^2 u}{\partial x^2}+\frac{\partial^2 u}{\partial y^2}=0$ subject to the conditions $u(x,0)=f(x), \frac{\partial u}{\partial y}=0$ at $y=0$.
3. An infinite string is initially at rest and that the initial displacement is $f(x), (-\infty<x<\infty)$. Determine the displacement $y(x,t)$ of the string.
4. Solve $\frac{\partial u}{\partial t}=\frac{\partial^2 u}{\partial x^2}$, $t>0$ subject to $u(x,0)=e^{-x^2}$.
5. The temperature distribution $u(x,t)$ in a thin, homogeneous semi-infinite bar can be modelled by the initial boundary value problem $\frac{\partial u}{\partial t}=c^2\frac{\partial^2 u}{\partial x^2}$, $0<x<\infty, t>0$; $u(x,0)=f(x), x>0,\ u(0,t)=0,\ t>0.$ Find the temperature distribution $u(x,t)$.
6. Find the solution of the Laplace equation $\frac{\partial^2 u}{\partial x^2}+\frac{\partial^2 u}{\partial y^2}=0$ inside the semi-infinite strip $x>0$, $0<y<b$ such that
$$u=\begin{cases}f(x); & y=0,\ 0<x<\infty\\ 0 & ; y=b,\ 0<x<\infty\\ 0 & ; x=0,\ 0<y<b\end{cases}$$
7. Solve $\frac{\partial u}{\partial t}=k\frac{\partial^2 u}{\partial x^2}$, if $u(0,t)=0$, $u(x,0)=e^{-x}$, $(x>0)$, $u(x,t)$ is bounded where $x>0$, $t>0$.
8. Use Fourier transform to solve the equation $\frac{\partial u}{\partial t}=\frac{\partial^2 u}{\partial x^2}$, $0<x<\infty, t>0$ where $u(x,t)$ satisfies the conditions

   (i) $\left(\frac{\partial u}{\partial x}\right)_{x=0}=0,\ t>0$

   (ii) $u(x,0)=\begin{cases}x; & 0<x<1\\ 0; & x>1\end{cases}$

   (iii) $|u(x,t)|<M$, i.e., bounded

9. Solve $\dfrac{\partial u}{\partial t}=k\dfrac{\partial^2 u}{\partial x^2}$ for $0\le x<\infty,\ t>0$ given the conditions
   (i) $u(x,0)=0$ for $x\ge 0$
   (ii) $\dfrac{\partial u}{\partial x}(0,t)=-a(\text{constant})$
   (iii) $u(x,t)$ is bounded.

10. The steady state temperature distribution $u(x,y)$ in a thin homogeneous semi-infinite plate is governed by the boundary value problem $\dfrac{\partial^2 u}{\partial x^2}+\dfrac{\partial^2 u}{\partial y^2}=0,\ 0<x<l,$ $0<y<\infty$; $u(0,y)=e^{-2y}$, $u(l,y)=0,\ y>0$; $\left(\dfrac{\partial u}{\partial y}\right)(x,0)=0,\ 0<x<l$. Find the temperature distribution $u(x,y)$, $0<x<l$, $y>0$.

11. Solve $\dfrac{dy}{dt}+3y=\cos 3t,\ y(0)=0;\ t\ge 0$

12. Solve $\dfrac{d^2y}{dt^2}+5\dfrac{dy}{dt}+6y=2\sin t,\ t\ge 0$ subject to the conditions $y'(0)=0,\ y(0)=0$.

13. Solve $\dfrac{d^2y}{dt}+3\dfrac{dy}{dt}+2y=H(t)\sin\omega t$ for $t>0$ satisfying $\lim\limits_{t\to 0+}y(t)=0$ and $\lim\limits_{t\to 0+}y'(t)=0$.

ANSWERS 3.6

1. $\theta(x,t)=\dfrac{\theta_0}{2}\left\{erf\dfrac{(a+x)}{2c\sqrt{t}}+erf\dfrac{(a-x)}{2c\sqrt{t}}\right\}$ where c^2 is thermal diffusivity.

2. $u(x,y)=\dfrac{1}{2}\left[f(x-iy)+f(x+iy)\right]$

3. $y(x,t)=\dfrac{1}{2}\left[f(x-ct)+f(x+ct)\right]$ where c^2 is diffusivity of string.

4. $u(x,t)=\dfrac{1}{(1+4t)^{1/2}}e^{-x^2/(1+4t)}$

5. $u(x,t)=\dfrac{2}{\pi}\int_0^{\infty}\left[\int_0^{\infty}f(\xi)\sin(\omega\xi)\,d\xi\right]e^{-c^2\omega^2 t}\sin(\omega x)\,d\omega$

6. $u(x,y)=\dfrac{2}{\pi}\int_0^{\infty}\int_0^{\infty}f(\xi)\sin(\omega\xi)\left[\cosh(\omega y)-\coth(\omega b)\sinh(\omega y)\right]\sin(\omega x)\,d\xi\,d\omega$

7. $u(x,t)=\dfrac{2}{\pi}\int_0^{\infty}\dfrac{\omega e^{-k\omega^2 t}}{1+\omega^2}\sin(\omega x)\,d\omega$

8. $u(x,t)=\dfrac{2}{\pi}\int_0^{\infty}\left[\dfrac{1}{\omega}\sin\omega-\dfrac{1}{\omega^2}(1-\cos\omega)\right]\cos(\omega x)e^{-\omega^2 t}\,d\omega$

9. $u(x,t)=\dfrac{2a}{\pi}\int_0^{\infty}\dfrac{1-e^{-k\omega^2 t}}{\omega^2}\cos(\omega x)\,d\omega$

10. $u(x,y)=\dfrac{4}{\pi}\int_0^{\infty}\dfrac{1}{(4+\omega^2)}\left[\cosh(\omega x)-\coth(\omega l)\sinh(\omega x)\right]\cos(\omega y)\,d\omega$

11. $y(t) = \frac{1}{6}(\cos 3t + \sin 3t) + c_1 e^{-3t}$

12. $y(t) = \frac{1}{5}\left(2e^{-2t} - e^{-3t} + \sin t - \cos t\right)$

13. $y(t) = -\frac{\omega e^{-2t}}{\omega^2 + 4} + \frac{\omega e^{-t}}{\omega^2 + 1} + \frac{1}{\left(1+\omega^2\right)\left(4+\omega^2\right)}\left[\left(2-\omega^2\right)\sin(\omega t) - 3\omega\cos(\omega t)\right]$

Partial Differential Equations 4

4.1 INTRODUCTION

When the functions involve only one variable then the system can be modelled by ordinary differential equations which we have already studied. Many problems in fluid mechanics, electricity, heat transfer, electromagnetic theory, quantum mechanics and other fields involve functions depending upon more than one variable usually one variable time and other one or more variables and hence these systems can be modelled by partial differential equations. The range of applications of partial differential equations is enormous compared to that of ordinary differential equations.

We discuss the formation of partial differential equations when either an equation containing the arbitrary constants or arbitrary functions is given or when a physical or geometrical system with certain conditions is provided.

We define the linear and non-linear, homogenous and non-homogenous partial differential equations. We also define partial differential equations linear in partial derivatives. Solutions of first order partial differential equations linear in partial derivatives are discussed by Lagrange's method and non-linear in partial derivatives is discussed by Charpit's method. Solutions of some particular types of first order partial differential equations are also discussed. Solutions of second order partial differential equations linear in second order partial derivatives are discussed by Monge's method. General solutions of homogenous and non-homogenous partial differential equations of higher orders with constant coefficients which are either linear homogenous in their order or linear of any type are discussed.

As applications of partial differential equations one dimensional heat equation, one dimensional wave equation governing the motion of a vibrating string, two dimensional wave equations in steady state governing vibrating membranes which are Laplace equations and transmission lines are considered.

4.2 FORMATION OF PARTIAL DIFFERENTIAL EQUATIONS

Partial differential equations (p.d.e.) are formed either by eliminating arbitrary constants or arbitrary functions from family of surfaces.

4.2.1 Elimination of Arbitrary Constants

Let
$$F(x, y, z, a, b) = 0 \tag{4.1}$$

represents two parameter family of surfaces. Differentiate (4.1) partially w.r.t. x and w.r.t. y

$$\frac{\partial F}{\partial x} + p\frac{\partial F}{\partial z} = 0 \tag{4.2}$$

$$\frac{\partial F}{\partial y}+q\frac{\partial F}{\partial z}=0 \tag{4.3}$$

Eliminating a and b from equations (4.1) to (4.3), we get a relation between x, y, z, p and q, i.e.

$$\phi(x,y,z,p,q)=0.$$

This will be first order partial differential equation of family of surfaces.

Remark 4.1: It may not be possible to eliminate a and b from equations (4.1) to (4.3). Then we find the second order partial derivatives and eliminate the arbitrary constants a and b. However, the higher order partial differential equation is not unique.

Remark 4.2: If we are to eliminate arbitrary constants $c_1, c_2, ..., c_n$ from the equation

$$F(x,y,z,c_1,c_2,...,c_n)=0$$

then it may be possible to obtain partial differential equations of order $n-1$ or less or of order more than $n-1$.

4.2.2 Elimination of Arbitrary Functions

Suppose we have family of surfaces containing arbitrary function f of x, y, z. We differentiate partially w.r.t. x and w.r.t. y Eliminate f' from these two equations formed from partial differentiation. We shall hence get the partial differential equation. If given equation contains two arbitrary functions then the equation obtained after differentiation partially w.r.t. x(or y) will have to be differentiated again partially w.r.t. x and y and then eliminate two arbitrary functions to obtain partial differential equation.

4.3 DEFINITIONS

4.3.1 Linear and Non-linear Partial Differential Equations

A partial differential equation is linear if it is of the first degree in the dependent variable and its partial derivatives, otherwise the equation is non-linear.

4.3.2 Homogenous and Non-homogenous Partial Differential Equations

In a partial differential equation, if each term contains either the dependent variable or one of its derivatives then the equation is called homogeneous, otherwise non-homogeneous.

4.3.3 Partial Differential Equations Linear in Partial Derivatives

A partial differential equation is linear in its partial derivatives if it is of the first degree in its partial derivatives. If nth partial derivatives are linear (others may or may not be linear) then partial differential equation is called linear in nth order partial derivatives.

4.3.4 Linear Homogenous in their Order Partial Differential Equations

Partial differential equations of nth order which have linear nth order partial derivatives and have no other partial derivative or dependent variable are called linear homogenous in nth order partial differential equations.

4.3.5 Solution of Partial Differential Equations

A solution of a partial differential equation in some region R is a function that has all the partial derivatives appearing in the equation in some domain containing R and satisfies the equation everywhere in R. In general, the totality of solutions of a partial differential equation is very large. For example, the functions $u = x^2 - y^2$, $u = e^x \cos y$, $u = \ln\left(x^2 + y^2\right)$ which are entirely different from each other are solutions of partial differential equation $\frac{\partial^2 u}{\partial x^2} + \frac{\partial^2 u}{\partial y^2} = 0$. These solutions may be derived from complete solutions or general solutions. Solution of differential equation containing arbitrary constants is called ***complete solution*** and that which contains arbitrary functions is called ***general solution***. Unique solution of a partial differential equation corresponding to a given physical problem is obtained by use of additional conditions arising from the problem. Such a solution is called a ***particular solution*** under these conditions. If additional conditions that the solution assume are given on the boundary of region R then these conditions are called boundary conditions and the problem is called ***boundary value problem***. If time t is one of the independent variable and conditions are prescribed at $t = 0$ then these are called initial conditions and the problem is called ***initial value problem***. There are solutions which cannot be derived from complete solution or general solution, and such solutions are called ***singular solutions***.

Example 4.1: Form partial differential equations from the following equations by eliminating the arbitrary constants:

(a) $z = ax + by + ab$

(b) $z = ax + a^2 y^2 + b$

(c) $\frac{x^2}{a^2} + \frac{y^2}{b^2} + \frac{z^2}{c^2} = 1$

(d) $x^2 + y^2 = (z - c)^2 \tan^2 \alpha$

(e) $z = ax^2 + 2bxy + cy^2$

(f) $(x - h)^2 + (y - k)^2 + z^2 = c^2$

(g) $z = ce^{\omega t} \cos(\omega x)$

Solution: (a)

$$z = ax + by + ab \tag{1}$$

Differentiating (1) partially w.r.t. x and y respectively, we have

$$p = a \tag{2}$$

$$q = b \tag{3}$$

Substituting for a and b from (2) and (3) in (1), we have

$$z = px + qy + pq$$

which is a required p.d.e.

(b) $$z = ax + a^2y^2 + b \quad (1)$$

Differentiating (1) partially w.r.t. x and y respectively, we have

$$p = a \quad (2)$$

$$q = 2a^2 y \quad (3)$$

Eliminating a from (2) and (3)

$$q = 2p^2 y$$

which is a required p.d.e.

(c) $$\frac{x^2}{a^2} + \frac{y^2}{b^2} + \frac{z^2}{c^2} = 1 \quad (1)$$

Differentiating (1) partially w.r.t. x and y respectively, we have

$$\frac{2x}{a^2} + \frac{2z}{c^2} p = 0 \quad (2)$$

$$\frac{2y}{b^2} + \frac{2z}{c^2} q = 0 \quad (3)$$

Differentiating (2) partially w.r.t. x again, we get

$$\frac{2}{a^2} + \frac{2}{c^2}\left(p^2 + zr\right) = 0 \quad (4)$$

Eliminating $\frac{2}{a^2}$ and $\frac{2}{c^2}$ from (2) and (4), we get

$$\begin{vmatrix} x & zp \\ 1 & \left(p^2 + zr\right) \end{vmatrix} = 0$$

i.e., $$\left(p^2 + zr\right)x - zp = 0 \quad (5)$$

which is a second-order p.d.e.

Note: The p.d.e. (5) obtained is not unique. One can also obtain the p.d.e.

$$\left(q^2 + zt\right)y - zq = 0 \quad \text{or} \quad pq + zs = 0.$$

(d) $$x^2 + y^2 = \left(z - c\right)^2 \tan^2\alpha \quad (1)$$

Differentiating (1) partially w.r.t. x and y respectively, we have

$$2x = 2\left(z - c\right)\tan^2\alpha \;.\; p \quad (2)$$

$$2y = 2\left(z - c\right)\tan^2\alpha \;.\; q \quad (3)$$

Dividing (2) by (3), we have

$$\frac{x}{y}=\frac{p}{q}$$

or $$qx-py=0$$

which is a required p.d.e.

(e) $$z=ax^2+2bxy+cy^2 \qquad (1)$$

Differentiating (1) partially w.r.t. x and y respectively, we have

$$p=2ax+2by \qquad (2)$$

$$q=2bx+2cy \qquad (3)$$

Differentiating (2) partially w.r.t. x and y respectively, we have

$$r=2a \qquad (4)$$

$$s=2b \qquad (5)$$

Differentiating (3) partially w.r.t. y, we have

$$t=2c \qquad (6)$$

Substituting for a, b and c from (4), (5) and (6) in (1), we have

$$2z=rx^2+2sxy+ty^2$$

which is a required p.d.e.

(f) $$(x-h)^2+(y-k)^2+z^2=c^2 \qquad (1)$$

Differentiating (1) partially w.r.t. x and y respectively, we have

$$2(x-h)+2zp=0 \qquad (2)$$

$$2(y-k)+2zq=0 \qquad (3)$$

Differentiating (2) again partially w.r.t. x, we have

$$2+2(p^2+zr)=0$$

i.e., $$zr+p^2+1=0$$

which is a second-order p.d.e.

(g) $$z=ce^{\omega t}\cos(\omega x) \qquad (1)$$

Differentiating (1) partially w.r.t. x and t respectively, we have

$$z_x=-c\omega\, e^{\omega t}\sin(\omega x) \qquad (2)$$

$$z_t=c\omega\, e^{\omega t}\cos(\omega x) \qquad (3)$$

Differentiate (2) partially w.r.t. x

$$z_{xx} = -c\omega^2 e^{\omega t} \cos(\omega x) \tag{4}$$

Differentiate (3) partially w.r.t. t

$$z_{tt} = c\omega^2 e^{\omega t} \cos(\omega x) \tag{5}$$

Add (4) and (5),

$$z_{xx} + z_{tt} = 0$$

which is a p.d.e.

Example 4.2: Find the differential equation of all spheres whose centres lie on the z-axis.
Solution: Equation of any sphere having its centre on z-axis say at $(0,0,c)$ and radius r is

$$x^2 + y^2 + (z-c)^2 = r^2 \tag{1}$$

Differentiating (1) partially w.r.t. x and y respectively, we have

$$2x + 2(z-c)p = 0 \tag{2}$$

$$2y + 2(z-c)q = 0 \tag{3}$$

Eliminating $z - c$ from (2) and (3), we have

$$qx - py = 0$$

which is a required p.d.e.

Example 4.3: Find the differential equation of all planes which are at a constant distance a from the origin.
Solution: The equation of family of planes in normal form is

$$lx + my + nz = a \tag{1}$$

where l, m and n are parameters of family which are the d.c's of the normal from the origin to the plane

$$\therefore \qquad l^2 + m^2 + n^2 = 1 \tag{2}$$

$\therefore$ Using (2), equation (1) becomes

$$lx + my \pm \sqrt{(1 - l^2 - m^2)}z = a \tag{3}$$

Differentiating (3) partially w.r.t. x and y respectively, we have

$$l \pm \sqrt{(1 - l^2 - m^2)}\ p = 0$$

$$\Rightarrow \qquad l = \mp\sqrt{(1 - l^2 - m^2)}\ p$$

$$m \pm \sqrt{(1-l^2-m^2)}\ q = 0$$

$$\Rightarrow \qquad m = \mp\sqrt{(1-l^2-m^2)}\ q$$

$$\therefore \qquad l^2+m^2 = (1-l^2-m^2)(p^2+q^2)$$

$$\Rightarrow \qquad (1-l^2-m^2)(p^2+q^2)+(1-l^2-m^2) = 1$$

$$\Rightarrow \qquad (1-l^2-m^2) = \frac{1}{(1+p^2+q^2)}$$

$$\therefore \qquad l = \frac{\mp p}{\sqrt{(p^2+q^2+1)}},\ m = \frac{\mp q}{\sqrt{p^2+q^2+1}}$$

$\therefore$ from equation (3)

$$\frac{-px-qy+z}{\sqrt{(p^2+q^2+1)}} = \pm a$$

or

$$z = px+qy \pm a\sqrt{(p^2+q^2+1)}$$

is a required p.d.e.

Example 4.4: Form a p.d.e. by eliminating the function f from the relation $z = y^2 + 2f\left(\frac{1}{x}+\log y\right)$.

Solution:

$$z = y^2 + 2f\left(\frac{1}{x}+\log y\right) \tag{1}$$

Differentiating (1) partially w.r.t. x and y respectively, we have

$$p = \frac{-2}{x^2}f'\left(\frac{1}{x}+\log y\right) \tag{2}$$

$$q = 2y + \frac{2}{y}f'\left(\frac{1}{x}+\log y\right) \tag{3}$$

Multiply (2) by x^2, (3) by y and add, we have

$$px^2 + qy = 2y^2$$

which is a required p.d.e.

Example 4.5: Form a p.d.e. by eliminating the arbitrary function f from the relation $z = f\left(\frac{xy}{z}\right)$.

Solution:

$$z = f\left(\frac{xy}{z}\right) \tag{1}$$

Differentiating (1) partially w.r.t. x and y respectively, we have

$$p = y\left(\frac{z - xp}{z^2}\right)f'\left(\frac{xy}{z}\right) \tag{2}$$

$$q = x\left(\frac{z - yq}{z^2}\right)f'\left(\frac{xy}{z}\right) \tag{3}$$

Divide (2) by (3)

$$\frac{p}{q} = \frac{y(z - xp)}{x(z - yq)}$$

or

$$px(z - yq) = qy(z - xp)$$

or

$$px = qy$$

which is a required p.d.e.

Example 4.6: Form a partial differential equation by eliminating the arbitrary functions f and ϕ from the equation $z = f(y/x) + \phi(xy)$.

Solution:

$$z = f(y/x) + \phi(xy) \tag{1}$$

Differentiating (1) partially w.r.t. x and y respectively, we have

$$p = \left(\frac{-y}{x^2}\right)f'\left(\frac{y}{x}\right) + y\phi'(xy) \tag{2}$$

$$q = \left(\frac{1}{x}\right)f'\left(\frac{y}{x}\right) + x\phi'(xy) \tag{3}$$

Multiply equation (2) by x, equation (3) by y and add

$\therefore$

$$px + qy = 2xy\phi'(xy) \tag{4}$$

Differentiating (4) partially w.r.t. x and y respectively, we have

$$p + xr + ys = 2y\left[\phi'(xy) + xy\phi''(xy)\right] \tag{5}$$

$$xs + q + yt = 2x\left[\phi'(xy) + xy\phi''(xy)\right] \tag{6}$$

Divide (5) by (6)

$$\frac{p + xr + ys}{xs + q + yt} = \frac{y}{x}$$

$\therefore$

$$xp + x^2r + xys = xys + yq + y^2t$$

or

$$px - qy = y^2t - x^2r$$

which is a required p.d.e.

Example 4.7: Form a partial differential equation by eliminating the arbitrary functions from $z = yf(x) + xg(y)$

Solution:

$$z = yf(x) + xg(y) \tag{1}$$

Differentiating (1) partially w.r.t. x and y respectively, we have

$$p = yf'(x) + g(y) \tag{2}$$

$$q = f(x) + xg'(y) \tag{3}$$

Differentiating (2) partially w.r.t. y, we have

$$s = f'(x) + g'(y)$$

$$\therefore \qquad xys = x\left(yf'(x)\right) + y\left[xg'(y)\right] \tag{4}$$

Substituting the values of $y\,f'(x)$ and $x\,g'(y)$ from equations (2) and (3), we have

$$xys = x\left[p - g(y)\right] + y\left[q - f(x)\right]$$

$$= px + qy - \left[yf(x) + xg(y)\right]$$

or $$xys = px + qy - z \qquad \text{(from equation (1))}$$

which is a required p.d.e.

Example 4.8: Form a partial differential equation by eliminating the arbitrary function ϕ from $lx + my + nz = \phi\left(x^2 + y^2 + z^2\right)$.

Solution:

$$lx + my + nz = \phi\left(x^2 + y^2 + z^2\right) \tag{1}$$

Differentiating (1) partially *w.r.t.* x and y respectively, we have

$$l + np = (2x + 2zp)\phi'\left(x^2 + y^2 + z^2\right) \tag{2}$$

$$m + nq = (2y + 2zq)\phi'\left(x^2 + y^2 + z^2\right) \tag{3}$$

Divide (2) by (3)

$$\frac{l + np}{m + nq} = \frac{x + zp}{y + zq}$$

or $$y(l + np) + zq(l + np) = x(m + nq) + zp(m + nq)$$

or $$y(l + np) - x(m + nq) = z(mp - ql)$$

which is a required p.d.e.

Example 4.9: Form partial differential equation from the following equation by eliminating the arbitrary function: $z = f(x^2 - y) + g(x^2 + y)$

Solution:

$$z = f(x^2 - y) + g(x^2 + y) \tag{1}$$

Differentiating (1) partially w.r.t. x and y respectively, we have

$$p = 2xf'(x^2 - y) + 2xg'(x^2 + y) \tag{2}$$

$$q = -f'(x^2 - y) + g'(x^2 + y) \tag{3}$$

Multiply equation (3) by $2x$ and add to equation (2)

$$\therefore \qquad p + 2xq = 4xg'(x^2 + y) \tag{4}$$

Differentiating (4) partially w.r.t. x and y respectively, we have

$$r + 2q + 2xs = 4g'(x^2 + y) + 8x^2 g''(x^2 + y) \tag{5}$$

$$s + 2xt = 4xg''(x^2 + y) \tag{6}$$

Multiply (6) by $2x$ and subtract from (5)

$$r + 2q + 2xs - 2x(s + 2xt) = 4g'(x^2 + y)$$

$$\therefore \qquad r + 2q - 4x^2 t = 4g'(x^2 + y)$$

$$\therefore \qquad x(r + 2q - 4x^2 t) = 4xg'(x^2 + y) = p + 2xq \qquad \text{(from equation (4))}$$

or

$$xr - 4x^3 t = p$$

which is a required p.d.e.

Example 4.10: Form a partial differential equation by eliminating the arbitrary function f from $f(xy + z^2, x + y + z) = 0$.

Solution: Let

$$u = xy + z^2, \; v = x + y + z \tag{1}$$

Then,

$$f(u, v) = 0 \tag{2}$$

Differentiating equation (2) partially w.r.t. x

$$\frac{\partial f}{\partial u}\left(\frac{\partial u}{\partial x} + \frac{\partial u}{\partial z}p\right) + \frac{\partial f}{\partial v}\left(\frac{\partial v}{\partial x} + \frac{\partial v}{\partial z}p\right) = 0$$

$$\therefore \qquad \frac{\partial f}{\partial u}(y + 2zp) + \frac{\partial f}{\partial v}(1 + p) = 0 \qquad \text{(from (1))} \tag{3}$$

Similarly differentiating (2) partially w.r.t. y, we have

$$\frac{\partial f}{\partial u}\left(\frac{\partial u}{\partial y}+\frac{\partial u}{\partial z}q\right)+\frac{\partial f}{\partial v}\left(\frac{\partial v}{\partial y}+\frac{\partial v}{\partial z}q\right)=0$$

$\therefore$ $$\frac{\partial f}{\partial u}(x+2zq)+\frac{\partial f}{\partial v}(1+q)=0 \qquad \text{(from (1))} \tag{4}$$

Eliminating $\frac{\partial f}{\partial u}$ and $\frac{\partial f}{\partial v}$ from (3) and (4), we get

$$\begin{vmatrix} y+2zp & 1+p \\ x+2zq & 1+q \end{vmatrix}=0$$

or $$(y+2zp)(1+q)-(x+2zq)(1+p)=0$$

or $$(y-x)+2z(p-q)+qy-px=0$$

or $$p(x-2z)+q(2z-y)=y-x$$

which is a required p.d.e.

Example 4.11: Form a partial differential equation by eliminating the arbitrary function f from $f\left(x^2+y^2+z^2, z^2-2xy\right)=0.$

Solution: Let

$$u=x^2+y^2+z^2, v=z^2-2xy \tag{1}$$

Then, $$f(u,v)=0 \tag{2}$$

Differentiate (2) partially w.r.t. x

$$\frac{\partial f}{\partial u}\left(\frac{\partial u}{\partial x}+\frac{\partial u}{\partial z}p\right)+\frac{\partial f}{\partial v}\left(\frac{\partial v}{\partial x}+\frac{\partial v}{\partial z}p\right)=0$$

$\therefore$ $$\frac{\partial f}{\partial u}(2x+2zp)+\frac{\partial f}{\partial v}(-2y+2zp)=0 \qquad \text{(from (1))} \tag{3}$$

Similarly differentiate (2) partially w.r.t. y

$$\frac{\partial f}{\partial u}\left(\frac{\partial u}{\partial y}+\frac{\partial u}{\partial z}q\right)+\frac{\partial f}{\partial v}\left(\frac{\partial v}{\partial y}+\frac{\partial v}{\partial z}q\right)=0$$

$\therefore$ $$\frac{\partial f}{\partial u}(2y+2zq)+\frac{\partial f}{\partial v}(-2x+2zq)=0 \qquad \text{(from (1))} \tag{4}$$

Eliminating $2\frac{\partial f}{\partial u}$ and $2\frac{\partial f}{\partial v}$ from (3) and (4), we get

$$\begin{vmatrix} x+zp & -y+zp \\ y+zq & -x+zq \end{vmatrix}=0$$

or $$(x+zp)(-x+zq)-(y+zq)(-y+zp)=0$$

or $$-x^2+xzq-xzp+z^2pq+y^2-yzp+yzq-z^2pq=0$$

or $$(x+y)zp-(x+y)zq=y^2-x^2$$

or $$zp-zq=y-x$$

is a required p.d.e.

4.4 DIRECT INTEGRATION METHOD FOR SOLUTIONS

There are partial differential equations which can be solved by directly integrating partially w.r.t. one variable keeping other constant.

Example 4.12: Solve $\dfrac{\partial^3 z}{\partial x^2 \partial y}-18x^2y+\sin(x-2y)=0.$

Solution: The given equation is

$$\frac{\partial^3 z}{\partial x^2 \partial y}=18x^2y-\sin(x-2y) \tag{1}$$

Integrate (1) w.r.t. x keeping y constant

$$\frac{\partial^2 z}{\partial x \partial y}=6x^3y+\cos(x-2y)+f(y)$$

Again, integrate w.r.t. x keeping y constant

$$\frac{\partial z}{\partial y}=\frac{3}{2}x^4y+\sin(x-2y)+x\,f(y)+g(y)$$

Now, integrate w.r.t. y keeping x constant

$$z=\frac{3}{4}x^4y^2+\frac{1}{2}\cos(x-2y)+x\,F(y)+G(y)+h(x)$$

where $$F(y)=\int f(y)\,dy,\ G(y)=\int g(y)\,dy$$

$\therefore$ $$z=\frac{3}{4}x^4y^2+\frac{1}{2}\cos(x-2y)+x\,F(y)+G(y)+h(x)$$

is the solution, where $h(x)$ is arbitrary function of x and $F(y)$, $G(y)$ are arbitrary functions of y.

Example 4.13: Solve $\dfrac{\partial^2 z}{\partial x^2}+z=0$, given that when $x=0, z=e^y$ and $\dfrac{\partial z}{\partial x}=1.$

Solution: The given equation is

$$\frac{\partial^2 z}{\partial x^2}+z=0 \tag{1}$$

If z were function of x only, equation would be $\dfrac{d^2 z}{dx^2}+z=0$ and its solution would have been $z=c_1\cos x+c_2\sin x$.

But here z is a function of x and y, therefore c_1 and c_2 are functions of y.

$\therefore$ the solution of (1) is

$$z = f(y)\cos x + g(y)\sin x$$

$\therefore$
$$\frac{\partial z}{\partial x} = -f(y)\sin x + g(y)\cos x$$

As $z = e^y$ and $\dfrac{\partial z}{\partial x} = 1$ when $x = 0$

$\therefore$
$$e^y = f(y)$$

and
$$1 = g(y)$$

Hence, the solution is $z = e^y \cos x + \sin x$.

Example 4.14: Solve $\dfrac{\partial^2 z}{\partial x^2} = a^2 z$ given that when $x = 0, \dfrac{\partial z}{\partial x} = a\sin y$ and $\dfrac{\partial z}{\partial y} = 0.$

Solution: The given equation is

$$\frac{\partial^2 z}{\partial x^2} = a^2 z \tag{1}$$

If z were function of x only, the solution of (1) would have been $z = c_1 e^{ax} + c_2 e^{-ax}$. But here z is a function of x and y, therefore c_1 and c_2 are functions of y.

$\therefore$ the solution of (1) is

$$z = f(y)e^{ax} + g(y)e^{-ax} \tag{2}$$

$$\frac{\partial z}{\partial x} = af(y)e^{ax} - ag(y)e^{-ax}$$

Given that $\dfrac{\partial z}{\partial x} = a\sin y$ when $x = 0$

$\therefore$
$$a\sin y = a\left(f(y) - g(y)\right)$$

$\therefore$
$$f(y) = \sin y + g(y)$$

$\therefore$ from (2)

$$z = \sin y e^{ax} + g(y)\left(e^{ax} + e^{-ax}\right) \tag{3}$$

or
$$z = \sin y e^{ax} + 2g(y)\cosh ax$$

$\therefore$
$$\frac{\partial z}{\partial y} = \cos y e^{ax} + 2g'(y)\cosh ax$$

Given that $\dfrac{\partial z}{\partial y} = 0$ when $x = 0$

$\therefore$
$$0 = \cos y + 2g'(y)$$

$\therefore$
$$g'(y) = -\frac{1}{2}\cos y$$

Integrate
$$g(y) = -\frac{1}{2}\sin y + c_1$$

$\therefore$ from (3), the solution is

$$z = \sin y\left(e^{ax} - \frac{1}{2}e^{ax} - \frac{1}{2}e^{-ax}\right) + c_1\left(e^{ax} + e^{-ax}\right)$$

$$= \sin y\left(\frac{e^{ax} - e^{-ax}}{2}\right) + 2c_1\left(\frac{e^{ax} + e^{-ax}}{2}\right)$$

Hence,

$$z = \sin y \sinh ax + c\cosh ax$$

where c is an arbitrary constant.

Example 4.15: Solve $z_x = 6x + 3y$ and $z_y = 3x - 4y$.

Solution: The given equations are

$$z_x = 6x + 3y \tag{1}$$

$$z_y = 3x - 4y \tag{2}$$

Integrate (1) w.r.t. x keeping y constant

$$z = 3x^2 + 3xy + f(y) \tag{3}$$

$$\therefore \quad z_y = 3x + f'(y) \tag{4}$$

From (2) and (4)

$$f'(y) = -4y$$

$$\therefore \quad f(y) = -2y^2 + c$$

$\therefore$ from (3), the solution is

$$z = 3x^2 + 3xy - 2y^2 + c$$

where c is an arbitrary constant.

EXERCISE 4.1

1. Form partial differential equations by eliminating the arbitrary constants:

   (a) $z = ax^2 + by^2$

   (b) $z = (x-a)^2 + (y-b)^2$

   (c) $2z = \dfrac{x^2}{a^2} + \dfrac{y^2}{b^2}$

   (d) $ax + by + cz = 1$

   (e) $(x-a)^2 + (y-b)^2 + z^2 = 1$

   (f) $x^2 + y^2 + (z-c)^2 = a^2$

   (g) $(x-a)^2 + (y-b)^2 = z^2\cot^2\alpha$ where α is a parameter.

2. Find the differential equation of all spheres of fixed radius r having their centres in the x–y plane.

3. Form a partial differential equation by eliminating the arbitrary constants a, b and c from $z = ax + by + cxy$.

4. Form partial differential equations by eliminating the arbitrary functions from
   (a) $z = xy + f(x^2 + y^2)$
   (b) $xyz = \phi(x + y + z)$
   (c) $z = (x + y)\phi(x^2 - y^2)$
   (d) $z = f(x^2 - y^2)$
   (e) $z = f(x + it) + g(x - it)$ where $i = \sqrt{-1}$
   (f) $z = f(y + 2x) + g(y - 3x)$
   (g) $u = f(x + at) + g(x - at)$
   (h) $z = xf(x + t) + g(x + t)$
   (i) $z = f(x)g(y)$
   (j) $z = xf_1(ax + by) + f_2(ax + by)$
5. Form partial differential equations by eliminating the arbitrary function from
   (a) $f(x^2 + y^2, x^2 - z^2) = 0$
   (b) $f(x + y + z, x^2 + y^2 + z^2) = 0$
   (c) $f(x^2 + y^2, z - xy) = 0$
   (d) $\phi(xyz, x + y + z) = 0$
   (e) $f\left(\dfrac{z}{x^3}, \dfrac{y}{x}\right) = 0$
6. Solve the following partial differential equations by direct integration
   (a) $\dfrac{\partial^2 z}{\partial x \partial y} = \dfrac{y}{x} + 2$
   (b) $\dfrac{\partial^2 z}{\partial x \partial y} = x^2 + y^2$
   (c) $\dfrac{\partial^2 z}{\partial x \partial y} = 4x \sin 3xy$
   (d) $\dfrac{\partial^3 z}{\partial x^2 \partial y} + 18xy^2 + \sin(2x - y) = 0$
   (e) $\dfrac{\partial^3 z}{\partial x^2 \partial y} = \cos(2x + 3y)$
7. Solve $\dfrac{\partial^2 z}{\partial y^2} = z$, given that when $y = 0$, $z = e^x$ and $\dfrac{\partial z}{\partial y} = e^{-x}$.
8. Solve $\dfrac{\partial^2 z}{\partial x \partial y} = x^2 y$ subject to the conditions $z(x, 0) = x^2$ and $z(1, y) = \cos y$.
9. Solve $\dfrac{\partial^2 z}{\partial x \partial y} = \sin x \sin y$, given that $\dfrac{\partial z}{\partial y} = -2\sin y$ when $x = 0$ and $z = 0$ when y is an odd multiple of $\dfrac{\pi}{2}$.

ANSWERS 4.1

1. (a) $2z = px + qy$ (b) $p^2 + q^2 = 4z$ (c) $px + qy = 2z$
   (d) $r = 0$ or $s = 0$ or $t = 0$ (e) $(p^2 + q^2 + 1)z^2 = 1$
   (f) $qx - py = 0$ (g) $p^2 + q^2 = \tan^2 \alpha$
2. $(p^2 + q^2 + 1)z^2 = r^2$
3. $r = 0$ or $t = 0$ or $z = px + qy - sxy$
4. (a) $py - qx = y^2 - x^2$ (b) $x(y - z)p + y(z - x)q = z(x - y)$
   (c) $yp + xq = z$ (d) $py + qx = 0$
   (e) $z_{xx} + z_{tt} = 0$ (f) $r + s = 6t$
   (g) $u_{tt} = a^2 u_{xx}$ (h) $z_{xx} - 2z_{xt} + z_{tt} = 0$
   (i) $pq = zs$ (j) $b^2 r - 2abs + a^2 t = 0$

5. (a) $yzp - xzq = xy$ (b) $(y-z)p+(z-x)q = x-y$
   (c) $py - qx = y^2 - x^2$ (d) $x(y-z)p+y(z-x)q = z(x-y)$
   (e) $px + qy = 3z$
6. (a) $z = \frac{y^2}{2}\log|x| + 2xy + f(x) + g(y)$
   (b) $z = \frac{1}{3}x^3y + \frac{1}{3}y^3x + f(x) + g(y)$
   (c) $z = \frac{-4}{9y}\sin(3xy) + f(x) + g(y)$
   (d) $z = \frac{1}{4}\cos(2x-y) - x^3y^3 + x\,f(y) + g(y) + h(x)$
   (e) $z = \frac{-1}{12}\sin(2x+3y) + f(x) + x\,g(y) + h(y)$
   where *f*, *g* and *h* are arbitrary constants.
7. $z = \cosh x\, e^y + \sinh x\, e^{-y}$
8. $z = \frac{1}{6}x^3y^2 + x^2 - \frac{1}{6}y^2 - 1 + \cos y$
9. $z = (1+\cos x)\cos y$

4.5 PARTIAL DIFFERENTIAL EQUATIONS OF THE FIRST ORDER

Partial differential equation of the form $F(x,y,z,p,q) = 0$ is first order partial differential equation where $p = \frac{\partial z}{\partial x}, q = \frac{\partial z}{\partial y}$.

We shall be dealing with the methods to solve some particular types of partial differential equations of first order.

4.5.1 Lagrange's Method

This method is used to solve partial differential equations linear in first order partial derivatives, i.e., equation of the form $Pp + Qq = R$ where *P*, *Q* and *R* are functions of *x*, *y* and *z*, respectively. This equation is called Lagrange equation.

Theorem 4.1: The general solution of the equation $Pp + Qq = R$ is given by $\phi(u,v) = 0$ where ϕ is an arbitrary function of *u* and *v* and $u(x,y,z) = c_1, v(x,y,z) = c_2$ are two linearly independent solutions of equations $\frac{dx}{P} = \frac{dy}{Q} = \frac{dz}{R}$.

Proof: $u(x,y,z) = c_1$ is a solution of equation $\frac{dx}{P} = \frac{dy}{Q} = \frac{dz}{R}$ (4.4)

We have $$du = \frac{\partial u}{\partial x}dx + \frac{\partial u}{\partial y}dy + \frac{\partial u}{\partial z}dz = 0 \tag{4.5}$$

Put each term of (4.4) equal to k and put values of dx, dy and dz from (4.4) in (4.5). Hence we now have

$$P\frac{\partial u}{\partial x} + Q\frac{\partial u}{\partial y} + R\frac{\partial u}{\partial z} = 0 \tag{4.6}$$

Similarly, $v(x, y, z) = c_2$ is solution of (4.4)

$$\therefore \quad P\frac{\partial v}{\partial x} + Q\frac{\partial v}{\partial y} + R\frac{\partial v}{\partial z} = 0 \tag{4.7}$$

Now, differentiate partially w.r.t. x and y

$$\phi(u, v) = 0 \tag{4.8}$$

Thus, we have

$$\frac{\partial \phi}{\partial u}\left[\frac{\partial u}{\partial x} + \frac{\partial u}{\partial z}p\right] + \frac{\partial \phi}{\partial v}\left[\frac{\partial v}{\partial x} + \frac{\partial v}{\partial z}p\right] = 0 \tag{4.9}$$

$$\frac{\partial \phi}{\partial u}\left[\frac{\partial u}{\partial y} + \frac{\partial u}{\partial z}q\right] + \frac{\partial \phi}{\partial v}\left[\frac{\partial v}{\partial y} + \frac{\partial v}{\partial z}q\right] = 0 \tag{4.10}$$

Since both $\frac{\partial \phi}{\partial u}$ and $\frac{\partial \phi}{\partial v}$ cannot be zero

$\therefore$ non-trial solution exist

$$\therefore \quad \begin{vmatrix} \frac{\partial u}{\partial x} + \frac{\partial u}{\partial z}p & \frac{\partial v}{\partial x} + \frac{\partial v}{\partial z}p \\ \frac{\partial u}{\partial y} + \frac{\partial u}{\partial z}q & \frac{\partial v}{\partial y} + \frac{\partial v}{\partial z}q \end{vmatrix} = 0$$

$$\therefore \quad \left[\frac{\partial u}{\partial x} + \frac{\partial u}{\partial z}p\right]\left[\frac{\partial v}{\partial y} + \frac{\partial v}{\partial z}q\right] - \left[\frac{\partial u}{\partial y} + \frac{\partial u}{\partial z}q\right]\left[\frac{\partial v}{\partial x} + \frac{\partial v}{\partial z}p\right] = 0$$

or $$\left[\frac{\partial u}{\partial z}\frac{\partial v}{\partial y} - \frac{\partial v}{\partial z}\frac{\partial u}{\partial y}\right]p + \left[\frac{\partial v}{\partial z}\frac{\partial u}{\partial x} - \frac{\partial u}{\partial z}\frac{\partial v}{\partial x}\right]q = \frac{\partial u}{\partial y}\frac{\partial v}{\partial x} - \frac{\partial u}{\partial x}\frac{\partial v}{\partial y} \tag{4.11}$$

Solving (4.6) and (4.7)

$$\frac{P}{\frac{\partial u}{\partial y}\frac{\partial v}{\partial z} - \frac{\partial u}{\partial z}\frac{\partial v}{\partial y}} = \frac{Q}{\frac{\partial u}{\partial z}\frac{\partial v}{\partial x} - \frac{\partial u}{\partial x}\frac{\partial v}{\partial z}} = \frac{R}{\frac{\partial u}{\partial x}\frac{\partial v}{\partial y} - \frac{\partial u}{\partial y}\frac{\partial v}{\partial x}} = c \text{ (say)} \tag{4.12}$$

Putting values from (4.12) in (4.11)

$$-\frac{P}{c}p - \frac{Q}{c}q = -\frac{R}{c}$$

or $$Pp + Qq = R \tag{4.13}$$

Thus, $\phi(u,v)=0$, where $u(x,y,z)=c_1$ and $v(x,y,z)=c_2$ are two independent solutions of $\frac{dx}{P}=\frac{dy}{Q}=\frac{dz}{R}$, which is general solution of $Pp+Qq=R$.

Remark 4.3: Equations $\frac{dx}{P}=\frac{dy}{Q}=\frac{dz}{R}$ are called Lagrange's auxiliary equations (A.E.) or Lagrange's subsidiary equations.

4.5.2 Geometrical Interpretation of Lagrange's Method

From
$$\frac{dx}{P}=\frac{dy}{Q}=\frac{dz}{R}$$

we observe that $P\hat{i}+Q\hat{j}+R\hat{k}$ is tangent to the integral curves represented by it and let $z=f(x,y)$ be integral surface represented by (4.13) then $\nabla f=p\hat{i}+q\hat{j}-\hat{k}$ is normal to this integral surface. Now,

$$\left(P\hat{i}+Q\hat{j}+R\hat{k}\right).\left(p\hat{i}+q\hat{j}-\hat{k}\right)=Pp+Qq-R=0 \text{ when } Pp+Qq=R.$$

Thus, the integral surfaces represented by solution of (4.13) contain integral curves represented by solution of (4.4).

Example 4.16: Solve the differential equation $y^2p-xyq=x(z-2y)$

Solution: Lagrange's auxiliary equations are

$$\frac{dx}{y^2}=\frac{dy}{-xy}=\frac{dz}{x(z-2y)} \tag{1}$$

From first and second members, $xdx+ydy=0$

Taking integral $x^2+y^2=c_1$ (2)

from second and third members of (1)

$$zdy-2ydy=-ydz$$

or $ydz+zdy-2ydy=0$

or $d\left(yz-y^2\right)=0$

Taking integral $y(z-y)=c_2$ (3)

Hence, general solution is $f\left(x^2+y^2,y(z-y)\right)=0$

where f is an arbitrary function.

Example 4.17: Find the general solution of the partial differential equation $\left(x+y^2\right)p+yq=z+x^2$

Solution: Lagrange's auxiliary equations are

$$\frac{dx}{x+y^2}=\frac{dy}{y}=\frac{dz}{z+x^2} \tag{1}$$

from first and second members

$$y\,dx - x\,dy - y^2 dy = 0$$

or
$$\frac{y\,dx - x\,dy}{y^2} - dy = 0$$

or
$$d\left(\frac{x}{y} - y\right) = 0$$

Taking integral

$$\frac{x}{y} - y = c_1$$

or
$$x = y(c_1 + y) \tag{2}$$

from second and third members of (1)

$$zdy + x^2 dy = ydz$$

or
$$zdy + y^2(c_1 + y)^2 dy = ydz \quad \text{(from (2))}$$

or
$$\frac{ydz - zdy}{y^2} - (c_1 + y)^2 dy = 0$$

or
$$d\left[\left(\frac{z}{y}\right) - \frac{1}{3}(c_1 + y)^3\right] = 0$$

Take integral

$$\frac{z}{y} - \frac{1}{3}(c_1 + y)^3 = c_2$$

or
$$3y^2 z - y^3(c_1 + y)^3 = 3c_2 y^3$$

or
$$3y^2 z - x^3 = 3c_2 y^3 \quad \text{(from (2))}$$

or
$$\frac{z}{y} - \frac{1}{3}\frac{x^3}{y^3} = c_2$$

∴ general solution is

$$f\left(\frac{x}{y} - y, \frac{z}{y} - \frac{x^3}{3y^3}\right) = 0$$

where f is an arbitrary function.

Example 4.18: Solve the following differential equation

$$(mz - ny)\,p + (nx - lz)\,q = ly - mx$$

Solution: Lagrange's auxiliary equations are

$$\frac{dx}{mz - ny} = \frac{dy}{nx - lz} = \frac{dz}{ly - mx} \tag{1}$$

From (1)

$$ldx + mdy + ndz = 0 \text{ and } xdx + ydy + zdz = 0$$

Taking integrals

$$lx + my + nz = c_1$$

$$x^2 + y^2 + z^2 = c_2$$

$\therefore$ general solution is

$$f\left(lx + my + nz, x^2 + y^2 + z^2\right) = 0$$

where f is an arbitrary function.

Example 4.19: Solve $p - x^2 = q + y^2$

Solution: Differential equation can be written as

$$p - q = x^2 + y^2$$

Lagrange's auxiliary equations are

$$\frac{dx}{1} = \frac{dy}{-1} = \frac{dz}{x^2 + y^2} \tag{1}$$

From first two members, $dx + dy = 0$

Take integrals

$$x + y = c_1 \tag{2}$$

from (1)

$$3x^2dx \quad 3y^2dy - 3dz = 0$$

Take integrals

$$x^3 - y^3 - 3z = c_2 \tag{3}$$

from (2) and (3)

$$x^3 + y^3 - 3z = f\left(x + y\right)$$

where f is an arbitrary function.

Example 4.20: Solve the following differential equation

$$\left(x^2 - yz\right)p + \left(y^2 - zx\right)q = z^2 - xy$$

Solution: Lagrange's auxiliary equations are

$$\frac{dx}{x^2 - yz} = \frac{dy}{y^2 - zx} = \frac{dz}{z^2 - xy} \tag{1}$$

From (1)

$$\frac{dx - dy}{(x - y)(x + y + z)} = \frac{dy - dz}{(y - z)(x + y + z)} = \frac{dz - dx}{(z - x)(x + y + z)}$$

$$\therefore \qquad \frac{dx-dy}{x-y}=\frac{dy-dz}{y-z}=\frac{dz-dx}{z-x} \qquad (2)$$

From first two members after integrating

$$\ln|x-y|-\ln|y-z|=\ln|c|$$

$$\therefore \qquad \ln\left|\frac{x-y}{y-z}\right|=\ln|c|$$

$$\therefore \qquad \frac{x-y}{y-z}=\pm c=c_1$$

Similarly, from last two members of (2)

$$\frac{y-z}{z-x}=c_2$$

$$\therefore \qquad f\left(\frac{x-y}{y-z},\frac{y-z}{z-x}\right)=0$$

is the general solution where f is an arbitrary function.

Example 4.21: Solve $(x^2-y^2-yz)p+(x^2-y^2-zx)q=z(x-y)$

Solution: Lagrange's auxiliary equations are

$$\frac{dx}{x^2-y^2-yz}=\frac{dy}{x^2-y^2-zx}=\frac{dz}{z(x-y)} \qquad (1)$$

From (1),

$$\frac{dx}{x^2-y^2-yz}=\frac{dy}{x^2-y^2-zx}=\frac{dz}{z(x-y)}=\frac{dx-dy}{z(x-y)}=\frac{xdx-ydy}{(x^2-y^2)(x-y)} \qquad (2)$$

From third and fourth members,

$$dz=dx-dy$$

or

$$dx-dy-dz=0$$

Take integrals

$$x-y-z=c_1 \qquad (3)$$

From third and fifth members of (2),

$$\frac{2dz}{z}=\frac{2(xdx-ydy)}{x^2-y^2}$$

Take integrals

$$\ln z^2=\ln|x^2-y^2|+\ln|c|$$

$$\therefore \qquad \frac{z^2}{x^2-y^2}=c_2 \qquad (4)$$

Hence, general solution is

$$f\left(x-y-z, \frac{z^2}{x^2-y^2}\right)=0$$

where f is an arbitrary function.

Example 4.22: Solve the following differential equation

$$x^2(y-z)p+y^2(z-x)q=z^2(x-y)$$

Solution: Lagrange's auxiliary equations are

$$\frac{dx}{x^2(y-z)}=\frac{dy}{y^2(z-x)}=\frac{dz}{z^2(x-y)} \tag{1}$$

From (1)

$$\frac{1}{x}dx+\frac{1}{y}dy+\frac{1}{z}dz=0$$

and

$$\frac{1}{x^2}dx+\frac{1}{y^2}dy+\frac{1}{z^2}dz=0$$

Taking their integrals

$$\ln|x|+\ln|y|+\ln|z|=\ln|c|$$

$\therefore$

$$xyz=c_1$$

and

$$-\left(\frac{1}{x}+\frac{1}{y}+\frac{1}{z}\right)=-c_2$$

$\therefore$

$$\frac{1}{x}+\frac{1}{y}+\frac{1}{z}=c_2$$

$\therefore$ general solution is

$$xyz=f\left(\frac{1}{x}+\frac{1}{y}+\frac{1}{z}\right)$$

where f is an arbitrary function.

Example 4.23: Solve $\dfrac{y-z}{yz}p+\dfrac{z-x}{zx}q=\dfrac{x-y}{xy}$

Solution: Differential equation is

$$\left(\frac{1}{z}-\frac{1}{y}\right)p+\left(\frac{1}{x}-\frac{1}{z}\right)q=\left(\frac{1}{y}-\frac{1}{x}\right)$$

Lagrange's auxiliary equations are

$$\frac{dx}{\left(\frac{1}{z}-\frac{1}{y}\right)}=\frac{dy}{\left(\frac{1}{x}-\frac{1}{z}\right)}=\frac{dz}{\left(\frac{1}{y}-\frac{1}{x}\right)} \tag{1}$$

From (1)

$$dx + dy + dz = 0$$

$$\text{and } \frac{1}{x}dx + \frac{1}{y}dy + \frac{1}{z}dz = 0$$

Taking their integrals

$$x + y + z = c_1$$

$$\log|x| + \log|y| + \log|z| = \log|c|$$

$\therefore$

$$xyz = c_2$$

$\therefore$ General solution is

$$f\left(x + y + z, xyz\right) = 0$$

where f is an arbitrary function.

Example 4.24: Find the general solution of the p.d.e. $\left(2y^2 + z\right)p + \left(y + 2x\right)q = 4xy - z$. Hence obtain the particular solution which passes through

(a) the curve $z = 1, \left(x + \frac{1}{2}\right)^2 - \left(y + \frac{1}{2}\right)^2 = 1$

(b) the straight line $z = 1, y = x$

Solution: Lagrange's auxiliary equations are

$$\frac{dx}{2y^2 + z} = \frac{dy}{y + 2x} = \frac{dz}{4xy - z} \tag{1}$$

From (1)

$$\frac{dx}{2y^2 + z} = \frac{dy}{y + 2x} = \frac{dz}{4xy - z} = \frac{dx + dz}{2y\left(y + 2x\right)} = \frac{zdy + ydz}{2x\left(z + 2y^2\right)} \tag{2}$$

From first and last members,

$$\frac{dx}{1} = \frac{zdy + ydz}{2x}$$

$\therefore$

$$2xdx - \left(ydz + zdy\right) = 0$$

Take integral

$$x^2 - yz = c_1 \tag{3}$$

From second and fourth members of (2)

$$dx - 2ydy + dz = 0$$

Take integral

$$x - y^2 + z = c_2 \tag{4}$$

∴ general solution is

$$f\left(x^2 - yz,\ x - y^2 + z\right) = 0$$

where f is an arbitrary function.

(a) When surface passes through curve

$$z = 1, \left(x + \frac{1}{2}\right)^2 - \left(y + \frac{1}{2}\right)^2 = 1 \tag{5}$$

From (3) and (4)

$$x^2 - y = c_1$$

$$x - y^2 = c_2 - 1$$

Add

$$x^2 + x - \left(y^2 + y\right) = c_1 + c_2 - 1$$

or
$$\left(x + \frac{1}{2}\right)^2 - \left(y + \frac{1}{2}\right)^2 = c_1 + c_2 - 1 \tag{6}$$

∴ from (5)

$$c_1 + c_2 = 2$$

∴ from (3) and (4), particular solution is

$$x^2 - yz + x - y^2 + z = c_1 + c_2 = 2$$

or
$$x^2 - y^2 - yz + x + z = 2$$

When surface passes through line $z = 1, y = x$ then from (3) and (4)

$$x^2 - x = c_1$$

$$x - x^2 = c_2 - 1$$

∴
$$c_1 + c_2 = 1$$

Add (3) and (4)

$$x^2 - yz + x - y^2 + z = c_1 + c_2 = 1$$

∴ particular solution is

$$x^2 - y^2 - yz + x + z = 1$$

Example 4.25: Find the equation of the surface which cuts orthogonally the family of spheres $x^2 + y^2 + z^2 = cy, c \neq 0$ and passes through the circle $z = 1, x^2 + y^2 = 4$.

Solution: Equation of given family of surfaces is

$$\phi(x, y, z) = \frac{x^2}{y} + y + \frac{z^2}{y} = c$$

$$\nabla\phi = \frac{\partial\phi}{\partial x}\hat{i} + \frac{\partial\phi}{\partial y}\hat{j} + \frac{\partial\phi}{\partial z}\hat{k}$$

$$= \frac{2x}{y}\hat{i} + \left(-\frac{x^2}{y^2} + 1 - \frac{z^2}{y^2}\right)\hat{j} + \left(\frac{2z}{y}\right)\hat{k}$$

is normal to surface.

Let surface $z = f(x, y)$ is normal to given family of spheres, then $\frac{\partial z}{\partial x}\hat{i} + \frac{\partial z}{\partial y}\hat{j} - \hat{k}$ is normal to this surface.

But it is orthogonal to given system of spheres.

$\therefore$
$$\nabla\phi \cdot \left(p\hat{i} + q\hat{j} - \hat{k}\right) = 0$$

$\therefore$ differential equation of surface orthogonal to given system of spheres is

$$\frac{2x}{y}p + \left(-\frac{x^2}{y^2} + 1 - \frac{z^2}{y^2}\right)q - \frac{2z}{y} = 0$$

or
$$2xyp + \left(-x^2 + y^2 - z^2\right)q = 2yz$$

It is Lagrange's equation.

Lagrange's auxiliary equations are

$$\frac{dx}{2xy} = \frac{dy}{-x^2 + y^2 - z^2} = \frac{dz}{2yz} \tag{1}$$

from first and third members

$$\frac{dx}{x} = \frac{dz}{z}$$

or
$$\frac{zdx - xdz}{z^2} = 0$$

$\therefore$
$$d\left(\frac{x}{z}\right) = 0$$

$\therefore$
$$\frac{x}{z} = c_1 \tag{2}$$

from second and third member of (1) taking $x = c_1 z$

we have

$$2yzdy = \left(-c_1^{\,2}z^2 + y^2 - z^2\right)dz$$

or
$$\frac{2yzdy - y^2dz}{z^2} + \left(1 + c_1^{\,2}\right)dz = 0$$

or
$$d\left(\frac{y^2}{z}\right) + d\left(\left(1 + c_1^{\,2}\right)z\right) = 0$$

$\therefore$ its solution is

$$\frac{y^2}{z} + \left(1 + c_1^{\,2}\right)z = c_2$$

Put value of c_1

$$\frac{y^2}{z}+\left(1+\frac{x^2}{z^2}\right)z=c_2$$

or
$$\frac{x^2+y^2+z^2}{z}=c_2 \tag{3}$$

$\therefore$ general solution is

$$F\left(\frac{x}{z},\ \frac{x^2+y^2+z^2}{z}\right)=0$$

where F is an arbitrary function.

When the surface passes through circle $z=1, x^2+y^2=4$

then from (2)

$$c_1=x$$

and from (3)

$$x^2+y^2+1=c_2$$

$$\Rightarrow \quad 4+1=c_2 \qquad (\because x^2+y^2=4)$$

$$\therefore \quad c_2=5$$

Put in (3) equation of surface passing through circle is $x^2+y^2+z^2=5z$.

Example 4.26: Solve $(t+y+z)\dfrac{\partial t}{\partial x}+(t+z+x)\dfrac{\partial t}{\partial y}+(t+x+y)\dfrac{\partial t}{\partial z}=x+y+z.$

Solution: Lagrange's auxiliary equations are

$$\frac{dx}{t+y+z}=\frac{dy}{t+z+x}=\frac{dz}{t+x+y}=\frac{dt}{x+y+z} \tag{1}$$

From (1)

$$\frac{dx+dy+dz+dt}{3(x+y+z+t)}=\frac{dx-dy}{-(x-y)}=\frac{dy-dz}{-(y-z)}=\frac{dz-dx}{-(z-x)} \tag{2}$$

From first and second members

$$(x-y)(dx+dy+dz+dt)+3(x+y+z+t)(dx-dy)=0$$

or
$$(x-y)^3(dx+dy+dz+dt)+3(x-y)^2(x+y+z+t)(dx-dy)=0$$

or
$$d\left((x-y)^3(x+y+z+t)\right)=0$$

$\therefore$ its solution is

$$(x-y)^3(x+y+z+t)=c_1$$

Similarly, from first and third members of (2) solution is

$$(y-z)^3(x+y+z+t)=c_2$$

and from first and fourth member of (2) solution is

$$(z-x)^3(x+y+z+t)=c_3$$

Hence, general solution is

$$f\left((x-y)^3(x+y+z+t),(y-z)^3(x+y+z+t),(z-x)^3(x+y+z+t)\right)=0$$

where f is an arbitrary function.

EXERCISE 4.2

Solve the following partial differential equations:

1. $yq-xp=z$
2. $yzp-xzq=xy$
3. $\frac{y^2z}{x}p+xzq=y^2$
4. $px+qy=az, a\neq 0$
5. $p\tan x+q\tan y=\tan z$
6. $y^2zp-x^2zq=x^2y$
7. $pz-qz=z^2+(x+y)^2$
8. $p-q=\ln(x+y)$
9. $zp+yq=x$
10. $z(z^2+xy)(px-qy)=x^4$
11. $px(z-2y^2)=(z-qy)(z-y^2-2x^3)$
12. $xzp+yzq=xy$
13. $(x^2-y^2-z^2)p+2xyq=2xz$
14. $z(xp-yq)=y^2-x^2$
15. $x^2p+y^2q=(x+y)z$
16. $2xzp+2yzq=z^2-x^2-y^2$
17. $(y^2+z^2)\frac{\partial z}{\partial x}-xy\frac{\partial z}{\partial y}+zx=0$
18. $x(y-z)p+y(z-x)q=z(x-y)$
19. $px(x+y)-qy(x+y)$
$+(2x+2y+z)(x-y)=0$
20. $x\frac{\partial u}{\partial x}+y\frac{\partial u}{\partial y}+z\frac{\partial u}{\partial z}=xyz$
21. $x(y^2-z^2)p+y(z^2-x^2)q=z(x^2-y^2)$
22. $(z-y)p+(x-z)q=y-x$
23. $(y+z)p-(x+z)q=x-y$
24. $(y+z)p+(z+x)q=x+y$
25. $(y+zx)p-(x+yz)q=x^2-y^2$
26. Find the general solution of the partial differential equation $y(x-z)\frac{\partial z}{\partial x}+(z^2-xz-x^2)\frac{\partial z}{\partial y}-y(2x-z)=0$. Hence obtain the particular solution which passes through the ellipse $z=0, 2x^2+4y^2=1$.
27. Find the equation of the surface which cuts orthogonally the system of surfaces $2xz+3yz=c(z+2)$, where c is an arbitrary constant and passes through the circle $z=0, x^2+y^2=9$.
28. Find the equation of the system of surfaces which cut orthogonally the family of cones $z^2=c(x^2+y^2)$. Obtain the particular surface which passes through the circle $z=3, x^2+y^2=9$.

ANSWERS 4.2

1. $f\left(xy, \dfrac{y}{z}\right) = 0$
2. $f\left(x^2 + y^2, y^2 + z^2\right) = 0$
3. $f\left(x^3 - y^3, x^2 - z^2\right) = 0$
4. $f\left(\dfrac{x}{y}, \dfrac{y^a}{z}\right) = 0$
5. $f\left(\dfrac{\sin x}{\sin y}, \dfrac{\sin x}{\sin z}\right) = 0$
6. $f\left(x^3 + y^3, y^2 + z^2\right) = 0$
7. $f\left(x + y, 2x - \log\left|x^2 + y^2 + z^2 + 2xy\right|\right) = 0$
8. $f\left(x + y, x\ln(x + y) - z\right) = 0$
9. $\ln|y| \pm \sinh^{-1}\dfrac{z}{\sqrt{x^2 - z^2}} = f\left(x^2 - z^2\right)$
10. $f\left(xy,\ x^4 - z^4 - 2xyz^2\right) = 0$
11. $\dfrac{z - y^2}{x} + x^2 = f\left(\dfrac{y}{z}\right)$
12. $f\left(\dfrac{x}{y}\right) = xy - z^2$
13. $f\left(\dfrac{y}{z}, \dfrac{x^2 + y^2 + z^2}{z}\right) = 0$
14. $f\left(xy,\ x^2 + y^2 + z^2\right) = 0$
15. $f\left(\dfrac{1}{x} - \dfrac{1}{y}, \dfrac{x - y}{z}\right) = 0$
16. $\left(x^2 + y^2 + z^2\right) = x\,f\left(\dfrac{x}{y}\right)$
17. $x^2 + y^2 + z^2 = f\left(\dfrac{y}{z}\right)$
18. $x + y + z = f(xyz)$
19. $(x + y)(x + y + z) = f(xy)$
20. $f\left(\dfrac{x}{y}, \dfrac{x}{z}, xyz - 3u\right) = 0$
21. $f\left(x^2 + y^2 + z^2, xyz\right) = 0$
22. $f\left(x + y + z, x^2 + y^2 + z^2\right) = 0$
23. $f\left(x + y + z, x^2 + y^2 - z^2\right) = 0$
24. $f\left(\dfrac{x - y}{y - z}\right) = (x - y)^2(x + y + z)$
25. $f\left(x^2 + y^2 - z^2, xy + z\right) = 0$
26. $f\left(x^2 + y^2 + z^2, x^2 - xz + \dfrac{z^2}{2}\right) = 0;$

    $2x^2 + 4y^2 + 3z^2 + 2xz = 1$
27. $3\left(x^2 + y^2 - z^2\right) - z^3 = 27$
28. $f\left(\dfrac{x}{y}, x^2 + y^2 + z^2\right) = 0;$

    $x^2 + y^2 + z^2 = 18$

We now explain a general method for finding the complete solution of general first order partial differential equations.

4.5.3 Charpit's Method

Consider the equation

$$f(x, y, z, p, q) = 0 \tag{4.14}$$

Since z depends on x and y

$$dz = \frac{\partial z}{\partial x}dx + \frac{\partial z}{\partial y}dy = pdx + qdy \tag{4.15}$$

Now, if we can find

$$\phi(x, y, z, p, q) = 0 \tag{4.16}$$

such that its solution is also solution of (4.14) then we can solve (4.14) and (4.16) for p and q and substitute in (4.15) and then solving it we can find complete solution.

To obtain ϕ, we differentiate (4.14) and (4.16) partially w.r.t. x and y

$$f_x + pf_z + f_p p_x + f_q q_x = 0 \tag{4.17}$$

$$\phi_x + p\phi_z + \phi_p p_x + \phi_q q_x = 0 \tag{4.18}$$

$$f_y + qf_z + f_p p_y + f_q q_y = 0 \tag{4.19}$$

$$\phi_y + q\phi_z + \phi_p p_y + \phi_q q_y = 0 \tag{4.20}$$

Use $q_x = \frac{\partial}{\partial x}\left(\frac{\partial z}{\partial y}\right) = \frac{\partial}{\partial y}\left(\frac{\partial z}{\partial x}\right) = p_y$ and eliminate p_x, p_y and q_y from (4.17) to (4.20)

$$\begin{vmatrix} f_p & f_q & 0 & f_x + pf_z \\ \phi_p & \phi_q & 0 & \phi_x + p\phi_z \\ 0 & f_p & f_q & f_y + qf_z \\ 0 & \phi_p & \phi_q & \phi_y + q\phi_z \end{vmatrix} = 0$$

Expand in terms of elements of last column

$$-[f_x + pf_z]\phi_p\left[f_p\phi_q - f_q\phi_p\right] + [\phi_x + p\phi_z]f_p\left[f_p\phi_q - f_q\phi_p\right] - \left[f_y + qf_z\right]\phi_q\left[f_p\phi_q - f_q\phi_p\right]$$
$$+\left[\phi_y + q\phi_z\right]f_q\left[f_p\phi_q - f_q\phi_p\right] = 0$$

Divide by $f_p\phi_q - f_q\phi_p \neq 0$ $\quad$ ($\because$ solution exists)

$$\phi_p(f_x + pf_z) - f_p(\phi_x + p\phi_z) + \phi_q(f_y + qf_z) - f_q(\phi_y + q\phi_z) = 0$$

or

$$-f_p\phi_x - f_q\phi_y + (-pf_p - qf_q)\phi_z + (f_x + pf_z)\phi_p + (f_y + qf_z)\phi_q = 0$$

It is Lagrange's equation for function ϕ with x, y, z, p and q as independent variables.

Lagrange's auxiliary equations are

$$\frac{dx}{-f_p} = \frac{dy}{-f_q} = \frac{dz}{-pf_p - qf_q} = \frac{dp}{f_x + pf_z} = \frac{dq}{f_y + qf_z}$$

These equations are called Charpit's auxiliary equations.

An integral of these equations involving p or q or both can be taken as the required relation (4.16). Then, solve (4.14) and (4.16) for p and q and put values of p and q in (4.15). After taking its integral, we get the complete solution.

We now discuss four standard forms of $f(x, y, z, p, q) = 0$ so that finding the solution be easier.

4.5.4 Standard Form $f(p,q) = 0$

In this case, given partial differential equation does not contain x, y, z explicitly.

$\therefore$ $$f_x + pf_z = 0, f_y + qf_z = 0$$

$\therefore$ Charpit's auxiliary equations give solutions $p = a, q = b$

Now, $$f(p,q) = 0$$

$\Rightarrow$ $$f(a,b) = 0$$

which gives $$b = \phi(a)$$

Now, $$dz = pdx + qdy = adx + bdy$$

Take integrals

$$z = ax + by + c$$

$\therefore$ complete solution of $f(p,q) = 0$ is

$$z = ax + by + c$$

where $$f(a,b) = 0$$

i.e., $$z = ax + \phi(a)y + c$$

where a and c are arbitrary constants.

4.5.5 Standard Form $f(z, p, q) = 0$

In this case, given partial differential equation does not contain x and y explicitly.

We assume

$$z = f(X),$$

where $$X = x + ay.$$

Then,

$$p = \frac{\partial z}{\partial x} = \frac{dz}{dX}\frac{\partial X}{\partial x} = \frac{dz}{dX}$$

$$q = \frac{\partial z}{\partial y} = \frac{dz}{dX}\frac{\partial X}{\partial y} = a\frac{dz}{dX}$$

$\therefore$ partial differential equation becomes

$$f\left(z, \frac{dz}{dX}, a\frac{dz}{dX}\right) = 0$$

It is ordinary differential equation of first order. After solving it, put $X = x + ay$; hence, we shall get the complete solution.

4.5.6 Standard Form $f(x,p) = \phi(y,q)$

In this form, partial differential equation does not contain z explicitly and terms of x and p can be separated from terms of y and q.

Let

$$f(x,p)=\phi(y,q)=a.$$

Solving $f(x,p)=a$ for p, we have $p=F(x,a)$

Solving $\phi(y,q)=a$ for q, we have $q=\psi(y,a)$

$$\therefore \qquad dz = pdx+qdy = F(x,a)dx+\psi(y,a)dy$$

Take integrals

$$z=\int F(x,a)dx+\int\psi(y,a)dy+b$$

which is the complete solution where a and b are arbitrary constants.

4.5.7 Clairut's Equation

Any first order partial differential equation of the form $z = px+qy+f(p,q)$ is called Clairut's equation.

This can be written as

$$f = z-px-qy$$

$$\therefore \qquad f_x=-p, f_y=-q, f_z=1$$

$$\therefore \qquad f_x+pf_z=0, f_y+qf_z=0$$

$\therefore$ two solutions from Charpit's equations will be $p=a$ and $q=b$.

$\therefore$ complete solution is

$$z=ax+by+f(a,b)$$

where a and b are arbitrary constants.

Example 4.27: Solve by Charpit's method

(i) $2zx-px^2-2qxy+pq=0$

(ii) $(p^2+q^2)y=qz$

(iii) $z^2=pqxy$

(iv) $q+xp=p^2$

Solution:

(i) Let $\qquad f\equiv 2zx-px^2-2qxy+pq=0 \qquad (1)$

Charpit's auxiliary equations are

$$\frac{dp}{f_x+pf_z}=\frac{dq}{f_y+qf_z}=\frac{dz}{-pf_p-qf_q}=\frac{dx}{-f_p}=\frac{dy}{-f_q}$$

Now,

$$f_y + qf_z = -2qx + 2qx = 0$$

$\therefore$ $$dq = 0$$

$\therefore$ $$q = a$$

$\therefore$ from (1),

$$2zx - px^2 - 2axy + ap = 0$$

$\therefore$ $$p = \frac{2zx - 2axy}{x^2 - a}$$

$$dz = pdx + qdy = \frac{2x(z - ay)}{x^2 - a}dx + ady$$

or $$\frac{dz - ady}{z - ay} = \frac{2x}{x^2 - a}dx$$

Take integrals

$$\log|z - ay| = \log|x^2 - a| + \log|c|$$

$\therefore$ $$\frac{z - ay}{x^2 - a} = \pm c = b$$

$\therefore$ $$z = ay + b(x^2 - a)$$

where a and b are arbitrary constants.

(ii) $$f \equiv (p^2 + q^2)y - qz = 0 \quad (1)$$

$$f_x = 0, f_y = p^2 + q^2, f_z = -q, f_p = 2py, f_q = 2qy - z$$

$\therefore$ Charpit's auxiliary equations

$$\frac{dx}{-f_p} = \frac{dy}{-f_q} = \frac{dz}{-pf_p - qf_q} = \frac{dp}{f_x + pf_z} = \frac{dq}{f_y + qf_z}$$

are

$$\frac{dx}{-2py} = \frac{dy}{z - 2qy} = \frac{dz}{-2p^2y + qz - 2q^2y} = \frac{dp}{-pq} = \frac{dq}{p^2}$$

from last two members

$$pdp + qdq = 0$$

Take integrals

$$p^2 + q^2 = a^2 \quad (2)$$

$\therefore$ from (1),

$$q = \frac{a^2 y}{z}$$

$\therefore$ from (2),

$$p = \pm a\sqrt{1 - \frac{a^2 y^2}{z^2}} = \pm\frac{a}{z}\sqrt{z^2 - a^2 y^2}$$

Now,

$$dz = pdx + qdy = \pm\frac{a\sqrt{z^2 - a^2 y^2}}{z}dx + \frac{a^2 ydy}{z}$$

or

$$\pm\frac{zdz - a^2 ydy}{\sqrt{z^2 - a^2 y^2}} = adx$$

Take integrals

$$\pm\sqrt{z^2 - a^2 y^2} = ax + b$$

or

$$z^2 - a^2 y^2 = (ax + b)^2$$

or

$$z^2 = a^2 y^2 + (ax + b)^2$$

where a and b are arbitrary constants.

(iii)

$$f \equiv z^2 - pqxy = 0 \tag{1}$$

$$f_x = -pqy, \quad f_y = -pqx, \quad f_z = 2z, \quad f_p = -qxy, \quad f_q = -pxy$$

$\therefore$ Charpit's auxiliary equations

$$\frac{dx}{-f_p} = \frac{dy}{-f_q} = \frac{dz}{-pf_p - qf_q} = \frac{dp}{f_x + pf_z} = \frac{dq}{f_y + qf_z}$$

are

$$\frac{dx}{qxy} = \frac{dy}{pxy} = \frac{dz}{2pqxy} = \frac{dp}{-pqy + 2pz} = \frac{dq}{-pqx + 2qz}$$

From these equations, we have

$$\frac{\frac{1}{x}dx - \frac{1}{y}dy}{qy - px} = \frac{\frac{1}{q}dq - \frac{1}{p}dp}{qy - px}$$

$$\therefore \quad \frac{dx}{x} - \frac{dy}{y} = \frac{dq}{q} - \frac{dp}{p}$$

Take integrals

$$\log|x| - \log|y| = \log|q| - \log|p| + \log|c_1|$$

$$\therefore \quad \frac{px}{qy} = \pm c_1 = a$$

$$\therefore \quad px = aqy \tag{2}$$

$\therefore$ from (1),

$$z^2 = aq^2 y^2$$

$\therefore \qquad z = cqy,$

where $\qquad c = \pm\sqrt{a}$

from (2),

$$px = c^2 qy$$

$\therefore \qquad q = \dfrac{z}{cy}, \; p = \dfrac{cz}{x}$

$\therefore \qquad dz = pdx + qdy = \dfrac{cz}{x}dx + \dfrac{z}{cy}dy$

$\therefore \qquad \dfrac{dz}{z} = c\dfrac{dx}{x} + \dfrac{1}{c}\dfrac{dy}{y}$

Take integrals

$$\log|z| = c\log|x| + \frac{1}{c}\log|y| + \log|c_1|$$

$$= \log\left|x^c\right| + \log\left|y^{1/c}\right| + \log|c_1|$$

$\therefore \qquad |z| = \left|c_1 x^c y^{1/c}\right|$

$\therefore \qquad z = \pm c_1 x^c y^{1/c} = b x^c y^{1/c} \qquad (b = \pm c_1)$

where b and c are arbitrary constants, $c \neq 0$.

(iv) $\qquad q + xp = p^2 \qquad (1)$

or $\qquad f \equiv q + xp - p^2 = 0$

$$f_x = p, f_y = 0, f_z = 0, f_p = x - 2p, f_q = 1.$$

$\therefore$ Charpit's auxiliary equations

$$\frac{dx}{-f_p} = \frac{dy}{-f_q} = \frac{dz}{-pf_p - qf_q} = \frac{dp}{f_x + pf_z} = \frac{dq}{f_y + qf_z},$$

become

$$\frac{dx}{2p - x} = \frac{dy}{-1} = \frac{dz}{2p^2 - px - q} = \frac{dp}{p} = \frac{dq}{0},$$

from $\qquad \dfrac{dy}{-1} = \dfrac{dp}{p}$

Taking integrals

$$\log|p| = -y + c_1$$

$\therefore \qquad |p| = a_1 e^{-y} \qquad$ where $a_1 = e^{c_1}$

$\therefore \qquad p = \pm a_1 e^{-y} = ae^{-y}$

$\therefore$ from (1),

$$q = a^2e^{-2y} - axe^{-y}$$

$\therefore$
$$dz = pdx + qdy = ae^{-y}dx + a^2e^{-2y}dy - axe^{-y}dy$$
$$= d\left(axe^{-y} - \frac{a^2}{2}e^{-2y}\right)$$

Take integrals

$$z = axe^{-y} - \frac{a^2}{2}e^{-2y} + b$$

where a and b are arbitrary constants.

Example 4.28: Find a complete integral of the p.d.e. $ypq + xp^2 = 1$. Hence, find a particular solution passing through the curve $x = 0, y - z = 0$.

Solution: p.d.e. is

$$f \equiv ypq + xp^2 - 1 = 0 \tag{1}$$
$$f_x = p^2, f_y = pq, f_z = 0, f_p = yq + 2xp, f_q = yp$$

$\therefore$ Charpit's auxiliary equations are

$$\frac{dx}{-f_p} = \frac{dy}{-f_q} = \frac{dz}{-pf_p - qf_q} = \frac{dp}{f_x + pf_z} = \frac{dq}{f_y + qf_z}$$

are

$$\frac{dx}{-yq - 2xp} = \frac{dy}{-py} = \frac{dz}{p(-yq - 2xp) - pqy} = \frac{dp}{p^2} = \frac{dq}{pq}$$

from last two members

$$pdq - qdp = 0$$

or
$$\frac{pdq - qdp}{p^2} = 0$$

or
$$d\left(\frac{q}{p}\right) = 0$$

$\therefore$
$$\frac{q}{p} = a$$

$\Rightarrow$
$$q = ap$$

$\therefore$ from (1)

$$ayp^2 + xp^2 = 1$$

$\therefore$
$$p = \pm\frac{1}{\sqrt{ay + x}}$$

∴ $$p = \pm\frac{1}{\sqrt{ay+x}},\ q = \pm\frac{a}{\sqrt{x+ay}}$$

∴ $$dz = pdx + qdy = \pm\frac{dx + ady}{\sqrt{x+ay}} = d\left(\pm 2\sqrt{x+ay}\right)$$

Integrate

$$z + b = \pm 2\sqrt{x+ay}$$

or $$(z+b)^2 = 4(x+ay) \quad (2)$$

where a and b are arbitrary constants.

(2) is a complete integral.

Any point on curve $x = 0, y - z = 0$ is $(0, t, t)$.

It lies on (2) when $(t+b)^2 = 4at$

i.e., $$t^2 + 2(b-2a)t + b^2 = 0$$

It has equal roots

∴ $$4(b-2a)^2 - 4b^2 = 0$$

∴ $$b - 2a = -b$$

⇒ $$a = b$$

Put value of b in (2)

$$(z+a)^2 = 4(x+ay) \quad (3)$$

Differentiating partially w.r.t. 'a'

$$2(z+a) = 4y$$

∴ $$a = 2y - z$$

Put this value in (3), particular solution passing through given curve is

$$(z + 2y - z)^2 = 4\left[x + (2y-z)y\right]$$

or $$4y^2 = 4\left(x + 2y^2 - yz\right)$$

or $$x + y^2 - yz = 0.$$

Example 4.29: Solve the following partial differential equations

(i) $\sqrt{p} + \sqrt{q} = 1$

(ii) $x^2p^2 + y^2q^2 = z^2$

(iii) $(y-x)(qy-px) = (p-q)^2$

(iv) $(x+y)(p+q)^2 + (x-y)(p-q)^2 = 1$

Solution: (i) Equation $\sqrt{p}+\sqrt{q}=1$ is $f(p,q)=\sqrt{p}+\sqrt{q}-1=0$.
Its solutions is $z=ax+by+c$ where

$$f(a,b)=\sqrt{a}+\sqrt{b}-1=0$$

or $$b=\left(1-\sqrt{a}\right)^2$$

$\therefore$ complete solution is

$$z=ax+\left(1-\sqrt{a}\right)^2 y+c$$

where a and c are arbitrary constants.

(ii) $$x^2p^2+y^2q^2=z^2$$

Let $$Z=\log|z|, Y=\log|y|, X=\log|x|.$$

$\therefore$ $$p=\frac{\partial z}{\partial x}=\frac{dz}{dZ}\frac{\partial Z}{\partial X}\frac{dX}{dx}=\frac{dz}{dZ}P\frac{dX}{dx}$$

$$=zP\frac{1}{x}=\frac{z}{x}P \qquad \text{where } P=\frac{\partial Z}{\partial X}$$

or $$x^2p^2=z^2P^2$$

Similarly, $$y^2q^2=z^2Q^2 \qquad \text{where } Q=\frac{\partial Z}{\partial Y}$$

$\therefore$ differential equation becomes

$$z^2P^2+z^2Q^2=z^2$$

or $$f(P,Q)=P^2+Q^2-1=0$$

Its solution is

$$Z=aX+bY+c$$

where $$f(a,b)=a^2+b^2-1=0$$

$\therefore$ $$b=\pm\sqrt{1-a^2}$$

$\therefore$ complete solution is

$$Z=aX\pm\sqrt{1-a^2}\,Y+c$$

or $$\log|z|=a\log|x|\pm\sqrt{1-a^2}\log|y|+c$$

where a and c are arbitrary constants.

(iii) $(y-x)(qy-px)=(p-q)^2$

Put $X=x+y, Y=xy$

$$\therefore \qquad p=\frac{\partial z}{\partial x}=\frac{\partial z}{\partial X}\frac{\partial X}{\partial x}+\frac{\partial z}{\partial Y}\frac{\partial Y}{\partial x}=P+yQ$$

$$q=\frac{\partial z}{\partial y}=\frac{\partial z}{\partial X}\frac{\partial X}{\partial y}+\frac{\partial z}{\partial Y}\frac{\partial Y}{\partial y}=P+xQ$$

where $\qquad P=\frac{\partial z}{\partial X}, Q=\frac{\partial z}{\partial Y}$

$$\therefore \qquad (y-x)(qy-px)=(y-x)[yP+xyQ-xP-xyQ]=(y-x)^2 P$$

and $\qquad (p-q)^2=(y-x)^2 Q^2$

$\therefore$ differential equation becomes

$$P=Q^2$$

or $\qquad f(P,Q)=P-Q^2=0$

Its solution is

$$z=aX+bY+c=a(x+y)+bxy+c$$

where $\quad f(a,b)=a-b^2=0$

$$\Rightarrow \qquad a=b^2$$

$\therefore$ complete solution is

$$z=b^2(x+y)+bxy+c$$

where b and c are arbitrary constants.

(iv) $(x+y)(p+q)^2+(x-y)(p-q)^2=1$

Put $X^2=x+y,\ Y^2=x-y$

then, $\qquad p=\frac{\partial z}{\partial x}=\frac{\partial z}{\partial X}\frac{\partial X}{\partial x}+\frac{\partial z}{\partial Y}\frac{\partial Y}{\partial x}=\frac{1}{2X}P+\frac{1}{2Y}Q$

$$q=\frac{\partial z}{\partial y}=\frac{\partial z}{\partial X}\frac{\partial X}{\partial y}+\frac{\partial z}{\partial Y}\frac{\partial Y}{\partial y}=\frac{1}{2X}P-\frac{1}{2Y}Q$$

where $\qquad \frac{\partial z}{\partial X}=P, \frac{\partial z}{\partial Y}=Q$

$\therefore$ equation becomes

$$X^2\left(\frac{1}{X}P\right)^2+Y^2\left(\frac{1}{Y}Q\right)^2=1$$

or $\qquad P^2+Q^2-1=0$

or $\qquad f(P,Q)=P^2+Q^2-1=0$

Its solution is

$$z = aX + bY + c$$
$$= \pm a\sqrt{x+y} \pm b\sqrt{x-y} + c$$

where
$$f(a,b) = a^2 + b^2 - 1 = 0$$

$\therefore$
$$b = \pm\sqrt{1-a^2}$$

$\therefore$ solution is

$$z = \pm a\sqrt{x+y} \pm \sqrt{1-a^2}\,\sqrt{x-y} + c$$

where a and c are arbitrary constants.

Example 4.30: Find the complete solutions of the following partial differential equations

(i) $z^2\left(p^2+q^2+1\right) = a^2$

(ii) $z^2\left(p^2x^2+q^2\right) = 1$

(iii) $p^2x^2 = z(z-qy)$

(iv) $z^2\left(p^2+q^2+1\right) = 0$

Solution: (i) Let

$$z = f(X)$$

where
$$X = x + by$$

then,
$$p = \frac{\partial z}{\partial x} = \frac{dz}{dX}\frac{\partial X}{\partial x} = \frac{dz}{dX}$$

$$q = \frac{\partial z}{\partial y} = \frac{dz}{dX}\frac{\partial X}{\partial y} = b\frac{dz}{dX}$$

$\therefore$ differential equation reduces to

$$z^2\left[\left(\frac{dz}{dX}\right)^2 + b^2\left(\frac{dz}{dX}\right)^2 + 1\right] = a^2$$

or
$$\left(\frac{dz}{dX}\right)^2\left(1+b^2\right) = \frac{a^2-z^2}{z^2}$$

or
$$\pm\sqrt{1+b^2}\,\frac{dz}{dX} = \frac{\sqrt{a^2-z^2}}{z}$$

$\therefore$
$$dX = \pm\frac{\sqrt{1+b^2}\,z}{\sqrt{a^2-z^2}}dz$$

Take integrals

$$c + X = \mp\sqrt{a^2-z^2}\cdot\sqrt{1+b^2}$$

or
$$(x+by+c)^2 = \left(1+b^2\right)\left(a^2-z^2\right)$$

or
$$\left(1+b^2\right)\left(a^2-z^2\right) = (x+by+c)^2$$

where b and c are arbitrary constants.

(ii) $$z^2\left(p^2x^2+q^2\right)=1$$

Let $$z=f(X)$$

where $$X=\ln|x|+ay$$

then, $$p=\frac{\partial z}{\partial x}=\frac{dz}{dX}\frac{\partial X}{\partial x}=\frac{1}{x}\frac{dz}{dX} \quad \Rightarrow \quad px=\frac{dz}{dX}$$

$$q=\frac{\partial z}{\partial y}=\frac{dz}{dX}\frac{\partial X}{\partial y}=a\frac{dz}{dX}$$

$\therefore$ differential equation reduces to

$$z^2\left[\left(\frac{dz}{dX}\right)^2+a^2\left(\frac{dz}{dX}\right)^2\right]=1$$

or $$z^2\left(1+a^2\right)\left(\frac{dz}{dX}\right)^2=1$$

or $$\left(\frac{dX}{dz}\right)^2=z^2\left(1+a^2\right)$$

$\therefore$ $$\frac{dX}{dz}=\pm\sqrt{1+a^2}\;z$$

$\therefore$ $$2\,dX=\pm 2\sqrt{1+a^2}\;z\,dz$$

Take integrals

$$b+2X=\pm\sqrt{1+a^2}\,z^2$$

or $$\left(b+2\ln|x|+2ay\right)^2=\left(1+a^2\right)z^2$$

where a and b are arbitrary constants.

(iii) $$p^2x^2=z(z-qy)$$

Let $$z=f(X)$$

where $$X=\ln|x|+a\ln|y|.$$

$\therefore$ $$p=\frac{\partial z}{\partial x}=\frac{dz}{dX}\frac{\partial X}{\partial x}=\frac{1}{x}\frac{dz}{dX} \quad \Rightarrow \quad px=\frac{dz}{dX}$$

$$q=\frac{\partial z}{\partial y}=\frac{dz}{dX}\frac{\partial X}{\partial y}=\frac{a}{y}\frac{dz}{dX} \qquad \Rightarrow \; qy=a\frac{dz}{dX}$$

$\therefore$ differential equation reduces to

$$\left(\frac{dz}{dX}\right)^2=z\left(z-a\frac{dz}{dX}\right)$$

$$\therefore \qquad \left(\frac{dz}{dX}\right)^2+az\frac{dz}{dX}-z^2=0$$

$$\therefore \qquad \frac{dz}{dX}=\frac{-az\pm\sqrt{a^2z^2+4z^2}}{2}=\frac{\left[-a\pm\sqrt{a^2+4}\right]z}{2}=kz \quad \text{where } k=\frac{-a\pm\sqrt{a^2+4}}{2}$$

$$\therefore \qquad \frac{dX}{dz}=\frac{1}{kz}$$

$$\Rightarrow \qquad dX=\frac{1}{kz}dz$$

Take integrals

$$b+X=\frac{1}{k}\ln|z|$$

or $$k\left(b+\ln|x|+a\ln|y|\right)=\ln|z|$$

or $$\frac{-a\pm\sqrt{a^2+4}}{2}\left[b+\ln|x|+a\ln|y|\right]=\ln|z|$$

where a and b are arbitrary constants.

(iv) $$z^2\left(p^2+q^2+1\right)=0$$

Let $$z=f(X)$$

where $$X=x+ay$$

$$\therefore \qquad p=\frac{\partial z}{\partial x}=\frac{dz}{dX}\frac{\partial X}{\partial x}=\frac{dz}{dX}$$

$$q=\frac{\partial z}{\partial y}=\frac{dz}{dX}\frac{\partial X}{\partial y}=a\frac{dz}{dX}$$

$\therefore$ differential equation reduces to

$$z^2\left[\left(1+a^2\right)\left(\frac{dz}{dX}\right)^2+1\right]=0$$

$z=0$ is the singular solution.

for complete solution

$$\left(1+a^2\right)\left(\frac{dz}{dX}\right)^2+1=0$$

$$\Rightarrow \qquad \frac{dz}{dX}=\pm\sqrt{-\frac{1}{1+a^2}}$$

$$\Rightarrow \qquad dz=\pm\sqrt{-\frac{1}{1+a^2}}\,dX$$

Take integrals

$$\pm\sqrt{-\left(1+a^2\right)}\,z=X+b=x+ay+b$$

or
$$-\left(1+a^2\right)z^2=\left(x+ay+b\right)^2$$

or
$$\left(x+ay+b\right)^2+\left(1+a^2\right)z^2=0$$

where a and b are arbitrary constants.

Example 4.31: Find the complete solution of the following partial differential equations

(i) $p^2-q^2=x-y$

(ii) $z^2\left(p^2+q^2\right)=x^2+y^2$

(iii) $p-q=x^2+y^2$

(iv) $p^2q^2=9p^2y^2\left(x^2+y^2\right)-9x^2y^2$

Solution: (i) Given differential equation can be written as

$$p^2-x=q^2-y=a \qquad \text{(say)}$$

$$\therefore \qquad p=\pm\sqrt{a+x},\ q=\pm\sqrt{a+y}$$

Now,
$$dz=pdx+qdy=\pm\left[\sqrt{a+x}dx\pm\sqrt{a+y}dy\right]$$

Take integrals

$$z=\pm\frac{2}{3}\left[\left(a+x\right)^{3/2}\pm\left(a+y\right)^{3/2}\right]+b$$

where a and b are arbitrary constants.

(ii) $z^2\left(p^2+q^2\right)=x^2+y^2$

Put
$$Z=\frac{z^2}{2}$$

$$\therefore \qquad p=\frac{\partial z}{\partial x}=\frac{dz}{dZ}\frac{\partial Z}{\partial x}=\frac{1}{z}\frac{\partial Z}{\partial x}=\frac{P}{z} \qquad \text{where } P=\frac{\partial Z}{\partial x}$$

$$q = \frac{\partial z}{\partial y} = \frac{dz}{dZ}\frac{\partial Z}{\partial y} = \frac{1}{z}\frac{\partial Z}{\partial y} = \frac{Q}{z} \qquad \text{where } Q = \frac{\partial Z}{\partial y}$$

$$\therefore \qquad p^2z^2 + q^2z^2 = P^2 + Q^2$$

$\therefore$ differential equation reduces to

$$P^2 - x^2 = y^2 - Q^2 = a \qquad \text{(say)}$$

$$\therefore \qquad P = \pm\sqrt{x^2+a},\ Q = \pm\sqrt{y^2-a}$$

$$\therefore \qquad dZ = Pdx + Qdy = \pm\left(\sqrt{x^2+a}\,dx \pm \sqrt{y^2-a}\,dy\right)$$

Take integrals

$$Z = \pm\left[\left\{\frac{x\sqrt{x^2+a}}{2} + \frac{a}{2}\log\left(x+\sqrt{x^2+a}\right)\right\} \pm \left\{\frac{y\sqrt{y^2-a}}{2} - \frac{a}{2}\log\left(y+\sqrt{y^2-a}\right)\right\}\right] + b_1$$

$$z^2 = \pm\left[\left\{x\sqrt{x^2+a} + a\log\left(x+\sqrt{x^2+a}\right)\right\} \pm \left\{y\sqrt{y^2-a} - a\log\left(y+\sqrt{y^2-a}\right)\right\}\right] + b \qquad (2b_1 = b)$$

where a and b are arbitrary constants.

(iii) $\quad p - q = x^2 + y^2$

$$\Rightarrow \qquad p - x^2 = q + y^2 = a \qquad \text{(say)}$$

$$\therefore \qquad p = x^2 + a,\ q = a - y^2$$

$$dz = pdx + qdy = \left(x^2+a\right)dx + \left(a-y^2\right)dy$$

Take integrals

$$z = \frac{x^3}{3} + ax + ay - \frac{y^3}{3} + b_1$$

or $\quad 3z = x^3 - y^3 + 3a(x+y) + b \qquad (b = 3b_1)$

where a and b are arbitrary constants.

(iv) $\ p^2q^2 = 9p^2y^2\left(x^2+y^2\right) - 9x^2y^2$

Divide by p^2y^2

$$\frac{q^2}{y^2} = 9\left(x^2+y^2\right) - 9\frac{x^2}{p^2}$$

$$\Rightarrow \qquad 9x^2\left(-1+\frac{1}{p^2}\right) = 9y^2 - \frac{q^2}{y^2} = a \qquad \text{(say)}$$

$$\therefore \qquad \frac{1}{p^2} = 1 + \frac{a}{9x^2},\ q^2 = y^2\left(9y^2 - a\right)$$

$$\therefore \qquad p = \pm\frac{3x}{\sqrt{9x^2 + a}}$$

$$q = \pm y\sqrt{9y^2 - a}$$

$$dz = pdx + qdy = \pm\left(\frac{3x}{\sqrt{9x^2 + a}}dx \pm y\sqrt{9y^2 - a}\,dy\right)$$

Take integrals

$$z = \pm\left(\frac{1}{3}\sqrt{9x^2 + a} \pm \frac{1}{27}\left(9y^2 - a\right)^{\frac{3}{2}}\right) + b$$

where a and b are arbitrary constants.

Example 4.32: Solve the partial differential equations

(i) $z = px + qy + \sqrt{1 + p^2 + q^2}$

(ii) $4xyz = pq + 2px^2y + 2qxy^2$

(iii) $pqz = p^2\left(xq + p^2\right) + q^2\left(yp + q^2\right)$

Solution:

(i) $\qquad z = px + qy\sqrt{1 + p^2 + q^2}$

This equation is in Clairut's form.

$\therefore$ solution is

$$z = ax + by + \sqrt{1 + a^2 + b^2}$$

where a and b are arbitrary constants.

(ii) $\qquad 4xyz = 2px^2y + 2qxy^2 + pq$

Divide by $4xy$

$$z = \frac{x}{2}p + \frac{y}{2}q + \frac{pq}{4xy}$$

Put $\qquad x^2 = X,\ y^2 = Y$

$$\therefore \qquad p = \frac{\partial z}{\partial x} = \frac{\partial z}{\partial X}\frac{dX}{dx} = 2xP \qquad \text{where } \frac{\partial z}{\partial X} = P$$

$$q = \frac{\partial z}{\partial y} = \frac{\partial z}{\partial Y}\frac{dY}{dy} = 2yQ \qquad \text{where } \frac{\partial z}{\partial Y} = Q$$

∴ equation reduces to

$$z = x^2P + y^2Q + PQ = PX + QY + PQ$$

which is in Clairut's form

∴ solution is

$$z = aX + bY + ab = ax^2 + by^2 + ab$$

where a and b are arbitrary constants.

(iii) $pqz = p^2\left(xq + p^2\right) + q^2\left(yp + q^2\right)$

$$\therefore \quad z = px + qy + \frac{p^4 + q^4}{pq}$$

which is in Clairut's form

∴ solution is

$$z = ax + by + \frac{a^4 + b^4}{ab}$$

where a and b are arbitrary constants.

EXERCISE 4.3

1. Solve the following partial differential equations by Charpit's method to obtain complete integrals
   (a) $px + qy = pq$
   (b) $2z + p^2 + qy + 2y^2 = 0$
   (c) $\left(p^2 + q^2\right)x = pz$
   (d) $2(z + xp + yq) = yp^2$
   (e) $pxy + pq + qy = yz$
   (f) $p = (z + qy)^2$
   (g) $xp + 3yq = 2\left(z - x^2q^2\right)$
   (h) $z = p^2x + q^2y$
2. Find the complete integral of the partial differential equation $px + q^2y = z$. Hence find a particular solution which passes through the curve $x = 1, y + z = 0$.
3. Find the complete integrals of the following partial differential equations
   (a) $pq = p + q$
   (b) $pq + p + q = 0$
   (c) $p^2 - q^2 = 1$
   (d) $p^2 - 3q^2 = 5$
   (e) $p^2 + q^2 = npq$
4. Find the complete solutions of the following partial differential equations
   (a) $p(1 + q) = qz$
   (b) $p(3 + q) = 2qz$
   (c) $zpq = p + q$
   (d) $p^2z^2 + q^2 = p^2q$
   (e) $z = p^2 + q^2$
   (f) $p\left(1 + q^2\right) = q(z - c)$
   (g) $q^2 = z^2p^2\left(1 - p^2\right)$
   (h) $p\left(1 - q^2\right) = q(1 - z)$
5. Solve the following partial differential equations to obtain complete integrals
   (a) $p^2 + q^2 = x + y$
   (b) $2\sqrt{p} + 3\sqrt{q} = 6x + 2y$

(c) $yp + xq + pq = 0$

(d) $yp = 2yx + \log q$

(e) $z(p^2 - q^2) = x - y$

(f) $\sqrt{p} - \sqrt{q} + 3x = 0$

(g) $z = a(x - y) - (\cos x + \cos y) + b$

(h) $p^2q^2 + x^2y^2 = x^2q^2(x^2 + y^2)$

(i) $p^2 + q^2 = z^2(x + y)$

(j) $zpy^2 = x(y^2 + z^2q^2)$

6. Find the complete integrals of the following partial differential equations

(a) $z = px + qy + 2\sqrt{pq}$

(b) $px + qy = z - 3p^{\frac{1}{3}}q^{\frac{1}{3}}$

(c) $z = px + qy + \ln pq$

(d) $(pq - p - q)(z - px - qy) = pq$

(e) $p^2q^2(px + qy - z) = 2$

(f) $pq(px + qy - z)^3 = 1$

(g) $(p - q)(z - px - qy) = 1$

(h) $(px + qy - z)^2 = 1 + p^2 + q^2$

ANSWERS 4.3

Here, a and b are arbitrary constants.

1. (a) $2az = (ax + y)^2 + b$

(b) $y^2\left[(x - a)^2 + y^2 + 2z\right] = b$

(c) $z^2 = a^2x^2 + (ay + b)^2$

(d) $4y^3z = 4y(ax + by) - a^2$

(e) $\log|z - ax| = y - a\log|a + y| + b$

(f) $(x - b)(a + yz) + y = 0$

(g) $zx = a(ax + y) + bx^3$

(h) $\sqrt{(1 + a)z} = \pm\left(\sqrt{ax} \pm \sqrt{y} + b\right)$

2. $\left(\sqrt{y} + b\right)^2 = z - ax$; $xy = z(x - 2)$

3. (a) $z = ax + \dfrac{a}{(a - 1)}y + b$

(b) $z = ax - \dfrac{a}{(a + 1)}y + b$

(c) $z = ax \pm \sqrt{a^2 - 1}\, y + b$

(d) $z = \pm\sqrt{(3a^2 + 5)}\, x + ay + b$

(e) $z = ax + \dfrac{a}{2}\left(n \pm \sqrt{n^2 - 4}\right)y + b$

4. (a) $\log|az - 1| = x + ay + b$

(b) $2az - 3 = be^{2(x + ay)}$

(c) $z^2 = \dfrac{2}{a}(a + 1)x + 2(a + 1)y + b$

(d) $z = a\tan(x + ay + b)$

(e) $4(1 + a^2)z = (x + ay + b)^2$

(f) $4(az - ac - 1) = (x + ay + b)^2$

(g) $z^2 - a^2 = (x + ay + b)^2$

(h) $4(1 - a + az) = (x + ay + b)^2$

5. (a) $z = \pm\dfrac{2}{3}\left[(x + a)^{\frac{3}{2}} \pm (y - a)^{\frac{3}{2}}\right] + b$

(b) $z = \dfrac{1}{72}(6x + a)^3 + \dfrac{1}{54}(2y - a)^3 + b$

(c) $z = \dfrac{ax^2}{2} - \dfrac{a}{(a + 1)}\dfrac{y^2}{2} + b$

(d) $z = x^2 + ax + \dfrac{1}{a}e^{ay} + b$

(e) $z^{\frac{3}{2}} = \pm\left[(x + a)^{\frac{3}{2}} \pm (y + a)^{\frac{3}{2}}\right] + b$

(f) $z = -\dfrac{1}{9}(a - 3x)^3 + a^2y + b$

(g) $z = a(x - y) - (\cos x + \cos y) + b$

(h) $z = \pm\left[\dfrac{1}{3}(x^2 + a^2)^{\frac{3}{2}} \pm (y^2 - a^2)^{\frac{1}{2}}\right] + b$

(i) $\log|z| = \pm\dfrac{2}{3}\left[(x + a)^{\frac{3}{2}} \pm (y - a)^{\frac{3}{2}}\right] + b$

(j) $z^2 = ax^2 \pm \sqrt{(a - 1)}\, y^2 + b$

6. (a) $z = ax + by + 2\sqrt{ab}$

(b) $z = ax + by + 3a^{\frac{1}{3}}b^{\frac{1}{3}}$

(c) $z = ax + by + \ln ab$

(d) $z = ax + by + \dfrac{ab}{ab - a - b}$

(e) $z = ax + by - \dfrac{2}{a^2b^2}$

(f) $z = ax + by - \dfrac{1}{(ab)^{\frac{1}{3}}}$

(g) $z = ax + by + \dfrac{1}{(a-b)}$

(h) $z = ax + by \pm \sqrt{1 + a^2 + b^2}$.

4.6 LINEAR IN SECOND ORDER PARTIAL DERIVATIVES DIFFERENTIAL EQUATIONS: MONGE'S METHOD

Monge's method gives the solution of partial differential equation

$$Rr + Ss + Tt = V \tag{4.21}$$

where R, S, T and V are functions of x, y, z, p, q.

This method reduces the equation (4.21) into an equivalent system of two equations from which we determine p or q or both p and q. If p or q is determined then solution is obtained following the procedure of solving Lagrange equation and if p and q both are determined then solution is obtained by integrating $dz = pdx + qdy$.

We have

$$dp = \frac{\partial p}{\partial x}dx + \frac{\partial p}{\partial y}dy = rdx + sdy$$

$$dq = \frac{\partial q}{\partial x}dx + \frac{\partial q}{\partial y}dy = sdx + tdy$$

From these equations,

$$r = \frac{dp - sdy}{dx},\ t = \frac{dq - sdx}{dy}$$

Substitute these values in (4.21)

$$R\frac{dp - sdy}{dx} + Ss + T\frac{dq - sdx}{dy} = V$$

or

$$\left(Rdpdy + Tdqdx - Vdxdy\right) - s\left(R(dy)^2 - Sdydx + T(dx)^2\right) = 0$$

This equation holds for arbitrary values of s and hence

$$R(dy)^2 - Sdydx + T(dx)^2 = 0 \tag{4.22}$$

and

$$Rdpdy + Tdqdx - Vdxdy = 0 \tag{4.23}$$

Equations (4.22) and (4.23) are called Monge's auxiliary equations or Monge's subsidiary equations.

From equation (4.22)

$$dy = \frac{1}{2R}\left[S \pm \sqrt{S^2 - 4RT}\right]dx \tag{4.24}$$

Now, dy has two non-zero values when $S^2 - 4RT \neq 0, R \neq 0, T \neq 0$.

Using these two values of dy and equation (4.23), we try to find two relations between x, y, z, p, q, each relation containing an arbitrary function. These relations are called intermediate integrals. We solve these relations for p and q and then we integrate

$$dz = pdx + qdy$$

to obtain the solution of given partial differential equation. We may also use only one of the values of dy from two values obtained and then use the method of solving the first order partial differential equations like Lagrange's equation to obtain the solution. For this purpose, we may use $dz = pdx + qdy$.

In equation (4.22)

if $S^2 - 4RT = 0$ or $R = 0$ or $T = 0$

then dy will have only one non-zero value.

In this case, we use the method of solving the first order partial differential equation like Lagrange's equation to obtain the solution. We may use $dz = pdx + qdy$.

We illustrate the above procedure through the following examples.

Example 4.33: Obtain the general solutions of the following partial differential equations by Monge's method

(i) $r - t\cos^2 x + p\tan x = 0$

(ii) $y^2 r - 2ys + t = p + 6y$

(iii) $q\left(q^2 + s\right) = pt$

(iv) $rq^2 - 2psq + tp^2 = qr - ps$

Solution: (i) Compare the given differential equations with $Rr + Ss + Tt = V$.

We have

$$R = 1,\ S = 0,\ T = -\cos^2 x,\ V = -p\tan x$$

$\therefore$ Monge's auxiliary equations

$$R(dy)^2 - Sdydx + T(dx)^2 = 0$$

and

$$Rdpdy + Tdqdx - Vdxdy = 0$$

are

$$(dy)^2 - \cos^2 x(dx)^2 = 0 \tag{1}$$

$$dpdy - \cos^2 x\, dqdx + p\tan x\, dxdy = 0 \tag{2}$$

From (1),

$$\left(dy + \cos x\, dx\right)\left(dy - \cos x\, dx\right) = 0$$

$\therefore \qquad dy = \cos x\, dx \qquad (3)$

or $\qquad dy = -\cos x\, dx \qquad (4)$

From (2) and (3)

$$\cos x\, dp\, dx - \cos^2 x\, dq dx + p \sin x (dx)^2 = 0$$

or

$$\cos x\, dp - \cos^2 x\, dq + p \sin x\, dx = 0$$

Divide by $\cos^2 x$

$$p \sec x \tan x\, dx + \sec x\, dp - dq = 0$$

or

$$d\left(p \sec x - q\right) = 0$$

$\therefore$

$$p \sec x - q = a \qquad (5)$$

Take integrals of (3)

$$y - \sin x = b \qquad (6)$$

From (5) and (6)

$$p \sec x - q = \phi_1\left(y - \sin x\right) \qquad (7)$$

Similarly, from (2) and (4)

$$p \sec x + q = \phi_2\left(y + \sin x\right) \qquad (8)$$

From (7) and (8)

$$p = \frac{1}{2} \cos x \left[\phi_1(y - \sin x) + \phi_2\left(y + \sin x\right)\right]$$

$$q = \frac{1}{2}\left[\phi_2\left(y + \sin x\right) - \phi_1\left(y - \sin x\right)\right]$$

Put these values in $dz = pdx + qdy$, hence we have

$$dz = \frac{1}{2} \cos x \left[\phi_1\left(y - \sin x\right) + \phi_2\left(y + \sin x\right)\right] dx + \frac{1}{2}\left[\phi_2\left(y + \sin x\right) - \phi_1\left(y - \sin x\right)\right] dy$$

$$= -\frac{1}{2}\phi_1\left(y - \sin x\right)\left(dy - \cos x\, dx\right) + \frac{1}{2}\phi_2\left(y + \sin x\right)\left(dy + \cos x\, dx\right)$$

$$= d\left[\frac{1}{2}\psi_2\left(y + \sin x\right) - \frac{1}{2}\psi_1\left(y - \sin x\right)\right]$$

where $\int \phi_1(t)\, dt = \psi_1(t)$ and $\int \phi_2(t)\, dt = \psi_2(t)$

$\therefore$

$$z = \frac{1}{2}\psi_2\left(y + \sin x\right) - \frac{1}{2}\psi_1\left(y - \sin x\right) + c$$

$$= f_1\left(y + \sin x\right) + f_2\left(y - \sin x\right)$$

where f_1 and f_2 are arbitrary functions.

(ii) Compare the given differential equation $y^2r-2ys+t=p+6y$ with $Rr+Ss+Tt=V$, hence we have

$$R=y^2,\ S=-2y,\ T=1,\ V=p+6y$$

∴ Monge's auxiliary equations

$$R(dy)^2-Sdydx+T(dx)^2=0$$

and
$$Rdpdy+Tdqdx-Vdxdy=0$$

are
$$y^2(dy)^2+2ydydx+(dx)^2=0 \qquad (1)$$

$$y^2dpdy+dqdx-(p+6y)dxdy=0 \qquad (2)$$

From (1),

$$(ydy+dx)^2=0$$

∴
$$ydy+dx=0$$

∴
$$dx=-ydy$$

Put in (2)

$$y^2dpdy-ydydq+y(p+6y)(dy)^2=0$$

⇒
$$ydp-dq+(p+6y)dy=0$$

or
$$ydp+pdy-dq+6ydy=0$$

or
$$d(py-q+3y^2)=0$$

∴
$$py-q+3y^2=a$$

or
$$py-q=-3y^2+a$$

It is Lagrange's equation.

Lagrange's auxiliary equations are

$$\frac{dx}{y}=\frac{dy}{-1}=\frac{dz}{-3y^2+a} \qquad (3)$$

From second and third members

$$dz=(3y^2-a)dy$$

Take integrals

$$z=y^3-ay+b \qquad (4)$$

From first and second members of (3)

$$2dx+2ydy=0$$

Take integrals

$$2x + y^2 = c \tag{5}$$

From (4) and (5)

$$z = y^3 - y\phi\left(y^2 + 2x\right) + f\left(y^2 + 2x\right)$$

where f and ϕ are arbitrary functions.

(iii) Compare the given differential equation $qs - pt = -q^3$ with $Rr + Ss + Tt = V$, hence we have

$$R = 0,\ S = q,\ T = -p,\ V = -q^3$$

Therefore, Monge's auxiliary equations

$$R(dy)^2 - Sdydx + T(dx)^2 = 0$$

and

$$Rdpdy + Tdqdx - Vdxdy = 0$$

are

$$-q\,dydx - p(dx)^2 = 0 \tag{1}$$

$$-p\,dqdx + q^3 dxdy = 0 \tag{2}$$

from (1),

$$pdx + qdy = 0$$

∴ $$dz = pdx + qdy = 0$$

∴ $$z = a \tag{3}$$

Using $pdx = -qdy$ in (2)

$$qdydq + q^3 dxdy = 0$$

or $$dq + q^2 dx = 0$$

or $$\frac{1}{q^2} dq + dx = 0$$

⇒ $$d\left(-\frac{1}{q} + x\right) = 0$$

∴ $$-\frac{1}{q} + x = b$$

⇒ $$q = \frac{1}{x - b}$$

∴ $$\frac{\partial z}{\partial y} = \frac{1}{x - b}$$

Integrating w.r.t. y keeping x constant

$$z = \frac{y}{x - b} + f(x)$$

$$\therefore \qquad y=(x-b)(z-f(x))=(x-\phi(z))(z-f(x)) \qquad \text{(from (3))}$$

$\therefore$ general solution is

$$y=xz-z\phi(z)-xf(x)+f(x)\phi(z)$$

where f and ϕ are arbitrary functions.

(iv) Compare the given differential equation $rq^2-2pqs+p^2t=qr-ps$

i.e.,
$$(q^2-q)r+(p-2pq)s+p^2t=0$$

with
$$Rr+Ss+Tt=V,$$

hence we have

$$R=q^2-q,\ S=p-2pq,\ T=p^2,\ V=0$$

$\therefore$ Monge's auxiliary equations

$$R(dy)^2-S\,dydx+T(dx)^2=0$$

and
$$Rdpdy+Tdqdx-Vdxdy=0$$

are
$$(q^2-q)(dy)^2-(p-2pq)\,dydx+p^2(dx)^2=0 \qquad (1)$$

$$(q^2-q)\,dp\,dy+p^2dq\,dx=0 \qquad (2)$$

from (1)

$$(q^2-q)(dy)^2+p(q+q-1)\,dydx+p^2(dx)^2=0$$

or
$$[(q-1)\,dy+pdx][qdy+pdx]=0$$

$\therefore$
$$pdx+qdy=0 \qquad (3)$$

or
$$pdx+(q-1)\,dy=0 \qquad (4)$$

from (3),

$$dz=pdx+qdy=0$$

$\therefore$
$$z=a \qquad (5)$$

Use (3) in (2)

$$(q^2-q)\,dpdy-pq\,dydq=0$$

or
$$(q-1)\,dp-pdq=0$$

or
$$\frac{pdq-qdp}{p^2}+\frac{1}{p^2}dp=0$$

or
$$d\left(\frac{q}{p}-\frac{1}{p}\right)=0$$

∴
$$\frac{q}{p}-\frac{1}{p}=b=f(z) \qquad \text{from (5)}$$

or
$$p\,f(z)-q=-1$$

which is Lagrange's equation.
Lagrange's auxiliary equations are

$$\frac{dx}{f(z)}=\frac{dy}{-1}=\frac{dz}{-1} \tag{5}$$

From first and third members

$$dx+f(z)\,dz=0$$

Integrate

$$x+g(z)=b, \qquad \text{where } g(z)=\int f(z)\,dz \tag{6}$$

From second and third members of (5)

$$dy-dz=0$$

Take integrals

$$y-z=c \tag{7}$$

From (6) and (7), general solution is

$$x+g(z)=\phi(y-z)$$

where g and ϕ are arbitrary functions.

EXERCISE 4.4

Obtain the general solutions of the following partial differential equations by Monge's method.

1. $r=a^2t$
2. $r-3s-10t=-3$
3. $(r-s)y+(s-t)x+q-p=0$
4. $r+(a+b)s+abt=xy$
5. $(x-y)(xr-xs-ys+yt)=(x+y)(p-q)$
6. $q^2r-2pqs+p^2t=pt-qs$
7. $y^2r+2sxy+x^2t+px+qy=0$
8. $x^2r-2xs+t+q=0$
9. $x^2r+2xys+y^2t=0$
10. $(e^x-1)(qr-ps)=pqe^x$
11. $qr-ps=p^3$
12. $(1-q)^2r-2(2-p-2q+pq)s$ $+(2-p)^2t=0$
13. $r-t\sin^2 x-p\cot x=0.$

ANSWERS 4.4

Here, f_1 and f_2 are arbitrary functions.

1. $z = f_1(y+ax) + f_2(y-ax)$
2. $z = f_1(y+5x) + f_2(y-2x) - \frac{3}{2}x^2$
3. $z = f_1(y+x) + f_2(y^2 - x^2)$
4. $z = f_1(y-ax) + f_2(y-bx) + \frac{1}{6}x^3y - \frac{1}{24}(a+b)x^4$
5. $z = f_1(x+y) - f_2(xy)$
6. $y + f_1(z) = f_2(x-z)$
7. $z = \log(y+x) f_1(y^2 - x^2) + f_2(y^2 - x^2)$
8. $z + f_1(\ln|x| + y) = f_2(\ln|x| + y)$
9. $z = f_1\left(\frac{x}{y}\right) + y f_2\left(\frac{x}{y}\right)$
10. $x = f_1(y) - f_2(z) + e^x$
11. $x = yz - f_1(z) + f_2(y)$
12. $x f_1(z-y-2x) + y = f_2(z-y-2x)$
13. $z = f_1(y+\cos x) + f_2(y-\cos x)$.

4.7 PARTIAL DIFFERENTIAL EQUATIONS LINEAR AND HOMOGENOUS IN PARTIAL DERIVATIVES WITH CONSTANT COEFFICIENTS

An equation of the form $\frac{\partial^n z}{\partial x^n} + k_1 \frac{\partial^n z}{\partial x^{n-1}\partial y} + k_2 \frac{\partial^n z}{\partial x^{n-2}\partial y^2} + \cdots + k_n \frac{\partial^n z}{\partial y^n} = F(x,y)$ is linear and homogenous in partial derivatives of order n and has constant coefficients.

On writing $\frac{\partial^r}{\partial x^r} \equiv D^r$ and $\frac{\partial^r}{\partial y^r} \equiv D'^r$ the equation can be written as

$$f(D,D')z = F(x,y)$$

where $$F(D,D') = D^n + k_1 D^{n-1}D' + k_2 D^{n-2}D'^2 + \cdots + k_n D'^n.$$

As in the case of ordinary linear equations with constant coefficients, the complete solution consist of two parts namely the complementary function (C.F.) and particular integral (P.I.)

The complementary function is the general solution or complete solution of $f(D,D')z = 0$ which must contain n arbitrary functions or n arbitrary constants.

The particular integral is the particular solution of $f(D,D')z = F(x,y)$.

4.7.1 Superposition or Linearity Principle

If u_1, u_2 are any solutions of $f(D, D')z = 0$ (4.25)

then $u = c_1u_1 + c_2u_2$, where c_1 and c_2 are arbitrary constants is solution of (4.25). This principle can be proved on the same line as the proof of Theorem (8.7) in ordinary differential equations in volume 1.

From this theorem, we observe that if $\phi_i(x, y),\ i = 1, 2, \ldots n$ are solutions of (4.25), where $\phi_1, \phi_2, \ldots$ are arbitrary functions then $\sum_{i=0}^{n} \phi_i(x, y)$ is C.F. of equation

$$f(D, D')z = F(x, y) \tag{4.26}$$

4.7.2 Rules for Finding the Complementary Function

Partial differential equation is

$$\frac{\partial^n z}{\partial x^n} + k_1 \frac{\partial^n z}{\partial x^{n-1}\partial y} + k_2 \frac{\partial^n z}{\partial x^{n-2}\partial y^2} + \cdots + k_n \frac{\partial^n z}{\partial y^n} = F(x, y).$$

We are to find its C.F. which is general solution or complete solution of

$$\frac{\partial^n z}{\partial x^n} + k_1 \frac{\partial^n z}{\partial x^{n-1}\partial y} + k_2 \frac{\partial^n z}{\partial x^{n-2}\partial y^2} + \cdots + k_n \frac{\partial^2 z}{\partial y^n} = 0. \tag{4.27}$$

This equation in symbolic form can be written as

$$\left[D^n + k_1 D^{n-1}D' + k_2 D^{n-2}D'^2 + \cdots + k_n D'^n\right]z = 0. \tag{4.28}$$

Replacing D by m and D' by 1, auxiliary equation (A.E.) is

$$m^n + k_1 m^{n-1} + k_2 m^{n-2} + \cdots + k_n = 0. \tag{4.29}$$

Let roots of this equation are $m_1, m_2, \ldots, m_n$.

Case I: When all roots are different

In this case, equation (4.28) can be written as

$$(D - m_1 D')(D - m_2 D')\ldots(D - m_n D')z = 0. \tag{4.30}$$

Equation (4.30) will be satisfied by solution of $(D - m_n D')z = 0$

or
$$p - m_n q = 0 \tag{4.31}$$

It is Lagrange's equation.

Lagrange's auxiliary equations are

$$\frac{dx}{1} = \frac{dy}{-m_n} = \frac{dz}{0}$$

Hence solutions are $y + m_n x = a,\ z = b$ where a and b are arbitrary constants.

$\therefore$
$$z = \phi_n(y + m_n x)$$

is solution of (4.31) where ϕ_n is an arbitrary function.

Now, the factors of equation (4.30) can be written in any order and hence it will be satisfied by solutions of $(D-m_1D')z=0,(D-m_2D')z=0,\ldots,(D-m_nD')z=0$
which are $z=\phi_1(y+m_1x),\ z=\phi_2(y+m_2x),\ldots,z=\phi_n(y+m_nx)$
where $\phi_1,\phi_2,\ldots,\phi_n$ are arbitrary functions.
Thus, by principle of superposition

$$\text{C.F.}=\sum_{i=0}^{n}\phi_i(y+m_ix).$$

Case II: If two roots of equation (4.29) are equal say $m_1=m_2$ and all others are different

In this case $\phi_1(y+m_1x),\ \phi_2(y+m_2x)$ will give only one arbitrary function of $y+m_1x=y+m_2x$. The part of C.F. corresponding to repeated root is solution of

$$(D-m_1D')(D-m_1D')z=0 \tag{4.32}$$

Let $$(D-m_1D')z=u \tag{4.33}$$

Now, we have $$(D-m_1D')u=0$$

Its solution is

$$u=\phi_1(y+m_1x)$$

$\therefore$ from equation (4.33)

$$(D-m_1D')z=\phi_1(y+m_1x)$$

or $$p-m_1q=\phi_1(y+m_1x)$$

It is Lagrange's equation.
Lagrange's auxiliary equations are

$$\frac{dx}{1}=\frac{dy}{-m_1}=\frac{dz}{\phi_1(y+m_1x)} \tag{4.34}$$

From first two members

$$y+m_1x=a$$

$\therefore$ from first and third members of (4.34)

$$dx=\frac{dz}{\phi_1(a)}$$

$\therefore$ $$z=x\phi_1(a)+b$$

$\therefore$ Complete solution of (4.32) is

$$z=x\phi_1(y+m_1x)+\phi_2(y+m_1x)$$

where ϕ_1 and ϕ_2 are arbitrary functions.
Hence

$$\text{C.F.}=x\phi_1(y+m_1x)+\phi_2(y+m_1x)+\phi_3(y+m_3x)+\cdots+\phi_n(y+m_nx).$$

Similarly, if $m_1 = m_2 = m_3$ and all other roots of (4.29) are different then

$$\text{C.F.} = \phi_1(y+m_1x) + x\phi_2(y+m_1x) + x^2\phi_3(y+m_1x) + \phi_4(y+m_4x) + \cdots + \phi_n(y+m_nx)$$

and so on.

4.7.3 Inverse Operator

Definition: If $f(D,D')z = F(x,y)$ then $\dfrac{1}{f(D,D')}F(x,y)$ is that function of x and y, not containing arbitrary constants, which when operated by $f(D,D')$ gives $F(x,y)$

i.e.,
$$f(D,D')\frac{1}{f(D,D')}F(x,y) = F(x,y).$$

Thus, $\dfrac{1}{f(D,D')}F(x,y)$ satisfies the differential equation $f(D,D')z = F(x,y)$ and hence is its particular integral. Clearly, $f(D,D')$ and $\dfrac{1}{f(D,D')}$ are inverse operators.

Theorem 4.2: $\dfrac{1}{D-mD'}F(x,\, y) = \int F(x,\, a-mx)\,dx$

where we put $a = y+mx$ after integration.

Proof: Let
$$\frac{1}{D-mD'}F(x,y) = G(x,y)$$

$\Rightarrow$
$$(D-mD')\frac{1}{D-mD'}F(x,y) = (D-mD')G(x,y)$$

$\therefore$
$$\frac{\partial G}{\partial x} - m\frac{\partial G}{\partial y} = F(x,y)$$

It is Lagrange's equation.

Lagrange's auxiliary equations are

$$\frac{dx}{1} = \frac{dy}{-m} = \frac{dG}{F(x,y)} \tag{4.35}$$

From first and second members

$$y + mx = a \tag{4.36}$$

From first and third members

$$dG = F(x,a-mx)\,dx \qquad \text{(from (4.36))}$$

$\therefore$
$$G(x,y) = \int F(x,a-mx)\,dx$$

$\therefore$
$$\frac{1}{D-mD'}F(x,y) = \int F(x,a-mx)\,dx \tag{4.37}$$

where we put $a = y+mx$ after integration.

4.7.4 Operator Methods for Finding Particular Integrals

Consider the equation

$$f(D,D')z = F(x,y)$$

then,
$$\text{P.I.} = \frac{1}{f(D,D')}F(x,y)$$

Case I: $F(x,y) = e^{ax+by}$

We have

$$D^r F(x,y) = a^r F(x,y)$$

and
$$D'^r F(x,y) = b^r F(x,y)$$

$$\therefore \qquad f(D,D')e^{ax+by} = f(a,b)e^{ax+by}$$

$$\therefore \qquad f(a,b)\frac{1}{f(D,D')}e^{ax+by} = \frac{1}{f(D,D')}f(D,D')e^{ax+by} = e^{ax+by}$$

$$\therefore \qquad \frac{1}{f(D,D')}e^{ax+by} = \frac{1}{f(a,b)}e^{ax+by} \quad \text{if } f(a,b) \neq 0$$

$$\therefore \qquad \text{P.I.} = \frac{1}{f(D,D')}e^{ax+by} = \frac{1}{f(a,b)}e^{ax+by} \quad \text{if } f(a,b) \neq 0$$

If $f(a,b) = 0$, i.e., $D = a$, $D' = b$ satisfies $f(D,D') = 0$

then
$$\left(D - \frac{a}{b}D'\right)^k \text{ is factor of } f(D,D') \text{ for some } k \in N$$

such that $f(D,D') = \left(D - \frac{a}{b}D'\right)^k \phi(D,D')$ where $\phi(a,b) \neq 0$.

$$\therefore \qquad \text{P.I.} = \frac{1}{f(D,D')}e^{ax+by} = \frac{1}{\left(D - \frac{a}{b}D'\right)^k \phi(D,D')}e^{ax+by}$$

$$= \frac{1}{\phi(a,b)} \frac{1}{\left(D - \frac{a}{b}D'\right)^{k-1}} \cdot \frac{1}{\left(D - \frac{a}{b}D'\right)}e^{ax+by}$$

$$= \frac{1}{\phi(a,b)} \frac{1}{\left(D - \frac{a}{b}D'\right)^{k-1}} \int e^{ax+b\left(c - \frac{a}{b}x\right)}dx \quad \text{where } c = y + \frac{a}{b}x \qquad \text{(from (4.37))}$$

$$= \frac{1}{\phi(a,b)} \frac{1}{\left(D - \frac{a}{b}D'\right)^{k-1}} \cdot xe^{bc}$$

$$= \frac{1}{\phi(a,b)} \frac{1}{\left(D - \frac{a}{b}D'\right)^{k-1}} \cdot xe^{ax+by} \qquad \left(\because c = y + \frac{a}{b}x\right)$$

$$= \frac{1}{\phi(a,b)} \frac{1}{\left(D - \frac{a}{b}D'\right)^{k-2}} \cdot \frac{1}{\left(D - \frac{a}{b}D'\right)} xe^{ax+by}$$

$$= \frac{1}{\phi(a,b)} \frac{1}{\left(D - \frac{a}{b}D'\right)^{k-2}} \int x\, e^{ax+b\left(c-\frac{a}{b}x\right)} dx; \ c = y + \frac{ax}{b}$$

$$= \frac{1}{\phi(a,b)} \frac{1}{\left(D - \frac{a}{b}D'\right)^{k-2}} \frac{x^2}{2} e^{ax+by}$$

Proceeding in this way

$$\text{P.I.} = \frac{1}{f(D,D')} e^{ax+by} = \frac{x^k}{k!\,\phi(a,b)} e^{ax+by}$$

But $f(D,D') = \left(D - \frac{a}{b}D'\right)^k \phi(D,D')$

$$\therefore \qquad \left[\frac{d^k}{dD^k} f(D,D')\right]_{D=a,\,D'=b} = k!\,\phi(a,b)$$

$$\therefore \qquad \text{P.I.} = \frac{x^k e^{ax+by}}{\left[\frac{d^k}{dD^k} f(D,D')\right]_{D=a,\,D'=b}}$$

Case II:

$$F(x,y) = \sin(ax + by + c) \quad \text{or} \quad \cos(ax + by + c)$$

We have $\quad D^2 F(x,y) = -a^2 F(x,y)$

$$D'^2 F(x,y) = -b^2 F(x,y)$$

$$DD' F(x,y) = -ab F(x,y)$$

Using these, we have for some *l, m, n*

$$f(D,D')\sin(ax+by+c)=(lD+mD'+n)\sin(ax+by+c) \tag{4.38}$$

$$=(la+mb)\cos(ax+by+c)+n\sin(ax+by+c)$$

$$f(D,D')\cos(ax+by+c)=(lD+mD'+n)\cos(ax+by+c)$$

$$=-(al+mb)\sin(ax+by+c)+n\cos(ax+by+c)$$

Thus, if $al+mb\neq 0$, we have

$$(la+mb)\frac{1}{f(D,D')}\cos(ax+by+c)+n\frac{1}{f(D,D')}\sin(ax+by+c)=\sin(ax+by+c)$$

$$n\frac{1}{f(D,D')}\cos(ax+by+c)-(la+mb)\frac{1}{f(D,D')}\sin(ax+by+c)=\cos(ax+by+c)$$

Solving these equations, we have

$$\frac{1}{f(D,D')}\cos(ax+by+c)=\frac{(la+mb)\sin(ax+by+c)+n\cos(ax+by+c)}{(la+mb)^2+n^2}$$

$$\frac{1}{f(D,D')}\sin(ax+by+c)=\frac{-(la+mb)\cos(ax+by+c)+n\sin(ax+by+c)}{(la+mb)^2+n^2}$$

Combining these, we can write

$$\frac{1}{f(D,D')}F(x,y)=\frac{-(lD+mD')+n}{-\left(l^2D^2+m^2D'^2+2lmDD'\right)_{D^2=-a^2,\ D'^2=-b^2,\ DD'=-ab}+n^2}F(x,y)$$

$$=\frac{n-(lD+mD')}{\left[n^2-(lD+mD')^2\right]_{D^2=-a^2,\ D'^2=-b^2,\ DD'=-ab}}F(x,y)$$

$$=\frac{1}{\left[lD+mD'+n\right]_{D^2=-a^2,\ D'^2=-b^2,\ DD'=-ab}}F(x,y)$$

$$\therefore\quad \frac{1}{f(D,D')}F(x,y)=\frac{1}{\left[f(D,D')\right]_{D^2=-a^2,\ D'^2=-b^2,\ DD'=-ab}}F(x,y) \qquad \text{(from (4.38))}$$

If $al+bm=0, n\neq 0$

then $\quad f(D,D')F(x,y)=nF(x,y)$

$$\therefore\quad \text{P.I.}=\frac{1}{f(D,D')}F(x,y)=\frac{1}{n}F(x,y)$$

If $al+bm=0, n=0$ i.e. $f(D,D')=0$ for $D^2=-a^2, D'^2=-b^2, DD'=-ab$

then $$\text{P.I.} = \frac{1}{f(D,D')}\sin(ax+by+c)$$

$$= \text{Im. part of } \frac{1}{f(D,D')}e^{i(ax+by+c)}$$

$$= \text{Im. part of } x\frac{1}{\frac{d}{dD}f(D,D')}e^{i(ax+by+c)} \qquad \text{(from case I, } \because f(ia,ib)=0\text{)}$$

$$= x\frac{1}{\frac{d}{dD}f(D,D')}\sin(ax+by+c)$$

After this, proceed as above.

Similar is the case, when $F(x,y)=\cos(ax+by+c)$

Case III:

$$F(x,y)=\sinh(ax+by+c) \text{ or } \cosh(ax+by+c)$$

Here, $$D^2F(x,y)=a^2F(x,y)$$
$$D'^2F(x,y)=b^2F(x,y)$$
$$DD'F(x,y)=ab\,F(x,y)$$

Hence proceed as in Case II, replacing D^2 by a^2, D'^2 by b^2 and DD' by ab when $f(D, D') \neq 0$ for $D^2 = a^2$, $D'^2 = b^2$, $DD' = ab$.

If $f(D, D') = 0$ for $D^2 = a^2$, $D'^2 = b^2$, $DD' = ab$, then convert $F(x, y)$ in exponential forms and solve it.

Case IV:

$$F(x,y)= \text{polynomial in } x \text{ and } y$$
$$\text{P.I.} = \left[f(D,D')\right]^{-1}F(x,y)$$

Expand $\left[f(D,D')\right]^{-1}$ in ascending powers of D or D' by Binomial theorem and then operate on $F(x,y)$.

Case V:

$$F(x,y)=e^{ax+by}\,V(x,y)$$

for any $g(x,y)$, by Leibnitz theorem

$$\begin{aligned}D^n\left(e^{ax+by}g(x,y)\right) &= e^{ax+by}D^n g(x,y) + {}^nC_1\,a e^{ax+by}D^{n-1}g(x,y)\\ &\quad + {}^nC_2\,a^2\,e^{ax+by}D^{n-2}g(x,y)+\cdots+a^n e^{ax+by}g(x,y)\\ &= e^{ax+by}\left[D^n + {}^nC_1\,a\,D^{n-1} + {}^nC_2\,a^2D^{n-2}+\cdots a^n\right]g(x,y)\\ &= e^{ax+by}(D+a)^n\,g(x,y)\end{aligned}$$

Similarly, $$D'^n\left(e^{ax+by}g(x,y)\right)=e^{ax+by}(D'+b)^n\,g(x,y)$$

$$\therefore \quad f(D,D')\left(e^{ax+by}g(x,y)\right)=e^{ax+by}f(D+a,\ D'+b)g(x,y)$$

$$\therefore \quad \frac{1}{f(D,D')}\left[e^{ax+by}f(D+a,\ D'+b)g(x,y)\right]=e^{ax+by}g(x,y)$$

Taking $g(x,y)=\dfrac{1}{f(D+a,D'+b)}V(x,y)$

we have $$\frac{1}{f(D,D')}e^{ax+by}V(x,y)=e^{ax+by}\frac{1}{f(D+a,D'+b)}V(x,y)$$

Case VI:

$$F(x,y)=\phi(ax+by).$$

Now, for any $G(ax+by)$

$$D^rG(ax+by)=a^rG^{(r)}(ax+by)$$

$$D'^rG(ax+by)=b^rG^{(r)}(ax+by)$$

$$\therefore \quad f(D,D')G(ax+by)=f(a,b)G^{(n)}(ax+by)$$

($\because f(D,D')$ is homogeneous of degree n)

If $f(a,b)\neq 0,$ then

$$\frac{1}{f(D,D')}G^{(n)}(ax+by)=\frac{1}{f(a,b)}G(ax+by)$$

Taking $G^{(n)}(ax+by)=\phi(ax+by)$

$$\frac{1}{f(D,D')}\phi(ax+by)=\frac{1}{f(a,b)}G(ax+by)$$

where $G(ax+by)$ is obtained after integrating $\phi(z)$ w.r.t. z, n times and then taking $z=ax+by$
If $f(a,b)=0,$ then this case can be dealt as in Case I replacing e^{ax+by} by $\phi(ax+by)$

and then $$\text{P.I.}=\frac{x^k\phi(ax+by)}{\left[\dfrac{d^k}{dD^k}f(D,D')\right]_{D=a,D'=b}}$$

when $$f(D,D')=\left(D-\frac{a}{b}D'\right)^k h(D,D');\ \left[h(D,D')\right]_{D=a,D'=b}\neq 0$$

In particular, if

$$f(D,D')=\left(D-\frac{a}{b}D'\right)^n$$

then $$\text{P.I.}=\frac{x^n}{n!}\phi(ax+by)$$

Hence, $$\frac{1}{(bD-aD')^n}\phi(ax+by)=\frac{1}{b^n}\frac{1}{\left(D-\frac{a}{b}D'\right)^n}\phi(ax+by)$$

$$=\frac{x^n}{n!b^n}\phi(ax+by)$$

Note: We can also use the symbols D_x and D_y at the places of D and D' respectively.

Example 4.34: Solve the following partial differential equations

(i) $\frac{\partial^4 z}{\partial x^4}-\frac{\partial^4 z}{\partial y^4}=0$ (ii) $\left(D_x^{\,3}-3D_x^{\,2}D_y+2D_y^{\,2}D_x\right)z=0$

Solution: (i) Differential equation written symbolically is

$$\left(D_x^{\,4}-D_y^{\,4}\right)z=0$$

A.E. is $m^4-1=0$

$\Rightarrow$ $(m^2-1)(m^2+1)=0$

$\therefore$ $m=\pm 1, \pm i$

$\therefore$ general solution is

$$z=\phi_1(y+x)+\phi_2(y-x)+\phi_3(y-ix)+\phi_4(y+ix)$$

where ϕ_1, ϕ_2, ϕ_3 and ϕ_4 are arbitrary functions.

(ii) Differential equation is

$$\left(D_x^{\,3}-3D_x^{\,2}D_y+2D_y^{\,2}D_x\right)z=0$$

A.E. is $m^3-3m^2+2m=0$

$\Rightarrow$ $m(m-1)(m-2)=0$

$\therefore$ $m=0, 1, 2$

$\therefore$ general solution is

$$z=\phi_1(y)+\phi_2(y+x)+\phi_3(y+2x)$$

where ϕ_1, ϕ_2 and ϕ_3 are arbitrary functions.

Example 4.35: Solve the following partial differential equations

(i) $\left(D^3-7DD'^2-6D'^3\right)z=\sin(x+2y)+e^{2x+y}$

(ii) $\left(2D^2-5DD'+2D'^2\right)z=5\sin(2x+y)$

(iii) $\left(D^2+DD'-6D'^2\right)z=y\cos x$

(iv) $\left(D^2-6DD'+9D'^2\right)z=12x^2+36xy$

Solution: (i) A.E. is

$$m^3 - 7m - 6 = 0$$

$m = -1$ satisfies it, by synthetic division

–1)	1	0	–7	–6
		–1	1	6
	1	–1	–6	0

$\therefore \quad m^3 - 7m - 6 = (m+1)(m^2 - m - 6) = (m+1)(m-3)(m+2) = 0$

$\therefore \quad m = -2, -1, 3$

$\therefore \quad \text{C.F.} = \phi_1(y-2x) + \phi_2(y-x) + \phi_3(y+3x)$

$$\text{P.I.} = \frac{1}{D^3 - 7DD'^2 - 6D'^3}\left[\sin(x+2y) + e^{2x+y}\right]$$

$$= \frac{1}{D^3 - 7DD'^2 - 6D'^3}\sin(x+2y) + \frac{1}{D^3 - 7DD'^2 - 6D'^3}e^{2x+y}$$

To find $\frac{1}{D^3 - 7DD'^2 - 6D'^3}\sin(x+2y)$

integrate $\sin z$ w.r.t. z three times and put $z = x+2y, D = 1, D' = 2$

$\therefore \quad \text{P.I.} = \frac{1}{1^3 - 7(1)(2)^2 - 6(2)^3}\cos(x+2y) + \frac{1}{2^3 - 7(2)(1) - 6(1)^3}e^{2x+y}$

$$= -\frac{1}{75}\cos(x+2y) - \frac{1}{12}e^{2x+y}$$

$\therefore$ general solution is

$$z = \phi_1(y-2x) + \phi_2(y-x) + \phi_3(y+3x) - \frac{1}{75}\cos(x+2y) - \frac{1}{12}e^{2x+y}$$

where ϕ_1, ϕ_2 and ϕ_3 are arbitrary functions.

(ii) A.E. is $2m^2 - 5m + 2 = 0$

$\therefore \quad m = \frac{5 \pm \sqrt{25-16}}{4} = \frac{1}{2}, 2$

$\therefore \quad \text{C.F.} = \phi_1(2y+x) + \phi_2(y+2x)$

$$\text{P.I.} = \frac{1}{2D^2 - 5DD' + 2D'^2}5\sin(2x+y)$$

$$= 5x\frac{1}{4D - 5D'}\sin(2x+y) \quad \left(\because 2(-2^2) + 5(2)(1) + 2(-1^2) = -8 + 10 - 2 = 0\right)$$

$$= 5x\frac{1}{4(2) - 5(1)}\left[-\cos(2x+y)\right]$$

$$= -\frac{5x}{3}\cos(2x+y)$$

∴ general solution is

$$z=\phi_1(2y+x)+\phi_2(y+2x)-\frac{5x}{3}\cos(2x+y)$$

where ϕ_1 and ϕ_2 are arbitrary functions.

(iii) $\left(D^2+DD'-6D'^2\right)z=y\cos x$

A.E. is

$$m^2+m-6=0$$

$$\Rightarrow \quad (m+3)(m-2)=0$$

$$\therefore \quad m=-3,2$$

$$\therefore \quad \text{C.F.}=\phi_1(y-3x)+\phi_2(y+2x)$$

$$\text{P.I.}=\frac{1}{D^2+DD'-6D'^2}y\cos x=\frac{1}{(D+3D')(D-2D')}y\cos x$$

$$=\frac{1}{D+3D'}\int(c-2x)\cos x\,dx\ ;\ c=2x+y$$

$$=\frac{1}{D+3D'}\left[(c-2x)\sin x-(-2)(-\cos x)\right]$$

$$=\frac{1}{D+3D'}[y\sin x-2\cos x]$$

$$=\int\left[(a+3x)\sin x-2\cos x\right]dx\ ;\ a=y-3x$$

$$=\left[(a+3x)(-\cos x)-(3)(-\sin x)-2\sin x\right]$$

$$=-y\cos x+3\sin x-2\sin x=-y\cos x+\sin x \qquad (\because a=y-3x)$$

∴ general solution is

$$z=\phi_1(y-3x)+\phi_2(y+2x)+\sin x-y\cos x$$

where ϕ_1 and ϕ_2 are arbitrary functions.

(iv) $\left[D^2-6DD'+9D'^2\right]z=12x^2+36xy$

A.E. is $m^2-6m+9=0$

$$\Rightarrow \quad (m-3)^2=0$$

$$\therefore \quad m=3,3$$

$$\therefore \quad \text{C.F.}=\phi_1(y+3x)+x\phi_2(y+3x)$$

$$\text{P.I.} = \frac{1}{D^2 - 6DD' + 9D'^2}\left(12x^2 + 36xy\right)$$

$$= \frac{1}{D^2}\left[1 - \frac{6DD' - 9D'^2}{D^2}\right]^{-1}\left(12x^2 + 36xy\right)$$

$$= \frac{1}{D^2}\left[1 + \left(\frac{6D'}{D} - \frac{9D'^2}{D^2}\right) + \cdots\cdots\cdots + \cdots\right]\left(12x^2 + 36xy\right)$$

$$= \frac{1}{D^2}\left[1 + \frac{6D'}{D} + \cdots + \cdots\right]\left(12x^2 + 36xy\right)$$

$$= \frac{1}{D^2}\left(12x^2 + 36xy + \frac{6}{D}(36x)\right)$$

$$= \frac{1}{D^2}\left[12x^2 + 36xy + 108x^2\right] = \frac{1}{D^2}\left(120x^2 + 36xy\right)$$

$$= 120\frac{x^4}{12} + 36\frac{x^3}{6}y = 10x^4 + 6x^3 y$$

$\therefore$ general solution is

$$z = \phi_1(y + 3x) + x\phi_2(y + 3x) + 10x^4 + 6x^3 y$$

where ϕ_1 and ϕ_2 are arbitrary functions.

Example 4.36: Find the general solutions of partial differential equations

(i) $\left(D^2 + 2DD' - 8D'^2\right)z = \sqrt{2x + 3y}$

(ii) $\left(D^3 + D^2D' - DD'^2 - D'^3\right)z = e^x \cos 2y$

(iii) $\left(D^2 - DD' - 2D'^2\right)z = (y - 1)e^x$

(iv) $\left(D^2 + DD' - 6D'^2\right)z = x^2 \sin(x + y)$

Solution: A.E. is

(i) $\quad m^2 + 2m - 8 = 0$

$\Rightarrow \quad (m + 4)(m - 2) = 0$

$\therefore \quad m = -4,\ 2$

$\therefore \quad \text{C.F.} = \phi_1(y - 4x) + \phi_2(y + 2x)$

$$\text{P.I.} = \frac{1}{D^2 + 2DD' - 8D'^2}\sqrt{2x + 3y}$$

Integrate $\sqrt{z}$ two times and then put $z = 2x + 3y,\ D = 2,\ D' = 3$

$$\therefore \quad \text{P.I.} = \frac{1}{\left(2^2 + 2(2)(3) - 8(3)^2\right)}\frac{(2x + 3y)^{5/2}}{(3/2)(5/2)} = -\frac{1}{210}(2x + 3y)^{5/2}$$

$\therefore$ general solution is

$$z = \phi_1(y-4x) + \phi_2(y+2x) - \frac{1}{210}(2x+3y)^{5/2}$$

where ϕ_1 and ϕ_2 are arbitrary functions.

(ii) $\left(D^3 + D^2D' - DD'^2 - D'^3\right)z = e^x \cos 2y$

A.E. is $\qquad m^3 + m^2 - m - 1 = 0$

$m = 1$ satisfies it.

By synthetic division

1)	1	1	−1	−1
		1	2	1
	1	2	1	0

$$\therefore \quad m^3 + m^2 - m - 1 = (m-1)(m^2+2m+1) = (m-1)(m+1)^2 = 0$$

$$\therefore \quad m = -1, -1, 1$$

$$\therefore \quad \text{C.F.} = \phi_1(y-x) + x\phi_2(y-x) + \phi_3(y+x)$$

$$\text{P.I.} = \frac{1}{D^3 + D^2D' - DD'^2 - D'^3} e^x \cos 2y$$

$$= e^x \frac{1}{(D+1)^3 + (D+1)^2 D' - (D+1)D'^2 - D'^3} \cos 2y$$

$$= e^x \frac{1}{1 + D' - D'^2 - D'^3} \cos 2y$$

$$= e^x \frac{1}{1 + D' - (-2^2) - (-2^2)D'} \cos 2y$$

$$= \frac{e^x}{5} \frac{1}{D'+1} \cos 2y = \frac{e^x}{5} \frac{D'-1}{D'^2-1} \cos 2y$$

$$= \frac{e^x}{5} \frac{1}{(-4-1)} (D'-1)\cos 2y = -\frac{e^x}{25}(-2\sin 2y - \cos 2y)$$

$$= \frac{e^x}{25}(2\sin 2y + \cos 2y)$$

$\therefore$ general solution is

$$z = \phi_1(y-x) + x\phi_2(y-x) + \phi_3(y+x) + \frac{e^x}{25}(2\sin 2y + \cos 2y)$$

where ϕ_1, ϕ_2 and ϕ_3 are arbitrary functions.

(iii) $\left(D^2 - DD' - 2D'^2\right)z = (y-1)e^x$

A.E. is $m^2 - m - 2 = 0$

$\Rightarrow \quad (m-2)(m+1) = 0$

$\therefore \quad m = -1, 2$

$\therefore \quad \text{C.F.} = \phi_1(y-x) + \phi_2(y+2x)$

$$\text{P.I.} = \frac{1}{D^2 - DD' - 2D'^2}(y-1)e^x$$

$$= e^x \frac{1}{(D+1)^2 - (D+1)D' - 2D'^2}(y-1)$$

$$= e^x \frac{1}{1 - D' - 2D'^2}(y-1)$$

$$= e^x \left(1 - D' - 2D'^2\right)^{-1}(y-1)$$

$$= e^x (1 + D' + \cdots)(y-1) = e^x (y-1+1) = ye^x$$

$\therefore$ general solution is

$$z = \phi_1(y-x) + \phi_2(y+2x) + ye^x$$

where ϕ_1 and ϕ_2 are arbitrary functions.

(iv) $\left(D^2 + DD' - 6D'^2\right)z = x^2 \sin(x+y)$

A.E. is $m^2 + m - 6 = 0$

$\Rightarrow \quad (m+3)(m-2) = 0$

$\therefore \quad m = -3, 2$

$\therefore \quad \text{C.F.} = \phi_1(y-3x) + \phi_2(y+2x)$

$$\text{P.I.} = \frac{1}{(D+3D')(D-2D')}x^2 \sin(x+y)$$

$$= \frac{1}{(D+3D')}\int x^2 \sin(x+c-2x)\,dx \text{ where } c = y+2x$$

$$= \frac{1}{(D+3D')}\left[x^2 \cos(c-x) - (2x)(-\sin(c-x)) + (2)(-\cos(c-x))\right]$$

$$= \frac{1}{(D+3D')}\left[x^2 \cos(x+y) + 2x\sin(x+y) - 2\cos(x+y)\right] \quad (\because\ c = y+2x)$$

$$= \frac{1}{(D+3D')}\left[\left(x^2-2\right)\cos(x+y) + 2x\sin(x+y)\right]$$

$$= \int \left[\left(x^2-2\right)\cos\left(x+c+3x\right)+2x\sin\left(x+c+3x\right) \right]dx \qquad \text{where } c = y-3x$$

$$= \int \left[\left(x^2-2\right)\cos\left(4x+c\right)+2x\sin\left(4x+c\right) \right]dx$$

$$= \left[\left(x^2-2\right)\left(\frac{1}{4}\sin\left(4x+c\right)\right)-\left(2x\right)\left(-\frac{1}{16}\cos\left(4x+c\right)\right)+2\left(-\frac{1}{64}\sin\left(4x+c\right)\right) \right.$$

$$\left. +\left(2x\right)\left(-\frac{1}{4}\cos\left(4x+c\right)\right)-2\left(-\frac{1}{16}\sin\left(4x+c\right)\right) \right]$$

$$= \left[\left(\frac{x^2-2}{4}-\frac{1}{32}+\frac{1}{8}\right)\sin\left(x+y\right)+\left(\frac{x}{8}-\frac{x}{2}\right)\cos\left(x+y\right) \right] \qquad (\because\ c = y-3x)$$

$$= \frac{1}{32}\left(8x^2-13\right)\sin\left(x+y\right)-\frac{3}{8}x\cos\left(x+y\right)$$

$\therefore$ general solution is

$$z = \phi_1\left(y-3x\right)+\phi_2\left(y+2x\right)+\frac{1}{32}\left(8x^2-13\right)\sin\left(x+y\right)-\frac{3x}{8}\cos\left(x+y\right)$$

where ϕ_1 and ϕ_2 are arbitrary functions.

Another method to find P.I.

$$\text{P.I.} = \text{Im. part of } \frac{1}{D^2+DD'-6D'^2}x^2e^{i(x+y)}$$

$$= \text{Im. part of } e^{i(x+y)}\frac{1}{\left(D+i\right)^2+\left(D+i\right)\left(D'+i\right)-6\left(D'+i\right)^2}x^2$$

$$= \text{Im. part of } e^{i(x+y)}\frac{1}{D^2+2iD-1+iD-1+6}x^2$$

$$= \text{Im. part of } e^{i(x+y)}\frac{1}{D^2+3iD+4}x^2$$

$$= \text{Im. part of } e^{i(x+y)}\frac{1}{4}\left(1+\frac{1}{4}\left(3iD+D^2\right)\right)^{-1}x^2$$

$$= \text{Im. part of } \frac{e^{i(x+y)}}{4}\left[1-\frac{1}{4}\left(3iD+D^2\right)-\frac{9}{16}D^2+\cdots\right]x^2$$

$$= \text{Im. part of } \frac{e^{i(x+y)}}{4}\left[1-\frac{3}{4}iD-\frac{13}{16}D^2+\cdots\right]x^2$$

$$= \text{Im. part of } \frac{1}{4}\left[\cos\left(x+y\right)+i\sin\left(x+y\right)\right]\left[x^2-\frac{3}{2}ix-\frac{13}{8}\right]$$

$$= \frac{1}{32}\left(8x^2-13\right)\sin\left(x+y\right)-\frac{3}{8}x\cos\left(x+y\right)$$

Example 4.37: Find the general solution of partial differential equations

(i) $4r+12s+9t=e^{3x-2y}$

(ii) $r-2s+t=2x\cos y$

(iii) $r+s-2t=8\ln(x+5y)$

Solution: (i) Differential equation in symbolic form is

$$\left(4D^2+12DD'+9D'^2\right)z=e^{3x-2y}$$

A.E. is

$$4m^2+12m+9=0$$

$$\Rightarrow \quad (2m+3)^2=0$$

$$\therefore \quad m=-\frac{3}{2},\frac{-3}{2}$$

$$\therefore \quad \text{C.F.}=\phi_1(2y-3x)+x\phi_2(2y-3x)$$

$$\text{P.I.}=\frac{1}{(2D+3D')^2}e^{3x-2y}$$

$$=x^2\frac{1}{\frac{d^2}{dD^2}(2D+3D')^2}e^{3x-2y} \qquad (\because 2(3)+3(-2)=0)$$

$$=x^2\frac{1}{8}e^{3x-2y}$$

$$=\frac{x^2}{8}e^{3x-2y}$$

$\therefore$ general solution is

$$z=\phi_1(2y-3x)+x\phi_2(2y-3x)+\frac{x^2}{8}e^{3x-2y}$$

where ϕ_1 and ϕ_2 are arbitrary functions.

(ii) Differential equation in symbolic form is

$$\left(D_x^2-2D_xD_y+D_y^2\right)z=2x\cos y$$

A.E. is

$$m^2-2m+1=0 \quad \text{or} \quad (m-1)^2=0 \qquad \Rightarrow m=1,1$$

$$\therefore \quad \text{C.F.}=\phi_1(y+x)+x\phi_2(y+x)$$

$$\text{P.I.}=\frac{1}{(D_x-D_y)^2}2x\cos y$$

$$= \text{Re part of } \frac{1}{(D_x - D_y)^2} 2xe^{iy}$$

$$= 2\text{Re part of } e^{iy} \frac{1}{(D_x - D_y - i)^2} x$$

$$= 2\text{Re part of } e^{iy}(-1)\left[1 + i(D_x - D_y)\right]^{-2} x$$

$$= 2\text{Re part of } (-e^{iy})[1 - 2iD_x + \cdots]x$$

$$= 2\text{Re part of } (-1)(\cos y + i\sin y)(x - 2i)$$

$$= -2[x\cos y + 2\sin y]$$

$\therefore$ general solution is

$$z = \phi_1(y + x) + x\phi_2(y + x) - 2(x\cos y + 2\sin y)$$

where ϕ_1 and ϕ_2 are arbitrary functions.

(iii) Differential equation in symbolic form is

$$\left(D_x^2 + D_x D_y - 2D_y^2\right)z = 8\ln(x + 5y)$$

A.E. is $\quad m^2 + m - 2 = 0$

$\Rightarrow \quad (m+2)(m-1) = 0$

$\therefore \quad m = -2, 1$

$\therefore \quad \text{C.F.} = \phi_1(y - 2x) + \phi_2(y + x)$

$$\text{P.I.} = \frac{1}{\left(D_x^2 + D_x D_y - 2D_y^2\right)} 8\ln(x + 5y)$$

$$= \frac{1}{(D_x + 2D_y)(D_x - D_y)} 8\ln(x + 5y)$$

$$= 8\frac{1}{(D_x + 2D_y)} \int \ln(x + 5(c - x))\,dx, \qquad \text{where } c = x + y$$

$$= 8\frac{1}{(D_x + 2D_y)}\left[x\ln(5c - 4x) + \int \frac{4x}{5c - 4x}\,dx\right] \qquad \text{(integrating by parts)}$$

$$= 8\frac{1}{(D_x + 2D_y)}\left[x\ln(5c - 4x) - \int\left(1 - \frac{5c}{5c - 4x}\right)dx\right]$$

$$= 8\frac{1}{(D_x + 2D_y)}\left[x\ln(5c - 4x) - x - \frac{5c}{4}\ln(5c - 4x)\right]$$

$$= 8\frac{1}{(D_x + 2D_y)}\left[\frac{-x-5y}{4}\ln(x+5y) - x\right] \qquad (\because\ c = x+y)$$

$$= -2\frac{1}{D_x + 2D_y}\left[(x+5y)\ln(x+5y) + 4x\right]$$

$$= -2\int\left[(5c+11x)\ln(5c+11x) + 4x\right]dx, \qquad \text{where } c = y-2x$$

$$= -2\left[\frac{(5c+11x)^2}{22}\ln(5c+11x) - \int\frac{11(5c+11x)^2}{22(5c+11x)}dx + 2x^2\right] \quad \text{(integrating by parts)}$$

$$= -\frac{1}{11}\left[(5c+11x)^2\ln(5c+11x) - 11\frac{(5c+11x)^2}{22} + 44x^2\right]$$

$$= -\frac{1}{22}\left[2(x+5y)^2\ln(x+5y) - (x+5y)^2 + 88x^2\right] \qquad (\because\ c = y-2x)$$

$$= -\frac{1}{22}\left[(x+5y)^2\left(2\ln(x+5y) - 1\right) + 88x^2\right]$$

∴ general solution is

$$z = \phi_1(y-2x) + \phi_2(y+x) - \frac{1}{22}(x+5y)^2\left(2\ln(x+5y) - 1\right) - 4x^2$$

where ϕ_1 and ϕ_2 are arbitrary functions.

EXERCISE 4.5

1. Solve the following partial differential equations:
   (a) $\left(D^3 - 7DD'^2 + 6D'^3\right)z = 0$
   (b) $25r - 40s + 16t = 0$
   (c) $2\frac{\partial^2 z}{\partial x^2} + 5\frac{\partial^2 z}{\partial x \partial y} + 2\frac{\partial^2 z}{\partial y^2} = 0$
   (d) $\left(D^2 - DD' - 6D'^2\right)z = 0$
   (e) $\left(D^3 - 4D^2D' + 3DD'^2\right)z = 0$
   (f) $\left(D^3 - 6D^2D' + 11DD'^2 - 6D'^3\right)z = 0$
   (g) $\left(D^3 - 2D^2D' + DD'^2\right)z = 0$
   (h) $r + 6s + 9t = 0$
   (i) $(D+2D')(D-3D')^2 z = 0$
   (j) $\frac{\partial^2 z}{\partial x^2} - 4\frac{\partial^2 z}{\partial x \partial y} + 4\frac{\partial^2 z}{\partial y^2} = 0$
   (k) $\left(D^4 - 2D^2D'^2 + D'^4\right)z = 0$
   (l) $\left(D^4 - 2D^3D' + 2DD'^3 - D'^4\right)z = 0$
2. Find the general solutions of the following partial differential equations
   (a) $\left(D^2 + 5DD' + 6D'^2\right)z = e^{x-y}$
   (b) $\left(D^3 - 3D^2D' + 4D'^3\right)z = e^{x+2y}$

(c) $r-4s+4t=e^{2x+y}$

(d) $r+2s+t=e^{2x+3y}$

(e) $\left(D^3-4D^2D'+5DD'^2-2D'^3\right)z$
$=e^{y-2x}+e^{y+2x}+e^{y+x}$

3. Obtain the general solutions of the following partial differential equations

(a) $\dfrac{\partial^2 z}{\partial x^2}+3\dfrac{\partial^2 z}{\partial x\partial y}+2\dfrac{\partial^2 z}{\partial y^2}=x+y$

(b) $\left(D^2-DD'-6D'^2\right)z=x+y$

(c) $\left(D^2+DD'-6D'^2\right)z=x+y$

(d) $\dfrac{\partial^2 z}{\partial x^2}-\dfrac{\partial^2 z}{\partial y^2}=x^2+y^2$

(e) $\left(D^2+D'^2\right)z=x^2y^2$

(f) $\left(D^2+2DD'+D'^2\right)z=3x+2y$

(g) $\dfrac{\partial^3 z}{\partial x^3}-\dfrac{\partial^3 z}{\partial y^3}=x^3y^3$

(h) $\left(4D^3-3DD'^2+D'^3\right)z=6x^2y^2$

(i) $\dfrac{\partial^3 z}{\partial x^3}-2\dfrac{\partial^3 z}{\partial x^2\partial y}=2e^{2x}+3x^2y$

4. Find the general solutions of the following partial differential equations

(a) $\dfrac{\partial^2 z}{\partial x^2}-\dfrac{\partial^2 z}{\partial x\partial y}=\sin x\cos 2y$

(b) $\left(D^3-4D^2D'+4DD'^2\right)z$
$=2\sin(3x+2y)$

(c) $\left(2D^2-3DD'+D'^2\right)z=\sin(x-2y)$

(d) $\left(D^2-2DD'+D'^2\right)z=\sin(2x+3y)$

(e) $\left(D^2+DD'-6D'^2\right)z=\cos(2x+y)$

(f) $\left(D^2-DD'\right)z=\cos x\cos 2y$

(g) $\dfrac{\partial^2 z}{\partial x^2}+\dfrac{\partial^2 z}{\partial y^2}=\cos mx\cos ny+30(2x+y)$

(h) $p-2q=\sin(x+2y)$

(i) $\dfrac{\partial^3 z}{\partial x^3}-7\dfrac{\partial^3 z}{\partial x\partial y^2}-6\dfrac{\partial^3 z}{\partial y^3}$
$=\sin(x+2y)+x^2y$

(j) $\left(D^2-2DD'+D'^2\right)z=\sin x$

5. Find the general solutions of the following differential equations

(a) $\left(2D^2+5DD'+3D'^2\right)z=ye^x$

(b) $4r-4s+t=16\log(x+2y)$

ANSWERS 4.5

Here, ϕ_1,ϕ_2,ϕ_3 and ϕ_4 are arbitrary functions.

1. (a) $z=\phi_1(y+x)+\phi_2(y+2x)+\phi_3(y-3x)$
   (b) $z=\phi_1(5y+4x)+x\phi_2(5y+4x)$
   (c) $z=\phi_1(y-2x)+\phi_2(2y-x)$
   (d) $z=\phi_1(y+3x)+\phi_2(y-2x)$
   (e) $z=\phi_1(y)+\phi_2(y+x)+\phi_3(y+3x)$
   (f) $z=\phi_1(y+x)+\phi_2(y+2x)+\phi_3(y+3x)$
   (g) $z=\phi_1(y)+\phi_2(y+x)+x\phi_3(y+x)$

(h) $z=\phi_1(y-3x)+x\phi_2(y-3x)$

(i) $z=\phi_1(y-2x)+\phi_2(y+3x)+x\phi_3(y+3x)$

(j) $z=\phi_1(y+2x)+x\phi_2(y+2x)$

(k) $z=\phi_1(y+x)+x\phi_2(y+x)+\phi_3(y-x)+x\phi_4(y-x)$

(l) $z=\phi_1(y+x)+x\phi_2(y+x)+x^2\phi_3(y+x)+\phi_4(y-x)$

2. (a) $z=\phi_1(y-2x)+\phi_2(y-3x)+\frac{1}{2}e^{x-y}$

(b) $z=\phi_1(y-x)+\phi_2(y+2x)+x\phi_3(y+2x)+\frac{1}{27}e^{x+2y}$

(c) $z=\phi_1(y+2x)+x\phi_2(y+2x)+\frac{1}{2}x^2e^{2x+y}$

(d) $z=\phi_1(y-x)+x\phi_2(y-x)+\frac{1}{25}e^{2x+3y}$

(e) $z=\phi_1(y+x)+x\phi_2(y+x)+\phi_3(y+2x)-\frac{1}{36}e^{y-2x}+xe^{y+2x}-\frac{x^2}{2}e^{y+x}$

3. (a) $z=\phi_1(y-x)+\phi_2(y-2x)+\frac{x^2y}{2}-\frac{x^3}{3}$

(b) $z=\phi_1(y-2x)+\phi_2(y+3x)+\frac{x^3}{3}+\frac{x^2}{2}y$

(c) $z=\phi_1(y+2x)+\phi_2(y-3x)+\frac{x^2y}{2}$

(d) $z=\phi_1(y+x)+\phi_2(y-x)+\frac{x^2}{6}(x^2+3y^2)$

(e) $z=\phi_1(y+ix)+\phi_2(y-ix)+\frac{1}{180}(15x^4y^2-x^6)$

(f) $z=\phi_1(y-x)+x\phi_2(y-x)+x^2y-\frac{x^3}{6}$

(g) $z=\phi_1(y+x)+\phi_2(y+\omega x)+\phi_3(y+\omega^2x)+\frac{x^6y^3}{120}+\frac{x^9}{10080}$; ω is cube root of unity.

(h) $z=\phi_1(y-x)+\phi_2(2y+x)+x\phi_3(2y+x)+\frac{1}{40}x^5y^2+\frac{x^7}{1120}$

(i) $z=\phi_1(y)+x\phi_2(y)+\phi_3(y+2x)+\frac{1}{60}(15e^{2x}+3x^5y+x^6)$

4. (a) $z=\phi_1(y)+\phi_2(y+x)+\frac{1}{2}\sin(x+2y)-\frac{1}{6}\sin(x-2y)$

(b) $z=\phi_1(y)+\phi_2(y+2x)+x\phi_3(y+2x)+\frac{2}{3}\cos(3x+2y)$

(c) $z=\phi_1(2y+x)+\phi_2(y+x)-\frac{1}{12}\sin(x-2y)$

(d) $z=\phi_1(y+x)+x\phi_2(y+x)-\sin(2x+3y)$

(e) $z=\phi_1(y+2x)+\phi_2(y-3x)+\frac{x}{5}\sin(2x+y)$

(f) $z=\phi_1(y)+\phi_2(y+x)+\frac{1}{6}[3\cos(x+2y)-\cos(x-2y)]$

(g) $z=\phi_1(y+ix)+\phi_2(y-ix)-\frac{\cos mx\cos ny}{(m^2+n^2)}+(2x+y)^3$

(h) $z=\phi_1(y+x)+\frac{1}{3}\cos(x+2y)$

(i) $z=\phi_1(y-x)+\phi_2(y-2x)+\phi_3(y+3x)-\frac{1}{75}\cos(x+2y)+\frac{x^5y}{60}$

(j) $z=\phi_1(y+x)+x\phi_2(y+x)-\sin x$

5. (a) $z=\phi_1(2y-3x)+\phi_2(y-x)+\frac{1}{4}(2y-5)e^x$

(b) $z=\phi_1(2y+x)+x\phi_2(2y+x)+2x^2\log(x+2y)$

4.8 LINEAR PARTIAL DIFFERENTIAL EQUATIONS WITH CONSTANT COEFFICIENTS, NON-HOMOGENEOUS IN PARTIAL DERIVATIVES

Differential equation is

$$f(D,D')z=F(x,y)$$

where $f(D,D')$ is not homogeneous, i.e., sum of powers of D and D' in terms may not be equal.

Here also general solution is sum of complementary function (C.F.) and particular integral (P.I.).

4.8.1 Rules for Finding Complementary Function

Case I: When $f(D,D')$ can be factorized into linear factors of the type $D+mD'+c$ or $D'+c$ where m, c be any constant (may be zero) and factors are not repeated.

To find the solution of

$$(D+mD'+c)z=0 \tag{4.39}$$

or

$$p+mq+cz=0$$

It is Lagrange's equation.

Lagrange's auxiliary equations are

$$\frac{dx}{1}=\frac{dy}{m}=\frac{dz}{-cz}$$

Solution from first two members is $y-mx=k_1$

and solution from first and third members is $z=k_2e^{-cx}=\phi(k_1)e^{-cx}$

Thus, solution of (4.39) is

$$z=e^{-cx}\phi(y-mx).$$

Now, we find solution of

$$(D'+c)z=0 \quad \text{or} \quad q=-cz \tag{4.40}$$

Lagrange's equations are

$$\frac{dx}{0}=\frac{dy}{1}=\frac{dz}{-cz}$$

Its two solutions are

$$x=k_1,\ z=k_2e^{-cy}.$$

Hence, in this case solution of (4.40) is

$$z=e^{-cy}\phi(x).$$

Hence, solution corresponding to each factor will be found if factors are not repeated.

Then by superposition principle complementary function will be sum of all these solutions.

Case II: When factors are repeated

Suppose $D+mD'+c$ is repeated twice.

Now, we find solution of

$$(D+mD'+c)^2 z=0 \tag{4.41}$$

Let

$$(D+mD'+c)z=u \tag{4.42}$$

$\therefore$ from (4.41)

$$(D+mD'+c)u=0$$

Its solution as in Case I is

$$u=e^{-cx}\phi(y-mx)$$

$\therefore$ from (4.42)

$$p+mq+cz=e^{-cx}\phi(y-mx)$$

or

$$p+mq=-cz+e^{-cx}\phi(y-mx)$$

It is Lagrange's equation.

Lagrange's auxiliary equations are

$$\frac{dx}{1}=\frac{dy}{m}=\frac{dz}{-cz+e^{-cx}\phi(y-mx)} \tag{4.43}$$

From first two members

$$y-mx=k_1$$

From first and third members

$$\frac{dx}{1}=\frac{dz}{-cz+e^{-cx}\phi(k_1)}$$

$\therefore$

$$\frac{dz}{dx}+cz=e^{-cx}\phi(k_1)$$

It is Leibnitz linear equation.

$$\text{I.F.}=e^{cx}$$

∴ solution is

$$ze^{cx} = \int e^{cx} e^{-cx} \phi(k_1)\, dx + k_2$$

$$= x\phi(k_1) + k_2$$

∴ $$z = \left[x\phi(k_1) + k_2\right] e^{-cx}$$

∴ $$z = e^{-cx}\left[x\phi(y - mx) + \psi(y - mx)\right]$$

will be solution corresponding to $(D + mD' + c)^2 z = 0$

Similarly $z = e^{-cx}\left[x^2\phi_1(y - mx) + x\phi_2(y - mx) + \phi_3(y - mx)\right]$ will be a solution corresponding to $(D + mD' + c)^3 z = 0$.

Case III: If $f(D, D')$ cannot be factorized into linear factors. For example $D^2 + D'$ cannot be factorized into linear factors.

As $$D^r e^{hx+ky} = h^r e^{hx+ky}$$

and $$D'^r e^{hx+ky} = k^r e^{hx+ky}$$

∴ $$f(D, D') e^{hx+ky} = f(h, k) e^{hx+ky}$$

Thus, $z = e^{hx+ky}$ will be solution of

$$f(D, D') z = 0 \text{ when } f(h, k) = 0$$

Hence, by principle of superposition

$$\text{C.F.} = \sum_{i=1}^{\infty} c_i e^{h_i x + k_i y} \quad \text{where } f(h_i, k_i) = 0$$

4.8.2 Operator Methods for Finding Particular Integral

Consider the equation

$$f(D, D') z = F(x, y).$$

Then, $$\text{P.I.} = \frac{1}{f(D, D')} F(x, y).$$

Cases I to V in (4.7.4) are proved for general $f(D, D')$ and hence particular integral can be found using them and for the general method, we prove the following theorem.

Theorem 4.3: $\dfrac{1}{D + mD' + c} F(x, y) = e^{-cx} \int e^{cx} F(x, a + mx)\, dx$ where we put $a = y - mx$ after integration.

Proof: Let $$\frac{1}{D + mD' + c} F(x, y) = G(x, y)$$

∴ $$(D + mD' + c) G(x, y) = F(x, y)$$

$$\therefore \qquad \frac{\partial G}{\partial x}+m\frac{\partial G}{\partial y}=-cG+F$$

It is Lagrange's equation.

Lagrange's auxiliary equations are

$$\frac{dx}{1}=\frac{dy}{m}=\frac{dG}{-cG+F}$$

From first and second members

$$y-mx=a$$

From first and third members

$$\frac{dG}{dx}+cG=F(x,y)=F(x,a+mx)$$

It is Leibnitz linear equation.

$$\text{I.F.}=e^{cx}$$

$\therefore$ solution is

$$Ge^{cx}=\int e^{cx}F(x,a+mx)\,dx$$

$$\therefore \qquad G=e^{-cx}\int e^{cx}F(x,a+mx)\,dx$$

4.9 PARTIAL DIFFERENTIAL EQUATIONS WITH VARIABLE COEFFICIENTS REDUCIBLE TO PARTIAL DIFFERENTIAL EQUATIONS WITH CONSTANT COEFFICIENTS

Consider the partial differential equation

$$f(xD,yD')z=F(x,y)$$

Put $\quad |x|=e^{X},|y|=e^{Y}$

ie., $\quad X=\log|x|,Y=\log|y|,$

then $\quad \dfrac{\partial z}{\partial x}=\dfrac{\partial z}{\partial X}\dfrac{dX}{dx}=\dfrac{1}{x}\dfrac{\partial z}{\partial X} \qquad \Rightarrow xD\equiv D_X$

$$\frac{\partial^2 z}{\partial x^2}=-\frac{1}{x^2}\frac{\partial z}{\partial X}+\frac{1}{x}\frac{\partial^2 z}{\partial X^2}\frac{dX}{dx}$$

$$=-\frac{1}{x^2}\frac{\partial z}{\partial X}+\frac{1}{x^2}\frac{\partial^2 z}{\partial X^2}$$

$$\Rightarrow \qquad x^2D^2\equiv D_X(D_X-1)$$

Similarly,

$$x^3D^3\equiv D_X(D_X-1)(D_X-2)$$

$$yD'\equiv D_Y, y^2D'^2\equiv D_Y(D_Y-1), y^3D'^3=D_Y(D_Y-1)(D_Y-2)$$

and so on.

Hence the differential equation will be reduced to partial differential equation with constant coefficients and be solved by methods already discussed.

Example 4.38: Find the solutions of the following partial differential equations

(i) $3r+7s+2t+7p+4q+2z=0$

(ii) $2r-s-t-p+q=0$

Solution: (i) Partial differential equation in symbolic form is

$$\left(3D^2+7DD'+2D'^2+7D+4D'+2\right)z=0$$

or
$$(D+2D'+2)(3D+D'+1)z=0$$

or
$$(D+2D'+2)\left(D+\frac{1}{3}D'+\frac{1}{3}\right)z=0$$

∴ General solution is

$$z=e^{-2x}\phi_1(y-2x)+e^{-x/3}\phi_2(3y-x)$$

where ϕ_1 and ϕ_2 are arbitrary functions.

(ii) Partial differential equation in symbolic form is

$$\left(2D^2-DD'-D'^2-D+D'\right)z=0$$

or
$$(D-D')(2D+D'-1)z=0$$

or
$$(D-D')\left(D+\frac{1}{2}D'-\frac{1}{2}\right)z=0$$

∴ general solution is

$$z=\phi_1(y+x)+e^{x/2}\phi_2(2y-x)$$

where ϕ_1 and ϕ_2 are arbitrary functions.

Example 4.39: Find the solutions of the following partial differential equations

(i) $\left(D^2-D'^2+3D'-3D\right)z=e^{x+2y}+xy$

(ii) $\left(4D^2+3DD'-D'^2-D-D'\right)z=3e^{\frac{1}{2}(x+2y)}$

(iii) $\left(D^2-D'\right)z=xe^{ax+a^2y}$

(iv) $\left(D^2-DD'-2D'^2+2D+2D'\right)z=e^{2x+3y}+\sin(2x+y)+xy$

(v) $(D+D'+1)(D'+2)z=e^{x-3y}$

(vi) $(D+D'+2)z=\dfrac{1}{\left(1+e^{x+y}\right)}$

Solution:

(i) $\left(D^2 - D'^2 + 3D' - 3D\right)z = e^{x+2y} + xy$

or $(D - D')(D + D' - 3)z = e^{x+2y} + xy$

$\therefore \qquad \text{C.F.} = \phi_1(y + x) + e^{3x}\phi_2(y - x)$

$$\text{P.I.} = \frac{1}{D^2 - D'^2 - 3D + 3D'}e^{x+2y} + xy$$

$$= x\frac{1}{\frac{d}{dD}\left(D^2 - D'^2 - 3D + 3D'\right)}e^{x+2y} + \frac{1}{(D - D')(D + D' - 3)}xy \quad (\because 1 - 4 - 3 + 6 = 0)$$

$$= x\frac{1}{2D - 3}e^{x+2y} - \frac{1}{3D}\left(1 - \frac{D'}{D}\right)^{-1}\left[1 - \left(\frac{D}{3} + \frac{D'}{3}\right)\right]^{-1}(xy)$$

$$= -xe^{x+2y} - \frac{1}{3D}\left(1 + \frac{D'}{D} + \cdots\right)\left(1 + \frac{D}{3} + \frac{D'}{3} + \frac{2}{9}DD' + \cdots\right)(xy)$$

$$= -xe^{x+2y} - \frac{1}{3D}\left(1 + \frac{D}{3} + \frac{2D'}{3} + \frac{D'}{D} + \frac{2}{9}DD' + \cdots\right)(xy)$$

$$= -xe^{x+2y} - \frac{1}{3D}\left(xy + \frac{y}{3} + \frac{2x}{3} + \frac{1}{D}x + \frac{2}{9}\right)$$

$$= -xe^{x+2y} - \frac{1}{3}\left(\frac{x^2}{2}y + \frac{xy}{3} + \frac{x^2}{3} + \frac{x^3}{6} + \frac{2}{9}x\right)$$

$\therefore$ general solution is

$$z = \phi_1(y + x) + e^{3x}\phi_2(y - x) - xe^{x+2y} - \frac{1}{54}\left(9x^2y + 6xy + 6x^2 + 3x^3 + 4x\right)$$

where ϕ_1 and ϕ_2 are arbitrary functions.

(ii) $\left(4D^2 + 3DD' - D'^2 - D - D'\right)z = 3e^{\frac{1}{2}(x+2y)}$

or $\qquad (4D - D' - 1)(D + D')z = 3e^{\frac{1}{2}(x+2y)}$

or $\qquad \left(D - \frac{1}{4}D' - \frac{1}{4}\right)(D + D')z = 3e^{\frac{1}{2}(x+2y)}$

$\therefore \qquad \text{C.F.} = \phi_1(y - x) + e^{x/4}\phi_2(4y + x)$

$$\text{P.I.} = \frac{1}{4D^2 + 3DD' - D'^2 - D - D'}3e^{(x+2y)/2}$$

$$= 3x\frac{1}{\frac{d}{dD}\left(4D^2+3DD'-D'^2-D-D'\right)}e^{(x+2y)/2} \qquad \left(\because\ 4\left(\frac{1}{4}\right)+3\left(\frac{1}{2}\right)(1)-1-\frac{1}{2}-1=0\right)$$

$$= 3x\frac{1}{8D+3D'-1}e^{(x+2y)/2}$$

$$= 3x\frac{1}{4+3-1}e^{(x+2y)/2} = \frac{x}{2}e^{(x+2y)/2}$$

$\therefore$ general solution is

$$z = \phi_1(y-x)+e^{x/4}\phi_2(4y+x)+\frac{x}{2}e^{(x+2y)/2}$$

where ϕ_1 and ϕ_2 are arbitrary functions.

(iii) $(D^2-D')z = xe^{ax+a^2y}$

Here, D^2-D' cannot be factorized into linear factors in D and D'.

$\therefore$ $\quad \text{C.F.} = \sum_{n=1}^{\infty} c_n e^{h_n x+k_n y} \quad$ where $h_n^2-k_n=0 \quad \therefore\ k_n=h_n^2$

$$= \sum_{n=1}^{\infty} c_n e^{h_n(x+h_n y)}$$

$$\text{P.I.} = \frac{1}{D^2-D'}xe^{ax+a^2y} = e^{ax+a^2y}\frac{1}{(D+a)^2-(D'+a^2)}x$$

$$= e^{ax+a^2y}\frac{1}{D^2+2aD}x$$

$$= e^{ax+a^2y}\frac{1}{2aD}\left(1+\frac{D}{2a}\right)^{-1}x = e^{ax+a^2y}\frac{1}{2aD}\left(1-\frac{D}{2a}+\cdots\right)x$$

$$= e^{ax+a^2y}\frac{1}{2aD}\left(x-\frac{1}{2a}\right) = e^{ax+a^2y}\frac{1}{2a}\left(\frac{x^2}{2}-\frac{x}{2a}\right)$$

$$= \frac{x}{4a^2}(ax-1)e^{ax+a^2y}$$

$\therefore$ complete solution is

$$z = \sum_{n=1}^{\infty} c_n e^{h_n(x+h_n y)} + \frac{x}{4a^2}(ax-1)e^{ax+a^2y}$$

where $c_n, h_n, n=1,2,3,\ldots$ are arbitrary constants.

(iv) $(D^2-DD'-2D'^2+2D+2D')z = e^{2x+3y}+\sin(2x+y)+xy$

or $\qquad (D-2D'+2)(D+D')z = e^{2x+3y}+\sin(2x+y)+xy$

$$\text{C.F.} = e^{-2x}\phi_1(y+2x)+\phi_2(y-x)$$

$$\text{P.I.} = \frac{1}{(D+D')(D-2D'+2)}e^{2x+3y} + \frac{1}{D^2 - DD' - 2D'^2 + 2D + 2D'}\sin(2x+y)$$

$$+\frac{1}{(D+D')(D-2D'+2)}xy$$

$$= \frac{1}{(2+3)(2-6+2)}e^{2x+3y} + \frac{1}{-4+2+2+2D+2D'}\sin(2x+y)$$

$$+\frac{1}{2D}\left(1+\frac{D'}{D}\right)^{-1}\left(1+\frac{D}{2}-D'\right)^{-1} xy$$

$$= -\frac{1}{10}e^{2x+3y} + \frac{1}{2}\frac{1}{D+D'}\sin(2x+y) + \frac{1}{2D}\left(1-\frac{D'}{D}+\cdots\right)\left(1-\frac{D}{2}+D'-DD'+\cdots\right)xy$$

$$= -\frac{1}{10}e^{2x+3y} - \frac{1}{2}\frac{1}{2+1}\cos(2x+y) + \frac{1}{2D}\left(1-\frac{D}{2}+D'-DD'-\frac{D'}{D}+\frac{D'}{2}+\cdots\right)xy$$

$$\left(\because \frac{1}{D+D'}\sin(2x+y) = \frac{1}{2+1}\int \sin z\, dz,\ z = 2x+y\right)$$

$$= -\frac{1}{10}e^{2x+3y} - \frac{1}{6}\cos(2x+y) + \frac{1}{2D}\left(xy - \frac{y}{2} + \frac{3}{2}x - 1 - \frac{1}{D}x\right)$$

$$= -\frac{1}{10}e^{2x+3y} - \frac{1}{6}\cos(2x+y) + \frac{1}{2}\left(\frac{x^2}{2}y - \frac{xy}{2} + \frac{3x^2}{4} - x - \frac{x^3}{6}\right)$$

$$= -\frac{1}{10}e^{2x+3y} - \frac{1}{6}\cos(2x+y) + \frac{1}{24}\left(6x^2y - 6xy + 9x^2 - 12x - 2x^3\right)$$

$\therefore$ general solution is

$$z = e^{-2x}\phi_1(y+2x) + \phi_2(y-x) - \frac{1}{10}e^{2x+3y} - \frac{1}{6}\cos(2x+y)$$

$$-\frac{1}{24}\left(2x^3 - 6x^2y - 9x^2 + 6xy + 12x\right)$$

where ϕ_1 and ϕ_2 are arbitrary functions.

(v) $(D+D'+1)(D'+2)z = e^{x-3y}$

$$\text{C.F.} = e^{-x}\phi_1(y-x) + e^{-2y}\phi_2(x)$$

$$\text{P.I.} = \frac{1}{(D+D'+1)(D'+2)}e^{x-3y}$$

$$= \frac{1}{(1-3+1)(-3+2)}e^{x-3y} = e^{x-3y}$$

$\therefore$ general solution is

$$z = e^{-x}\phi_1(y-x) + e^{-2y}\phi_2(x) + e^{x-3y}$$

where ϕ_1 and ϕ_2 are arbitrary functions.

(vi) $(D+D'+2)z = \dfrac{1}{(1+e^{x+y})}$

$\therefore$ $\quad \text{C.F.} = e^{-2x}\phi_1(y-x)$

$$\text{P.I.} = \frac{1}{(D+D'+2)}\cdot\frac{1}{(1+e^{x+y})}$$

$$= e^{-2x}\int e^{2x}\cdot\frac{1}{(1+e^{x}\cdot e^{a+x})}dx \qquad \text{where } a = y-x$$

(using formula given in (Theorem 4.3))

$$= e^{-2x}\int\frac{e^{2x}}{(1+e^{a}\cdot e^{2x})}dx$$

$$= e^{-2x}\cdot\frac{1}{2e^{a}}\ln\left(1+e^{a+2x}\right)$$

$$= e^{-2x}\cdot\frac{1}{2e^{y-x}}\ln\left(1+e^{x+y}\right) \qquad (\because a = y-x)$$

$$= \frac{1}{2e^{x+y}}\ln\left(1+e^{x+y}\right)$$

$\therefore$ general solution is

$$z = e^{-2x}\phi_1(y-x) + \frac{1}{2e^{x+y}}\ln\left(1+e^{x+y}\right)$$

where ϕ_1 is an arbitrary function.

Example 4.40: Find the general solutions of the following partial differential equations

(i) $x^2r - 3xys + 2y^2t + px + 2qy = x + 2y$

(ii) $\left(x^2D^2 - y^2D'^2 + xD - yD'\right)z = \log x$

(iii) $\dfrac{1}{x^2}\dfrac{\partial^2 z}{\partial x^2} - \dfrac{1}{x^3}\dfrac{\partial z}{\partial x} = \dfrac{1}{y^2}\dfrac{\partial^2 z}{\partial y^2} - \dfrac{1}{y^3}\dfrac{\partial z}{\partial y}$

Solution: (i) Differential equation in symbolic form is

$$\left(x^2D_x^2 - 3xyD_xD_y + 2y^2D_y^2 + xD_x + 2yD_y\right)z = x + 2y$$

Put $\quad |x| = e^X, |y| = e^Y$, i.e., $X = \log|x|$, $Y = \log|y|$

We have

$$xD_x \equiv D_X,\ yD_y \equiv D_Y,\ x^2D_x^2 \equiv D_X\left(D_X - 1\right),$$

$$xyD_xD_y \equiv D_XD_Y,\ y^2D_y^2 \equiv D_Y\left(D_Y - 1\right)$$

$\therefore$ differential equation becomes

$$\left[D_X\left(D_X-1\right)-3D_XD_Y+2D_Y\left(D_Y-1\right)+D_X+2D_Y\right]z$$

$$= e^X + 2e^Y, \quad \text{if } x > 0,\ y > 0$$

$$= -e^X + 2e^Y, \text{ if } x < 0,\ y > 0$$

$$= -e^X - 2e^Y, \text{ if } x < 0,\ y < 0$$

$$= e^X - 2e^Y, \quad \text{if } x > 0,\ y < 0$$

$\therefore$ $$\left[D_X^2 - 3D_XD_Y + 2D_Y^2\right]z = e^X + 2e^Y, \quad \text{if } x > 0,\ y > 0$$

or $$\left(D_X - D_Y\right)\left(D_X - 2D_Y\right)z = e^X + 2e^Y$$

A.E. is

$$(m-1)(m-2) = 0$$

$\therefore$ $$m = 1, 2$$

$\therefore$ $$\text{C.F.} = \phi_1\left(Y + 2X\right) + \phi_2\left(Y + X\right)$$

$$= \phi_1\left(\log|y| + 2\log|x|\right) + \phi_2\left(\log|y| + \log|x|\right)$$

$$= \phi_1\left(\log\left|x^2y\right|\right) + \phi_2\left(\log|xy|\right)$$

$$= f_1\left(x^2y\right) + f_2\left(xy\right) \text{ where } f_1 \equiv \phi_1 \circ \ln,\ f_2 \equiv \phi_2 \circ \ln$$

$$\text{P.I.} = \frac{1}{D_X^2 - 3D_XD_Y + 2D_Y^2}\left(e^X + 2e^Y\right)$$

$$= \frac{1}{1-0+0}e^X + 2\cdot\frac{1}{0-0+2}e^Y = e^X + e^Y$$

$\therefore$ $$\text{P.I.} = e^X + e^Y = x + y \quad \text{if } x > 0, y > 0$$

$$= -e^X + e^Y = x + y \text{ if } x < 0, y > 0$$

$$= e^X - e^Y = x + y \quad \text{if } x > 0, y < 0$$

$$= -e^X - e^Y = x + y \text{ if } x < 0, y < 0$$

Hence in each case, $\text{P.I.} = x + y$

∴ general solution is

$$z = f_1\left(x^2y\right) + f_2\left(xy\right) + x + y$$

where f_1 and f_2 are arbitrary functions.

(ii) $\left(x^2D^2 - y^2D'^2 + xD - yD'\right)z = \log x$

Put $x = e^X$, $|y| = e^Y$, i.e., $X = \log x$, $Y = \log|y|$, as $x > 0$ ($\because \log x$ is defined)

We have $xD \equiv D_X$, $yD' \equiv D_Y$, $x^2D^2 \equiv D_X\left(D_X - 1\right)$,

$$xyDD' \equiv D_XD_Y,\ y^2D'^2 \equiv D_Y\left(D_Y - 1\right)$$

∴ differential equation becomes

$$\left[D_X\left(D_X - 1\right) - D_Y\left(D_Y - 1\right) + D_X - D_Y\right]z = \log x = X$$

or $$\left[D_X^2 - D_Y^2\right]z = X$$

∴ A.E. is $m^2 - 1 = 0$

$\Rightarrow$ $m = \pm 1$

∴ $$\text{C.F.} = \phi_1\left(Y + X\right) + \phi_2\left(Y - X\right) = \phi_1\left(\log|y| + \log x\right) + \phi_2\left(\log|y| - \log x\right)$$

$$= \phi_1\left(\log|xy|\right) + \phi_2\left(\log\left|\frac{y}{x}\right|\right) = f_1\left(xy\right) + f_2\left(\frac{y}{x}\right) \text{ where } f_1 \equiv \phi_1 \text{ o} \ln,\ f_2 \equiv \phi_2 \text{ o} \ln$$

$$\text{P.I.} = \frac{1}{D_X^2 - D_Y^2}X = \frac{X^3}{2.3} = \frac{X^3}{6} = \frac{1}{6}\left(\log x\right)^3$$

∴ general solution is

$$z = f_1\left(xy\right) + f_2\left(\frac{y}{x}\right) + \frac{1}{6}\left(\log x\right)^3$$

where f_1 and f_2 are arbitrary functions.

(iii) $\dfrac{1}{x^2}\dfrac{\partial^2 z}{\partial x^2} - \dfrac{1}{x^3}\dfrac{\partial z}{\partial x} = \dfrac{1}{y^2}\dfrac{\partial^2 z}{\partial y^2} - \dfrac{1}{y^3}\dfrac{\partial z}{\partial y}$

Put $\dfrac{x^2}{2} = X$, $\dfrac{y^2}{2} = Y$

∴ $$\frac{\partial z}{\partial x} = \frac{\partial z}{\partial X}\frac{dX}{dx} = x\frac{\partial z}{\partial X}$$

$$\frac{\partial^2 z}{\partial x^2} = x\frac{\partial^2 z}{\partial X^2}\frac{dX}{dx} + \frac{\partial z}{\partial X}$$

$$= x^2\frac{\partial^2 z}{\partial X^2} + \frac{\partial z}{\partial X}$$

$$\therefore \qquad \frac{1}{x^2}\frac{\partial^2 z}{\partial x^2}-\frac{1}{x^3}\frac{\partial z}{\partial x}=\frac{\partial^2 z}{\partial X^2}=D_X^2 z$$

Similarly,

$$\frac{1}{y^2}\frac{\partial^2 z}{\partial y^2}-\frac{1}{y^3}\frac{\partial z}{\partial y}=D_Y^2 z$$

$\therefore$ differential equation reduces to

$$\left(D_X^2-D_Y^2\right)z=0$$

A.E. is

$$m^2-1=0$$

$$\therefore \qquad m=\pm 1$$

$$\therefore \qquad \text{C.F.}=\phi_1\left(Y+X\right)+\phi_2\left(Y-X\right)$$

$$=\phi_1\left(\frac{x^2+y^2}{2}\right)+\phi_2\left(\frac{y^2-x^2}{2}\right)=f_1\left(x^2+y^2\right)+f_2\left(x^2-y^2\right)$$

$\therefore$ general solution is

$$z=f_1\left(x^2+y^2\right)+f_2\left(x^2-y^2\right)$$

where f_1 and f_2 are arbitrary functions.

EXERCISE 4.6

1. Solve the following partial differential equations
   (a) $r-t+p-q=0$
   (b) $(D+2D'-3)(D+D'-1)z=0$
   (c) $\left(D_x+3D_y+4\right)^2 z=0$
   (d) $\left(D+2D'\right)\left(D+3D'+1\right)\left(D+2D'+2\right)^2 z=0$
   (e) $\left(D^2+DD'-D'^2+D-D'\right)z=0$
2. Solve $(D-D'-2)(D-D'-3)z=e^{3x-2y}$ to find general solution.
3. Find the general solutions of the following partial differential equations
   (a) $\left(2D^2-DD'-D'^2+D-D'\right)z=e^{2x+3y}$
   (b) $\left(D^2-DD'+D'-1\right)z=\cos\left(x+2y\right)+e^y$
   (c) $\left(2D^2+3DD'+D'^2+D+D'\right)z=x-y$
   (d) $\left(D^2-D'^2+D+3D'-2\right)z=x^2y$

(e) $(D_x - D_y - 1)(D_x - D_y - 2)z = e^{2x-y} + x$

(f) $(D^3 - 3DD' + D' + 4)z = e^{2x+y}$

(g) $(D - 3D' - 2)^2 z = 2e^{2x}\sin(y + 3x)$

(h) $(D^2 + 2DD' + D'^2 - 2D - 2D')z = \sin(x + 2y)$

(i) $(D + 1)(D + D' - 1)z = \sin(x + 2y)$

(j) $(D + D' + 1)z = e^{-x}\tan(x + 2y)$

4. Obtain the general solutions of the following partial differential equations

(a) $(x^2D^2 + 2xyDD' + y^2D'^2)z = x^m y^n$

(b) $(x^2D_x^2 - y^2D_y^2)z = x^2 y$

(c) $(x^2D^2 - 4xyDD' + 4y^2D'^2 + 6yD')z = x^3y^4$

(d) $x^2\dfrac{\partial^2 z}{\partial x^2} + 2xy\dfrac{\partial^2 z}{\partial x\partial y} + y^2\dfrac{\partial^2 z}{\partial y^2} - nx\dfrac{\partial z}{\partial x} - ny\dfrac{\partial z}{\partial y} + nz = x^2 + y^2$

(e) $(xD^3D'^2 - yD^2D'^3)z = 0$

(f) $(x^2D^2 - 2xyDD' - 3y^2D'^2 + xD - 3yD')z = x^2 y$

ANSWERS 4.6

Here, ϕ_1, ϕ_2, ϕ_3 and ϕ_4 are arbitrary functions.

1. (a) $z = \phi_1(y + x) + e^{-x}\phi_2(y - x)$

(b) $z = e^x\phi_1(y - x) + e^{3x}\phi_2(y - 2x)$

(c) $z = e^{-4x}\phi_1(y - 3x) + xe^{-4x}\phi_2(y - 3x)$

(d) $z = \phi_1(y - 2x) + e^{-x}\phi_2(y - 3x) + e^{-2x}[\phi_3(y - 2x) + x\phi_4(y - 2x)]$

(e) $z = \sum_{n=1}^{\infty} c_n e^{a_n x + b_n y}$ where $a_n^2 + a_n b_n - b_n^2 + a_n - b_n = 0$; a_n, b_n and c_n are arbitrary constants.

2. $z = e^{2x}\phi_1(y + x) + e^{3x}\phi_2(y + x) + \dfrac{1}{6}e^{3x-2y}$

3. (a) $z = \phi_1(y + x) + e^{-x/2}\phi_2(2y - x) - \dfrac{1}{8}e^{2x+3y}$

(b) $z = e^x\phi_1(y) + e^{-x}\phi_2(y + x) + \dfrac{1}{2}\sin(x + 2y) - xe^y$

(c) $z = \phi_1(y - x) + e^{-x/2}\phi_2(2y - x) + x^2 - x(y + 1)$

(d) $z = e^{-2x}\phi_1(y + x) + e^x\phi_2(y - x) - \dfrac{1}{2}\left(x^2y + \dfrac{3}{2}x^2 + xy + \dfrac{3}{2}y + 3x + \dfrac{21}{4}\right)$

(e) $z=e^{x}\phi_1(y+x)+e^{2x}\phi_2(y+x)+\frac{1}{2}e^{2x-y}+\frac{x}{2}+\frac{3}{4}$

(f) $z=\sum_{n=1}^{\infty}c_n e^{a_n x+b_n y}+\frac{1}{7}e^{2x+y}$ where $a_n^3-3a_n b_n+b_n+4=0; a_n, b_n$ and c_n are arbitrary constants.

(g) $z=e^{2x}\phi_1(y+3x)+xe^{2x}\phi_2(y+3x)+x^2e^{2x}\sin(y+3x)$

(h) $z=\phi_1(y-x)+e^{2x}\phi_2(y-x)+\frac{1}{39}\left[2\cos(x+2y)-3\sin(x+2y)\right]$

(i) $z=e^{-x}\phi_1(y)+e^{x}\phi_2(y-x)-\frac{1}{10}\left[\cos(x+2y)+2\sin(x+2y)\right]$

(j) $z=e^{-x}\phi(y-x)+\frac{1}{3}e^{-x}\ln\left|\sec(x+2y)\right|$

4. (a) $z=\phi_1\left(\frac{y}{x}\right)+x\phi_2\left(\frac{y}{x}\right)+\frac{x^m y^n}{(m+n)(m+n-1)}$

(b) $z=\phi_1(xy)+x\phi_2\left(\frac{y}{x}\right)+\frac{1}{2}x^2y$

(c) $z=\phi_1(x^2y)+x\phi_2(x^2y)+\frac{1}{30}x^3y^4$

(d) $z=x\phi_1\left(\frac{y}{x}\right)+x^n\phi_2\left(\frac{y}{x}\right)+\frac{x^2+y^2}{2-n}$

(e) $z=\phi_1(x)+\phi_2(y)+x\phi_3(y)+y\phi_4(x)+\phi_5(xy)$

(f) $z=\phi_1(x^3y)+\phi_2\left(\frac{y}{x}\right)-\frac{1}{3}x^2y$

4.10 APPLICATIONS OF PARTIAL DIFFERENTIAL EQUATIONS

In most of engineering applications, second order partial differential equations are formed. First of all, we shall classify second order partial differential equations.

A general second order linear partial differential equation is

$$A\frac{\partial^2 u}{\partial x^2}+B\frac{\partial^2 u}{\partial x\partial y}+C\frac{\partial^2 u}{\partial y^2}+D\frac{\partial u}{\partial x}+E\frac{\partial u}{\partial y}+Fu=f(x,y)$$

where A, B, C, D, E and F are functions of x and y.

These differential equations are classified into three parts.

Differential equation is called parabolic in a region if $B^2-4AC=0$ in that region.

For example, one dimensional heat equation $\frac{\partial u}{\partial t}=c^2\frac{\partial^2 u}{\partial x^2}$ is parabolic.

The differential equation is called hyperbolic in a region if $B^2-4AC>0$ in that region.

For example, one dimensional wave equation $\frac{\partial^2 u}{\partial t^2}=c^2\frac{\partial^2 u}{\partial x^2}$ is hyperbolic.

The differential equation is called elliptic in a region if $B^2 - 4AC < 0$ in that region. For example, two dimensional Laplace equation $\frac{\partial^2 u}{\partial x^2} + \frac{\partial^2 u}{\partial y^2} = 0$ is elliptic. Nature of p.d.e. depends only on coefficients of second-order derivatives.

Example 4.41: Classify the following p.d.e.

$$u_{xx} + 4u_{xy} + \left(x^2 + 4y^2\right)u_{yy} = \sin(x + y)$$

Solution: Coeff of $u_{xx} = A = 1$, Coeff of $u_{xy} = B = 4$

Coeff of $u_{yy} = C = x^2 + 4y^2$

$$\therefore \quad B^2 - 4AC = 16 - 4\left(x^2 + 4y^2\right) = 16\left[1 - \left(\frac{x^2}{4} + y^2\right)\right]$$

If region R is the ellipse $\frac{x^2}{4} + y^2 = 1$

then outside R, $\quad \frac{x^2}{4} + y^2 - 1 > 0$

$\Rightarrow \quad B^2 - 4AC < 0$

and inside R, $\quad \frac{x^2}{4} + y^2 - 1 < 0$

$\Rightarrow \quad B^2 - 4AC > 0$

on the ellipse R, $\quad \frac{x^2}{4} + y^2 - 1 = 0$

$\Rightarrow \quad B^2 - 4AC = 0$

Hence, the partial differential equation is elliptic outside this ellipse, hyperbolic inside this ellipse and parabolic on this ellipse.

Method of Separation of Variables

Suppose, we are given partial differential equation in $u(x, y)$ and its partial derivatives.

We suppose

$$u(x, y) = X(x)Y(y)$$

where X is function of x only and Y is function of y only.

Substitute it in given differential equation. Take terms of X and its derivatives on one side and terms of Y and its derivatives on other side. Since X and Y are independent, so both these sides will be constant.

Thus, two ordinary differential equations will be formed. Solve them for X and Y and then $u = XY$ will be solution of the given p.d.e.

Example 4.42: Using the method of separation of variables, solve

$$\frac{\partial u}{\partial x} = 2\frac{\partial u}{\partial t} + u$$

where $$u(x,0) = 6e^{-3x};\ x > 0,\ t > 0$$

Solution: Let $$u(x,t) = X(x)T(t)$$

$\therefore$ $$\frac{\partial u}{\partial x} = X'T,\ \frac{\partial u}{\partial t} = XT'$$

where dashes denote derivatives w.r.t. their variables.

$\therefore$ differential equation becomes

$$X'T = 2XT' + XT$$

$\Rightarrow$ $$\frac{X'}{X} = 2\frac{T'}{T} + 1$$

Now L.H.S. is function of x and R.H.S. is function of t and hence each must be constant

$\therefore$ $$\frac{X'}{X} = 2\frac{T'}{T} + 1 = \lambda$$

$\therefore$ $$X^1 = \lambda X,\ T' = \frac{\lambda - 1}{2}T$$

$\therefore$ $$X = Ae^{\lambda x},\ T = Be^{\frac{\lambda-1}{2}t}$$

$\therefore$ $$u = XT = AB\,e^{\lambda x}\,e^{\frac{\lambda-1}{2}t} = c\,e^{\lambda x}\,e^{\frac{\lambda-1}{2}t}$$

$$u(x,0) = c\,e^{\lambda x} = 6e^{-3x}$$

$\Rightarrow$ $$c = 6, \lambda = -3$$

$\therefore$ solution is

$$u(x,t) = 6e^{-(3x+2t)}$$

Example 4.43: Use the method of separation of variables to solve the p.d.e.

$$3\frac{\partial u}{\partial x} + 2\frac{\partial u}{\partial y} = 0,\ u(x,0) = 4e^{-x}$$

Solution: Let $$u(x,y) = X(x)Y(y)$$

$\therefore$ $$\frac{\partial u}{\partial x} = X'Y,\ \frac{\partial u}{\partial y} = XY'$$

where dashes denote derivatives w.r.t their variables.

$\therefore$ differential equation becomes

$$3X'Y + 2XY' = 0$$

$$\Rightarrow \qquad 3\frac{X'}{X} = -2\frac{Y'}{Y}$$

L.H.S. is function of x and R.H.S. is function of y and hence each is constant

$$\therefore \qquad 3\frac{X'}{X} = -2\frac{Y'}{Y} = \lambda$$

$$\therefore \qquad X' = \frac{1}{3}\lambda X,\ Y' = -\frac{\lambda}{2}Y$$

$$\therefore \qquad X = Ae^{\frac{\lambda}{3}x}, Y = Be^{-\frac{\lambda}{2}y}$$

$$\therefore \qquad u(x,y) = AB\,e^{\frac{\lambda}{3}x}e^{-\frac{\lambda}{2}y} = c\,e^{\frac{\lambda}{3}x}e^{-\frac{\lambda}{2}y}$$

$$u(x,0) = c\,e^{\frac{\lambda}{3}x} = 4e^{-x}$$

$$\Rightarrow \qquad c = 4,\ \lambda = -3$$

$\therefore$ solution is

$$u(x,y) = 4e^{\frac{1}{2}(3y-2x)}$$

Example 4.44: Solve the equation $4\frac{\partial u}{\partial x} + \frac{\partial u}{\partial y} = 3u$ given $u = 3e^{-y} - e^{-5y}$ when $x = 0$.

Solution: Let

$$u(x,y) = X(x)Y(y)$$

$$\therefore \qquad \frac{\partial u}{\partial x} = X'Y,\ \frac{\partial u}{\partial y} = XY'$$

where dashes denote derivatives w.r.t. their variables

$\therefore$ differential equation becomes

$$4X'Y + XY' = 3XY$$

$$\Rightarrow \qquad \frac{4X'}{X} = \frac{-Y'}{Y} + 3$$

L.H.S. is function of x and R.H.S. is function of y and hence each is constant.

$$\therefore \qquad 4\frac{X'}{X} = \frac{-Y'}{Y} + 3 = \lambda$$

$$\therefore \qquad X' = \frac{\lambda}{4}X,\ Y' = (3-\lambda)Y$$

$$\therefore \qquad X = Ae^{\frac{\lambda}{4}x};\ Y = Be^{(3-\lambda)y}$$

$$\therefore \qquad u(x,y) = AB\,e^{\frac{\lambda}{4}x}e^{(3-\lambda)y} = c\,e^{\frac{\lambda}{4}x}e^{(3-\lambda)y}$$

Now, $u(0,y)=3e^{-y}-e^{-5y}$

$\therefore$ $u(x,y)$ is sum of two solutions $c_1 e^{\frac{\lambda_1}{4}x} e^{(3-\lambda_1)y}$ and $c_2 e^{\frac{\lambda_2}{4}x} e^{(3-\lambda_2)y}$

$$\therefore \qquad u(x,y)=c_1 e^{\frac{\lambda_1}{4}x} e^{(3-\lambda_1)y}+c_2 e^{\frac{\lambda_2}{4}x} e^{(3-\lambda_2)y}$$

$$u(0,y)=c_1 e^{(3-\lambda_1)y}+c_2 e^{(3-\lambda_2)y}=3e^{-y}-e^{-5y}$$

$\therefore$ either $c_1=3,\ \lambda_1=4,\ c_2=-1,\ \lambda_2=8$ or $c_1=-1,\ \lambda_1=8,\ c_2=3,\ \lambda_2=4$

In both cases, solution is

$$u(x,y)=3e^{x-y}-e^{2x-5y}$$

Example 4.45: Use the method of separation of variables to solve the equation $\frac{\partial^2 V}{\partial x^2}=\frac{\partial V}{\partial t}$ given that $V=0$ when $t\to\infty$ as well as $V=0$ at $x=0$ and $x=l$.

Solution: Let $$V=X(x)T(t)$$

$$\therefore \qquad \frac{\partial^2 V}{\partial x^2}=X''T,\ \frac{\partial V}{\partial t}=XT'$$

where dashes denote derivatives with respect to their variables.

$\therefore$ differential equation becomes

$$X''T=XT'$$

$$\Rightarrow \qquad \frac{X''}{X}=\frac{T'}{T}$$

Now L.H.S. is function of x and R.H.S. is function of t and hence each must be constant.

$$\therefore \qquad \frac{X''}{X}=\frac{T'}{T}=\lambda$$

$$\therefore \qquad X''=\lambda X,\ T'=\lambda T$$

Now, $$T'=\lambda T$$

$$\Rightarrow \qquad T=Ae^{\lambda t}$$

As $V=XT=0$ when $t\to\infty$, so λ must be negative

Take $$\lambda=-p^2,\ p>0$$

$$\therefore \qquad T=Ae^{-p^2 t}$$

Now $$X''=\lambda X,$$

A.E. is

$$m^2 = \lambda = -p^2$$

$\therefore$ $$m = \pm ip$$

Its solution is

$$X = B\cos px + C\sin px$$

$\therefore$ $$V = Ae^{-p^2t}\left[B\cos px + C\sin px\right]$$

$$= e^{-p^2t}\left[c_1\cos px + c_2\sin px\right]$$

$$V(0,t) = c_1 e^{-p^2t} = 0$$

$\Rightarrow$ $$c_1 = 0$$

$\therefore$ $$V(x,t) = c_2 e^{-p^2t}\sin px$$

$$V(l,t) = c_2 e^{-p^2t}\sin pl = 0$$

$\Rightarrow$ $$pl = n\pi, n = 1,2,\ldots \qquad (\because p > 0)$$

$\therefore$ solutions are

$$V(x,t) = b_n e^{-\frac{n^2\pi^2}{l^2}t}\sin\frac{n\pi x}{l};\ n = 1,2,\ldots \qquad (b_n = c_2)$$

$\therefore$ By principle of superposition, complete solution is

$$V(x,t) = \sum_{n=1}^{\infty} b_n e^{-\frac{n^2\pi^2}{l^2}t}\sin\frac{n\pi x}{l}$$

Example 4.46: Using the method of separation of variables, solve the parabolic partial differential equation $\frac{\partial^2 u}{\partial x^2} = 16\frac{\partial u}{\partial y}$

Solution: Let $$u(x,y) = X(x)Y(y)$$

$\therefore$ $$\frac{\partial^2 u}{\partial x^2} = X'', \frac{\partial u}{\partial y} = XY'$$

where dashes denote derivatives w.r.t. their variables.

$\therefore$ differential equation becomes

$$X''Y = 16XY'$$

or $$\frac{X''}{X} = \frac{16Y'}{Y}$$

L.H.S. is function of x and R.H.S. is function of y and hence each is constant.

$\therefore$ $$\frac{X''}{X} = \frac{16Y'}{Y} = \lambda$$

$$X'' = \lambda X, Y' = \frac{1}{16}\lambda Y$$

Now, three cases arise as discussed hereunder.

Case I: $\lambda = 0$ then

$$X'' = 0, Y' = 0$$

$\therefore$ $$X = Ax + B,\ Y = C$$

$\therefore$ $$u(x, y) = C(Ax + B) = c_1 x + c_2$$

Case II: $\lambda > 0$, let $\lambda = p^2,\ p > 0$

$\therefore$ $$X'' = p^2 X,\ Y' = \frac{p^2}{16} Y$$

A.E. of $X'' = p^2 X$ is $m^2 - p^2 = 0$ $\quad\therefore\ m = \pm p$

$\therefore$ $$X = Ae^{-px} + Be^{px},\ Y = C e^{\frac{p^2}{16} y}$$

$\therefore$ $$u = C e^{\frac{p^2}{16} y}\left(Ae^{-px} + Be^{px}\right)$$

$$= e^{k^2 y}\left(c_3 e^{-4kx} + c_4 e^{4kx}\right) \qquad (p = 4k)$$

Case III: $\lambda < 0$, let $\lambda = -16k_1^2,\ k_1 > 0$

$$X'' = -16k_1^2 X,\ Y' = -k_1^2 Y$$

A.E. of $X'' = -16k_1^2 X$ is $m^2 + 16k_1^2 = 0$ $\quad\Rightarrow m = \pm 4i\, k_1$

$\therefore$ $$X = A\cos 4k_1 x + B\sin 4k_1 x$$

and solution of $Y' = -k_1^2 Y$ is

$$Y = C e^{-k_1^2 y}$$

$\therefore$ solution is

$$u = XY = e^{-k_1^2 y}\left(c_5 \cos 4k_1 x + c_6 \sin 4k_1 x\right)$$

$\therefore$ solutions are

$$u(x, y) = c_1 x + c_2,\ u(x, y) = e^{k^2 y}\left(c_3 e^{-4kx} + c_4 e^{4kx}\right),$$

$$u(x, y) = e^{-k_1^2 y}\left(c_5 \cos 4k_1 x + c_6 \sin 4k_1 x\right)$$

$\therefore$ general solution is

$$u(x, y) = c_1 x + c_2 + e^{k^2 y}\left(c_3 e^{-4kx} + c_4 e^{4kx}\right) + e^{-k_1^2 y}\left(c_5 \cos 4k_1 x + c_6 \sin 4k_1 x\right)$$

where $c_1, c_2, c_3, c_4, c_5, c_6, k > 0, k_1 > 0$ are arbitrary constants.

4.11 VIBRATIONS OF A STRETCHED STRING (ONE DIMENSIONAL WAVE EQUATION)

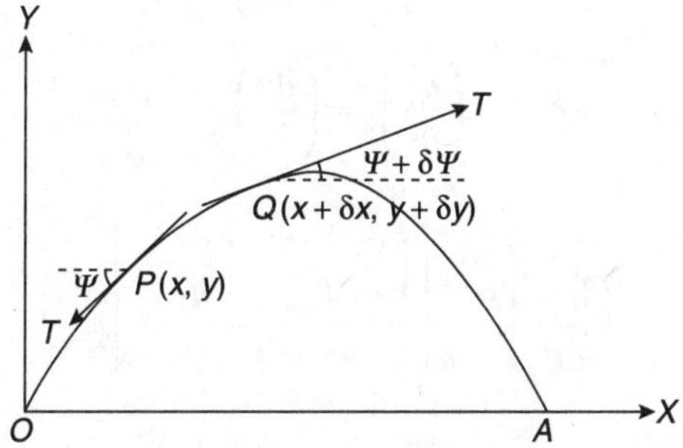

Figure 4.1

Consider a uniform elastic string of length l stretched tightly between two points O and A.

Suppose that the string is displaced slightly from its position OA. Taking O as origin, x-axis along OA and y-axis $\perp$ to OA at O, displacement y of any point will depend upon x and t where x is x-co-ordinate of point and t is time. We shall consider vibration of string under the following assumptions.

(i) Each point of string moves perpendicular to the equilibrium position OA in x–y plane.
(ii) String is flexible and does not offer resistance of bending.
(iii) The displacement y and slope $\frac{\partial y}{\partial x}$ are small so that their higher powers can be neglected.
(iv) Tension in string is large and constant throughout the string and weight of string is negligible in comparison to tension.

Let m be mass per unit length of string. Let $P(x, y)$ and $Q(x+\delta x, y+\delta y)$ be position of two points on the string at time t. Tension T acts at P and Q as shown in the figure.

Since there is no motion horizontally

$$\therefore \qquad T\cos(\psi+\delta\psi) = T\cos\psi = T \qquad (\because \psi \text{ and } \psi+\delta\psi \text{ are small}) \tag{4.44}$$

Let $\overset{\frown}{PQ} = \delta s$ then $m\,\delta s$ is mass of portion PQ of the string. By Newton's second law of motion, equation of vertical motion is

$$m\,\delta s \frac{\partial^2 y}{\partial t^2} = T\sin(\psi+\delta\psi) - T\sin\psi$$

$$\therefore \qquad \frac{m\,\delta s}{T}\frac{\partial^2 y}{\partial t^2} = \frac{T\sin(\psi+\delta\psi)}{T\cos(\psi+\delta\psi)} - \frac{T\sin\psi}{T\cos\psi} \qquad (\text{from}\,(4.44))$$

or

$$\frac{m\,\delta s}{T}\frac{\partial^2 y}{\partial t^2} = \tan(\psi+\delta\psi) - \tan\psi$$

But $\tan\psi$ and $\tan(\psi+\delta\psi)$ are slopes of tangents at P and Q, respectively

Hence,

$$\tan\psi = \left(\frac{\partial y}{\partial x}\right)_P = \left(\frac{\partial y}{\partial x}\right)_{(x,y)}$$

$$\tan(\psi + \delta\psi) = \left(\frac{\partial y}{\partial x}\right)_Q = \left(\frac{\partial y}{\partial x}\right)_{(x+\delta x,\, y+\delta y)}$$

$\therefore$
$$\left(\frac{\partial^2 y}{\partial t^2}\right) = \frac{T}{m}\left[\frac{\left(\frac{\partial y}{\partial x}\right)_{x+\delta x} - \left(\frac{\partial y}{\partial x}\right)_x}{\delta x}\right] \qquad (\because\ \delta s \cong \delta x)$$

Taking limit as $\delta x \to 0$

$$\frac{\partial^2 y}{\partial t^2} = \frac{T}{m}\frac{\partial^2 y}{\partial x^2}$$

or
$$\frac{\partial^2 y}{\partial t^2} = c^2 \frac{\partial^2 y}{\partial x^2}$$

where $c^2 = \dfrac{T}{m}$ is called the diffusivity of string.

It is the partial differential equation giving the vertical displacement of the points of the string. Differential equation of vibrations of a stretched string is also called one dimensional wave equation.

4.11.1 Solution of the Wave Equation

Wave equation is
$$\frac{\partial^2 y}{\partial t^2} = c^2 \frac{\partial^2 y}{\partial x^2}$$

Let
$$y(x,t) = X(x)T(t)$$

where $X(x)$ is function of x only and $T(t)$ is function of t only

$\therefore$
$$\frac{\partial^2 y}{\partial t^2} = X(x)T''(t),\ \frac{\partial^2 y}{\partial x^2} = X''(x)T(t)$$

$\therefore$ differential equation becomes

$$X(x)T''(t) = c^2 X''(x)T(t)$$

or
$$\frac{X''(x)}{X(x)} = \frac{1}{c^2}\frac{T''(t)}{T(t)} \qquad (4.45)$$

L.H.S. is function of x and R.H.S. is function of t and hence each must be constant. This constant λ may be zero, positive or negative.

Case I: $\lambda = 0$

In this case, we have

$$X''(x) = 0,\ T''(t) = 0$$

$$\therefore \quad X(x) = Ax + B,\ T(t) = Ct + D$$

$$\therefore \quad y(x,t) = (Ax + B)(Ct + D) \tag{4.46}$$

Case II: λ is positive, i.e., $\lambda = p^2;\ p > 0$

We have

$$X''(x) - p^2 X(x) = 0$$

$$T''(t) - p^2c^2\,T(t) = 0$$

Their solutions are

$$X = Ae^{px} + Be^{-px},\ T = C\,e^{pct} + De^{-pct}$$

$$\therefore \quad y(x,t) = \left(Ae^{px} + Be^{-px}\right)\left(Ce^{pct} + De^{-pct}\right) \tag{4.47}$$

Case III: λ is negative, i.e., $\lambda = -p^2;\ p > 0$

We have

$$X''(x) + p^2 X(x) = 0$$

$$T''(t) + p^2c^2 T(t) = 0$$

Their solutions are

$$X = A\cos(px) + B\sin(px),\ T = C\cos(pct) + D\sin(pct)$$

$$\therefore \quad y(x,t) = \left[A\cos(px) + B\sin(px)\right]\left[C\cos(pct) + D\sin(pct)\right] \tag{4.48}$$

There can be one or combination of these solutions.

But when the ends O and A are fixed then $y(0,t) = y(l,t) = 0$ for all t and hence solutions (4.46) and (4.47) are invalid.

$$\therefore \quad y(x,t) = \left[A\cos(px) + B\sin(px)\right]\left[C\cos(pct) + D\sin(pct)\right]$$

$$y(0,t) = 0 \quad \Rightarrow A = 0$$

and $\quad y(l,t) = 0$

$\Rightarrow \quad \sin pl = 0$

$$\Rightarrow \quad p = \frac{\pi n}{l};\ n \in N$$

∴ solutions are

$$y(x,t)=\left[b_n\cos\left(\frac{\pi nct}{l}\right)+e_n\sin\left(\frac{\pi nct}{l}\right)\right]\sin\frac{\pi nx}{l};\ n=1,2,3,\ldots$$

By principle of superposition, solution is

$$y(x,t)=\sum_{n=1}^{\infty}\left[b_n\cos\left(\frac{\pi nct}{l}\right)+e_n\sin\left(\frac{\pi nct}{l}\right)\right]\sin\frac{\pi nx}{l}$$

Constants b_n and e_n will be found from initial conditions.

4.11.2 D'Alembert's Method of Solving Wave Equation

Wave equation is

$$\frac{\partial^2 y}{\partial t^2}=c^2\frac{\partial^2 y}{\partial x^2}$$

Let us introduce new variables

$$\xi=x+ct,\ \eta=x-ct$$

so that y becomes function of ξ and η

$$\frac{\partial y}{\partial t}=\frac{\partial y}{\partial \xi}\frac{\partial \xi}{\partial t}+\frac{\partial y}{\partial \eta}\frac{\partial \eta}{\partial t}=c\left(\frac{\partial y}{\partial \xi}-\frac{\partial y}{\partial \eta}\right)$$

$$\frac{\partial^2 y}{\partial t^2}=c\left[\frac{\partial}{\partial \xi}\left(\frac{\partial y}{\partial \xi}-\frac{\partial y}{\partial \eta}\right)\frac{\partial \xi}{\partial t}+\frac{\partial}{\partial \eta}\left(\frac{\partial y}{\partial \xi}-\frac{\partial y}{\partial \eta}\right)\frac{\partial \eta}{\partial t}\right]$$

$$=c^2\left[\left(\frac{\partial^2 y}{\partial \xi^2}-\frac{\partial^2 y}{\partial \xi\partial \eta}-\frac{\partial^2 y}{\partial \eta\partial \xi}+\frac{\partial^2 y}{\partial \eta^2}\right)\right]$$

$$=c^2\left[\left(\frac{\partial^2 y}{\partial \xi^2}-2\frac{\partial^2 y}{\partial \xi\partial \eta}+\frac{\partial^2 y}{\partial \eta^2}\right)\right]$$

$$\frac{\partial y}{\partial x}=\frac{\partial y}{\partial \xi}\frac{\partial \xi}{\partial x}+\frac{\partial y}{\partial \eta}\frac{\partial \eta}{\partial x}=\frac{\partial y}{\partial \xi}+\frac{\partial y}{\partial \eta}$$

$$\frac{\partial^2 y}{\partial x^2}=\frac{\partial}{\partial \xi}\left(\frac{\partial y}{\partial \xi}+\frac{\partial y}{\partial \eta}\right)\frac{\partial \xi}{\partial x}+\frac{\partial}{\partial \eta}\left(\frac{\partial y}{\partial \xi}+\frac{\partial y}{\partial \eta}\right)\frac{\partial \eta}{\partial x}$$

$$=\frac{\partial^2 y}{\partial \xi^2}+\frac{\partial^2 y}{\partial \xi\partial \eta}+\frac{\partial^2 y}{\partial \eta\partial \xi}+\frac{\partial^2 y}{\partial \eta^2}$$

$$=\frac{\partial^2 y}{\partial \xi^2}+2\frac{\partial^2 y}{\partial \xi\partial \eta}+\frac{\partial^2 y}{\partial \eta^2}$$

$\therefore$ wave equation becomes

$$c^2\left[\frac{\partial^2 y}{\partial \xi^2}-2\frac{\partial^2 y}{\partial \xi \partial \eta}+\frac{\partial^2 y}{\partial \eta^2}\right]=c^2\left[\frac{\partial^2 y}{\partial \xi^2}+2\frac{\partial^2 y}{\partial \xi \partial \eta}+\frac{\partial^2 y}{\partial \eta^2}\right]$$

or
$$\frac{\partial^2 y}{\partial \xi \partial \eta}=0$$

Integrate w.r.t. ξ, keeping η constant

$$\frac{\partial y}{\partial \eta}=f(\eta)$$

Integrate w.r.t. η, keeping ξ constant

$$y(x,t)=\int f(\eta)\,d\eta+\phi(\xi)=\phi(\xi)+\psi(\eta) \text{ where } \psi(\eta)=\int f(\eta)\,d\eta$$

$\therefore$ solution is

$$y(x,t)=\phi(x+ct)+\psi(x-ct)$$

where ϕ and ψ are arbitrary functions which will be determined from initial conditions. This solution can be obtained directly also as wave equation:

$$\left(D_t^2-c^2D_x^2\right)y=0$$

or
$$\left(D_t+cD_x\right)\left(D_t-cD_x\right)y=0$$

Hence, solution is

$$y(x,t)=\phi(x+ct)+\psi(x-ct)$$

Suppose initial conditions are

$$y(x,0)=f(x) \text{ and } \left(\frac{\partial y}{\partial t}\right)_{(x,0)}=g(x)$$

then,
$$y(x,0)=\phi(x)+\psi(x)=f(x) \tag{4.49}$$

$$\frac{\partial y}{\partial t}=c\left[\phi'(x+ct)-\psi'(x-ct)\right]$$

$\therefore$
$$\left(\frac{\partial y}{\partial t}\right)_{(x,0)}=c\left[\phi'(x)-\psi'(x)\right]=g(x)$$

$\Rightarrow$
$$\phi'(x)-\psi'(x)=\frac{1}{c}g(x)$$

Integrate w.r.t. x

$$\left[\phi(x)-\psi(x)\right]=\frac{1}{c}\int_{x_0}^{x}g(x)\,dx+k \tag{4.50}$$

where x_0 and k are arbitrary constants.

Solving (4.49) and (4.50)

$$\phi(x)=\frac{1}{2}\left[f(x)+\frac{1}{c}\int_{x_0}^{x}g(x)\,dx+k\right]$$

$$\psi(x)=\frac{1}{2}\left[f(x)-\frac{1}{c}\int_{x_0}^{x}g(x)\,dx-k\right]$$

$\therefore$
$$y(x,t)=\phi(x+ct)+\psi(x-ct)$$

$$=\frac{1}{2}\left[f(x+ct)+\frac{1}{c}\int_{x_0}^{x+ct}g(x)\,dx+k\right]+\frac{1}{2}\left[f(x-ct)-\frac{1}{c}\int_{x_0}^{x-ct}g(x)\,dx+k\right]$$

$$=\frac{1}{2}\left[f(x+ct)+f(x-ct)+\frac{1}{c}\left\{\int_{x_0}^{x+ct}g(x)\,dx+\int_{x-ct}^{x_0}g(x)\,dx\right\}\right]$$

$$=\frac{1}{2}\left[f(x+ct)+f(x-ct)+\frac{1}{c}\int_{x-ct}^{x+ct}g(x)\,dx\right]$$

Remark 4.4:

(i) If we write half range Fourier sine series of $f(x)$ and $g(x)$ in $(0, l)$ we shall get the solution obtained in article (4.11.1).

(ii) If stretched string is infinite then method of solution in article (4.11.1) cannot be applied and solution will be obtained by D' Alembert's method.

Example 4.47: Solve the p.d.e. $\frac{\partial^2 y}{\partial t^2}=a^2\frac{\partial^2 y}{\partial x^2}$ representing the vibrations of a string of length l, fixed at both ends, subject to the boundary conditions $y(0,t)=y(l,t)=0$ and initial conditions

$$y=y_0\sin\frac{\pi x}{l},\ \frac{\partial y}{\partial t}=0 \text{ at } t=0$$

Solution: Here, the boundary conditions are

$$y(0,t)=y(l,t)=0$$

$\therefore$ solution of wave equation is

$$y(x,t)=\sum_{n=1}^{\infty}\left[b_n\cos\left(\frac{\pi nat}{l}\right)+e_n\sin\left(\frac{\pi nat}{l}\right)\right]\sin\frac{n\pi x}{l}$$

$$y(x,0)=y_0\sin\frac{\pi x}{l}=\sum_{n=1}^{\infty}b_n\sin\frac{n\pi x}{l}$$

$\therefore$
$$b_1=y_0,\ b_n=0;\ n=2,3,\ldots$$

$\therefore$
$$y(x,t)=y_0\cos\left(\frac{\pi at}{l}\right)\sin\frac{\pi x}{l}+\sum_{n=1}^{\infty}e_n\sin\frac{n\pi at}{l}\sin\frac{n\pi x}{l}$$

$$\frac{\partial y}{\partial t} = -y_0 \frac{\pi a}{l} \sin\left(\frac{\pi at}{l}\right) \sin\frac{\pi x}{l} + \sum_{n=1}^{\infty} e_n \frac{n\pi a}{l} \cos\frac{n\pi at}{l} \sin\frac{n\pi x}{l}$$

$\therefore$ $$\left(\frac{\partial y}{\partial t}\right)_{t=0} = 0 = \sum_{n=1}^{\infty} e_n \frac{\pi na}{l} \sin\frac{n\pi x}{l}$$

$\therefore$ $$e_n = 0;\ n = 1, 2, 3, \ldots$$

$\therefore$ solution is

$$y(x,t) = y_0 \cos\left(\frac{\pi at}{l}\right) \sin\left(\frac{\pi x}{l}\right)$$

Example 4.48: A thin uniform tightly stretched vibrating string fixed at the points $x = 0$ and $x = l$ satisfy the equation $\frac{\partial^2 y}{\partial t^2} = c^2 \frac{\partial^2 y}{\partial x^2}; y(x,0) = y_0 \sin^3 \frac{\pi x}{l}$ and released from rest from this position. Find the displacement $y(x,t)$ at any x and any time t.

Solution: Since the ends are fixed

$\therefore$ $$y(0,t) = y(l,t) = 0$$

$\therefore$ solution of wave equation $\frac{\partial^2 y}{\partial t^2} = c^2 \frac{\partial^2 y}{\partial x^2}$ is

$$y(x,t) = \sum_{n=1}^{\infty} \left[b_n \cos\left(\frac{n\pi ct}{l}\right) + e_n \sin\left(\frac{n\pi ct}{l}\right) \right] \sin\frac{n\pi x}{l}$$

$$y(x,0) = y_0 \sin^3 \frac{\pi x}{l} = y_0 \left[\frac{1}{4}\left(3\sin\frac{\pi x}{l} - \sin\frac{3\pi x}{l} \right) \right]$$

$\therefore$ $$\frac{y_0}{4}\left[3\sin\frac{\pi x}{l} - \sin\frac{3\pi x}{l} \right] = \sum_{n=1}^{\infty} b_n \sin\frac{n\pi x}{l}$$

$\therefore$ $$b_1 = \frac{3y_0}{4}, b_3 = \frac{-y_0}{4};\ b_n = 0;\ n = 2, 4, 5, \ldots$$

$\therefore$ $$y(x,t) = \frac{3y_0}{4} \cos\left(\frac{\pi ct}{l}\right) \sin\frac{\pi x}{l} - \frac{y_0}{4} \cos\left(\frac{3\pi ct}{l}\right) \sin\frac{3\pi x}{l} + \sum_{n=1}^{\infty} e_n \sin\left(\frac{n\pi ct}{l}\right) \sin\frac{n\pi x}{l}$$

$\therefore$ $$\frac{\partial y}{\partial t} = -\frac{3y_0 \pi c}{4l} \sin\left(\frac{\pi ct}{l}\right) \sin\frac{\pi x}{l} + \frac{3y_0 \pi c}{4l} \sin\left(\frac{3\pi ct}{l}\right) \sin\frac{3\pi x}{l} + \sum_{n=1}^{\infty} \frac{n\pi c}{l} e_n \cos\left(\frac{n\pi ct}{l}\right) \sin\frac{n\pi x}{l}$$

$$\therefore \qquad \left(\frac{\partial y}{\partial t}\right)_{t=0} = 0 = \sum_{n=1}^{\infty} \frac{n\pi c}{l} e_n \sin\frac{n\pi x}{l}$$

$$\therefore \qquad e_n = 0;\ n = 1, 2, 3, \ldots$$

$\therefore$ solution is

$$y(x,t) = \frac{y_0}{4}\left[3\cos\left(\frac{\pi ct}{l}\right)\sin\frac{\pi x}{l} - \cos\left(\frac{3\pi ct}{l}\right)\sin\frac{3\pi x}{l}\right]$$

Example 4.49: The vibrations of an elastic string is governed by the p.d.e. $\frac{\partial^2 u}{\partial t^2} = \frac{\partial^2 u}{\partial x^2}$.

The length of the string is π and the ends are fixed. The initial velocity is zero and the initial deflection is $u(x,0) = 2(\sin x + \sin 3x)$. Find the deflection $u(x,t)$ of the vibrating string for $t > 0$.
Solution: Since the ends are fixed

$$\therefore \qquad u(0,t) = u(\pi,t) = 0$$

$\therefore$ solution of wave equation $\frac{\partial^2 y}{\partial t^2} = \frac{\partial^2 u}{\partial x^2}$ with diffusivity unity is

$$u(x,t) = \sum_{n=1}^{\infty}\left[b_n \cos\left(\frac{n\pi t}{\pi}\right) + e_n \sin\left(\frac{n\pi t}{\pi}\right)\right]\sin\frac{n\pi x}{\pi}$$

$$= \sum_{n=1}^{\infty}\left[b_n \cos(nt) + e_n \sin(nt)\right]\sin(nx)$$

$$u(x,0) = 2(\sin x + \sin 3x) = \sum_{n=1}^{\infty} b_n \sin(nx)$$

$$\therefore \qquad b_1 = 2, b_3 = 2;\ b_n = 0, n = 2, 4, 5, 6 \ldots$$

$$\therefore \qquad u(x,t) = 2\cos t \sin x + 2\cos 3t \sin 3x + \sum_{n=1}^{\infty} e_n \sin(nt)\sin(nx)$$

$$\frac{\partial u}{\partial t} = -2\sin t \sin x - 6\sin 3t \sin 3x + \sum_{n=1}^{\infty} ne_n \cos(nt)\sin(nx)$$

$$\left(\frac{\partial u}{\partial t}\right)_{t=0} = 0 = \sum_{n=1}^{\infty} ne_n \sin(nx)$$

$$\therefore \qquad e_n = 0;\ n = 1, 2, 3 \ldots$$

$\therefore$ solution is

$$u(x,t) = 2(\cos t \sin x + \cos 3t \sin 3x).$$

Example 4.50: Solve the boundary value problem $\frac{\partial^2 y}{\partial t^2} = 4\frac{\partial^2 y}{\partial x^2}$, given that

$$y(0,t) = y(5,t) = 0,$$

$$\left(\frac{\partial y}{\partial t}\right)_{t=0} = 5\sin \pi x,\ y(x,0) = 0$$

Solution: Since $y(0,t) = y(5,t) = 0$

$\therefore$ solution of wave equation $\dfrac{\partial^2 y}{\partial t^2} = 4\dfrac{\partial^2 y}{\partial x^2}$ with diffusivity $= c^2 = 4$ is

$$y(x,t) = \sum_{n=1}^{\infty}\left[b_n \cos\frac{2n\pi t}{5} + e_n \sin\frac{2n\pi t}{5}\right]\sin\frac{n\pi x}{5}$$

$$y(x,0) = 0 = \sum_{n=1}^{\infty} b_n \sin\frac{n\pi x}{5}$$

$$\therefore \qquad b_n = 0;\ n = 1,2,3\ldots$$

$$\therefore \qquad y(x,t) = \sum_{n=1}^{\infty} e_n \sin\frac{2n\pi t}{5}\sin\frac{n\pi x}{5}$$

$$\frac{\partial y}{\partial t} = \sum_{n=1}^{\infty}\frac{2n\pi}{5} e_n \cos\frac{2n\pi t}{5}\sin\frac{n\pi x}{5}$$

$$\therefore \qquad \left(\frac{\partial y}{\partial t}\right)_{t=0} = 5\sin\pi x = \sum_{n=1}^{\infty}\frac{2n\pi}{5} e_n \sin\frac{n\pi x}{5}$$

$$\therefore \qquad e_5 = \frac{5}{2\pi};\ e_n = 0;\ n = 1,2,3\ldots;\ n \neq 5$$

$\therefore$ solution is

$$y(x,t) = \frac{5}{2\pi}\sin(2\pi t)\sin(\pi x)$$

Example 4.51: Solve the wave equation $\dfrac{\partial^2 u}{\partial t^2} = c^2\dfrac{\partial^2 u}{\partial x^2}$; $0 < x < l,\ 0 < t < 4$ with the boundary conditions, $u(0,t) = u(l,t) = 0$ and initial conditions $u(x,0) = f(x)$ and $\left(\dfrac{\partial u}{\partial t}\right)_{t=0} = g(x);\ 0 < x < l$

Solution: As

$$u(0,t) = u(l,t) = 0$$

$\therefore$ solution of wave equation $\dfrac{\partial^2 u}{\partial t^2} = c^2\dfrac{\partial^2 u}{\partial x^2}$; $0 < x < l,\ 0 < t < 4$ is

$$u(x,t) = \sum_{n=1}^{\infty}\left[b_n \cos\frac{n\pi ct}{l} + e_n \sin\frac{n\pi ct}{l}\right]\sin\frac{n\pi x}{l};\ 0 < x < l,\ 0 < t < 4$$

$$u(x,0) = f(x) = \sum_{n=1}^{\infty} b_n \sin\frac{n\pi x}{l};\ 0 < x < l$$

It is Fourier half range sine series of $f(x)$ in $(0,l)$

$$\therefore \qquad b_n = \frac{2}{l}\int_0^l f(x)\sin\frac{n\pi x}{l}\,dx$$

$$\frac{\partial u}{\partial t}=\sum_{n=1}^{\infty}\frac{n\pi c}{l}\left[-b_n\sin\frac{n\pi ct}{l}+e_n\cos\frac{n\pi ct}{l}\right]\sin\frac{n\pi x}{l}$$

$\therefore$ $$\left(\frac{\partial u}{\partial t}\right)_{t=0}=g(x)=\sum_{n=1}^{\infty}\frac{n\pi c}{l}e_n\sin\frac{n\pi x}{l};\ 0<x<l$$

It is Fourier half range sine series of $g(x)$ in $(0,l)$

$\therefore$ $$\frac{n\pi c}{l}e_n=\frac{2}{l}\int_0^l g(x)\sin\frac{n\pi x}{l}dx$$

$\therefore$ $$e_n=\frac{2}{n\pi c}\int_0^l g(x)\sin\frac{n\pi x}{l}dx$$

$\therefore$ solution is

$$u(x,t)=\sum_{n=1}^{\infty}\left[b_n\cos\frac{n\pi ct}{l}+e_n\sin\frac{n\pi ct}{l}\right]\sin\frac{n\pi x}{l},\ 0\le x\le l;\ 0\le t<4$$

where $$b_n=\frac{2}{l}\int_0^l f(x)\sin\frac{n\pi x}{l}dx,\ e_n=\frac{2}{n\pi c}\int_0^l g(x)\sin\frac{n\pi x}{l}dx$$

Example 4.52: The points of trisection of a string of length l are pulled aside through the same distance d on opposite sides of the position of equilibrium and the string is released from rest. Derive an expression for the displacement of the string at subsequent time and show that the midpoint of the string always remains at rest.

Solution: Let OA be string of length l. B and C are points of trisection and P is midpoint of string. Then initially at $t = 0$, the string is in position $OB'PC'A$.

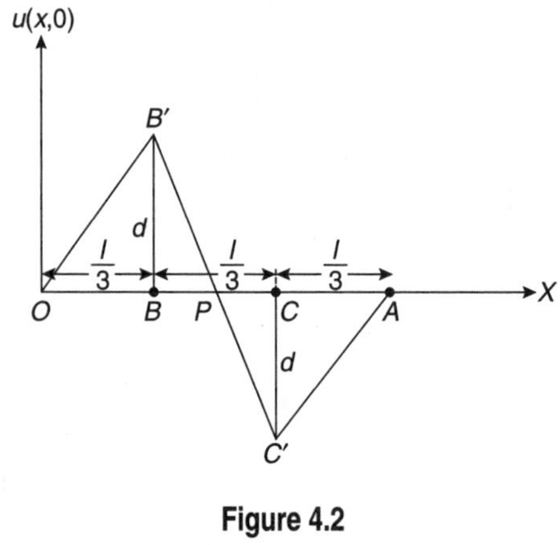

Figure 4.2

We have

$$O(0,0),B'\left(\frac{l}{3},d\right),P\left(\frac{l}{2},0\right),C'\left(\frac{2l}{3},-d\right),A(l,0)$$

Let $y(x,t)$ be displacement of string at distance x from O at time t.

Equation of portion OB' is

$$y(x,0)-0=\frac{d-0}{\frac{l}{3}-0}(x-0)$$

$$\Rightarrow \qquad y(x,0)=\frac{3d}{l}x$$

Equation of portion $B'C'$ is

$$y(x,0)-d=\frac{-d-d}{\frac{2l}{3}-\frac{l}{3}}\left(x-\frac{l}{3}\right)$$

$$\Rightarrow \qquad y(x,0)=-\frac{6d}{l}x+3d$$

Equation of portion $C'A$ is

$$y(x,0)-0=\frac{0+d}{l-\frac{2l}{3}}(x-l)$$

$$\Rightarrow \qquad y(x,0)=\frac{3d}{l}x-3d$$

$$\therefore \qquad y(x,0)=\frac{3d}{l}x;\ 0\le x\le\frac{l}{3}$$

$$=-\frac{6d}{l}x+3d;\ \frac{l}{3}\le x\le\frac{2l}{3}$$

$$=\frac{3d}{l}x-3d;\ \frac{2l}{3}\le x\le l$$

As ends O and A are fixed

$$\therefore \qquad y(0,t)=y(l,t)=0$$

Now, deflection $y(x,t)$ follow wave equation

$$\frac{\partial^2 y}{\partial t^2}=c^2\frac{\partial^2 y}{\partial x^2}$$

where c^2 is diffusivity of string.

Its solution is

$$y(x,t)=\sum_{n=1}^{\infty}\left(b_n\cos\frac{n\pi ct}{l}+e_n\sin\frac{n\pi ct}{l}\right)\sin\frac{n\pi x}{l}$$

$$\frac{\partial y}{\partial t}=\sum_{n=1}^{\infty}\frac{n\pi c}{l}\left(-b_n\sin\frac{n\pi ct}{l}+e_n\cos\frac{n\pi ct}{l}\right)\sin\frac{n\pi x}{l}$$

$$\therefore \quad \left(\frac{\partial y}{\partial t}\right)_{t=0} = 0 = \sum_{n=1}^{\infty} \frac{n\pi c}{l} e_n \sin\frac{n\pi x}{l}$$

$$\therefore \quad e_n = 0;\ n = 1,2,3,\ldots$$

$$\therefore \quad y(x,t) = \sum_{n=1}^{\infty} b_n \cos\frac{n\pi ct}{l} \sin\frac{n\pi x}{l}$$

$$y(x,0) = \sum_{n=1}^{\infty} b_n \sin\frac{n\pi x}{l}$$

It is half range Fourier sine series of $y(x,0)$ in $0 \le x \le l$

$$\therefore \quad b_n = \frac{2}{l}\int_0^l y(x,0)\sin\frac{n\pi x}{l}\,dx$$

$$= \frac{2}{l}\left[\int_0^{\frac{l}{3}} \frac{3d}{l}x\sin\frac{n\pi x}{l}\,dx + \int_{\frac{l}{3}}^{\frac{2l}{3}}\left(3d - \frac{6d}{l}x\right)\sin\frac{n\pi x}{l}\,dx + \int_{\frac{2l}{3}}^{l}\left(\frac{3d}{l}x - 3d\right)\sin\frac{n\pi x}{l}\,dx\right]$$

$$= \frac{6d}{l^2}\left[\left\{x\left(-\frac{l}{n\pi}\cos\frac{n\pi x}{l}\right) - \left(-\frac{l^2}{n^2\pi^2}\sin\frac{n\pi x}{l}\right)\right\}_0^{l/3}\right.$$

$$+\left\{(l-2x)\left(-\frac{l}{n\pi}\cos\frac{n\pi x}{l}\right) - (-2)\left(-\frac{l^2}{n^2\pi^2}\sin\frac{n\pi x}{l}\right)\right\}_{l/3}^{2l/3}$$

$$\left.+\left\{(x-l)\left(-\frac{l}{n\pi}\cos\frac{n\pi x}{l}\right) - \left(-\frac{l^2}{n^2\pi^2}\sin\frac{n\pi x}{l}\right)\right\}_{2l/3}^{l}\right]$$

$$= \frac{6d}{l^2}\left[-\frac{l^2}{3n\pi}\cos\frac{n\pi}{3} + \frac{l^2}{n^2\pi^2}\sin\frac{n\pi}{3} + \frac{l^2}{3n\pi}\cos\frac{2n\pi}{3} + \frac{l^2}{3n\pi}\cos\frac{n\pi}{3} - \frac{2l^2}{n^2\pi^2}\sin\frac{2n\pi}{3}\right.$$

$$\left.+\frac{2l^2}{n^2\pi^2}\sin\frac{n\pi}{3} - \frac{l^2}{3n\pi}\cos\frac{2n\pi}{3} - \frac{l^2}{n^2\pi^2}\sin\frac{2n\pi}{3}\right]$$

$$= \frac{6d}{l^2}\left[\frac{3l^2}{n^2\pi^2}\sin\frac{n\pi}{3} - \frac{3l^2}{n^2\pi^2}\sin\left(n\pi - \frac{n\pi}{3}\right)\right]$$

$$= \frac{18d}{n^2\pi^2}\left[\sin\frac{n\pi}{3} - (-1)^{n+1}\sin\frac{n\pi}{3}\right]$$

$$= \frac{18d}{n^2\pi^2}\left[1 + (-1)^n\right]\sin\frac{n\pi}{3}$$

$$\therefore \quad b_{2n} = \frac{36d}{(2n)^2\pi^2}\sin\frac{2n\pi}{3} = \frac{9d}{n^2\pi^2}\sin\frac{2n\pi}{3};\ b_{2n-1} = 0;\ n = 1,2,3,\ldots$$

$$\therefore \qquad y(x,t)=\frac{9d}{\pi^2}\sum_{n=1}^{\infty}\frac{1}{n^2}\sin\frac{2n\pi}{3}\cos\frac{2n\pi ct}{l}\sin\frac{2n\pi x}{l}$$

which gives the displacement of string at time *t*.

$$\frac{\partial y}{\partial t}=\frac{9d}{\pi^2}\sum_{n=1}^{\infty}\frac{1}{n^2}\sin\frac{2n\pi}{3}\cdot\left(\frac{-2n\pi c}{l}\right)\sin\frac{2n\pi ct}{l}\sin\frac{2n\pi x}{l}$$

$$\therefore \qquad \left(\frac{\partial y}{\partial t}\right)_{x=\frac{l}{2}}=0 \qquad (\because \sin n\pi=0;\ n=1,2\ldots)$$

$\therefore$ midpoint of string always remain at rest.

Example 4.53: Solve the wave equation $\frac{\partial^2 u}{\partial t^2}=a^2\frac{\partial^2 u}{\partial x^2}$ under the conditions $u=0$ when $x=0$ and $x=\pi$, $\frac{\partial u}{\partial t}=0$ when $t=0$ and $u(x,0)=x;\ 0<x<\pi$.

Solution: As $u(0,t)=u(\pi,t)=0$

$\therefore$ solution of wave equation $\frac{\partial^2 u}{\partial t^2}=a^2\frac{\partial^2 u}{\partial x^2}$ is

$$u(x,t)=\sum_{n=1}^{\infty}\left[b_n\cos(nat)+e_n\sin(nat)\right]\sin(nx)$$

$$\frac{\partial u}{\partial t}=\sum_{n=1}^{\infty}na\left(-b_n\sin(nat)+e_n\cos(nat)\right)\sin(nx)$$

$$\left(\frac{\partial u}{\partial t}\right)_{t=0}=0=\sum_{n=1}^{\infty}nae_n\sin(nx)$$

$$\Rightarrow \qquad e_n=0;\ n=1,2,\ldots$$

$$\therefore \qquad u(x,t)=\sum_{n=1}^{\infty}b_n\cos(nat)\sin(nx)$$

$$u(x,0)=x=\sum_{n=1}^{\infty}b_n\sin(nx)$$

It is half range Fourier sine series of x in $[0,\pi]$

$$\therefore \qquad b_n=\frac{2}{\pi}\int_0^{\pi}x\sin(nx)\,dx$$

$$=\frac{2}{\pi}\left[x\left(-\frac{1}{n}\cos nx\right)-\left(-\frac{1}{n^2}\sin nx\right)\right]_0^{\pi}$$

$$=-\frac{2}{n}\cos n\pi=\frac{2}{n}(-1)^{n+1}$$

$$\therefore \qquad u(x,t)=2\sum_{n=1}^{\infty}\frac{(-1)^{n+1}}{n}\cos(nat)\sin(nx)$$

Example 4.54: An elastic string of length l which is fastened at its ends $x=0$ and $x=l$ is released from its horizontal position (zero initial displacement) with initial velocity $g(x)$ given as

$$g(x)=x;\ 0\le x\le \frac{l}{3}$$
$$=0;\ \frac{l}{3}<x<l$$

Find the displacement of the string at any instant of time.

Solution: Let $y(x,t)$ be displacement at distant x from end $x=0$ at time t and c^2 is diffusivity of string then $\frac{\partial^2 y}{\partial t^2}=c^2\frac{\partial^2 y}{\partial x^2}$.

As ends $x=0$, $x=l$ are fastened

$$\therefore \quad y(0,t)=y(l,t)=0$$

$\therefore$ solution of wave equation is

$$y(x,t)=\sum_{n=1}^{\infty}\left[b_n\cos\frac{n\pi ct}{l}+e_n\sin\frac{n\pi ct}{l}\right]\sin\frac{n\pi x}{l}$$

$$y(x,0)=0=\sum_{n=1}^{\infty}b_n\sin\frac{n\pi x}{l}$$

$$\Rightarrow \quad b_n=0;\ n=1,2,3,\ldots$$

$$\therefore \quad y(x,t)=\sum_{n=1}^{\infty}e_n\sin\frac{n\pi ct}{l}\sin\frac{n\pi x}{l}$$

$$\frac{\partial y}{\partial t}=\sum_{n=1}^{\infty}\frac{n\pi c}{l}e_n\cos\frac{n\pi ct}{l}\sin\frac{n\pi x}{l}$$

$$\therefore \quad \left(\frac{\partial y}{\partial t}\right)_{t=0}=g(x)=\sum_{n=1}^{\infty}\frac{n\pi c}{l}e_n\sin\frac{n\pi x}{l}$$

It is half range Fourier sine series of $g(x)$ in $[0, l]$.

$$\therefore \quad \frac{n\pi c}{l}e_n=\frac{2}{l}\int_0^l g(x)\sin\frac{n\pi x}{l}dx$$

$$=\frac{2}{l}\int_0^{\frac{l}{3}}x\sin\frac{n\pi x}{l}dx \qquad \text{(by definition of } g(x)\text{)}$$

$$=\frac{2}{l}\left[x\left(-\frac{l}{n\pi}\cos\frac{n\pi x}{l}\right)-\left(-\frac{l^2}{n^2\pi^2}\sin\frac{n\pi x}{l}\right)\right]_0^{\frac{l}{3}}$$

$$=\frac{2}{l}\left[-\frac{l^2}{3n\pi}\cos\frac{n\pi}{3}+\frac{l^2}{n^2\pi^2}\sin\frac{n\pi}{3}\right]$$

$$\therefore \qquad e_n = \frac{2l^2}{\pi^2 c}\left[\frac{1}{n^3\pi}\sin\frac{n\pi}{3} - \frac{1}{3n^2}\cos\frac{n\pi}{3}\right]$$

$$\therefore \qquad y(x,t) = \frac{2l^2}{\pi^2 c}\sum_{n=1}^{\infty}\left(\frac{1}{n^3\pi}\sin\frac{n\pi}{3} - \frac{1}{3n^2}\cos\frac{n\pi}{3}\right)\sin\frac{n\pi ct}{l}\sin\frac{n\pi x}{l}$$

Example 4.55: Solve the vibrating string problem with

(i) $y(0,t) = 0, y(L,t) = 0$

(ii) $y(x,0) = \begin{cases} x \quad ; \; 0 < x < \dfrac{L}{2} \\ L - x; \; \dfrac{L}{2} < x < L \end{cases}$

(iii) $y_t(x,0) = x(x-L); 0 < x < L$

Solution: As

$$y(0,t) = y(L,t) = 0$$

$\therefore$ wave equation $\dfrac{\partial^2 y}{\partial t^2} = c^2\dfrac{\partial^2 y}{\partial x^2}$ where c^2 is diffusivity of string has solution

$$y(x,t) = \sum_{n=1}^{\infty}\left[b_n\cos\frac{n\pi ct}{L} + e_n\sin\frac{n\pi ct}{L}\right]\sin\frac{n\pi x}{L}$$

$$y_t(x,t) = \sum_{n=1}^{\infty}\frac{n\pi c}{L}\left(-b_n\sin\frac{n\pi ct}{L} + e_n\cos\frac{n\pi ct}{L}\right)\sin\frac{n\pi x}{L}$$

$$\therefore \qquad y_t(x,0) = x(x-L) = \sum_{n=1}^{\infty}\frac{n\pi c}{L}e_n\sin\frac{n\pi x}{L}$$

It is half range Fourier sine series of $x\,(x-L)$ in $0 < x < L$

$$\therefore \qquad \frac{n\pi c}{L}e_n = \frac{2}{L}\int_0^L (x^2 - Lx)\sin\frac{n\pi x}{L}\,dx$$

$$\therefore \qquad e_n = \frac{2}{n\pi c}\left[(x^2 - Lx)\left(-\frac{L}{n\pi}\cos\frac{n\pi x}{L}\right) - (2x - L)\left(-\frac{L^2}{n^2\pi^2}\sin\frac{n\pi x}{L}\right) + 2\left(\frac{L^3}{n^3\pi^3}\cos\frac{n\pi x}{L}\right)\right]_0^L$$

$$= \frac{2}{n\pi c}\left[\frac{2L^3}{n^3\pi^3}\left((-1)^n - 1\right)\right]$$

$$\therefore \qquad e_{2n} = 0,\; e_{2n-1} = -\frac{8L^3}{(2n-1)^4 c\pi^4};\; n = 1,2,3,\ldots\ldots$$

$$y(x,0)=\sum_{n=1}^{\infty} b_n \sin\frac{n\pi x}{L};\ 0<x<\frac{L}{2},\ \frac{L}{2}<x<L$$

It is half range Fourier sine series of $y\,(x,\,0)$ in $0<x<L$

$$\therefore \qquad b_n=\frac{2}{L}\int_0^L y(x,0)\sin\frac{n\pi x}{L}\,dx$$

$$=\frac{2}{L}\left[\int_0^{L/2} x\sin\frac{n\pi x}{L}\,dx+\int_{L/2}^{L}(L-x)\sin\frac{n\pi x}{L}\,dx\right] \qquad \text{(by definition of } y(x,0))$$

$$=\frac{2}{L}\left[\left\{x\left(-\frac{L}{n\pi}\cos\frac{n\pi x}{L}\right)-\left(-\frac{L^2}{n^2\pi^2}\sin\frac{n\pi x}{L}\right)\right\}_0^{L/2}+\left\{(L-x)\left(-\frac{L}{n\pi}\cos\frac{n\pi x}{L}\right)+\left(-\frac{L^2}{n^2\pi^2}\sin\frac{n\pi x}{L}\right)\right\}_{L/2}^{L}\right]$$

$$=\frac{2}{L}\left[-\frac{L^2}{2n\pi}\cos\frac{n\pi}{2}+\frac{L^2}{n^2\pi^2}\sin\frac{n\pi}{2}+\frac{L^2}{2n\pi}\cos\frac{n\pi}{2}+\frac{L^2}{n^2\pi^2}\sin\frac{n\pi}{2}\right]$$

$$=\frac{4L}{n^2\pi^2}\sin\frac{n\pi}{2}$$

$$\therefore \qquad b_{2n}=0,\ b_{2n-1}=\frac{4L(-1)^{n+1}}{\pi^2(2n-1)^2};n=1,2,3,\ldots$$

$$\therefore \quad y(x,t)=\frac{4L}{\pi^2}\sum_{n=1}^{\infty}\left[\frac{(-1)^{n+1}}{(2n-1)^2}\cos\frac{(2n-1)\pi ct}{L}-\frac{2L^2}{c\pi^2(2n-1)^4}\sin\frac{(2n-1)\pi ct}{L}\right]\sin\frac{(2n-1)\pi x}{L}$$

Example 4.56: Solve the differential equation $\frac{\partial^2 u}{\partial t^2}=4\frac{\partial^2 u}{\partial x^2}$ subject to the conditions $u=\sin t$ at $x=0$ and $\frac{\partial u}{\partial x}=\sin t$ at $x=0$.

Solution: Since boundary conditions at $x=0$ are trigonometric functions, so solution of differential equation $\frac{\partial^2 u}{\partial t^2}=4\frac{\partial^2 u}{\partial x^2}$ is

$$u(x,t)=(A\cos px+B\sin px)(E\cos(2pt)+F\sin(2pt))$$

$$u(0,t)=\sin t=A[E\cos(2pt)+F\sin(2pt)]$$

$$\therefore \quad AE = 0, AF = 1, p = \frac{1}{2}$$

$$\Rightarrow \quad E = 0, AF = 1, p = \frac{1}{2}$$

$$\therefore \quad u(x,t) = \cos\frac{x}{2}\sin t + (BF)\sin\frac{x}{2}\sin t$$

$$\frac{\partial u}{\partial x} = -\frac{1}{2}\sin\left(\frac{1}{2}x\right)\sin t + \frac{1}{2}(BF)\cos\frac{x}{2}\sin t$$

$$\therefore \quad \left(\frac{\partial u}{\partial x}\right)_{x=0} = \sin t = \frac{1}{2}(BF)\sin t$$

$$\therefore \quad BF = 2$$

$$\therefore \quad u(x,t) = \cos\frac{x}{2}\sin t + 2\sin\frac{x}{2}\sin t$$

$$= \sin t\left(\cos\frac{x}{2} + 2\sin\frac{x}{2}\right)$$

Example 4.57: Find the D'Alembert solution of a one dimensional wave equation. Use it to find the deflection of a vibrating string of unit length having fixed ends with initial velocity zero and initial deflection $f(x) = k(\sin x - \sin 2x)$.

Solution: One dimensional wave equation is $\frac{\partial^2 y}{\partial t^2} = c^2\frac{\partial^2 y}{\partial x^2}$.

In the theory, we have already found its general solution

$$y(x,t) = \phi(x+ct) + \psi(x-ct)$$

where ϕ and ψ are arbitrary functions

Initial deflection $= y(x,0) = \phi(x) + \psi(x) = f(x) = k(\sin x - \sin 2x)$ (1)

$$\frac{\partial y}{\partial t} = c\phi'(x+ct) - c\psi'(x-ct)$$

$\therefore$ initial velocity $= \left(\frac{\partial y}{\partial t}\right)_{t=0} = c(\phi'(x) - \psi'(x)) = 0$

which on integration gives

$$\phi(x) - \psi(x) = \text{constant} = k_1 \quad (2)$$

From (1) and (2)

$$\phi(x) = \frac{k}{2}(\sin x - \sin 2x) + \frac{1}{2}k_1$$

$$\psi(x) = \frac{k}{2}(\sin x - \sin 2x) - \frac{1}{2}k_1$$

$$\therefore \quad y(x,t) = \phi(x+ct) + \psi(x-ct)$$

$$= \frac{k}{2}\left[\sin(x+ct) - \sin 2(x+ct)\right] + \frac{k_1}{2} + \frac{k}{2}\left[\sin(x-ct) - \sin 2(x-ct)\right] - \frac{k_1}{2}$$

$$= \frac{k}{2}\left[\{\sin(x+ct) + \sin(x-ct)\} - \{\sin 2(x+ct) + \sin 2(x-ct)\}\right]$$

$$= k\left[\sin x \cos ct - \sin 2x \cos 2ct\right]$$

Example 4.58: Use D'Alembert's solution to find the solution of the initial value problem defining the vibrations of an infinitely long elastic string

$$\frac{\partial^2 u}{\partial t^2} = c^2 \frac{\partial^2 u}{\partial x^2}, \ -\infty < x < \infty, \ t > 0$$

$$u(x,0) = f(x), \frac{\partial u}{\partial t}(x,0) = g(x)$$

where

(i) $f(x) = \sin 2x, \ g(x) = \cos 2x$

(ii) $f(x) = a\sin^2 \pi x, \ g(x) = 0$

Solution:

$$\frac{\partial^2 u}{\partial t^2} = c^2 \frac{\partial^2 u}{\partial x^2}$$

By D'Alembert method of solution, its general solution is

$$u(x,t) = \phi(x+ct) + \psi(x-ct) \tag{1}$$

$$u(x,0) = \phi(x) + \psi(x) = f(x) \tag{2}$$

$$\frac{\partial u}{\partial t} = c\phi'(x+ct) - c\psi'(x-ct)$$

$$\therefore \quad \frac{\partial u}{\partial t}(x,0) = c\left(\phi'(x) - \psi'(x)\right) = g(x) \tag{3}$$

(i) $f(x) = \sin 2x, \ g(x) = \cos 2x$

$\therefore$ from (2) and (3)

$$\phi(x) + \psi(x) = \sin 2x \tag{4}$$

$$\phi'(x) - \psi'(x) = \frac{1}{c}\cos 2x$$

which on integration gives

$$\phi(x) - \psi(x) = \frac{1}{2c}\sin 2x + k \tag{5}$$

Solving (4) and (5) we have

$$\phi(x) = \frac{1}{2}\left(1 + \frac{1}{2c}\right)\sin 2x + \frac{1}{2}k$$

$$\psi(x) = \frac{1}{2}\left(1 - \frac{1}{2c}\right)\sin 2x - \frac{1}{2}k$$

$\therefore$ from (1)

$$u(x,t)=\left[\frac{1}{2}\left(1+\frac{1}{2c}\right)\sin 2(x+ct)+\frac{1}{2}k\right]+\left[\frac{1}{2}\left(1-\frac{1}{2c}\right)\sin 2(x-ct)-\frac{1}{2}k\right]$$

$$=\frac{1}{2}\left[\sin 2(x+ct)+\sin 2(x-ct)+\frac{1}{2c}\{\sin 2(x+ct)-\sin 2(x-ct)\}\right]$$

$$=\frac{1}{2}\left[2\sin 2x\cos 2ct+\frac{1}{c}\cos 2x\sin 2ct\right]$$

$$=\sin 2x\cos 2ct+\frac{1}{2c}\cos 2x\sin 2ct$$

(ii) $f(x)=a\sin^2\pi x,\ g(x)=0$

$\therefore$ from (2) and (3)

$$\phi(x)+\psi(x)=a\sin^2\pi x \tag{6}$$

$$\phi'(x)-\psi'(x)=0$$

which on integration gives

$$\phi(x)-\psi(x)=k \tag{7}$$

Solving (6) and (7), we have

$$\phi(x)=\frac{1}{2}a\sin^2\pi x+\frac{1}{2}k$$

$$\psi(x)=\frac{1}{2}a\sin^2\pi x-\frac{1}{2}k$$

$\therefore$ from (1)

$$u(x,t)=\phi(x+ct)+\psi(x-ct)$$

$$=\left[\frac{1}{2}a\sin^2\pi(x+ct)+\frac{1}{2}k\right]+\left[\frac{1}{2}a\sin^2\pi(x-ct)-\frac{1}{2}k\right]$$

$$=\frac{a}{2}\left[\sin^2\pi(x+ct)+\sin^2\pi(x-ct)\right]$$

$$=\frac{a}{2}\left[\frac{1-\cos 2\pi(x+ct)}{2}+\frac{1-\cos 2\pi(x-ct)}{2}\right]$$

$$=\frac{a}{2}\left[1-\frac{1}{2}\{\cos 2\pi(x+ct)+\cos 2\pi(x-ct)\}\right]$$

$$=\frac{a}{2}[1-\cos 2\pi x\cos 2\pi ct]$$

4.12 ONE DIMENSIONAL HEAT FLOW

Consider the flow of heat by conduction in a uniform bar. We assume that the sides of the bar are insulated and the loss of heat from the sides by conduction or radiation is negligible.

Take one end of the bar as origin and x-axis along the direction of flow of heat. Let $u(x,t)$ be temperature at a point distance x from origin at time t and the temperature at all points of a cross section is same.

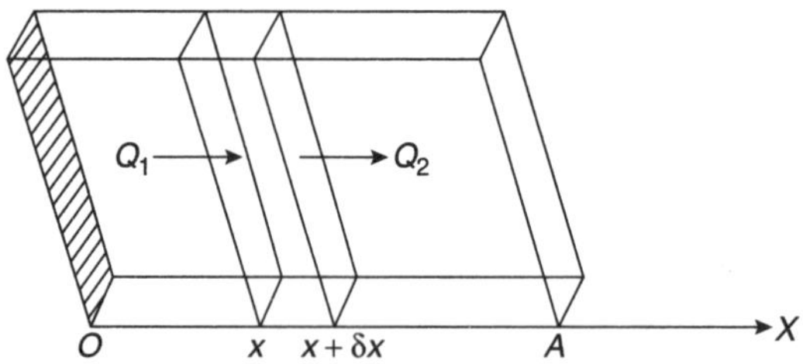

Figure 4.3

We know that in a body heat flows in a direction of decreasing temperature. Physical experiments show that the rate of flow is proportional to $\frac{\partial u}{\partial x}$ and area of cross section A.

Thus, quantity of heat Q_1 flowing into the section at distance x is $-KA\left(\frac{\partial u}{\partial x}\right)_x$ per second, the negative sign indicates that u decreases as x increases and K is constant called thermal conductivity of body.

The quantity of heat Q_2 flowing out at section at distance $x+\delta x$ is $-KA\left(\frac{\partial u}{\partial x}\right)_{x+\delta x}$ per second.

Hence, amount of heat retained by bar of thickness δx at distance x is

$$= Q_1 - Q_2 = KA\left[\left(\frac{\partial u}{\partial x}\right)_{x+\delta x} - \left(\frac{\partial u}{\partial x}\right)_x\right] \text{ per second}$$

But rate of increase of heat in this bar of thickness $\delta x = \sigma\rho A\delta x \frac{\partial u}{\partial t}$

where σ is specific heat and ρ is density of material.

$$\therefore \qquad \sigma\rho A\delta x \frac{\partial u}{\partial t} = KA\left[\left(\frac{\partial u}{\partial x}\right)_{x+\delta x} - \left(\frac{\partial u}{\partial x}\right)_x\right]$$

$$\therefore \qquad \frac{\partial u}{\partial t} = \frac{K}{\sigma\rho}\left[\frac{\left(\frac{\partial u}{\partial x}\right)_{x+\delta x} - \left(\frac{\partial u}{\partial x}\right)_x}{\delta x}\right]$$

Taking limit as $\delta x \to 0$

$$\frac{\partial u}{\partial t} = c^2 \frac{\partial^2 u}{\partial x^2} \tag{4.51}$$

where $c^2 = \dfrac{K}{\sigma\rho}$ is called thermal diffusivity.

Equation (4.51) is called heat equation in one dimension.

4.12.1 Solution of the Heat Equation

Heat equation is

$$\frac{\partial u}{\partial t} = c^2 \frac{\partial^2 u}{\partial x^2}$$

Let
$$u(x,t) = X(x)T(t)$$

$\therefore$
$$\frac{\partial u}{\partial t} = XT' \quad \text{and} \quad \frac{\partial^2 u}{\partial x^2} = X''T$$

$\therefore$ differential equation becomes

$$XT' = c^2 X''T$$

$\therefore$
$$\frac{X''}{X} = \frac{1}{c^2}\frac{T'}{T}$$

L.H.S. is function of x and R.H.S. is function of t and hence both are constant say λ.

If $\lambda = 0$, then $X'' = 0$, $T' = 0$

$\therefore$
$$X = Ax + B,\ T = C$$

$\therefore$
$$u(x,t) = C(Ax + B) = A_1 x + B_1$$

If $\lambda > 0$, i.e., $\lambda = p^2$, $p > 0$ then $X'' - p^2 X = 0$, $T' = c^2 p^2 T$

$\therefore$
$$X = \left(Ae^{px} + Be^{-px}\right),\ T = C\,e^{c^2 p^2 t}$$

$\therefore$
$$u(x,t) = \left(Ae^{px} + Be^{-px}\right)C\,e^{p^2c^2t} = \left(A_1e^{px} + B_1e^{-px}\right)e^{p^2c^2t}$$

Similarly, if $\lambda < 0$, i.e., $\lambda = -p^2$, $p > 0$ then solution is $u(x,t) = e^{-p^2c^2t}\left(A_1 \cos px + B_1 \sin px\right)$.

Of these solutions, we are to choose solutions satisfying initial and boundary conditions.

The solution $Ax + B$ is the solution of $\dfrac{\partial^2 u}{\partial x^2} = 0$ and hence $\dfrac{\partial u}{\partial t} = 0$ in heat equation.

Thus, solution $Ax + B$ is a steady-state solution.

Case I: If $u(0,t) = u(l,t) = 0$ then there is no steady-state solution and $u(x,t)$ decreases as time increases and hence in this case solution will be

$$u(x,t) = e^{-p^2c^2t}\left(A_1 \cos px + B_1 \sin px\right)$$

$$u(0,t) = e^{-p^2c^2t}A_1 = 0$$

$\Rightarrow$
$$A_1 = 0$$

$\therefore$
$$u(x,t) = e^{-p^2c^2t}B_1 \sin px$$

$$u(l,t) = e^{-p^2c^2t}B_1 \sin pl = 0$$

$$\Rightarrow \qquad p = \frac{n\pi}{l};\ n \in N$$

$$\therefore \qquad u(x,t) = b_n \sin\frac{n\pi x}{l} e^{\frac{-n^2\pi^2c^2t}{l^2}};\ n \in N$$

By principle of superposition

$$u(x,t) = \sum_{n=1}^{\infty} b_n \sin\frac{n\pi x}{l} e^{\frac{-n^2\pi^2c^2t}{l^2}}$$

Case II: If either $u(0,t)$ or $u(l,t)$ or both are non-zero and given, then $u(x,t)$ will consist of steady-state solution $u_s(x) = Ax + B$ and transient-state solution $u_t(x,t)$.

$$\therefore \qquad u(x,t) = u_s(x) + u_t(x,t)$$

$$= Ax + B + \sum_{n=1}^{\infty} b_n \sin\frac{n\pi x}{l} e^{\frac{-n^2\pi^2c^2t}{l^2}}$$

Case III: If at least one of $y(0,t)$ and $y(l,t)$ is not given, then

$$u(x,t) = Ax + B + (A_1 \cos px + B_1 \sin px)e^{-p^2c^2t}$$

Example 4.59: A rod of length l with insulated sides is initially at a uniform temperature u_0. Its ends are suddenly cooled to $0°\text{C}$ and are kept at that temperature. Prove that the temperature function $u(x,t)$ is given by $u(x,t) = \sum_{n=1}^{\infty} b_n \sin\frac{n\pi x}{l} e^{-\frac{c^2\pi^2n^2}{l^2}t}$ where b_n is determined from $u(x,0) = u_0$. Find the value of b_n.

Solution:

The temperature function $u(x,t)$ satisfies heat equation $c^2\frac{\partial^2 u}{\partial x^2} = \frac{\partial u}{\partial t}$ under the conditions $u(x,0) = u_0$, $u(0,t) = 0$, $u(l,t) = 0$.

Let $\qquad u(x,t) = X(x)T(t)$

$$\therefore \qquad \frac{\partial^2 u}{\partial x^2} = X''T,\ \frac{\partial u}{\partial t} = XT';$$

where dashes denote derivatives with respect to their variables.

$\therefore$ differential equation becomes

$$c^2X''T = XT',\ X(x)T(0) = u_0,\ X(0) = 0,\ X(l) = 0$$

$$\therefore \qquad \frac{c^2X''}{X} = \frac{T'}{T}$$

L.H.S. is function of x and R.H.S. is function of t and hence both are constant.

$$\therefore \qquad \frac{c^2X''}{X} = \frac{T'}{T} = \lambda$$

$$\therefore \qquad c^2X'' - \lambda X = 0,\ T' = \lambda T$$

Case I: $\lambda = 0$, then $c^2X'' = 0$

$$\therefore \quad X'' = 0$$

$$\Rightarrow \quad X = Ax + B$$

$$X(0) = B = 0$$

then $\quad X(l) = Al = 0 \quad \Rightarrow A = 0$

$\therefore \; X = 0$ for all t which is not valid

Case II: $\lambda > 0$. Let $\lambda = p^2,\; p > 0$,

then $$c^2X'' - \lambda X = 0$$

$$\Rightarrow \quad c^2X'' - p^2X = 0$$

Its solution is $$X(x) = Ae^{\frac{-p}{c}x} + Be^{\frac{p}{c}x}$$

$$X(0) = A + B = 0 \qquad \therefore \; B = -A$$

$$\therefore \quad X(x) = A\left(e^{\frac{-p}{c}x} - e^{\frac{p}{c}x}\right)$$

$$X(l) = A\left(e^{\frac{-p}{c}l} - e^{\frac{p}{c}l}\right) = 0 \qquad \Rightarrow A = 0$$

$\therefore A = B = 0$ which is invalid

Case III: $\lambda < 0$. Let $\lambda = -p^2;\; p > 0$,
then solutions of $c^2X'' + p^2X = 0$ and $T' = -p^2T$ are

$$X = A\cos\frac{px}{c} + B\sin\frac{px}{c},\; T = Ce^{-p^2t}$$

$$\therefore \quad u(x,t) = e^{-p^2t}\left[c_1\cos\frac{px}{c} + c_2\sin\frac{px}{c}\right] \qquad \text{where } AC = c_1,\; BC = c_2$$

$$u(0,t) = e^{-p^2t}c_1 = 0 \qquad \Rightarrow c_1 = 0$$

$$\therefore \quad u(x,t) = c_2e^{-p^2t}\sin\frac{px}{c}$$

$$u(l,t) = c_2e^{-p^2t}\sin\frac{pl}{c} = 0$$

$$\therefore \quad \frac{pl}{c} = n\pi;\; n = 1,2,3\ldots$$

$$\therefore \quad p = \frac{cn\pi}{l};\; n = 1,2,3\ldots$$

$$\therefore \qquad u(x,t) = b_n e^{-\frac{c^2 n^2 \pi^2}{l^2} t} \sin \frac{n\pi x}{l};\ n = 1,2,3\ldots$$

$\therefore$ By principle of superposition, solution is

$$u(x,t) = \sum_{n=1}^{\infty} b_n \sin \frac{n\pi x}{l} e^{-\frac{c^2 n^2 \pi^2}{l^2} t}$$

where b_n are determined from $u(x,0) = u_0 = \sum_{n=1}^{\infty} b_n \sin \frac{n\pi x}{l};\ 0 \le x \le l$

This is Fourier half range sine series of u_0 in $[0,l]$, hence b_n are Fourier coefficients given by

$$b_n = \frac{2}{l}\int_0^l u(x,0)\sin \frac{n\pi x}{l}dx = \frac{2u_0}{l}\int_0^l \sin \frac{n\pi x}{l}dx$$

$$= \frac{2u_0}{l}\cdot\frac{l}{n\pi}\left(-\cos\frac{n\pi x}{l}\right)_0^l = \frac{2u_0}{n\pi}\left(1-(-1)^n\right) \qquad \left(\because \cos n\pi = (-1)^n\right)$$

$$\therefore \qquad b_{2n} = 0,\ b_{2n-1} = \frac{4u_0}{(2n-1)\pi};\ n = 1,2,3,\ldots$$

$\therefore$ solution is

$$u(x,t) = \frac{4u_0}{\pi}\sum_{n=1}^{\infty}\frac{1}{2n-1}\sin\frac{(2n-1)\pi x}{l} e^{-\frac{c^2 (2n-1)^2 \pi^2 t}{l^2}}$$

Example 4.60: The equation for heat conduction along a bar of length l is $\frac{\partial u}{\partial t} = a^2 \frac{\partial^2 u}{\partial x^2}$ neglecting the radiation. Find an expression for $u(x,t)$ if the ends of the bar are maintained at zero temperature and if initially the temperature is T at the centre of the bar and falls uniformally to zero with time.

Solution: Ends of the bar are maintained at zero temperature and hence there is no steady-state solution

$$\therefore \qquad u(x,t) = \sum_{n=1}^{\infty} b_n \sin \frac{n\pi x}{l} e^{-\frac{n^2 \pi^2 a^2 t}{l^2}}$$

Initially, $\qquad u(x,0) = Ax + B$

$$u(0,0) = B = 0$$

$$\therefore \qquad u\left(\frac{l}{2},0\right) = A\frac{l}{2} + B = A\frac{l}{2} = T$$

$$\therefore \qquad A = \frac{2T}{l}$$

$$\therefore \quad u(x,0)=\frac{2T}{l}x;\ 0\le x\le\frac{l}{2}$$

for $\frac{l}{2}\le x\le l,$

$$u\left(\frac{l}{2},0\right)=T$$

$$u(l,0)=0$$

$$\therefore \quad A\frac{l}{2}+B=T$$

$$Al+B=0$$

$$\therefore \quad A=\frac{-2T}{l},B=2T$$

$$\therefore \quad u(x,0)=\frac{2T}{l}x;\ 0\le x\le\frac{l}{2}$$

$$=\frac{-2T}{l}x+2T;\ \frac{l}{2}\le x\le l$$

$$u(x,0)=\sum_{n=1}^{\infty}b_n\sin\frac{n\pi x}{l}$$

It is half range Fourier sine series of $u(x,0)$ in $[0,l]$.

$$\therefore \quad b_n=\frac{2}{l}\int_0^l u(x,0)\sin\frac{n\pi x}{l}dx$$

$$=\frac{2}{l}\left[\int_0^{l/2}\frac{2T}{l}x\sin\frac{n\pi x}{l}dx+\int_{l/2}^{l}\frac{2T}{l}(l-x)\sin\frac{n\pi x}{l}dx\right]$$

$$=\frac{4T}{l^2}\left[\left\{x\left(-\frac{l}{n\pi}\cos\frac{n\pi x}{l}\right)-\left(-\frac{l^2}{n^2\pi^2}\sin\frac{n\pi x}{l}\right)\right\}_0^{l/2}\right.$$

$$\left.+\left\{(l-x)\left(-\frac{l}{n\pi}\cos\frac{n\pi x}{l}\right)+\left(-\frac{l^2}{n^2\pi^2}\sin\frac{n\pi x}{l}\right)\right\}_{l/2}^{l}\right]$$

$$=\frac{4T}{l^2}\left[-\frac{l^2}{2n\pi}\cos\frac{n\pi}{2}+\frac{l^2}{n^2\pi^2}\sin\frac{n\pi}{2}+\frac{l^2}{2n\pi}\cos\frac{n\pi}{2}+\frac{l^2}{n^2\pi^2}\sin\frac{n\pi}{2}\right]$$

$$=\frac{8T}{n^2\pi^2}\sin\frac{n\pi}{2}$$

$$\therefore \quad b_{2n}=0,\ b_{2n-1}=\frac{8T(-1)^{n+1}}{\pi^2(2n-1)^2};\ n=1,2,3,\ldots$$

∴ solution is

$$u(x,t)=\frac{8T}{\pi^2}\sum_{n=1}^{\infty}\frac{(-1)^{n+1}}{(2n-1)^2}\sin\frac{(2n-1)\pi x}{l}e^{-\frac{(2n-1)^2\pi^2a^2t}{l^2}}$$

Example 4.61: (a) An insulated rod of length *l* has its ends *A* and *B* maintained at $0°$ and $100°$C respectively until steady state conditions prevail. If *B* is suddenly reduced to $0°$C and maintained at $0°$C, find the temperature at a distance *x* from *A* at time *t*.

(b) Find also the temperature if the change consists of raising the temperature of *A* to $20°$C and reducing that of *B* to $80°$C.

Solution: (a) Let $u(x,t)$ be temperature at distance *x* from *A* at time *t* then $\frac{\partial u}{\partial t}=c^2\frac{\partial^2 u}{\partial x^2}$ where c^2 is diffusivity of rod. As

$$u(0,t)=u(l,t)=0$$

∴ solution is

$$u(x,t)=\sum_{n=1}^{\infty}b_n\sin\frac{n\pi x}{l}e^{-\frac{n^2\pi^2c^2t}{l^2}}$$

$$u(x,0)=Ax+B;\ u(0,0)=0,\ u(l,0)=100$$

∴ $$B=0,\ A=\frac{100}{l}$$

∴ $$u(x,0)=\frac{100x}{l}$$

$$u(x,0)=\frac{100x}{l}=\sum_{n=1}^{\infty}b_n\sin\frac{n\pi x}{l}$$

It is half range Fourier sine series of $\frac{100x}{l}$ in $[0,l]$.

∴ $$b_n=\frac{2}{l}\int_0^l\frac{100x}{l}\sin\frac{n\pi x}{l}dx$$

$$=\frac{200}{l^2}\left[x\left(-\frac{l}{n\pi}\cos\frac{n\pi x}{l}\right)-\left(-\frac{l^2}{n^2\pi^2}\sin\frac{n\pi x}{l}\right)\right]_0^l$$

$$=\frac{200}{l^2}\left[-\frac{l^2}{n\pi}(-1)^n\right]=\frac{200}{n\pi}(-1)^{n+1}$$

∴ solution is

$$u(x,t)=\frac{200}{\pi}\sum_{n=1}^{\infty}\frac{(-1)^{n+1}}{n}\sin\frac{n\pi x}{l}e^{-\frac{n^2\pi^2c^2t}{l^2}}$$

(b) Here, temperatures at end points are not zero.

∴ $$u(x,t)=u_s(x)+u_t(x,t)$$

where $$u_s(x) = Ax + B$$

$$u_s(0) = B = 20$$

$$u_s(l) = Al + B = 80$$

$$\therefore \quad A = \frac{60}{l}, B = 20$$

$$\therefore \quad u_s(x) = \frac{60x}{l} + 20$$

$$\therefore \quad u(x,t) = \frac{60x}{l} + 20 + \sum_{n=1}^{\infty} b_n \sin\frac{n\pi x}{l} e^{-\frac{n^2\pi^2c^2t}{l^2}}$$

$$u(x,0) = \frac{100x}{l} \qquad \text{(from part (a))}$$

$$\therefore \quad \frac{100x}{l} = \frac{60x}{l} + 20 + \sum_{n=1}^{\infty} b_n \sin\frac{n\pi x}{l}$$

$$\therefore \quad \frac{40x}{l} - 20 = \sum_{n=1}^{\infty} b_n \sin\frac{n\pi x}{l}$$

It is half range Fourier sine series of $\frac{40x}{l} - 20$ in $[0,l]$.

$$\therefore \quad b_n = \frac{2}{l}\int_0^l \left(\frac{40x}{l} - 20\right)\sin\frac{n\pi x}{l}\,dx$$

$$= \frac{40}{l^2}\left[(2x-l)\left(-\frac{l}{n\pi}\cos\frac{n\pi x}{l}\right) - 2\left(-\frac{l^2}{n^2\pi^2}\sin\frac{n\pi x}{l}\right)\right]_0^l$$

$$= \frac{40}{l^2}\left[-\frac{l^2}{n\pi}(-1)^n - \frac{l^2}{n\pi}\right]$$

$$= -\frac{40}{n\pi}\left(1+(-1)^n\right)$$

$$\therefore \quad b_{2n} = -\frac{40}{n\pi}, \; b_{2n-1} = 0; \; n = 1, 2, 3, \ldots$$

∴ solution is

$$u(x,t) = \frac{60x}{l} + 20 - \frac{40}{\pi}\sum_{n=1}^{\infty} \frac{1}{n} \sin\frac{2n\pi x}{l} e^{-\frac{4n^2\pi^2c^2t}{l^2}}$$

Example 4.62: A bar of length l with insulated sides is initially at 0°C temperature throughout. The end $x=0$ is kept at 0° C for all time and heat is suddenly applied such that $\frac{\partial u}{\partial x}=10$ at $x=l$ for all time. Find the temperature function $u(x,t)$.

Solution: Let $u(x,t)$ be temperature at distance x from end at 0° C at time t

then, $$\frac{\partial u}{\partial t}=c^2\frac{\partial^2 u}{\partial x^2}$$

where c^2 is diffusivity of bar.

As both ends are not at 0° C

$$\therefore \quad u(x,t)=u_s(x)+u_t(x,t)$$

$$u_s(x)=Ax+B$$

$$u_s(0)=B=0$$

$$\left(\frac{\partial u_s}{\partial x}\right)_{x=l}=A=10$$

$$\therefore \quad u_s(x)=10x$$

$$\therefore \quad u(x,t)=10x+(c_1\cos px+c_2\sin px)e^{-p^2c^2t}$$

($\because$ temperature at end $x=l$ is not given)

$$u(0,t)=c_1e^{-p^2c^2t}=0$$

$$\Rightarrow \quad c_1=0$$

$$\therefore \quad u(x,t)=10x+c_2\sin px\, e^{-p^2c^2t}$$

$$\frac{\partial u}{\partial x}=10+pc_2\cos px\, e^{-p^2c^2t}$$

$$\left(\frac{\partial u}{\partial x}\right)_{(l,t)}=10=10+pc_2\cos(pl)e^{-p^2c^2t}$$

$$\therefore \quad \cos pl=0$$

$$\Rightarrow \quad p=\frac{(2n-1)\pi}{2l};\ n\in N$$

$\therefore$ By principle of superposition

$$u(x,t)=10x+\sum_{n=1}^{\infty}c_n\sin\frac{(2n-1)\pi x}{2l}e^{-\frac{(2n-1)^2\pi^2c^2t}{4l^2}}$$

$$u(x,0)=0,\text{ for all }x$$

$$\therefore \qquad 0 = 10x + \sum_{n=1}^{\infty} c_n \sin\frac{(2n-1)\pi x}{2l}$$

$$\text{or} \qquad -10x = \sum_{n=1}^{\infty} c_n \sin\frac{(2n-1)\pi x}{2l}$$

It is half range Fourier sine series of $-10x$ in $[0, l]$.

$$\therefore \qquad c_n = \frac{2}{l}\int_0^l (-10x)\sin\frac{(2n-1)\pi x}{2l}\,dx$$

$$= \frac{-20}{l}\left[x\left(-\frac{2l}{(2n-1)\pi}\cos\frac{(2n-1)\pi x}{2l}\right) - \left(-\frac{4l^2}{(2n-1)^2\pi^2}\sin\frac{(2n-1)\pi x}{2l}\right)\right]_0^l$$

$$= -\frac{20}{l}\left[\frac{4l^2}{(2n-1)^2\pi^2}\sin\left(n\pi - \frac{\pi}{2}\right)\right]$$

$$= \frac{80l}{(2n-1)^2\pi^2}(-1)^n$$

$\therefore$ solution is

$$u(x,t) = 10x + \frac{80l}{\pi^2}\sum_{n=1}^{\infty}\frac{(-1)^n}{(2n-1)^2}\sin\frac{(2n-1)\pi x}{2l}\,e^{-\frac{(2n-1)^2\pi^2c^2t}{4l^2}}$$

Example 4.63: A bar 100 cm long with insulated sides, has its ends kept at 0°C and 100°C until steady-state conditions prevail. The two ends are then suddenly insulated and kept so. Find the temperature distribution.

Solution: Let AB be the rod of length 100 cm and $u(x,t)$ be temperature at distance x from A at time t.

Heat equation is $c^2\frac{\partial^2 u}{\partial x^2} = \frac{\partial u}{\partial t}$ where c^2 is diffusivity of bar.

Since two ends are insulated

$$\therefore \qquad \left(\frac{\partial u}{\partial x}\right)_{x=0} = \left(\frac{\partial u}{\partial x}\right)_{x=100} = 0$$

$\therefore$ steady-state temp. is a constant say $\frac{a_0}{2}$

$\therefore$ solution is

$$u(x,t) = \frac{a_0}{2} + (A_1\cos px + B_1\sin px)e^{-p^2c^2t}$$

($\because$ temperatures at end points are not given)

$$\text{Now,} \qquad \frac{\partial u}{\partial x} = (-pA_1\sin px + pB_1\cos px)e^{-p^2c^2t}$$

$$\left(\frac{\partial u}{\partial x}\right)_{x=0} = pB_1 e^{-p^2c^2t} = 0$$

$\Rightarrow$ $$B_1 = 0$$

and $$\left(\frac{\partial u}{\partial x}\right)_{x=l} = -pA_1 \sin pl\, e^{-p^2c^2t} = 0 \qquad (\because B_1 = 0)$$

$\therefore$ $$\sin pl = 0$$

$\Rightarrow$ $$p = \frac{n\pi}{l};\ n = 1,2,3,\ldots$$

$\therefore$ $$u(x,t) = \frac{a_0}{2} + a_n \cos\frac{n\pi x}{l} e^{-\frac{n^2\pi^2c^2t}{l^2}};\ n = 1,2,3,\ldots$$

are solutions.

$\therefore$ By principle of superposition, solution is

$$u(x,t) = \frac{a_0}{2} + \sum_{n=1}^{\infty} a_n \cos\frac{n\pi x}{l} e^{-\frac{n^2\pi^2c^2t}{l^2}}$$

Now, $$u(0,0) = 0,\ u(l,0) = 100$$

$\therefore$ $$u(x,0) = \frac{100}{l}x$$

$\therefore$ $$\frac{100x}{l} = \frac{a_0}{2} + \sum_{n=1}^{\infty} a_n \cos\frac{n\pi x}{l}$$

It is half range Fourier cosine series of $\frac{100x}{l}$ in $[0,l]$.

$\therefore$ $$a_0 = \frac{2}{l}\int_0^l \frac{100x}{l}dx = \frac{200}{l^2}\left(\frac{x^2}{2}\right)_0^l = 100$$

$$a_n = \frac{2}{l}\int_0^l \frac{100x}{l}\cos\frac{n\pi x}{l}dx$$

$$= \frac{200}{l^2}\left[x\left(\frac{l}{n\pi}\sin\frac{n\pi x}{l}\right) - \left(-\frac{l^2}{n^2\pi^2}\cos\frac{n\pi x}{l}\right)\right]_0^l$$

$$= \frac{200}{l^2}\cdot\frac{l^2}{n^2\pi^2}\left[(-1)^n - 1\right] = \frac{200}{n^2\pi^2}\left[(-1)^n - 1\right]$$

$\therefore$ $$a_{2n} = 0,\ a_{2n-1} = -\frac{400}{\pi^2(2n-1)^2};\ n = 1,2,3,\ldots$$

$\therefore$ $$u(x,t) = 50 - \frac{400}{\pi^2}\sum_{n=1}^{\infty}\frac{1}{(2n-1)^2}\cos\frac{(2n-1)\pi x}{l} e^{-\frac{(2n-1)^2\pi^2c^2t}{l^2}}$$

But $l = 100$ cm

$$\therefore \quad u(x,t) = 50 - \frac{400}{\pi^2}\sum_{n=1}^{\infty}\frac{1}{(2n-1)^2}\cos\frac{(2n-1)\pi x}{100}\, e^{-\frac{(2n-1)^2\pi^2 c^2 t}{10000}}$$

Example 4.64: Solve $u_t = c^2 u_{xx}$ when

(i) $u \neq \infty$ as $t \to \infty$

(ii) $u_x = 0$ when $x = 0$ for all t

(iii) $u = 0$ when $x = l$ for all t

(iv) $u = u_0 =$ constant when $t = 0$ for all $0 < x < l$.

Solution: Here temperature at both ends is not zero for all t and is also not given at end $x = 0$.

$\therefore$ solution is

$$u(x,t) = ax + b + (c_1 \cos px + c_2 \sin px)e^{-p^2c^2t}$$

For steady-state solution

$$u_s(x) = ax + b$$

$$\frac{\partial u_s}{\partial x} = a = 0 \quad \text{for} \quad x = 0$$

$$\therefore \quad a = 0$$

$$\therefore \quad u_s(x) = b$$

Also, $$u_s(l) = 0$$

$$\Rightarrow \quad b = 0$$

$$\therefore \quad ax + b = 0$$

$$\therefore \quad u(x,t) = (c_1 \cos px + c_2 \sin px)e^{-p^2c^2t}$$

$$\frac{\partial u}{\partial x} = (-pc_1 \sin px + pc_2 \cos px)e^{-p^2c^2t}$$

$$\left(\frac{\partial u}{\partial x}\right)_{x=0} = pc_2 e^{-p^2c^2t} = 0 \text{ for all } t$$

$$\Rightarrow \quad c_2 = 0$$

$$\therefore \quad u(x,t) = c_1 \cos px\, e^{-p^2c^2t}$$

$$\therefore \quad u(l,t) = c_1 \cos pl\, e^{-p^2c^2t} = 0 \text{ for all } t$$

$$\therefore \quad p = \frac{(2n-1)\pi}{2l};\ n = 1, 2, 3, \ldots$$

$$\therefore \quad u(x,t) = c_n \cos\frac{(2n-1)\pi x}{2l}\, e^{-\frac{(2n-1)^2\pi^2c^2t}{4l^2}};\ n = 1, 2, \ldots$$

By principle of superposition

$$u(x,t) = \sum_{n=1}^{\infty} c_n \cos\frac{(2n-1)\pi x}{2l} e^{-\frac{(2n-1)^2\pi^2c^2t}{4l^2}}$$

$$u(x,0) = u_0 = \sum_{n=1}^{\infty} c_n \cos\frac{(2n-1)\pi x}{2l}$$

It is half range Fourier cosine series of u_0 in [0, l].

$$\therefore \qquad c_n = \frac{2}{l}\int_0^l u_0 \cos\frac{(2n-1)\pi x}{2l}\,dx$$

$$= \frac{2u_0}{l}\cdot\frac{2l}{(2n-1)\pi}\left[\sin\frac{(2n-1)\pi x}{2l}\right]_0^l$$

$$= \frac{4u_0}{(2n-1)\pi}(-1)^{n+1}$$

$\therefore$ solution is

$$u(x,t) = \frac{4u_0}{\pi}\sum_{n=1}^{\infty}\frac{(-1)^{n+1}}{2n-1}\cos\frac{(2n-1)\pi x}{2l} e^{-\frac{(2n-1)^2\pi^2c^2t}{4l^2}}$$

4.13 TRANSMISSION LINE EQUATIONS

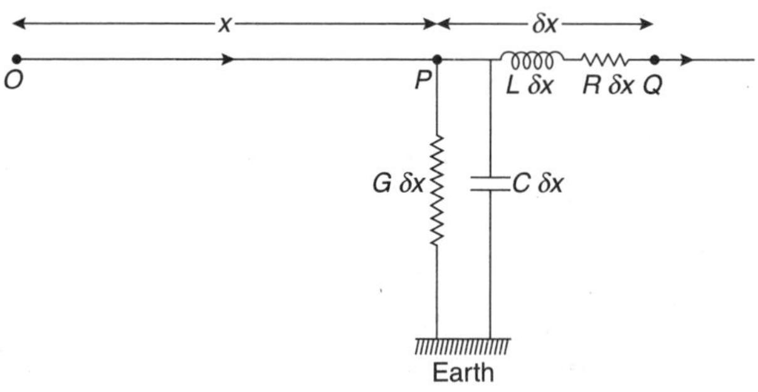

Figure 4.4

Consider the flow of electricity in an insulated cable. Let P, Q be points on the cable at distance x and $x+\delta x$ from starting point O. Let R and L respectively be resistance and inductance established per unit length of the cable. Let C and G be capacitance and leakance respectively to the ground per unit length. We assume that R, L, C, G are constants.

Let I, $I+\delta I$ be currents and V, $V+\delta V$ be potentials at P and Q respectively at time t. Here, δI and δV will be negative.

Potential drop across segment PQ = potential drop due to resistance + potential drop due to inductance

$$\therefore \quad -\delta V = (R\delta x)I + (L\delta x)\frac{\partial I}{\partial t} \tag{4.52}$$

Decrease in current in crossing segment PQ = decrease in current due to leakance + decrease in current due to capacitance

$$\therefore \quad -\delta I = (G\delta x)V + (C\delta x)\frac{\partial V}{\partial t} \tag{4.53}$$

Divide (4.52) and (4.53) by δx and take limit as $\delta x \to 0$

$$-\frac{\partial V}{\partial x} = RI + L\frac{\partial I}{\partial t} \tag{4.54}$$

$$-\frac{\partial I}{\partial x} = GV + C\frac{\partial V}{\partial t} \tag{4.55}$$

We shall find the partial differential equations eliminating V and I.

Differentiate both sides of (4.54) partially w.r.t. x

$$-\frac{\partial^2 V}{\partial x^2} = R\frac{\partial I}{\partial x} + L\frac{\partial^2 I}{\partial x \partial t} = R\frac{\partial I}{\partial x} + L\frac{\partial^2 I}{\partial t \partial x}$$

$$= \left(R + L\frac{\partial}{\partial t}\right)\frac{\partial I}{\partial x}$$

$$= \left(R + L\frac{\partial}{\partial t}\right)\left(-GV - C\frac{\partial V}{\partial t}\right) \qquad \text{(from (4.55))}$$

$$\therefore \quad \frac{\partial^2 V}{\partial x^2} = RGV + RC\frac{\partial V}{\partial t} + LG\frac{\partial V}{\partial t} + LC\frac{\partial^2 V}{\partial t^2}$$

or

$$\frac{\partial^2 V}{\partial x^2} = LC\frac{\partial^2 V}{\partial t^2} + (RC + LG)\frac{\partial V}{\partial t} + RGV \tag{4.56}$$

Similarly differentiating both side of (4.55) partially w.r.t. x and using (4.54) will give

$$\frac{\partial^2 I}{\partial x^2} = LC\frac{\partial^2 I}{\partial t^2} + (RC + LG)\frac{\partial I}{\partial t} + RGI \tag{4.57}$$

Equations (4.56) and (4.57) are called telephone equations.

Remark 4.5:

(i) If L and G are negligible as in telegraph lines then (4.56) and (4.57) become

$$\frac{\partial^2 V}{\partial x^2} = RC\frac{\partial V}{\partial t} \quad \text{and} \quad \frac{\partial^2 I}{\partial x^2} = RC\frac{\partial I}{\partial t}$$

These are telegraph equations. These are similar to one dimensional heat flow.

(ii) If R and G are negligible as in radio then (4.56) and (4.57) become

$$\frac{\partial^2 V}{\partial x^2} = LC\frac{\partial^2 V}{\partial t^2} \text{ and } \frac{\partial^2 I}{\partial x^2} = LC\frac{\partial^2 I}{\partial t^2}$$

These are called radio equations. These are similar to one dimensional wave equations.

Example 4.65: A transmission line 1000 km long is initially under steady-state conditions with potential 1300 volts at the sending end $(x = 0)$ and 1200 volts at the receiving end $(x = 1000)$. The terminal end of the line is suddenly grounded, but the potential at the source is kept at 1300 volts. Assuming that the inductance and leakance to be negligible, find the potential $V(x,t)$.

Solution: As the inductance and leakance are negligible, the line is a telegraph line.

The equation of telegraph line is

$$\frac{\partial^2 V}{\partial x^2} = RC\frac{\partial V}{\partial t} \tag{1}$$

In steady state,

$$\frac{\partial V}{\partial t} = 0$$

$$\therefore \quad \frac{\partial^2 V}{\partial x^2} = 0$$

$\therefore$ steady-state voltage is $V = Ax + B$

But $V = 1300$ when $x = 0$

$V = 1200$ when $x = 1000$

$$\therefore \quad V = 1300 - \frac{x}{10}$$

$$\therefore \quad V(x,0) = 1300 - \frac{x}{10}$$

After grounding the terminal end $x = 1000$ steady-state voltage at $x = 0$ is 1300 volts and steady-state voltage at $x = 1000$ is zero.

$\therefore$ steady-state voltage after grounding

$$V_s(x) = 1300 - 1.3x$$

Now, $V(x,t)$ is sum of steady-state voltage and transient-state voltage.

$\therefore$ solution of (1) considering as one dimensional heat equation is

$$V(x,t) = 1300 - 1.3x + \sum_{n=1}^{\infty} E_n\, e^{-\frac{n^2\pi^2 t}{10^6 RC}} \sin\frac{n\pi x}{1000} \qquad (\because\ l = 1000 \text{ km})$$

$$\therefore \quad 1300 - \frac{x}{10} = V(x,0) = 1300 - 1.3x + \sum_{n=1}^{\infty} E_n \sin\frac{n\pi x}{1000}$$

$$\therefore \qquad 1.2x = \sum_{n=1}^{\infty} E_n \sin\frac{n\pi x}{1000};\ 0 \le x \le 1000$$

It is Fourier half range sine series in $[0,1000]$.

$$\therefore \qquad E_n = \frac{2}{1000}\int_0^{1000} 1.2x \sin\frac{n\pi x}{1000}\,dx$$

$$= \frac{24}{10^4}\left[x\left(-\frac{1000}{n\pi}\cos\frac{n\pi x}{1000}\right)-(1)\left(\frac{-10^6}{n^2\pi^2}\sin\frac{n\pi x}{1000}\right)\right]_0^{1000}$$

$$= \frac{24}{10^4}\left[\frac{10^6}{n\pi}(-1)^{n+1}\right] = \frac{2400}{n\pi}(-1)^{n+1}$$

$\therefore$ solution is

$$V(x,t) = 1300 - \frac{13}{10}x + \frac{2400}{\pi}\sum_{n=1}^{\infty}\frac{(-1)^{n+1}}{n}e^{-\frac{n^2\pi^2 t}{10^6 RC}}\sin\frac{n\pi x}{1000}$$

Example 4.66: Neglecting R and G, find the e.m.f. $V(x,t)$ in a line of length l, t seconds after the ends were suddenly grounded, given that $i(x,0) = i_0$ and $V(x,0) = e_1 \sin\frac{\pi x}{l} + e_5 \sin\frac{5\pi x}{l}$.

Solution: As R and G are neglected, we use radio equation $\frac{\partial^2 V}{\partial x^2} = LC\frac{\partial^2 V}{\partial t^2}$.

Let $\qquad V = X(x)T(t).$

$\therefore$ differential equation becomes

$$X''T = LCXT''$$

$$\therefore \qquad \frac{X''}{X} = \frac{LCT''}{T} = -p^2 \text{ (say)} \qquad (\because V(0,t) = V(l,t) = 0)$$

$$\therefore \qquad X'' + p^2 X = 0,\ LCT'' + p^2 T = 0$$

Solution of $X'' + p^2 X = 0$ is

$$X = c_1 \cos px + c_2 \sin px$$

$$X(0) = c_1 = 0$$

$$\therefore \qquad X = c_2 \sin px$$

$$X(l) = c_2 \sin pl = 0$$

$$\Rightarrow \qquad p = \frac{n\pi}{l};\ n = 1,2,\ldots$$

$$\therefore \qquad X = c_n \sin\frac{n\pi x}{l};\ n = 1,2,3\ldots$$

Solution of $LCT'' + p^2T = 0$ is

$$T = b\cos\frac{pt}{\sqrt{LC}} + e\sin\frac{pt}{\sqrt{LC}}$$

$$= b_n \cos\frac{n\pi t}{l\sqrt{LC}} + e_n \sin\frac{n\pi t}{l\sqrt{LC}};\ n = 1, 2, \ldots$$

By principle of superposition, general solution is

$$V(x,t) = \sum_{n=1}^{\infty}\left(B_n \cos\frac{n\pi t}{l\sqrt{LC}} + E_n \sin\frac{n\pi t}{l\sqrt{LC}}\right)\sin\frac{n\pi x}{l} \tag{1}$$

Now, $\quad i(x,0) = i_0$

$\therefore \quad \left(\frac{\partial i}{\partial x}\right)_{t=0} = 0$

But $\quad \frac{\partial i}{\partial x} = -C\frac{\partial V}{\partial t} \quad$ (from (4.55))

$\therefore \quad \left(\frac{\partial V}{\partial t}\right)_{t=0} = 0$

$\therefore$ from (1)

$$0 = \left(\frac{\partial V}{\partial t}\right)_{t=0} = \sum_{n=1}^{\infty} E_n \frac{n\pi}{l\sqrt{LC}} \sin\frac{n\pi x}{l}$$

$\therefore \quad E_n = 0;\ n = 1, 2, 3, \ldots$

$$\therefore \quad V(x,t) = \sum_{n=1}^{\infty} B_n \cos\frac{n\pi t}{l\sqrt{LC}} \sin\frac{n\pi x}{l} \tag{2}$$

$$\therefore \quad e_1 \sin\frac{\pi x}{l} + e_5 \sin\frac{5\pi x}{l} = V(x,0) = \sum_{n=1}^{\infty} B_n \sin\frac{n\pi x}{l}$$

$\therefore \quad B_1 = e_1, B_5 = e_5;\ B_n = 0;\ n \in N,\ n \neq 1, 5$

$\therefore$ from (2)

$$V(x,t) = e_1 \cos\frac{\pi t}{l\sqrt{LC}} \sin\frac{\pi x}{l} + e_5 \cos\frac{5\pi t}{l\sqrt{LC}} \sin\frac{5\pi x}{l}.$$

EXERCISE 4.7

1. Classify the following partial differential equations

   (i) $\frac{\partial^2 u}{\partial x^2} + \frac{\partial^2 u}{\partial y^2} = 0$

   (ii) $\frac{\partial^2 u}{\partial x^2} + \frac{\partial^2 u}{\partial y^2} = f(x,y)$

   (iii) $c^2\frac{\partial^2 u}{\partial x^2} = \frac{\partial u}{\partial t}$

(iv) $c^2 \frac{\partial^2 u}{\partial x^2} = \frac{\partial^2 u}{\partial t^2}$

(v) $4(x+1)z_{xx} - 4(x+2)z_{xy} + (x+3)z_{yy} = 0$

(vi) $(x-2)z_{xx} + 4(x+1)z_{xy} - 4x\, z_{yy} + z_x = \cos x$

2. Solve the equation $\frac{\partial u}{\partial x} - 2\frac{\partial u}{\partial y} = u$, $u(x,0) = 3e^{-5x} + 2e^{-3x}$ by the method of separation of variables.
3. Solve the following equation by the method of separation of variables:
$\frac{\partial u}{\partial x} = 4\frac{\partial u}{\partial y}$ where $u(0,y) = 8e^{-3y}$
4. Solve the equation $\frac{\partial^2 z}{\partial x^2} - 2\frac{\partial z}{\partial x} + \frac{\partial z}{\partial y} = 0$ by the method of separation of variables.
5. Solve one dimensional heat equation $\frac{\partial u}{\partial t} = k\frac{\partial^2 u}{\partial x^2}$, $x \in [0, l]$ with initial condition $u(x,0) = f(x)$ and the boundary conditions $u(0,t) = 0$, $u(l,t) = 0$, $t \geq 0$.
6. Find the solution of $\frac{\partial^2 u}{\partial x^2} = c^2 \frac{\partial u}{\partial t}$ for which $u(0,t) = u(l,t) = 0, u(x,0) = \sin\frac{\pi x}{l}$ by the method of separation of variables.
7. Find the temperature in a bar of length 2 whose ends are kept at zero and lateral surface insulated if the initial temperature is $\sin\frac{\pi x}{2} + 3\sin\frac{5\pi x}{2}$.
8. Solve $\frac{\partial u}{\partial t} = k\frac{\partial^2 u}{\partial x^2}, 0 < x < 2\pi$ with the conditions $u(x,0) = x^2, u(0,t) = u(2\pi, t) = 0$.
9. The ends A and B of a rod 20 cm long have the temperatures at 30°C and at 80°C until steady state prevails. The temperatures of the ends are changed to 40°C and 60°C, respectively. Find the temperature distribution in the rod at time t.
10. Find the temperature in a thin metal rod of length l, with both the ends insulated and with initial temperature in the rod $\sin(\pi x / l)$.
11. Solve $\frac{\partial u}{\partial t} = k\frac{\partial^2 u}{\partial x^2}$, $0 < x < l, t > 0$ with the boundary conditions $u_x(0,t) = 0$, $u_x(l,t) = 0$, $t \geq 0$ and the initial condition $u(x,0) = f(x)$, $0 \leq x \leq 1$.
12. Find the temperature $u(x,t)$ in a homogenous bar of heat conducting material of length l cm with its ends kept at zero temperature and initial temperature given by $ax(l-x)/l^2$.
13. A bar 10 cm long with insulated sides has its ends A and B maintained at temperature 50°C and 100°C respectively, until steady-state condition prevails. The temperature at A is suddenly raised to 90°C and at the same time lowered to 60°C at B. Find the temperature distribution in the bar at time t.
14. A bar of length l, laterally insulated, has its ends A and B kept at 0° and u_0° respectively, until steady-state conditions prevail. If the temperature at B is then suddenly reduced to 0° and kept so while that of A is maintained at 0° find the temperature in the bar at any subsequent time.
15. Find the temperature $u(x,t)$ in a bar which is perfectly insulated laterally, whose ends are kept at temperature 0°C and whose initial temperature is $f(x) = x(10-x)$ given that its length is 10 cm, constant cross section of area 1 cm², density 10.6 g/cm³, thermal conductivity 1.04 cal/cm deg sec and specific heat 0.056 call/gm deg.

16. A bar AB of length 10 cm has its ends A and B kept at 30° and 100° temperature respectively until steady-state condition is reached. Then the temperature at A is lowered to 20° and that at B to 40° and these temperatures are maintained. Find the subsequent temperature distribution in the bar.
17. Solve the differential equation $\frac{\partial u}{\partial t}=c^2\frac{\partial^2 u}{\partial x^2}$ for the conduction of heat along a rod of length l without radiation, subject to the following conditions:
    (i) u is not infinite for $t \to \infty$
    (ii) $\frac{\partial u}{\partial x}=0$ for $x=0$ and $x=l$
    (iii) $u=lx-x^2$ for $t=0$, between $x=0$ and $x=l$
18. A string is stretched and fastened to two points l apart. Motion is started by displacing the string in the form $y=a\sin\frac{\pi x}{l}$ from which it is released at time $t=0$. Show that the displacement of any point at a distance x from one end at time t is given by

    $y(x,t)=a\sin\frac{\pi x}{l}\cos\frac{\pi ct}{l}$, where c^2 is diffusivity of string.
19. Find the deflection of vibrating string of unit length having fixed ends with initial velocity zero and initial deflection $f(x)=k(\sin x-\sin 2x)$ using the D' Alembert solution.
20. Find the displacement of a string stretched between the fixed points $(0,0)$ and $(1,0)$ and released from rest from the position $A\sin\pi x+B\sin 2\pi x$.
21. A tightly stretched string of length l with fixed ends is initially in equilibrium position. It is set vibrating by giving each point a velocity $v_0\sin^3\pi x/l$. Find the displacement $y(x,t)$.
22. Solve the wave equation $\frac{\partial^2 u}{\partial t^2}=a^2\frac{\partial^2 u}{\partial x^2}$, $0<x<l$, $t>0$, where a is a constant related to tension in the vibrating string of length l having fixed ends. The boundary conditions and initial conditions are

    $$u(0,t)=u(l,t)=0, t\geq 0$$
    $$u(x,0)=f(x), 0\leq x\leq l$$
    $$u_t(x,0)=0, 0\leq x\leq l$$
23. A string of length l is fastened at both ends A and C. At a distance 'b' from the end A, the string is transversely displaced to distance 'd' and is released from rest when it is in this position. Find the equation of the subsequent motion.
24. A tightly stretched string of length l is attached at $x=0$ and at $x=l$ and released from rest at $t=0$. Find the expression for y the displacement of the string at a distance x, given that $y=A\sin\frac{2\pi x}{l}$ at $t=0$.
25. A tightly stretched flexible string has its ends fixed at $x=0$ and $x=l$. At time $t=0$, the string is given a shape defined by $f(x)=\mu x(l-x)$, where μ is a constant and then released. Find the displacement of any point x of the string at any time $t>0$.
26. A string is stretched between the fixed points $(0,0)$ and $(l,0)$ and released at rest from the initial deflection given by

    $$f(x)=\begin{cases}\frac{2kx}{l} & \text{when } 0<x<\frac{l}{2}\\ \frac{2k}{l}(l-x) & \text{when } \frac{l}{2}<x<l\end{cases}.$$ Find the deflection of the string at any time t.
27. An elastic string of length l which is fastened at its ends $x=0$ and $x=l$ is picked up at its centre point $x=\frac{l}{2}$ to a height of $\frac{l}{2}$ and released from rest. Find the displacement of the string at any instant of time.

28. Find the displacement of a string stretched between two fixed points at a distance $2c$ apart when the string is initially at rest in equilibrium position and points of the string are given initial velocities v where

$$v = \begin{cases} x/c, 0 < x < c \\ (2c-x)/c, c < x < 2c \end{cases}; \; x \text{ being the}$$

distance measured from one end.

29. Show that the solution of the wave equation

$\frac{\partial^2 y}{\partial t^2} = c^2 \frac{\partial^2 y}{\partial x^2}$ can be expressed in the form $y(x,t) = \phi(x+ct) + \psi(x-ct)$

If $y(x,0) = f(x)$ and $\frac{\partial y}{\partial t}(x,0) = 0$, show that $y(x,t) = \frac{1}{2}\left[f(x+ct) + f(x-ct)\right]$

30. Use D' Alembert solution to find the solution of the initial value problem defining the vibrations of an infinitely long elastic string $\frac{\partial^2 y}{\partial t^2} = c^2 \frac{\partial^2 y}{\partial x^2}, -\infty < x < \infty, t > 0$

$$y(x,0) = f(x), \frac{\partial y}{\partial t}(x,0) = g(x)$$

when

(i) $f(x) = \sin x, g(x) = k$ where k is a constant

(ii) $f(x) = 0, \quad g(x) = \sin 3x$

(iii) $f(x) = k(x - x^2), g(x) = 0$

31. Find the current i and voltage v in a transmission line of length l, t seconds after the ends are suddenly grounded given that $i(x,0) = i_0$, $v(x,0) = v_0 \sin\frac{\pi x}{l}$. Also, R and G are negligible.

32. A telephone line 3000 km long has a resistance of 4 ohms/km and a capacitance of 5×10^{-7} farad/km. Initially, both the ends are grounded so that the line is uncharged. At time $t = 0$, a constant e.m.f. E is applied to one end, while the other end is left grounded. Assuming the inductance and leakance to be negligible, show that steady-state current of the grounded end at the end of 1 sec. is 5.3 per cent.

33. In a telephone wire of length l, a steady voltage distribution of 20 volts at the source end and 12 volts at the terminal end is maintained. At time $t = 0$, the terminal end is grounded. Determine the voltage and current. Assume that $L = 0$ and $G = 0$.

ANSWERS 4.7

1. (i) elliptic (ii) elliptic (iii) parabolic (iv) hyperbolic
   (v) hyperbolic (vi) hyperbolic

2. $u(x,y) = 3e^{-(5x+3y)} + 2e^{-(3x+2y)}$

3. $u(x,y) = 8e^{-3(4x+y)}$

4. $z(x,y) = (c_1 x + c_2)e^{x+y} + e^{x+(1-k^2)y}\left(c_3 e^{-kx} + c_4 e^{-kx}\right).+ e^{x+(1+k_1^2)y}\left(c_5 \cos k_1 x + c_6 \sin k_1 x\right)$

   where $c_1, c_2, c_3, c_4, c_5, c_6, k > 0, k_1 > 0$ are arbitrary constants.

5. $u(x,t) = \sum_{n=1}^{\infty} b_n e^{\frac{-n^2\pi^2 k}{l^2}t} \sin\left(\frac{n\pi x}{l}\right)$, where $b_n = \frac{2}{l}\int_0^l f(x)\sin\left(\frac{n\pi x}{l}\right)dx, n = 1,2,3,\ldots$

6. $u(x,t) = \sin\frac{\pi x}{l} e^{\frac{-\pi^2}{c^2 l^2}t}$.

7. $u(x,t)=\sin\frac{\pi x}{2}e^{-\frac{c^2\pi^2}{4}t}+3\sin\frac{5\pi x}{2}e^{-\frac{25c^2\pi^2}{4}t}$, where c^2 is the thermal diffusivity.

8. $u(x,t)=\sum_{n=1}^{\infty}\frac{8}{n}\left[(-1)^n\left(\frac{2}{\pi n^2}-\pi\right)-\frac{2}{8n^2}\right]\sin\frac{nx}{2}e^{-\frac{n^2k}{4}t}$

9. $u(x,t)=x+40-\frac{20}{\pi}\sum_{n=1}^{\infty}\frac{1-2(-1)^n}{n}\sin\frac{n\pi x}{20}e^{\frac{-c^2n^2\pi^2}{400}t}$, where c^2 is the thermal diffusivity.

10 $u(x,t)=\frac{2}{\pi}-\frac{4}{\pi}\sum_{n=1}^{\infty}\frac{1}{(4n^2-1)}\cos\left(\frac{2n\pi x}{l}\right)e^{-\frac{4n^2\pi^2c^2}{l^2}t}$, where c^2 is the thermal diffusivity.

11. $u(x,t)=\frac{a_0}{2}+\sum_{n=1}^{\infty}a_n\cos\frac{n\pi x}{l}e^{\frac{-n^2\pi^2k}{l^2}t}$, where $a_n=\frac{2}{l}\int_0^l f(x)\cos\frac{n\pi x}{l}dx;\ n=0,1,2,\ldots$

12. $u(x,t)=\frac{8a}{\pi^3}\sum_{n=1}^{\infty}\frac{1}{(2n-1)^3}\sin\frac{(2n-1)\pi x}{l}e^{-\frac{c^2(2n-1)^2\pi^2}{l^2}t}$, where c^2 is the thermal diffusivity.

13. $u(x,t)=90-3x-\frac{80}{\pi}\sum_{n=1}^{\infty}\frac{1}{n}\sin\frac{n\pi x}{5}e^{-\frac{c^2\pi^2n^2}{25}t}$, where c^2 is the thermal diffusivity.

14. $u(x,t)=\frac{2u_0}{\pi}\sum_{n=1}^{\infty}\frac{(-1)^{n+1}}{n}\sin\frac{n\pi x}{l}e^{\frac{-n^2\pi^2c^2}{l^2}t}$, where c^2 is the thermal diffusivity.

15. $u(x,t)=\frac{800}{\pi^3}\sum_{n=1}^{\infty}\frac{1}{(2n-1)^3}e^{-0.01752\pi^2(2n-1)^2t}\sin\frac{(2n-1)\pi x}{10}$

16. $u(x,t)=2x+20+\frac{20}{\pi}\sum_{n=1}^{\infty}\left\{\frac{1-6(-1)^n}{n}\right\}\sin\frac{n\pi x}{10}e^{-\frac{c^2n^2\pi^2}{100}t}$, where c^2 is the thermal diffusivity.

17. $u(x,t)=\frac{l^2}{6}-\frac{l^2}{\pi^2}\sum_{n=1}^{\infty}\frac{1}{n^2}\cos\left(\frac{2n\pi x}{l}\right)e^{\frac{-4n^2\pi^2c^2}{l^2}t}$

19. $y(x,t)=k[\sin x\cos ct-\sin 2x\cos 2ct]$, where c^2 is diffusivity of string.

20. $y(x,t)=A\sin\pi x\cos\pi ct+B\sin 2\pi x\cos(2\pi ct)$, where c^2 is diffusivity of string.

21. $y(x,t)=\frac{lv_0}{12\pi c}\left(9\sin\frac{\pi x}{l}\sin\frac{\pi ct}{l}-\sin\frac{3\pi x}{l}\sin\frac{3\pi ct}{l}\right)$, where c^2 is diffusivity of string.

22. $u(x,t)=\sum_{n=1}^{\infty}b_n\cos\frac{n\pi at}{l}\sin\frac{n\pi x}{l}$, where $b_n=\frac{2}{l}\int_0^l f(x)\sin\frac{n\pi x}{l}dx$

23. $y(x,t)=\frac{2dl^2}{b(l-b)\pi^2}\sum_{n=1}^{\infty}\frac{1}{n^2}\sin\frac{n\pi b}{l}\sin\frac{n\pi x}{l}\cos\frac{n\pi ct}{l}$, where c^2 is diffusivity of string.

24. $y(x,t)=A\sin\frac{2\pi x}{l}\cos\frac{2c\pi t}{l}$, where c^2 is diffusivity of string.

25. $y(x,t)=\frac{8\mu l^2}{\pi^3}\sum_{n=1}^{\infty}\frac{1}{(2n-1)^3}\cos\frac{(2n-1)c\pi t}{l}\sin\frac{(2n-1)\pi x}{l}$, where c^2 is diffusivity of string.

26. $y(x,t)=\frac{8k}{\pi^2}\sum_{n=1}^{\infty}\frac{(-1)^{n+1}}{(2n-1)^2}\cos\frac{(2n-1)c\pi t}{l}\sin\frac{(2n-1)\pi x}{l}$, where c^2 is diffusivity of string.

27. $y(x,t)=\frac{4l}{\pi^2}\sum_{n=1}^{\infty}\frac{(-1)^{n+1}}{(2n-1)^2}\cos\frac{(2n-1)c\pi t}{l}\sin\frac{(2n-1)\pi x}{l}$, where c^2 is diffusivity of string.

28. $y(x,t)=\frac{16c}{a\pi^3}\sum_{n=1}^{\infty}\frac{(-1)^{n+1}}{(2n-1)^3}\sin\frac{(2n-1)\pi x}{2c}\sin\frac{(2n-1)\pi at}{2c}$, where a^2 is diffusivity of string.

30. (i) $y(x,t)=\sin x\cos(ct)+kt$

(ii) $y(x,t)=\frac{1}{3c}\sin(3x)\sin(3ct)$

(iii) $y(x,t)=k\left(x-x^2-c^2t^2\right)$

31. $v(x,t)=v_0\cos\left(\frac{\pi t}{l\sqrt{LC}}\right)\sin\left(\frac{\pi x}{l}\right)$

$i(x,t)=i_0-v_0\sqrt{\frac{C}{L}}\sin\left(\frac{\pi t}{l\sqrt{LC}}\right)\cos\left(\frac{\pi x}{l}\right)$

33. $v(x,t)=\frac{20(l-x)}{l}+\frac{24}{\pi}\sum_{n=1}^{\infty}\frac{(-1)^{n+1}}{n}\sin\left(\frac{n\pi x}{l}\right)e^{\frac{-n^2\pi^2}{l^2RC}t}$

$i(x,t)=\frac{20}{Rl}+\frac{24}{Rl}\sum_{n=1}^{\infty}(-1)^n\cos\left(\frac{n\pi x}{l}\right)e^{\frac{-n^2\pi^2}{l^2RC}t}$

4.14 TWO DIMENSIONAL HEAT FLOW

Consider the flow of heat in a metal plate in the x–y plane. We assume that the sides of plate are insulated. Let $u(x,y,t)$ be temperature at any point (x,y) of plate at time t.

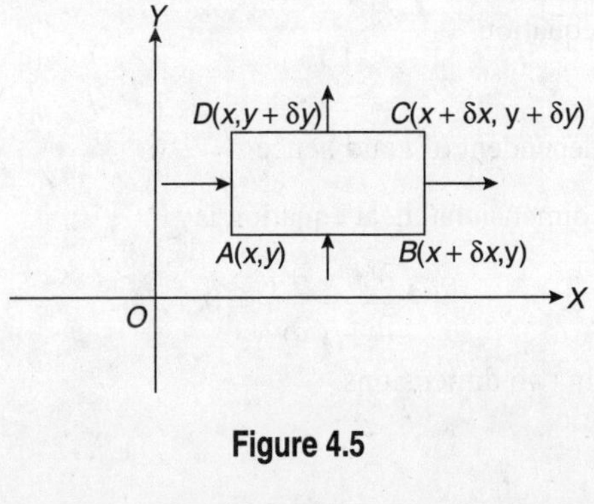

Figure 4.5

Consider a portion of plate of thickness α as shown in the figure.
As discussed in one dimensional heat flow

Amount of heat entering AD per second $= -K\alpha\delta y\left(\frac{\partial u}{\partial x}\right)_x$

Amount of heat entering AB per second $= -K\alpha\delta x\left(\frac{\partial u}{\partial y}\right)_y$

Amount of heat leaving BC per second $= -K\alpha\delta y\left(\frac{\partial u}{\partial x}\right)_{x+\delta x}$

Amount of heat leaving DC per second $= -K\alpha\delta x\left(\frac{\partial u}{\partial y}\right)_{y+\delta y}$

Hence, total amount of heat retained by this element of plate per second

$$= K\alpha\left[\delta y\left\{\left(\frac{\partial u}{\partial x}\right)_{x+\delta x} - \left(\frac{\partial u}{\partial x}\right)_x\right\} + \delta x\left\{\left(\frac{\partial u}{\partial y}\right)_{y+\delta y} - \left(\frac{\partial u}{\partial y}\right)_y\right\}\right]$$

But rate of increase of heat in this element $= \sigma\rho\alpha\delta x\delta y\frac{\partial u}{\partial t}$

where σ is specific heat and ρ is density of material of plate.

$$\therefore \qquad \sigma\rho\alpha\delta x\delta y\frac{\partial u}{\partial t} = K\alpha\left[\delta y\left\{\left(\frac{\partial u}{\partial x}\right)_{x+\delta x} - \left(\frac{\partial u}{\partial x}\right)_x\right\} + \delta x\left\{\left(\frac{\delta u}{\delta y}\right)_{y+\delta y} - \left(\frac{\partial u}{\partial y}\right)_y\right\}\right]$$

$$\therefore \qquad \frac{\partial u}{\partial t} = \frac{K}{\sigma\rho}\left[\frac{\left(\frac{\partial u}{\partial x}\right)_{x+\delta x} - \left(\frac{\partial u}{\partial x}\right)_x}{\delta x} + \frac{\left(\frac{\partial u}{\partial y}\right)_{y+\delta y} - \left(\frac{\partial u}{\partial y}\right)_y}{\delta y}\right]$$

Take limit as $\delta x \to 0, \delta y \to 0$

$$\frac{\partial u}{\partial t} = c^2\left[\frac{\partial^2 u}{\partial x^2} + \frac{\partial^2 u}{\partial y^2}\right]$$

where $c^2 = \frac{K}{\sigma\rho}$ is diffusivity of material of plate

It is two dimensional heat equation.

Remark 4.6:

(i) In steady state, u is independent of t and hence $\frac{\partial u}{\partial t} = 0$.

$\therefore$ In steady state, two dimensional heat equation is

$$\frac{\partial^2 u}{\partial x^2} + \frac{\partial^2 u}{\partial y^2} = 0$$

It is Laplace equation in two dimensions.

(ii) In three dimensions, heat equation is

$$\frac{\partial u}{\partial t} = c^2\left(\frac{\partial^2 u}{\partial x^2} + \frac{\partial^2 u}{\partial y^2} + \frac{\partial^2 u}{\partial z^2}\right)$$

In steady state, it becomes

$$\frac{\partial^2 u}{\partial x^2} + \frac{\partial^2 u}{\partial y^2} + \frac{\partial^2 u}{\partial z^2} = 0$$

It is Laplace equation in three dimensions.

4.15 SOLUTION OF TWO DIMENSIONAL LAPLACE EQUATION

Two dimensional Laplace equation is

$$\frac{\partial^2 u}{\partial x^2} + \frac{\partial^2 u}{\partial y^2} = 0$$

Let $$u(x, y) = X(x)Y(y)$$

$\therefore$ $$\frac{\partial^2 u}{\partial x^2} = X''Y, \quad \frac{\partial^2 u}{\partial y^2} = XY''$$

$\therefore$ differential equation becomes

$$X''Y + XY'' = 0$$

or $$\frac{X''}{X} = -\frac{Y''}{Y}$$

L.H.S. is a function of x and R.H.S. is a function of y and hence each is constant say λ.

Case I: $\lambda = 0$, then $X'' = 0,\ Y'' = 0$

$\therefore$ solutions are $X = Ax + B,\ Y = Cy + D$

$\therefore$ $$u(x, y) = (Ax + B)(Cy + D)$$

Case II: λ is positive, i.e., $\lambda = p^2, p > 0$ then

$$X'' - p^2X = 0,\ Y'' + p^2Y = 0.$$

Solutions are $X = Ae^{px} + Be^{-px}, Y = C\cos py + D\sin py$

$\therefore$ $$u(x, y) = \left(Ae^{px} + Be^{-px}\right)(C\cos py + D\sin py)$$

Case III: λ is negative, i.e., $\lambda = -p^2, p > 0$

$\therefore$ $$X'' + p^2X = 0,\ Y'' - p^2Y = 0$$

$\therefore$ solutions are $X = A\cos px + B\sin px,\ Y = Ce^{py} + De^{-py}$

$\therefore$ $$u(x, y) = \left(Ce^{py} + De^{-py}\right)(A\cos px + B\sin px)$$

Among these solutions, we are to find those solutions which satisfy initial and boundary conditions consistent with the physical nature.

In particular, if $u \to 0$ as $y \to \infty$ for all x

Then solution must be

$$u(x,y) = e^{-py}(A_1 \cos px + B_1 \sin py); p > 0$$

and if $u \to 0$ as $x \to \infty$ for all y

then solution must be

$$u(x,y) = e^{-px}(A_1 \cos py + B_1 \sin py), p > 0$$

Example 4.67: Find the solution of the Laplace equation

$$\frac{\partial^2 u}{\partial x^2} + \frac{\partial^2 u}{\partial y^2} = 0$$

which satisfies the conditions

(i) $u \to 0$ as $y \to \infty$ for all x (ii) $u = 0$ at $x = 0$ for all y

(iii) $u = 0$ at $x = l$ for all y (iv) $u = lx - x^2$ if $y = 0$ for all $x \in (0, l)$

Solution: Since $u \to 0$ as $y \to \infty$ for all x

Solution is of the form

$$u(x,y) = e^{-py}(C_1 \cos px + C_2 \sin px); \ p > 0$$

$$u(0,y) = C_1 e^{-py} = 0 \text{ for all } y$$

$$\Rightarrow \quad C_1 = 0$$

$$\therefore \quad u(x,y) = C_2 e^{-py} \sin px$$

$$u(l,y) = C_2 e^{-py} \sin pl = 0$$

$$\Rightarrow \quad p = \frac{n\pi}{l}; \ n = 1, 2, 3, \ldots$$

$$\therefore \quad u(x, y) = C_n e^{-\frac{n\pi y}{l}} \sin \frac{n\pi x}{l}; \ n = 1, 2, 3, \ldots$$

$\therefore$ By principle of superposition, solution is

$$u(x,y) = \sum_{n=1}^{\infty} C_n e^{-\frac{n\pi y}{l}} \sin \frac{n\pi x}{l} \tag{1}$$

$$lx - x^2 = u(x,0) = \sum_{n=1}^{\infty} C_n \sin \frac{n\pi x}{l}$$

It is Fourier half range sine-series in $[0, l]$

$$\therefore \quad C_n = \frac{2}{l} \int_0^l (lx - x^2) \sin \frac{n\pi x}{l} \, dx$$

$$= \frac{2}{l} \left[(lx - x^2)\left(-\frac{l}{n\pi} \cos \frac{n\pi x}{l} \right) - (l - 2x)\left(-\frac{l^2}{n^2\pi^2} \sin \frac{n\pi x}{l} \right) + (-2)\left(\frac{l^3}{n^3\pi^3} \cos \frac{n\pi x}{l} \right) \right]_0^l$$

$$= -\frac{4l^2}{n^3\pi^3}\left((-1)^n - 1\right)$$

$$\therefore \qquad C_{2n} = 0,\ C_{2n-1} = \frac{8l^2}{(2n-1)^3\pi^3}\ ; n = 1,2,3,\ldots$$

$\therefore$ from (1), solution is

$$u(x,y) = \frac{8l^2}{\pi^3}\sum_{n=1}^{\infty}\frac{1}{(2n-1)^3}e^{-\frac{(2n-1)\pi y}{l}}\sin\frac{(2n-1)\pi x}{l}$$

Example 4.68: Solve $\dfrac{\partial^2 u}{\partial x^2} + \dfrac{\partial^2 u}{\partial y^2} = 0;\quad 0 \le x \le a,\ 0 \le y \le b$ subject to the conditions

$$u(0,y) = u(a,y) = u(x,b) = 0;\ u(x,0) = x(a-x).$$

Solution: Solutions can be of type

(i) $(C_1x + C_2)(C_3y + C_4)$

(ii) $\left(C_1e^{px} + C_2e^{-px}\right)(C_3\cos py + C_4\sin py)$

(iii) $\left(C_1e^{py} + C_2e^{-py}\right)(C_3\cos px + C_4\sin px)$

Since $u(0,y) = u(a,y) = 0$

$\therefore$ solution must be of the type

$$u(x,y) = \left(C_1e^{py} + C_2e^{-py}\right)\sin px$$

$$u(x,b) = \left(C_1e^{pb} + C_2e^{-pb}\right)\sin px = 0$$

$$\therefore \qquad C_2 = -C_1e^{2pb}$$

$$\therefore \qquad u(x,y) = C_1\left(e^{py} - e^{p(2b-y)}\right)\sin px$$

$$= C_1e^{pb}\left(e^{p(y-b)} - e^{p(b-y)}\right)\sin px$$

$$= -2C_1e^{pb}\left(\frac{e^{p(b-y)} - e^{-p(b-y)}}{2}\right)\sin px$$

$$= Ce^{pb}\sinh p(b-y)\sin px \qquad (C = -2C_1)$$

$$u(a,y) = Ce^{pb}\sinh p(b-y)\sin pa = 0$$

$$\Rightarrow \qquad p = \frac{n\pi}{a};\ n = 1,2,\ldots$$

$$\therefore \qquad u(x,y) = Ce^{\frac{n\pi b}{a}}\sinh\frac{n\pi}{a}(b-y)\sin\frac{n\pi x}{a};\ n = 1,2,\ldots$$

$$= C_n\sinh\frac{n\pi(b-y)}{a}\sin\frac{n\pi x}{a};\ n = 1,2,\ldots \text{ where } C_n = Ce^{\frac{n\pi b}{a}}.$$

∴ By principle of superposition, solution is

$$u(x,y)=\sum_{n=1}^{\infty}C_n \sinh\frac{n\pi(b-y)}{a}\sin\frac{n\pi x}{a} \tag{1}$$

∴ $$x(a-x)=u(x,0)=\sum_{n=1}^{\infty}C_n \sinh\frac{n\pi b}{a}\sin\frac{n\pi x}{a};\quad 0\le x\le a$$

It is Fourier half range sine series of $x(a-x)$ in $[0,a]$.

∴ $$C_n \sinh\frac{n\pi b}{a}=\frac{2}{a}\int_0^a (ax-x^2)\sin\frac{n\pi x}{a}dx$$

∴ $$C_n=\frac{2}{a\sinh\frac{n\pi b}{a}}\left[(ax-x^2)\left(-\frac{a}{n\pi}\cos\frac{n\pi x}{a}\right)-(a-2x)\left(-\frac{a^2}{n^2\pi^2}\sin\frac{n\pi x}{a}\right)\right.$$
$$\left.+(-2)\left(\frac{a^3}{n^3\pi^3}\cos\frac{n\pi x}{a}\right)\right]_0^a$$
$$=\frac{-4a^2\left[(-1)^n-1\right]}{n^3\pi^3\sinh\frac{n\pi b}{a}}$$

$$C_{2n}=0,\quad C_{2n-1}=\frac{8a^2}{(2n-1)^3\pi^3\sinh\frac{(2n-1)\pi b}{a}};\ n=1,2,3,...$$

∴ from (1), solution is

$$u(x,y)=\frac{8a^2}{\pi^3}\sum_{n=1}^{\infty}\frac{\sinh\frac{(2n-1)\pi(b-y)}{a}\sin\frac{(2n-1)\pi x}{a}}{(2n-1)^3\sinh\frac{(2n-1)\pi b}{a}}$$

Example 4.69: Solve Laplace equation $\frac{\partial^2 u}{\partial x^2}+\frac{\partial^2 u}{\partial y^2}=0$ in rectangle with $u(0,y)=u(a,y)=u(x,b)=0$, $u(x,0)=f(x)$.
Also find the solution in a square of side π and $f(x)=\sin^2 x;\ 0<x<\pi$.

Solution: As $u(0,y)=u(a,y)=0$

∴ solution is of the form

$$u(x,y)=\left(C_1e^{py}+C_2e^{-py}\right)\sin px;\ p>0$$
$$u(x,b)=\left(C_1e^{pb}+C_2e^{-pb}\right)\sin px=0$$

∴ $$C_2=-C_1e^{2pb}$$

$$\therefore \qquad u(x,y) = C_1\left(e^{py} - e^{2pb}e^{-py}\right)\sin px$$

$$= -C_1 e^{pb}\left(e^{p(b-y)} - e^{-p(b-y)}\right)\sin px$$

$$= Ce^{pb}\sinh p(b-y)\sin px \qquad (-2C_1 = C)$$

$$u(a,y) = Ce^{pb}\sinh p(b-y)\sin pa = 0$$

$$\Rightarrow \qquad p = \frac{n\pi}{a};\; n = 1,2,3,\ldots$$

$$\therefore \qquad u(x,y) = C_n \sinh\frac{n\pi(b-y)}{a}\sin\frac{n\pi x}{a};\; n = 1,2,3,\ldots. \qquad \left(Ce^{\frac{n\pi b}{a}} = C_n\right)$$

$\therefore$ By principle of superposition, solution is

$$u(x,y) = \sum_{n=1}^{\infty} C_n \sinh\frac{n\pi(b-y)}{a}\sin\frac{n\pi x}{a}$$

$$\therefore \qquad f(x) = u(x,0) = \sum_{n=1}^{\infty} C_n \sinh\frac{n\pi b}{a}\sin\frac{n\pi x}{a}$$

It is Fourier half range sin-series of $f(x)$ in $[0,a]$.

$$\therefore \qquad C_n \sinh\frac{n\pi b}{a} = \frac{2}{a}\int_0^a f(x)\sin\frac{n\pi x}{a}\,dx$$

$$\therefore \qquad C_n = \frac{2}{a\sinh\frac{n\pi b}{a}}\int_0^a f(x)\sin\frac{n\pi x}{a}\,dx$$

$\therefore$ solution is

$$u(x,y) = \sum_{n=1}^{\infty} C_n \sinh\frac{n\pi(b-y)}{a}\sin\frac{n\pi x}{a}$$

where
$$C_n = \frac{2}{a\sinh\frac{n\pi b}{a}}\int_0^a f(x)\sin\frac{n\pi x}{a}\,dx$$

In square of side π, $a = b = \pi$ and when $f(x) = \sin^2 x$, $0 < x < \pi$

we have

$$u(x,y) = \sum_{n=1}^{\infty} C_n \sinh n(\pi - y)\sin nx$$

where
$$C_n = \frac{2}{\pi\sinh n\pi}\int_0^{\pi}\sin^2 x\sin nx\,dx$$

$$= \frac{1}{\pi\sinh n\pi}\int_0^{\pi}(1-\cos 2x)\sin nx\,dx$$

$$= \frac{1}{\pi \sinh n\pi}\left[-\frac{1}{n}\cos nx\Big|_0^{\pi} - \frac{1}{2}\int_0^{\pi}\left[\sin(n+2)x + \sin(n-2)x\right]dx \right]$$

$$= \frac{1}{\pi \sinh n\pi}\left[-\frac{1}{n}\left((-1)^n - 1\right) - \frac{1}{2}\left\{ -\frac{1}{n+2}\cos(n+2)x - \frac{1}{n-2}\cos(n-2)x \right\}_0^{\pi} \right]$$

$$= \frac{1}{\pi \sinh n\pi}\left[-\frac{1}{n}\left((-1)^n - 1\right) - \frac{1}{2}\left\{ -\frac{(-1)^n - 1}{n+2} - \frac{(-1)^n - 1}{n-2} \right\} \right]$$

$\therefore$ $\quad C_{2n} = 0;\ n = 1, 2, \ldots$

$$C_{2n-1} = \frac{1}{\pi \sinh(2n-1)\pi}\left[\frac{2}{2n-1} - \frac{1}{2}\left\{ \frac{2}{2n+1} + \frac{2}{2n-3} \right\} \right]$$

$$= \frac{1}{\pi \sinh(2n-1)\pi}\left[\frac{2}{2n-1} - \frac{1}{2n+1} - \frac{1}{2n-3} \right]$$

$$= \frac{1}{\pi \sinh(2n-1)\pi}\left[\frac{4n+2-2n+1}{4n^2-1} - \frac{1}{2n-3} \right]$$

$$= \frac{1}{\pi \sinh(2n-1)\pi}\left[\frac{2n+3}{4n^2-1} - \frac{1}{(2n-3)} \right] = \frac{1}{\pi \sinh(2n-1)\pi}\left[\frac{4n^2 - 9 - 4n^2 + 1}{(4n^2-1)(2n-3)} \right]$$

$$= \frac{-8}{\pi(4n^2-1)(2n-3)\sinh(2n-1)\pi};\ n = 1, 2, 3, \ldots$$

$\therefore$ solution is

$$u(x,y) = \frac{-8}{\pi}\sum_{n=1}^{\infty}\frac{\sinh[(2n-1)(\pi - y)]\sin(2n-1)x}{(4n^2-1)(2n-3)\sinh(2n-1)\pi}$$

Example 4.70: A rectangular plate with insulated surface is 10 cm wide and so long compared to its width that it may be considered infinite in length without introducing an appreciable error. If the temperature of the short edge $y = 0$ is given by

$$u = \begin{cases} 20x & ;\ 0 \le x \le 5 \\ 20(10-x); & 5 \le x \le 10 \end{cases}$$

and the two long edges $x = 0$, $x = 10$ as well as the other short edges are kept at 0°C. Prove that the temperature u at any point (x, y) is given by

$$u = \left(\frac{800}{\pi^2}\right)\sum_{n=1}^{\infty}\frac{(-1)^{n+1}}{(2n-1)^2}\sin\frac{(2n-1)\pi x}{10}\cdot e^{-\frac{(2n-1)\pi y}{10}}.$$

Solution: Temperature function $u(x,y)$ at any point (x,y) satisfies Laplace equation $\frac{\partial^2 u}{\partial x^2} + \frac{\partial^2 u}{\partial y^2} = 0$ subject to the conditions $u(0,y) = u(10,y) = 0$, $u(x,y) \to 0$ as $y \to \infty$ and

$$u(x,0)=\begin{cases}20x & ;\ 0\le x\le 5\\ 20(10-x); & 5\le x\le 10\end{cases}$$

As $u(x,y)\to 0$ when $y\to\infty$

$\therefore$ solution is of the form

$$u(x,y)=e^{-py}\left(c_1\cos px+c_2\sin px\right);\ p>0$$

$$u(0,y)=c_1e^{-py}=0$$

$$\Rightarrow\qquad c_1=0$$

$$\therefore\qquad u(x,y)=c\,e^{-py}\sin px\qquad (c=c_2)$$

$$u(10,y)=ce^{-py}\sin 10p=0$$

$$\Rightarrow\qquad p=\frac{n\pi}{10};\ n=1,2,\ldots$$

$$\therefore\qquad u(x,y)=c_ne^{-\frac{n\pi y}{10}}\sin\frac{n\pi x}{10};\ n=1,2,3,\ldots\qquad (c_n=c)$$

$\therefore$ By principle of superposition, solution is

$$u(x,y)=\sum_{n=1}^{\infty}c_ne^{-\frac{n\pi y}{10}}\sin\frac{n\pi x}{10}\tag{1}$$

$$\therefore\qquad u(x,0)=\sum_{n=1}^{\infty}c_n\sin\frac{n\pi x}{10};\ 0\le x\le 10$$

It is Fourier half range sine series of $u(x,0)$ in [0,10].

$$\therefore\qquad c_n=\frac{2}{10}\int_0^{10}u(x,0)\sin\frac{n\pi x}{10}dx$$

$$=\frac{1}{5}\left[\int_0^{5}20x\sin\frac{n\pi x}{10}dx+\int_5^{10}(200-20x)\sin\frac{n\pi x}{10}dx\right]\quad \text{(by definition of } u(x,0))$$

$$=4\left[\left\{x\left(-\frac{10}{n\pi}\cos\frac{n\pi x}{10}\right)-(1)\left(-\frac{10^2}{n^2\pi^2}\sin\frac{n\pi x}{10}\right)\right\}_0^5\right.$$

$$\left.+\left\{(10-x)\left(-\frac{10}{n\pi}\cos\frac{n\pi x}{10}\right)-(-1)\left(-\frac{10^2}{n^2\pi^2}\sin\frac{n\pi x}{10}\right)\right\}_5^{10}\right]$$

$$=4\left[-\frac{50}{n\pi}\cos\frac{n\pi}{2}+\frac{100}{n^2\pi^2}\sin\frac{n\pi}{2}+\frac{50}{n\pi}\cos\frac{n\pi}{2}+\frac{100}{n^2\pi^2}\sin\frac{n\pi}{2}\right]$$

$$=\frac{800}{n^2\pi^2}\sin\frac{n\pi}{2}$$

$$\therefore \quad c_{2n} = 0; \quad c_{2n-1} = \frac{800}{(2n-1)^2 \pi^2} \sin\frac{(2n-1)\pi}{2} = \frac{800(-1)^{n+1}}{(2n-1)^2 \pi^2}; \; n = 1, 2, \ldots$$

$\therefore$ from (1), solution is

$$u(x, y) = \frac{800}{\pi^2} \sum_{n=1}^{\infty} \frac{(-1)^{n+1}}{(2n-1)^2} e^{-\frac{(2n-1)\pi y}{10}} \sin\frac{(2n-1)\pi x}{10}$$

4.16 TWO DIMENSIONAL WAVE EQUATION

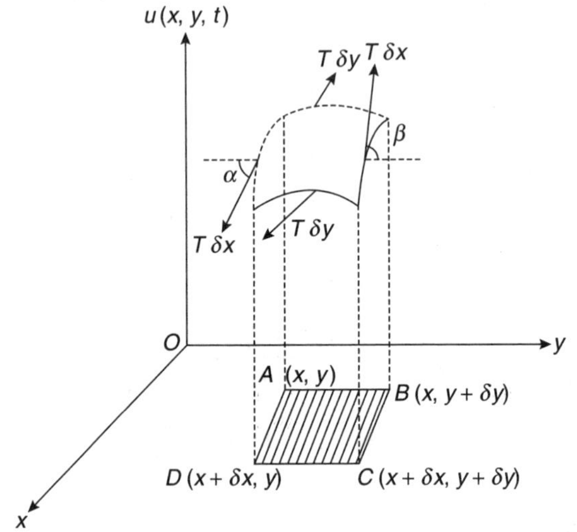

Figure 4.6

We shall obtain the partial differential equation for the vibrations of a tightly stretched membrane (such as the membrane of a drum). We shall assume

(i) Membrane is homogeneous and hence the mass per unit area m is constant.
(ii) The membrane is perfectly flexible and offers no resistance to bending.
(iii) The membrane is stretched and then fixed along its entire boundary in the x–y plane. The tension per unit length T caused by stretching the membrane is same at all points and in all directions and does not change during motion and it is so large that weight of membrane is negligible in its comparison.
(iv) The deflection $u(x, y, t)$ of the membrane during the motion is small compared to size of membrane and all angles of inclinations are small.

Consider a small portion $ABCD$ of the membrane. Forces acting on sides are $T\delta x$, $T\delta y$.

Since motion is vertical, so horizontal components of tensions cancel.

Vertical components of tensions on deflected portion of AD and BC are $-T\delta x \sin\alpha$ and $T\delta x \sin\beta$.

Since angles are small, we can replace their sines by their tangents. Hence, resultant of these two components is $T\delta x[\tan\beta-\tan\alpha]$ which is equal to $T\delta x\left[u_y\left(x_1,y+\delta y\right)-u_y\left(x_2,y\right)\right]$

where x_1,x_2 lie in $[x,x+\delta x]$.

Similarly, the resultant vertical component of tensions on deflected portions of *AB* and *DC* is

$$T\delta y\left[u_x\left(x+\delta x,y_1\right)-u_x\left(x,y_2\right)\right]$$

where y_1,y_2 lie in $[y,y+\delta y]$.

If ρ is mass per unit area of membrane, then by Newton's second law of motion (vertically upward)

$$\rho\delta x\delta y\frac{\partial^2 u}{\partial t^2}=T\delta x\left[u_y\left(x_1,y+\delta y\right)-u_y\left(x_2,y\right)\right]+T\delta y\left[u_x\left(x+\delta x,y_1\right)-u_x\left(x,y_2\right)\right]$$

$$\therefore \quad \frac{\partial^2 u}{\partial t^2}=\frac{T}{\rho}\left[\left\{\frac{u_y\left(x_1,y+\delta y\right)-u_y\left(x_2,y\right)}{\delta y}\right\}+\left\{\frac{u_x\left(x+\delta x,y_1\right)-u_x\left(x,y_2\right)}{\delta x}\right\}\right]$$

Take limit as $\delta x\to 0,\delta y\to 0$

$$\frac{\partial^2 u}{\partial t^2}=c^2\left(\frac{\partial^2 u}{\partial y^2}+\frac{\partial^2 u}{\partial x^2}\right)$$

where $c^2=\dfrac{T}{\rho}$ is diffusivity of membrane.

which is the partial differential equation of two dimensional wave motion.

4.16.1 Solution of Two Dimensional Wave Equation

Two dimensional wave equation is

$$\frac{\partial^2 u}{\partial t^2}=c^2\left(\frac{\partial^2 u}{\partial x^2}+\frac{\partial^2 u}{\partial y^2}\right)$$

Let

$$u(x,y,t)=F(x,y)\,T(t)$$

$$\therefore \quad \frac{\partial^2 u}{\partial t^2}=FT'',\ \frac{\partial^2 u}{\partial x^2}=\frac{\partial^2 F}{\partial x^2}T,\ \frac{\partial^2 u}{\partial y^2}=\frac{\partial^2 F}{\partial y^2}T$$

$\therefore$ differential equation becomes

$$FT''=c^2\left(\frac{\partial^2 F}{\partial x^2}+\frac{\partial^2 F}{\partial y^2}\right)T$$

or

$$\frac{1}{F}\left(\frac{\partial^2 F}{\partial x^2}+\frac{\partial^2 F}{\partial y^2}\right)=\frac{1}{c^2}\frac{T''}{T}$$

L.H.S. is function of *x* and *y* and R.H.S. is function of *t* and hence each is constant, say λ.

As boundaries are fixed, i.e., $u(x,y,t)=0$ for all t and all points of boundary. Hence, λ must be negative say $-p^2$.

$$\therefore \qquad \frac{\partial^2 F}{\partial x^2}+\frac{\partial^2 F}{\partial y^2}+p^2F=0 \tag{4.58}$$

and $\qquad T''=-p^2c^2T$

Its solution is

$$T=k_1\cos cpt+k_2\sin cpt \tag{4.59}$$

Now, let $\qquad F(x,y)=X(x)Y(y)$

$$\therefore \qquad \frac{\partial^2 F}{\partial x^2}=X''Y,\ \frac{\partial^2 F}{\partial y^2}=XY''$$

$\therefore$ from equation (4.58)

$$X''Y+XY''+p^2XY=0$$

$$\therefore \qquad \frac{X''}{X}=-\frac{Y''+p^2Y}{Y}$$

L.H.S. is function of x and R.H.S. is function of y and hence each is constant. As boundaries are fixed, so constant must be negative, say $-q^2$.

$$\therefore \qquad X''+q^2X=0,\ Y''+s^2Y=0 \text{ where } p^2-q^2=s^2 \tag{4.60}$$

Solutions are

$$X(x)=A\cos qx+B\sin qx \tag{4.61}$$

$$Y(y)=C\cos sy+E\sin sy \tag{4.62}$$

Now, $\qquad u(x,y,t)=X(x)Y(y)T(t)$

and on boundary, i.e., $0\le x\le a,\ 0\le y\le b$

$$u(x,y,t)=0 \text{ for all } t\ge 0$$

$$\therefore \qquad X(0)=0, X(a)=0, Y(0)=0, Y(b)=0$$

From (4.61),

$$X(0)=0 \qquad \Rightarrow \text{A}=0$$

From (4.62),

$$Y(0)=0 \qquad \Rightarrow C=0$$

$$\therefore \qquad X(x)=B\sin qx,\ Y(y)=E\sin sy$$

$$X(a) = B \sin qa = 0$$

$\Rightarrow$
$$q = \frac{m\pi}{a};\ m \in N \tag{4.63}$$

and
$$Y(b) = E \sin sb = 0$$

$\Rightarrow$
$$s = \frac{n\pi}{b};\ n \in N \tag{4.64}$$

$\therefore$
$$X(x) = B_m \sin\frac{m\pi x}{a},\ Y(y) = E_n \sin\frac{n\pi y}{b}$$

$\therefore$
$$F(x,y) = B_{m,n} \sin\frac{m\pi x}{a} \sin\frac{n\pi y}{b};\ B_{m,n} = B_m E_n;\ \begin{array}{l} m = 1,2,.... \\ n = 1,2,.... \end{array}$$

From (4.60),

$$\begin{aligned} p^2 &= q^2 + s^2 \\ &= \frac{m^2\pi^2}{a^2} + \frac{n^2\pi^2}{b^2} \qquad \text{from (4.63) and (4.64)} \\ &= \pi^2\left(\frac{m^2}{a^2} + \frac{n^2}{b^2}\right) = p_{m,n}^2 \text{ (say)} \end{aligned}$$

$\therefore$ from (4.59)

$$T = k_{m,n} \cos c\, p_{m,n} t + l_{m,n} \sin c\, p_{m,n} t$$

$\therefore$
$$\begin{aligned} u(x,y,t) &= X(x)Y(y)T(t) \\ &= F(x,y)T(t) \\ &= \sin\frac{m\pi x}{a} \sin\frac{n\pi y}{b}\left[K_{m,n} \cos c\, p_{m,n} t + L_{m,n} \sin c\, p_{m,n} t\right] \end{aligned}$$

where $K_{m,n} = B_{m,n} k_{m,n},\ L_{m,n} = B_{m,n} l_{m,n}$

By principle of superposition

$$u(x,y,t) = \sum_{m=1}^{\infty}\sum_{n=1}^{\infty}\left[K_{m,n} \cos\left(cp_{m,n}t\right) + L_{m,n} \sin\left(cp_{m,n}t\right)\right] \sin\frac{m\pi x}{a} \sin\frac{n\pi y}{b} \tag{4.65}$$

where
$$p_{m,n} = \pi\sqrt{\frac{m^2}{a^2} + \frac{n^2}{b^2}}$$

which gives us the solution of two dimension wave equation. Constants $K_{m,n}$ and $L_{m,n}$ will be obtained from initial conditions.

Suppose, the initial conditions are

$$u(x,y,0) = f(x,y),\ \frac{\partial u}{\partial t}(x,y,0) = g(x,y)$$

then $$u(x,y,0)=f(x,y)=\sum_{m=1}^{\infty}\sum_{n=1}^{\infty}K_{m,n}\sin\frac{m\pi x}{a}\sin\frac{n\pi y}{b}$$

$$=\sum_{m=1}^{\infty}\left(\sum_{n=1}^{\infty}K_{m,n}\sin\frac{n\pi y}{b}\right)\sin\frac{m\pi x}{a}$$

Let $$l_m(y)=\sum_{n=1}^{\infty}K_{m,n}\sin\frac{n\pi y}{b} \tag{4.66}$$

$\therefore$ $$f(x,y)=\sum_{m=1}^{\infty}l_m(y)\sin\frac{m\pi x}{a}$$

It is half range Fourier sine series of $f(x, y)$ in $x \in [0, a]$ where y is taken as constant

$\therefore$ $$l_m(y)=\frac{2}{a}\int_0^a f(x,y)\sin\frac{m\pi x}{a}dx \text{ (taking } y \text{ constant)} \tag{4.67}$$

Now, (4.66) is Fourier sine series of $l_m(y)$ in $y\in[0,b]$.

$\therefore$ $$K_{m,n}=\frac{2}{b}\int_0^b l_m(y)\sin\frac{n\pi y}{b}dy$$

Put value of $l_m(y)$ from (4.67)

$$K_{m,n}=\frac{4}{ab}\int_0^b\int_0^a f(x,y)\sin\frac{m\pi x}{a}\sin\frac{n\pi y}{b}dxdy \tag{4.68}$$

From (4.65)

$$\frac{\partial}{\partial t}u(x,y,t)=\sum_{m=1}^{\infty}\sum_{n=1}^{\infty}c\,p_{m,n}\left[-K_{m,n}\sin(c\,p_{m,n}t)+L_{m,n}\cos(c\,p_{m,n}t)\right]\cdot\sin\left(\frac{m\pi x}{a}\right)\sin\left(\frac{n\pi y}{b}\right)$$

$$\therefore \quad \frac{\partial u}{\partial t}(x,y,0)=\sum_{m=1}^{\infty}\sum_{n=1}^{\infty}c\,p_{m,n}L_{m,n}\sin\frac{m\pi x}{a}\sin\frac{n\pi y}{b}=g(x,y)$$

Proceeding as above

$$L_{m,n}=\frac{4}{cp_{m,n}ab}\int_0^b\int_0^a g(x,y)\sin\frac{m\pi x}{a}\sin\frac{n\pi y}{b}dxdy \tag{4.69}$$

Substituting the value of $K_{m,n}$ and $L_{m,n}$ in (4.65), we obtain the solution.

Example 4.71: Find the deflection $u(x,y,t)$ of a rectangular membrane $(0\le x\le a, 0\le y\le b)$ whose boundary is fixed, given that it starts from rest and $u(x,y,0)=xy(a-x)(b-y)$.

Solution: Let $u(x,y,t)$ be deflection of point (x,y) at time t.

Then, wave equation is

$$\frac{\partial^2 u}{\partial t^2}=c^2\left(\frac{\partial^2 u}{\partial x^2}+\frac{\partial^2 u}{\partial y^2}\right)$$

where c^2 is diffusivity of membrane.

When boundaries are fixed then on the boundaries $u(x,y,t)=0$ for all t.

Under this boundary condition solution is

$$u(x,y,t)=\sum_{m=1}^{\infty}\sum_{n=1}^{\infty}\left[k_{m,n}\cos\left(p_{m,n}ct\right)+l_{m,n}\sin\left(p_{m,n}ct\right)\right]\sin\frac{m\pi x}{a}\sin\frac{n\pi y}{b}, \tag{1}$$

where $$p_{m,n}^2=\pi^2\left(\frac{m^2}{a^2}+\frac{n^2}{b^2}\right)$$

$$xy(a-x)(b-y)=u(x,y,0)=\sum_{m=1}^{\infty}\sum_{n=1}^{\infty}k_{m,n}\sin\frac{m\pi x}{a}\sin\frac{n\pi y}{b}$$

$$=\sum_{m=1}^{\infty}\left[\sum_{n=1}^{\infty}k_{m,n}\sin\frac{n\pi y}{b}\right]\sin\frac{m\pi x}{a}$$

Considering y constant, it is Fourier half range sine series of $xy(a-x)(b-y)$ in $[0,a]$.

$$\therefore \quad \sum_{n=1}^{\infty}k_{m,n}\sin\frac{n\pi y}{b}=\frac{2}{a}\int_0^a xy(a-x)(b-y)\sin\frac{m\pi x}{a}dx$$

$$=\frac{2y(b-y)}{a}\left[\left(ax-x^2\right)\left(-\frac{a}{m\pi}\cos\frac{m\pi x}{a}\right)-(a-2x)\left(-\frac{a^2}{m^2\pi^2}\sin\frac{m\pi x}{a}\right)+(-2)\left(\frac{a^3}{m^3\pi^3}\cos\frac{m\pi x}{a}\right)\right]_0^a$$

$$=\frac{2y(b-y)}{a}\left[-\frac{2a^3}{m^3\pi^3}\left((-1)^m-1\right)\right]$$

$$=\frac{-4a^2y(b-y)}{m^3\pi^3}\left[(-1)^m-1\right]$$

$$\therefore \quad k_{2m,n}=0;\ m=1,2,3,\ldots$$

$$\sum_{n=1}^{\infty}k_{2m-1,n}\sin\frac{n\pi y}{b}=\frac{8a^2y(b-y)}{(2m-1)^3\pi^3};\ m=1,2,3,\ldots$$

It is, again, Fourier half range sine series of $\frac{8a^2y(b-y)}{(2m-1)^3\pi^3}$ in $[0,b]$.

$$\therefore \quad k_{2m-1,n}=\frac{2}{b}\int_0^b\frac{8a^2y(b-y)}{(2m-1)^3\pi^3}\sin\frac{n\pi y}{b}dy$$

$$=\frac{16a^2}{(2m-1)^3b\pi^3}\left[\left(by-y^2\right)\left(-\frac{b}{n\pi}\cos\frac{n\pi y}{b}\right)-(b-2y)\left(-\frac{b^2}{n^2\pi^2}\sin\frac{n\pi y}{b}\right)+(-2)\left(\frac{b^3}{n^3\pi^3}\cos\frac{n\pi y}{b}\right)\right]_0^b$$

$$= \frac{-32a^2b^2}{(2m-1)^3\pi^6 n^3}\left[(-1)^n - 1\right]$$

$$\therefore \qquad k_{2m-1,\,2n-1} = \frac{64a^2b^2}{(2m-1)^3(2n-1)^3\pi^6} \text{ and all other } k_{m,n} = 0$$

From (1)

$$\frac{\partial}{\partial t}u(x,y,t) = \sum_{m=1}^{\infty}\sum_{n=1}^{\infty} c\,p_{m,n}\left[-k_{m,n}\sin(p_{m,n}ct) + l_{m,n}\cos(p_{m,n}ct)\right]\sin\left(\frac{m\pi x}{a}\right)\sin\left(\frac{n\pi y}{b}\right)$$

$$\therefore \qquad 0 = \frac{\partial}{\partial t}u(x,y,0) = \sum_{m=1}^{\infty}\sum_{n=1}^{\infty} l_{m,n}p_{m,n}\,c\sin\frac{m\pi x}{a}\sin\frac{n\pi y}{b}$$

$$\therefore \qquad l_{m,n} = 0 \text{ for all } m, n.$$

Put values of $k_{m,n}$ and $l_{m,n}$ in (1)

$$u(x,y,t) = \sum_{m=1}^{\infty}\sum_{n=1}^{\infty}\frac{64a^2b^2}{(2m-1)^3(2n-1)^3\pi^6}\cos(p_{2m-1,2n-1}ct)\sin\frac{(2m-1)\pi x}{a}\sin\frac{(2n-1)\pi y}{b}$$

where $\quad p^2_{2m-1,2n-1} = \pi^2\left[\frac{(2m-1)^2}{a^2} + \frac{(2n-1)^2}{b^2}\right]$

Example 4.72: Find the deflection $u(x,y,t)$ of a square membrane of one unit side whose boundaries are fixed, if the initial velocity is zero and the initial deflection is given by $f(x,y) = \sin(3\pi x)\ \sin(4\pi y)$. Assume $c = 1$ in the differential equation.

Solution: Let $u(x,y,t)$ be deflection of point (x,y) at time t.

Then, wave equation is

$$\frac{\partial^2 u}{\partial t^2} = \frac{\partial^2 u}{\partial x^2} + \frac{\partial^2 u}{\partial y^2};\ 0 \le x \le 1,\ 0 \le y \le 1.$$

When boundaries are fixed then on the boundaries $u(x,y,t) = 0$ and solution then will be

$$u(x,y,t) = \sum_{m=1}^{\infty}\sum_{n=1}^{\infty}\left[k_{m,n}\cos(p_{m,n}t) + l_{m,n}\sin(p_{m,n}t)\right]\sin(m\pi x)\sin(n\pi y),$$

where $\qquad p^2_{m,n} = \pi^2(m^2+n^2)$

$$\frac{\partial}{\partial t}u(x,y,t) = \sum_{m=1}^{\infty}\sum_{n=1}^{\infty} p_{m,n}\left[-k_{m,n}\sin(p_{m,n}t) + l_{m,n}\cos(p_{m,n}t)\right]\sin(m\pi x)\sin(n\pi y)$$

$$\therefore \qquad 0 = \left(\frac{\partial u}{\partial t}\right)_{t=0} = \sum_{m=1}^{\infty}\sum_{n=1}^{\infty} l_{m,n}p_{m,n}\sin(m\pi x)\sin(n\pi y)$$

$$\Rightarrow \qquad l_{m,n} = 0 \text{ for all } m, n.$$

$$\therefore \quad u(x,y,t)=\sum_{m=1}^{\infty}\sum_{n=1}^{\infty}k_{m,n}\cos\left(p_{m,n}t\right)\sin(m\pi x)\sin(n\pi y)$$

$$\therefore \quad \sin(3\pi x)\sin(4\pi y)=u(x,y,0)=\sum_{m=1}^{\infty}\sum_{n=1}^{\infty}k_{m,n}\sin(m\pi x)\sin(n\pi y)$$

$$\therefore \quad k_{3,4}=1 \text{ and all other } k_{m,n}=0$$

$\therefore$ solution is

$$u(x,y,t)=\cos\left(p_{3,4}t\right)\sin(3\pi x)\sin(4\pi y)$$

where $$p_{3,4}=\pi\sqrt{3^2+4^2}=5\pi$$

$$\therefore \quad u(x,y,t)=\cos(5\pi t)\sin(3\pi x)\sin(4\pi y).$$

EXERCISE 4.8

1. Solve $\dfrac{\partial^2 u}{\partial x^2}+\dfrac{\partial^2 u}{\partial y^2}=0$ subject to boundary conditions $u(0,y)=\sin y$ and $u\to 0$ as $x\to\infty$.

2. Solve $\dfrac{\partial^2 u}{\partial x^2}+\dfrac{\partial^2 u}{\partial y^2}=0$ in a rectangle in the x–y plane, $0<x<a$ and $0<y<b$ satisfying the following boundary conditions $u(x,0)=u(x,b)=u(0,y)=0$ and $u(a,y)=f(y)$. Also, find the solution when $f(y)=ky(b-y), 0<y<b$.

3. Solve $\dfrac{\partial^2 u}{\partial x^2}+\dfrac{\partial^2 u}{\partial y^2}=0$ in the interval $0\le x\le\pi$, subject to the boundary conditions: $u(0,y)=0, u(\pi,y)=0, u(x,0)=1$ and $u(x,y)\to 0$ as $y\to\infty$ for all x.

4. Solve $\dfrac{\partial^2 u}{\partial x^2}+\dfrac{\partial^2 u}{\partial y^2}=0$, subject to the conditions $u(x,0)=0, u(x,a)=0, u(x,y)\to 0$ as $x\to\infty$ when $x\ge 0$ and $0\le y\le a$.

5. An infinitely long metal plate of width 1 with insulated surfaces has its temperature zero along both the long edges $y=0$ and $y=1$ and the short edge at infinity. If the edge $x=0$ is kept at fixed temperature T_0, find the temperature T at any point (x,y) of the plate in steady state.

6. Solve $\dfrac{\partial^2 u}{\partial x^2}+\dfrac{\partial^2 u}{\partial y^2}=0$ which satisfies the conditions: $u(0,y)=u(l,y)=u(x,0)=0$ and $u(x,a)=\dfrac{\sin n\pi x}{l}$.

7. An infinitely long plane uniform plate is bounded by two parallel edges and an end at right angles to them. The breadth is π. This end is maintained at temperature u_0 at all points and the other edges are at zero temperature. Determine the temperature at any point of the plate in the steady state.

8. A tightly stretched unit square membrane starts vibrating from rest and its initial displacement is $k\sin 2\pi x\sin\pi y$. Show that the deflection at any instant is $k\sin 2\pi x\sin\pi y\cos\left(\sqrt{5}\pi ct\right)$.

9. Find the deflection $u(x,y,t)$ of the square membrane of side unity and boundaries fixed with $c=1$, if the initial velocity is zero and the initial deflection $f(x,y)=A\sin\pi x\sin 2\pi y$.

10. Determine the displacement function $u(x,y,t)$ of a rectangular membrane $0<x<l_1,\ 0<y<l_2$ with the entire boundary fixed and with initial conditions $u(x,y,0)=0$ and $u_t(x,y,0)=1$.

ANSWERS 4.8

1. $u(x,y)=e^{-x}\sin y$.

2. $u(x,y)=\sum_{n=1}^{\infty} b_n \sinh\left(\frac{n\pi x}{b}\right)\sin\left(\frac{n\pi y}{b}\right)$, where $b_n=\frac{2}{b\sinh\left(\frac{n\pi a}{b}\right)}\int_0^b f(y)\sin\left(\frac{n\pi y}{b}\right)dy$

and if $f(y)=ky(b-y)$, $u(x,y)=\frac{8kb^2}{\pi^3}\sum_{n=1}^{\infty}\frac{\sinh\frac{(2n-1)\pi x}{b}\sin\frac{(2n-1)\pi y}{b}}{(2n-1)^3\sinh\frac{(2n-1)\pi a}{b}}$.

3. $u(x,y)=\frac{4}{\pi}\sum_{n=1}^{\infty}\frac{e^{-(2n-1)y}}{(2n-1)}\sin(2n-1)x$.

4. $u(x,y)=\sum_{n=1}^{\infty} c_n e^{\frac{-n\pi x}{a}}\sin\frac{n\pi y}{a}$, where $c_n, n=1,2,3\ldots$ are arbitrary constants.

5. $T(x,y)=\frac{4T_0}{\pi}\sum_{n=1}^{\infty}\frac{1}{(2n-1)}e^{-(2n-1)\pi x}\sin(2n-1)\pi y$.

6. $u(x,y)=\sin\frac{n\pi x}{l}\left[\frac{\sinh\frac{n\pi y}{l}}{\sinh\frac{n\pi a}{l}}\right]$.

7. $u(x,y)=\frac{4u_0}{\pi}\sum_{n=1}^{\infty}\frac{1}{(2n-1)}e^{-(2n-1)y}\sin(2n-1)x$.

9. $u(x,y,t)=A\cos\sqrt{5}\pi t\sin\pi x\sin 2\pi y$.

10. $u(x,y,t)=\frac{16}{\pi^2 c}\sum_{m=1}^{\infty}\sum_{n=1}^{\infty}\frac{\sin\left(c\,p_{2m-1,2n-1}t\right)}{(2m-1)(2n-1)\,p_{2m-1,2n-1}}\sin\left\{\frac{(2m-1)\pi x}{l_1}\right\}\sin\left\{\frac{(2n-1)\pi y}{l_2}\right\}$

where $p^2_{2m-1,2n-1}=\pi^2\left\{\frac{(2m-1)^2}{l_1^2}+\frac{(2n-1)^2}{l_2^2}\right\}$.

Numerical Methods in General and Linear Algebra

5

5.1 INTRODUCTION

Numerical methods are referred to the methods for solving problems numerically on a computer or a calculator, or in older times by hand. Computers have changed the field as a whole as well as many individual methods. Numerical methods are necessary because for many problems, there is no solution formula as for many algebraic and transcendental equations, or in some cases a solution formula is practically useless. Ideas of round-off errors and various types of errors occurring in numerical methods are discussed. Methods to find roots of algebraic and transcendental equations and various methods to solve linear system of equations and eigenvalue problems are also discussed. Interpolation means to find approximate values of a function $f(x)$ for x between given x values, and extrapolation means to find $f(x)$ out of the range of given x values. Various interpolation formulae are studied in detail.

5.2 ERRORS IN NUMERICAL COMPUTATIONS

Usually, the following three different types of errors occur in numerical computations.

(i) Gross errors
(ii) Round-off errors
(iii) Truncation errors

We shall deal them one by one.

Gross Errors: These errors are due to human mistakes or due to malfunctioning of a calculator or a computer as the case may be. Although these errors occur quite frequently in numerical computations, yet we shall not treat them here.

Round-off Errors: A number is rounded to position n by making all digits to the right of this position zero. The digit in position n is unchanged, if the truncated part is less than half a unit of the position value of n th place. The digit in position n is increased by one, if the truncated part is more than half a unit of the position value of the n th place. If the truncated part is exactly half a unit of the position value of n th place, then n th place is unchanged if n th place has even digit (0 is considered as even), and n th place is increased by one unit, if n th place has odd digit. It should also be noted that the significant digits in a number start with the first non-zero digit and end with the non-zero digit if it does not contain decimal part, and if it contains decimal part, then zeros to the right in decimal part are also significant.

For example, 895472603 rounded to 4 significant digits is 895500000.
5.2435904 rounded to 5 decimal places is 5.24359
8.73500 rounded to two decimal places is 8.74
7.24500 rounded to two decimal places is 7.24
11.34576523 rounded to five decimal places is 11.34577.

Truncation Errors: Truncation error arises when an infinite process is approximated by a finite one. Suppose, we approximate e^x by some polynomial, then truncation error arises. Truncation error also arises when some differential equation is solved by difference method.

In practical computations, we start with some initial data. Even if the intermediate calculations are exact, the final result may give large deviations. Such type of problem is called ill-conditioned.

In numerical computations, each step introduces a round-off error. The error in each step will influence the error in subsequent steps. We must be careful that error propagation may give the stable results.

5.3 ALGEBRAIC AND TRANSCENDENTAL EQUATIONS

An equation $f(x) = 0$ is called an algebraic equation of degree n, if $f(x)$ is a polynomial of degree n. If $f(x)$ contains some other functions such as trigonometric, logarithmic, exponential, etc., then $f(x) = 0$ is called a transcendental equation.

We shall now consider the concept of a root of an equation $f(x) = 0$. By definition, α is a root of $f(x) = 0$ if $f(\alpha) = 0$. However, in numerical applications, the equation usually cannot be satisfied exactly. Therefore, we modify the mathematical definition of root here.

α will be called a root of $f(x) = 0$, if $|f(\alpha)| < \varepsilon$, where ε is given tolerance. According to this definition, $f(x) = 0$ and $M f(x) = 0$ for some constant M do not have same roots. Before dealing the methods to find root of an equation, we state a property.

Intermediate Value Property

If $f(x)$ is continuous in $[a, b]$ and $f(a) \cdot f(b) < 0$, i.e., $f(a)$ and $f(b)$ have opposite signs, then the equation $f(x) = 0$ has at least one real root in (a, b). Further, if $|f(a)| < |f(b)|$, then, in general, root is near a, as compared to b.

Iterative Process

Suppose one or more approximations of the root of an equation are known and we find the next approximation using the formula containing known approximations, then this process is known as iterative process. Suppose x_1 is an approximation of root of an equation and we find x_2 using formula containing x_1, then approximation x_3 from x_2 and so on, then this process is called iterative process.

Rate of Convergence

The fastness of convergence in any method is represented by its rate of convergence.

Let $x_1, x_2, x_3, \cdots, x_n, x_{n+1}$ be successive approximations of root of an equation having errors $\varepsilon_1, \varepsilon_2, \varepsilon_3, \cdots, \varepsilon_n, \varepsilon_{n+1}$. If $\varepsilon_{n+1} = K\varepsilon_n^m$ then convergence is said to be of order *m*. If *m* = 1, then convergence is called linear and if *m* = 2, then convergence is called quadratic. When the convergence is quadratic, then the number of correct decimals is approximately doubled at every iteration, at least if the factor *K* is not too large. In case of linear convergence $\varepsilon_{n+1} = K\varepsilon_n$ and hence $\varepsilon_{n+1} = K\varepsilon_n = K^2\varepsilon_{n-1} = K^3\varepsilon_{n-2} = \cdots = K^n\varepsilon_1$.

Thus, the error multiplies by *K* at each step and hence this convergence is also called as geometric convergence.

Now, we shall be dealing some methods to find roots of a given equation.

5.3.1 Bisection Method or Bolzano Method or Halving Method

This method is based on the repeated application of intermediate value property.

Suppose we are to find real root of the equation $f(x) = 0$, where $f(x)$ is a continuous function. Let *a* and *b* be real numbers such that $f(a)$ and $f(b)$ have opposite signs, then the first approximation to the root is $x_1 = \frac{1}{2}(a+b)$. If $f(x_1) = 0$, then x_1 is root. If $f(x_1) \neq 0$, then either $f(a)$ and $f(x_1)$ have opposite signs in which case second approximation will be $x_2 = \frac{a+x_1}{2}$ or $f(x_1)$ and $f(b)$ have opposite signs in which case second approximation will be $x_2 = \frac{x_1+b}{2}$.

Now, replace *a* or *b* by x_1 as the case be, then next approximation will be $x_3 = \frac{x_1+x_2}{2}$ and so on.

Number of Iterations Required to Reach Accuracy ε

Suppose *M* is the length of interval (*a*, *b*), then after first approximation x_1 the root will lie in $\left(a, \frac{a+b}{2}\right)$ or in $\left(\frac{a+b}{2}, b\right)$ or $x_1 = \frac{a+b}{2}$ is root and thus root will lie in an interval of length $\frac{M}{2}$. Thus, at every step, the new interval containing the root is exactly half the length of the previous one. At the end of *n* steps when we obtain x_n, the root will lie in an interval of length $\frac{b-a}{2^n}$. Thus, the number of iterations *n* required to reach accuracy ε must satisfy

$$\frac{b-a}{2^n} \leq \varepsilon$$

or

$$\log(b-a) - n\log 2 \leq \log\varepsilon$$

or

$$n \geq \frac{\log(b-a) - \log\varepsilon}{\log 2}$$

Smallest natural number *n* satisfying this inequality gives the number of iterations required to reach accuracy ε.

As the length of interval at each step is $\frac{1}{2}$ the length of interval in the previous step in which root lies, so if ε_{n+1} is error in x_{n+1} and ε_n is error in x_n then

$$\varepsilon_{n+1} = \frac{1}{2}\varepsilon_n$$

Hence, convergence is linear. Also, the convergence is geometric with common ratio $\frac{1}{2} < 1$ and thus, the process must converge to root. Hence, the process is slow but must converge.

5.3.2 Direct Iteration Method

To find the real root of the equation $f(x) = 0$ by successive approximations, we write this equation as $x = \phi(x)$. Starting with suitable x_0 we find $x_1 = \phi(x_0)$, $x_2 = \phi(x_1)\cdots$, $x_{n+1} = \phi(x_n)$. If the sequence $\{x_1, x_2, \cdots, x_n, x_{n+1}\}$ has a limit ξ, then ξ is a root of $f(x) = 0$. The root is same as the point of intersection of the straight line $y = x$ and the curve $y = \phi(x)$. Figures (5.1) to (5.4) illustrate the working of the iterative method.

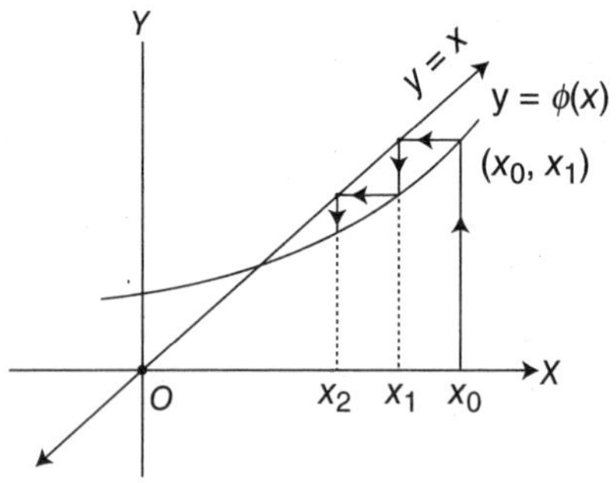

Figrue 5.1

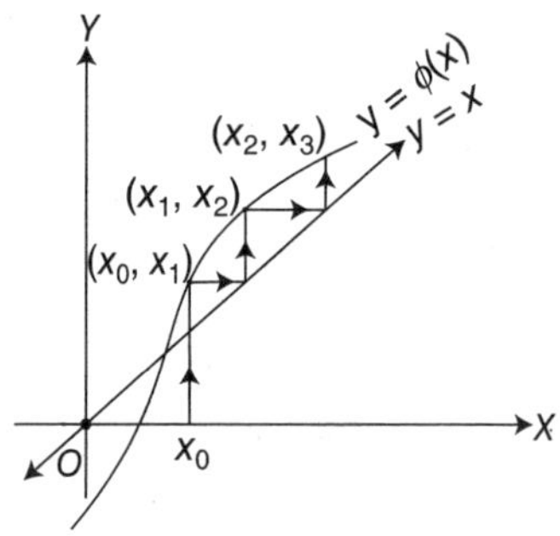

Figure 5.2

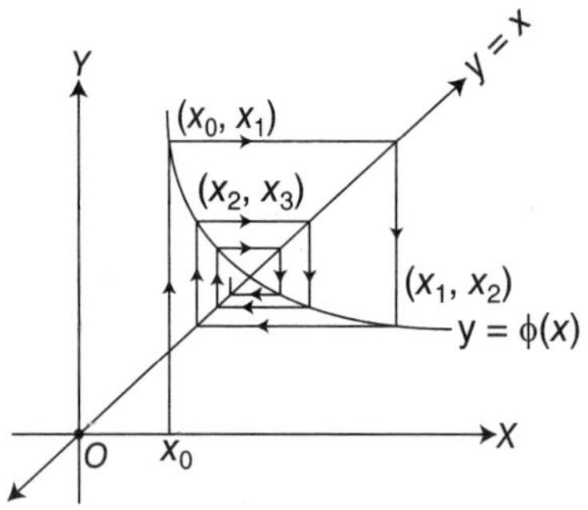

Figure 5.3

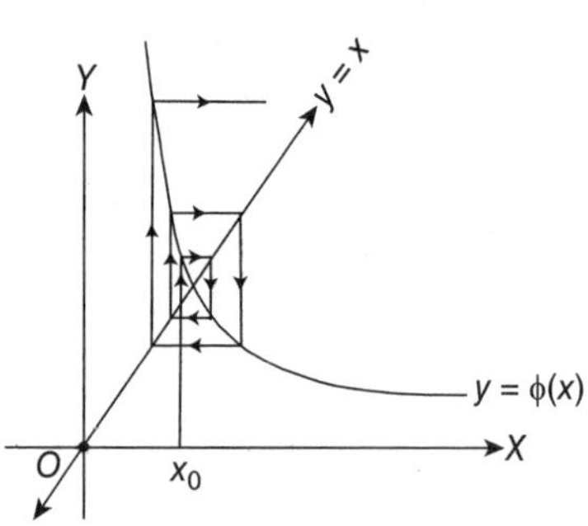

Figure 5.4

From the figures, it is clear that the convergence depends upon the form of $\phi(x)$. In Figures (5.1) and (5.3) the iterations converge to the root and in Figures (5.2) and (5.4), the iterations do not converge to the root. The equation $f(x) = 0$ can be written in the form $x = \phi(x)$ in an infinite number of ways. For example, the equation $x^3 - 5 = 0$ can be written as $x = x^3 + x - 5$ or $x = \frac{5+4x-x^3}{4}$ or in other ways. Thus, now we derive the condition of convergence so that the equation $f(x) = 0$ be changed to the form $x = \phi(x)$ and the process of iteration converges to the root.

Let ξ be exact root of $f(x) = 0$ which is changed to $x = \phi(x)$

then $\xi = \phi(\xi)$.

Let ε_{n+1} be error in x_{n+1}

$\therefore \quad \xi + \varepsilon_{n+1} = x_{n+1}$

Now, by Cauchy mean value theorem,

$\varepsilon_1 = x_1 - \xi = \phi(x_0) - \phi(\xi) = (x_0 - \xi)\phi'(\xi_0)$; where ξ_0 lies in the interval formed by x_0 and ξ.

Similarly $x_2 - \xi = (x_1 - \xi)\phi'(\xi_1)$; where ξ_1 lies in the interval formed by x_1 & ξ

and so on.

$x_{n+1} - \xi = (x_n - \xi)\phi'(\xi_n)$; where ξ_n lies in the interval formed by x_n & ξ

Multiplying these, we have

$x_{n+1} - \xi = (x_0 - \xi)\phi'(\xi_0)\phi'(\xi_1)\cdots\phi'(\xi_n)$; $\xi_0, \xi_1, \cdots, \xi_n$ lie in the interval formed by x_0, x_1, $x_2, \cdots, x_n$ & ξ.

$$\therefore \quad \varepsilon_{n+1} = \varepsilon_0 \, \phi'(\xi_0)\phi'(\xi_1)\cdots\phi'(\xi_n)$$

Let

$$|\phi'(\xi_i)| \le m \text{ for } i = 0, 1, 2, \cdots, n$$

then $|\varepsilon_{n+1}| \le m^{n+1} |\varepsilon_0|$

Thus, convergence is linear, and iteration converges if $m < 1$

$\therefore$ Iteration converges if $|\phi'(x)| < 1$ where x lies in the interval formed by iteration values.

Now, we shall be taking some examples.

Example 5.1: Find a root of the equation $x^3 - 4x - 9 = 0$ using the bisection method in four stages.

Solution:

$$f(x) = x^3 - 4x - 9 = 0$$

$$f(2) = -9, f(3) = 6$$

$$|f(2)| > |f(3)|$$

$$f(2.7) = 2.7^3 - 4(2.7) - 9 = -0.117$$

$$f(2.8) = 2.8^3 - 4(2.8) - 9 = 1.752$$

$\therefore$ root lies between 2.7 and 2.8

$$\therefore \quad x_1 = \frac{2.7+2.8}{2} = 2.75$$

Approximate root x	$f(x)$	Root between	Next approximation
$x_1 = 2.75$	+ive	2.7 and x_1	$\frac{2.7+2.75}{2} = 2.725$
$x_2 = 2.725$	+ive	2.7 and x_2	$\frac{2.7+2.725}{2} = 2.7125$
$x_3 = 2.7125$	+ive	2.7 and x_3	$\frac{2.7+2.7125}{2} = 2.70625$
$x_4 = 2.70625$	–ive	x_3 and x_4	$\frac{2.7125+2.70625}{2} = 2.709375$

∴ approximate root = 2.71

Example 5.2: (a) Find a root of the equation $x^3 - x - 11 = 0$ and correct to four decimal places using bisection method.

(b) Using bisection method, find a negative root of $x^3 - x + 11 = 0$.

Solution: (a) $f(x) = x^3 - x - 11 = 0$

$$f(0) = -11,\ f(1) = -11,\ f(2) = -5,\ f(3) = 13$$

∴ Root lies between 2 and 3

∵ $$|f(2)| < |f(3)|$$

∴ Root is near 2

$$f(2.3) = (2.3)^3 - (2.3) - 11 = -1.133 < 0$$

$$f(2.4) = (2.4)^3 - (2.4) - 11 = 0.424 > 0$$

$$|f(2.4)| < |f(2.3)|$$

∴ Root lies near 2.4

$$f(2.37) = (2.37)^3 - (2.37) - 11 = -0.058 < 0$$

$$f(2.38) = (2.38)^3 - (2.38) - 11 = 0.101 > 0$$

∴ Root lies between 2.37 and 2.38

$$x_1 = \frac{2.37 + 2.38}{2} = 2.375$$

Approximate root *x*	*f*(*x*)	Root between	Next approximation
$x_1 = 2.375$	+ive	2.37 and x_1	$\dfrac{2.37+2.375}{2} = 2.3725$
$x_2 = 2.3725$	–ive	x_1 and x_2	$\dfrac{2.375+2.3725}{2} = 2.37375$
$x_3 = 2.37375$	+ive	x_2 and x_3	$\dfrac{2.3725+2.37375}{2} \simeq 2.37313$
$x_4 = 2.37313$	–ive	x_3 and x_4	$\dfrac{2.37375+2.37313}{2} = 2.37344$
$x_5 = 2.37344$	–ive	x_3 and x_5	$\dfrac{2.37375+2.37344}{2} \simeq 2.37360$
$x_6 = 2.37360$	–ive	x_3 and x_6	$\dfrac{2.37375+2.37360}{2} \simeq 2.37368$
$x_7 = 2.37368$	+ive	x_6 and x_7	$\dfrac{2.37360+2.37368}{2} \simeq 2.37364$
$x_8 = 2.37364$	–ive	x_7 and x_8	$\dfrac{2.37368+2.37364}{2} \simeq 2.37366$
$x_9 = 2.37366$	+ive	x_8 and x_9	$\dfrac{2.37364+2.37366}{2} = 2.37365$
$x_{10} = 2.37365$	+ive	x_8 and x_{10}	$\dfrac{2.37364+2.37365}{2} \simeq 2.373645$

$\therefore$ Approximate root corrected upto four decimal places = 2.3736

(b) If α is negative root of $x^3 - x + 11 = 0$

then $-\alpha$ is positive root of $(-x)^3 - (-x) + 11 = 0$, i.e. $x^3 - x - 11 = 0$

from part (a), 2.3736 is positive root of $x^3 - x - 11 = 0$

$$\therefore \qquad -\alpha = 2.3736$$

$$\therefore \qquad \alpha = -2.3736$$

$\therefore$ $\alpha = -2.3736$ is a negative root of $x^3 - x + 11 = 0$

Example 5.3: Find a root of the equation $2x = \cos x + 3$ and correct to three decimal places using direct iteration method.

Solution: Equation is

$$f(x) = 2x - \cos x - 3 = 0$$

$$f(0) = -4,\ f\left(\frac{\pi}{2}\right) = \pi - 3 \simeq 0.1416 \text{ and } \left|f\left(\frac{\pi}{2}\right)\right| < |f(0)|$$

$\therefore$ Root lies between 0 and $\dfrac{\pi}{2}$ and is near $\dfrac{\pi}{2}$.

Given equation can be written as

$$x = \frac{1}{2}(\cos x + 3) = \phi(x)$$

$$\therefore \quad \phi'(x) = -\frac{1}{2}\sin x$$

$$\therefore \quad |\phi'(x)| < 1 \text{ for all } x.$$

$\therefore$ Iteration will converge. Take $x_0 = \dfrac{\pi}{2}$

x	$\phi(x)$
$x_0 = \pi/2$	1.5
$x_1 = 1.5$	1.5354
$x_2 = 1.5354$	1.5177
$x_3 = 1.5177$	1.5265
$x_4 = 1.5265$	1.5221
$x_5 = 1.5221$	1.5243
$x_6 = 1.5243$	1.5232
$x_7 = 1.5232$	1.5238
$x_8 = 1.5238$	1.5235
$x_9 = 1.5235$	1.5236
$x_{10} = 1.5236$	1.5236

$\therefore$ upto three decimal places root = 1.524

Example 5.4: Solve by iteration $2x - \log_{10} x = 7$

Solution: Equation is

$$f(x) = 2x - \log_{10} x - 7 = 0$$

$$f(1) = -5,\ f(2) \simeq -3.30,\ f(3) \simeq -1.477,\ f(4) \simeq .398$$

$$f(3) < 0,\ f(4) > 0$$

$\therefore$ Root lies between 3 and 4 and is near 4

Given equation can be written as

$$x = \frac{1}{2}(7 + \log_{10} x) = \phi(x)$$

$$\phi'(x) = \frac{1}{2x}\log_{10} e$$

$$\therefore \quad |\phi'(x)| < 1 \quad \text{for } x \in (3,4)$$

$\therefore$ Iteration will converge

Take $x_0 = 3.7$

x	ϕ(x)
3.7	3.78410
3.78410	3.78898
3.78898	3.78926
3.78926	3.78928
3.78928	3.78928

$\therefore$ Root upto four decimal places = 3.7893

Example 5.5: Find the smallest root of the equation $f(x) = 1 - x + \frac{x^2}{(2!)^2} - \frac{x^3}{(3!)^2} + \frac{x^4}{(4!)^2} - \cdots = 0.$

Solution: $f(x) = 1 - x + \frac{x^2}{(2!)^2} - \frac{x^3}{(3!)^2} + \frac{x^4}{(4!)^2} - \cdots = 0$ (1)

Here $f(1) > 0, f(2) < 0$

$\therefore$ Smallest root lies between 1 and 2

we take $x_0 = 1.5$

(1) can be written as

$$x = 1 + \frac{x^2}{(2!)^2} - \frac{x^3}{(3!)^2} + \frac{x^4}{(4!)^2} - \cdots = \phi(x)$$

$$\phi(1.5) = 1 + \frac{(1.5)^2}{(2!)^2} - \frac{(1.5)^3}{(3!)^2} + \frac{(1.5)^4}{(4!)^2} - \cdots$$

$$= 1 + 0.5625 - 0.0938 + .0088 - .0005 + \ldots \simeq 1.4770$$

$\phi(1.4770) = 1 + 0.5454 - 0.0895 + 0.0083 - 0.0005 + \ldots \simeq 1.4637$

$\phi(1.4637) \simeq 1 + 0.5356 - 0.0871 + 0.0080 - 0.0005 = 1.4560$

$\phi(1.4560) \simeq 1 + 0.5300 - 0.0857 + 0.0078 - 0.0005 = 1.4516$

$\phi(1.4516) \simeq 1 + 0.5268 - 0.0850 + 0.0077 - 0.0004 = 1.4491$

$\phi(1.4491) \simeq 1 + 0.5250 - 0.0845 + 0.0077 - 0.0004 = 1.4478$

$\therefore$

x	ϕ(x)
1.5	1.4770
1.4770	1.4637
1.4637	1.4560
1.4560	1.4516
1.4516	1.4491
1.4491	1.4478

$\therefore$ upto two decimal places $x = 1.45$

5.3.3 Secant and Regula-falsi Methods

Secant Method

Let x_0, x_1 be two approximations of root of $y = f(x) = 0$. Then $P(x_0, y_0)$ and $Q(x_1, y_1)$ are two points on the curve $y = f(x)$ where $y_0 = f(x_0)$, $y_1 = f(x_1)$. Join PQ. We approximate the curve by secant (chord) PQ and take the point of intersection of PQ with x-axis as the next approximation x_2 of the root. Then, we take secant joining $Q(x_1, y_1)$ and $R(x_2, y_2)$ as approximation of the curve and point of intersection of QR with x-axis as the next approximation x_3 of the root. Proceeding in this way, curve is approximated by secant joining (x_{n-1}, y_{n-1}) and (x_n, y_n) and its point of intersection with x-axis as the approximation x_{n+1} of the root. Equation of secant joining (x_{n-1}, y_{n-1}) and (x_n, y_n) is $y - y_n = \dfrac{y_n - y_{n-1}}{x_n - x_{n-1}}(x - x_n)$. For the point where it meets x-axis, we have $y = 0$ and x be x_{n+1}

$$\therefore \qquad -y_n = \frac{y_n - y_{n-1}}{x_n - x_{n-1}}(x_{n+1} - x_n)$$

$$\therefore \qquad x_{n+1} = x_n - \frac{x_n - x_{n-1}}{y_n - y_{n-1}} y_n = x_n - \frac{x_n - x_{n-1}}{f(x_n) - f(x_{n-1})} f(x_n)$$

Equation of secant joining (x_{n-1}, y_{n-1}) and (x_n, y_n) can also be written as

$$y - y_{n-1} = \frac{y_n - y_{n-1}}{x_n - x_{n-1}}(x - x_{n-1})$$

$\therefore$ Interchanging x_n and x_{n-1} above, iteration can also be written as

$$x_{n+1} = x_{n-1} - \frac{x_n - x_{n-1}}{f(x_n) - f(x_{n-1})} f(x_{n-1})$$

which is the iterative formula to find the approximations. Secant method is shown in Figures (5.5) and (5.6).

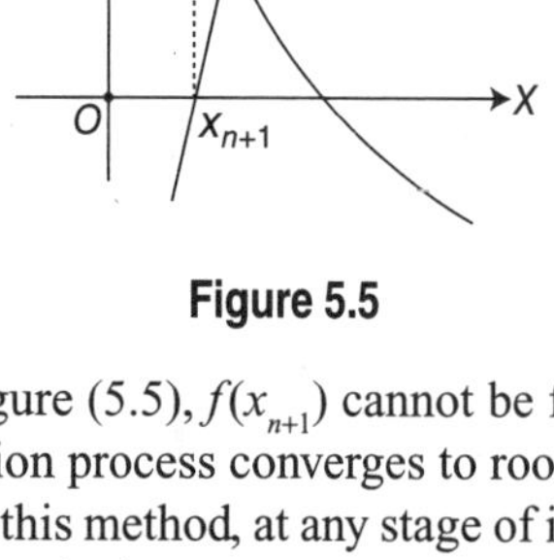

Figure 5.5

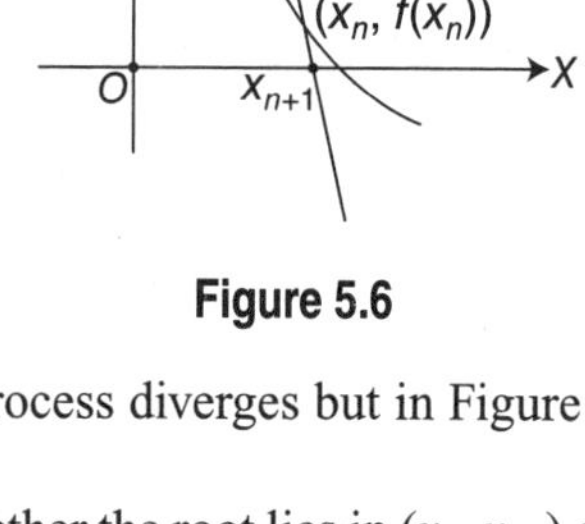

Figure 5.6

In Figure (5.5), $f(x_{n+1})$ cannot be found and hence iteration process diverges but in Figure (5.6), iteration process converges to root.

In this method, at any stage of iterations, we do not test whether the root lies in (x_i, x_{i+1}) or not. We use the last two approximations to obtain the next approximation. This is a draw-back of this method that the iteration process may not converge.

Let ξ be exact root and ε_{n+1} be error in x_{n+1}

$\therefore$ from the iterative formula

$$x_{n+1} = x_n - \frac{x_n - x_{n-1}}{f(x_n) - f(x_{n-1})} f(x_n)$$

$$\xi + \varepsilon_{n+1} = \xi + \varepsilon_n - \frac{(\xi + \varepsilon_n) - (\xi + \varepsilon_{n-1})}{f(\xi + \varepsilon_n) - f(\xi + \varepsilon_{n-1})} f(\xi + \varepsilon_n)$$

$$\text{or } \varepsilon_{n+1} = \frac{\varepsilon_{n-1} f(\xi + \varepsilon_n) - \varepsilon_n f(\xi + \varepsilon_{n-1})}{f(\xi + \varepsilon_n) - f(\xi + \varepsilon_{n-1})}$$

$$= \frac{\varepsilon_{n-1}\left[f(\xi) + \varepsilon_n f'(\xi) + \frac{\varepsilon_n^2}{2} f''(\xi) + \cdots\right] - \varepsilon_n \left[f(\xi) + \varepsilon_{n-1} f'(\xi) + \frac{\varepsilon_{n-1}^2}{2} f''(\xi) + \cdots\right]}{\left[f(\xi) + \varepsilon_n f'(\xi) + \cdots\right] - \left[f(\xi) + \varepsilon_{n-1} f'(\xi) + \cdots\right]}$$

But $\quad f(\xi) = 0$

$$\therefore \quad \varepsilon_{n+1} = \frac{\frac{\varepsilon_n \varepsilon_{n-1}}{2}(\varepsilon_n - \varepsilon_{n-1}) f''(\xi) + \cdots}{(\varepsilon_n - \varepsilon_{n-1}) f'(\xi) + \cdots}$$

$$\therefore \quad \varepsilon_{n+1} = \varepsilon_n \, \varepsilon_{n-1} \frac{f''(\xi)}{2f'(\xi)} + \cdots$$

If the remaining terms are neglected, then $\varepsilon_{n+1} = A\varepsilon_n \, \varepsilon_{n-1}$ where $A = \dfrac{f''(\xi)}{2f'(\xi)}$

Let m be order of convergence then $\varepsilon_{n+1} = k\varepsilon_n^m$, $\varepsilon_n = k\varepsilon_{n-1}^m$ for some k

$$\therefore \quad \varepsilon_{n-1} = \left(\frac{1}{k}\varepsilon_n\right)^{\frac{1}{m}}$$

$\therefore$ from $\quad \varepsilon_{n+1} = A\varepsilon_n \, \varepsilon_{n-1}$

we have $\quad \varepsilon_{n+1} = A\varepsilon_n . \dfrac{1}{k^{1/m}} \varepsilon_n^{\frac{1}{m}} = \dfrac{A}{k^{1/m}} \varepsilon_n^{1+1/m}$

But order of convergence is m

$\therefore$ From this, we get

$$m = 1 + \frac{1}{m}$$

or $\quad m^2 - m - 1 = 0$

$$\therefore \quad m = \frac{1 \pm \sqrt{5}}{2}$$

But $\quad m > 0$

$$\therefore \quad m = \frac{1 + \sqrt{5}}{2} \simeq 1.62$$

$\therefore$ order of convergence is 1.62

Regula-falsi Method or Method of False Position

In this method, x_0 and x_1 are two approximations of the root of $y = f(x) = 0$ with the condition $f(x_0)$ and $f(x_1)$ have opposite signs so that the root lies between x_0 and x_1. Then $P(x_0, y_0)$ and $Q(x_1, y_1)$ are two points on the curve $y = f(x)$ where $y_0 = f(x_0)$ and $y_1 = f(x_1)$. Here, also we approximate the curve by secant PQ and take the point of intersection of PQ with x-axis as the next approximation x_2 of the root. We find $f(x_2)$. If $f(x_0)$ and $f(x_2)$ have opposite signs, then root lies between x_0 and x_2 and if $f(x_1)$ and $f(x_2)$ have opposite signs, then root lies between x_1 and x_2. If root lies between x_1 and x_2, then we interchange x_0 and x_1 so that $P(x_0, y_0)$ is fixed point for every secant. Figure (5.7) shows the case when $P(x_0, y_0)$ is fixed point for secants. Figure (5.8) shows the case when root lies between x_1 and x_2, and hence $P(x_0, y_0)$ and $Q(x_1, y_1)$ will be named as $P(x_1, y_1)$ and $Q(x_2, y_2)$ for finding the next approximations. Now the root lies between x_0 and x_2. Again, we draw secant through $P(x_0, y_0)$ and $R(x_2, y_2)$. Let x_3 be point of intersection of this secant with x – axis, then x_3 will be next approximation. When secant will be drawn joining (x_0, y_0) and (x_n, y_n), then its point of intersection with x-axis will be x_{n+1}. Equation of secant joining (x_0, y_0) and (x_n, y_n) is

$$y - y_n = \frac{y_n - y_0}{x_n - x_0}(x - x_n)$$

For the point where it meets x-axis, we have $y = 0$ and $x = x_{n+1}$

$$\therefore \qquad -y_n = \frac{y_n - y_0}{x_n - x_0}(x_{n+1} - x_n)$$

$$\text{or} \qquad x_{n+1} - x_n = -\frac{x_n - x_0}{y_n - y_0} y_n$$

$$\therefore \qquad x_{n+1} = x_n - \frac{x_n - x_0}{y_n - y_0} y_n = \frac{x_0 y_n - x_n y_0}{y_n - y_0}$$

$$= \frac{x_0 f(x_n) - x_n f(x_0)}{f(x_n) - f(x_0)} = \frac{x_0 f_n - x_n f_0}{f_n - f_0}$$

where $f_n = f(x_n)$, $f_0 = f(x_0)$.

It is the iterative formula to find the approximations.

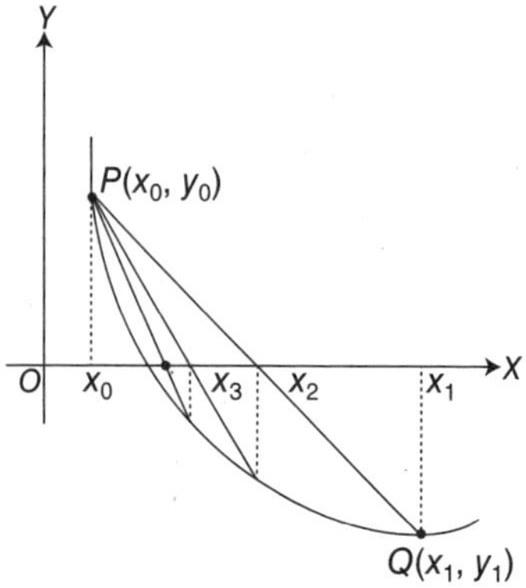

Figure 5.7

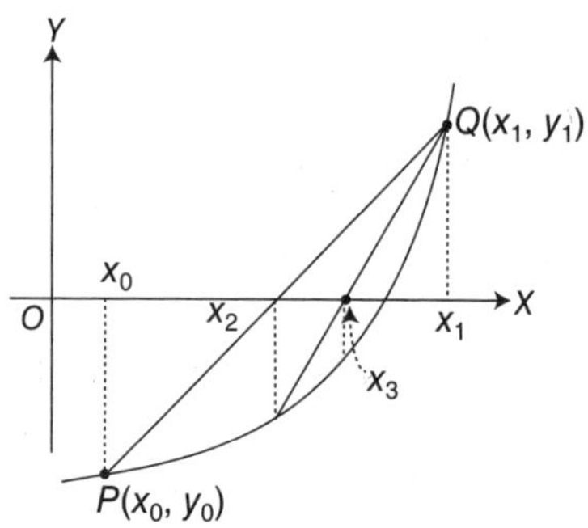

Figure 5.8

This method always converges to the root as $f(x_0)$ and $f(x_n)$ for all n are of opposite signs.

Order of Convergence

Iterative formula is

$$x_{n+1} = \frac{x_0 f_n - x_n f_0}{f_n - f_0} = \frac{x_0 f(x_n) - x_n f(x_0)}{f(x_n) - f(x_0)}$$

$$= x_n - \frac{x_n - x_0}{f(x_n) - f(x_0)} f(x_n)$$

Let ξ be exact root of $f(x) = 0$ and $\varepsilon_0, \varepsilon_n, \varepsilon_{n+1}$ be error in x_0, x_n, x_{n+1}, respectively.

$\therefore$ From iterative formula

$$\xi + \varepsilon_{n+1} = \xi + \varepsilon_n - \frac{(\xi + \varepsilon_n) - (\xi + \varepsilon_0)}{f(\xi + \varepsilon_n) - f(\xi + \varepsilon_0)} f(\xi + \varepsilon_n)$$

$$\therefore \qquad \varepsilon_{n+1} = \frac{\varepsilon_0 f(\xi + \varepsilon_n) - \varepsilon_n f(\xi + \varepsilon_0)}{f(\xi + \varepsilon_n) - f(\xi + \varepsilon_0)}$$

$$= \frac{\varepsilon_0 \left[f(\xi) + \varepsilon_n f'(\xi) + \frac{\varepsilon_n^2}{2} f''(\xi) + \cdots \right] - \varepsilon_n \left[f(\xi) + \varepsilon_0 f'(\xi) + \frac{\varepsilon_0^2}{2} f''(\xi) + \cdots \right]}{\left[f(\xi) + \varepsilon_n f'(\xi) + \cdots \right] - \left[f(\xi) + \varepsilon_0 f'(\xi) + \cdots \right]}$$

But $f(\xi) = 0$

$$\therefore \qquad \varepsilon_{n+1} = \frac{\frac{\varepsilon_0 \varepsilon_n}{2} (\varepsilon_n - \varepsilon_0) f''(\xi) + \cdots}{(\varepsilon_n - \varepsilon_0) f'(\xi) + \cdots}$$

$$= \frac{\varepsilon_0}{2} \varepsilon_n \frac{f''(\xi)}{f'(\xi)} + \cdots$$

If the remaining terms are neglected, then $\varepsilon_{n+1} = K\varepsilon_n$, where $K = \frac{\varepsilon_0}{2} \frac{f''(\xi)}{f'(\xi)}$ and hence the convergence is linear.

Though the method always converges to the root, yet the convergence is very slow. In the secant method, the convergence is faster but we are not sure whether the process will converge. Thus, we shall be dealing in the examples only the regula-falsi method.

Example 5.6: Using regula-falsi method, find the root of $x \log_{10} x = 1.2$

(a) correct to three decimal places

(b) correct to five decimal places

Solution:

$$f(x) = x \log_{10} x - 1.2 = 0$$

$$f(2) = -0.5979, \; f(3) = 0.2314$$

Since $f(2)$ and $f(3)$ have opposite signs, root lies between 2 and 3

Taking $x_0 = 2$ and $x_1 = 3$

we have

$$x_2 = \frac{x_0 f_1 - x_1 f_0}{f_1 - f_0} = \frac{2(0.2314) - 3(-0.5979)}{0.2314 + 0.5979} = 2.721$$

$$f(x_2) = -0.0171$$

Since $f(x_1)$ and $f(x_2)$ have opposite signs, so root lies between x_1 and x_2. To keep x_0 fixed, we interchange x_0 and x_1

i.e. $\quad x_0 = 3, x_1 = 2$

$$f_0 = 0.2314$$

Iterative formula becomes

$$x_{n+1} = \frac{x_0 f_n - x_n f_0}{f_n - f_0} = \frac{3f_n - (0.2314)x_n}{f_n - 0.2314}$$

n	x_n	f_n	$x_{n+1} = \frac{3f_n - (0.2314)x_n}{f_n - 0.2314}$
2	2.721	–0.0171	$\frac{0.6809}{0.2485} = 2.74$
3	2.74	-5.6346×10^{-4}	$\frac{6357.2638}{2319.6346} = 2.7406$
4	2.7406	-0.4020×10^{-4}	$\frac{6342.9544}{2314.402} = 2.7406451$
5	2.7406451	-0.8686×10^{-6}	$\frac{634187.8819}{231400.8686} = 2.7406461$
6	2.7406461	3.51×10^{-9}	

∴ Root lies between 2.7406451 and 2.7406461

∴ Root upto three decimals = 2.741

Root upto five decimals = 2.74065

Example 5.7: Determine the root of $xe^x - 2 = 0$ by method of false position correct to four decimal places.

Solution:

$$f(x) = xe^x - 2 = 0$$

$$f(0.8) = -0.2196, \quad f(0.9) = 0.2136$$

Since $f(0.8)$ and $f(0.9)$ have opposite signs, root lies between 0.8 and 0.9.

Taking $x_0 = 0.8$ and $x_1 = 0.9$, we have

$$x_2 = \frac{x_0 f_1 - x_1 f_0}{f_1 - f_0} = \frac{0.8(0.2136) - 0.9(-0.2196)}{0.2136 + 0.2196} = 0.851$$

$$f(x_2) = -6.9685 \times 10^{-3}$$

Since $f(x_1)$ and $f(x_2)$ have opposite signs, so root lies between x_1 and x_2. To keep x_0 fixed, we interchange x_0 and x_1

i.e. $\quad x_0 = 0.9, x_1 = 0.8$

$$f_0 = 0.2136$$

Iterative formula becomes

$$x_{n+1} = \frac{x_0 f_n - x_n f_0}{f_n - f_0} = \frac{0.9 f_n - (0.2136) x_n}{f_n - 0.2136}$$

n	x_n	f_n	$x_{n+1} = \frac{0.9f_n - (0.2136)x_n}{f_n - 0.2136}$
2	0.851	-6.9685×10^{-3}	$\frac{188.04525}{220.5685} = 0.852548$
3	0.852548	-2.4988×10^{-4}	$\frac{1823.2914}{2138.4988} = 0.8526034$
4	0.8526034	-9.1348×10^{-6}	$\frac{182124.3076}{213609.1348} = 0.8526054$
5	0.8526054	-4.4333×10^{-7}	

As $f(0.8526054)$ is very small in magnitude, root is near it.

$\therefore$ Root corrected to four decimal places = 0.8526

Example 5.8: Use regula-falsi method to find the real root of the equation $x^3 - 5x + 1 = 0$ and correct to four decimals.

Solution:

$$f(x) = x^3 - 5x + 1 = 0$$

$$f(0) = 1, \; f(1) = -3$$

Since $f(0)$ and $f(1)$ are of opposite signs, the root lies between $x_0 = 0$ and $x_1 = 1$.

$$x_2 = \frac{x_0 f_1 - x_1 f_0}{f_1 - f_0} = \frac{0-1}{-4} = 0.25$$

$$f(x_2) = -0.2344$$

$f(x_0)$ and $f(x_2)$ are of opposite signs, so root lies between x_0 and x_2 in which x_0 is fixed.

Iterative formula becomes

$$x_{n+1} = \frac{x_0 f_n - x_n f_0}{f_n - f_0} = \frac{-x_n}{f_n - 1} = \frac{x_n}{1 - f_n}$$

n	x_n	f_n	$x_{n+1} = \frac{x_n}{1-f_n}$
2	0.25	–0.2344	0.2025275
3	0.2025275	-4.3303×10^{-3}	0.2016543
4	0.2016543	-7.1337×10^{-5}	0.2016399
5	0.2016399	-1.0940×10^{-6}	0.2016397
6	0.2016397	-1.1842×10^{-7}	

As $f(0.2016397)$ is very small in magnitude, root is near it.

$\therefore$ Root corrected to four decimal places = 0.2016

5.3.4 Newton–Raphson Method (or Newton's Iteration Method or Method of Tangents)

In secant method, we replace the curve by variable secants and in regula-falsi method we replace the curve by secants in which one end point of secants is fixed. In Newton–Raphson method, the curve is replaced by variable tangents. Let x_0 is an approximation to the root of $f(x) = 0$. We find the equation of tangent at (x_0, y_0) to the graph of curve $y = f(x)$ where $y_0 = f(x_0)$. Let this tangent meets x-axis at x_1 then x_1 will be next approximation and we find (x_1, y_1) on the graph and draw tangent at (x_1, y_1) to the curve $y = f(x)$. Its intersection with x-axis will be x_2. Proceeding in this way, when approximation x_n is found then intersection of tangent at (x_n, y_n) to $y = f(x)$ with x-axis will give next approximation x_{n+1}.

Now, equation of tangent at (x_n, y_n) to $y = f(x)$ is

$$y - y_n = f'(x_n).(x - x_n)$$

For its intersection with x-axis, we have $y = 0$ and $x = x_{n+1}$

$$\therefore \quad -y_n = f'(x_n).(x_{n+1} - x_n)$$

or

$$x_{n+1} - x_n = -\frac{y_n}{f'(x_n)} = -\frac{f(x_n)}{f'(x_n)}$$

or

$$x_{n+1} = x_n - \frac{f(x_n)}{f'(x_n)}$$

It is Newton's iterative formula to obtain the approximations. Geometrically, the process is shown in Figure (5.9)

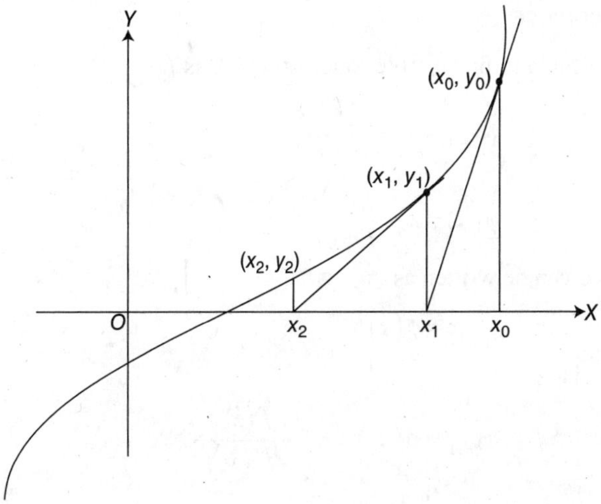

Figure 5.9

Newton–Raphson formula can also be obtained analytically. Let ε_n is error in approximation x_n to root ξ of $f(x) = 0$,

then $$x_n = \xi + \varepsilon_n$$

$\therefore$ $$\xi = x_n - \varepsilon_n$$

$\therefore$ $$f(\xi) = f(x_n - \varepsilon_n) = 0$$

By Taylor-series expansion

$$f(x_n) - \varepsilon_n f'(x_n) + \frac{\varepsilon_n^2}{2!} f''(x_n) - \cdots = 0$$

As ε_n is small

$\therefore$ $$f(x_n) - \varepsilon_n f'(x_n) \simeq 0$$

$\therefore$ $$\varepsilon_n \simeq \frac{f(x_n)}{f'(x_n)}$$

$\therefore$ $$\xi \simeq x_n - \frac{f(x_n)}{f'(x_n)}$$

$\therefore$ next approximation to the root will be

$$x_{n+1} = x_n - \frac{f(x_n)}{f'(x_n)}$$

which gives the same iterative formula we have derived geometrically.

Condition of Convergence

Newton's iterative formula to find simple root of $f(x) = 0$ is

$$x_{n+1} = x_n - \frac{f(x_n)}{f'(x_n)}$$

If we take
$$\phi(x) = x - \frac{f(x)}{f'(x)}$$

then equation $f(x) = 0$ can be written as

$$x = \phi(x)$$

and iterative formula is

$$x_{n+1} = \phi(x_n) = x_n - \frac{f(x_n)}{f'(x_n)}$$

But iteration

$$x_{n+1} = \phi(x_n)$$

converges if $|\phi'(x)| < 1$ when x is near the root of equation.

Here
$$\phi'(x) = 1 - \frac{f'(x).f'(x) - f(x)f''(x)}{(f'(x))^2}$$

$\therefore$ Newton–Raphson iterative formula converges if

$$\left|1 - \frac{(f'(x))^2 - f(x)f''(x)}{(f'(x))^2}\right| < 1$$

or
$$|f(x)f''(x)| < (f'(x))^2$$

It is the sufficient condition for convergence of Newton–Raphson formula.

Order of Convergence

Let ξ be exact root of $f(x) = 0$ and approximations x_n and x_{n+1} have errors ε_n and ε_{n+1}, respectively.
Thus, from Newton–Raphson iterative formula

$$x_{n+1} = x_n - \frac{f(x_n)}{f'(x_n)}$$

we have

$$\xi + \varepsilon_{n+1} = \xi + \varepsilon_n - \frac{f(\xi + \varepsilon_n)}{f'(\xi + \varepsilon_n)}$$

$$\therefore \qquad \varepsilon_{n+1} = \frac{\varepsilon_n f'(\xi+\varepsilon_n) - f(\xi+\varepsilon_n)}{f'(\xi+\varepsilon_n)}$$

$$= \frac{\varepsilon_n\left[f'(\xi)+\varepsilon_n f''(\xi)+\cdots\right]-\left[f(\xi)+\varepsilon_n f'(\xi)+\frac{\varepsilon_n^{\,2}}{2}f''(\xi)+\cdots\right]}{f'(\xi)+\varepsilon_n f''(\xi)+\cdots}$$

But $f(\xi) = 0$

$$\therefore \qquad \varepsilon_{n+1} = \frac{\frac{\varepsilon_n^{\,2}}{2}f''(\xi)+\cdots}{f'(\xi)+\varepsilon_n f''(\xi)}$$

$$= \frac{1}{2}\varepsilon_n^2 \frac{f''(\xi)}{f'(\xi)} + \cdots$$

If the remaining terms are neglected, then

$$\varepsilon_{n+1} = k\varepsilon_n^2 \quad \text{where } k = \frac{1}{2}\frac{f''(\xi)}{f'(\xi)}$$

Hence, the convergence is of order 2, i.e. convergence is quadratic.

This fact also means that the number of correct decimals is approximately doubled at every iteration, at least if the factor $\frac{1}{2}\frac{f''(\xi)}{f'(\xi)}$ is not too large.

Remark 5.1: Newton–Raphson formula also works when coefficients or roots are complex. It should be noted that if the algebraic equation has real coefficients, then a complex root cannot be reached if we take initial approximation real. Thus, initial approximation will have to be taken complex with imaginary part non-zero.

Modification of Newton's Iteration Formula for Multiple Roots

If the equation $f(x) = 0$ has root ξ of multiplicity $m > 1$ then Newton–Raphson iteration formula

$$x_{n+1} = x_n - \frac{f(x_n)}{f'(x_n)}$$

possesses linear convergence which we prove below.

Let ε_n and ε_{n+1} are errors in x_n and x_{n+1}, respectively,

then $$\xi + \varepsilon_{n+1} = \xi + \varepsilon_n - \frac{f(\xi+\varepsilon_n)}{f'(\xi+\varepsilon_n)}$$

$$\therefore \qquad \varepsilon_{n+1} = \frac{\varepsilon_n f'(\xi+\varepsilon_n) - f(\xi+\varepsilon_n)}{f'(\xi+\varepsilon_n)}$$

$$= \frac{\varepsilon_n\left[f'(\xi)+\varepsilon_n f''(\xi)+\cdots+\frac{\varepsilon_n^{\,m-1}}{(m-1)!}f^{(m)}(\xi)+\cdots\right]-\left[f(\xi)+\varepsilon_n f'(\xi)+\cdots+\frac{\varepsilon_n^{\,m}}{m!}f^{(m)}(\xi)+\cdots\right]}{f'(\xi)+\varepsilon_n f''(\xi)+\cdots+\frac{\varepsilon_n^{\,m-1}}{(m-1)!}f^{(m)}(\xi)+\cdots}$$

But ξ is root of $f(x) = 0$ of multiplicity m

$$\therefore \qquad f(\xi) = f'(\xi) = \cdots = f^{(m-1)}(\xi) = 0$$

$$\therefore \qquad \varepsilon_{n+1} = \frac{\left(\frac{1}{(m-1)!}-\frac{1}{m!}\right)\varepsilon_n^m f^{(m)}(\xi)+\cdots}{\frac{\varepsilon_n^{m-1}}{(m-1)!}f^{(m)}(\xi)+\cdots}$$

$$= \left(1-\frac{1}{m}\right)\varepsilon_n + \cdots$$

If the remaining terms are neglected

$$\varepsilon_{n+1} = K\,\varepsilon_n \text{ where } K = \left(1-\frac{1}{m}\right) \neq 0 \quad (\because m > 1)$$

$\therefore$ Convergence is linear.

Thus, Newton's iterative formula is slow for multiple roots and thus, this formula requires modification.

If equation $f(x) = 0$ has root ξ of multiplicity m, then equation $\left[f(x)\right]^{\frac{1}{m}} = 0$ has ξ as simple root.

i.e. $F(x) = \left(f(x)\right)^{\frac{1}{m}} = 0$ has simple root ξ

$$F'(x) = \frac{1}{m}\left(f(x)\right)^{\frac{1}{m}-1} f'(x)$$

$\therefore$ Iterative formula

$$x_{n+1} = x_n - \frac{F(x_n)}{F'(x_n)} = x_n - \frac{\left(f(x_n)\right)^{\frac{1}{m}}}{\frac{1}{m}\left(f(x_n)\right)^{\frac{1}{m}-1} f'(x_n)}$$

$$= x_n - \frac{m\, f(x_n)}{f'(x_n)}$$

will have quadratic convergence. This iterative formula is called modified (or sometime generalised) iterative formula for root of multiplicity m.

We, explicitly show that this iterative formula has quadratic convergence.

Let ε_n and ε_{n+1} be errors in x_n and x_{n+1}, respectively,

$\therefore$ from iterative formula

$$\xi + \varepsilon_{n+1} = \xi + \varepsilon_n - \frac{m\, f(\xi+\varepsilon_n)}{f'(\xi+\varepsilon_n)}$$

$$\therefore \quad \varepsilon_{n+1} = \frac{\varepsilon_n f'(\xi+\varepsilon_n) - m f(\xi+\varepsilon_n)}{f'(\xi+\varepsilon_n)}$$

$$\therefore \quad \varepsilon_{n+1} = \frac{\varepsilon_n\left[f'(\xi)+\varepsilon_n f''(\xi)+\cdots+\dfrac{\varepsilon_n^{m-1}}{(m-1)!}f^{(m)}(\xi)+\cdots\right] - m\left[f(\xi)+\varepsilon_n f'(\xi)+\cdots+\dfrac{\varepsilon_n^{m}}{m!}f^{(m)}(\xi)+\cdots\right]}{f'(\xi)+\varepsilon_n f''(\xi)+\cdots+\dfrac{\varepsilon_n^{m-1}}{(m-1)!}f^{(m)}(\xi)+\cdots}$$

But ξ is root of multiplicity m of $f(x) = 0$

$$\therefore \quad f(\xi) = f'(\xi) = \cdots = f^{(m-1)}(\xi) = 0$$

$$\therefore \quad \varepsilon_{n+1} = \frac{\left(\dfrac{1}{m!} - \dfrac{m}{(m+1)!}\right)\varepsilon_n^{m+1} f^{(m+1)}(\xi)+\cdots}{\dfrac{\varepsilon_n^{m-1}}{(m-1)!}f^{(m)}(\xi)+\cdots}$$

$$= \left(\frac{1}{m} - \frac{1}{m+1}\right)\varepsilon_n^2 \frac{f^{(m+1)}(\xi)}{f^{(m)}(\xi)} + \cdots$$

If the remaining terms are neglected, then

$$\varepsilon_{n+1} = K\varepsilon_n^2 \text{ where } K = \frac{f^{m+1}(\xi)}{m(m+1)f^{(m)}(\xi)}$$

Hence, the convergence is quadratic.

Remark 5.2: Newton–Raphson technique is widely used on automatic computers for calculation of various simple functions as inverse, square root, cube root, etc.

The quantity $\frac{1}{a}$; $a \neq 0$ can be interpreted as a root of the equation $\frac{1}{x} - a = 0$. Thus, iterative formula will be

$$x_{n+1} = x_n + \frac{\dfrac{1}{x_n} - a}{\dfrac{1}{x_n^2}} = x_n(2 - a x_n)$$

This can also be written as

$$1 - a x_{n+1} = \left(1 - 2a x_n + a^2 x_n^2\right) = \left(1 - a x_n\right)^2$$

It clearly shows the quadratic convergence.

From above, we conclude that the iterative formula for finding $\frac{1}{a}$ can also be written as

$$1 - a x_{n+1} = \left(1 - a x_n\right)^3$$

i.e

$$x_{n+1} = x_n\left(3 - 3a x_n + a^2 x_n^2\right)$$

which is still faster as its order of convergence is 3, but in this formula more computational work will be required at each iteration and hence there is no real advantage.

If we want to compute $\sqrt{a}$; $a > 0$, then we start with the equation $x^2 - a = 0$ and thus, iterative formula will be

$$x_{n+1} = x_n - \frac{x_n^2 - a}{2x_n} = \frac{1}{2}\left(x_n + \frac{a}{x_n}\right)$$

A corresponding formula for N th root of 'a' from the starting equation $f(x) = x^N - a = 0$ is

$$x_{n+1} = x_n - \frac{x_n^N - a}{Nx_n^{N-1}} = \frac{(N-1)x_n^N + a}{Nx_n^{N-1}}$$

and thus, for cube root of 'a' we have

$$x_{n+1} = \frac{1}{3x_n^2}\left(2x_n^3 + a\right) = \frac{1}{3}\left(2x_n + \frac{a}{x_n^2}\right)$$

and for $\frac{1}{\sqrt{a}}$, the iterative formula is

$$x_{n+1} = -\frac{1}{2x_n^{-3}}\left(-3x_n^{-2} + a\right) = \frac{x_n}{2}\left(3 - a x_n^2\right)$$

This formula does not make use of division.

Method of Finding Initial Approximations if Two Roots are Given Close to a Number

Let two roots of equation $f(x) = 0$ are close to ξ then for small ε, $\xi + \varepsilon$ and $\xi - \varepsilon$ are initial approximations to roots of $f(x) = 0$

$$\therefore \qquad f(\xi+\varepsilon) = f(\xi) + \varepsilon f'(\xi) + \frac{\varepsilon^2}{2!} f''(\xi) + \cdots = 0$$

$$f(\xi-\varepsilon) = f(\xi) - \varepsilon f'(\xi) + \frac{\varepsilon^2}{2!} f''(\xi) - \cdots = 0$$

Since ε is small so the terms of ε^3 and higher powers of ε can be neglected and adding we have

$$2f(\xi) + \varepsilon^2 f''(\xi) = 0$$

$$\therefore \qquad \varepsilon^2 = -\frac{2f(\xi)}{f''(\xi)}$$

$$\therefore \qquad \varepsilon = \pm\sqrt{-\frac{2f(\xi)}{f''(\xi)}}$$

Thus, the first approximations for two roots can be taken near to $\xi + \sqrt{-\frac{2f(\xi)}{f''(\xi)}}$ and $\xi - \sqrt{-\frac{2f(\xi)}{f''(\xi)}}$

Example 5.9: Using Newton–Raphson method, derive a formula to find $\sqrt[3]{N}$ where N is a real number. Hence evaluate $\sqrt[3]{41}$ correct to four places of decimals.

Solution: $\sqrt[3]{N}$ is root of equation

$$f(x) = x^3 - N = 0$$

$$\therefore \qquad f'(x) = 3x^2$$

$\therefore$ Newton–Raphson iterative formula

$$x_{n+1} = x_n - \frac{f(x_n)}{f'(x_n)}$$

to evaluate $\sqrt[3]{N}$ is

$$x_{n+1} = x_n - \frac{x_n^3 - N}{3x_n^2} = \frac{1}{3}\left(2x_n + \frac{N}{x_n^2}\right)$$

Take $N = 41$;

Since, $\qquad 3^3 = 27, 4^3 = 64$

$$\therefore \qquad \sqrt[3]{27} = 3, \sqrt[3]{64} = 4$$

Take $x_0 = 3.4$

n	x_n	$x_{n+1} = \frac{1}{3}\left(2x_n + \frac{41}{x_n^2}\right)$
0	3.4	3.4489
1	3.4489	3.44822
2	3.44822	3.44822

$\therefore$ $\sqrt[3]{41}$ correct to four places of decimals = 3.4482

Example 5.10: Apply Newton–Raphson method to compute

(i) $\sqrt{5}$ (ii) $30^{-\frac{1}{5}}$ (iii) $\sqrt[3]{24}$

Solution: $\sqrt[k]{N}$ is root of equation

$$f(x) = x^k - N = 0$$

$$f'(x) = k\,x^{k-1}$$

∴ Newton–Raphson iterative formula

$$x_{n+1} = x_n - \frac{f(x_n)}{f'(x_n)}$$

to evaluate $\sqrt[k]{N}$ is

$$x_{n+1} = x_n - \frac{x_n^k - N}{k\,x_n^{k-1}} = \frac{1}{k}\left[(k-1)x_n + \frac{N}{x_n^{k-1}}\right] \tag{1}$$

(i) Take $N = 5$, $k = 2$ in (1)

$$x_{n+1} = \frac{1}{2}\left[x_n + \frac{5}{x_n}\right]$$

Take $x_0 = 2.2$

n	x_n	$x_{n+1} = \frac{1}{2}\left(x_n + \frac{5}{x_n}\right)$
0	2.2	2.236
1	2.236	2.236068
2	2.236068	2.236068

∴ $\sqrt{5}$ correct up to five decimal places = 2.23607

(ii) Take $N = 30$, $k = -5$ in (1)

$$x_{n+1} = -\frac{1}{5}\left[-6x_n + \frac{30}{x_n^{-6}}\right]$$

$$= 1.2\,x_n - 6x_n^6$$

We have $2^5 = 32$

∴ $(32)^{-\frac{1}{5}} = \frac{1}{2} = 0.5$

∴ we take $x_0 = 0.5$

n	x_n	$x_{n+1} = 1.2x_n - 6x_n^6$
0	0.5	0.50625
1	0.50625	0.50650
2	0.50650	0.506496

∴ $(30)^{-\frac{1}{5}}$ correct up to four decimal places = 0.5065

(iii) Take $N = 24$, $k = 3$ in (1)

Iterative formula is

$$x_{n+1} = \frac{1}{3}\left(2x_n + \frac{24}{x_n^2}\right)$$

We take $x_0 = 3$

n	x_n	$x_{n+1} = \frac{1}{3}\left(2x_n + \frac{24}{x_n^2}\right)$
0	3	2.889
1	2.889	2.88451
2	2.88451	2.88450

$\therefore$ $\sqrt[3]{24}$ correct up to four decimal places = 2.8845

Example 5.11: Use Newton–Raphson method to solve the equation $3x - \cos x - 1 = 0$
Solution: $f(x) = 3x - \cos x - 1 = 0$
$\therefore \quad f'(x) = 3 + \sin x$
$\therefore$ Newton–Raphson iteration formula is

$$x_{n+1} = x_n - \frac{f(x_n)}{f'(x_n)} = x_n - \frac{3x_n - \cos x_n - 1}{3 + \sin x_n}$$

$$= \frac{x_n \sin x_n + \cos x_n + 1}{3 + \sin x_n}$$

$$f(0) = -2, \quad f(1) = 1.4597 \quad \therefore \text{ we take } x_0 = 0.6$$

n	x_n	$x_n \sin x_n + \cos x_n + 1$	$3 + \sin x_n$	x_{n+1}
0	0.6	2.1641	3.5646	0.6071
1	0.6071	2.1676	3.5705	0.607099

$\therefore$ Root to four decimal places = 0.6071

Example 5.12: Compute to four decimal places the non-zero real root of $x^2 + 4\sin x = 0$
Solution: $f(x) = x^2 + 4\sin x = 0,$

$f'(x) = 2x + 4\cos x$

We have $\quad -4 \le 4\sin x \le 4$

Thus, positive root must satisfy $0 < x \le 2$

for $0 < x \le 2, \; f(x) > 0$

$\therefore$ There is no positive real root.

$f(-1) = -2.3659$

$f(-2) = 0.3628$

$\therefore$ we take $\quad x_0 = -1.9$

Newton–Raphson iterative formula is

$$x_{n+1} = x_n - \frac{f(x_n)}{f'(x_n)} = x_n - \frac{x_n^2 + 4\sin x_n}{2x_n + 4\cos x_n}$$

$$= \frac{x_n^2 + 4x_n \cos x_n - 4\sin x_n}{2x_n + 4\cos x_n}$$

n	x_n	$x_n^2 + 4x_n \cos x_n - 4\sin x_n$	$2x_n + 4\cos x_n$	x_{n+1}
0	−1.9	9.8522	−5.0932	−1.934
1	−1.934	10.2278	−5.2891	−1.93375
2	−1.93375	10.2250	−5.2876	−1.93377

∴ Non-zero real root to four decimal places = −1.9338

Example 5.13: Determine the root of the equation $\cos x - xe^x = 0$ using Newton–Raphson method.

Solution: $f(x) = \cos x - xe^x = 0$

∴ $f'(x) = -\sin x - (x+1)e^x$

Newton–Raphson iterative formula is

$$x_{n+1} = x_n - \frac{f(x_n)}{f'(x_n)}$$

$$= x_n - \frac{\cos x_n - x_n e^{x_n}}{-\sin x_n - (x_n+1)e^{x_n}}$$

$$= x_n + \frac{\cos x_n - x_n e^{x_n}}{\sin x_n + (x_n+1)e^{x_n}}$$

$$= \frac{x_n^2 e^{x_n} + x_n \sin x_n + \cos x_n}{\sin x_n + (x_n+1)e^{x_n}}$$

$$f(0) = 1, f(1) = -2.178$$

We take $x_0 = 0.5$

n	x_n	$x_n^2 e^{x_n} + x_n \sin x_n + \cos x_n$	$\sin x_n + (x_n+1)e^{x_n}$	x_{n+1}
0	0.5	1.52948	2.9525	0.5180
1	0.5180	1.57572	3.0434	0.51775
2	0.51775	1.5750625	3.0421	0.5177550
3	0.5177550	1.5750757	3.0421033	0.5177588

∴ Positive real root up to five decimal places = 0.51776

Example 5.14: Find the double root of the equation $x^3 - x^2 - x + 1 = 0$ near 0.9

Solution: $$f(x) = x^3 - x^2 - x + 1 = 0$$

$$\therefore \quad f'(x) = 3x^2 - 2x - 1$$

Newton's iterative formula for double root is

$$x_{n+1} = x_n - 2\frac{f(x_n)}{f'(x_n)}$$

$$= x_n - 2.\frac{x_n^3 - x_n^2 - x_n + 1}{3x_n^2 - 2x_n - 1} = \frac{x_n^3 + x_n - 2}{3x_n^2 - 2x_n - 1}$$

$x_0 = 0.9$

n	x_n	$x_n^3 + x_n - 2$	$3x_n^2 - 2x_n - 1$	x_{n+1}
0	0.9	−0.371	−0.37	1.003
1	1.003	0.012027027	0.012027	1.0000022

$\therefore$ Double root = 1.

Example 5.15: The equation $f(x) = x^3 - 7x^2 + 16x - 12 = 0$ has a double root. Starting with initial approximation $x_0 = 1$, find the double root correct to three decimal places using

(i) Newton–Raphson method

(ii) Modified Newton–Raphson method

Solution: $$f(x) = x^3 - 7x^2 + 16x - 12 = 0$$

$$f'(x) = 3x^2 - 14x + 16$$

(i) **By Newton–Raphson method**

Newton's iterative formula is

$$x_{n+1} = x_n - \frac{f(x_n)}{f'(x_n)} = x_n - \frac{x_n^3 - 7x_n^2 + 16x_n - 12}{3x_n^2 - 14x_n + 16} = \frac{2x_n^3 - 7x_n^2 + 12}{3x_n^2 - 14x_n + 16}$$

$x_0 = 1$

n	x_n	$2x_n^3 - 7x_n^2 + 12$	$3x_n^2 - 14x_n + 16$	x_{n+1}
0	1	7	5	1.4
1	1.4	3.768	2.28	1.65
2	1.65	1.92675	1.0675	1.80
3	1.80	0.984	0.52	1.89
4	1.89	0.497838	0.2563	1.94
5	1.94	0.257568	0.1308	1.97
6	1.97	0.124446	0.0627	1.98

n	x_n	$2x_n^3 - 7x_n^2 + 12$	$3x_n^2 - 14x_n + 16$	x_{n+1}
7	1.98	0.081984	0.0412	1.990
8	1.990	0.040498	0.0203	1.9950
9	1.9950	0.0201247	0.010075	1.9975
10	1.9975	0.0100312	5.018749×10^{-3}	1.9987
11	1.9987	5.208446×10^{-3}	2.605069×10^{-3}	1.9994
12	1.9994	2.401801×10^{-3}	1.201079×10^{-3}	1.9997
13	1.9997	1.200451×10^{-3}	6.00269×10^{-4}	1.99986

$\therefore$ Double root corrected to three decimal places = 2.000

(ii) **By Modified Newton–Raphson method**

Modified Newton–Raphson iterative formula is

$$x_{n+1} = x_n - 2\frac{f(x_n)}{f'(x_n)} = x_n - 2\frac{x_n^3 - 7x_n^2 + 16x_n - 12}{3x_n^2 - 14x_n + 16}$$

$$= \frac{x_n^3 - 16x_n + 24}{3x_n^2 - 14x_n + 16}$$

$x_0 = 1$

n	x_n	$x_n^3 - 16x_n + 24$	$3x_n^2 - 14x_n + 16$	x_{n+1}
0	1	9	5	1.8
1	1.8	1.032	0.52	1.985
2	1.985	0.0613466	0.030675	1.99989
3	1.99989	4.40073×10^{-4}	2.20036×10^{-4}	2.0000045

$\therefore$ Double root corrected to three decimal places = 2.000

Example 5.16: The equation $x^4 - 5x^3 - 12x^2 + 76x - 79 = 0$ has two roots close to $x = 2$. Find these roots to four decimals.

Solution:

$$f(x) = x^4 - 5x^3 - 12x^2 + 76x - 79 = 0$$

Let approximations of two roots close to 2 are $2 + \varepsilon$ and $2 - \varepsilon$

$\therefore$

$$f(2+\varepsilon) = f(2) + \varepsilon f'(2) + \frac{\varepsilon^2}{2!} f''(2) + \cdots = 0 \tag{1}$$

$$f(2-\varepsilon) = f(2) - \varepsilon f'(2) + \frac{\varepsilon^2}{2!} f''(2) - \cdots = 0 \tag{2}$$

Since ε is small so neglect terms of ε^3 and higher powers of ε

Add (1) and (2)

$$2f(2) + \varepsilon^2 f''(2) = 0$$

$\therefore \qquad \varepsilon^2 = -\dfrac{2f(2)}{f''(2)}$

Now, $\qquad f(x) = x^4 - 5x^3 - 12x^2 + 76x - 79$

$$f'(x) = 4x^3 - 15x^2 - 24x + 76$$

$$f''(x) = 12x^2 - 30x - 24$$

$\therefore \qquad f(2) = 1, f''(2) = -36$

$\therefore \qquad \varepsilon^2 = -\dfrac{2(1)}{-36} = \dfrac{1}{18}$

$\therefore \qquad \varepsilon = \pm\dfrac{1}{\sqrt{18}} \simeq \pm 0.24$

Initial approximations to two roots close to 2 are 2.24 and 1.76

Newton's iterative formula is

$$x_{n+1} = x_n - \frac{f(x_n)}{f'(x_n)} = x_n - \frac{x_n^4 - 5x_n^3 - 12x_n^2 + 76x_n - 79}{4x_n^3 - 15x_n^2 - 24x_n + 76}$$

$$= \frac{3x_n^4 - 10x_n^3 - 12x_n^2 + 79}{4x_n^3 - 15x_n^2 - 24x_n + 76}$$

Taking $x_0 = 2.24$

n	x_n	$3x_n^4 - 10x_n^3 - 12x_n^2 + 79$	$4x_n^4 - 15x_n^2 - 24x_n + 76$	x_{n+1}
0	2.24	−18.076511	−8.066304	2.2410
1	2.2410	−18.145914	−8.0972809	2.2410

Taking $x_0 = 1.76$

n	x_n	$3x_n^4 - 10x_n^3 - 12x_n^2 + 79$	$4x_n^4 - 15x_n^2 - 24x_n + 76$	x_{n+1}
0	1.76	16.096417	9.103104	1.7682
1	1.7682	15.523893	8.7785616	1.7684
2	1.7684	15.509914	8.7706563	1.7684

$\therefore$ Two roots close to 2 up to 4 decimals are 2.2410 and 1.7684

EXERCISE 5.1

1. Perform five iterations of the bisection method to obtain the smallest positive root of the equation $f(x) = x^3 - 5x + 1 = 0$
2. Find a real root of the equation $f(x) = x^3 - x - 1 = 0$ using bisection method.

3. Perform five iterations of the bisection method to obtain a root of the equation $f(x) = \cos x - x e^x = 0$
4. Isolate the roots of the equation $x^3 - 4x + 1 = 0$. Find all the roots using bisection method.
5. Evaluate $\sqrt{5}$ by direct iteration method.
6. Find a real root of the equation $x^3 - x - 1 = 0$ correct to two decimal places by iterative method.
7. Find a real root of the equation $x^3 + x^2 - 1 = 0$ by iteration method.
8. Solve the equation by direct iteration, $e^x - 3x = 0$
9. Starting with $x = 0.12$, solve $x = 0.21 \sin(0.5 + x)$ by iteration method.
10. Use the method of iteration to solve the equation $x = \exp(-x)$, starting with $x = 1.00$. Perform four iterations, taking the readings up to four decimal places.
11. Find a real root of the equation $x^3 - 2x - 5 = 0$ by the method of false position correct to three decimal places.
12. Find an approximate value of the root of the equation $x^3 + x - 1 = 0$ near $x = 1$, using the method of false position.
13. Using regula-falsi method, find the real root of the equation $f(x) = x^3 - 5x + 1 = 0$ which lies in the interval (0,1). Perform iterations to obtain this root.
14. The equation $x^6 - x^4 - x^3 - 1 = 0$ has one real root between 1.4 and 1.5. Find this root to four decimal places by false position method.
15. The negative root of the equation $3x^3 + 8x^2 + 8x + 5 = 0$ is to be determined. Find the root by regula-falsi method. Stop iteration when $f(x_2) < 0.02$.
16. Find the root of $x\, e^x = 3$ by regula-falsi method correct to three decimal places.
17. Using false position method, solve the equation $3x - \cos x - 1 = 0$
18. Using regula–falsi method, find the root of the equation $x^3 + x^2 - 3x - 3 = 0$ lying between 1 and 2.
19. Show that the initial approximation x_0 for finding $1/N$, where N is a positive integer, by the Newton–Raphson method must satisfy $0 < x_0 < 2/N$ for convergence.
20. Use Newton's method to find a root of the equation $x^3 - 3x - 5 = 0$.
21. Perform four iterations of the Newton–Raphson method to find the smallest positive root of the equation $f(x) = x^3 - 5x + 1 = 0$.
22. Find a positive root of $x^4 - x = 10$ using Newton–Raphson method.
23. Find the largest root of $x^2 - 5x + 2 = 0$ correct to five decimal places by Newton–Raphson method.
24. Find the positive root of the equation $x^4 - 3x^3 + 2x^2 + 2x - 7 = 0$ by Newton–Raphson method.
25. Find the root of the equation $x \log_{10} x = 1.2$ by Newton–Raphson method correct to six decimal places.
26. Find the smallest root of the equation $e^{-x} - \sin x = 0$ correct to four decimal places.
27. Use Newton–Raphson method to solve the transcendental equation $e^x = 5x$
28. Find an interval of length 1, in which the root of $f(x) = 3x^3 - 4x^2 - 4x - 7 = 0$ lies. Take the middle point of this interval as the starting approximation and iterate two times, using the Newton–Raphson method.
29. Find a double root of the equation $x^3 - 5x^2 + 8x - 4 = 0$ near 1.8.
30. Find the double root of $x^3 - x^2 - x + 1 = 0$ close to 0.8.

ANSWERS 5.1

1. 0.20	2. 1.323	3. 0.5	4. – 2.115, 0.254, 1.86	5. 2.23607
6. 1.32	7. 0.7549	8. 0.6191	9. 0.1224	10. 0.6062
11. 2.095	12. 0.6823	13. 0.2016	14. 1.4036	15. –1.66
16. 1.050	17. 0.607	18. 1.732	19. 2.2790	20. 0.201640
21. 1.8556	22. 4.56155	23. 2.32672	24. 2.740646	25. 0.5885
26. 0.25917	27. 2.3	28. 2	29. 1	

5.4 SYSTEM OF LINEAR EQUATIONS

In this unit, we shall be solving n linear equations in n unknowns having unique solution. We shall distinguish mainly between direct and iterative methods. Firstly, we consider the direct methods.

5.4.1 Gauss Elimination Method

Let n linear equations in n unknowns be

$$
\begin{aligned}
a_{11}x_1 + a_{12}x_2 + \ldots + a_{1n}x_n &= y_1 \\
a_{21}x_1 + a_{22}x_2 + \ldots + a_{2n}x_n &= y_2 \\
&\vdots \\
a_{n1}x_1 + a_{n2}x_2 + \ldots + a_{nn}x_n &= y_n
\end{aligned}
$$

If $a_{11} = 0$, then equations are permuted in a suitable way so that coefficient a_{11} of x_1 in first equation is not zero. Then a_{11} is called pivot element. Divide first equation by a_{11} and then subtract this equation multiplied by $a_{21}, a_{31}, \ldots, a_{n1}$ from the second, third, …, n^{th} equation. Thus, x_1 will remain in the first equation only. Next the coefficient a'_{22} of the variable x_2 if not zero will be pivot element (if it is zero, then permute second equation with any of third, fourth, …, n^{th} equation in which coefficient of x_2 is non zero). We divide the second equation by this pivot element and then x_2 is eliminated in a similar way from the third, fourth, …, n^{th} equation. The procedure is continued until $x_1, x_2, \ldots, x_{n-1}$ are eliminated from the last equation. The coefficient of x_n in last equation is pivot element and divide this equation by this pivot element. When elimination is completed the system has the following form

$$
\begin{aligned}
x_1 + a'_{12}x_2 + a'_{13}x_3 + \ldots + a'_{1n}x_n &= z_1 \\
x_2 + a'_{23}x_3 + \ldots + a'_{2n}x_n &= z_2 \\
&\vdots \\
x_n &= z_n
\end{aligned}
$$

The new coefficient matrix is upper triangular matrix with unit elements in the diagonal. Then $x_n, x_{n-1}, \ldots, x_2, x_1$ are obtained from the nth, $(n-1)^{\text{th}}$, …, second and first equations by back substitutions.

The above process can also be performed by writing the given equations in matrix form $AX = Y$ where augmented matrix is $[A:Y] = \begin{bmatrix} a_{11} & a_{12} & \cdots & a_{1n} & y_1 \\ a_{21} & a_{22} & \cdots & a_{2n} & y_2 \\ \vdots & & & & \\ a_{n1} & a_{n2} & \cdots & a_{nn} & y_n \end{bmatrix}, X = \begin{bmatrix} x_1 \\ x_2 \\ \vdots \\ x_n \end{bmatrix}$ and performing the row transformations as explained above to reduce augmented matrix $[A:Y]$ to the form

$$\begin{bmatrix} 1 & a'_{12} & a'_{13} & \cdots & a'_{1n} & z_1 \\ 0 & 1 & a'_{23} & \cdots & a'_{2n} & z_2 \\ \vdots & & & & & \\ 0 & 0 & & \cdots & 1 & z_n \end{bmatrix} = [A' : Z]$$

Remark 5.3:

(i) It may occur that the pivot element be different from zero but very small in magnitude. It will give too large errors. The reason is that the small coefficient is usually formed as the difference between two almost equal numbers. This difficulty is overcome by suitable permutation of this equation with an equation below it which has the largest coefficient in magnitude. This process is called **partial pivoting.**

(ii) In practical work, it is recommended that an extra column be carried which contains the sum of the coefficients in one row. Elements in this column are treated like the other elements in the row in which they lie and same transformations be applied to these elements. In this way, we get an easy check throughout the computation in each step by comparing the row sums with the numbers in this extra column.

(iii) In the reduction of AY to $A'Z$ if A' contains zero row, then there will be no solution or infinite number of solutions.

5.4.2 Gauss–Jordan Method

Jordan's modification to Gauss elimination method is that the elimination is performed not only in the equations below but also in the equations above. In this way, we shall finally obtain unit coefficient matrix, i.e. identity matrix and we have the solution without further computations.

Remark 5.4: In Gauss elimination method, the number of operations of addition, subtraction, multiplication and division comes out to be $\frac{n^3}{3}$ (this includes the operations during back substitutions) and in Gauss Jordan method, number of operations are $\frac{n^3}{2}$. Hence the number of operations in Gauss elimination method are less than the number of operations required in Gauss Jordan method and hence Gauss elimination method is preferred as compared to Gauss Jordan method.

Example 5.17: Solve the following equations by Gauss elimination method with partial pivoting.

$$2x_2 + x_4 = 0,\ 2x_1 + 2x_2 + 3x_3 + 2x_4 = -2,\ 4x_1 - 3x_2 + x_4 = -7 \text{ and } 6x_1 + x_2 - 6x_3 - 5x_4 = 6$$

Solution: With initial partial pivoting, augmented matrix is

$$\begin{bmatrix} 6 & 1 & -6 & -5 & 6 \\ 0 & 2 & 0 & 1 & 0 \\ 2 & 2 & 3 & 2 & -2 \\ 4 & -3 & 0 & 1 & -7 \end{bmatrix} \qquad \begin{matrix} \text{check} \\ 2 \\ 3 \\ 7 \\ -5 \end{matrix}$$

Operate $\dfrac{1}{6}R_1$

$$\begin{bmatrix} 1 & .16667 & -1 & -.83333 & 1 \\ 0 & 2 & 0 & 1 & 0 \\ 2 & 2 & 3 & 2 & -2 \\ 4 & -3 & 0 & 1 & -7 \end{bmatrix} \qquad \begin{matrix} .33334 \\ 3 \\ 7 \\ -5 \end{matrix}$$

Operate $R_3 - 2R_1,\ R_4 - 4R_1$

$$\begin{bmatrix} 1 & .16667 & -1 & -.83333 & 1 \\ 0 & 2 & 0 & 1 & 0 \\ 0 & 1.66666 & 5 & 3.66666 & -4 \\ 0 & -3.66668 & 4 & 4.33332 & -11 \end{bmatrix} \qquad \begin{matrix} .33334 \\ 3 \\ 6.33332 \\ -6.33336 \end{matrix}$$

Operate R_{24}

$$\begin{bmatrix} 1 & .16667 & -1 & -.83333 & 1 \\ 0 & -3.66668 & 4 & 4.33332 & -11 \\ 0 & 1.66666 & 5 & 3.66666 & -4 \\ 0 & 2 & 0 & 1 & 0 \end{bmatrix} \qquad \begin{matrix} .33334 \\ -6.33336 \\ 6.33332 \\ 3 \end{matrix}$$

$$\left(-\frac{1}{3.66668}\right)R_2$$

$$\begin{bmatrix} 1 & .16667 & -1 & -.83333 & 1 \\ 0 & 1 & -1.09091 & -1.18181 & 2.99999 \\ 0 & 1.66666 & 5 & 3.66666 & -4 \\ 0 & 2 & 0 & 1 & 0 \end{bmatrix} \qquad \begin{matrix} .33334 \\ 1.72727 \\ 6.33332 \\ 3 \end{matrix}$$

$$R_3 - 1.66666R_2,\ R_4 - 2R_2$$

$$\begin{bmatrix} 1 & .16667 & -1 & -.83333 & 1 \\ 0 & 1 & -1.09091 & -1.18181 & 2.99999 \\ 0 & 0 & 6.81818 & 5.63634 & -8.99996 \\ 0 & 0 & 2.18182 & 3.36362 & -5.99998 \end{bmatrix} \qquad \begin{matrix} .33334 \\ 1.72727 \\ 3.45456 \\ -0.45454 \end{matrix}$$

$$\frac{1}{6.81818}R_3$$

$$\left[\begin{array}{ccccc|c} 1 & .16667 & -1 & -.83333 & 1 & .33334 \\ 0 & 1 & -1.09091 & -1.18181 & 2.99999 & 1.72727 \\ 0 & 0 & 1 & .82666 & -1.31999 & .50667 \\ 0 & 0 & 2.18182 & 3.36362 & -5.99998 & -0.45454 \end{array}\right]$$

$$R_4 - 2.18182\,R_3$$

$$\left[\begin{array}{ccccc|c} 1 & .16667 & -1 & -.83333 & 1 & .33334 \\ 0 & 1 & -1.09091 & -1.18181 & 2.99999 & 1.72727 \\ 0 & 0 & 1 & .82666 & -1.31999 & .50667 \\ 0 & 0 & 0 & 1.56000 & -3.12000 & -1.56000 \end{array}\right]$$

$$\frac{1}{1.56000}R_4$$

$$\begin{bmatrix} 1 & .16667 & -1 & -.83333 & 1 \\ 0 & 1 & -1.09091 & -1.18181 & 2.99999 \\ 0 & 0 & 1 & .82666 & -1.31999 \\ 0 & 0 & 0 & 1 & -2 \end{bmatrix}$$

$\therefore$ $$x_4 = -2$$

$$x_3 = -1.31999 - .82666(-2) = 0.3333$$

$$x_2 = 3 + 1.09091(.33333) + 1.18181(-2) = 1$$

$$x_1 = 1 - .16667(1) + 1(.33333) + .83333(-2) = -0.5$$

$\therefore$ $$x_1 = -0.5,\ x_2 = 1,\ x_3 = 0.3333,\ x_4 = -2$$

Example 5.18: Solve the following equations using Gauss elimination method. Find the solution correct up to three decimal places.

$$5x_1 - x_2 + x_3 = 10,\ 2x_1 + 4x_2 = 12,\ x_1 + x_2 + 5x_3 = -1$$

Solution: Given equations are

$$x_1 + x_2 + 5x_3 = -1$$
$$2x_1 + 4x_2 = 12$$
$$5x_1 - x_2 + x_3 = 10$$

Augmented matrix is

$$\begin{bmatrix} 1 & 1 & 5 & -1 \\ 2 & 4 & 0 & 12 \\ 5 & -1 & 1 & 10 \end{bmatrix} \quad \begin{matrix} \text{check} \\ 6 \\ 18 \\ 15 \end{matrix}$$

$$R_2 - 2R_1,\ R_3 - 5R_1$$

$$\begin{bmatrix} 1 & 1 & 5 & -1 \\ 0 & 2 & -10 & 14 \\ 0 & -6 & -24 & 15 \end{bmatrix} \quad \begin{matrix} 6 \\ 6 \\ -15 \end{matrix}$$

$$\frac{1}{2}R_2$$

$$\begin{bmatrix} 1 & 1 & 5 & -1 \\ 0 & 1 & -5 & 7 \\ 0 & -6 & -24 & 15 \end{bmatrix} \quad \begin{matrix} 6 \\ 3 \\ -15 \end{matrix}$$

$$R_3 + 6R_2$$

$$\begin{bmatrix} 1 & 1 & 5 & -1 \\ 0 & 1 & -5 & 7 \\ 0 & 0 & -54 & 57 \end{bmatrix} \quad \begin{matrix} 6 \\ 3 \\ 3 \end{matrix}$$

$$\left(-\frac{1}{54}\right)R_3$$

$$\begin{bmatrix} 1 & 1 & 5 & -1 \\ 0 & 1 & -5 & 7 \\ 0 & 0 & 1 & -1.0556 \end{bmatrix} \quad \begin{matrix} 6 \\ 3 \\ -.0556 \end{matrix}$$

$\therefore$
$$x_3 = -1.056$$
$$x_2 = 7 + 5(-1.0556) = 1.722$$
$$x_1 = -1 - 1(1.722) - 5(-1.0556) = 2.556$$

$\therefore$
$$x_1 = 2.556,\ x_2 = 1.722,\ x_3 = -1.056$$

Example 5.19: Solve the following equations using Gauss–Jordan method up to three decimal places

$$2x_1 + 2x_2 + x_3 = 12,\ 3x_1 + 2x_2 + 2x_3 = 8,\ 5x_1 + 10x_2 - 8x_3 = 10$$

Solution: Given equations are

$$5x_1 + 10x_2 - 8x_3 = 10$$
$$3x_1 + 2x_2 + 2x_3 = 8$$
$$2x_1 + 2x_2 + x_3 = 12$$

Augmented matrix is

$$\begin{bmatrix} 5 & 10 & -8 & 10 \\ 3 & 2 & 2 & 8 \\ 2 & 2 & 1 & 12 \end{bmatrix} \qquad \begin{matrix} \text{check} \\ 17 \\ 15 \\ 17 \end{matrix}$$

$$\frac{1}{5}R_1$$

$$\begin{bmatrix} 1 & 2 & -1.6 & 2 \\ 3 & 2 & 2 & 8 \\ 2 & 2 & 1 & 12 \end{bmatrix} \qquad \begin{matrix} 3.4 \\ 15 \\ 17 \end{matrix}$$

$$R_2 - 3R_1,\ R_3 - 2R_1$$

$$\begin{bmatrix} 1 & 2 & -1.6 & 2 \\ 0 & -4 & 6.8 & 2 \\ 0 & -2 & 4.2 & 8 \end{bmatrix} \qquad \begin{matrix} 3.4 \\ 4.8 \\ 10.2 \end{matrix}$$

$$\left(-\frac{1}{4}\right)R_2$$

$$\begin{bmatrix} 1 & 2 & -1.6 & 2 \\ 0 & 1 & -1.7 & -.5 \\ 0 & -2 & 4.2 & 8 \end{bmatrix} \qquad \begin{matrix} 3.4 \\ -1.2 \\ 10.2 \end{matrix}$$

$$R_1 - 2R_2,\ R_3 + 2R_2$$

$$\begin{bmatrix} 1 & 0 & 1.8 & 3 \\ 0 & 1 & -1.7 & -.5 \\ 0 & 0 & 0.8 & 7 \end{bmatrix} \qquad \begin{matrix} 5.8 \\ -1.2 \\ 7.8 \end{matrix}$$

$$\frac{1}{0.8}R_3$$

$$\begin{bmatrix} 1 & 0 & 1.8 & 3 \\ 0 & 1 & -1.7 & -.5 \\ 0 & 0 & 1 & 8.75 \end{bmatrix} \qquad \begin{matrix} 5.8 \\ -1.2 \\ 9.75 \end{matrix}$$

$$R_1 - 1.8R_3,\ R_2 + 1.7R_3$$

$$\begin{bmatrix} 1 & 0 & 0 & -12.75 \\ 0 & 1 & 0 & 14.375 \\ 0 & 0 & 1 & 8.75 \end{bmatrix} \qquad \begin{matrix} -11.75 \\ 15.375 \\ 9.75 \end{matrix}$$

$\therefore$ $$x_1 = -12.75,\ x_2 = 14.375,\ x_3 = 8.75$$

Example 5.20: Solve the following system of equations by

(i) Gauss elimination method
(ii) Gauss–Jordan method

$$10x-7y+3z+5u=6,\ -6x+8y-z-4u=5$$
$$3x+y+4z+11u=2,\ 5x-9y-2z+4u=7$$

Solution: (i) **By Gauss elimination method**

Augmented matrix is

$$\begin{bmatrix} 10 & -7 & 3 & 5 & 6 \\ -6 & 8 & -1 & -4 & 5 \\ 3 & 1 & 4 & 11 & 2 \\ 5 & -9 & -2 & 4 & 7 \end{bmatrix} \qquad \begin{matrix} \text{check} \\ 17 \\ 2 \\ 21 \\ 5 \end{matrix}$$

$$\frac{1}{10}R_1$$

$$\begin{bmatrix} 1 & -0.7 & 0.3 & 0.5 & 0.6 \\ -6 & 8 & -1 & -4 & 5 \\ 3 & 1 & 4 & 11 & 2 \\ 5 & -9 & -2 & 4 & 7 \end{bmatrix} \qquad \begin{matrix} 1.7 \\ 2 \\ 21 \\ 5 \end{matrix}$$

$$R_2+6R_1,\ R_3-3R_1,\ R_4-5R_1$$

$$\begin{bmatrix} 1 & -0.7 & 0.3 & 0.5 & 0.6 \\ 0 & 3.8 & 0.8 & -1 & 8.6 \\ 0 & 3.1 & 3.1 & 9.5 & 0.2 \\ 0 & -5.5 & -3.5 & 1.5 & 4 \end{bmatrix} \qquad \begin{matrix} 1.7 \\ 12.2 \\ 15.9 \\ -3.5 \end{matrix}$$

$$\frac{1}{3.8}R_2$$

$$\begin{bmatrix} 1 & -0.7 & 0.3 & 0.5 & 0.6 \\ 0 & 1 & .2105 & -.2632 & 2.2632 \\ 0 & 3.1 & 3.1 & 9.5 & 0.2 \\ 0 & -5.5 & -3.5 & 1.5 & 4 \end{bmatrix} \qquad \begin{matrix} 1.7 \\ 3.2105 \\ 15.9 \\ -3.5 \end{matrix} \qquad (1)$$

$$R_3-3.1R_2,\ R_4+5.5R_2$$

$$\begin{bmatrix} 1 & -0.7 & 0.3 & 0.5 & 0.6 \\ 0 & 1 & .2105 & -.2632 & 2.2632 \\ 0 & 0 & 2.4474 & 10.3159 & -6.8159 \\ 0 & 0 & -2.3422 & 0.0524 & 16.4476 \end{bmatrix} \qquad \begin{matrix} 1.7 \\ 3.2105 \\ 5.9474 \\ 14.1578 \end{matrix}$$

$$\frac{1}{2.4474}R_3$$

$$\begin{bmatrix} 1 & -0.7 & 0.3 & 0.5 & 0.6 \\ 0 & 1 & .2105 & -.2632 & 2.2632 \\ 0 & 0 & 1 & 4.2150 & -2.7850 \\ 0 & 0 & -2.3422 & 0.0524 & 16.4476 \end{bmatrix} \quad \begin{matrix} 1.7 \\ 3.2105 \\ 2.4300 \\ 14.1578 \end{matrix}$$

$$R_4 + 2.3422R_3$$

$$\begin{bmatrix} 1 & -0.7 & 0.3 & 0.5 & 0.6 \\ 0 & 1 & .2105 & -.2632 & 2.2632 \\ 0 & 0 & 1 & 4.2150 & -2.7850 \\ 0 & 0 & 0 & 9.9248 & 9.9246 \end{bmatrix} \quad \begin{matrix} 1.7 \\ 3.2105 \\ 2.4300 \\ 19.8494 \end{matrix}$$

$$\frac{1}{9.9248}R_4$$

$$\begin{bmatrix} 1 & -0.7 & 0.3 & 0.5 & 0.6 \\ 0 & 1 & .2105 & -.2632 & 2.2632 \\ 0 & 0 & 1 & 4.2150 & -2.7850 \\ 0 & 0 & 0 & 1 & 1 \end{bmatrix} \quad \begin{matrix} 1.7 \\ 3.2105 \\ 2.4300 \\ 2 \end{matrix}$$

$\therefore$

$$u = 1$$

$$z = -2.7850 - 4.2150(1) = -7$$

$$y = 2.2632 - .2105(-7) + .2632(1) = 4$$

$$x = .6 + .7(4) - .3(-7) - .5(1) = 5$$

$\therefore$ $x = 5,\ y = 4,\ z = -7,\ u = 1$

(ii) **By Gauss–Jordan Method**

Solution up to (1) is same

$$\begin{bmatrix} 1 & -0.7 & 0.3 & 0.5 & 0.6 \\ 0 & 1 & 0.2105 & -.2632 & 2.2632 \\ 0 & 3.1 & 3.1 & 9.5 & 0.2 \\ 0 & -5.5 & -3.5 & 1.5 & 4 \end{bmatrix} \quad \begin{matrix} 1.7 \\ 3.2105 \\ 15.9 \\ -3.5 \end{matrix}$$

$$R_1 + 0.7R_2,\ R_3 - 3.1R_2,\ R_4 + 5.5R_2$$

$$\begin{bmatrix} 1 & 0 & .4474 & .3158 & 2.1842 \\ 0 & 1 & .2105 & -.2632 & 2.2632 \\ 0 & 0 & 2.4474 & 10.3159 & -6.8159 \\ 0 & 0 & -2.3422 & .0524 & 16.4476 \end{bmatrix} \quad \begin{matrix} 3.9474 \\ 3.2105 \\ 5.9474 \\ 14.1578 \end{matrix}$$

$$\frac{1}{2.4474}R_3$$

$$\left[\begin{array}{ccccc|c} 1 & 0 & .4474 & .3158 & 2.1842 & 3.9474 \\ 0 & 1 & .2105 & -.2632 & 2.2632 & 3.2105 \\ 0 & 0 & 1 & 4.2150 & -2.7850 & 2.4300 \\ 0 & 0 & -2.3422 & 0.0524 & 16.4476 & 14.1578 \end{array}\right]$$

$$R_1 - .4474R_3,\ R_2 - .2105R_3,\ R_4 + 2.3422R_3$$

$$\left[\begin{array}{ccccc|c} 1 & 0 & 0 & -1.5700 & 3.4302 & 2.8602 \\ 0 & 1 & 0 & -1.1505 & 2.8494 & 2.6989 \\ 0 & 0 & 1 & 4.2150 & -2.7850 & 2.43 \\ 0 & 0 & 0 & 9.9248 & 9.9246 & 19.8494 \end{array}\right]$$

$$\frac{1}{9.9248}R_4$$

$$\left[\begin{array}{ccccc|c} 1 & 0 & 0 & -1.5700 & 3.4302 & 2.8602 \\ 0 & 1 & 0 & -1.1505 & 2.8494 & 2.6989 \\ 0 & 0 & 1 & 4.2150 & -2.7850 & 2.43 \\ 0 & 0 & 0 & 1 & 1 & 2 \end{array}\right]$$

$$R_1 + 1.57R_4,\ R_2 + 1.1505R_4,\ R_3 - 4.2150R_4$$

$$\begin{bmatrix} 1 & 0 & 0 & 0 & 5.0002 \\ 0 & 1 & 0 & 0 & 3.9999 \\ 0 & 0 & 1 & 0 & -7 \\ 0 & 0 & 0 & 1 & 1 \end{bmatrix}$$

$\therefore$ $x = 5,\ y = 4,\ z = -7,\ u = 1$

5.4.3 Triangularisation Method

This method is also called ***LU decomposition method*** or ***factorisation method.***

In Gauss elimination method, we have seen that the coefficient matrix A is reduced to upper triangular matrix with diagonal elements unity. In fact, the elimination can be interpreted as the premultiplication of A by a lower triangular matrix. Hence, in three dimensions we have

$$\begin{bmatrix} l_{11} & 0 & 0 \\ l_{21} & l_{22} & 0 \\ l_{31} & l_{32} & l_{33} \end{bmatrix}\begin{bmatrix} a_{11} & a_{12} & a_{13} \\ a_{21} & a_{22} & a_{23} \\ a_{31} & a_{32} & a_{33} \end{bmatrix} = \begin{bmatrix} 1 & u_{12} & u_{13} \\ 0 & 1 & u_{23} \\ 0 & 0 & 1 \end{bmatrix}$$

If the lower and upper triangular are denoted by L and U

then $$LA = U$$

or $$A = L^{-1} U$$

Now L^{-1} is also a lower triangular matrix, we can find a factorisation of A as a product of one lower triangular matrix and one upper triangular matrix. Replacing L^{-1} by L, we have $LU = A$. Thus, any $n \times n$ matrix can be factorised as $A = LU$ where

$$L = \begin{bmatrix} l_{11} & 0 & 0 \cdots 0 \\ l_{21} & l_{22} & 0 \cdots 0 \\ \vdots & & \\ l_{n1} & l_{n2} & l_{n3} \cdots l_{nn} \end{bmatrix}$$

and

$$U = \begin{bmatrix} u_{11} & u_{12} & u_{13} \cdots u_{1n} \\ 0 & u_{22} & u_{23} \cdots u_{2n} \\ \vdots & & \\ 0 & 0 & 0 \cdots u_{nn} \end{bmatrix}$$

Now, each of L and U will have $\dfrac{n(n+1)}{2}$ unknowns and hence number of unknowns will be $n(n + 1)$ and number of equations will be n^2 which will be obtained by multiplying each row of L with each column of U. Thus, we shall have to take n unknowns ourselves.

We take these n unknowns either all $l_{ii} = 1$ or all $u_{ii} = 1$

Thus, we have

$$LUX = Y$$

Suppose, $UX = Z$

$$\therefore \quad LZ = Y$$

From $LZ = Y$ by back substitutions, we can find Z and then from $UX = Z$ by back substitutions, we can find X and equations will be solved.

Now, we explain the method of factorisation A to LU. In three dimensions, if L has diagonal elements unity we have

$$\begin{bmatrix} 1 & 0 & 0 \\ l_{21} & 1 & 0 \\ l_{31} & l_{32} & 1 \end{bmatrix} \begin{bmatrix} u_{11} & u_{12} & u_{13} \\ 0 & u_{22} & u_{23} \\ 0 & 0 & u_{33} \end{bmatrix} = \begin{bmatrix} a_{11} & a_{12} & a_{13} \\ a_{21} & a_{22} & a_{23} \\ a_{31} & a_{32} & a_{33} \end{bmatrix}$$

The following equations in order will be obtained

$$u_{11} = a_{11}$$

$$u_{12} = a_{12}$$

$$u_{13} = a_{13}$$

$$l_{21}u_{11} = a_{21}$$

$$\Rightarrow \qquad l_{21} = \frac{a_{21}}{u_{11}}$$

$$l_{21}u_{12} + u_{22} = a_{22}$$

$$\Rightarrow \qquad u_{22} = a_{22} - l_{21}u_{12}$$

$$l_{21}u_{13} + u_{23} = a_{23}$$

$$\Rightarrow \qquad u_{23} = a_{23} - l_{21}u_{13}$$

$$l_{31}u_{11} = a_{31}$$

$$\Rightarrow \qquad l_{31} = \frac{a_{31}}{u_{11}}$$

$$l_{31}u_{12} + l_{32}u_{22} = a_{32}$$

$$\Rightarrow \qquad l_{32} = \frac{a_{32} - l_{31}u_{12}}{u_{22}}$$

$$l_{31}u_{13} + l_{32}u_{23} + u_{33} = a_{33}$$

$$\Rightarrow \qquad u_{33} = a_{33} - l_{31}u_{13} - l_{32}u_{23}$$

Thus, all the unknowns will be obtained even without writing these equations if we find them in this order.

Similarly if unity diagonal elements are taken in U, we can write L and U in $LU = A$ without writing the equations.

Remark 5.5: In determining elements in L and U in the above equations

$$u_{11} = a_{11} \text{ and } u_{22} = a_{22} - l_{21}u_{12} = a_{22} - \frac{a_{21}a_{12}}{a_{11}} = \frac{a_{11}a_{22} - a_{21}a_{12}}{a_{11}}$$

are in the denominator and hence unknowns can be found only if

$$a_{11} \neq 0, \; a_{11}a_{22} - a_{21}a_{12} = \begin{vmatrix} a_{11} & a_{12} \\ a_{21} & a_{22} \end{vmatrix} \neq 0$$

Thus, LU decomposition is possible only when leading minors in A are non-zero.

5.4.4 Doolittle Method

In this method, augmented matrix $[A : Y]$ is factorised into $L[U : Z]$ where diagonal elements of L are unity. The elements of L and $[U : Z]$ are obtained in the same way as in the triangularisation. In $[U : Z]$, Z will be obtained in the same way as elements of U as A is augmented with Y. As in three dimensions

$$\begin{bmatrix} 1 & 0 & 0 \\ l_{21} & 1 & 0 \\ l_{31} & l_{32} & 1 \end{bmatrix} \begin{bmatrix} u_{11} & u_{12} & u_{13} & z_1 \\ 0 & u_{22} & u_{23} & z_2 \\ 0 & 0 & u_{33} & z_3 \end{bmatrix} = \begin{bmatrix} a_{11} & a_{12} & a_{13} & y_1 \\ a_{21} & a_{22} & a_{23} & y_2 \\ a_{31} & a_{32} & a_{33} & y_3 \end{bmatrix}$$

z_1, z_2, z_3 will be obtained from

$$z_1 = y_1,\ l_{21}z_1 + z_2 = y_2,\ l_{31}z_1 + l_{32}z_2 + z_3 = y_3$$

Then, from the equations

$$\begin{aligned} u_{11}x_1 + u_{12}x_2 + u_{13}x_3 &= z_1 \\ u_{22}x_2 + u_{23}x_3 &= z_2 \\ u_{33}x_3 &= z_3 \end{aligned}$$

By back substitution, we find x_1, x_2, x_3.

5.4.5 Crout's Method

Crout suggested a technique to determine systematically the entries of the lower and upper triangular matrix augmented in which diagonal elements of upper triangular matrix are unity.

Augmented matrix from given equations is

$$\begin{bmatrix} a_{11} & a_{12} & \cdots & a_{1n} & y_1 \\ a_{21} & a_{22} & \cdots & a_{2n} & y_2 \\ \vdots & & & & \\ a_{n1} & a_{n2} & \cdots & a_{nn} & y_n \end{bmatrix}.$$

Unknown elements of lower triangular matrix L and upper triangular matrix (with diagonal elements unity) U augmented with z is written in one matrix

$$(A' : Z) = \begin{bmatrix} l_{11} & u_{12} & u_{13} & \cdots u_{1n} & z_1 \\ l_{21} & l_{22} & u_{23} & \cdots u_{2n} & z_2 \\ \vdots & & & & \\ l_{n1} & l_{n2} & l_{n3} & \cdots l_{nn} & z_n \end{bmatrix}$$

This matrix is called the derived matrix or auxiliary matrix. Without writing any equations the elements of auxiliary matrix are found in the order elements of first column, remaining elements of first row, remaining elements of second column, remaining elements of second row, remaining elements of third column, remaining elements of third row, ... These elements are

$$\begin{array}{llllll} l_{11} = a_{11} & u_{12} = \dfrac{a_{12}}{l_{11}} & u_{13} = \dfrac{a_{13}}{l_{11}} & \cdots & u_{1n} = \dfrac{a_{1n}}{l_{11}} & z_1 = \dfrac{y_1}{l_{11}} \\ l_{21} = a_{21} & l_{22} = a_{22} - l_{21}u_{12} & u_{23} = \dfrac{a_{23} - l_{21}u_{13}}{l_{22}} & \cdots & u_{2n} = \dfrac{a_{2n} - l_{21}u_{1n}}{l_{22}} & z_2 = \dfrac{y_2 - l_{21}z_1}{l_{22}} \\ l_{31} = a_{31} & l_{32} = a_{32} - l_{31}u_{12} & l_{33} = a_{33} - l_{31}u_{13} - l_{32}u_{23} & \cdots & u_{3n} = \dfrac{a_{3n} - l_{31}u_{1n} - l_{32}u_{2n}}{l_{33}} & z_3 = \dfrac{y_3 - l_{31}z_1 - l_{32}z_2}{l_{33}} \\ \vdots & \vdots & \vdots & & & \vdots \\ l_{n1} = a_{n1} & l_{n2} = a_{n2} - l_{n1}u_{12} & l_{n3} = a_{n3} - l_{n1}u_{13} - l_{n2}u_{23} & \cdots & l_{nn} = a_{nn} - \sum_{j=1}^{n-1} l_{nj}u_{jn} & z_n = \dfrac{y_n - \sum_{j=1}^{n-1} l_{nj}z_j}{l_{nn}} \end{array}$$

After finding $[A' : Z]$, we have

$$\begin{bmatrix} 1 & u_{12} & u_{13} & \cdots & u_{1n} \\ 0 & 1 & u_{23} & \cdots & u_{2n} \\ \vdots & & & & \\ 0 & 0 & 0 & \cdots & 1 \end{bmatrix} \begin{bmatrix} x_1 \\ x_2 \\ \vdots \\ x_n \end{bmatrix} = \begin{bmatrix} z_1 \\ z_2 \\ \vdots \\ z_n \end{bmatrix}$$

By back substitution, we can find $x_1, x_2, \ldots, x_n$.

Example 5.21: Show that the *LU* decomposition method fails to solve the system of equations

$$x_1 + x_2 - x_3 = 2$$
$$2x_1 + 2x_2 + 5x_3 = -3$$
$$3x_1 + 2x_2 - 3x_3 = 6$$

in this order. After writing the equations in other order, solve the equations.
Solution: Given equations in matrix form can be written as

$$\begin{bmatrix} 1 & 1 & -1 \\ 2 & 2 & 5 \\ 3 & 2 & -3 \end{bmatrix} \begin{bmatrix} x \\ y \\ z \end{bmatrix} = \begin{bmatrix} 2 \\ -3 \\ 6 \end{bmatrix}$$

The leading minor $\begin{vmatrix} 1 & 1 \\ 2 & 2 \end{vmatrix} = 2 - 2 = 0$

Hence, *LU* decomposition method fails to solve the equations written in this order.
If we write the given equations in the order

$$x_1 + x_2 - x_3 = 2$$
$$3x_1 + 2x_2 - 3x_3 = 6$$
$$2x_1 + 2x_2 + 5x_3 = -3$$

then $$AX = Y$$

where $$A = \begin{bmatrix} 1 & 1 & -1 \\ 3 & 2 & -3 \\ 2 & 2 & 5 \end{bmatrix}, \quad X = \begin{bmatrix} x_1 \\ x_2 \\ x_3 \end{bmatrix}, \quad Y = \begin{bmatrix} 2 \\ 6 \\ -3 \end{bmatrix}$$

Let $$\begin{bmatrix} l_{11} & 0 & 0 \\ l_{21} & l_{22} & 0 \\ l_{31} & l_{32} & l_{33} \end{bmatrix} \begin{bmatrix} 1 & u_{12} & u_{13} \\ 0 & 1 & u_{23} \\ 0 & 0 & 1 \end{bmatrix} = \begin{bmatrix} 1 & 1 & -1 \\ 3 & 2 & -3 \\ 2 & 2 & 5 \end{bmatrix}$$

$\therefore$ $$l_{11} = 1 \qquad u_{12} = \frac{1}{1} = 1 \qquad u_{13} = \frac{-1}{1} = -1$$

$$l_{21} = 3 \qquad l_{22} = 2-(3)(1) = -1 \qquad u_{23} = \frac{-3+3}{-1} = 0$$

$$l_{31} = 2 \qquad l_{32} = 2-(2)(1) = 0 \qquad l_{33} = 5-2(-1)-0(0) = 7$$

$\therefore$ $$\begin{bmatrix} 1 & 0 & 0 \\ 3 & -1 & 0 \\ 2 & 0 & 7 \end{bmatrix}\begin{bmatrix} 1 & 1 & -1 \\ 0 & 1 & 0 \\ 0 & 0 & 1 \end{bmatrix} = \begin{bmatrix} 1 & 1 & -1 \\ 3 & 2 & -3 \\ 2 & 2 & 5 \end{bmatrix}$$

$\therefore$ $$\begin{bmatrix} 1 & 0 & 0 \\ 3 & -1 & 0 \\ 2 & 0 & 7 \end{bmatrix}\begin{bmatrix} 1 & 1 & -1 \\ 0 & 1 & 0 \\ 0 & 0 & 1 \end{bmatrix}\begin{bmatrix} x_1 \\ x_2 \\ x_3 \end{bmatrix} = \begin{bmatrix} 2 \\ 6 \\ -3 \end{bmatrix}$$

Let $$\begin{bmatrix} 1 & 1 & -1 \\ 0 & 1 & 0 \\ 0 & 0 & 1 \end{bmatrix}\begin{bmatrix} x_1 \\ x_2 \\ x_3 \end{bmatrix} = \begin{bmatrix} z_1 \\ z_2 \\ z_3 \end{bmatrix} \tag{1}$$

$\therefore$ $$\begin{bmatrix} 1 & 0 & 0 \\ 3 & -1 & 0 \\ 2 & 0 & 7 \end{bmatrix}\begin{bmatrix} z_1 \\ z_2 \\ z_3 \end{bmatrix} = \begin{bmatrix} 2 \\ 6 \\ -3 \end{bmatrix}$$

$\therefore$ $$z_1 = 2$$

$$3z_1 - z_2 = 6$$

$\therefore$ $$z_2 = 3(2) - 6 = 0$$

$$2z_1 + 7z_3 = -3$$

$\therefore$ $$z_3 = \frac{-3-2(2)}{7} = -1$$

$\therefore$ from (1)

$$\begin{bmatrix} 1 & 1 & -1 \\ 0 & 1 & 0 \\ 0 & 0 & 1 \end{bmatrix}\begin{bmatrix} x_1 \\ x_2 \\ x_3 \end{bmatrix} = \begin{bmatrix} 2 \\ 0 \\ -1 \end{bmatrix}$$

$\therefore$ $$x_3 = -1$$

$$x_2 = 0$$

$$x_1 + x_2 - x_3 = 2$$

$\Rightarrow$ $$x_1 = 2-(0)+(-1) = 1$$

$\therefore$ $$x_1 = 1,\ x_2 = 0,\ x_3 = -1$$

Example 5.22: Solve the system of equations $AX = B$ by triangularisation method where

$$A = \begin{bmatrix} 2 & 1 & 2 \\ 8 & 5 & 13 \\ 6 & 3 & 12 \end{bmatrix}, X = \begin{bmatrix} x_1 \\ x_2 \\ x_3 \end{bmatrix}, B = \begin{bmatrix} 5.6 \\ 20.9 \\ 11.4 \end{bmatrix}$$

Solution: Let
$$\begin{bmatrix} l_{11} & 0 & 0 \\ l_{21} & l_{22} & 0 \\ l_{31} & l_{32} & l_{33} \end{bmatrix} \begin{bmatrix} 1 & u_{12} & u_{13} \\ 0 & 1 & u_{23} \\ 0 & 0 & 1 \end{bmatrix} = \begin{bmatrix} 2 & 1 & 2 \\ 8 & 5 & 13 \\ 6 & 3 & 12 \end{bmatrix}$$

$\therefore$
$$l_{11} = 2 \quad u_{12} = \frac{1}{2} = 0.5 \quad u_{13} = \frac{2}{2} = 1$$

$$l_{21} = 8 \quad l_{22} = 5 - 4 = 1 \quad u_{23} = \frac{13 - 8}{1} = 5$$

$$l_{31} = 6 \quad l_{32} = 3 - 3 = 0 \quad l_{33} = 12 - 6 - 0 = 6$$

$\therefore$
$$\begin{bmatrix} 2 & 0 & 0 \\ 8 & 1 & 0 \\ 6 & 0 & 6 \end{bmatrix} \begin{bmatrix} 1 & 0.5 & 1 \\ 0 & 1 & 5 \\ 0 & 0 & 1 \end{bmatrix} = \begin{bmatrix} 2 & 1 & 2 \\ 8 & 5 & 13 \\ 6 & 3 & 12 \end{bmatrix}$$

$\therefore$
$$\begin{bmatrix} 2 & 0 & 0 \\ 8 & 1 & 0 \\ 6 & 0 & 6 \end{bmatrix} \begin{bmatrix} 1 & 0.5 & 1 \\ 0 & 1 & 5 \\ 0 & 0 & 1 \end{bmatrix} \begin{bmatrix} x_1 \\ x_2 \\ x_3 \end{bmatrix} = \begin{bmatrix} 5.6 \\ 20.9 \\ 11.4 \end{bmatrix}$$

Let
$$\begin{bmatrix} 1 & 0.5 & 1 \\ 0 & 1 & 5 \\ 0 & 0 & 1 \end{bmatrix} \begin{bmatrix} x_1 \\ x_2 \\ x_3 \end{bmatrix} = \begin{bmatrix} z_1 \\ z_2 \\ z_3 \end{bmatrix} \tag{1}$$

$\therefore$
$$\begin{bmatrix} 2 & 0 & 0 \\ 8 & 1 & 0 \\ 6 & 0 & 6 \end{bmatrix} \begin{bmatrix} z_1 \\ z_2 \\ z_3 \end{bmatrix} = \begin{bmatrix} 5.6 \\ 20.9 \\ 11.4 \end{bmatrix}$$

$\therefore$
$$2z_1 = 5.6$$

$\Rightarrow$
$$z_1 = 2.8$$

$$8z_1 + z_2 = 20.9$$

$\Rightarrow$
$$z_2 = 20.9 - 8(2.8) = -1.5$$

$$6z_1 + 6z_3 = 11.4$$

$\Rightarrow$
$$z_3 = \frac{11.4 - 6(2.8)}{6} = -0.9$$

∴ from (1)

$$\begin{bmatrix} 1 & 0.5 & 1 \\ 0 & 1 & 5 \\ 0 & 0 & 1 \end{bmatrix}\begin{bmatrix} x_1 \\ x_2 \\ x_3 \end{bmatrix} = \begin{bmatrix} 2.8 \\ -1.5 \\ -0.9 \end{bmatrix}$$

$$\therefore \qquad x_3 = -0.9$$

$$x_2 + 5x_3 = -1.5$$

$$\Rightarrow \qquad x_2 = -1.5 - 5(-0.9) = 3$$

$$x_1 + 0.5x_2 + x_3 = 2.8$$

$$\Rightarrow \qquad x_1 = 2.8 - 0.5(3) - (-0.9) = 2.2$$

$$\therefore \qquad x_1 = 2.2,\ x_2 = 3,\ x_3 = -0.9$$

Example 5.23: Solve the following system of equations by Doolittle method

(i)

$$2x + y + z - 2u = -10$$
$$4x + 2z + u = 8$$
$$3x + 2y + 2z = 7$$
$$x + 3y + 2z - u = -5$$

(ii)

$$5x_1 + 4x_2 + x_3 = 3.4$$
$$10x_1 + 9x_2 + 4x_3 = 8.8$$
$$10x_1 + 13x_2 + 15x_3 = 19.2$$

Solution: (i) Augmented matrix is

$$[A:Y] = \begin{bmatrix} 2 & 1 & 1 & -2 & -10 \\ 4 & 0 & 2 & 1 & 8 \\ 3 & 2 & 2 & 0 & 7 \\ 1 & 3 & 2 & -1 & -5 \end{bmatrix}$$

Let

$$\begin{bmatrix} 1 & 0 & 0 & 0 \\ l_{21} & 1 & 0 & 0 \\ l_{31} & l_{32} & 1 & 0 \\ l_{41} & l_{42} & l_{43} & 1 \end{bmatrix}\begin{bmatrix} u_{11} & u_{12} & u_{13} & u_{14} & z_1 \\ 0 & u_{22} & u_{23} & u_{24} & z_2 \\ 0 & 0 & u_{33} & u_{34} & z_3 \\ 0 & 0 & 0 & u_{44} & z_4 \end{bmatrix} = \begin{bmatrix} 2 & 1 & 1 & -2 & -10 \\ 4 & 0 & 2 & 1 & 8 \\ 3 & 2 & 2 & 0 & 7 \\ 1 & 3 & 2 & -1 & -5 \end{bmatrix}$$

$$u_{11} = 2 \qquad u_{12} = 1 \qquad u_{13} = 1 \qquad u_{14} = -2 \qquad z_1 = -10$$

$$l_{21} = \frac{4}{2} = 2 \qquad u_{22} = 0 - 2 = -2 \qquad u_{23} = 2 - 2 = 0 \qquad u_{24} = 1 + 4 = 5 \qquad z_2 = 8 + 20 = 28$$

$l_{31}=\dfrac{3}{2}=1.5 \quad l_{32}=\dfrac{2-1.5}{-2}=-.25 \quad u_{33}=2-1.5-0=0.5 \quad u_{34}=0+3+1.25=4.25 \quad z_3=7+15+7=29$

$l_{41}=\dfrac{1}{2}=0.5 \quad l_{42}=\dfrac{3-0.5}{-2}=-1.25 \quad l_{43}=\dfrac{2-0.5-0}{0.5}=3 \quad u_{44}=-1+1+6.25-12.75=-6.5 \quad z_4=-5+5+35-87=-52$

$$\therefore \quad \begin{bmatrix} 1 & 0 & 0 & 0 \\ 2 & 1 & 0 & 0 \\ 1.5 & -0.25 & 1 & 0 \\ 0.5 & -1.25 & 3 & 1 \end{bmatrix}\begin{bmatrix} 2 & 1 & 1 & -2 & -10 \\ 0 & -2 & 0 & 5 & 28 \\ 0 & 0 & 0.5 & 4.25 & 29 \\ 0 & 0 & 0 & -6.5 & -52 \end{bmatrix}=\begin{bmatrix} 2 & 1 & 1 & -2 & -10 \\ 4 & 0 & 2 & 1 & 8 \\ 3 & 2 & 2 & 0 & 7 \\ 1 & 3 & 2 & -1 & -5 \end{bmatrix}$$

$$\therefore \quad -6.5u=-52$$

$$\Rightarrow \quad u=8$$

$$0.5z+4.25u=29$$

$$\Rightarrow \quad z=\frac{29-4.25(8)}{0.5}=-10$$

$$-2y+5u=28$$

$$\Rightarrow \quad y=\frac{28-5(8)}{-2}=6$$

$$2x+y+z-2u=-10$$

$$\Rightarrow \quad x=\frac{-10-6+10+2(8)}{2}=5$$

$$\therefore \quad x=5,\ y=6,\ z=-10,\ u=8$$

(ii) Augmented matrix is

$$[A:Y]=\begin{bmatrix} 5 & 4 & 1 & 3.4 \\ 10 & 9 & 4 & 8.8 \\ 10 & 13 & 15 & 19.2 \end{bmatrix}$$

Let $\begin{bmatrix} 1 & 0 & 0 \\ l_{21} & 1 & 0 \\ l_{31} & l_{32} & 1 \end{bmatrix}\begin{bmatrix} u_{11} & u_{12} & u_{13} & z_1 \\ 0 & u_{22} & u_{23} & z_2 \\ 0 & 0 & u_{33} & z_3 \end{bmatrix}=\begin{bmatrix} 5 & 4 & 1 & 3.4 \\ 10 & 9 & 4 & 8.8 \\ 10 & 13 & 15 & 19.2 \end{bmatrix}$

$\therefore \quad u_{11}=5 \quad u_{12}=4 \quad u_{13}=1 \quad z_1=3.4$

$l_{21}=\dfrac{10}{5}=2 \quad u_{22}=9-8=1 \quad u_{23}=4-2=2 \quad z_2=8.8-6.8=2.0$

$l_{31}=\dfrac{10}{5}=2 \quad l_{32}=\dfrac{13-8}{1}=5 \quad u_{33}=15-2-10=3 \quad z_3=19.2-6.8-10=2.4$

$$\therefore \quad \begin{bmatrix} 1 & 0 & 0 \\ 2 & 1 & 0 \\ 2 & 5 & 1 \end{bmatrix} \begin{bmatrix} 5 & 4 & 1 & 3.4 \\ 0 & 1 & 2 & 2.0 \\ 0 & 0 & 3 & 2.4 \end{bmatrix} = \begin{bmatrix} 5 & 4 & 1 & 3.4 \\ 10 & 9 & 4 & 8.8 \\ 10 & 13 & 15 & 19.2 \end{bmatrix}$$

$$\therefore \quad 3x_3 = 2.4$$

$$\Rightarrow \quad x_3 = 0.8$$

$$x_2 + 2x_3 = 2$$

$$\Rightarrow \quad x_2 = 2 - 2(0.8) = 0.4$$

$$5x_1 + 4x_2 + x_3 = 3.4$$

$$\Rightarrow \quad x_1 = \frac{3.4 - 4(0.4) - 0.8}{5} = 0.2$$

$$\therefore \quad x_1 = 0.2,\ x_2 = 0.4,\ x_3 = 0.8$$

Example 5.24: Solve the following system of equations by Crout's method

(i)
$$\begin{aligned} 2x + 3y + z &= 9 \\ x + 2y + 3z &= 6 \\ 3x + y + 2z &= 8 \end{aligned}$$

(ii)
$$\begin{aligned} 10x_1 - 7x_2 + 3x_3 + 5x_4 &= 6 \\ -6x_1 + 8x_2 - x_3 - 4x_4 &= 5 \\ 3x_1 + x_2 + 4x_3 + 11x_4 &= 2 \\ 5x_1 - 9x_2 - 2x_3 + 4x_4 &= 7 \end{aligned}$$

Solution: (i) Augmented matrix of the given system is

$$[A:Y] = \begin{bmatrix} 2 & 3 & 1 & 9 \\ 1 & 2 & 3 & 6 \\ 3 & 1 & 2 & 8 \end{bmatrix}$$

Let derived matrix be

$$[A':Z] = \begin{bmatrix} l_{11} & u_{12} & u_{13} & z_1 \\ l_{21} & l_{22} & u_{23} & z_2 \\ l_{31} & l_{32} & l_{33} & z_3 \end{bmatrix}$$

$$\therefore\ l_{11} = 2 \qquad u_{12} = \frac{3}{2} = 1.5 \qquad u_{13} = \frac{1}{2} = 0.5 \qquad z_1 = \frac{9}{2} = 4.5$$

$$l_{21} = 1 \qquad l_{22} = 2 - 1.5 = 0.5 \qquad u_{23} = \frac{3 - 0.5}{0.5} = 5 \qquad z_2 = \frac{6 - 4.5}{0.5} = 3$$

$$l_{31} = 3 \qquad l_{32} = 1 - 4.5 = -3.5 \qquad l_{33} = 2 - 1.5 + 17.5 = 18 \qquad z_3 = \frac{8 - 13.5 + 10.5}{18} = 0.277778$$

$\therefore$ $UX = Z$ gives

$$\begin{bmatrix} 1 & 1.5 & 0.5 \\ 0 & 1 & 5 \\ 0 & 0 & 1 \end{bmatrix}\begin{bmatrix} x \\ y \\ z \end{bmatrix} = \begin{bmatrix} 4.5 \\ 3 \\ 0.277778 \end{bmatrix}$$

$$\therefore \qquad z = 0.277778$$

$$y + 5z = 3$$

$$\Rightarrow \qquad y = 3 - 5(0.277778) = 1.611110$$

$$x + 1.5y + 0.5z = 4.5$$

$$\Rightarrow \qquad x = 4.5 - 1.5(1.611110) - 0.5(0.277778) = 1.944446$$

To four decimals

$$\therefore \qquad x = 1.9444,\ y = 1.6111,\ z = 0.2778$$

(ii) Augmented matrix of the given system is

$$[A:Y] = \begin{bmatrix} 10 & -7 & 3 & 5 & 6 \\ -6 & 8 & -1 & -4 & 5 \\ 3 & 1 & 4 & 11 & 2 \\ 5 & -9 & -2 & 4 & 7 \end{bmatrix}$$

Let derived matrix be

$$[A':Z] = \begin{bmatrix} l_{11} & u_{12} & u_{13} & u_{14} & z_1 \\ l_{21} & l_{22} & u_{23} & u_{24} & z_2 \\ l_{31} & l_{32} & l_{33} & u_{34} & z_3 \\ l_{41} & l_{42} & l_{43} & l_{44} & z_4 \end{bmatrix}$$

$\therefore\ l_{11} = 10 \quad u_{12} = \dfrac{-7}{10} = -0.7 \quad u_{13} = \dfrac{3}{10} = 0.3 \quad u_{14} = \dfrac{5}{10} = 0.5 \quad z_1 = \dfrac{6}{10} = 0.6$

$l_{21} = -6 \quad l_{22} = 8 - 4.2 = 3.8 \quad u_{23} = \dfrac{-1 + 1.8}{3.8} = 0.21053 \quad u_{24} = \dfrac{-4 + 3}{3.8} = -.26316 \quad z_2 = \dfrac{5 + 3.6}{3.8} = 2.26316$

$l_{31} = 3 \quad l_{32} = 1 + 2.1 = 3.1 \quad l_{33} = 4 - .9 - .65264 = 2.44736 \quad u_{34} = \dfrac{11 - 1.5 - .81580}{2.44736} = 4.21507 \quad z_3 = \dfrac{2 - 1.8 - 7.01580}{2.44736} = -2.78496$

$l_{41} = 5 \quad l_{42} = -9 + 3.5 = -5.5 \quad l_{43} = -2 - 1.5 + 1.15792 = -2.34208 \quad l_{44} = 4 - 2.5 - 1.44378 + 9.87203 = 9.92465 \quad z_4 = \dfrac{7 - 3 + 12.44378 - 6.52260}{9.92465} = 1.00001$

$\therefore$ $UX = Z$ gives

$$\begin{bmatrix} 1 & -0.7 & 0.3 & 0.5 \\ 0 & 1 & 0.21053 & -0.26316 \\ 0 & 0 & 1 & 4.21507 \\ 0 & 0 & 0 & 1 \end{bmatrix} \begin{bmatrix} x_1 \\ x_2 \\ x_3 \\ x_4 \end{bmatrix} = \begin{bmatrix} 0.6 \\ 2.26316 \\ -2.78496 \\ 1.00001 \end{bmatrix}$$

$$\therefore \quad x_4 = 1.00001 \simeq 1$$

$$x_3 + 4.21507x_4 = -2.78496$$

$$\therefore \quad x_3 = -2.78496 - 4.21507(1) = -7.00003 \simeq -7$$

$$x_2 + 0.21053x_3 - 0.26316x_4 = 2.26316$$

$$\therefore \quad x_2 = 2.26316 - 0.21053(-7) + 0.26316(1) = 4.00003 \simeq 4$$

$$x_1 - 0.7x_2 + 0.3x_3 + 0.5x_4 = 0.6$$

$$\therefore \quad x_1 = 0.6 + 0.7(4) - 0.3(-7) - 0.5(1) = 5$$

$$\therefore \quad x_1 = 5,\ x_2 = 4,\ x_3 = -7,\ x_4 = 1$$

5.5 ITERATIVE METHODS FOR SOLVING SYSTEM OF LINEAR EQUATIONS

The direct methods for the solution of the system of linear equations provide the solution after a certain amount of fixed computation. In iterative methods (also called indirect methods), we start from an approximation to the true solution and, if convergent, we form a sequence of closer approximations till the required accuracy is achieved. Thus, the difference between direct and iterative methods is that in direct methods the amount of computation is fixed while in iterative methods, the amount of computation is not fixed and depends upon the accuracy required.

In general, we prefer a direct method for solving system of linear equations but for large systems, iterative methods may be faster than the direct methods. Round-off errors in iterative methods are smaller. If any error is committed at any stage of computation, it will get corrected at subsequent stages. An iterative process may or may not converge. If the coefficient matrix A of given system is diagonally dominant, i.e. $|a_{ii}| \ge \sum_{\substack{j=1 \\ j\ne i}}^{n} |a_{ij}|$ for all i then the iterative process is sure to converge.

We shall consider two iterative processes.

5.5.1 Jacobi's Iterative Method

Let the system of linear equations be

$$a_{11}x_1 + a_{12}x_2 + \cdots + a_{1n}x_n = y_1$$

$$a_{21}x_1 + a_{22}x_2 + \cdots + a_{2n}x_n = y_2$$

$$\vdots$$

$$a_{n1}x_1 + a_{n2}x_2 + \cdots + a_{nn}x_n = y_n$$

Permute the equations such that diagonal elements of coefficient matrix have large magnitude in comparison to other elements of the row.

Let $\left(x_1^{(0)}, x_2^{(0)}, \cdots, x_n^{(0)}\right)^{\mathrm{T}}$ be initial approximation.

In the absence of better estimate for initial approximation, we may take

$$\left(x_1^{(0)}, x_2^{(0)}, \cdots, x_n^{(0)}\right)^{\mathrm{T}} = (0,0,0,\cdots,0)^{\mathrm{T}}$$

Now the equations can be written as

$$\left.\begin{aligned}
x_1 &= \frac{1}{a_{11}}(y_1 - a_{12}x_2 - a_{13}x_3 \cdots - a_{1n}x_n) \\
x_2 &= \frac{1}{a_{22}}(y_2 - a_{21}x_1 - a_{23}x_3 - \cdots - a_{2n}x_n) \\
x_3 &= \frac{1}{a_{33}}(y_3 - a_{31}x_1 - a_{32}x_2 - a_{34}x_4 - \cdots - a_{3n}x_n) \\
&\vdots \\
x_n &= \frac{1}{a_{nn}}(y_n - a_{n1}x_1 - a_{n2}x_2 - \cdots - a_{n,n-1}x_{n-1})
\end{aligned}\right] \tag{5.1}$$

Substituting $x_1^{(0)}, x_2^{(0)}, \ldots, x_n^{(0)}$ on *R.H.S.* of equations (5.1), we get first approximations

$$\begin{aligned}
x_1^{(1)} &= \frac{1}{a_{11}}\left(y_1 - a_{12}x_2^{(0)} - a_{13}x_3^{(0)} \cdots - a_{1n}x_n^{(0)}\right) \\
x_2^{(1)} &= \frac{1}{a_{22}}\left(y_2 - a_{21}x_1^{(0)} - a_{23}x_3^{(0)} - \cdots - a_{2n}x_n^{(0)}\right) \\
&\vdots \\
x_n^{(1)} &= \frac{1}{a_{nn}}\left(y_n - a_{n1}x_1^{(0)} - a_{n2}x_2^{(0)} - \cdots - a_{n,n-1}x_{n-1}^{(0)}\right)
\end{aligned}$$

Again, substituting $x_1^{(1)}, x_2^{(1)}, \ldots, x_n^{(1)}$ on *R.H.S.* of system (5.1) we obtain next approximations $x_1^{(2)}, x_2^{(2)}, \ldots, x_n^{(2)}$.

The process is repeated till the last two approximations are equal up to desired accuracy.

5.5.2 Gauss–Seidel Iteration Method

This method is modification of Jacobi's method. As in Jacobi's method, the system of linear equations

$$\begin{aligned}
a_{11}x_1 + a_{12}x_2 + \cdots + a_{1n}x_n &= y_1 \\
a_{21}x_1 + a_{22}x_2 + \cdots + a_{2n}x_n &= y_2 \\
&\vdots \\
a_{n1}x_1 + a_{n2}x_2 + \cdots + a_{nn}x_n &= y_n
\end{aligned}$$

is written as

$$\left.\begin{aligned} x_1 &= \frac{1}{a_{11}}\left(y_1 - a_{12}x_2 - a_{13}x_3 - \cdots - a_{1n}x_n\right) \\ x_2 &= \frac{1}{a_{22}}\left(y_2 - a_{21}x_1 - a_{23}x_3 - \cdots - a_{2n}x_n\right) \\ &\vdots \\ x_n &= \frac{1}{a_{nn}}\left(y_n - a_{n1}x_1 - a_{n2}x_2 - \cdots - a_{n,n-1}x_{n-1}\right) \end{aligned}\right] \tag{5.2}$$

Let $\left(x_1^{(0)}, x_2^{(0)}, \ldots, x_n^{(0)}\right)^{\mathrm{T}}$ be initial approximation.

Substituting $x_2^{(0)}, x_3^{(0)}, \ldots, x_n^{(0)}$ for $x_2, x_3, \ldots, x_n$ in first equation of system (5.2), we get

$$x_1^{(1)} = \frac{1}{a_{11}}\left(y_1 - a_{12}x_2^{(0)} - a_{13}x_3^{(0)} \ldots - a_{1n}x_n^{(0)}\right)$$

Now, substituting $x_1^{(1)}, x_3^{(0)}, x_4^{(0)}, \ldots, x_n^{(0)}$ for $x_1, x_3, x_4, \ldots, x_n$ in *R.H.S.* of second equation of system (5.2), we get

$$x_2^{(1)} = \frac{1}{a_{22}}\left(y_2 - a_{21}x_1^{(1)} - a_{23}x_3^{(0)} - \cdots - a_{2n}x_n^{(0)}\right)$$

next substituting $x_1^{(1)}, x_2^{(1)}, x_4^{(0)}, \ldots, x_n^{(0)}$ for $x_1, x_2, x_4, \ldots, x_n$ in *R.H.S.* of third equation of system (5.2), we get

$$x_3^{(1)} = \frac{1}{a_{33}}\left(y_3 - a_{31}x_1^{(1)} - a_{32}x_2^{(1)} - a_{34}x_4^{(0)} - \cdots - a_{3n}x_n^{(0)}\right)$$

and so on. We shall have

$$x_n^{(1)} = \frac{1}{a_{nn}}\left(y_n - a_{n1}x_1^{(1)} - a_{n2}x_2^{(1)} - \cdots - a_{n,n-1}x_{n-1}^{(1)}\right)$$

In the substitutions as soon as a new approximation for an unknown is found, it is immediately used in the next step.

The process of iteration is repeated till $\left(x_1^{(n+1)}, x_2^{(n+1)}, \ldots, x_n^{(n+1)}\right)^{\mathrm{T}} = \left(x_1^{(n)}, x_2^{(n)}, \ldots, x_n^{(n)}\right)$ up to the desired accuracy.

Remark 5.6: Since the most recent approximations of the unknowns are used in the next step, the convergence in the Gauss–Seidel method is twice as fast as in Jacobi's method.

Example 5.25: Starting with $(x_0, y_0, z_0) = (0, 0, 0)$ and using Jacobi's method, find the next five iterations for the system

$$5x - y + z = 10,\ 2x + 8y - z = 11,\ -x + y + 4z = 3$$

Solution: From the given equations

$$x = \frac{1}{5}(10 + y - z)$$

$$y = \frac{1}{8}(11 - 2x + z)$$

$$z = \frac{1}{4}(3 + x - y)$$

We have $(x_0, y_0, z_0) = (0,0,0)$

$\therefore$

$$x_1 = \frac{1}{5}(10) = 2,\ y_1 = \frac{11}{8} = 1.375,\ z_1 = \frac{3}{4} = 0.75$$

$$x_2 = \frac{1}{5}(10 + 1.375 - 0.75) = 2.125$$

$$y_2 = \frac{1}{8}(11 - 4 + 0.75) = 0.9688$$

$$z_2 = \frac{1}{4}(3 + 2 - 1.375) = 0.9063$$

$$x_3 = \frac{1}{5}(10 + 0.9688 - 0.9063) = 2.0125$$

$$y_3 = \frac{1}{8}(11 - 2(2.125) + 0.9063) = 0.9570$$

$$z_3 = \frac{1}{4}(3 + 2.125 - 0.9688) = 1.0390$$

$$x_4 = \frac{1}{5}(10 + 0.9570 - 1.0390) = 1.9836$$

$$y_4 = \frac{1}{8}(11 - 2(2.0125) + 1.0390) = 1.0018$$

$$z_4 = \frac{1}{4}(3 + 2.0125 - 0.9570) = 1.0139$$

$$x_5 = \frac{1}{5}(10 + 1.0018 - 1.0139) = 1.9976$$

$$y_5 = \frac{1}{8}(11 - 2(1.9836) + 1.0139) = 1.0058$$

$$z_5 = \frac{1}{4}(3 + 1.9836 - 1.0018) = 0.9954$$

$\therefore$

$$(x_5, y_5, z_5) = (1.9976, 1.0058, 0.9954)$$

The iteration converges to (2, 1, 1)

Example 5.26: Use Gauss–Seidel procedure to solve

$$5x + 2y + z = 12,\ x + 4y + 2z = 15,\ x + 2y + 5z = 20$$

Solution: From the given equations

$$x = \frac{1}{5}(12 - 2y - z)$$

$$y = \frac{1}{4}(15 - x - 2z)$$

$$z = \frac{1}{5}(20 - x - 2y)$$

Taking $(x_0, y_0, z_0) = (0, 0, 0)$

we have

$$x_1 = \frac{12}{5} = 2.4,$$

$$y_1 = \frac{1}{4}(15 - 2.4) = 3.15$$

$$z_1 = \frac{1}{5}(20 - 2.4 - 2(3.15)) = 2.26$$

$$x_2 = \frac{1}{5}(12 - 2(3.15) - 2.26) = 0.688$$

$$y_2 = \frac{1}{4}(15 - 0.688 - 2(2.26)) = 2.448$$

$$z_2 = \frac{1}{5}(20 - 0.688 - 2(2.448)) = 2.8832$$

$$x_3 = \frac{1}{5}(12 - 2(2.448) - 2.8832) = 0.8442$$

$$y_3 = \frac{1}{4}(15 - 0.8442 - 2(2.8832)) = 2.0974$$

$$z_3 = \frac{1}{5}(20 - 0.8442 - 2(2.0974)) = 2.9922$$

$$x_4 = \frac{1}{5}(12 - 2(2.0974) - 2.9922) = 0.9626$$

$$y_4 = \frac{1}{4}(15 - 0.9626 - 2(2.9922)) = 2.0132$$

$$z_4 = \frac{1}{5}(20 - 0.9626 - 2(2.0132)) = 3.0022$$

$$x_5 = \frac{1}{5}(12 - 2(2.0132) - 3.0022) = 0.9943$$

$$y_5 = \frac{1}{4}(15 - 0.9943 - 2(3.0022)) = 2.0003$$

$$z_5 = \frac{1}{5}(20 - 0.9943 - 2(2.0003)) = 3.0010$$

$$x_6 = \frac{1}{5}(12 - 2(2.0003) - 3.001) = 0.9997$$

$$y_6 = \frac{1}{4}(15 - 0.9997 - 2(3.001)) = 1.9996$$

$$z_6 = \frac{1}{5}(20 - 0.9997 - 2(1.9996)) = 3.0002$$

$$\therefore \quad (x_5, y_5, z_5) = (0.9943, 2.0003, 3.0010)$$

$$(x_6, y_6, z_6) = (0.9997, 1.9996, 3.0002)$$

$\therefore$ Iteration converges to

$$(x, y, z) = (1, 2, 3)$$

Example 5.27: Using Gauss–Seidel method, solve the following system of equations $10x + y - z = 17$, $2x + 20y + z = 28$, $3x - y + 25z = 105$ starting with initial approximation $(1, 1, 1)^T$ and perform three iterations.

Solution: From given equations

$$x = 1.7 - 0.1y + 0.1z$$

$$y = 1.4 - 0.1x - 0.05z$$

$$z = 4.2 - 0.12x + 0.04y$$

Initial approximation is

$$(x_0, y_0, z_0)^T = (1, 1, 1)^T$$

$$x_1 = 1.7$$

$$y_1 = 1.4 - 0.1(1.7) - 0.05 = 1.18$$

$$z_1 = 4.2 - 0.12(1.7) + 0.04(1.18) = 4.043$$

$$x_2 = 1.7 - 0.1(1.18) + 0.1(4.043) = 1.986$$

$$y_2 = 1.4 - 0.1(1.986) - 0.05(4.043) = 0.999$$

$$z_2 = 4.2 - 0.12(1.986) + 0.04(0.999) = 4.002$$

$$x_3 = 1.7 - 0.1(0.999) + 0.1(4.002) = 2.000$$

$$y_3 = 1.4 - 0.1(2) - 0.05(4.002) = 1.000$$

$$z_3 = 4.2 - 0.12(2) + 0.04(1) = 4.000$$

After three iterations, we observe that iteration converges to solution $(x, y, z)^T = (2, 1, 4)^T$

Example 5.28: Solve the equations

$$10x_1 - 2x_2 - x_3 - x_4 = 3$$
$$-2x_1 + 10x_2 - x_3 - x_4 = 15$$
$$-x_1 - x_2 + 10x_3 - 2x_4 = 27$$
$$-x_1 - x_2 - 2x_3 + 10x_4 = -9$$

by (i) Gauss–Jacobi method (ii) Gauss–Seidel method.

Solution: From given equations

$$x_1 = 0.3 + 0.2x_2 + 0.1x_3 + 0.1x_4$$
$$x_2 = 1.5 + 0.2x_1 + 0.1x_3 + 0.1x_4$$
$$x_3 = 2.7 + 0.1x_1 + 0.1x_2 + 0.2x_4$$
$$x_4 = -0.9 + 0.1x_1 + 0.1x_2 + 0.2x_3$$

Take initial solution $\left(x_1^{(0)}, x_2^{(0)}, x_3^{(0)}, x_4^{(0)}\right)^T = (0,0,0,0)^T$

$$x_1^{(1)} = 0.3$$
$$x_2^{(1)} = 1.5$$
$$x_3^{(1)} = 2.7$$
$$x_4^{(1)} = -0.9$$
$$x_1^{(2)} = 0.3 + 0.2(1.5) + 0.1(2.7) + 0.1(-0.9) = 0.78$$
$$x_2^{(2)} = 1.5 + 0.2(0.3) + 0.1(2.7) + 0.1(-0.9) = 1.74$$
$$x_3^{(2)} = 2.7 + 0.1(0.3) + 0.1(1.5) + 0.2(-0.9) = 2.7$$
$$x_4^{(2)} = -0.9 + 0.1(0.3) + 0.1(1.5) + 0.2(2.7) = -0.18$$
$$x_1^{(3)} = 0.3 + 0.2(1.74) + 0.1(2.7) + 0.1(-0.18) = 0.90$$
$$x_2^{(3)} = 1.5 + 0.2(0.78) + 0.1(2.7) + 0.1(-0.18) = 1.908$$
$$x_3^{(3)} = 2.7 + 0.1(0.78) + 0.1(1.74) + 0.2(-0.18) = 2.916$$
$$x_4^{(3)} = -0.9 + 0.1(0.78) + 0.1(1.74) + 0.2(2.7) = -0.108$$
$$x_1^{(4)} = 0.3 + 0.2(1.908) + 0.1(2.916) + 0.1(-0.108) = 0.9624$$
$$x_2^{(4)} = 1.5 + 0.2(0.90) + 0.1(2.916) + 0.1(-0.108) = 1.9608$$

$$x_3^{(4)} = 2.7 + 0.1(0.90) + 0.1(1.908) + 0.2(-0.108) = 2.9592$$

$$x_4^{(4)} = -0.9 + 0.1(0.90) + 0.1(1.908) + 0.2(2.916) = -0.036$$

$$x_1^{(5)} = 0.3 + 0.2(1.9608) + 0.1(2.9592) + 0.1(-0.036) = 0.9845$$

$$x_2^{(5)} = 1.5 + 0.2(0.9624) + 0.1(2.9592) + 0.1(-0.036) = 1.9848$$

$$x_3^{(5)} = 2.7 + 0.1(0.9624) + 0.1(1.9608) + 0.2(-0.036) = 2.9851$$

$$x_4^{(5)} = -0.9 + 0.1(0.9624) + 0.1(1.9608) + 0.2(2.9592) = -0.0158$$

$\therefore$ up to one decimal, the root converges to

$$(x_1, x_2, x_3, x_4)^T = (1.0, 2.0, 3.0, 0.0)$$

(ii) **By Gauss–Seidel Method**

$$x_1^{(1)} = 0.3$$

$$x_2^{(1)} = 1.5 + 0.2(0.3) = 1.56$$

$$x_3^{(1)} = 2.7 + 0.1(0.3) + 0.1(1.56) = 2.886$$

$$x_4^{(1)} = -0.9 + 0.1(0.3) + 0.1(1.56) + 0.2(2.886) = -0.137$$

$$x_1^{(2)} = 0.3 + 0.2(1.56) + 0.1(2.886) + 0.1(-0.137) = 0.887$$

$$x_2^{(2)} = 1.5 + 0.2(0.887) + 0.1(2.886) + 0.1(-0.137) = 1.952$$

$$x_3^{(2)} = 2.7 + 0.1(0.887) + 0.1(1.952) + 0.2(-0.137) = 2.956$$

$$x_4^{(2)} = -0.9 + 0.1(0.887) + 0.1(1.952) + 0.2(2.956) = -0.025$$

$$x_1^{(3)} = 0.3 + 0.2(1.952) + 0.1(2.956) + 0.1(-0.025) = 0.984$$

$$x_2^{(3)} = 1.5 + 0.2(0.984) + 0.1(2.956) + 0.1(-0.025) = 1.990$$

$$x_3^{(3)} = 2.7 + 0.1(0.984) + 0.1(1.990) + 0.2(-0.025) = 2.992$$

$$x_4^{(3)} = -0.9 + 0.1(0.984) + 0.1(1.990) + 0.2(2.992) = -.004$$

$$x_1^{(4)} = 0.3 + 0.2(1.990) + 0.1(2.992) + 0.1(-.004) = 0.997$$

$$x_2^{(4)} = 1.5 + 0.2(0.997) + 0.1(2.992) + 0.1(-0.004) = 1.998$$

$$x_3^{(4)} = 2.7 + 0.1(0.997) + 0.1(1.998) + 0.2(-0.004) = 2.999$$

$$x_4^{(4)} = -0.9 + 0.1(0.997) + 0.1(1.998) + 0.2(2.999) = -0.001$$

$\therefore$ up to one decimal place iteration converges to solution $(x_1, x_2, x_3, x_4)^T = (1.0, 2.0, 3.0, 0.0)^T$

5.6 ALGEBRAIC EIGENVALUE PROBLEMS

Suppose we are given a matrix

$$A = \begin{bmatrix} a_{11} & a_{12} \cdots & a_{1n} \\ a_{21} & a_{22} \cdots & a_{2n} \\ \vdots & & \\ a_{n1} & a_{n2} \cdots & a_{nn} \end{bmatrix}$$

n eigenvalues of A are the solutions of n th degree polynomial equations in λ which is

$$\begin{vmatrix} a_{11} - \lambda & a_{12} & \cdots & a_{1n} \\ a_{21} & a_{22} - \lambda & \cdots & a_{2n} \\ \vdots & & & \\ a_{n1} & a_{n2} & \cdots & a_{nn} - \lambda \end{vmatrix} = 0$$

and corresponding to eigenvalue λ, eigenvector is non-zero solution of n linear homogeneous system of equations

$$\begin{bmatrix} a_{11} - \lambda & a_{12} & \cdots & a_{1n} \\ a_{21} & a_{22} - \lambda & \cdots & a_{2n} \\ \vdots & & & \\ a_{n1} & a_{n2} & \cdots & a_{nn} - \lambda \end{bmatrix} \begin{bmatrix} x_1 \\ x_2 \\ \vdots \\ x_n \end{bmatrix} = 0$$

Hence, theoretically the problem of finding eigenvalues and eigenvectors is reduced to finding the roots of an algebraic equation and to solving n linear homogeneous system of equations. In practical computation, this method is unsuitable and better method must be applied. If only the numerically largest eigenvalue and its corresponding eigenvector is required, then the power method is quite suitable.

5.6.1 Power Method

Suppose $\lambda_1, \lambda_2, \ldots, \lambda_n$ are eigenvalues and $X_1, X_2, \ldots, X_n$ are corresponding eigenvectors of $n \times n$ matrix A.

Let $X = c_1X_1 + c_2X_2 + \cdots + c_nX_n$ is a linear combination of eigenvectors.

Then
$$AX = c_1AX_1 + c_2AX_2 + \cdots + c_nAX_n$$
$$= c_1\lambda_1X_1 + c_2\lambda_2X_2 + \cdots + c_n\lambda_nX_n$$
$$= \lambda_1\left(c_1X_1 + c_2\frac{\lambda_2}{\lambda_1}X_2 + \cdots + c_n\frac{\lambda_n}{\lambda_1}X_n\right) \text{ provided } \lambda_1 \neq 0$$

$$A^2X = \lambda_1\left(c_1AX_1 + c_2\frac{\lambda_2}{\lambda_1}AX_2 + \cdots + c_n\frac{\lambda_n}{\lambda_1}AX_n\right)$$

$$= \lambda_1\left(c_1\lambda_1X_1 + c_2\frac{\lambda_2}{\lambda_1}\cdot\lambda_2X_2 + \cdots + c_n\frac{\lambda_n}{\lambda_1}\lambda_nX_n\right)$$

$$= \lambda_1^2\left(c_1X_1 + c_2\left(\frac{\lambda_2}{\lambda_1}\right)^2X_2 + \cdots + c_n\left(\frac{\lambda_n}{\lambda_1}\right)^2X_n\right)$$

Proceeding in this way, we have

$$A^pX = \lambda_1^p\left(c_1X_1 + c_2\left(\frac{\lambda_2}{\lambda_1}\right)^pX_2 + \cdots + c_n\left(\frac{\lambda_n}{\lambda_1}\right)^pX_n\right)$$

Now, for large values of p, the vector

$$c_1X_1 + c_2\left(\frac{\lambda_2}{\lambda_1}\right)^pX_2 + \cdots + c_n\left(\frac{\lambda_n}{\lambda_1}\right)^pX_n$$

will converge to c_1X_1 if $|\lambda_1| > |\lambda_2| \geq |\lambda_3| \cdots \geq |\lambda_n|$ and c_1X_1 is eigenvector corresponding to eigenvalue λ_1.

$\therefore$ The eigenvalue λ_1 of largest magnitude is obtained as

$$\lambda = \lim_{p\to\infty}\frac{\left(A^{p+1}X\right)_r}{\left(A^pX\right)_r};\ r = 1, 2, \ldots, n$$

where $\left(A^{p+1}X\right)_r$ and $\left(A^pX\right)_r$ are r th components of $A^{p+1}X$ and A^pX, respectively.

The rate of convergence is determined by the quotient $\left|\frac{\lambda_2}{\lambda_1}\right|$. Convergence is faster if $\left|\frac{\lambda_2}{\lambda_1}\right|$ is smaller.

For numerical purposes, the algorithm just described can be formulated in the following way:

$$Y^{(k+1)} = AX^{(k)}$$

$$\lambda^{(k+1)} = \max_r\left(\left|\left(Y^{k+1}\right)_r\right|\right)$$

$$X^{(k+1)} = \frac{Y^{(k+1)}}{\lambda^{(k+1)}}$$

i.e., $$\lambda^{(k+1)}X^{(k+1)} = Y^{(k+1)}$$

The initial value $X^{(0)}$ can be chosen in a convenient way, either we may take $\begin{bmatrix}1\\0\\0\end{bmatrix}$ or $\begin{bmatrix}1\\1\\1\end{bmatrix}$ or component 1 be taken as r th component if r th row of A has largest magnitude element and other components zero. Iteration is stopped when $AX^{(n)} \simeq \lambda^{(n)}X^{(n)}$ or $AX^{(n)} \simeq -\lambda^{(n)}X^{(n)}$.

In the first case $\lambda^{(n)}$ will be eigenvalue and $X^{(n)}$ will be eigenvector. In the second case $-\lambda^{(n)}$ will be eigenvalue and $X^{(n)}$ will be eigenvector.

5.6.2 Modification of Power Method

The power method can be modified in such a way that other eigenvalues are obtained.

We know that if A has eigenvalue λ, then $A-qI$ has eigenvalue $\lambda-q$. Using this principle, we can find the next dominant eigenvalue. We will now discuss how the next absolutely largest eigenvalue can be calculated if we know the absolutely largest eigenvalue λ_1 and the corresponding eigenvector X_1.

Let a_1^T be the first row vector of A and X_1 is normalised in such a way that its first component is 1.

Form $A_1 = A - X_1 a_1^T$. Here, first row of A_1 will be zero.

Let λ_2 and X_2 be other eigenvalue and corresponding eigenvector of A, where first component of X_2 is unity. As first components of X_1 and X_2 are unity

$$\therefore \quad a_1^T X_1 = \lambda_1 \qquad (\because AX_1 = \lambda_1 X_1)$$

$$a_1^T X_2 = \lambda_2 \qquad (\because AX_2 = \lambda_2 X_2)$$

$$\therefore \quad A_1(X_1 - X_2) = (A - X_1 a_1^T)(X_1 - X_2)$$

$$= AX_1 - AX_2 - X_1 a_1^T X_1 + X_1 a_1^T X_2$$

$$= \lambda_1 X_1 - \lambda_2 X_2 - \lambda_1 X_1 + \lambda_2 X_1$$

$$= \lambda_2 (X_1 - X_2)$$

Thus, λ_2 is an eigenvalue of A_1 with corresponding eigenvector $X_1 - X_2$.

As first component of $X_1 - X_2$ is zero, so the first column of A_1 is irrelevant in $A_1 (X_1 - X_2) = \lambda_2 (X_1 - X_2)$. Thus, remove first row and first column of A_1. Let the remaining matrix is B_1.

Determine eigenvalue λ_2 of B_1 and corresponding eigenvector.

By adding 0 as first component in this eigenvector, we get a vector Z which is an eigenvector of A_1 and we shall have

$$X_2 - X_1 = cZ$$

$$\Rightarrow \quad a_1^T (X_2 - X_1) = c a_1^T Z$$

$$\Rightarrow \quad \lambda_2 - \lambda_1 = c a_1^T Z$$

$$\therefore \quad c = \frac{\lambda_2 - \lambda_1}{a_1^T Z}$$

Hence,
$$X_2 - X_1 = \left(\frac{\lambda_2 - \lambda_1}{a_1^T z} \right) Z$$

$$\therefore \quad X_2 = X_1 + \left(\frac{\lambda_2 - \lambda_1}{a_1^T z} \right) Z$$

Thus, eigenvalue λ_2 and corresponding eigenvector X_2 will be determined.

Remark 5.7: (i) If λ_1 is dominant eigenvalue of A (i.e magnitude of λ_1 is maximum among the magnitudes of eigenvalues of the matrix A) and λ_2 is dominant eigenvalue of A_1, then λ_2 will be the next dominant eigenvalue of A.

(ii) If λ is dominant eigenvalue of A^{-1}, then $\frac{1}{\lambda}$ will be smallest magnitude eigenvalue of A.

Example 5.29: Determine the largest eigenvalue in magnitude and the corresponding eigenvector of the following matrices.

(i) $A=\begin{bmatrix}2 & -1 & 0\\ -1 & 2 & -1\\ 0 & -1 & 2\end{bmatrix}$ (ii) $A=\begin{bmatrix}-15 & 4 & 3\\ 10 & -12 & 6\\ 20 & -4 & 2\end{bmatrix}$

Solution: (i) Let $X^{(0)}=\begin{bmatrix}1\\0\\1\end{bmatrix}$

$$\therefore \quad AX^{(0)}=\begin{bmatrix}2 & -1 & 0\\ -1 & 2 & -1\\ 0 & -1 & 2\end{bmatrix}\begin{bmatrix}1\\0\\1\end{bmatrix}=\begin{bmatrix}2\\-2\\2\end{bmatrix}=2\begin{bmatrix}1\\-1\\1\end{bmatrix}=\lambda^{(1)}X^{(1)}$$

$$AX^{(1)}=\begin{bmatrix}2 & -1 & 0\\ -1 & 2 & -1\\ 0 & -1 & 2\end{bmatrix}\begin{bmatrix}1\\-1\\1\end{bmatrix}=\begin{bmatrix}3\\-4\\3\end{bmatrix}=4\begin{bmatrix}0.75\\-1\\0.75\end{bmatrix}=\lambda^{(2)}X^{(2)}$$

$$AX^{(2)}=\begin{bmatrix}2 & -1 & 0\\ -1 & 2 & -1\\ 0 & -1 & 2\end{bmatrix}\begin{bmatrix}0.75\\-1\\0.75\end{bmatrix}=\begin{bmatrix}2.5\\-3.5\\2.5\end{bmatrix}=3.5\begin{bmatrix}0.714\\-1\\0.714\end{bmatrix}=\lambda^{(3)}X^{(3)}$$

$$AX^{(3)}=\begin{bmatrix}2 & -1 & 0\\ -1 & 2 & -1\\ 0 & -1 & 2\end{bmatrix}\begin{bmatrix}0.714\\-1\\0.714\end{bmatrix}=\begin{bmatrix}2.428\\-3.428\\2.428\end{bmatrix}=3.428\begin{bmatrix}0.708\\-1\\0.708\end{bmatrix}=\lambda^{(4)}X^{(4)}$$

$$AX^{(4)}=\begin{bmatrix}2 & -1 & 0\\ -1 & 2 & -1\\ 0 & -1 & 2\end{bmatrix}\begin{bmatrix}0.708\\-1\\0.708\end{bmatrix}=\begin{bmatrix}2.416\\-3.416\\2.416\end{bmatrix}=3.416\begin{bmatrix}0.707\\-1\\0.707\end{bmatrix}=\lambda^{(5)}X^{(5)}$$

$$AX^{(5)}=\begin{bmatrix}2 & -1 & 0\\ -1 & 2 & -1\\ 0 & -1 & 2\end{bmatrix}\begin{bmatrix}0.707\\-1\\0.707\end{bmatrix}=\begin{bmatrix}2.414\\-3.414\\2.414\end{bmatrix}=3.414\begin{bmatrix}0.707\\-1\\0.707\end{bmatrix}$$

$\therefore$ eigenvalue of largest magnitude = 3.414

corresponding eigenvector = $\begin{bmatrix}0.707\\-1\\0.707\end{bmatrix}$

(ii) Let $X^{(0)} = \begin{bmatrix} 1 \\ 0 \\ 0 \end{bmatrix}$

$$AX^{(0)} = \begin{bmatrix} -15 & 4 & 3 \\ 10 & -12 & 6 \\ 20 & -4 & 2 \end{bmatrix}\begin{bmatrix} 1 \\ 0 \\ 0 \end{bmatrix} = \begin{bmatrix} -15 \\ 10 \\ 20 \end{bmatrix} = 20\begin{bmatrix} -0.75 \\ 0.50 \\ 1 \end{bmatrix} = \lambda^{(1)} X^{(1)}$$

$$AX^{(1)} = \begin{bmatrix} -15 & 4 & 3 \\ 10 & -12 & 6 \\ 20 & -4 & 2 \end{bmatrix}\begin{bmatrix} -0.75 \\ 0.50 \\ 1 \end{bmatrix} = \begin{bmatrix} 16.25 \\ -7.50 \\ -15.0 \end{bmatrix} = 16.25\begin{bmatrix} 1 \\ -0.461 \\ -0.923 \end{bmatrix} = \lambda^{(2)} X^{(2)}$$

$$AX^{(2)} = \begin{bmatrix} -15 & 4 & 3 \\ 10 & -12 & 6 \\ 20 & -4 & 2 \end{bmatrix}\begin{bmatrix} 1 \\ -0.461 \\ -0.923 \end{bmatrix} = \begin{bmatrix} -19.613 \\ 9.994 \\ 19.998 \end{bmatrix} = 19.998\begin{bmatrix} -0.981 \\ 0.500 \\ 1 \end{bmatrix} = \lambda^{(3)} X^{(3)}$$

$$AX^{(3)} = \begin{bmatrix} -15 & 4 & 3 \\ 10 & -12 & 6 \\ 20 & -4 & 2 \end{bmatrix}\begin{bmatrix} -0.981 \\ 0.500 \\ 1 \end{bmatrix} = \begin{bmatrix} 15.715 \\ -9.810 \\ -19.620 \end{bmatrix} = 19.620\begin{bmatrix} 0.801 \\ -0.500 \\ -1 \end{bmatrix} = \lambda^{(4)} X^{(4)}$$

$$AX^{(4)} = \begin{bmatrix} -15 & 4 & 3 \\ 10 & -12 & 6 \\ 20 & -4 & 2 \end{bmatrix}\begin{bmatrix} 0.801 \\ -0.500 \\ -1 \end{bmatrix} = \begin{bmatrix} -17.015 \\ 8.010 \\ 16.020 \end{bmatrix} = 17.015\begin{bmatrix} -1 \\ 0.471 \\ 0.942 \end{bmatrix} = \lambda^{(5)} X^{(5)}$$

$$AX^{(5)} = \begin{bmatrix} -15 & 4 & 3 \\ 10 & -12 & 6 \\ 20 & -4 & 2 \end{bmatrix}\begin{bmatrix} -1 \\ 0.471 \\ 0.942 \end{bmatrix} = \begin{bmatrix} 19.71 \\ -10.00 \\ -20.00 \end{bmatrix} = 20\begin{bmatrix} 0.986 \\ -0.500 \\ -1 \end{bmatrix} = \lambda^{(6)} X^{(6)}$$

$$AX^{(6)} = \begin{bmatrix} -15 & 4 & 3 \\ 10 & -12 & 6 \\ 20 & -4 & 2 \end{bmatrix}\begin{bmatrix} 0.986 \\ -0.500 \\ -1 \end{bmatrix} = \begin{bmatrix} -19.79 \\ 9.86 \\ 19.72 \end{bmatrix} = 19.79\begin{bmatrix} -1 \\ 0.498 \\ 0.996 \end{bmatrix} = \lambda^{(7)} X^{(7)}$$

$$AX^{(7)} = \begin{bmatrix} -15 & 4 & 3 \\ 10 & -12 & 6 \\ 20 & -4 & 2 \end{bmatrix}\begin{bmatrix} -1 \\ 0.498 \\ 0.996 \end{bmatrix} = \begin{bmatrix} 19.980 \\ -10.000 \\ -20.000 \end{bmatrix} = 20\begin{bmatrix} 0.999 \\ -0.5 \\ -1 \end{bmatrix} \simeq -20\begin{bmatrix} -1 \\ 0.5 \\ 1 \end{bmatrix}$$

$$AX^{(8)} = \begin{bmatrix} -15 & 4 & 3 \\ 10 & -12 & 6 \\ 20 & -4 & 2 \end{bmatrix}\begin{bmatrix} -1 \\ 0.5 \\ 1 \end{bmatrix} = \begin{bmatrix} 20 \\ -10 \\ -20 \end{bmatrix} = -20\begin{bmatrix} -1 \\ 0.5 \\ 1 \end{bmatrix}$$

$\therefore$ eigenvalue of largest magnitude = −20

$$\text{corresponding eigenvector} = \begin{bmatrix} -1 \\ 0.5 \\ 1 \end{bmatrix}$$

Example 5.30: Find the absolutely smallest eigenvalue of the matrix $A = \begin{bmatrix} 2 & -1 & 0 \\ -1 & 2 & -1 \\ 0 & -1 & 2 \end{bmatrix}$

Solution:

$$A = \begin{bmatrix} 2 & -1 & 0 \\ -1 & 2 & -1 \\ 0 & -1 & 2 \end{bmatrix}$$

$$|A| = 2(4-1) + 1(-2-0) = 4$$

$$\therefore \qquad A^{-1} = \frac{1}{4}\begin{bmatrix} 3 & 2 & 1 \\ 2 & 4 & 2 \\ 1 & 2 & 3 \end{bmatrix} \qquad \left(\text{from } A^{-1} = \frac{1}{|A|}\text{adj } A\right)$$

Let

$$B = \begin{bmatrix} 3 & 2 & 1 \\ 2 & 4 & 2 \\ 1 & 2 & 3 \end{bmatrix}$$

Let

$$X^{(0)} = \begin{bmatrix} 0 \\ 1 \\ 0 \end{bmatrix}$$

$$BX^{(0)} = \begin{bmatrix} 3 & 2 & 1 \\ 2 & 4 & 2 \\ 1 & 2 & 3 \end{bmatrix}\begin{bmatrix} 0 \\ 1 \\ 0 \end{bmatrix} = \begin{bmatrix} 2 \\ 4 \\ 2 \end{bmatrix} = 4\begin{bmatrix} 0.5 \\ 1 \\ 0.5 \end{bmatrix} = \lambda^{(1)} X^{(1)}$$

$$BX^{(1)} = \begin{bmatrix} 3 & 2 & 1 \\ 2 & 4 & 2 \\ 1 & 2 & 3 \end{bmatrix}\begin{bmatrix} 0.5 \\ 1 \\ 0.5 \end{bmatrix} = \begin{bmatrix} 4 \\ 6 \\ 4 \end{bmatrix} = 6\begin{bmatrix} 0.667 \\ 1 \\ 0.667 \end{bmatrix} = \lambda^{(2)} X^{(2)}$$

$$BX^{(2)} = \begin{bmatrix} 3 & 2 & 1 \\ 2 & 4 & 2 \\ 1 & 2 & 3 \end{bmatrix}\begin{bmatrix} 0.667 \\ 1 \\ 0.667 \end{bmatrix} = \begin{bmatrix} 4.668 \\ 6.668 \\ 4.668 \end{bmatrix} = 6.668\begin{bmatrix} 0.7 \\ 1 \\ 0.7 \end{bmatrix} = \lambda^{(3)} X^{(3)}$$

$$BX^{(3)} = \begin{bmatrix} 3 & 2 & 1 \\ 2 & 4 & 2 \\ 1 & 2 & 3 \end{bmatrix}\begin{bmatrix} 0.7 \\ 1 \\ 0.7 \end{bmatrix} = \begin{bmatrix} 4.8 \\ 6.8 \\ 4.8 \end{bmatrix} = 6.8\begin{bmatrix} 0.706 \\ 1 \\ 0.706 \end{bmatrix} = \lambda^{(4)} X^{(4)}$$

$$BX^{(4)} = \begin{bmatrix} 3 & 2 & 1 \\ 2 & 4 & 2 \\ 1 & 2 & 3 \end{bmatrix} \begin{bmatrix} 0.706 \\ 1 \\ 0.706 \end{bmatrix} = \begin{bmatrix} 4.824 \\ 6.824 \\ 4.824 \end{bmatrix} = 6.824 \begin{bmatrix} 0.707 \\ 1 \\ 0.707 \end{bmatrix} = \lambda^{(5)} X^{(5)}$$

$$BX^{(5)} = \begin{bmatrix} 3 & 2 & 1 \\ 2 & 4 & 2 \\ 1 & 2 & 3 \end{bmatrix} \begin{bmatrix} 0.707 \\ 1 \\ 0.707 \end{bmatrix} = \begin{bmatrix} 4.828 \\ 6.828 \\ 4.828 \end{bmatrix} = 6.828 \begin{bmatrix} 0.707 \\ 1 \\ 0.707 \end{bmatrix}$$

$\therefore$ eigenvalue of B of maximum magnitude = 6.828

$\therefore$ eigenvalue of A^{-1} of maximum magnitude $= \dfrac{6.828}{4} = 1.707$

$\therefore$ eigenvalue of A of minimum magnitude $= \dfrac{1}{1.707} \simeq 0.5858$

Example 5.31: The matrix $A = \begin{bmatrix} -306 & -198 & 426 \\ 104 & 67 & -147 \\ -176 & -114 & 244 \end{bmatrix}$

has an eigenvalue 6 and corresponding eigenvector $\begin{bmatrix} 2 \\ -1 \\ 1 \end{bmatrix}$. Find its other eigenvalues and corresponding eigenvectors.

Solution: Eigenvector corresponding to eigenvalue $\lambda_1 = 6$ with first component unity is

$$X_1 = \begin{bmatrix} 1 \\ -0.5 \\ 0.5 \end{bmatrix}$$

$$A_1 = A - X_1 a_1^T = \begin{bmatrix} -306 & -198 & 426 \\ 104 & 67 & -147 \\ -176 & -114 & 244 \end{bmatrix} - \begin{bmatrix} 1 \\ -0.5 \\ 0.5 \end{bmatrix} \begin{bmatrix} -306 & -198 & 426 \end{bmatrix}$$

$$= \begin{bmatrix} 0 & 0 & 0 \\ -49 & -32 & 66 \\ -23 & -15 & 31 \end{bmatrix}$$

After leaving the first row and first column,

$$B_1 = \begin{bmatrix} -32 & 66 \\ -15 & 31 \end{bmatrix}$$

Sum of eigenvalues = Trace of $B_1 = -32 + 31 = -1$

Product of eigenvalues = $|B_1| = -992 + 990 = -2$

$\therefore$ eigenvalues are roots of

$$t^2 + t - 2 = 0$$

$\therefore$ eigenvalues are –2 and 1

$\therefore$ $\lambda_2 = -2$, $\lambda_3 = 1$ are other eigenvalues of A

for $\lambda_2 = -2$, two-dimensional eigenvector is the solution of

$$\begin{bmatrix} -32+2 & 66 \\ -15 & 31+2 \end{bmatrix}\begin{bmatrix} x_1 \\ x_2 \end{bmatrix} = \begin{bmatrix} 0 \\ 0 \end{bmatrix}$$

$\therefore$ $-30x_1 + 66x_2 = 0$, we can take eigenvector $\begin{bmatrix} 11 \\ 5 \end{bmatrix}$

for $\lambda_3 = 1$, two-dimensional eigenvector is the solution of

$$\begin{bmatrix} -32-1 & 66 \\ -15 & 31-1 \end{bmatrix}\begin{bmatrix} x_1 \\ x_2 \end{bmatrix} = \begin{bmatrix} 0 \\ 0 \end{bmatrix}$$

$\therefore$ $-33x_1 + 66x_2 = 0$, we can take eigenvector $\begin{bmatrix} 2 \\ 1 \end{bmatrix}$

$\therefore$ for $\lambda_2 = -2$, eigenvector of A_1 is $Z_2 = \begin{bmatrix} 0 \\ 11 \\ 5 \end{bmatrix}$

Now, $a_1^T Z_2 = \begin{bmatrix} -306 & -198 & 426 \end{bmatrix}\begin{bmatrix} 0 \\ 11 \\ 5 \end{bmatrix} = -48$

$\therefore$ eigenvector of A is $\begin{bmatrix} 1 \\ -0.5 \\ 0.5 \end{bmatrix} + \dfrac{-2-6}{-48}\begin{bmatrix} 0 \\ 11 \\ 5 \end{bmatrix}$ $\left(\because X_2 = X_1 + cZ \;\&\; c = \dfrac{\lambda_2 - \lambda_1}{a_1^T Z}\right)$

$$= \frac{1}{6}\left[\begin{bmatrix} 6 \\ -3 \\ 3 \end{bmatrix} + \begin{bmatrix} 0 \\ 11 \\ 5 \end{bmatrix}\right] = \frac{1}{6}\begin{bmatrix} 6 \\ 8 \\ 8 \end{bmatrix}$$

$\therefore$ for $\lambda_2 = -2$, we can take eigenvector $\begin{bmatrix} 3 \\ 4 \\ 4 \end{bmatrix}$

for $\lambda_3 = 1$, eigenvector of A_1 is $Z_3 = \begin{bmatrix} 0 \\ 2 \\ 1 \end{bmatrix}$

Now, $a_1^T Z_3 = [-306 - 198\ 426]\begin{bmatrix}0\\2\\1\end{bmatrix} = 30$

∴ eigenvector of A is

$$\begin{bmatrix}1\\-0.5\\0.5\end{bmatrix} + \frac{1-6}{30}\begin{bmatrix}0\\2\\1\end{bmatrix} = \frac{1}{6}\left[\begin{bmatrix}6\\-3\\3\end{bmatrix} - \begin{bmatrix}0\\2\\1\end{bmatrix}\right] = \frac{1}{6}\begin{bmatrix}6\\-5\\2\end{bmatrix}$$

∴ we can take eigenvector $\begin{bmatrix}6\\-5\\2\end{bmatrix}$

∴ other eigenvalues of A are –2 and 1 and their corresponding eigenvectors are, respectively,

$$\begin{bmatrix}3\\4\\4\end{bmatrix} \text{and} \begin{bmatrix}6\\-5\\2\end{bmatrix}$$

Example 5.32: Using power method, find the dominant eigenvalue and corresponding eigenvector of the matrix

$$A = \begin{bmatrix}3 & -1 & 0\\-1 & 2 & -1\\0 & -1 & 3\end{bmatrix}$$

and then using deflation method, find the other eigenvalues and corresponding eigenvectors.

Solution: Let $X^{(0)} = \begin{bmatrix}1\\0\\1\end{bmatrix}$

∴ $$AX^{(0)} = \begin{bmatrix}3 & -1 & 0\\-1 & 2 & -1\\0 & -1 & 3\end{bmatrix}\begin{bmatrix}1\\0\\1\end{bmatrix} = \begin{bmatrix}3\\-2\\3\end{bmatrix} = 3\begin{bmatrix}1\\-0.6667\\1\end{bmatrix} = \lambda^{(1)}X^{(1)}$$

$$AX^{(1)} = \begin{bmatrix}3 & -1 & 0\\-1 & 2 & -1\\0 & -1 & 3\end{bmatrix}\begin{bmatrix}1\\-0.6667\\1\end{bmatrix} = \begin{bmatrix}3.6667\\-3.3334\\3.6667\end{bmatrix} = 3.6667\begin{bmatrix}1\\-0.9091\\1\end{bmatrix} = \lambda^{(2)}X^{(2)}$$

$$AX^{(2)} = \begin{bmatrix}3 & -1 & 0\\-1 & 2 & -1\\0 & -1 & 3\end{bmatrix}\begin{bmatrix}1\\-0.9091\\1\end{bmatrix} = \begin{bmatrix}3.9091\\-3.8182\\3.9091\end{bmatrix} = 3.9091\begin{bmatrix}1\\-0.9767\\1\end{bmatrix} = \lambda^{(3)}X^{(3)}$$

$$AX^{(3)} = \begin{bmatrix} 3 & -1 & 0 \\ -1 & 2 & -1 \\ 0 & -1 & 3 \end{bmatrix}\begin{bmatrix} 1 \\ -0.9767 \\ 1 \end{bmatrix} = \begin{bmatrix} 3.9767 \\ -3.9534 \\ 3.9767 \end{bmatrix} = 3.9767\begin{bmatrix} 1 \\ -0.9941 \\ 1 \end{bmatrix} = \lambda^{(4)} X^{(4)}$$

$$AX^{(4)} = \begin{bmatrix} 3 & -1 & 0 \\ -1 & 2 & -1 \\ 0 & -1 & 3 \end{bmatrix}\begin{bmatrix} 1 \\ -0.9941 \\ 1 \end{bmatrix} = \begin{bmatrix} 3.9941 \\ -3.9882 \\ 3.9941 \end{bmatrix} = 3.9941\begin{bmatrix} 1 \\ -0.9985 \\ 1 \end{bmatrix} = \lambda^{(5)} X^{(5)}$$

$$AX^{(5)} = \begin{bmatrix} 3 & -1 & 0 \\ -1 & 2 & -1 \\ 0 & -1 & 3 \end{bmatrix}\begin{bmatrix} 1 \\ -0.9985 \\ 1 \end{bmatrix} = \begin{bmatrix} 3.9985 \\ -3.9970 \\ 3.9985 \end{bmatrix} = 3.9985\begin{bmatrix} 1 \\ -0.9996 \\ 1 \end{bmatrix} = \lambda^{(6)} X^{(6)}$$

$$AX^{(6)} = \begin{bmatrix} 3 & -1 & 0 \\ -1 & 2 & -1 \\ 0 & -1 & 3 \end{bmatrix}\begin{bmatrix} 1 \\ -0.9996 \\ 1 \end{bmatrix} = \begin{bmatrix} 3.9996 \\ -3.9992 \\ 3.9996 \end{bmatrix} = 3.9996\begin{bmatrix} 1 \\ -0.9999 \\ 1 \end{bmatrix}$$

$\therefore$ Iteration converges to $\lambda = 4$ and corresponding eigenvector $\begin{bmatrix} 1 \\ -1 \\ 1 \end{bmatrix}$

$\therefore$ Dominant eigenvalue $= \lambda_1 = 4$

and corresponding eigenvector with first component unity $= X_1 = \begin{bmatrix} 1 \\ -1 \\ 1 \end{bmatrix}$

$$A_1 = A - X_1 a_1^T = \begin{bmatrix} 3 & -1 & 0 \\ -1 & 2 & -1 \\ 0 & -1 & 3 \end{bmatrix} - \begin{bmatrix} 1 \\ -1 \\ 1 \end{bmatrix}[3\ -1\ 0] = \begin{bmatrix} 0 & 0 & 0 \\ 2 & 1 & -1 \\ -3 & 0 & 3 \end{bmatrix}$$

Leaving its first row and first column

$$B_1 = \begin{bmatrix} 1 & -1 \\ 0 & 3 \end{bmatrix}$$

Its eigenvalues are $\lambda_2 = 3$, $\lambda_3 = 1$

Two-dimensional eigenvector corresponding to $\lambda_2 = 3$ is given by $\begin{bmatrix} 1-3 & -1 \\ 0 & 3-3 \end{bmatrix}\begin{bmatrix} x_1 \\ x_2 \end{bmatrix} = \begin{bmatrix} 0 \\ 0 \end{bmatrix}$ which can be taken as $\begin{bmatrix} 1 \\ -2 \end{bmatrix}$

$\therefore$ Eigenvector of A_1 is $Z_2 = \begin{bmatrix} 0 \\ 1 \\ -2 \end{bmatrix}$

Now, $a_1^T Z_2 = [3\ \ -1\ \ 0]\begin{bmatrix} 0 \\ 1 \\ -2 \end{bmatrix} = -1$

$\therefore$ Eigenvector of A is

$$X_1 + \frac{\lambda_2 - \lambda_1}{a_1^T Z_2} Z_2 = \begin{bmatrix} 1 \\ -1 \\ 1 \end{bmatrix} + \frac{3-4}{-1} \begin{bmatrix} 0 \\ 1 \\ -2 \end{bmatrix} = \begin{bmatrix} 1 \\ 0 \\ -1 \end{bmatrix}$$

Two-dimensional eigenvector corresponding to $\lambda_3 = 1$ is given by $\begin{bmatrix} 1-1 & -1 \\ 0 & 3-1 \end{bmatrix} \begin{bmatrix} x_1 \\ x_2 \end{bmatrix} = \begin{bmatrix} 0 \\ 0 \end{bmatrix}$

i.e., $\begin{bmatrix} 0 & -1 \\ 0 & 2 \end{bmatrix} \begin{bmatrix} x_1 \\ x_2 \end{bmatrix} = \begin{bmatrix} 0 \\ 0 \end{bmatrix}$

we can take as $\begin{bmatrix} 1 \\ 0 \end{bmatrix} \Rightarrow$ eigenvectors of A_1 is $Z_3 = \begin{bmatrix} 0 \\ 1 \\ 0 \end{bmatrix}$

Now $a_1^T Z_3 = \begin{bmatrix} 3 & -1 & 0 \end{bmatrix} \begin{bmatrix} 0 \\ 1 \\ 0 \end{bmatrix} = -1$

Eigenvector of A is

$$X_1 + \frac{\lambda_3 - \lambda_1}{a_1^T Z_3} Z_3 = \begin{bmatrix} 1 \\ -1 \\ 1 \end{bmatrix} + \frac{1-4}{-1} \begin{bmatrix} 0 \\ 1 \\ 0 \end{bmatrix} = \begin{bmatrix} 1 \\ 2 \\ 1 \end{bmatrix}$$

Dominant eigenvalue is 4 with eigenvector $\begin{bmatrix} 1 \\ -1 \\ 1 \end{bmatrix}$ and other eigenvalues are 3 and 1 with corresponding eigenvectors $\begin{bmatrix} 1 \\ 0 \\ -1 \end{bmatrix}$ and $\begin{bmatrix} 1 \\ 2 \\ 1 \end{bmatrix}$, respectively.

EXERCISE 5.2

1. Solve the system of equations

$$\begin{bmatrix} 2 & 1 & 1 & -2 \\ 4 & 0 & 2 & 1 \\ 3 & 2 & 2 & 0 \\ 1 & 3 & 2 & -1 \end{bmatrix} \begin{bmatrix} x_1 \\ x_2 \\ x_3 \\ x_4 \end{bmatrix} = \begin{bmatrix} -10 \\ 8 \\ 7 \\ -5 \end{bmatrix}$$

using the Gauss elimination method with partial pivoting.

2. Solve the following system of equations

$$2x_1 + 3x_2 + x_3 = 9$$
$$x_1 + 2x_2 + 3x_3 = 6$$
$$3x_1 + x_2 + 2x_3 = 8$$

using Gauss elimination method.

3. Solve the following system of equations by Gauss elimination method

(a) $2x+3y+z=13$

$x-y-2z=-1$

$3x+y+4z=15$

(b) $x+y+z=9$

$2x+5y+7z=52$

$2x+y-z=0$

(c) $3x+y-z=3$

$2x-8y+z=-5$

$x-2y+9z=8$

(d) $8y+2z=-7$

$3x+5y+2z=8$

$6x+2y+8z=26$

4. Solve the following linear system by Gauss elimination, with partial pivoting if necessary (but without scaling). Check the result by substitution. If no solution or more than one solution exists, give a reason.

(a) $1.5x_1+2.3x_2=16$

$-4.5x_1-6.9x_2=48$

(b) $2x_1+5x_2+7x_3=25$

$-5x_1+7x_2+2x_3=-4$

$x_1+22x_2+23x_3=71$

(c) $3.2x_1+1.6x_2=-0.8$

$1.6x_1-0.8x_2+2.4x_3=16.0$

$2.4x_2-4.8x_3+3.6x_4=-39.0$

$3.6x_3+2.4x_4=10.2$

5. Apply Gauss–Jordan method to solve the following equations

$$x+y+z=9$$

$$2x-3y+4z=13$$

$$3x+4y+5z=40$$

6. Solve the following system of equations by Gauss–Jordan method

(a) $x+2y+z=8$

$2x+3y+4z=20$

$4x+3y+2z=16$

(b) $2x_1-7x_2+4x_3=9$

$x_1+9x_2-6x_3=1$

$-3x_1+8x_2+5x_3=6$

(c) $10x+y+z=12$

$x+10y+z=12$

$x+y+10z=12$

7. Solve the following equations by Gauss–Jordan method

$$x+2y+z-w=-2,\ 2x+3y-z+2w=7$$

$$x+y+3z-2w=-6,\ x+y+z+w=2$$

8. Solve the following system of equations by Gauss–Jordan method

$$\begin{bmatrix} 1 & 1 & 1 \\ 4 & 3 & -1 \\ 3 & 5 & 3 \end{bmatrix}\begin{bmatrix} x \\ y \\ z \end{bmatrix}=\begin{bmatrix} 1 \\ 6 \\ 4 \end{bmatrix}$$

9. Solve the following system of equations by

(a) Gauss elimination method

(b) Gauss–Jordan method

$$2x+y+z=10$$

$$3x+2y+3z=18$$

$$x+4y+9z=16$$

10. Solve the following system of equations by method of factorisation

$$2x-3y+10z=3$$

$$-x+4y+2z=20$$

$$5x+2y+z=-12$$

11. Solve the equations

$$2x+3y+z=9$$

$$x+2y+3z=6$$

$$3x+y+2z=8$$

by the factorisation method.

12. Solve the following system of equations by Doolittle's method

(a)
$$4x_1 + 3x_2 - x_3 = 6$$
$$3x_1 + 5x_2 + 3x_3 = 4$$
$$x_1 + x_2 + x_3 = 1$$

(b)
$$x_1 + 2x_2 + 3x_3 = 14$$
$$2x_1 + 5x_2 + 2x_3 = 18$$
$$3x_1 + x_2 + 5x_3 = 20$$

(c)
$$3x + 2y + 7z = 4$$
$$2x + 3y + z = 5$$
$$3x + 4y + z = 7$$

13. Solve the following system of equations by Crout's method

(a)
$$2x + y + 4z = 12$$
$$8x - 3y + 2z = 20$$
$$4x + 11y - z = 33$$

(b)
$$2x_1 + 3x_2 - 4x_3 + 2x_4 = -4$$
$$x_1 + 2x_2 + 3x_3 - 4x_4 = 7$$
$$4x_1 - x_2 + 2x_3 - 2x_4 = 7$$
$$3x_1 + 5x_2 - x_3 + 6x_4 = 5$$

(c)
$$x_1 + x_2 + x_3 = 1$$
$$3x_1 + x_2 - 3x_3 = 5$$
$$x_1 - 2x_2 - 5x_3 = 10$$

(d)
$$4x + 3y + z - w = 14$$
$$2x + 5y + 2z + w = 17$$
$$x + 4y + 4z + 6w = 20$$
$$3x + y - z + 5w = 12$$

(e)
$$2x - 6y + 8z = 24$$
$$5x + 4y - 3z = 2$$
$$3x + y + 2z = 16$$

(f)
$$4x + y - z = 13$$
$$3x + 5y + 2z = 21$$
$$2x + y + 6z = 14$$

14. Solve the system of equations
$$4x_1 + x_2 + x_3 = 2$$
$$x_1 + 5x_2 + 2x_3 = -6$$
$$x_1 + 2x_2 + 3x_3 = -4$$
using Gauss-Jacobi method with initial approximation as $x^{(0)} = [0.5, -0.5, -0.5]^T$. Perform three iterations. The exact solution is $x_1 = 1,\ x_2 = -1,\ x_3 = -1$.

15. Solve by Jacobi's method
$$4x + y + 3z = 17$$
$$x + 5y + z = 14$$
$$2x - y + 8z = 12$$

16. Solve the system of equations
$$54x + y + z = 110$$
$$2x + 15y + 6z = 72$$
$$-x + 6y + 27z = 85$$
by Gauss–Seidel method.

17. Using Gauss–Seidel method, solve the following system of equations
$$2x_1 - x_2 = 7$$
$$-x_1 + 2x_2 - x_3 = 1$$
$$-x_2 + 2x_3 = 1$$
Starting with initial approximation $(0,0,0)^T$ perform three iterations.

18. Solve the linear system of equations by Gauss–Seidel method
$$13x_1 + 5x_2 - 3x_3 + x_4 = 18$$
$$2x_1 + 12x_2 + x_3 - 4x_4 = 13$$
$$3x_1 - 4x_2 + 10x_3 + x_4 = 29$$
$$x_2 + 3x_3 + 5x_4 = 31$$

19. Solve the following system of equations by
    (a) Gauss–Jacobi method
    (b) Gauss–Seidel method

$$20x + y - 2z = 17$$
$$3x + 20y - z = -18$$
$$2x - 3y + 20z = 25$$

20. Determine the largest eigenvalue in magnitude and the corresponding eigenvector of each of the given matrix by power method.

(a) $A = \begin{bmatrix} 1 & 6 & 1 \\ 1 & 2 & 0 \\ 0 & 0 & 3 \end{bmatrix}$

(b) $A = \begin{bmatrix} -1 & 1 & 2 \\ 0 & 1 & -1 \\ 4 & -2 & 9 \end{bmatrix}$

(c) $A = \begin{bmatrix} 1 & 3 & -1 \\ 3 & 2 & 4 \\ -1 & 4 & 10 \end{bmatrix}$

(d) $A = \begin{bmatrix} 5 & 2 & 1 & -2 \\ 2 & 6 & 3 & -4 \\ 1 & 3 & 19 & 2 \\ -2 & -4 & 2 & 1 \end{bmatrix}$

(e) $A = \begin{bmatrix} 10 & -2 & 1 \\ -2 & 10 & -2 \\ 1 & -2 & 10 \end{bmatrix}$

21. Find the absolutely smallest eigenvalue of the given matrix A by power method.

$$A = \begin{bmatrix} -6 & 18 & 2 \\ 3 & -3 & -1 \\ 0 & 0 & 4 \end{bmatrix}$$

ANSWERS 5.2

1. $x_1 = 5,\ x_2 = 6,\ x_3 = -10,\ x_4 = 8$
2. $x_1 = 1.9444,\ x_2 = 1.6111,\ x_3 = 0.2778$
3. (a) $x = 3,\ y = 2,\ z = 1$ (b) $x = 1,\ y = 3,\ z = 5$
   (c) $x = y = z = 1$ (d) $x = 4,\ y = -1,\ z = 0.5$
4. (a) No Solution
   (b) Infinite number of solutions; $x_1 = 2 + x_2,\ x_2$ arbitrary, $x_3 = 3 - x_2$
   (c) $x_1 = 1.5,\ x_2 = -3.5,\ x_3 = 4.5,\ x_4 = -2.5$
5. $x = 1,\ y = 3,\ z = 5$
6. (a) $x = 1, y = 2, z = 3$ (b) $x_1 = 4, x_2 = 1, x_3 = 2$ (c) $x = y = z = 1$
7. $x = 1, y = 0, z = -1, w = 2$
8. $x = 1,\ y = 0.5,\ z = -0.5$
9. $x = 7,\ y = -9,\ z = 5$
10. $x = -4,\ y = 3,\ z = 2$
11. $x = 1.9444,\ y = 1.6111,\ z = 0.2778$
12. (a) $x_1 = 1,\ x_2 = 0.5,\ x_3 = -0.5$ (b) $x_1 = 1,\ x_2 = 2,\ x_3 = 3$
    (c) $x = 0.875,\ y = 1.125,\ z = -0.125$

13. (a) $x = 3,\ y = 2,\ z = 1$ (b) $x_1 = 1.0,\ x_2 = 0.5,\ x_3 = 2.0,\ x_4 = 0.25$
(c) $x_1 = 6,\ x_2 = -7,\ x_3 = 2$ (d) $x = 2, y = 2, z = 1, w = 1$
(e) $x = 1,\ y = 3,\ z = 5$ (f) $x = 3, y = 2, z = 1$
14. $x^{(3)} = [0.9333, -1.0733, -1.1000]^T$
15. $(x_6, y_6, z_6) = (2.9773, 1.9863, 0.9917)$ and iteration converges to $(x, y, z) = (3, 2, 1)$.
16. $(x, y, z)^T = (1.926, 3.573, 2.425)^T$
17. $x^{(3)} = (5.3125, 4.3125, 2.6563)^T$
18. Iteration converges to $(x_1, x_2, x_3, x_4)^T = (1, 2, 3, 4)^T$
19. By both the methods, iteration converges to $(x, y, z)^T = (1, -1, 1)^T$.
20. (a) 4; $(2, 1, 0)^T$ (b) 9.916; $(0.173, -0.112, 1)^T$ (c) 11.66; $(0.025, 0.422, 1)^T$
(d) 19.837; $(0.090, 0.215, 1, 0.051)^T$ (e) 13.376; $(0.844, -1, 0.844)^T$
21. 3

5.7 LINEAR OPERATORS

Let a function $y = f(x)$ has values $f(x_0), f(x_0 + h), f(x_0 + 2h), \ldots$ at points $x_0,\ x_0 + h,\ x_0 + 2h, \ldots$, respectively. Then, the arguments are equispaced and h is called interval of differencing. We define the following operators when h is interval of differencing.

(i) E is shifting operator defined by

$$E\, f(x) = f(x + h)$$

(ii) Δ is forward difference operator defined by

$$\Delta f(x) = f(x + h) - f(x)$$

(iii) ∇ is backward difference operator defined by

$$\nabla f(x) = f(x) - f(x - h)$$

(iv) δ is central difference operator defined by

$$\delta f(x) = f\left(x + \frac{h}{2}\right) - f\left(x - \frac{h}{2}\right)$$

(v) μ is mean value operator defined by

$$\mu f(x) = \frac{1}{2}\left[f\left(x + \frac{h}{2}\right) + f\left(x - \frac{h}{2}\right)\right]$$

If $f(x)$ is an analytic function, then differentiation operator D is defined as

$$D\, f(x) = \frac{d}{dx} f(x) = f'(x)$$

Relation between Operators *E* and *D*

We have $E\,f(x) = f(x+h)$

and by Taylor series expansion

$$E\,f(x) = f(x+h) = f(x) + h\,f'(x) + \frac{h^2}{2!}f''(x) + \frac{h^3}{3!}f'''(x) + \cdots$$

$$= \left(1 + hD + \frac{(hD)^2}{2!} + \frac{(hD)^3}{3!} + \cdots\right) f(x) = e^{hD} f(x)$$

$$\therefore \quad E = e^{hD}$$

We define operator U by

$$U = hD$$

$$\therefore \quad E = e^U \tag{5.3}$$

Now,

$$\delta f(x) = f\left(x + \frac{h}{2}\right) - f\left(x - \frac{h}{2}\right)$$

$$= \left(E^{\frac{1}{2}} - E^{-\frac{1}{2}}\right) f(x)$$

$$\therefore \quad \delta = E^{\frac{1}{2}} - E^{-\frac{1}{2}} = e^{\frac{1}{2}hD} - e^{-\frac{1}{2}hD} = e^{\frac{U}{2}} - e^{-\frac{U}{2}}$$

$$= 2\sinh\frac{U}{2}$$

and

$$\mu f(x) = \frac{1}{2}\left[f\left(x + \frac{h}{2}\right) + f\left(x - \frac{h}{2}\right)\right]$$

$$= \frac{1}{2}\left(E^{\frac{1}{2}} + E^{-\frac{1}{2}}\right) f(x)$$

$$\therefore \quad \mu = \frac{1}{2}\left(E^{\frac{1}{2}} + E^{-\frac{1}{2}}\right)$$

$$\therefore \quad \mu = \frac{1}{2}\left(e^{U/2} + e^{-U/2}\right) \qquad \text{(from 5.3)}$$

$$= \cosh\frac{U}{2} \geq 1$$

Now,

$$\cosh^2\frac{U}{2} = 1 + \sinh^2\frac{U}{2}$$

$$\therefore \quad \mu^2 = 1 + \frac{1}{4}\delta^2$$

$$\therefore \quad \mu = \sqrt{1 + \frac{1}{4}\delta^2}$$

The formulae $\delta = 2\sinh\frac{U}{2}$ and $\mu = \sqrt{1+\frac{1}{4}\delta^2}$ are frequently used.

Now, we find the relation between the operators

(i) ***E* in terms of other operators**

$$\Delta f(x) = f(x+h) - f(x) = (E-1)f(x)$$

$$\therefore \quad \Delta = E - 1$$

$$\therefore \quad E = 1 + \Delta \tag{5.4}$$

$$\nabla f(x) = f(x) - f(x-h) = (1-E^{-1})f(x)$$

$$\therefore \quad \nabla = 1 - E^{-1}$$

$$\therefore \quad E = (1-\nabla)^{-1} \tag{5.5}$$

$$\delta f(x) = f\left(x+\frac{h}{2}\right) - f\left(x-\frac{h}{2}\right) = \left(E^{\frac{1}{2}} - E^{-\frac{1}{2}}\right)f(x)$$

$$\therefore \quad \delta = E^{\frac{1}{2}} - E^{-\frac{1}{2}}$$

$$\therefore \quad E^{\frac{1}{2}}\delta = E - 1$$

$$\therefore \quad E - \delta E^{\frac{1}{2}} - 1 = 0$$

$$\therefore \quad E^{\frac{1}{2}} = \frac{\delta + \sqrt{\delta^2+4}}{2}$$

$$\therefore \quad E = \frac{1}{4}\left[\delta^2 + (\delta^2+4) + 2\delta\sqrt{\delta^2+4}\right]$$

$$= 1 + \frac{\delta^2}{2} + \delta\sqrt{1+\frac{\delta^2}{4}} \tag{5.6}$$

(ii) **Δ in terms of other operators**

$$\Delta f(x) = f(x+h) - f(x) = (E-1)f(x)$$

$$\therefore \quad \Delta = E - 1 = (1-\nabla)^{-1} - 1 \qquad \text{from (5.5)}$$

$$= \frac{\delta^2}{2} + \delta\sqrt{1+\frac{\delta^2}{4}} \qquad \text{from (5.6)}$$

$$= e^U - 1 \qquad \text{from (5.3)}$$

(iii) **∇ in terms of other operators**

$$\nabla f(x) = f(x) - f(x-h) = (1-E^{-1})f(x)$$

$$\therefore \quad \nabla = 1 - E^{-1}$$

$$= 1 - (1+\Delta)^{-1} \qquad \text{from (5.4)}$$

Now,
$$\nabla = 1-(E+E^{-1}-2)+E-2$$
$$= E-1-\left(E^{\frac{1}{2}}-E^{-\frac{1}{2}}\right)^2$$
$$= E-1-\delta^2$$
$$= \left[1+\frac{\delta^2}{2}+\delta\sqrt{1+\frac{\delta^2}{4}}\right]-1-\delta^2 \qquad \text{from (5.6)}$$
$$= -\frac{\delta^2}{2}+\delta\sqrt{1+\frac{\delta^2}{4}}$$

and
$$\nabla = 1-E^{-1} = 1-e^{-U} \qquad \text{from (5.3)}$$

(iv) **δ in terms of other operators**

$$\delta f(x) = f\left(x+\frac{h}{2}\right)-f\left(x-\frac{h}{2}\right) = \left(E^{\frac{1}{2}}-E^{-\frac{1}{2}}\right)f(x)$$

$\therefore$
$$\delta = E^{\frac{1}{2}}-E^{-\frac{1}{2}}$$
$$= E^{-\frac{1}{2}}(E-1) = \Delta(1+\Delta)^{-\frac{1}{2}} \qquad \text{from (5.4)}$$

Also,
$$\delta = \left(1-E^{-1}\right)E^{\frac{1}{2}} = \nabla(1-\nabla)^{-\frac{1}{2}} \qquad \text{from (5.5)}$$

we have already proved
$$\delta = 2\sinh\frac{U}{2}$$

(v) **μ in terms of other operators**

$$\mu f(x) = \frac{1}{2}\left[f\left(x+\frac{h}{2}\right)+f\left(x-\frac{h}{2}\right)\right] = \frac{1}{2}\left(E^{\frac{1}{2}}+E^{-\frac{1}{2}}\right)f(x)$$

$\therefore$
$$\mu = \frac{1}{2}\left(E^{\frac{1}{2}}+E^{-\frac{1}{2}}\right)$$
$$= \frac{1}{2}(E+1)E^{-\frac{1}{2}}$$
$$= \left(1+\frac{\Delta}{2}\right)(1+\Delta)^{-\frac{1}{2}} \qquad \text{from (5.4)}$$

Also,
$$\mu = \frac{1}{2}\left(1+E^{-1}\right)E^{\frac{1}{2}}$$
$$= \frac{1}{2}(2-\nabla)(1-\nabla)^{-\frac{1}{2}} \qquad \text{from (5.5)}$$

$$= \left(1 - \frac{\nabla}{2}\right)(1-\nabla)^{-\frac{1}{2}}$$

$$\mu = \frac{1}{2}\left(E^{\frac{1}{2}} + E^{-\frac{1}{2}}\right) = \frac{1}{2}\left(e^{\frac{hD}{2}} + e^{-\frac{hD}{2}}\right) = \frac{1}{2}\left(e^{\frac{1}{2}U} + e^{-\frac{1}{2}U}\right)$$

$$= \cosh\frac{U}{2}$$

(vi) ***U* in terms of other operators**

$$E = e^U$$

$\therefore$ $$U = \log E$$

$$= \log(1+\Delta) \qquad \text{from (5.4)}$$

$$= \log(1-\nabla)^{-1} = -\log(1-\nabla) \qquad \text{from (5.5)}$$

As $$\delta = 2\sinh\frac{U}{2}$$

$$U = 2\sinh^{-1}\frac{\delta}{2}$$

We give the above relations in the table below.

Relations between shift operator, difference operators, mean value operator and differentiation operator

	E	Δ	∇	δ	*U*
E	E	$\Delta + 1$	$(1-\nabla)^{-1}$	$1+\frac{1}{2}\delta^2+\delta\sqrt{1+\frac{\delta^2}{4}}$	e^U
Δ	$E-1$	Δ	$(1-\nabla)^{-1}-1$	$\frac{1}{2}\delta^2+\delta\sqrt{1+\frac{\delta^2}{4}}$	e^U-1
∇	$1-E^{-1}$	$1-(1+\Delta)^{-1}$	∇	$-\frac{1}{2}\delta^2+\delta\sqrt{1+\frac{\delta^2}{4}}$	$1-e^{-U}$
δ	$E^{\frac{1}{2}}-E^{-\frac{1}{2}}$	$\Delta(1+\Delta)^{-\frac{1}{2}}$	$\nabla(1-\nabla)^{-\frac{1}{2}}$	δ	$2\sinh\frac{U}{2}$
μ	$\frac{1}{2}\left(E^{\frac{1}{2}}+E^{-\frac{1}{2}}\right)$	$\left(1+\frac{\Delta}{2}\right)(1+\Delta)^{-\frac{1}{2}}$	$\left(1-\frac{\nabla}{2}\right)(1-\nabla)^{-\frac{1}{2}}$	$\sqrt{1+\frac{\delta^2}{4}}$	$\cosh\frac{U}{2}$
U	$\log E$	$\log(1+\Delta)$	$-\log(1-\nabla)$	$2\sinh^{-1}\frac{\delta}{2}$	U

If P is any of the above operator, then

$$P\left(f(x)\pm g(x)\right)=P\left(f(x)\right)\pm P\left(g(x)\right)$$

and
$$P\left(c_1 f(x)\right)=c_1 P\left(f(x)\right)$$

and hence all the above operators are linear operators.

5.7.1 Forward Differences

The differences $y_1-y_0, y_2-y_1,\ldots,y_n-y_{n-1}$ are denoted by $\Delta y_0,\Delta y_1,\ldots,\Delta y_{n-1}$, respectively, and are called first-order forward differences.

Similarly, the second-order forward differences are defined by $\Delta^2 y_r=\Delta y_{r+1}-\Delta y_r$

In general, pth-order forward differences are defined as

$$\Delta^p y_r=\Delta^{p-1}y_{r+1}-\Delta^{p-1}y_r$$

These differences are shown below in forward difference table.

Forward difference table

x	y	Δy	$\Delta^2 y$	$\Delta^3 y$	$\Delta^4 y$	$\Delta^5 y$
x_0	y_0					
		Δy_0				
x_0+h	y_1		$\Delta^2 y_0$			
		Δy_1		$\Delta^3 y_0$		
x_0+2h	y_2		$\Delta^2 y_1$		$\Delta^4 y_0$	
		Δy_2		$\Delta^3 y_1$		$\Delta^5 y_0$
x_0+3h	y_3		$\Delta^2 y_2$		$\Delta^4 y_1$	
		Δy_3		$\Delta^3 y_2$		
x_0+4h	y_4		$\Delta^2 y_3$			
		Δy_4				
x_0+5h	y_5					

Remark 5.8: Any higher order forward difference can be expressed in terms of entries.

We have

$$\Delta^2 y_0=\Delta y_1-\Delta y_0=\left(y_2-y_1\right)-\left(y_1-y_0\right)=y_2-2y_1+y_0$$

$$\Delta^3 y_0=\Delta^2 y_1-\Delta^2 y_0=\left(y_3-2y_2+y_1\right)-\left(y_2-2y_1+y_0\right)=y_3-3y_2+3y_1-y_0$$

$$\Delta^4 y_0=\Delta^3 y_1-\Delta^3 y_0=\left(y_4-3y_3+3y_2-y_1\right)-\left(y_3-3y_2+3y_1-y_0\right)=y_4-4y_3+6y_2-4y_1+y_0$$

In general

$$\Delta^n y_0=y_n-{}^nC_1 y_{n-1}+{}^nC_2 y_{n-2}-{}^nC_3 y_{n-3}+\cdots+(-1)^n y_0$$

This result can be proved as follows

$$\Delta^n y_0 = (E-1)^n y_0 = \left[E^n - {}^nC_1E^{n-1} + {}^nC_2E^{n-2} - {}^nC_3E^{n-3} + \cdots + (-1)^n\right]y_0$$

$$= y_n - {}^nC_1\, y_{n-1} + {}^nC_2 y_{n-2} - {}^nC_3 y_{n-3} + \cdots + (-1)^n\, y_0$$

5.7.2 Backward Differences

The differences $y_1 - y_0, y_2 - y_1, y_3 - y_2, \ldots, y_n - y_{n-1}$ are denoted by ∇y_1, $\nabla y_2, \ldots, \nabla y_n$, respectively, and are called first-order backward differences.

Similarly, the second-order backward differences are defined by $\nabla^2 y_r = \nabla y_r - \nabla y_{r-1}$

In general, pth-order backward differences are defined as

$$\nabla^p y_r = \nabla^{p-1} y_r - \nabla^{p-1} y_{r-1}$$

These differences are shown below in backward difference table.

Backward difference table

x	y	∇y	∇²y	∇³y	∇⁴y	∇⁵y
x_0	y_0					
		∇y_1				
$x_0 + h$	y_1		$\nabla^2 y_2$			
		∇y_2		$\nabla^3 y_3$		
$x_0 + 2h$	y_2		$\nabla^2 y_3$		$\nabla^4 y_4$	
		∇y_3		$\nabla^3 y_4$		$\nabla^5 y_5$
$x_0 + 3h$	y_3		$\nabla^2 y_4$		$\nabla^4 y_5$	
		∇y_4		$\nabla^3 y_5$		
$x_0 + 4h$	y_4		$\nabla^2 y_5$			
		∇y_5				
$x_0 + 5h$	y_5					

Remark 5.9:

$$\nabla^n y_n = \left(1 - E^{-1}\right)^n y_n = \left(1 - {}^nC_1E^{-1} + {}^nC_2E^{-2} - {}^nC_3E^{-3} + \cdots + (-1)^n E^{-n}\right)y_n$$

$$= y_n - {}^nC_1\, y_{n-1} + {}^nC_2\, y_{n-2} - {}^nC_3\, y_{n-3} + \cdots + (-1)^n\, y_0$$

5.7.3 Central Differences

The differences $y_1 - y_0, y_2 - y_1, y_3 - y_2, \ldots, y_n - y_{n-1}$ are denoted by $\delta y_{1/2}, \delta y_{3/2}, \delta y_{5/2}, \ldots, \delta y_{n-\frac{1}{2}}$, respectively, and are called first-order central differences.

Similarly, the second-order central differences are defined by $\delta^2 y_r = \delta y_{r+\frac{1}{2}} - \delta y_{r-\frac{1}{2}}$

In general, *p*th-order central differences are defined by

$$\delta^p y_r = \delta^{p-1} y_{r+\frac{1}{2}} - \delta^{p-1} y_{r-\frac{1}{2}}$$

These differences are shown below in central difference table.

Central difference table

x	y	δy	$\delta^2 y$	$\delta^3 y$	$\delta^4 y$	$\delta^5 y$
x_0	y_0					
		$\delta y_{1/2}$				
$x_0 + h$	y_1		$\delta^2 y_1$			
		$\delta y_{3/2}$		$\delta^3 y_{3/2}$		
$x_0 + 2h$	y_2		$\delta^2 y_2$		$\delta^4 y_2$	
		$\delta y_{5/2}$		$\delta^3 y_{5/2}$		$\delta^5 y_{5/2}$
$x_0 + 3h$	y_3		$\delta^2 y_3$		$\delta^4 y_3$	
		$\delta y_{7/2}$		$\delta^3 y_{7/2}$		
$x_0 + 4h$	y_4		$\delta^2 y_4$			
		$\delta y_{9/2}$				
$x_0 + 5h$	y_5					

Remark 5.10: (i) In the central difference table, the central differences on the same horizontal line have the same suffix. The differences of odd order are known only for half values of suffixes and the differences of even order for integral values of suffixes.

(ii) The differences in each table are same but only the notation for that difference is changed.

5.7.4 Factorial Polynomials

The factorial polynomial is of special importance in the theory of finite differences. It helps us in finding the successive forward differences of a polynomial directly by simple rule of differentiation and also to find the polynomial by simple rule of integration if differences of a polynomial function are given in factorial notation. The factorial polynomial of degree n (n is a positive integer) is defined by the formula

$x^{(n)} = x(x-1)(x-2)\ldots(x-n+1)$ and for $n = 0$, $x^{(0)} = 1$ if the interval of differencing is 1 and if interval of differencing is h then

$$x^{(n)} = x(x-h)(x-2h)\ldots\left(x-\overline{n-1}h\right);\ x^{(0)} = 1$$

We have

$$\Delta x^{(n)} = (x+h)^{(n)} - x^{(n)}$$

$$= (x+h)x(x-h)\ldots\left(x-\overline{n-2}h\right) - x(x-h)\ldots\left(x-\overline{n-1}h\right)$$

$$= x(x-h)\cdots(x-\overline{n-2}\,h)\left[(x+h)-(x+h-nh)\right]$$

$$= nh\,x(x-h)\cdots(x-\overline{n-2}\,h) = nh\,x^{(n-1)}$$

Theorem 5.1 $\Delta^r x^{(n)} = \dfrac{n!}{(n-r)!} h^r x^{(n-r)}; \quad r \le n$

$$= 0 \qquad ; \quad r > n$$

Proof: We prove the result by principle of mathematical induction

For $r = 1$

$$\Delta x^{(n)} = \Delta\left[x(x-h)(x-2h)\cdots(x-\overline{n-1}\,h)\right]$$

$$= (x+h)x(x-h)\cdots(x-\overline{n-2}\,h) - x(x-h)(x-2h)\cdots\cdots(x-\overline{n-1}\,h)$$

$$= x(x-h)\cdots(x-\overline{n-2}\,h)\left[x+h-x+nh-h\right]$$

$$= nh\,x(x-h)\cdots(x-\overline{n-2}\,h)$$

$$= nh\,x^{(n-1)} = \frac{n!}{(n-1)!} h\,x^{(n-1)}$$

$\therefore$ Result is true for $r = 1$

Let the result is true for $r = p; p \le n-1$

$$\therefore\ \Delta^p (x+h)^{(n)} = \frac{n!}{(n-p)!} h^p (x+h)^{(n-p)}$$

$$\Delta^p x^{(n)} = \frac{n!}{(n-p)!} h^p x^{(n-p)}$$

$$\therefore \qquad \Delta^{p+1} x^{(n)} = \Delta^p (x+h)^{(n)} - \Delta^p x^{(n)}$$

$$= \frac{n!}{(n-p)!} h^p (x+h)^{(n-p)} - \frac{n!}{(n-p)!} h^p x^{(n-p)}$$

$$= \frac{n!}{(n-p)!} h^p \left[(x+h)x(x-h)\cdots(x+h-\overline{n-p-1}\,h) - x(x-h)\cdots(x-\overline{n-p-1}\,h)\right]$$

$$= \frac{n!}{(n-p)!} h^p x(x-h)\cdots(x-\overline{n-p-2}\,h)(x+h-x+\overline{n-p-1}\,h)$$

$$= \frac{n!}{(n-p)!} h^{p+1} (n-p) x^{(n-p-1)} = \frac{n!}{(n-p-1)!} h^{p+1} x^{(n-p-1)}$$

$\therefore$ Result holds for $p+1$

$\therefore$ By the principle of mathematical induction result holds for all $r = 1, 2, \ldots, n$

$\therefore$ For $r = n$ we have

$$\Delta^r x^{(n)} = \Delta^n x^{(n)} = \frac{n!}{0!} h^n x^{(0)} = n!\,h^n$$

which is constant.

$$\therefore \qquad \Delta^r x^{(n)} = 0 \qquad \text{for all } r > n$$

Remark 5.11: (i) The nth-order forward differences of a polynomial of nth degree $a_0x^n + a_1x^{n-1} + \ldots + a_{n-1}x + a_n$ are constant and its value is $a_0 n!\, h^n$ and all forward differences of greater than n are zero.

Proof: $a_0x^n + a_1x^{n-1} + \ldots + a_{n-1}x + a_n = a_0x^{(n)} +$ factorial polynomial of degree $n-1$

$\therefore$ By above theorem

$$\Delta^n (a_0x^n + a_1x^{n-1} + \ldots + a_{n-1}x + a_n) = a_0\, \Delta^n x^{(n)} + 0 = a_0\, n!\, h^n$$

and
$$\Delta^{n+p}\left(a_0x^n + a_1x^{n-1} + \ldots + a_n\right) = 0; \quad p = 1, 2, 3 \ldots$$

(ii) A polynomial can be changed to factorial polynomial by synthetic division which will be clear in the examples.

We have defined factorial $x^{(n)}$ for non-negative values of n.

When $n \geq 1$, we have $x^{(n)} = (x-n+1)x^{(n-1)}$ and requiring that this formula also hold for $n = 0$, we get $x^{(-1)} = \dfrac{1}{x+1}$. Using the formula repeatedly for $n = -1, -2, \ldots$ we obtain

$$x^{(-n)} = \frac{1}{(x+1)(x+2)\ldots(x+n)} = \frac{1}{(x+n)^{(n)}}; \qquad n = 1, 2, 3, \ldots$$

with this formula $x^{(n)} = (x-n+1)x^{(n-1)}$ and $\Delta x^{(n)} = nx^{(n-1)}$ hold for negative values of n also.

5.7.5 Error Propagation

Firstly, we demonstrate the propagation of round-off errors. We assume that in the given data, maximum round-off error is ε. Here, we shall be considering the maximum possible variations of errors which will be possible when magnitude of round-off error in each entry of the data is maximum and two consecutive entries have error of opposite signs.

ε					
	-2ε				
$-\varepsilon$		4ε			
	2ε		-8ε		
ε		-4ε		16ε	
	-2ε		8ε		-32ε
$-\varepsilon$		4ε		-16ε	
	2ε		-8ε		
ε		-4ε			
	-2ε				
$-\varepsilon$					

Thus, we observe that in the worst possible case, the error will be doubled for every new difference introduced.

Now, it may be possible that an entry is written wrong in the data. Then, this entry will have error say ε and no other entry will have error. Now, in the following table, we see the propagation of this error.

	0		0		0	
0		0		0		ε
	0		0		ε	
0		0		ε		-6ε
	0		ε		-5ε	
0		ε		-4ε		15ε
	ε		-3ε		10ε	
ε		-2ε		6ε		-20ε
	$-\varepsilon$		3ε		-10ε	
0		ε		-4ε		15ε
	0		$-\varepsilon$		5ε	
0		0		ε		-6ε
	0		0		$-\varepsilon$	
0		0		0		ε
	0		0		0	

We observe that the error propagation is in a triangular pattern and grows quickly. Excluding the signs, there are binomial coefficients ${}^nC_0, {}^nC_1, {}^nC_2, \ldots, {}^nC_n$ in the nth difference column. The error will be in that entry which has maximum value of binomial coefficient in its row.

5.7.6 Missing Entries in the Data

Suppose n entries in the data are given and some other entries are missing. Assuming these entries some unknowns, complete the difference table up to nth-order differences.

Now, all nth-order differences will be zero and we shall get number of equations equal to number of unknowns. Solving these equations, we can find values of unknowns.

Example 5.33: Show that

(i) $\nabla+\Delta=\frac{\Delta}{\nabla}-\frac{\nabla}{\Delta}$ (ii) $\Delta\nabla=\nabla\Delta=\Delta-\nabla=\delta^2$ (iii) $\mu\delta=\frac{1}{2}(\Delta+\nabla)=\frac{1}{2}\left(E-E^{-1}\right)$

(iv) $\mu=\left(1+\frac{\delta^2}{4}\right)^{\frac{1}{2}}$ (v) $\Delta f_k^2=\left(f_k+f_{k+1}\right)\Delta f_k$ (vi) $1+\mu^2\delta^2=\left(1+\frac{1}{2}\delta^2\right)^2$

(vii) $1+\Delta=e^{hD}$ (viii) $\nabla^2=h^2D^2-h^3D^3+\frac{7}{12}h^4D^4-\ldots$

Solution:

(i) $$\nabla+\Delta=\left(1-E^{-1}\right)+(E-1)=E-E^{-1} \quad (1)$$

and $$\frac{\Delta}{\nabla}-\frac{\nabla}{\Delta}=\frac{E-1}{1-E^{-1}}-\frac{1-E^{-1}}{E-1}=\frac{E\left(1-E^{-1}\right)}{1-E^{-1}}-\frac{1-E^{-1}}{E\left(1-E^{-1}\right)}$$

$$=E-\frac{1}{E}=E-E^{-1} \quad (2)$$

from (1) and (2)

$$\nabla+\Delta=\frac{\Delta}{\nabla}-\frac{\nabla}{\Delta}$$

(ii) $$\Delta\nabla=(E-1)\left(1-E^{-1}\right)=\left(E^{\frac{1}{2}}-E^{-\frac{1}{2}}\right)E^{\frac{1}{2}}\left(1-E^{-1}\right)=\left(E^{\frac{1}{2}}-E^{-\frac{1}{2}}\right)\left(E^{\frac{1}{2}}-E^{-\frac{1}{2}}\right)$$

$$=\delta\cdot\delta=\delta^2 \quad (1)$$

$$\nabla\Delta=\left(1-E^{-1}\right)(E-1)=\left(1-E^{-1}\right)E^{\frac{1}{2}}\left(E^{\frac{1}{2}}-E^{-\frac{1}{2}}\right)=\left(E^{\frac{1}{2}}-E^{-\frac{1}{2}}\right)\left(E^{\frac{1}{2}}-E^{-\frac{1}{2}}\right)$$

$$=\delta\cdot\delta=\delta^2 \quad (2)$$

$$\Delta-\nabla=(E-1)-\left(1-E^{-1}\right)=E-2+E^{-1}=\left(E^{\frac{1}{2}}-E^{-\frac{1}{2}}\right)^2=\delta^2 \quad (3)$$

from (1), (2) and (3)

$$\Delta\nabla=\nabla\Delta=\Delta-\nabla=\delta^2$$

(iii) $$\mu\delta=\frac{1}{2}\left(E^{\frac{1}{2}}+E^{-\frac{1}{2}}\right)\left(E^{\frac{1}{2}}-E^{-\frac{1}{2}}\right)=\frac{1}{2}\left(E-E^{-1}\right) \quad (1)$$

$$\frac{1}{2}(\Delta+\nabla)=\frac{1}{2}\left[(E-1)+\left(1-E^{-1}\right)\right]=\frac{1}{2}\left(E-E^{-1}\right) \quad (2)$$

from (1) and (2)

$$\mu\delta = \frac{1}{2}(\Delta + \nabla) = \frac{1}{2}\left(E - E^{-1}\right)$$

(iv) $$\left(1 + \frac{\delta^2}{4}\right)^{\frac{1}{2}} = \frac{1}{2}\left(4 + \delta^2\right)^{\frac{1}{2}} = \frac{1}{2}\left[4 + \left(E^{\frac{1}{2}} - E^{-\frac{1}{2}}\right)^2\right]^{\frac{1}{2}}$$

$$= \frac{1}{2}\left[4 + E - 2 + E^{-1}\right]^{\frac{1}{2}} = \frac{1}{2}\left[E + 2 + E^{-1}\right]^{\frac{1}{2}}$$

$$= \frac{1}{2}\left[\left(E^{\frac{1}{2}} + E^{-\frac{1}{2}}\right)^2\right]^{\frac{1}{2}}$$

$$= \frac{1}{2}\left(E^{\frac{1}{2}} + E^{-\frac{1}{2}}\right) = \mu.$$

(v) $$\Delta f_k^2 = f_{k+1}^2 - f_k^2 = \left(f_k + f_{k+1}\right)\left(f_{k+1} - f_k\right)$$

$$= \left(f_k + f_{k+1}\right)\Delta f_k$$

(vi) $$1 + \mu^2\delta^2 = 1 + \frac{1}{4}\left(E^{\frac{1}{2}} + E^{-\frac{1}{2}}\right)^2 \delta^2 = 1 + \frac{1}{4}\left[\left(E^{\frac{1}{2}} - E^{-\frac{1}{2}}\right)^2 + 4E^{\frac{1}{2}}E^{-\frac{1}{2}}\right]\delta^2$$

$$= 1 + \frac{1}{4}\left(\delta^2 + 4\right)\delta^2 = \left(1 + \delta^2 + \frac{1}{4}\delta^4\right) = \left(1 + \frac{1}{2}\delta^2\right)^2$$

(vii) $$(1 + \Delta)f(x) = E\,f(x) = f(x + h)$$

$$= f(x) + h\,f'(x) + \frac{h^2}{2!}f''(x) + \frac{h^3}{3!}f'''(x) + \ldots$$

$$= \left(1 + hD + \frac{(hD)^2}{2!} + \frac{(hD)^3}{3!} + \ldots\right)f(x)$$

$$= e^{hD}f(x)$$

$\therefore$ $$1 + \Delta = e^{hD}$$

(viii) $$\nabla^2 = \left(1 - E^{-1}\right)^2 = \left(1 - e^{-hD}\right)^2$$

$$= \left[1 - \left(1 - hD + \frac{h^2D^2}{2!} - \frac{h^3D^3}{3!} + \ldots\right)\right]^2$$

$$= \left(hD - \frac{h^2 D^2}{2!} + \frac{h^3 D^3}{3!} - \ldots \right)^2$$

$$= h^2 D^2 - 2\frac{h^3 D^3}{2!} + \left(\frac{1}{4} + \frac{2}{6} \right) h^4 D^4 - \ldots$$

$$= h^2 D^2 - h^3 D^3 + \frac{7}{12} h^4 D^4 - \ldots$$

Example 5.34: Prove that

(i) $hD = \log(1+\Delta) = -\log(1-\nabla) = \sinh^{-1}(\mu\delta)$

(ii) $\left(\frac{\Delta^2}{E} \right) e^x . \frac{E(e^x)}{\Delta^2 e^x} = e^x$

Solution: (i) $e^{hD} f(x) = \left(1 + hD + \frac{h^2 D^2}{2!} + \frac{h^3 D^3}{3!} + \ldots \right) f(x)$

$$= f(x) + h f'(x) + \frac{h^2}{2!} f''(x) + \frac{h^3}{3!} f'''(x) + \ldots = f(x+h) = E f(x)$$

$\therefore \quad e^{hD} = E = 1 + \Delta$

$\therefore \quad hD = \log(1+\Delta) \quad (1)$

Also, $\quad e^{hD} = E$

$\therefore \quad e^{-hD} = E^{-1} = 1 - \nabla$

$\therefore \quad -hD = \log(1-\nabla)$

$\therefore \quad hD = -\log(1-\nabla) \quad (2)$

and $\quad \sinh hD = \frac{e^{hD} - e^{-hD}}{2} = \frac{1}{2}(E - E^{-1})$

$$= \frac{1}{2}\left(E^{\frac{1}{2}} + E^{-\frac{1}{2}} \right)\left(E^{\frac{1}{2}} - E^{-\frac{1}{2}} \right) = \mu\delta$$

$\therefore \quad hD = \sinh^{-1} \mu\delta \quad (3)$

From (1), (2) and (3)

$$hD = \log(1+\Delta) = -\log(1-\nabla) = \sinh^{-1}(\mu\delta)$$

(ii) $\left(\frac{\Delta^2}{E} \right) e^x \frac{E(e^x)}{\Delta^2 e^x} = (\Delta^2 E^{-1}) e^x \frac{E(e^x)}{\Delta^2 (e^x)} = \Delta^2 (e)^{x-h} \frac{E(e^x)}{\Delta^2 (e^x)}$

$$= e^{-h}\Delta^2\left(e^x\right)\frac{E\left(e^x\right)}{\Delta^2\left(e^x\right)} = e^{-h}E\left(e^x\right)$$

$$= e^{-h}e^{x+h} = e^x$$

Example 5.35: Using the method of separation of symbols, show that

$$\Delta^n\ u_{x-n} = u_x - n u_{x-1} + \frac{n(n-1)}{2!}u_{x-2} - \frac{n(n-1)(n-2)}{3!}u_{x-3} + \ldots + (-1)^{n-1} n u_{x-n+1} + (-1)^n u_{x-n}$$

Solution:

$$\Delta^n\ u_{x-n} = (E-1)^n u_{x-n}$$

$$= \left[E^n - {}^nC_1\,E^{n-1} + {}^nC_2\,E^{n-2} - {}^nC_3\,E^{n-3} + \ldots + {}^nC_{n-1}(-1)^{n-1}E + {}^nC_n(-1)^n\right]u_{x-n}$$

$$= u_x - n u_{x-1} + \frac{n(n-1)}{2!}u_{x-2} - \frac{n(n-1)(n-2)}{3!}u_{x-3} + \ldots + (-1)^{n-1} n u_{x-n+1} + (-1)^n u_{x-n}$$

Example 5.36: Prove the following identities

(i) $u_0 + u_1 x + u_2 x^2 + \ldots \text{to } \infty = \dfrac{u_0}{1-x} + \dfrac{x\Delta u_0}{(1-x)^2} + \dfrac{x^2\Delta^2 u_0}{(1-x)^3} + \ldots \text{to } \infty$

(ii) $\Delta^n y_0 = y_n - {}^nC_1\,y_{n-1} + {}^nC_2\,y_{n-2} - \ldots + (-1)^n y_0$

Solution:

(i) R.H.S. $= \left[\dfrac{1}{1-x} + \dfrac{x}{(1-x)^2}(E-1) + \dfrac{x^2}{(1-x)^3}(E-1)^2 + \ldots\right]u_0$

$$= \frac{1}{1-x}\left[1 + \frac{x}{1-x}(E-1) + \left(\frac{x}{1-x}\right)^2(E-1)^2 + \ldots\right]u_0$$

$$= \frac{1}{1-x}\left[\frac{1}{1-\frac{x}{1-x}(E-1)}\right]u_0$$

$$= \frac{1}{1-x-x(E-1)}u_0 = \frac{1}{1-xE}u_0 = (1-xE)^{-1}u_0$$

$$= \left(1 + xE + x^2E^2 + x^3E^3 + \ldots\right)u_0$$

$$= u_0 + xu_1 + x^2u_2 + x^3u_3 + \ldots$$

$$= u_0 + u_1x + u_2x^2 + u_3x^3 + \ldots = \text{L.H.S.}$$

(ii) $\Delta^n y_0 = (E-1)^n y_0 = (1-E^{-1})^n E^n y_0 = (1-E^{-1})^n y_n$

$$= \left[1 - {}^nC_1 E^{-1} + {}^nC_2 E^{-2} - {}^nC_3 E^{-3} + \ldots + (-1)^n E^{-n}\right] y_n$$

$$= y_n - {}^nC_1 y_{n-1} + {}^nC_2 y_{n-2} - {}^nC_3 y_{n-3} + \ldots + (-1)^n y_0$$

Example 5.37: Given that $y_5 = 4,\ y_6 = 3,\ y_7 = 4,\ y_8 = 10$ and $y_9 = 24$. Find the value of $\Delta^4 y_5$ by using difference table.

Solution: Difference table for given data is

x	y	Δy	$\Delta^2 y$	$\Delta^3 y$	$\Delta^4 y$
5	4				
		−1			
6	3		2		
		1		3	
7	4		5		0
		6		3	
8	10		8		
		14			
9	24				

From the table

$$\Delta^4 y_5 = 0$$

Example 5.38: Find the missing entry in the following table

x	0	1	2	3	4
$y(x)$	1	3	9	–	81

Solution: Four entries are given

∴ 4th-order differences are zero

$$\therefore \quad \Delta^4 y_0 = 0$$

$$\therefore \quad (E-1)^4 y_0 = (E^4 - 4E^3 + 6E^2 - 4E + 1) y_0 = 0$$

$$\therefore \quad y_4 - 4y_3 + 6y_2 - 4y_1 + y_0 = 0$$

$$\therefore \quad y_3 = \frac{1}{4}(y_4 + 6y_2 - 4y_1 + y_0) = \frac{1}{4}\left(81 + 6(9) - 4(3) + 1\right) = 31$$

Other method

Let $y_3 = a$

Then, difference table is

x	y	Δy	$\Delta^2 y$	$\Delta^3 y$	$\Delta^4 y$
0	1				
		2			
1	3		4		
		6		$a - 19$	
2	9		$a - 15$		$124 - 4a$
		$a - 9$		$105 - 3a$	
3	a		$90 - 2a$		
		$81 - a$			
4	81				

Four entries are given

∴ 4th-order differences are zero

$$\therefore \quad 124 - 4a = 0$$

$$\therefore \quad a = 31$$

$$\therefore \quad y(3) = 31$$

Example 5.39: Find the missing values in the following table:

x	0	5	10	15	20	25
y	6	10	–	17	–	31

Solution: Let $y\,(10) = a$, $y\,(20) = b$

Then difference table is

x	y	Δy	$\Delta^2 y$	$\Delta^3 y$	$\Delta^4 y$
0	6				
		4			
5	10		$a - 14$		
		$a - 10$		$41 - 3a$	
10	a		$27 - 2a$		$6a + b - 102$
		$17 - a$		$3a + b - 61$	
15	17		$a + b - 34$		$143 - 4a - 4b$
		$b - 17$		$82 - a - 3b$	
20	b		$48 - 2b$		
		$31 - b$			
25	31				

As four entries are given, so 4th-order differences are zero

$$\therefore \quad 6a + b - 102 = 0$$
$$143 - 4a - 4b = 0$$
$$\therefore \quad 6a + b = 102$$
$$4a + 4b = 143$$

∴ By Cramer's rule

$$a = \frac{\begin{vmatrix} 102 & 1 \\ 143 & 4 \end{vmatrix}}{\begin{vmatrix} 6 & 1 \\ 4 & 4 \end{vmatrix}} = \frac{408-143}{24-4} = \frac{265}{20} = 13.25$$

$$b = \frac{\begin{vmatrix} 6 & 102 \\ 4 & 143 \end{vmatrix}}{\begin{vmatrix} 6 & 1 \\ 4 & 4 \end{vmatrix}} = \frac{858-408}{24-4} = \frac{450}{20} = 22.50$$

Other method

Four entries are given

$$\therefore \quad \Delta^4 y_0 = 0 \text{ and } \Delta^4 y_5 = 0$$

Now, $$\Delta^4 y_0 = 0$$

$$\Rightarrow \quad (E-1)^4 y_0 = 0$$
$$\Rightarrow \quad \left(E^4 - 4E^3 + 6E^2 - 4E + 1\right) y_0 = 0$$
$$\therefore \quad y_{20} - 4y_{15} + 6y_{10} - 4y_5 + y_0 = 0$$
$$\therefore \quad 6y_{10} + y_{20} = 4y_{15} + 4y_5 - y_0 = 4(17) + 4(10) - 6 = 102 \qquad (1)$$

Similarly, $\Delta^4 y_5 = 0$ gives

$$y_{25} - 4y_{20} + 6y_{15} - 4y_{10} + y_5 = 0$$
$$\therefore \quad 4y_{10} + 4y_{20} = y_{25} + 6y_{15} + y_5 = 31 + 6(17) + 10 = 143 \qquad (2)$$

Solving (1) and (2) by Cramer's rule as above, we have

$$y_{10} = 13.25, \; y_{20} = 22.50$$

Example 5.40: One entry in the following table of a polynomial of degree 4 is incorrect. Correct the entry by locating it.

x	1.0	1.1	1.2	1.3	1.4	1.5	1.6	1.7	1.8	1.9	2.0
y	1.0000	1.5191	2.0736	2.6611	3.2816	3.9375	4.6363	5.3771	6.1776	7.0471	8.0

Solution: Difference table for given data is

x	y	Δy	$\Delta^2 y$	$\Delta^3 y$	$\Delta^4 y$
1.0	1.0000				
		0.5191			
1.1	1.5191		0.0354		
		0.5545		–0.0024	
1.2	2.0736		0.0330		0.0024
		0.5875		0.0000	
1.3	2.6611		0.0330		0.0024
		0.6205		0.0024	
1.4	3.2816		0.0354		$0.0051+\varepsilon$
		0.6559		$0.0075+\varepsilon$	
1.5	3.9375		$0.0429+\varepsilon$		$-0.0084-4\varepsilon$
		$0.6988+\varepsilon$		$-0.0009-3\varepsilon$	
1.6	$4.6363+\varepsilon$		$0.0420-2\varepsilon$		$0.0186+6\varepsilon$
		$0.7408-\varepsilon$		$0.0177+3\varepsilon$	
1.7	5.3771		$0.0597+\varepsilon$		$-0.0084-4\varepsilon$
		0.8005		$0.0093-\varepsilon$	
1.8	6.1776		0.0690		$0.0051+\varepsilon$
		0.8695		0.0144	
1.9	7.0471		0.0834		
		0.9529			
2.0	8.0000				

Since the degree of polynomial is 4, the fourth differences must be constant. Numerically largest fourth difference is 0.0186 which is in the row of $x = 1.6$. Thus $y_{1.6}$ has error. Let error is ε then error propagation is shown in the table.

As 4th-order differences are constant

$$\therefore \quad 0.0186 + 6\varepsilon = 0.0024$$

$$\therefore \quad 6\varepsilon = -0.0162$$

$$\therefore \quad \varepsilon = -0.0027$$

$$\therefore \quad y_{1.6} = 4.6363 - 0.0027 = 4.6336$$

Example 5.41: Evaluate $\Delta^2\left(\dfrac{1}{x^2+5x+6}\right)$

Solution: $\Delta^2\left(\dfrac{1}{(x+2)(x+3)}\right)$

$$= \Delta^2 (x+1)^{(-2)}$$

$$= \Delta\left(-2(x+1)^{(-3)}\right)$$

$$= (-2)(-3)(x+1)^{(-4)}$$

$$= \frac{6}{(x+2)(x+3)(x+4)(x+5)} = \frac{6}{x^4+14x^3+71x^2+154x+120}$$

Example 5.42: Evaluate $\Delta^2\left(\dfrac{5x+12}{x^2+5x+6}\right)$

Solution:

$$\Delta^2\left(\frac{5x+12}{x^2+5x+6}\right) = \Delta^2\left[\frac{5x+12}{(x+2)(x+3)}\right] = \Delta^2\left[\frac{2}{x+2}+\frac{3}{x+3}\right] \quad \text{(By suppression method)}$$

$$= \Delta^2\left[2(x+1)^{(-1)}+3(x+2)^{(-1)}\right] = \Delta\left[-2(x+1)^{(-2)}-3(x+2)^{(-2)}\right]$$

$$= 4(x+1)^{(-3)}+6(x+2)^{(-3)} = \frac{4}{(x+2)(x+3)(x+4)}+\frac{6}{(x+3)(x+4)(x+5)}$$

$$= \frac{10x+32}{(x+2)(x+3)(x+4)(x+5)} = \frac{2(5x+16)}{x^4+14x^3+71x^2+154x+120}$$

Example 5.43: Evaluate $\Delta\left(\dfrac{1}{x(x+4)(x+8)}\right)$

Solution: $\Delta\left(\dfrac{1}{x(x+4)(x+8)}\right) = \Delta(x-4)^{(-3)} = (-3)(4)(x-4)^{(-4)}$

$$= -\frac{12}{x(x+4)(x+8)(x+12)} \qquad (\because h=4)$$

Example 5.44: Express the polynomial $f(x)=2x^3+3x^2-5x+4$ in factorial notation and find its successive differences. Also obtain a function whose first-order finite difference is $f(x)$

Solution: By synthetic division

$$\begin{array}{r|rrr|r}
0) & 2 & 3 & -5 & 4 \\
 & & 0 & 0 & 0 \\
\hline
1) & 2 & 3 & -5 & 4 \\
 & & 2 & 5 & \\
\hline
2) & 2 & 5 & 0 & \\
 & & 4 & & \\
\hline
 & 2 & 9 & &
\end{array}$$

$\therefore$ $$f(x)=2x^3+3x^2-5x+4=2x^{(3)}+9x^{(2)}+4$$

$\therefore$ $$\Delta f(x)=6x^{(2)}+18x^{(1)}=6x(x-1)+18x=6x^2+12x$$

$$\Delta^2 f(x)=12x^{(1)}+18=12x+18$$

$$\Delta^3 f(x)=12$$

$$\Delta^n f(x)=0;\ n\in N,\ n>3$$

Let $$\Delta F(x)=f(x)=2x^{(3)}+9x^{(2)}+4$$

$\therefore$ $$F(x)=\frac{2}{4}x^{(4)}+\frac{9}{3}x^{(3)}+4x^{(1)}+C$$

$$=\frac{1}{2}x(x-1)(x-2)(x-3)+3x(x-1)(x-2)+4x+C$$

$$=\frac{1}{2}\left[x^4-6x^3+11x^2-6x\right]+3\left[x^3-3x^2+2x\right]+4x+C$$

$$=\frac{1}{2}\left[x^4-7x^2+14x\right]+C$$

$\therefore$ function whose first-order finite difference is $f(x)$ is $\frac{1}{2}\left(x^4-7x^2+14x\right)+C$ where C is arbitrary constant.

Example 5.45: Express $f(x) = x^3 - 2x^2 + x - 1$ in factorial notation and show that $\Delta^4 f(x)=0$
Solution By synthetic division

$$\begin{array}{r|rrr|r}
0) & 1 & -2 & 1 & -1 \\
 & & 0 & 0 & 0 \\
\hline
1) & 1 & -2 & 1 & -1 \\
 & & 1 & -1 & \\
\hline
2) & 1 & -1 & 0 & \\
 & & 2 & & \\
\hline
 & 1 & 1 & &
\end{array}$$

$\therefore$ $$f(x)=x^3-2x^2+x-1=x^{(3)}+x^{(2)}-1$$

$\therefore$ $$\Delta f(x)=3x^{(2)}+2x^{(1)}$$

$$\Delta^2 f(x) = 6x^{(1)} + 2$$
$$\Delta^3 f(x) = 6$$
$$\Delta^4 f(x) = 0$$

Example 5.46: Evaluate $\Delta^3\left((1-x)(1-2x)(1-3x)\right)$
Solution: We have

$$f(x) = (1-x)(1-2x)(1-3x) = -6x^3 + 11x^2 - 6x + 1$$

By synthetic division

$$\begin{array}{r|rrrr}
0) & -6 & 11 & -6 & 1 \\
 & & 0 & 0 & 0 \\
\hline
1) & -6 & 11 & -6 & \multicolumn{1}{|r}{1} \\
 & & -6 & 5 & \\
\hline
2) & -6 & 5 & \multicolumn{1}{|r}{-1} & \\
 & & -12 & & \\
\hline
 & -6 & \multicolumn{1}{|r}{-7} & & \\
\end{array}$$

$\therefore$ $$f(x) = -6x^3 + 11x^2 - 6x + 1 = -6x^{(3)} - 7x^{(2)} - x^{(1)} + 1$$

$\therefore$ $$\Delta^3 f(x) = -6(3)(2)(1) = -36$$

Example 5.47: Find the second-order difference of the polynomial $x^4 - 12x^3 + 42x^2 - 30x + 9$ with interval of differencing $h = 2$
Solution: By synthetic division

$$\begin{array}{r|rrrrr}
0) & 1 & -12 & 42 & -30 & 9 \\
 & & 0 & 0 & 0 & 0 \\
\hline
2) & 1 & -12 & 42 & -30 & \multicolumn{1}{|r}{9} \\
 & & 2 & -20 & 44 & \\
\hline
4) & 1 & -10 & 22 & \multicolumn{1}{|r}{14} & \\
 & & 4 & -24 & & \\
\hline
6) & 1 & -6 & \multicolumn{1}{|r}{-2} & & \\
 & & 6 & & & \\
\hline
 & 1 & \multicolumn{1}{|r}{0} & & & \\
\end{array}$$

$\therefore$ $$f(x) = x^4 - 12x^3 + 42x^2 - 30x + 9 = x^{(4)} - 2x^{(2)} + 14x^{(1)} + 9$$

$\therefore$ $$\Delta f(x) = 4hx^{(3)} - 2(2)hx^{(1)} + 14h = 4(2)x^{(3)} - 4(2)x^{(1)} + 14(2) \qquad (\because h = 2)$$

$$= 8x^{(3)} - 8x^{(1)} + 28$$

$$\Delta^2 f(x) = 8(3)hx^{(2)} - 8h = 24(2)x^{(2)} - 8(2) = 48x^{(2)} - 16$$

$$= 48x(x-2) - 16 = 48x^2 - 96x - 16$$

EXERCISE 5.3

1. Evaluate

   (a) $\Delta \cos x$ (b) $\Delta \log f(x)$ (c) $\Delta \tan^{-1} x$

   (d) $\Delta^2 \cos 2x$ (e) $\Delta^2 \sin(ax+b)$ (f) $\Delta^n e^{ax+b}$

2. Show that

   (i) $\Delta[f(x)g(x)] = f(x+h)\,\Delta g(x) + g(x)\,\Delta f(x)$

   (ii) $\Delta\left[\dfrac{f(x)}{g(x)}\right] = \dfrac{g(x)\,\Delta f(x) - f(x)\,\Delta g(x)}{g(x+h)\,g(x)}$

3. Show that

   (a) $(\Delta+1)(1-\nabla) = 1$ (b) $\Delta = \delta E^{1/2}$ (c) $\delta = \nabla(1-\nabla)^{-1/2}$

   (d) $\mu = \dfrac{1}{2}\left(E^{1/2} + E^{-1/2}\right) = \left(1+\dfrac{\Delta}{2}\right)(1+\Delta)^{-1/2}$ (e) $E^r = \left(\mu + \dfrac{\delta}{2}\right)^{2r}$

   (f) $\Delta = \dfrac{1}{2}\delta^2 + \delta\sqrt{\left(1+\dfrac{\delta^2}{4}\right)}$ (g) $D = \dfrac{1}{h}\log E$

   (h) $\mu^{-1} = 1 - \dfrac{1}{8}\delta^2 + \dfrac{3}{128}\delta^4 - \dfrac{5}{1024}\delta^6 + \cdots\cdots$ (i) $\delta = 2\sinh\left(\dfrac{hD}{2}\right)$

4. Prove the following:

   (a) $\left(\dfrac{\Delta^2}{E}\right)u_x \neq \dfrac{\Delta^2 u_x}{Eu_x}$ (b) $\nabla^m f_n = \Delta^m f_{n\ m}$

5. Show that (a) $\dfrac{\Delta^2}{E}x^3 = 6xh^2$ (b) $(\Delta+\nabla)^2\left(x^2+x\right) = 8$ with $h=1$

6. If $f(x) = x^3 - 3x^2 + 5x + 7$, find $\Delta f(x)$, $\Delta^2 f(x)$, $\Delta^3 f(x)$, $\Delta^4 f(x)$ when $h = 1$.

7. Evaluate $\Delta^2\left(\dfrac{5x+12}{x^2+5x+6}\right)$ interval of differencing being unity.

8. Evaluate (a) $\Delta^2\left[\dfrac{1}{x(x+3)(x+6)}\right]$ (b) $\Delta^{10}(1-ax)(1-bx^2)(1-cx^3)(1-dx^4)$

9. If $y = (3x+1)(3x+4)\cdots\cdots(3x+22)$, prove that

   $\Delta^4 y = 136080(3x+13)(3x+16)(3x+19)(3x+22)$

10. Using the method of separation of symbols, prove that

    (i) $u_1 x + u_2 x^2 + u_3 x^3 + \cdots\cdots = \left(\dfrac{x}{1-x}\right)u_1 + \left(\dfrac{x}{1-x}\right)^2 \Delta u_1 + \left(\dfrac{x}{1-x}\right)^3 \Delta^2 u_1 + \cdots$

    (ii) $u_0 + u_1 + u_2 + \cdots\cdots + u_n = {}^{n+1}C_1\,u_0 + {}^{n+1}C_2\,\Delta u_0 + {}^{n+1}C_3\,\Delta^2 u_0 + \cdots + \Delta^n u_0$

11. Prove the following identities:

(i) $u_0 - u_1 + u_2 - u_3 + \cdots\cdots = \frac{1}{2}u_0 - \frac{1}{4}\Delta u_0 + \frac{1}{8}\Delta^2 u_0 + \cdots\cdots$

(ii) $u_0 + \frac{u_1}{1!}x + \frac{u_2}{2!}x^2 + \cdots\cdots = e^x\left[u_0 + x\frac{\Delta u_0}{1!} + x^2\frac{\Delta^2 u_0}{2!} + \cdots\cdots\right]$

12. If $u_x = ax^2 + bx + c$, show that

$u_{2x} - {}^xC_1.2.u_{2x-1} + {}^xC_2.2^2 u_{2x-2} + \cdots\cdots + (-2)^x u_x = (-1)^x (c - 2ax)$

13. Form the forward difference table for $y = x^3 - 2x^2 + 7$ for $x = 0, 1, 2, \ldots\ldots 6$.

14. Construct a forward difference table from the following:

x	0	1	2	3	4
y_x	1	1.5	2.2	3.1	4.6

Evaluate $\Delta^3 y_1$ and y_5.

15. If $f(x) = x^3 + 5x - 7$, then form the backward difference table for $x = -1, 0, 1, 2, 3, 4, 5$.

16. Form the backward difference table for the function $f(x) = x^3 - 3x^2 + 5x - 7$ for $x = -1$, 0, 1, 2, 3, 4 and 5.

17. Express y_4 in terms of successive forward finite differences of y_0.

18. A cubic polynomial takes the following values: $y(0) = 1$, $y(1) = 0$, $y(2) = 1$ and $y(3) = 10$. Obtain $y(4)$.

19. A cubic polynomial $f(x)$ takes the following values: $f_0 = -5, f_1 = 1, f_2 = 9, f_3 = 25, f_4 = 55$, $f_5 = 105$. Find f_{-1} and f_6.

20. Given $y_0 = -8, y_1 = -6, y_2 = 22, y_3 = 148, y_4 = 492$ and $y_5 = 1222$; find the value of y_6.

21. Find the missing value in the following table:

x	16	18	20	22	24	26
y	43	89	–	155	268	388

22. Following are the population data from the census of same district (in thousands). Estimate the missing population of year 1911.

Year	1881	1891	1901	1911	1921	1931
Population	363	391	421	–	461	501

23. Find the missing value in the following table:

x	45	50	55	60	65
y	3.0	–	2.0	–	–2.4

24. Assuming that the following values of y_x belong to a polynomial of degree four, compute the next two values.

x	2	4	6	8	10
y	2	3	5	8	9

25. Interpolate the missing entries in the following table:

x	0	1	2	3	4	5
$y(x)$	0	–	8	15	–	35

26. Find the missing values in the following table of values of x and y:

x	0	1	2	3	4	5	6
y	–4	–2	–	–	220	546	1148

27. Assuming that the following values of y belong to a polynomial of degree 4, compute the next three values:

x	0	1	2	3	4
y	1	–1	1	–1	1

28. Find and correct the error by means of differences, in the given data:

x	0	1	2	3	4	5	6	7	8	9	10
y	2	5	8	17	38	75	140	233	362	533	752

29. In the following table, one value of y is incorrect and that y is a cubic polynomial in x

x	0	1	2	3	4	5	6	7
y	25	21	18	18	27	45	76	123

Construct a difference table for y and use it to locate and correct the wrong value.

30. Express $f(x)=3x^3-2x^2+7x-6$ in factorial polynomial.
31. Express $f(u)=u^4-3u^2+2u+6$ in terms of factorial polynomial. Hence show that $\Delta^4 f(u)=24$.
32. Express $x^4+3x^3-5x^2+6x-7$ in factorial polynomial and find their successive forward differences.
33. Express $y=2x^3-3x^2+3x-10$ in factorial notation and hence show that $\Delta^3 y=12$.
34. Represent the function $f(x) = x^4 - 12x^3 + 42x^3 - 30x + 9$ and its successive differences in factorial notation in which the interval of differencing is one.
35. Obtain the function whose first difference is $9x^2 + 11x + 5$.

ANSWERS 5.3

1. (a) $-2\sin\left(x+\frac{h}{2}\right)\sin\frac{h}{2}$ (b) $\log\left(\frac{f(x+h)}{f(x)}\right)$ (c) $\tan^{-1}\frac{h}{1+x(x+h)}$

(b) $-4\sin^2 h\cos(2x+2h)$ (e) $-4\sin^2\frac{ah}{2}\sin(ax+b+ah)$ (f) $\left(e^{ah}-1\right)^n e^{ax+b}$

6. $3\left(x^2-x+1\right), 6x, 6, 0$

7. $\dfrac{2(5x+16)}{(x+2)(x+3)(x+4)(x+5)}$

8. (a) $\dfrac{108}{x(x+3)(x+6)(x+9)(x+12)}$ (b) $a\,b\,c\,d\,(10!)$

13.

x	y	Δy	$\Delta^2 y$	$\Delta^3 y$	$\Delta^4 y$	$\Delta^5 y$	$\Delta^6 y$
0	7						
		–1					
1	6		2				
		1		6			
2	7		8		0		
		9		6		0	
3	16		14		0		0
		23		6		0	
4	39		20		0		
		43		6			
5	82		26				
		69					
6	151						

14.

x	y_x	Δy_x	$\Delta^2 y_x$	$\Delta^3 y_x$	$\Delta^4 y_x$
0	1				
		0.5			
1	1.5		0.2		
		0.7		0	
2	2.2		0.2		0.4
		0.9		0.4	
3	3.1		0.6		
		1.5			
4	4.6				

$\Delta^3 y_1 = 0.4,\ y_5 = 7.5$

15.

x	y	∇y	$\nabla^2 y$	$\nabla^3 y$	$\nabla^4 y$	$\nabla^5 y$	$\nabla^6 y$
–1	–13						
		6					
0	–7		0				
		6		6			
1	–1		6		0		
		12		6		0	
2	11		12		0		0
		24		6		0	
3	35		18		0		
		42		6			
4	77		24				
		66					
5	143						

16.

x	y	∇y	$\nabla^2 y$	$\nabla^3 y$	$\nabla^4 y$	$\nabla^5 y$	$\nabla^6 y$
–1	–16						
		9					
0	–7		–6				
		3		6			
1	–4		0		0		
		3		6		0	
2	–1		6		0		0
		9		6		0	
3	8		12		0		
		21		6			
4	29		18				
		39					
5	68						

17. $y_4 = y_0 + 4\Delta y_0 + 6\Delta^2 y_0 + 4\Delta^3 y_0 + \Delta^4 y_0$

18. 33 19. $f_{-1} = -15,\ f_6 = 181$ 20. 2554 21. 100

22. 442.2 thousands 23. 2.925, 0.225 24. 2 and –22

25. 3.24 26. 12, 68 27. 31, 129, 351

28. The correct entry corresponding to $x = 5$ is 77

29. The correct entry corresponding to $x = 3$ is 19

30. $f(x) = 3x^{(3)} + 7x^{(2)} + 8x^{(1)} - 6$

31. $f(u) = u^{(4)} + 6u^{(3)} + 4u^{(2)} + 6$

32. $f(x) = x^{(4)} + 9x^{(3)} + 11x^{(2)} + 5x^{(1)} - 7$

$\Delta f(x) = 4x^{(3)} + 27x^{(2)} + 22x^{(1)} + 5$

$\Delta^2 f(x) = 12x^{(2)} + 54x^{(1)} + 22$

$\Delta^3 f(x) = 24x^{(1)} + 54$

$\Delta^4 f(x) = 24$

and $\Delta^n f(x) = 0;\ n > 4, n \in N$

33. $y = 2x^{(3)} + 3x^{(2)} + 2x^{(1)} - 10$

34. $f(x) = x^{(4)} - 6x^{(3)} + 13x^{(2)} + x^{(1)} + 9$

$\Delta f(x) = 4x^{(3)} - 18x^{(2)} + 26x^{(1)} + 1$

$\Delta^2 f(x) = 12x^{(2)} - 36x^{(1)} + 26$

$\Delta^3 f(x) = 24x^{(1)} - 36$

$\Delta^4 f(x) = 24$

and $\Delta^n f(x) = 0;\ n > 4,\ n \in N$

35. $f(x) = 3x^3 + x^2 + x + c$ where c is arbitrary constant.

5.8 INTERPOLATION

Suppose the values of a function $y = f(x)$ are given for some values of x called arguments. Let I be interval formed by these values of x. If the value of y is required for some $x \in I$, then it is called problem of interpolation and if value of y is required for some x outside the interval I, then it is called problem of extrapolation. We shall be treating the problem of extrapolation in the same way as the problem of interpolation. The following two methods will be used for interpolation when arguments may be equispaced or not equispaced. Other methods can be used only when the arguments are equispaced.

For finding the value of $f(x)$ for some x we shall have to approximate $f(x)$ by some function $\phi(x)$ which can be found. If $\phi(x)$ is taken as a polynomial taking the given values at given arguments, then this polynomial is called interpolating polynomial.

5.8.1 Lagrange's Interpolation Formula

Let $y = f(x)$ be the function whose values at $x_0, x_1, x_2, \ldots x_n$ is given to be $y_0, y_1, y_2, \ldots, y_n$, respectively. We approximate $f(x)$ by nth degree polynomial passing through $(x_0, y_0), (x_1, y_1), \ldots, (x_n, y_n)$ which is called Lagrange's polynomial $P(x)$. Let

$$P(x) = a_0(x-x_1)(x-x_2)\ldots(x-x_n) + a_1(x-x_0)(x-x_2)(x-x_3)\ldots(x-x_n)$$

$$+ a_2(x-x_0)(x-x_1)(x-x_3)\ldots(x-x_n) + \ldots + a_n(x-x_0)(x-x_1)(x-x_2)\ldots(x-x_{n-1}) \quad (5.7)$$

Now $\qquad P(x_k) = y_k;\ k = 0, 1, 2, \cdots n$

$\therefore \qquad y_k = a_k(x_k - x_0)(x_k - x_1)\cdots(x_k - x_{k-1})(x_k - x_{k+1})\cdots(x_k - x_n)$

$\therefore \qquad a_k = \dfrac{y_k}{(x_k - x_0)(x_k - x_1)\cdots(x_k - x_{k-1})(x_k - x_{k+1})\cdots(x_k - x_n)};\ k = 0, 1, 2, \ldots, n$

Substituting in (5.7)

$$f(x) \simeq P(x) = \sum_{k=0}^{n} l_k(x) y_k \quad (5.8)$$

where

$$l_k(x) = \frac{(x-x_0)(x-x_1)\cdots(x-x_{k-1})(x-x_{k+1})\cdots(x-x_n)}{(x_k-x_0)(x_k-x_1)\cdots(x_k-x_{k-1})(x_k-x_{k+1})\cdots(x_k-x_n)} \quad (5.9)$$

(5.8) is Lagrange's interpolation formula to approximate $f(x)$ at any x where $l_k(x)$ is given by (5.9)

Error in Lagrange-Interpolation Polynomial Approximation

Let $F(x) = \prod_{r=0}^{n}(x - x_r)$.

Let I be interval bounded by $x_0, x_1, x_2, \ldots, x_n$ and η is any point in I such that $\eta \neq x_i;\ i = 0, 1, 2, \ldots, n$.

We want to find error in $P(\eta)$ when exact value is $f(\eta)$.

The function $f(x) - P(x)$ vanishes for $x_0, x_1, \ldots, x_n$

Let $\qquad f(x) = P(x) + RF(x)$ at $x = \eta$.

Here, R can be determined as follows

Define $$G(x) = f(x) - P(x) - RF(x)$$

we have $$G(x) = 0;\ x = x_0, x_1, x_2, \ldots, x_n, \eta$$

Arranging the $(n + 2)$ points $x_0, x_1, x_2, \ldots, x_n, \eta$ in ascending order, we can form $(n + 1)$ subintervals of I. Applying Rolle's theorem to the function $G(x)$ in each of these subintervals we can find $\xi_i \in I$ such that $G'(\xi_i) = 0$, $i = 1, 2, \ldots, (n+1)$. Again n subintervals of I can be formed with the help of the points $\xi_1, \xi_2, \ldots \xi_{n+1}$ and we can apply Rolle's theorem to the function $G'(x)$ in each of these subintervals. Proceeding in this way applying Rolle's theorem repeatedly, we have

$$G^{(n+1)}(\xi) = 0 \text{ for some } \xi \in I$$

$$\therefore \quad G^{(n+1)}(\xi) = f^{(n+1)}(\xi) - R.(n+1)! = 0,$$

($\because$ $P(x)$ is polynomial of degree n and $F^{(n+1)}(x) = (n + 1)!$ for all x)

$$\therefore \quad R = \frac{f^{(n+1)}(\xi)}{(n+1)!}$$

Hence, $$\text{error} = RF(\eta) = \frac{f^{(n+1)}(\xi)}{(n+1)!}(x - x_0)(x - x_1)\cdots(x - x_n)$$

for some ξ lying in the interval bounded by $x_0, x_1, \ldots, x_n$.

If problem is of extrapolation, then ξ will lie in the interval bounded by $\eta, x_0, x_1, x_2, \ldots, x_n$.

Remark 5.12: (i) If we take $F(x) = \prod_{r=0}^{n}(x - x_r)$
then dividing (5.8) by $F(x)$ we have

$$\frac{f(x)}{F(x)} = \sum_{k=0}^{n} \frac{y_k}{(x - x_k)(x_k - x_0)(x_k - x_1)\cdots(x_k - x_{k-1})(x_k - x_{k+1})\cdots(x_k - x_n)}$$

It is the partial fraction of $\frac{f(x)}{F(x)}$ where $f(x)$ is taken as $P(x)$.

(ii) From (5.9)

$$\frac{l_k(x)}{F(x)} = \frac{1}{(x - x_k)(x_k - x_0)(x_k - x_1)\cdots(x_k - x_{k-1})(x_k - x_{k+1})\cdots(x_k - x_n)}$$

$$\therefore \quad \frac{1}{F(x)}\sum_{k=0}^{n} l_k(x) = \sum_{k=0}^{n} \frac{1}{(x - x_k)(x_k - x_0)(x_k - x_1)\cdots(x_k - x_{k-1})(x_k - x_{k+1})\cdots(x_k - x_n)}$$

But R.H.S. is partial fraction of $\frac{1}{F(x)}$

$$\therefore \quad \frac{1}{F(x)}\sum_{k=0}^{n} l_k(x) = \frac{1}{F(x)}$$

$$\therefore \quad \sum_{k=0}^{n} l_k(x) = 1 \text{ for all } x$$

This can be used as a check in the calculations.

5.8.2 Divided Differences

If the functional values are given for non-equispaced arguments, then Lagrange's interpolation formula require much labour. In this respect, divided differences offer better possibilities.

Let $x_0, x_1, x_2, \ldots, x_n$ be arguments, then first divided differences between x_0 and x_1, x_1 and x_2, x_2 and x_3, …, x_{n-1} and x_n are defined by

$$f(x_0, x_1) = \frac{f(x_1) - f(x_0)}{x_1 - x_0} = \frac{f(x_0) - f(x_1)}{x_0 - x_1} = f(x_1, x_0)$$

$$f(x_1, x_2) = \frac{f(x_2) - f(x_1)}{x_2 - x_1}$$

$$\vdots$$

$$f(x_{n-1}, x_n) = \frac{f(x_n) - f(x_{n-1})}{x_n - x_{n-1}}$$

The second divided difference between $x_0, x_1, x_2; x_1, x_2, x_3; \ldots; x_{n-2}, x_{n-1}, x_n$ are

$$f(x_0, x_1, x_2) = \frac{f(x_1, x_2) - f(x_0, x_1)}{x_2 - x_0}$$

$$f(x_1, x_2, x_3) = \frac{f(x_2, x_3) - f(x_1, x_2)}{x_3 - x_1}$$

$$\vdots$$

$$f(x_{n-2}, x_{n-1}, x_n) = \frac{f(x_{n-1}\, x_n) - f(x_{n-2}, x_{n-1})}{x_n - x_{n-2}}$$

Similarly, we define nth divided difference

$$f(x_0, x_1 \ldots, x_n) = \frac{f(x_1, x_2, \ldots, x_n) - f(x_0, x_1, \ldots, x_{n-1})}{x_n - x_0}.$$

Divided difference of order n of $f(x)$ will be denoted by $\Delta^n f(x)$

Theorem 5.2 $f(x_0, x_1, \ldots, x_n) = \sum_{k=0}^{n} \frac{f(x_k)}{\prod_{\substack{j=0 \\ j \neq k}}^{n} (x_k - x_j)}$

Proof: We prove this theorem by principle of mathematical induction

for $n = 1$,

$$f(x_0, x_1) = \frac{f(x_1) - f(x_0)}{x_1 - x_0} = \frac{f(x_0)}{x_0 - x_1} + \frac{f(x_1)}{x_1 - x_0}$$

$\therefore$ the result is true for $n = 1$.

Let the result is true for $n = m$

$$\therefore \qquad f(x_0, x_1, \ldots, x_m) = \sum_{k=0}^{m} \frac{f(x_k)}{\prod\limits_{\substack{j=0 \\ j\neq k}}^{m} (x_k - x_j)}$$

$$f(x_1, x_2, \ldots, x_{m+1}) = \sum_{k=1}^{m+1} \frac{f(x_k)}{\prod\limits_{\substack{j=1 \\ j\neq k}}^{m+1} (x_k - x_j)}$$

$\therefore \ f(x_0, x_1, \ldots, x_{m+1})$

$$= \frac{f(x_1, x_2, \ldots, x_{m+1}) - f(x_0, x_1, \ldots, x_m)}{x_{m+1} - x_0}$$

$$= \frac{1}{(x_{m+1} - x_0)} \left[\sum_{k=1}^{m+1} \frac{f(x_k)}{\prod\limits_{\substack{j=1 \\ j\neq k}}^{m+1} (x_k - x_j)} - \sum_{k=0}^{m} \frac{f(x_k)}{\prod\limits_{\substack{j=0 \\ j\neq k}}^{m} (x_k - x_j)} \right]$$

$$= \frac{1}{(x_{m+1} - x_0)} \left[-\frac{f(x_0)}{\prod\limits_{j=1}^{m} (x_0 - x_j)} + \frac{f(x_{m+1})}{\prod\limits_{j=1}^{m} (x_{m+1} - x_j)} + \sum_{k=1}^{m} f(x_k) \left\{ \frac{1}{\prod\limits_{\substack{j=1 \\ j\neq k}}^{m+1} (x_k - x_j)} - \frac{1}{\prod\limits_{\substack{j=0 \\ j\neq k}}^{m} (x_k - x_j)} \right\} \right]$$

$$= \frac{f(x_0)}{\prod\limits_{j=1}^{m+1} (x_0 - x_j)} + \frac{f(x_{m+1})}{\prod\limits_{j=0}^{m} (x_{m+1} - x_j)} + \sum_{k=1}^{m} \frac{f(x_k)}{(x_{m+1} - x_0) \prod\limits_{\substack{j=0 \\ j\neq k}}^{m+1} (x_k - x_j)} \left[x_k - x_0 - (x_k - x_{m+1}) \right]$$

$$= \frac{f(x_0)}{\prod\limits_{j=1}^{m+1} (x_0 - x_j)} + \frac{f(x_{m+1})}{\prod\limits_{j=0}^{m} (x_{m+1} - x_j)} + \sum_{k=1}^{m} \frac{f(x_k)}{\prod\limits_{\substack{j=0 \\ j\neq k}}^{m+1} (x_k - x_j)}$$

$$= \sum_{k=0}^{m+1} \frac{f(x_k)}{\prod\limits_{\substack{j=0 \\ j\neq k}}^{m+1} (x_k - x_j)}$$

Hence by principle of mathematical induction, the result holds for all $n \in N$.

Theorem 5.3 For equidistant arguments with interval of differencing h

$$f(x_0, x_1, \ldots, x_n) = \frac{1}{h^n n!} \Delta^n f_0$$

Proof: We shall prove this result by principle of mathematical induction

for $n = 1$,

$$f(x_0, x_1) = \frac{f(x_1) - f(x_0)}{x_1 - x_0} = \frac{\Delta f(x_0)}{h} = \frac{1}{h} \Delta f_0$$

$\therefore$ result holds for $n = 1$

Let the result is true for $n = m$

$$\therefore \quad f(x_0, x_1, x_2 \ldots, x_m) = \frac{1}{h^m\, m!}\Delta^m f_0$$

$$f(x_1, x_2, \ldots, x_{m+1}) = \frac{1}{h^m\, m!}\Delta^m f_1$$

$$\therefore \quad f(x_0, x_1, \ldots, x_{m+1}) = \frac{f(x_1, x_2, \ldots, x_{m+1}) - f(x_0, x_1, \ldots, x_m)}{x_{m+1} - x_0}$$

$$= \frac{\Delta^m f_1 - \Delta^m f_0}{h^m\, m!\,(m+1)h}$$

$$= \frac{\Delta^{m+1} f_0}{h^{m+1}(m+1)!}$$

$\therefore$ result is true for $n = m + 1$.

Hence by principle of mathematical induction, the result is true for all $n \in N$.

Remark 5.13: (i) Even if the arguments are equal, the divided differences still have a meaning

$$f(x_0, x_0) = \lim_{\varepsilon \to 0} f(x_0, x_0 + \varepsilon) = \lim_{\varepsilon \to 0} \frac{f(x_0 + \varepsilon) - f(x_0)}{\varepsilon}$$

$$= f'(x_0); \text{ provided } f(x) \text{ is differentiable at } x_0 \qquad (1)$$

$$f(x_0, x_0, x_0) = \lim_{\varepsilon \to 0} f(x_0, x_0 + \varepsilon, x_0 + \varepsilon)$$

$$= \lim_{\varepsilon \to 0} \frac{f(x_0 + \varepsilon, x_0 + \varepsilon) - f(x_0, x_0 + \varepsilon)}{\varepsilon}$$

$$= \lim_{\varepsilon \to 0} \frac{1}{\varepsilon}\left[f'(x_0 + \varepsilon) - \frac{1}{\varepsilon}\big(f(x_0 + \varepsilon) - f(x_0)\big)\right] \qquad \text{(from (1))}$$

$$= \lim_{\varepsilon \to 0} \frac{1}{\varepsilon}\left[f'(x_0) + \varepsilon f''(x_0) + \cdots - \frac{1}{\varepsilon}\left(\varepsilon f'(x_0) + \frac{\varepsilon^2}{2!} f''(x_0) + \cdots\right)\right]$$

$$= \frac{f''(x_0)}{2}$$

Similarly $\quad f(\underbrace{x_0, x_0, \ldots, x_0}_{(r+1)\text{arguments}}) = \dfrac{f^{(r)}(x_0)}{r!}$

(ii) $f(x, x, x_0, x_1, \ldots, x_n) = \dfrac{d}{dx} f(x, x_0, x_1, \ldots, x_n)$

5.8.3 Newton's Divided Difference Interpolation Formula

We have

$$f(x, x_0) = \frac{f(x) - f(x_0)}{x - x_0}$$

Thus, from defining equations

$$f(x)=f(x_0)+(x-x_0)f(x,x_0)$$
$$f(x,x_0)=f(x_0,x_1)+(x-x_1)f(x,x_0,x_1)$$
$$f(x,x_0,x_1)=f(x_0,x_1,x_2)+(x-x_2)f(x,x_0,x_1,x_2)$$
$$f(x,x_0,x_1,\ldots,x_{n-1})=f(x_0,x_1,\ldots,x_n)+(x-x_n)f(x,x_0,x_1,\ldots,x_n)$$

Multiply the second equation by $x - x_0$, the third by $(x-x_0)(x-x_1),\ldots$ and last equation by $(x-x_0)(x-x_1)\ldots(x-x_{n-1})$ and add

$$f(x)=f(x_0)+(x-x_0)f(x_0,x_1)+(x-x_0)(x-x_1)f(x_0,x_1,x_2)$$
$$+\ldots+(x-x_0)(x-x_1)\ldots(x-x_{n-1})f(x_0,x_1,\ldots,x_n)+R$$

where

$$R=(x-x_0)(x-x_1)\ldots(x-x_n)f(x,x_0,x_1,\ldots,x_n)$$

It is Newton's divided difference interpolation formula where R is the error.

We can write

$$f(x)=P(x)+R$$

where $P(x)$ is nth degree polynomial and $R=0$ for $x=x_0,x_1,\ldots,x_n$

As $f(x_k)=P(x_k);\ k=0,1,2,\ldots,n$

$\therefore$ $P(x)$ must be identical with Lagrange's interpolation polynomial. Hence

$$R=\frac{f^{(n+1)}(\xi)}{(n+1)!}(x-x_0)(x-x_1)\cdots(x-x_n)$$

Thus, $$f(x,x_0,x_1,\ldots,x_n)=\frac{f^{(n+1)}(\xi)}{(n+1)!},\ \xi\in(x_0,x_n)$$

Example 5.48: Find the interpolating polynomial for $(0,2),(1,3),(2,12)$ and $(5,147)$ using Lagrange's interpolation formula

Solution:

x	0	1	2	5
y	2	3	12	147

If $P(x)$ is Lagrange interpolating polynomial for $y(x)$ then

$$P(x)=\frac{(x-1)(x-2)(x-5)}{(0-1)(0-2)(0-5)}(2)+\frac{(x-0)(x-2)(x-5)}{(1-0)(1-2)(1-5)}(3)$$
$$+\frac{(x-0)(x-1)(x-5)}{(2-0)(2-1)(2-5)}(12)+\frac{(x-0)(x-1)(x-2)}{(5-0)(5-1)(5-2)}(147)$$

$$= -\frac{1}{5}\left(x^3 - 8x^2 + 17x - 10\right) + \frac{3}{4}\left(x^3 - 7x^2 + 10x\right)$$
$$-2\left(x^3 - 6x^2 + 5x\right) + \frac{49}{20}\left(x^3 - 3x^2 + 2x\right)$$
$$= \frac{1}{20}\left[(-4 + 15 - 40 + 49)x^3 + (32 - 105 + 240 - 147)x^2\right.$$
$$\left. + (-68 + 150 - 200 + 98)x + 40\right]$$
$$= x^3 + x^2 - x + 2$$

Note:

$$(x - x_1)(x - x_2)\cdots(x - x_n)$$
$$= x^n - (x_1 + x_2 + \cdots + x_n)x^{n-1} + \left(\sum x_i x_j\right)x^{n-2} - \left(\sum x_i x_j x_k\right)x^{n-3} + \cdots + (-1)^n x_1 x_2 \cdots x_n$$

Example 5.49: Obtain a unique polynomial $f(x)$ of degree 2 or less such that $f(1) = 1, f(3) = 27, f(4) = 64$ using Lagrange's interpolation concept.
Solution:

x	1	3	4
$f(x)$	1	27	64

Here, $f(x)$ itself is Lagrange interpolation polynomial

$$f(x) = \frac{(x-3)(x-4)}{(1-3)(1-4)}(1) + \frac{(x-1)(x-4)}{(3-1)(3-4)}(27) + \frac{(x-1)(x-3)}{(4-1)(4-3)}(64)$$
$$= \frac{1}{6}\left(x^2 - 7x + 12\right) - \frac{27}{2}\left(x^2 - 5x + 4\right) + \frac{64}{3}\left(x^2 - 4x + 3\right)$$
$$= \frac{1}{6}\left[(1 - 81 + 128)x^2 + (-7 + 405 - 512)x + 12 - 324 + 384\right]$$
$$= 8x^2 - 19x + 12$$

Example 5.50: Derive Lagrange's interpolation formula. Apply it to find interpolating polynomial to fit the following data:

x	0	1	2	3
$y = e^x - 1$	0	1.72	6.39	19.09

Solution: Lagrange's interpolation formula to approximate $f(x)$ to Lagrange polynomial $P(x)$ is already derived. In the given data

$$P(x) = \frac{(x-1)(x-2)(x-3)}{(0-1)(0-2)(0-3)}(0) + \frac{(x-0)(x-2)(x-3)}{(1-0)(1-2)(1-3)}(1.72)$$

$$+\frac{(x-0)(x-1)(x-3)}{(2-0)(2-1)(2-3)}(6.39)+\frac{(x-0)(x-1)(x-2)}{(3-0)(3-1)(3-2)}(19.09)$$

$$\therefore \qquad P(x)=0.86\left(x^3-5x^2+6x\right)-\frac{6.39}{2}\left(x^3-4x^2+3x\right)+\frac{19.09}{6}\left(x^3-3x^2+2x\right)$$

$$=\frac{1}{6}\Big[\big(6(0.86)-3(6.39)+19.09\big)x^3+\big(-30(0.86)+12(6.39)-3(19.09)\big)x^2$$

$$+\big(36(0.86)-9(6.39)+2(19.09)\big)x\Big]$$

$$=\frac{1}{6}\left(5.08x^3-6.39x^2+11.63x\right)$$

Example 5.51: Write the Lagrange's polynomial passing through the points $(x_0, f_0), (x_1, f_1)$ and (x_2, f_2) and hence resolve $\dfrac{3x^2+x+1}{(x-1)(x-2)(x-3)}$ into partial fractions.

Solution: Lagrange's polynomial passing through $(x_0, f_0), (x_1, f_1)$ and (x_2, f_2) is

$$P(x)=\frac{(x-x_1)(x-x_2)}{(x_0-x_1)(x_0-x_2)}f_0+\frac{(x-x_0)(x-x_2)}{(x_1-x_0)(x_1-x_2)}f_1+\frac{(x-x_0)(x-x_1)}{(x_2-x_0)(x_2-x_1)}f_2$$

Now $\quad P(x)=3x^2+x+1$ is the polynomial passing through $(1,5),(2,15),(3,31)$

$$\therefore \qquad 3x^2+x+1=\frac{(x-2)(x-3)}{(1-2)(1-3)}(5)+\frac{(x-1)(x-3)}{(2-1)(2-3)}(15)+\frac{(x-1)(x-2)}{(3-1)(3-2)}(31)$$

$$=\frac{5}{2}(x-2)(x-3)-15(x-1)(x-3)+\frac{31}{2}(x-1)(x-2)$$

Divide by $(x-1)(x-2)(x-3)$

$$\frac{3x^2+x+1}{(x-1)(x-2)(x-3)}=\frac{5}{2(x-1)}-\frac{15}{x-2}+\frac{31}{2(x-3)}$$

which is the partial fraction of $\dfrac{3x^2+x+1}{(x-1)(x-2)(x-3)}$

Example 5.52: Using the data sin (0.1) = 0.09983 and sin (0.2) = 0.19867, find an approximate value of sin (0.15) by Lagrange's interpolation. Obtain a bound on the truncation error.
Solution:

x	0.1	0.2
$y(x) = \sin x$	.09983	0.19867

Let $P(x)$ be Lagrange's polynomial, then

$$y(x) \simeq P(x) = \frac{x-0.2}{0.1-0.2}(.09983) + \frac{x-0.1}{0.2-0.1}(0.19867)$$

$\therefore$ $$\sin(0.15) = y(0.15) \simeq \frac{0.15-0.2}{0.1-0.2}(.09983) + \frac{0.15-0.1}{0.2-0.1}(0.19867)$$

$$= 0.5(.09983) + 0.5(0.19867)$$

$$= 0.5(.09983 + 0.19867) = 0.14925$$

$$\text{Truncation error } = (0.15-0.10)(0.15-0.20)\frac{f''(\xi)}{2!}$$

where $$f(x) = \sin x,\ \xi \in (0.10,\ 0.20)$$

$\therefore$ $$f''(x) = -\sin x$$

$\therefore$ $$\text{Max.}\left|f''(\xi)\right| = \underset{\xi\in(0.10,\ 0.20)}{\text{Max.}} \sin\xi < \sin(0.20) = 0.19867$$

$\therefore$ $$\left|\text{Truncation error}\right| < \frac{(0.05)(0.05)}{2}(0.19867)$$

$\therefore$ $$\left|\text{Truncation error}\right| < 0.0002484$$

$\therefore$ $$\sin(0.15) \simeq 0.14925$$

and $$\left|\text{Truncation error}\right| < 0.0002484$$

Example 5.53: Find the nth divided difference of $\frac{1}{x}$ based on the points $x_0, x_1, x_2, \cdots, x_n$

Solution: Let $f(x) = \frac{1}{x}$. We shall prove by principle of mathematical induction that

$$f[x_0, x_1, x_2, \ldots, x_n] = \frac{(-1)^n}{x_0\, x_1\, x_2 \ldots x_n}$$

for $n = 1$

$$f[x_0, x_1] = \frac{f(x_1) - f(x_0)}{x_1 - x_0} = \frac{\frac{1}{x_1} - \frac{1}{x_0}}{x_1 - x_0}$$

$$= \frac{x_0 - x_1}{x_0 x_1 (x_1 - x_0)} = -\frac{1}{x_0 x_1}$$

$\therefore$ the result is true for $n = 1$

Let the result is true for $n = m$

$$f[x_0, x_1, x_2, \cdots, x_m] = \frac{(-1)^m}{x_0 x_1 x_2 \cdots x_m}$$

$$f\left[x_1, x_2, x_3, \cdots, x_{m+1}\right] = \frac{(-1)^m}{x_1 x_2 x_3 \cdots x_{m+1}}$$

$$f\left[x_0, x_1, x_2, \cdots, x_{m+1}\right] = \frac{f\left[x_1, x_2, \cdots, x_{m+1}\right] - f\left[x_0, x_1, \cdots, x_m\right]}{x_{m+1} - x_0}$$

$$= \frac{\dfrac{(-1)^m}{x_1 x_2 \cdots x_{m+1}} - \dfrac{(-1)^m}{x_0 x_1 \cdots x_m}}{x_{m+1} - x_0}$$

$$= \frac{(-1)^m \left(x_0 - x_{m+1}\right)}{x_0 x_1 x_2 \cdots x_m \cdot x_{m+1}\left(x_{m+1} - x_0\right)}$$

$$= \frac{(-1)^{m+1}}{x_0 x_1 x_2 \cdots x_{m+1}}$$

$\therefore$ Result is true for $n = m \Rightarrow$ Result is true for $n = m + 1$

$\therefore$ By principle of mathematical induction, $f\left[x_0, x_1, x_2, \ldots, x_n\right] = \dfrac{(-1)^n}{x_0\, x_1\, x_2 \ldots x_n}$; for all $n \in N$

Example 5.54: If $f(x) = \dfrac{1}{x^2}$ then find $f\left[a, b, c\right]$.

Solution: $f\left[a,b\right] = \dfrac{f(b) - f(a)}{b - a} = \dfrac{\dfrac{1}{b^2} - \dfrac{1}{a^2}}{b - a} = \dfrac{a^2 - b^2}{a^2 b^2 (b - a)}$

$$= \frac{(a+b)(a-b)}{a^2 b^2 (b-a)} = \frac{-(a+b)}{a^2 b^2}$$

By Symmetry

$$f[b,c] = \frac{-(b+c)}{b^2 c^2}$$

$\therefore$
$$f\left[a,b,c\right] = \frac{f[b,c] - f[a,b]}{c - a} = \frac{-\dfrac{(b+c)}{b^2 c^2} + \dfrac{(a+b)}{a^2 b^2}}{c - a}$$

$$= \frac{-a^2 b - a^2 c + ac^2 + bc^2}{(c-a) a^2 b^2 c^2} = \frac{bc^2 - a^2 b + ac^2 - a^2 c}{(c-a) a^2 b^2 c^2}$$

$$= \frac{b(c-a)(c+a) + ac(c-a)}{(c-a) a^2 b^2 c^2} = \frac{bc + ab + ac}{a^2 b^2 c^2} = \frac{1}{abc}\left(\frac{1}{a} + \frac{1}{b} + \frac{1}{c}\right)$$

Example 5.55: Employing Newton's divided difference interpolation, estimate $f(x)$ from the following data:

x	0	1	2	4	5	6
$f(x)$	1	14	15	5	6	19

Solution: Divided difference table for given data is

x	f(x)	Δf(x)	Δ²f(x)	Δ³f(x)	Δ⁴f(x)	Δ⁵f(x)
0	1					
		13				
1	14		−6			
		1		1		
2	15		−2		0	
		−5		1		0
4	5		2		0	
		1		1		
5	6		6			
		13				
6	19					

By Newton's divided difference interpolation formula

$$f(x) \simeq f(x_0) + (x - x_0) f(x_0, x_1) + (x - x_0)(x - x_1) f(x_0, x_1, x_2)$$
$$+ (x - x_0)(x - x_1)(x - x_2) f(x_0, x_1, x_2, x_3) + \cdots$$
$$= 1 + 13(x - 0) + (-6)x(x - 1) + (1)x(x - 1)(x - 2) = x^3 - 9x^2 + 21x + 1$$

Example 5.56: Form the divided difference table for the following data

x	0.5	1.5	3.0	5.0	6.5	8.0
$f(x)$	1.625	5.875	31.0	131.0	282.125	521.0

and find the interpolating polynomial and estimate the value of $f(7)$

Solution: Divided difference table for given data is

x	f(x)	Δf(x)	Δ²f(x)	Δ³f(x)	Δ⁴f(x)	Δ⁵f(x)
0.5	1.625					
		4.25				
1.5	5.875		5.0			
		16.75		1		
3.0	31.000		9.5		0	
		50.00		1		0
5.0	131.000		14.5		0	
		100.75		1		
6.5	282.125		19.5			
		159.25				
8.0	521.000					

If $P(x)$ is interpolating polynomial then

$$P(x) = 1.625 + 4.25(x - 0.5) + 5.0(x - 0.5)(x - 1.5) + 1(x - 0.5)(x - 1.5)(x - 3.0)$$

$$= 1.625 + 4.25(x - 0.5) + 5.0(x^2 - 2x + 0.75) + x^3 - 5x^2 + 6.75x - 2.25 = x^3 + x + 1$$

$$f(7) \simeq P(7) = 7^3 + 7 + 1 = 351$$

Example 5.57: Using the following table find $f(x)$ as a polynomial in powers of $(x - 6)$

x	–1	0	2	3	7	10
$f(x)$	–11	1	1	1	141	561

Solution Divided difference table for given data is

x	f(x)	Δf(x)	Δ²f(x)	Δ³f(x)	Δ⁴f(x)	Δ⁵f(x)
–1	–11					
		12				
0	1		–4			
		0		1		
2	1		0		0	
		0		1		0
3	1		7		0	
		35		1		
7	141		15			
		140				
10	561					

$f(x)$ is 5th degree polynomial. By Newton's divided difference interpolation formula

$$f(x) = -11 + (x + 1)(12) + (x + 1)\,x(-4) + (x + 1)\,x\,(x - 2)\,(1)$$

$$= -11 + 12\,(x + 1) - 4x\,(x + 1) + (x - 2)\,x\,(x + 1)$$

For writing the polynomial in powers of $x - 6$

we take $\quad x - 6 = y$

i.e., $\quad x = y + 6$

$\therefore \quad f(x) = -11 + 12\,(y + 7) - 4(y + 6)\,(y + 7) + (y + 4)\,(y + 6)\,(y + 7)$

$$= y^3 + (17 - 4)y^2 + (94 - 52 + 12)y + 168 - 168 + 84 - 11$$

$$= (x - 6)^3 + 13(x - 6)^2 + 54(x - 6) + 73$$

Example 5.58: Given that $\log_{10} 654 = 2.8156$, $\log_{10} 658 = 2.8182$, $\log_{10} 659 = 2.8189$, $\log_{10} 661 = 2.8202$ find $\log_{10} 656$.

Solution: Divided difference table for given data is

x	$f(x) = \log_{10} x$	$\Delta f(x)$	$\Delta^2 f(x)$	$\Delta^3 f(x)$
654	2.8156			
		.00065		
658	2.8182		.00001	
		.00070		–.000004
659	2.8189		–.00002	
		.00065		
661	2.8202			

By Newton's divided difference formula

$$\log_{10} 656 = f(656) \simeq 2.8156 + (656 - 654)\,(.00065) + (656 - 654).(656 - 658)\,(.00001)$$
$$+ (656 - 654)\,(656 - 658).(656 - 659)\,(-.000004)$$

$$\therefore \quad \log_{10} 656 \simeq 2.8156 + 2\,(.00065) - 4(.00001) - 12(.000004)$$

$$\therefore \quad \log_{10} 656 \simeq 2.8168$$

Now, we shall discuss interpolation formulae which can be applied only when arguments are equally spaced.

5.8.4 Newton's Forward Difference Interpolation Formula

Let $y_0, y_1, y_2, \ldots, y_n$ be the values of $f(x)$ at arguments $x_0, x_0 + h, \ldots, x_0 + nh$ where h is the interval of differencing and we want to interpolate $f(x)$ at $x = x_0 + ph$

$$f(x_0 + ph) = E^p f(x_0) = (1 + \Delta)^p f(x_0) = (1 + \Delta)^p y_0$$

$$= \left[1 + p\Delta + \frac{p(p-1)}{2!}\Delta^2 + \frac{p(p-1)(p-2)}{3!}\Delta^3 + \ldots + \frac{p(p-1)(p-2)\ldots(p-\overline{n-1})}{n!}\Delta^n + \ldots\right] y_0$$

$$\simeq y_0 + p\Delta y_0 + \frac{p(p-1)}{2!}\Delta^2 y_0 + \ldots + \frac{p(p-1)(p-2)\ldots(p-\overline{n-1})}{n!}\Delta^n y_0$$

which is Newton's forward difference interpolation formula with error obtained below.

If we define

$$\phi(p) = y_0 + p\Delta y_0 + \binom{p}{2}\Delta^2 y_0 + \ldots + \binom{p}{n}\Delta^n y_0$$

then

$$\phi(k) = y_0 + k\Delta y_0 + \binom{k}{2}\Delta^2 y_0 + \ldots + \binom{k}{k}\Delta^k y_0$$
$$= (1+\Delta)^k y_0 = E^k y_0 = y_k \quad \text{for } k = 0,1,2,\ldots,n$$

Thus, $\phi(p)$ is identical with the Lagrangian interpolation polynomial and hence error is given by

$$\frac{f^{(n+1)}(\xi)}{(n+1)!}h^{n+1}p(p-1)\cdots(p-n) = \frac{f^{(n+1)}(\xi)}{(n+1)!}h^{n+1}p^{(n+1)},\ \xi \in (x_0, x_0+nh)$$

Remark 5.14: From the forward difference table (already described), we observe that the quantities $\Delta^k y_0$ lie on a straight line slopping down to the right from y_0. Now, error will be less if more of forward differences are used. Hence, it will be preferable to use forward difference interpolation formula, if we are to interpolate the value of the function near the beginning of set of tabulated value and $0 < p < 1$.

5.8.5 Newton's Backward Difference Interpolation Formula

Let $y_0, y_1, y_2, \ldots, y_n$ be the values of $f(x)$ at arguments $x_0, x_1, x_2, \ldots, x_n$ where $x_k = x_0 + kh;\ k = 1, 2, \ldots, n$ and we want to interpolate $f(x)$ at $x = x_n + ph$

$$f(x_n + ph) = E^p f(x_n) = (1-\nabla)^{-p} f(x_n) = (1-\nabla)^{-p} y_n$$
$$= \left[1 + \binom{p}{1}\nabla + \binom{p+1}{2}\nabla^2 + \binom{p+2}{3}\nabla^3 + \cdots + \binom{p+n-1}{n}\nabla^n + \cdots\right] y_n$$
$$\simeq y_n + \binom{p}{1}\nabla y_n + \binom{p+1}{2}\nabla^2 y_n + \binom{p+2}{3}\nabla^3 y_n + \cdots + \binom{p+n-1}{n}\nabla^n y_n$$

which is Newton's backward difference interpolation formula with error

$$\frac{f^{(n+1)}(\xi)}{(n+1)!}h^{n+1}p(p+1)(p+2)\cdots(p+n) = \frac{f^{(n+1)}(\xi)}{(n+1)!}h^{n+1}(p+n)^{(n+1)},\ \xi \in (x_0,\ x_0+nh)$$

The error is found in the similar way as in Newton's forward interpolation formula.

Remark 5.15: From the backward difference table (already described), we observe that the quantities $\nabla^k y_n$ lie on a straight line sloping up to the right from y_n. For less error, we use Newton's backward interpolation formula if we are to interpolate $f(x)$ near the end of set of tabulated values and $-1 < p < 0$.

Example 5.59: The population of a city in the decennial census was as given below. Estimate the population for the year 1895 and 1925.

years x	1891	1901	1911	1921	1931
Population y (in thousands)	46	66	81	93	101

Solution: Difference table for given data is

x	*y*				
1891	46				
		20			
1901	66		–5		
		15		2	
1911	81		–3		–3
		12		–1	
1921	93		–4		
		18			
1931	101				

For 1895, we shall use Newton's forward difference formula,

$x_0 = 1891, h = 10, x = 1895$

$$p = \frac{x - x_0}{h} = \frac{1895 - 1891}{10} = 0.4$$

$$y(1895) \simeq y(1891) + p\,\Delta y(1891) + \frac{p(p-1)}{2!}\Delta^2 y(1891)$$

$$+ \frac{p(p-1)(p-2)}{3!}\Delta^3 y(1891) + \frac{p(p-1)(p-2)(p-3)}{4!}\Delta^4 y(1891)$$

$$= 46 + (0.4)(20) + \frac{(0.4)(-0.6)}{2}(-5) + \frac{(0.4)(-0.6)(-1.6)}{6}(2) + \frac{(0.4)(-0.6)(-1.6)(-2.6)}{24}(-3)$$

$$= 54.8528 \simeq 55$$

$\therefore \quad y(1895) \simeq 55$ thousands

For $x = 1925$, we shall use Newton's backward difference formula.

$x = 1925, x_n = 1931, h = 10$

$$p = \frac{x - x_n}{h} = \frac{1925 - 1931}{10} = -0.6$$

$$y(1925) \simeq y(1931) + p\nabla y(1931) + \frac{(p+1)p}{2!}\nabla^2 y(1931)$$

$$+ \frac{(p+2)(p+1)p}{3!}\nabla^3 y(1931) + \frac{(p+3)(p+2)(p+1)p}{4!}\nabla^4 y(1931)$$

$$= 101 + (-0.6)(8) + \frac{(0.4)(-0.6)}{2}(-4) + \frac{(1.4)(0.4)(-0.6)}{6}(-1) + \frac{(2.4)(1.4)(0.4)(-0.6)}{24}(-3)$$

$$= 96.8368 \simeq 97$$

$\therefore \quad y(1925) \simeq 97$ thousands.

Example 5.60: For the following data obtain the forward and backward differences interpolation polynomials and estimate $f(x)$ at $x = 0.25$ and $x = 0.35$

x	0.1	0.2	0.3	0.4	0.5
$f(x)$	1.40	1.56	1.76	2.00	2.28

Solution: Difference table for given data is

x	**f(x)**		
0.1	1.40		
		0.16	
0.2	1.56		0.04
		0.20	
0.3	1.76		0.04
		0.24	
0.4	2.00		0.04
		0.28	
0.5	2.28		

Second order differences are constant and hence both forward and backward interpolation polynomials are second degree polynomials.

Here, $h = 0.1, x_0 = 0.1$

For forward difference interpolation polynomial $P(x)$

$$p = \frac{x - x_0}{h} = \frac{x - 0.1}{0.1} = 10x - 1$$

$$\therefore \qquad P(x) = f(0.1) + p\Delta f(0.1) + \frac{p(p-1)}{2!}\Delta^2 f(0.1)$$

$$= 1.40 + (10x-1)(0.16) + \frac{(10x-1)(10x-2)}{2}(0.04)$$

$$= 1.40 + 0.16\left(10x-1\right) + 0.02\left(100x^2 - 30x + 2\right)$$

$$= 2x^2 + x + 1.28$$

$$f(0.25) \simeq P(0.25) = 2(0.25)^2 + 0.25 + 1.28 = 1.655$$

For backward difference interpolation polynomial

$x_n = 0.5, h = 0.1$

$$p = \frac{x - x_n}{h} = \frac{x - 0.5}{0.1} = 10x - 5$$

∴ Backward difference interpolation polynomial $F(x)$ is

$$F(x) = f(0.5) + p\nabla f(0.5) + \frac{(p+1)p}{2!}\nabla^2 f(0.5)$$

$$= 2.28 + (10x-5)(0.28) + \frac{(10x-4)(10x-5)}{2}(0.04)$$

$$= 2.28 + 0.28(10x-5) + 0.02(100x^2 - 90x + 20)$$

$$= 2x^2 + x + 1.28$$

$$f(0.35) \simeq F(0.35) = 2(0.35)^2 + 0.35 + 1.28 = 1.875$$

Here $P(x)$ and $F(x)$ are same. These must be same as $f(x)$ is unique second degree polynomial.

Example 5.61: Given sin 45° = 0.7071, sin 50° = 0.7660, sin 55° = 0.8192 and sin 60° = 0.8660, find sin 52° using Newton's interpolation formula. Estimate the error.
Solution: Forward difference table is

x	y = sin x	Δy	Δ²y	Δ³y
45°	0.7071			
		0.0589		
50°	0.7660		–0.0057	
		0.0532		–0.0007
55°	0.8192		–0.0064	
		0.0468		
60°	0.8660			

We shall use Newton's forward interpolation formula to find approximate value of sin 52°
$x = 52°, x_0 = 50°, h = 5°$

$$\therefore \quad p = \frac{x - x_0}{h} = \frac{52-50}{5} = 0.4$$

$$\therefore \quad \sin 52° \simeq y(50°) + p\Delta y(50°) + \frac{p(p-1)}{2!}\Delta^2 y(50°)$$

$$= 0.7660 + (0.4)(0.0532) + \frac{(0.4)(-0.6)}{2}(-0.0064)$$

$$\therefore \quad \sin 52° \simeq 0.7880$$

Exact value of sin 52° to 4 decimal places = 0.7880. Thus, upto 4 decimal places, there is no error.

Example 5.62: Find the number of men getting wages between ₹10 and 15 from the following data:

Wages in ₹	0–10	10–20	20–30	30–40
Frequency	9	30	35	42

Solution: First of all we find cumulative frequency

Wages in ₹	Frequency f	Cumulative frequency
0–10	9	9
10–20	30	39
20–30	35	74
30–40	42	116

Let y denote number of men getting wages below ₹x. Then difference table is

x	y	Δy	$\Delta^2 y$	$\Delta^3 y$
10	9			
		30		
20	39		5	
		35		2
30	74		7	
		42		
40	116			

We estimate $y(15)$ using Newton's forward interpolation formula

$$x = 15, x_0 = 10, h = 10$$

$$\therefore \quad p = \frac{x - x_0}{h} = \frac{15-10}{10} = 0.5$$

$$y(15) \simeq y(10) + p\Delta y(10) + \frac{p(p-1)}{2!}\Delta^2 y(10) + \frac{p(p-1)(p-2)}{3!}\Delta^3 y(10)$$

$$= 9 + (0.5)(30) + \frac{(0.5)(-0.5)}{2}(5) + \frac{(0.5)(-0.5)(-1.5)(2)}{6}$$

$$\therefore \quad y(15) \simeq 24$$

$\therefore$ Number of men getting wages between ₹10 and 15

$$= y(15) - y(10) = 24 - 9 = 15$$

5.8.6 Gauss Forward Interpolation Formula

This formula will be derived in terms of following central differences in the central difference table.

$$\begin{array}{ccccccc} y_0 & & \delta^2 y_0 & & \delta^4 y_0 & & \delta^6 y_0 \\ & \delta y_{1/2} & & \delta^3 y_{1/2} & & \delta^5 y_{1/2} & \end{array}$$

Let the values of function $f(x)$ are given at $x_0-3h, x_0-2h, x_0-h, x_0, x_0+h, x_0+2h, x_0+3h$ and values are $y_{-3}, y_{-2}, y_{-1}, y_0, y_1, y_2, y_3$, respectively. We want to interpolate $f(x)$ at $x = x_0 + ph$; $-1 < p < 1$. Let value is y_p.

By Newton's forward interpolation formula

$$y_p = f(x) = f(x_0 + ph) = y_0 + \binom{p}{1}\Delta y_0 + \binom{p}{2}\Delta^2 y_0 + \binom{p}{3}\Delta^3 y_0 + \binom{p}{4}\Delta^4 y_0 + \binom{p}{5}\Delta^5 y_0 + \binom{p}{6}\Delta^6 y_0 + \cdots$$

$$= y_0 + \binom{p}{1}\Delta y_0 + \binom{p}{2}\Delta^2 E\, y_{-1} + \binom{p}{3}\Delta^3 E\, y_{-1} + \binom{p}{4}\Delta^4 E\, y_{-1} + \binom{p}{5}\Delta^5 E\, y_{-1} + \binom{p}{6}\Delta^6 E\, y_{-1} + \cdots$$

$$= y_0 + \binom{p}{1}\Delta y_0 + \binom{p}{2}\Delta^2 (1+\Delta) y_{-1} + \binom{p}{3}\Delta^3 (1+\Delta) y_{-1} + \binom{p}{4}\Delta^4 (1+\Delta) y_{-1}$$
$$+ \binom{p}{5}\Delta^5 (1+\Delta) y_{-1} + \binom{p}{6}\Delta^6 (1+\Delta) y_{-1} + \cdots$$

$$= y_0 + \binom{p}{1}\Delta y_0 + \binom{p}{2}\Delta^2 y_{-1} + \left[\binom{p}{2} + \binom{p}{3}\right]\Delta^3 y_{-1} + \left[\binom{p}{3} + \binom{p}{4}\right]\Delta^4 y_{-1}$$
$$+ \left[\binom{p}{4} + \binom{p}{5}\right]\Delta^5 y_{-1} + \left[\binom{p}{5} + \binom{p}{6}\right]\Delta^6 y_{-1} + \cdots$$

$$= y_0 + \binom{p}{1}\Delta y_0 + \binom{p}{2}\Delta^2 y_{-1} + \binom{p+1}{3}\Delta^3 y_{-1} + \binom{p+1}{4}\Delta^4 (1+\Delta) y_{-2}$$
$$+ \binom{p+1}{5}\Delta^5 (1+\Delta) y_{-2} + \binom{p+1}{6}\Delta^6 (1+\Delta) y_{-2} + \cdots$$

$$= y_0 + \binom{p}{1}\Delta y_0 + \binom{p}{2}\Delta^2 y_{-1} + \binom{p+1}{3}\Delta^3 y_{-1} + \binom{p+1}{4}\Delta^4 y_{-2}$$
$$+ \left[\binom{p+1}{4} + \binom{p+1}{5}\right]\Delta^5 y_{-2} + \left[\binom{p+1}{5} + \binom{p+1}{6}\right]\Delta^6 y_{-2} + \ldots$$

$$= y_0 + \binom{p}{1}\Delta y_0 + \binom{p}{2}\Delta^2 y_{-1} + \binom{p+1}{3}\Delta^3 y_{-1} + \binom{p+1}{4}\Delta^4 y_{-2}$$
$$+ \binom{p+2}{5}\Delta^5 y_{-2} + \binom{p+2}{6}\Delta^6 (1+\Delta) y_{-3} + \binom{p+2}{7}\Delta^7 (1+\Delta) y_{-3} + \ldots$$

$$= y_0 + \binom{p}{1}\Delta y_0 + \binom{p}{2}\Delta^2 y_{-1} + \binom{p+1}{3}\Delta^3 y_{-1} + \binom{p+1}{4}\Delta^4 y_{-2}$$
$$+ \binom{p+2}{5}\Delta^5 y_{-2} + \binom{p+2}{6}\Delta^6 y_{-3} + \cdots$$

But $\Delta^k y_{-\frac{k-1}{2}} = (\delta E^{1/2})^k y_{-\frac{k-1}{2}} = \delta^k y_{\frac{1}{2}}$; $k = 1, 3, 5, \ldots\ldots$

and $\Delta^k y_{-\frac{k}{2}} = (\delta E^{1/2})^k y_{-\frac{k}{2}} = \delta^k y_0$; $k = 2, 4, 6, \ldots$

$$\therefore \quad y_p = y_0 + \binom{p}{1}\delta y_{\frac{1}{2}} + \binom{p}{2}\delta^2 y_0 + \binom{p+1}{3}\delta^3 y_{\frac{1}{2}} + \binom{p+1}{4}\delta^4 y_0 + \binom{p+2}{5}\delta^5 y_{\frac{1}{2}} + \binom{p+2}{6}\delta^6 y_0 + \cdots$$

This is Gauss forward interpolation formula.

5.8.7 Gauss Backward Interpolation Formula

This formula will be derived in terms of following central differences in the central difference table.

$$
\begin{array}{ccccccc}
 & \delta y_{-\frac{1}{2}} & & \delta^3 y_{-\frac{1}{2}} & & \delta^5 y_{-\frac{1}{2}} & \\
y_0 & & \delta^2 y_0 & & \delta^4 y_0 & & \delta^6 y_0
\end{array}
$$

Let the values of the function $f(x)$ are given at $x_0 - 3h, x_0 - 2h, x_0 - h, x_0, x_0 + h, x_0 + 2h, x_0 + 3h$ and values are $y_{-3}, y_{-2}, y_{-1}, y_0, y_1, y_2, y_3$, respectively. We want to interpolate $f(x)$ at $x = x_0 + ph$; $-1 < p < 1$. Let value is y_p.

$\therefore$ By Newton's backward difference interpolation formula

$$
\begin{aligned}
y_p &= f(x_0 + ph) \\
&= y_0 + \binom{p}{1}\nabla y_0 + \binom{p+1}{2}\nabla^2 y_0 + \binom{p+2}{3}\nabla^3 y_0 \\
&\quad + \binom{p+3}{4}\nabla^4 y_0 + \binom{p+4}{5}\nabla^5 y_0 + \binom{p+5}{6}\nabla^6 y_0 + \ldots \\
&= y_0 + \binom{p}{1}\nabla y_0 + \binom{p+1}{2}\nabla^2 E^{-1} y_1 + \binom{p+2}{3}\nabla^3 E^{-1} y_1 \\
&\quad + \binom{p+3}{4}\nabla^4 E^{-1} y_1 + \binom{p+4}{5}\nabla^5 E^{-1} y_1 + \binom{p+5}{6}\nabla^6 E^{-1} y_1 + \cdots \\
&= y_0 + \binom{p}{1}\nabla y_0 + \binom{p+1}{2}\nabla^2 (1-\nabla) y_1 + \binom{p+2}{3}\nabla^3 (1-\nabla) y_1 + \binom{p+3}{4}\nabla^4 (1-\nabla) y_1 \\
&\quad + \binom{p+4}{5}\nabla^5 (1-\nabla) y_1 + \binom{p+5}{6}\nabla^6 (1-\nabla) y_1 + \cdots \\
&= y_0 + \binom{p}{1}\nabla y_0 + \binom{p+1}{2}\nabla^2 y_1 + \left[\binom{p+2}{3} - \binom{p+1}{2}\right]\nabla^3 y_1 + \left[\binom{p+3}{4} - \binom{p+2}{3}\right]\nabla^4 y_1 \\
&\quad + \left[\binom{p+4}{5} - \binom{p+3}{4}\right]\nabla^5 y_1 + \left[\binom{p+5}{6} - \binom{p+4}{5}\right]\nabla^6 y_1 + \cdots \\
&= y_0 + \binom{p}{1}\nabla y_0 + \binom{p+1}{2}\nabla^2 y_1 + \binom{p+1}{3}\nabla^3 y_1 + \binom{p+2}{4}\nabla^4 (1-\nabla) y_2 \\
&\quad + \binom{p+3}{5}\nabla^5 (1-\nabla) y_2 + \binom{p+4}{6}\nabla^6 (1-\nabla) y_2 + \cdots \\
&= y_0 + \binom{p}{1}\nabla y_0 + \binom{p+1}{2}\nabla^2 y_1 + \binom{p+1}{3}\nabla^3 y_1 + \binom{p+2}{4}\nabla^4 y_2 \\
&\quad + \left[\binom{p+3}{5} - \binom{p+2}{4}\right]\nabla^5 y_2 + \left[\binom{p+4}{6} - \binom{p+3}{5}\right]\nabla^6 y_2 + \cdots \\
&= y_0 + \binom{p}{1}\nabla y_0 + \binom{p+1}{2}\nabla^2 y_1 + \binom{p+1}{3}\nabla^3 y_1 + \binom{p+2}{4}\nabla^4 y_2 + \binom{p+2}{5}\nabla^5 y_2 + \binom{p+3}{6}\nabla^6 (1-\nabla) y_3 + \cdots \\
&= y_0 + \binom{p}{1}\nabla y_0 + \binom{p+1}{2}\nabla^2 y_1 + \binom{p+1}{3}\nabla^3 y_1 + \binom{p+2}{4}\nabla^4 y_2 + \binom{p+2}{5}\nabla^5 y_2 + \binom{p+3}{6}\nabla^6 y_3 + \cdots
\end{aligned}
$$

But $\quad \nabla^k y_{\frac{k-1}{2}} = (\delta E^{-\frac{1}{2}})^k y_{\frac{k-1}{2}} = \delta^k y_{-\frac{1}{2}};\ k = 1, 3, 5 \cdots$

and $\quad \nabla^k y_{\frac{k}{2}} = (\delta E^{-\frac{1}{2}})^k y_{\frac{k}{2}} = \delta^k y_0;\ k = 2, 4, 6 \ldots$

$$\therefore \quad y_p = y_0 + \binom{p}{1}\delta y_{-\frac{1}{2}} + \binom{p+1}{2}\delta^2 y_0 + \binom{p+1}{3}\delta^3 y_{-\frac{1}{2}} + \binom{p+2}{4}\delta^4 y_0 + \binom{p+2}{5}\delta^5 y_{-\frac{1}{2}} + \binom{p+3}{6}\delta^6 y_0 + \ldots$$

This is Gauss backward difference formula.

5.8.8 Stirling's Formula

This formula will be derived in terms of following central differences in the central difference table

$$\begin{array}{ccccccc} & \delta y_{-\frac{1}{2}} & & \delta^3 y_{-\frac{1}{2}} & & \delta^5 y_{-\frac{1}{2}} & \\ y_0 & & \delta^2 y_0 & & \delta^4 y_0 & & \delta^6 y_0 \\ & \delta y_{\frac{1}{2}} & & \delta^3 y_{\frac{1}{2}} & & \delta^5 y_{\frac{1}{2}} & \end{array}$$

Stirling formula uses the entries in the table of both Gauss forward and Gauss backward formula. This suggests us that we may take the mean of these values.

From Gauss forward formula

$$y_p = y_0 + \binom{p}{1}\delta y_{\frac{1}{2}} + \binom{p}{2}\delta^2 y_0 + \binom{p+1}{3}\delta^3 y_{\frac{1}{2}} + \binom{p+1}{4}\delta^4 y_0 + \binom{p+2}{5}\delta^5 y_{\frac{1}{2}} + \binom{p+2}{6}\delta^6 y_0 + \cdots$$

From Gauss backward formula

$$y_p = y_0 + \binom{p}{1}\delta y_{-\frac{1}{2}} + \binom{p+1}{2}\delta^2 y_0 + \binom{p+1}{3}\delta^3 y_{-\frac{1}{2}} + \binom{p+2}{4}\delta^4 y_0 + \binom{p+2}{5}\delta^5 y_{-\frac{1}{2}} + \binom{p+3}{6}\delta^6 y_0 + \cdots$$

Add these and divide by 2

$$y_p = y_0 + \binom{p}{1}\delta \frac{E^{\frac{1}{2}} + E^{-\frac{1}{2}}}{2} y_0 + \frac{1}{2}(p-1+p+1)\frac{p}{2}\delta^2 y_0 + \binom{p+1}{3}\delta^3 \frac{E^{\frac{1}{2}} + E^{-\frac{1}{2}}}{2} y_0$$

$$+\frac{1}{2}\cdot\frac{1}{4}(p-2+p+2)\binom{p+1}{3}\delta^4 y_0 + \binom{p+2}{5}\delta^5 \frac{E^{\frac{1}{2}} + E^{-\frac{1}{2}}}{2} y_0 + \frac{1}{2}\cdot\frac{1}{6}(p-3+p+3)\binom{p+2}{5}\delta^6 y_0 + \cdots$$

$$= y_0 + \binom{p}{1}\mu\delta y_0 + \frac{p^2}{2!}\delta^2 y_0 + \binom{p+1}{3}\mu\delta^3 y_0 + \frac{p^2(p^2-1)}{4!}\delta^4 y_0 + \binom{p+2}{5}\mu\delta^5 y_0$$

$$+\frac{p^2\left(p^2-1\right)\left(p^2-4\right)}{6!}\delta^6 y_0 + \cdots$$

which is Stirling's interpolation formula.

5.8.9 Bessel's Interpolation Formula

This formula will be derived in terms of following central differences in the central difference formula

$$\begin{array}{ccccccc} y_0 & & \delta^2 y_0 & & \delta^4 y_0 & & \delta^6 y_0 \\ & \delta y_{1/2} & & \delta^3 y_1 & & \delta^5 y_{1/2} & \\ y_1 & & \delta^2 y_1 & & \delta^4 y_1 & & \delta^6 y_1 \end{array}$$

If $\quad x = x_0 + ph$

then $\quad f(x) = f(x_0 + ph) = E^p f(x_0) = y_p$

Also, $f(x_0 + ph) = f(x_0 + h + (p-1)h) = E^{p-1} f(x_0 + h)$

Thus, Bessel interpolation will be mean of Gauss forward interpolation formula and Gauss backward interpolation in which y_0 is replaced by y_1 and p is replaced by $p - 1$.

By Gauss forward interpolation formula

$$y_p = y_0 + \binom{p}{1}\delta y_{\frac{1}{2}} + \binom{p}{2}\delta^2 y_0 + \binom{p+1}{3}\delta^3 y_{\frac{1}{2}} + \binom{p+1}{4}\delta^4 y_0 + \binom{p+2}{5}\delta^5 y_{\frac{1}{2}} + \binom{p+2}{6}\delta^6 y_0 + \cdots$$

By Gauss backward interpolation formula replacing p by $p - 1$ and y_0 by y_1

$$y_p = y_1 + \binom{p-1}{1}\delta y_{\frac{1}{2}} + \binom{p}{2}\delta^2 y_1 + \binom{p}{3}\delta^3 y_{\frac{1}{2}} + \binom{p+1}{4}\delta^4 y_1 + \binom{p+1}{5}\delta^5 y_{\frac{1}{2}} + \binom{p+2}{6}\delta^6 y_1 + \cdots$$

Add these and divide by 2

$$y_p = \frac{1}{2}(y_1 + y_0) + \left(p - \frac{1}{2}\right)\delta y_{\frac{1}{2}} + \binom{p}{2}\delta^2 \frac{y_1 + y_0}{2} + \frac{1}{2}\left(\frac{p+1+p-2}{3}\right)\binom{p}{2}\delta^3 y_{\frac{1}{2}}$$

$$+ \binom{p+1}{4}\delta^4 \frac{y_1 + y_0}{2} + \frac{1}{2}\left(\frac{p+2+p-3}{5}\right)\binom{p+1}{4}\delta^5 y_{\frac{1}{2}} + \binom{p+2}{6}\delta^6 \frac{y_1 + y_0}{2} + \cdots$$

$$= \mu y_{\frac{1}{2}} + \left(p - \frac{1}{2}\right)\delta y_{\frac{1}{2}} + \binom{p}{2}\mu\delta^2 y_{\frac{1}{2}} + \binom{p}{2}\frac{p - \frac{1}{2}}{3}\delta^3 y_{\frac{1}{2}}$$

$$+ \binom{p+1}{4}\mu\delta^4 y_{\frac{1}{2}} + \binom{p+1}{4}\frac{p - \frac{1}{2}}{5}\delta^5 y_{\frac{1}{2}} + \binom{p+2}{6}\mu\delta^6 y_{\frac{1}{2}} + \cdots$$

which is Bessel's formula.

5.8.10 Laplace–Everett's Interpolation Formula

This formula will be derived in terms of following central differences in the central difference table.

$$\begin{array}{ccccccc} y_0 & \cdots\cdots & \delta^2 y_0 & \cdots\cdots & \delta^4 y_0 & \cdots\cdots & \delta^6 y_0 \\ \cdot & & \cdot & & \cdot & & \\ y_1 & \cdots\cdots & \delta^2 y_1 & \cdots\cdots & \delta^4 y_1 & \cdots\cdots & \delta^6 y_1 \end{array}$$

Here, only even order central difference of y_0 and y_1 are required. Thus, we are to eliminate odd-order central differences in the Bessel formula.

Bessel interpolation formula is

$$y_p = \mu y_{\frac{1}{2}} + \left(p - \frac{1}{2}\right)\delta y_{\frac{1}{2}} + \binom{p}{2}\mu\delta^2 y_{\frac{1}{2}} + \binom{p}{2}\frac{p - \frac{1}{2}}{3}\delta^3 y_{\frac{1}{2}} + \binom{p+1}{4}\mu\delta^4 y_{\frac{1}{2}}$$

$$+\binom{p+1}{4}\frac{p-\frac{1}{2}}{5}\delta^5 y_{\frac{1}{2}}+\binom{p+2}{6}\mu\delta^6 y_{\frac{1}{2}}+\binom{p+2}{6}\frac{p-\frac{1}{2}}{7}\delta^7 y_{\frac{1}{2}}+\ldots$$

$$=\frac{1}{2}(y_1+y_0)+\left(p-\frac{1}{2}\right)(y_1-y_0)+\binom{p}{2}\frac{1}{2}\left(\delta^2 y_1+\delta^2 y_0\right)$$

$$+\binom{p}{2}\frac{p-\frac{1}{2}}{3}\left(\delta^2 y_1-\delta^2 y_0\right)+\binom{p+1}{4}\frac{1}{2}\left(\delta^4 y_1+\delta^4 y_0\right)$$

$$+\binom{p+1}{4}\frac{p-\frac{1}{2}}{5}\left(\delta^4 y_1-\delta^4 y_0\right)+\binom{p+2}{6}\frac{1}{2}\left(\delta^6 y_1+\delta^6 y_0\right)+\binom{p+2}{6}\frac{p-\frac{1}{2}}{7}\left(\delta^6 y_1-\delta^6 y_0\right)+\cdots$$

$$=(1-p)y_0+\binom{p}{2}\left(\frac{1}{2}-\frac{p-\frac{1}{2}}{3}\right)\delta^2 y_0+\binom{p+1}{4}\left(\frac{1}{2}-\frac{p-\frac{1}{2}}{5}\right)\delta^4 y_0+\binom{p+2}{6}\left(\frac{1}{2}-\frac{p-\frac{1}{2}}{7}\right)\delta^6 y_0+\cdots$$

$$+py_1+\binom{p}{2}\left(\frac{1}{2}+\frac{p-\frac{1}{2}}{3}\right)\delta^2 y_1+\binom{p+1}{4}\left(\frac{1}{2}+\frac{p-\frac{1}{2}}{5}\right)\delta^4 y_1+\binom{p+2}{6}\left(\frac{1}{2}+\frac{p-\frac{1}{2}}{7}\right)\delta^6 y_1+\cdots$$

$$=(1-p)y_0+\binom{p}{2}\frac{2-p}{3}\delta^2 y_0+\binom{p+1}{4}\frac{3-p}{5}\delta^4 y_0+\binom{p+2}{6}\frac{4-p}{7}\delta^6 y_0+\cdots$$

$$+py_1+\binom{p}{2}\frac{p+1}{3}\delta^2 y_1+\binom{p+1}{4}\frac{p+2}{5}\delta^4 y_1+\binom{p+2}{6}\frac{p+3}{7}\delta^6 y_1+\cdots$$

$$=qy_0+\binom{1-q}{2}\frac{q+1}{3}\delta^2 y_0+\binom{2-q}{4}\frac{q+2}{5}\delta^4 y_0+\binom{3-q}{6}\frac{q+3}{7}\delta^6 y_0+\ldots.$$

$$+py_1+\binom{p+1}{3}\delta^2 y_1+\binom{p+2}{5}\delta^4 y_1+\binom{p+3}{7}\delta^6 y_1+\cdots$$

$$=qy_0+\binom{q}{2}\frac{q+1}{3}\delta^2 y_0+\binom{q+1}{4}\frac{q+2}{5}\delta^4 y_0+\binom{q+2}{6}\frac{q+3}{7}\delta^6 y_0+\cdots$$

$$+py_1+\binom{p+1}{3}\delta^2 y_1+\binom{p+2}{5}\delta^4 y_1+\binom{p+3}{7}\delta^6 y_1+\ldots$$

where $q=1-p$

$$\therefore\quad y_p=qy_0+\binom{q+1}{3}\delta^2 y_0+\binom{q+2}{5}\delta^4 y_0+\binom{q+3}{7}\delta^6 y_0+\cdots$$

$$+py_1+\binom{p+1}{3}\delta^2 y_1+\binom{p+2}{5}\delta^4 y_1+\binom{p+3}{7}\delta^6 y_1+\cdots$$

where $q=1-p$

which is Laplace–Everett's formula.

Comparison of Central Difference Formulae

Gaussian interpolation formulae are of theoretical interest. Stirling's formula is suitable for small values of p, for example $-\frac{1}{4} \le p \le \frac{1}{4}$. Bessel's formula is suitable for values of p not too far from $\frac{1}{2}$, for example $\frac{1}{4} \le p \le \frac{3}{4}$. Everett's formula is most generally useful and in this formula even order differences are together placed in horizontal lines.

Example 5.63: Use Gauss forward interpolation formula to find the value of log 337.5 from the following table:

x	310	320	330	340	350	360
$y_x = \log x$	2.4914	2.5051	2.5185	2.5315	2.5441	2.5563

Solution: Central difference table for given data is

x	y	δy	$\delta^2 y$	$\delta^3 y$	$\delta^4 y$	$\delta^5 y$
310	2.4914					
		0.0137				
320	2.5051		−0.0003			
		0.0134		−0.0001		
330	2.5185		−0.0004		0.0001	
		↘ 0.0130 ↗		↘ 0 ↗		↘ −0.0001
340	2.5315		−0.0004		0	
		0.0126		0		
350	2.5441		−0.0004			
		0.0122				
360	2.5563					

$$x = 337.5,\ x_0 = 330,\ h = 10$$

$$\therefore \quad p = \frac{x - x_0}{h} = \frac{337.5 - 330}{10} = 0.75$$

By Gauss forward interpolation formula

$$\log 337.5 = y_p = y_0 + p\delta y_{\frac{1}{2}} + \frac{p(p-1)}{2!}\delta^2 y_0 + \frac{(p+1)p(p-1)}{3!}\delta^3 y_{\frac{1}{2}}$$

$$+ \frac{(p+1)p(p-1)(p-2)}{4!}\delta^4 y_0 + \frac{(p+2)(p+1)p(p-1)(p-2)}{5!}\delta^5 y_{\frac{1}{2}}$$

$$= 2.5185 + (0.75)(0.0130) + \frac{(0.75)(-0.25)}{2}(-0.0004) + \frac{(1.75)(0.75)(-0.25)}{6}(0)$$

$$+\frac{(1.75)(0.75)(-0.25)(-1.25)}{24}(0.0001)$$

$$+\frac{(2.75)(1.75)(0.75)(-0.25)(-1.25)}{120}(-0.0001)$$

$\therefore$ log 337.5 $\simeq$ 2.5283

Example 5.64: Apply Gauss's backward interpolation formula to find the sales for the year 1974 from the table

(year) x	1939	1949	1959	1969	1979	1989
Sales (y) (in lakhs)	12	15	20	27	39	52

Solution: Central difference table for given data is

x	y	δy	$\delta^2 y$	$\delta^3 y$	$\delta^4 y$	$\delta^5 y$
1939	12					
		3				
1949	15		2			
		5		0		
1959	20		2		3	
		7		3		−10
1969	27		5		−7	
		12		−4		
1979	39		1			
		13				
1989	52					

$x = 1974, \; x_0 = 1969, \; h = 10$

$$\therefore \quad p = \frac{x - x_0}{h} = \frac{1974 - 1969}{10} = 0.5$$

By Gauss backward interpolation formula

$$y_p \simeq y_0 + p\delta y_{-\frac{1}{2}} + \frac{(p+1)p}{2!}\delta^2 y_0 + \frac{(p+1)p(p-1)}{3!}\delta^3 y_{-\frac{1}{2}} + \frac{(p+2)(p+1)p(p-1)}{4!}\delta^4 y_0$$

$$+\frac{(p+2)(p+1)p(p-1)(p-2)}{5!}\delta^5 y_{-\frac{1}{2}}$$

$$\therefore \quad y(1974) \simeq 27 + (0.5)(7) + \frac{(1.5)(0.5)}{2}(5) + \frac{(1.5)(0.5)(-0.5)}{6}(3)$$
$$+ \frac{(2.5)(1.5)(0.5)(-0.5)}{24}(-7) + \frac{(2.5)(1.5)(0.5)(-0.5)(-1.5)}{120}(-10)$$

$$\therefore \quad y(1974) \simeq 32 \text{ lakhs}$$

Example 5.65: Compute $u_{12.2}$ from the following table using Stirling's formula

x°	10	11	12	13	14
$10^5 u_x$	23967	28060	31788	35209	38368

where $\quad u_x = 1 + \log_{10} \sin x$

Solution: Central difference table is

x (in degrees)	$10^5 u_x$				
10	23967				
		4093			
11	28060		−365		
		3728		58	
12	31788		−307		−13
		3421		45	
13	35209		−262		
		3159			
14	38368				

$$x = 12.2, \; x_0 = 12, \; h = 1$$

$$\therefore \quad p = \frac{x - x_0}{h} = \frac{12.2 - 12}{1} = 0.2$$

By Stirling's formula

$$10^5 u_p = y_0 + p\mu\delta y_0 + \frac{p^2}{2!}\delta^2 y_0 + \frac{(p+1)p(p-1)}{3!}\mu\delta^2 y_0 + \frac{(p+1)p^2(p-1)}{4!}\delta^4 y_0 + \cdots$$

$$\therefore \quad 10^5 u_{12.2} = 31788 + (0.2)\left(\frac{3728+3421}{2}\right) + \frac{(0.2)^2}{2}(-307) + \frac{(1.2)(0.2)(-0.8)}{6}\left(\frac{58+45}{2}\right)$$
$$+ \frac{(1.2)(0.2)^2(-0.8)}{24}(-13)$$

$$= 31788 + (0.1)(7149) - \frac{(0.2)^2}{2}(307) - \frac{(1.2)(0.2)(0.8)}{12}(103)$$

$$+ \frac{(1.2)(0.2)^2(0.8)}{24}(13) \simeq 32495$$

$\therefore \quad u_{12.2} \simeq 0.32495$

Example 5.66: Apply Bessel's formula to obtain y_{25} given that $y_{20} = 2854, y_{24} = 3162, y_{28} = 3544,$ $y_{32} = 3992.$

Solution: Central difference table for given data is

x	y	δy	$\delta^2 y$	$\delta^3 y$
20	2854			
		308		
24	3162		74	
		382		−8
28	3544		66	
		448		
32	3992			

To obtain y_{25}

$$x = 25,\ x_0 = 24,\ h = 4$$

$$\therefore \quad p = \frac{x - x_0}{h} = \frac{25 - 24}{4} = 0.25$$

By Bessel's interpolation formula

$$y_p = \mu y_{\frac{1}{2}} + \left(p - \frac{1}{2}\right)\delta y_{\frac{1}{2}} + \frac{p(p-1)}{2!}\mu\delta^2 y_{\frac{1}{2}} + \frac{p(p-1)}{2!}\cdot\frac{p-\frac{1}{2}}{3}\delta^3 y_{\frac{1}{2}} + \cdots$$

$$\therefore \quad y_{25} \simeq \frac{3544 + 3162}{2} + (-0.25)(382) + \frac{(0.25)(-0.75)}{2}\cdot\left(\frac{66 + 74}{2}\right)$$

$$+ \frac{(0.25)(-0.75)}{2}\cdot\frac{(-0.25)}{3}(-8)$$

$$= \frac{6706}{2} - (0.25)(382) - \frac{(0.25)(0.75)(140)}{4} - \frac{(0.25)(0.75)(0.25)(8)}{6}$$

$\therefore \quad y_{25} \simeq 3251$

Example 5.67: Given the table

x	310	320	330	340	350	360
$y = \log x$	2.49136	2.50515	2.51851	2.53148	2.54407	2.55630

find the value of log 337.5 by Everett's formula.
Solution: Central difference table for given data is

x	y	δy	$\delta^2 y$	$\delta^3 y$	$\delta^4 y$	$\delta^5 y$
310	2.49136					
		0.01379				
320	2.50515		–0.00043			
		0.01336		0.00004		
330	2.51851		–0.00039		–0.00003	
		0.01297		0.00001		0.00004
340	2.53148		–0.00038		0.00001	
		0.01259		0.00002		
350	2.54407		–0.00036			
		0.01223				
360	2.55630					

To obtain $y = \log 337.5$

$$x = 337.5,\ x_0 = 330,\ \ x_1 = 340,\ h = 10$$

$$p = \frac{x - x_0}{h} = \frac{337.5 - 330}{10} = 0.75, \quad q = 1 - p = 0.25$$

$$y_p = qy_0 + \frac{(q+1)q(q-1)}{3!}\delta^2 y_0 + \frac{(q+2)(q+1)q(q-1)(q-2)}{5!}\delta^4 y_0 + \cdots$$

$$+ py_1 + \frac{(p+1)p(p-1)}{3!}\delta^2 y_1 + \frac{(p+2)(p+1)p(p-1)(p-2)}{5!}\delta^4 y_1 + \cdots$$

$$\therefore \quad y(337.5) = \log 337.5 \simeq (0.25)(2.51851) + \frac{(1.25)(0.25)(-0.75)}{6}(-.00039)$$

$$+ \frac{(2.25)(1.25)(0.25)(-0.75)(-1.75)}{120}(-0.00003)$$

$$+ (0.75)(2.53148) + \frac{(1.75)(0.75)(-0.25)}{6}(-0.00038)$$

$$+ \frac{(2.75)(1.75)(0.75)(-0.25)(-1.25)}{120}(0.00001)$$

$$\therefore \quad \log 337.5 \simeq 2.52827$$

Example 5.68: The following table gives the value of e^x for certain equidistant values of x. Find the value of e^x when $x = 0.644$ using

(i) Stirling's formula (ii) Bessel's formula (iii) Everett's formula

x	0.61	0.62	0.63	0.64	0.65	0.66	0.67
$y = e^x$	1.840431	1.858928	1.877610	1.896481	1.915541	1.934792	1.954237

Solution: Central difference table for given data is

x	y	δy	$\delta^2 y$	$\delta^3 y$	$\delta^4 y$	$\delta^5 y$	$\delta^6 y$
0.61	1.840431						
		0.018497					
0.62	1.858928		0.000185				
		0.018682		0.000004			
0.63	1.877610		0.000189		–0.000004		
		0.018871		0		0.000006	
0.64	1.896481		0.000189		0.000002		–0.000007
		0.019060		0.000002		–0.000001	
0.65	1.915541		0.000191		0.000001		
		0.019251		0.000003			
0.66	1.934792		0.000194				
		0.019445					
0.67	1.954237						

To obtain $y(0.644)$

$$x = 0.644,\ x_0 = 0.64,\ x_1 = 0.65,\ h = 0.01$$

$$p = \frac{x - x_0}{h} = \frac{0.644 - 0.64}{0.01} = 0.4, \quad q = 1 - p = 0.6$$

(i) By Stirling's formula

$$y_p = y_0 + p\mu\delta y_0 + \frac{p^2}{2!}\delta^2 y_0 + \frac{(p+1)p(p-1)}{3!}\mu\delta^2 y_0 + \frac{(p+1)p^2(p-1)}{4!}\delta^4 y_0 + \cdots$$

$$\therefore \quad y(0.644) \simeq 1.896481 + (0.4)\left(\frac{0.019060 + 0.018671}{2}\right) + \frac{(0.4)^2}{2}(0.000189)$$

$$\therefore \quad y(0.644) \simeq 1.896481 + \frac{(0.4)(0.037931)}{2} + \frac{(0.4)^2}{2}(0.000189)$$

$$\therefore \quad y(0.644) \simeq 1.904082$$

(ii) By Bessel's formula

$$y_p = \mu y_{\frac{1}{2}} + \left(p - \frac{1}{2}\right)\delta y_{\frac{1}{2}} + \frac{p(p-1)}{2!}\mu\delta^2 y_{\frac{1}{2}} + \frac{p(p-1)}{2!}\cdot\frac{p-\frac{1}{2}}{3}\delta^3 y_{\frac{1}{2}} + \cdots$$

$$\therefore\ y(0.644) \simeq \frac{1.915541 + 1.896481}{2} + (-0.1)(0.019060) + \frac{(0.4)(-0.6)}{2}\cdot\frac{(0.000191 + 0.000189)}{2}$$

$$+ \frac{(0.4)(-0.6)}{2}\cdot\frac{(-0.1)}{3}(0.000002)$$

$$= \frac{3.812022}{2} - (0.1)(0.019060) - \frac{(0.4)(0.6)}{2}(0.000190) + \frac{(0.4)(0.6)(0.1)}{6}(0.000002)$$

$$\therefore\ y(0.644) \simeq 1.904082$$

(iii) By Everett's formula

$$y_p = qy_0 + \frac{(q+1)q(q-1)}{3!}\delta^2 y_0 + \frac{(q+2)(q+1)q(q-1)(q-2)}{5!}\delta^4 y_0 + \cdots$$

$$+ py_1 + \frac{(p+1)p(p-1)}{3!}\delta^2 y_1 + \frac{(p+2)(p+1)p(p-1)(p-2)}{5!}\delta^4 y_1 + \cdots$$

$$\therefore\quad y(0.644) \simeq (0.6)(1.896481) + \frac{(1.6)(0.6)(-0.4)}{6}(.000189)$$

$$+ (0.4)(1.915541) + \frac{(1.4)(0.4)(-0.6)}{6}(0.000191)$$

$$\therefore\quad y(0.644) \simeq 1.904082$$

5.9 INVERSE INTERPOLATION

Inverse interpolation is the process of finding the value of x for some given value of $f(x)$ when the given value of $f(x)$ is between two tabulated values of $f(x)$.

5.9.1 Lagrange's Method for Inverse Interpolation

Lagrange's polynomial is merely a relation between two variables where we take x as independent variable and $y = f(x)$ as dependent variable.

We can write the Lagrange inverse interpolation formula by taking $y = f(x)$ as independent variable and x as dependent variable.

Suppose, we are given that $y = f(x)$ has values $y_0, y_1, y_2, \ldots, y_n$ at values of arguments $x_0, x_1, x_2, \ldots, x_n$ and we are to find x for given value of y. We have

$$x = \sum_{k=0}^{n} l_k(y)\, x_k$$

where

$$l_k(y) = \frac{(y-y_0)(y-y_1)\cdots(y-y_{k-1})(y-y_{k+1})\cdots(y-y_n)}{(y_k-y_0)(y_k-y_1)\cdots(y_k-y_{k-1})(y_k-y_{k+1})\cdots(y_k-y_n)}$$

Iterative methods

Lagrange's method is used when arguments are not equispaced. When the arguments are equispaced, then any of the interpolation formula (not Lagrange's or divided difference) can be used and by iteration process we can find value of x for given y. We shall explain it by using forward difference interpolation formula and using Everett's formula.

5.9.2 Inverse Interpolation using Newton's Forward Interpolation Formula

From the given data, complete the forward difference table. Let y is given value for which x is required. We name this value of y as y_p then

$$y_p = f(x_0 + ph)$$

$$\therefore \quad x = x_0 + ph$$

After finding p, putting value of p in this equation, we get x.

Now Newton's forward difference interpolation formula is

$$y_p = y_0 + p\Delta y_0 + \binom{p}{2}\Delta^2 y_0 + \binom{p}{3}\Delta^3 y_0 + \binom{p}{4}\Delta^4 y_0 + \cdots$$

$$\therefore \quad p = \frac{1}{\Delta y_0}\left[y_p - y_0 - \binom{p}{2}\Delta^2 y_0 - \binom{p}{3}\Delta^3 y_0 - \binom{p}{4}\Delta^4 y_0 - \cdots\right]$$

First approximation of p is taken as

$$p_1 = \frac{1}{\Delta y_0}(y_p - y_0)$$

Second approximation p_2 of p is

$$p_2 = \frac{1}{\Delta y_0}\left[y_p - y_0 - \binom{p_1}{2}\Delta^2 y_0\right]$$

Third approximation p_3 of p is

$$p_3 = \frac{1}{\Delta y_0}\left[y_p - y_0 - \binom{p_2}{2}\Delta^2 y_0 - \binom{p_2}{3}\Delta^3 y_0\right]$$

The process is continued till two successive approximations of p are equal up to desired accuracy. After finding p, x is found from $x = x_0 + ph$.

5.9.3 Inverse Interpolation using Everett's Formula

Let f is tabulated with $\delta^2 f$ and $\delta^4 f$ from given data and we want to find x for $y = f(x) = y_p$

We have $x = x_0 + ph$ and by Everett's formula

$$y_p = py_1 + \binom{p+1}{3}\delta^2 y_1 + \binom{p+2}{5}\delta^4 y_1 + qy_0 + \binom{q+1}{3}\delta^2 y_0 + \binom{q+2}{5}\delta^4 y_0$$

where $q = 1 - p$

$$= py_1 + \binom{p+1}{3}\delta^2 y_1 + \binom{p+2}{5}\delta^4 y_1 + (1-p)y_0 + \binom{2-p}{3}\delta^2 y_0 + \binom{3-p}{5}\delta^4 y_0$$

First approximation p_1 of p is given by $p_1 y_1 + (1 - p_1)y_0 = y_p$

Second approximation p_2 of p is given by $p_2 y_1 + (1 - p_2)y_0 = y_p - \binom{p_1+1}{3}\delta^2 y_1 - \binom{2-p_1}{3}\delta^2 y_0$

Third approximation p_3 of p is given by

$$p_3 y_1 + (1-p_3)y_0 = y_p - \binom{p_2+1}{3}\delta^2 y_1 - \binom{2-p_2}{3}\delta^2 y_0 - \binom{p_2+2}{5}\delta^4 y_1 - \binom{3-p_2}{5}\delta^4 y_0$$

The process is continued till two successive approximation of p are equal up to desired accuracy.

After finding p, x is found from $x = x_0 + ph$.

Example 5.69: If $y_1 = 4, y_3 = 12, y_4 = 19$ and $y_x = 7$, find x using inverse interpolation.

Solution:

y	4	12	19
x	1	3	4

We are to find x for $y = 7$

By Lagrange's inverse interpolation formula

$$x \simeq \frac{(y-y_1)(y-y_2)}{(y_0-y_1)(y_0-y_2)}x_0 + \frac{(y-y_0)(y-y_2)}{(y_1-y_0)(y_1-y_2)}x_1 + \frac{(y-y_0)(y-y_1)}{(y_2-y_0)(y_2-y_1)}x_2$$

$$= \frac{(7-12)(7-19)}{(4-12)(4-19)}(1) + \frac{(7-4)(7-19)}{(12-4)(12-19)}(3) + \frac{(7-4)(7-12)}{(19-4)(19-12)}(4)$$

$$\therefore \quad x \simeq \frac{(-5)(-12)}{(-8)(-15)} + \frac{(3)(-12)}{(8)(-7)}(3) + \frac{(3)(-5)}{(15)(7)}(4)$$

$$= \frac{(5)(12)}{(8)(15)} + \frac{(3)(12)}{(8)(7)}(3) - \frac{(3)(5)(4)}{(15)(7)}$$

$$\therefore \quad x \simeq 2$$

Example 5.70: Following table gives the annuity value y for various ages x.

Age (x)	30	35	40	45	50
Annuity (y)	15.9	14.9	14.1	13.3	12.5

Find the age corresponding to annuity value 13.6

Solution:

Central difference table for given data is

x	y	δy	$\delta^2 y$	$\delta^3 y$	$\delta^4 y$
30	15.9				
		−1.0			
35	14.9		0.2		
		−0.8		−0.2	
40	14.1 ↘		↗ 0.0 ↘		↗ 0.2
		−0.8		0.0	
45	13.3 ↗		↘ 0.0 ↗		
		−0.8			
50	12.5				

Take $x_0 = 40,\ h = 5$

Let x be age for annuity 13.6

$$\therefore \quad p = \frac{x - x_0}{h} = \frac{x - 40}{5}\ ;\ y_p = 13.6$$

By Bessel's interpolation formula

$$y_p = \mu y_{\frac{1}{2}} + \left(p - \frac{1}{2}\right)\delta y_{\frac{1}{2}} + \frac{p(p-1)}{2}\mu\delta^2 y_{\frac{1}{2}} + \frac{p(p-1)}{2}\cdot\frac{p-\frac{1}{2}}{3}\delta^3 y_{\frac{1}{2}}$$

$$+\frac{(p+1)p(p-1)(p-2)}{4!}\mu\delta^4 y_{\frac{1}{2}} + \cdots$$

$$\therefore \quad p_1 - \frac{1}{2} \simeq \frac{y_p - \mu y_{\frac{1}{2}}}{\delta y_{\frac{1}{2}}} = \frac{13.6 - \frac{1}{2}(13.3 + 14.1)}{13.3 - 14.1}$$

$$= \frac{13.6 - 13.7}{-0.8} = 0.125$$

$$\therefore \quad p_1 \simeq \frac{1}{2} + 0.125 = 0.625$$

$$p_2 - \frac{1}{2} \simeq \frac{1}{\delta y_{\frac{1}{2}}}\left[y_p - \mu y_{\frac{1}{2}} - \frac{p_1(p_1 - 1)}{2}\mu\delta^2 y_{\frac{1}{2}}\right]$$

$$= \frac{1}{13.3 - 14.1}\left[13.6 - \frac{1}{2}(13.3 + 14.1) - \frac{(0.625)(-0.375)}{2}\cdot\frac{1}{2}(0 + 0)\right] = 0.125$$

$$p_2 \simeq \frac{1}{2} + 0.125 = 0.625$$

$$\therefore \qquad p \simeq 0.625$$

$$\frac{x-40}{5} = p \simeq 0.625$$

$$\therefore \qquad x \simeq 40 + 5(0.625) = 43.125 \simeq 43$$

$\therefore$ Age corresponding to annuity 13.6 $\simeq$ 43

Example 5.71: Using inverse interpolation, find the real root of the equation $x^3 + x - 3 = 0$ which is close to 1.2

Solution: Let $\qquad y = x^3 + x - 3$

We have

x	1	2	3	4
y	– 1	7	27	65

Forward difference table for this data is

x	y	Δy	Δ²y	Δ³y
1	– 1			
		8		
2	7		12	
		20		6
3	27		18	
		38		
4	65			

We are to find x close to 1.2 for $y = 0$

$$h = 1, x_0 = 1$$

$$p = \frac{x - x_0}{1} = x - 1$$

By Newton's forward interpolation formula

$$y_p \simeq y_0 + p\Delta y_0 + \frac{p(p-1)}{2!}\Delta^2 y_0 + \frac{p(p-1)(p-2)}{3!}\Delta^3 y_0$$

for $\quad y_p = 0$

$$p_1 \simeq \frac{y_p - y_0}{\Delta y_0} = \frac{0-(-1)}{8} = 0.125$$

$$p_2 \simeq \frac{1}{\Delta y_0}\left[y_p - y_0 - \frac{p_1(p_1-1)}{2}\Delta^2 y_0\right]$$

$$= \frac{1}{8}\left[0 + 1 - \frac{(0.125)(-0.875)}{2}(12)\right]$$

$$= \frac{1}{8}\left[1 + \frac{(0.125)(0.875)(12)}{2}\right] \simeq 0.207$$

$$p_3 \simeq \frac{1}{\Delta y_0}\left[y_p - y_0 - \frac{p_2(p_2-1)}{2}\Delta^2 y_0 - \frac{p_2(p_2-1)(p_2-2)}{6}\Delta^3 y_0\right]$$

$$= \frac{1}{8}\left[0 + 1 - \frac{(0.207)(-0.793)}{2}(12) - \frac{(0.207)(-0.793)(-1.793)}{6}(6)\right]$$

$$= \frac{1}{8}\left[1 + (0.207)(0.793)(6) - (0.207)(0.793)(1.793)\right]$$

$$\simeq 0.211$$

$$p_4 \simeq \frac{1}{\Delta y_0}\left[y_p - y_0 - \frac{p_3(p_3-1)}{2}\Delta^2 y_0 - \frac{p_3(p_3-1)(p_3-2)}{6}\Delta^3 y_0\right]$$

$$= \frac{1}{8}\left[0 + 1 - \frac{(0.211)(-0.789)}{2}(12) - \frac{(0.211)(-0.789)(-1.789)}{6}(6)\right]$$

$$= \frac{1}{8}\left[1 + (0.211)(0.789)(6) - (0.211)(0.789)(1.789)\right]$$

$$\simeq 0.2126$$

$$p_5 \simeq \frac{1}{\Delta y_0}\left[y_p - y_0 - \frac{p_4(p_4-1)}{2}\Delta^2 y_0 - \frac{p_4(p_4-1)(p_4-2)}{6}\Delta^3 y_0\right]$$

$$= \frac{1}{8}\left[0 + 1 - \frac{(0.2126)(-0.7874)}{2}(12) - (0.2126)(-0.7874)(-1.7874)\right]$$

$$= \frac{1}{8}\left[1 + (0.2126)(0.7874)(6) - (0.2126)(0.7874)(1.7874)\right]$$

$$\simeq 0.2131$$

$\therefore \quad p \simeq 0.213$

$\therefore \quad x - 1 = p \simeq 0.213$

$\therefore$ Root upto three decimal places = 1.213

EXERCISE 5.4

1. The following data give I, the indicated HP and V, the speed in knots developed by a ship

V	8	10	12	14	16
I	1000	1900	3250	5400	8950

Find I when $V = 9$, using Newton's forward interpolation formula.

2. Compute (a) $y(9)$ (b) $y(7)$ (c) $y(17)$ and (d) $y(19)$ from the following data:

x	8	10	12	14	16	18
y	10	19	32.5	54	89.5	154

3. In the table below, the values of y are consecutive terms of a series of which 23.6 is the sixth term. Find the first and tenth term of the series:

x	3	4	5	6	7	8	9
y	4.8	8.4	14.5	23.6	36.2	52.8	73.9

4. The following are data from the steam table:

Temp $C°$(t)	140	150	160	170	180
Pressure kg f/cm^2 (p)	3.685	4.854	6.302	8.076	10.225

Using Newton's formula, find the pressure of the steam for temperature 142° and 175°.

5. The table below gives the values of tan x for $0.10 \le x \le 0.30$

x	0.10	0.15	0.20	0.25	0.30
$y = \tan x$	0.1003	0.1511	0.2027	0.2553	0.3093

Find (i) tan 0.12 (ii) tan 0.26 (iii) tan 0.40 (iv) tan 0.50

6. The amount A of a substance remaining in a reacting system after an interval of time t in a certain chemical experiment is tabulated below:

t (min.)	2	5	8	11
A (gm.)	94.8	87.9	81.3	75.1

Obtain the value of A when $t = 9$ using Newton's backward interpolation formula.

7. By using Newton's forward difference formula, fit a polynomial of degree three which takes the following values:

x	3	4	5	6
y	6	24	60	120

8. Find the cubic polynomial $f(x)$ which takes the values $f(0) = -4$, $f(1) = -1$, $f(2) = 2$, $f(3) = 11$, $f(4) = 32$, $f(5) = 71$. Find $f(6)$ and $f(2.5)$.

9. In the table below the values of y are consecutive terms of a series of which the number 21.6 is the sixth term. Find the first and tenth terms of the series:

x	3	4	5	6	7	8	9
y	2.7	6.4	12.5	21.6	34.3	51.3	72.9

10. The following table gives corresponding values of x and y. Prepare a forward difference table and using it express y as a function of x. Also obtain y when $x = 2.5$.

x	0	1	2	3	4
y	7	10	13	22	43

11. Using Newton's formula, find a cubic polynomial $f(x)$ which takes the following set of values (0,1), (1, 2), (2, 1) and (3, 10). Hence or otherwise evaluate $f(4)$.
12. Calculate the approximate value of $\sin x$ for $x = 0.54$ and $x = 1.36$ using the following table:

x	0.5	0.7	0.9	1.1	1.3	1.5
$\sin x$	0.47943	0.64422	0.78333	0.89121	0.96356	0.99749

13. The table gives the distance in nautical miles of the visible horizon for the given heights in feet above the earth's surface:

x = height	100	150	200	250	300	350	400
y = distance	10.63	13.03	15.04	16.81	18.42	19.90	21.27

Find the value of y when $x = 218$ ft. and 410 ft. by using Newton's formulae.

14. The following table gives the values of density of saturated water for various temperatures of saturated steam:

Temp°c (= T)	100	150	200	250	300
Density hg/m^3(= d)	958	917	865	799	712

Find by interpolation, the densities when the temperatures are 130°c and 275°c, respectively.

15. By using Newton's forward difference formula, find a polynomial which takes the following values

x	1	3	5	7	9	11
y_x	3	14	19	21	23	28

and hence compute y_x at $x = 2, 12$.

16. Approximate cos 23° by interpolation from the following table correct to four places of decimal

x	10°	20°	30°	40°	50°	60°	70°	80°
$\cos x$	0.9848	0.9397	0.8660	0.7660	0.6428	0.5000	0.3420	0.1737

17. From the following data, estimate the number of persons having income in rupess in between

(i) 1000–1700 and (ii) 3500–4000

Income	Below 500	500–1000	1000–2000	2000–3000	3000–4000
No. of persons	6000	4250	3600	1500	650

18. Following table gives the grouped data for no. of students lying in various weight groups:

Wt. in lbs.	0–40	40–60	60–80	80–100	100–120
No. of students	250	120	100	70	50

Find the number of students having weight between 60 and 70 lbs.

19. Use Gauss's forward interpolation formula to find the value of y_9 if $y_0 = 14$, $y_4 = 24$, $y_8 = 32$, $y_{12} = 35$ and $y_{16} = 40$

20. Apply Gauss's forward formula to evaluate y_{30}, given that $y_{21} = 18.4708$, $y_{25} = 17.8144$, $y_{29} = 17.1070$, $y_{33} = 16.3432$ and $y_{37} = 15.5154$.

21. Apply Gauss's backward interpolation formula to find the population for the year 1976, given that

Year x	1940	1950	1960	1970	1980	1990
Population (y) in thousands	17	20	27	32	36	38

22. For the following table use Stirling's formula to compute $f(x)$ at $x = 11$

x	2	6	10	14	18
$y = f(x)$	21.857	21.025	20.132	19.145	18.057

23. Using Stirling's formula, compute $f(1.22)$ from the following data:

x	1.0	1.1	1.2	1.3	1.4
$f(x)$	0.841	0.891	0.932	0.963	0.985

24. Use Stirling's formula to find y_{35} given that $y_{10} = 600$, $y_{20} = 512$, $y_{30} = 439$, $y_{40} = 346$, $y_{50} = 243$

25. Use Bessel's formula to compute $f(1.95)$ from the following data:

x	1.7	1.8	1.9	2.0	2.1	2.2	2.3
$f(x)$	2.979	3.144	3.283	3.391	3.463	3.997	4.491

26. Use Laplace – Everett's formula to obtain $f(1.15)$ given that $f(1) = 1.000$, $f(1.10) = 1.049$, $f(1.20) = 1.096$ and $f(1.30) = 1.140$.

27. The following table gives the values of the probability integral $f(x) = \frac{2}{\sqrt{\pi}} \int_0^x e^{-x^2} dx$ for some values of x. Find the values of this integral when $x = 0.5437$ using

(i) Stirling's formula (ii) Bessel's formula

x	0.51	0.52	0.53	0.54	0.55	0.56	0.57
$f(x)$	0.5292437	0.5378987	0.5464641	0.5549392	0.5633233	0.5716157	0.5798158

28. Using Lagrange's interpolation formula, find the value of y corresponding to $x = 10$ from the following table:

x	5	6	9	11
y	12	13	14	16

29. Use Lagrange's interpolation formula to fit a polynomial to the following data. Hence find $y(-2)$, $y(1)$ and $y(4)$.

x	−1	0	2	3
y	−8	3	1	2

30. Use Lagrange's interpolation formula to express the function $\dfrac{x^2+x-3}{x^3-2x^2-x+2}$ as a sum of partial fractions.
31. Use Lagrange's formula to find the form of $f(x)$, given

x	0	2	3	6
$f(x)$	648	704	729	792

32. Express the function $\dfrac{x^2+6x-1}{(x^2-1)(x-4)(x-6)}$ as a sum of partial fractions using Lagrange's formula.
33. Using Lagrange's formula, prove that $32\,f(1) = -3\,f(-4) + 10\,f(-2) + 30\,f(2) - 5\,f(4)$
34. By means of Lagrange's formula, prove that

$$y_1 = y_3 - 0.3\,(y_5 - y_{-3}) + 0.2\,(y_{-3} - y_{-5}) \quad \text{(approximately)}$$

35. The function $y = f(x)$ is given in the points (7, 3), (8, 1), (9, 1) and (10, 9). Find y (9.5) using Lagrange's interpolation formula.
36. If a, b, c, d are the arguments of $f(x) = \dfrac{1}{x}$, show that $f(a, b, c, d) = -\dfrac{1}{abcd}$
37. Find a polynomial satisfied by the following table:

x	−4	−1	0	2	5
$f(x)$	1245	33	5	9	1335

38. Find Newton's divided difference polynomial for the data given in the table below, Also find $f(2.5)$

x	−3	−1	0	3	5
$f(x)$	−30	−22	−12	330	3458

39. By means of Newton's divided difference formula, find the value of $f(8)$ and $f(15)$ from the following table:

x	4	5	7	10	11	13
$f(x)$	48	100	294	900	1210	2028

40. Using divided difference, find the value of $f(8)$, given that $f(6) = 1.556$, $f(7) = 1.690$, $f(9) = 1.908$, $f(12) = 2.158$

41. Given that

x	5	7	11	13	17
$f(x)$	150	392	1452	2366	5202

Evaluate $f(9)$, using Lagrange's and Newton's divided difference formulae.

42. Given that

x	300	304	305	307
$\log_{10} x$	2.4771	2.4829	2.4843	2.4871

Calculate the approximate value of $\log_{10} 301$.

43. Solve $e^x = 3.14$ by inverse interpolation using $x = f(y)$ with $f(3.0) = 1.0986$, $f(3.2) = 1.1632$, $f(3.4) = 1.2238$

44. Apply Lagrange's formula inversely to find, to two decimal places, the value of x when $y = f(x) = 19$, given the following table:

x	0	1	2
$f(x)$	0	1	20

45. Compute the value of x, when $y = 8$ by inverse interpolation using Lagrange's formula

x	−2	−1	1	2
y	−7	2	0	11

46. The following values of $y = f(x)$ are given

x	10	15	20
y	1754	2648	3564

Find the value of x for $y = 3000$ by iterative method.

47. From the following data

x	1.8	2.0	2.2	2.4	2.6
y	2.9	3.6	4.4	5.5	6.7

find x when $y = 5$, using iterative method.

48. Find the root of the equation $10x^3 - 15x + 3 = 0$ which lies between 1 and 2.

ANSWERS 5.4

1. 1406 2. (a) 14.2 (b) 5.2 (c) 116.6 (d) 205.9

3. $y(1) = 3.1, y(10) = 100$ 4. 3.899, 9.1005

5. (i) 0.1205 (ii) 0.2660 (iii) 0.4241 (iv) 0.5543

6. 79.2 gm. 7. $y(x) = x^3 - 3x^2 + 2x$

8. $f(x) = x^3 - 3x^2 + 5x - 4, f(6) = 134, f(2.5) = 5.375$

9. $y(1) = 0.1, y(10) = 100$ 10. $y(x) = x^3 - 3x^2 + 5x + 7, y(2.5) = 16.375$

11. $f(x) = 2x^3 - 7x^2 + 6x + 1, f(4) = 41$ 12. 0.51414, 0.97786

13. 15.70 nautical miles, 21.54 nautical miles 14. 935 hg/m^3, 759 hg/m^3

15. $y_x = \frac{1}{16}(x^3 - 21x^2 + 159x - 91), y_2 = 9.4375, y_{12} = 32.5625$ 16. 0.9205

17. (i) 2797 (ii) 402 18. 54 19. 33 20. 16.9216

21. 34 thousands 22. 19.895 23. 0.939 24. 395

25. 3.347 26. 1.073 27. (i) 0.5580520 (ii) 0.5580520 28. 14.6667

29. $y = \frac{1}{6}(7x^3 - 31x^2 + 28x + 18); y(-2) = -36.3333, y(1) = 3.6667, y(4) = 13.6667,$

30. $-\dfrac{1}{2(x+1)} + \dfrac{1}{2(x-1)} + \dfrac{1}{(x-2)}$ 31. $f(x) = -x^2 + 30x + 648$

32. $\dfrac{3}{35(x+1)} + \dfrac{1}{5(x-1)} - \dfrac{13}{10(x-4)} + \dfrac{71}{70(x-6)}$ 35. 3.625

37. $f(x) = 3x^4 - 5x^3 + 6x^2 - 14x + 5$

38. $f(x) = 5x^4 + 9x^3 - 27x^2 - 21x - 12, f(2.5) = 102.6875$

39. $f(8) = 448, f(15) = 3150$ 40. 1.806 41. 810 42. 2.4786

43. 1.1442 44. 2.80 45. –3 46. 17 47. 2.3 48. 1.109

Numerical Methods for Differentiation, Integration and Ordinary Differential Equations 6

6.1 INTRODUCTION

Numerical differentiation is the computation of values of the derivative of a function from its given values. Derivative is the limit of the difference quotient in which we divide by small quantity which can create too much error. This is the reason that numerical differentiation is the weakest point of numerical methods and should be avoided wherever possible. However, the formulae obtained in numerical differentiation are basic in numerical solution of differential equations.

Numerical integration means the numerical evaluation of integral $\int_a^b f(x)dx$ where a and b are given and f is a function given analytically by a formula or empirically by a table of values. Numerical methods are applied when f is complicate to be integrated or only numerical values of f are given.

Numerical methods for differential equations are of great importance to the engineer and physicist because practical problems often lead to differential equations which cannot be solved by known methods or solution is very much complicated and thus numerical methods will be preferable.

6.2 NUMERICAL DIFFERENTIATION

In interpolation, $f(x)$ is approximated by a polynomial $P(x)$. Now if we plot $f(x)$ and $P(x)$, we may observe that even though the difference between $P(x)$ and $f(x)$ is small throughout the interval but the slopes of tangents to two curves may differ quite appreciably and hence the derivative will differ quite appreciably. It can also be possible that slopes may differ in sign and then even the round off error will affect the calculations. This is the reason that numerical differentiation is considered the weakest concept in numerical methods.

In the given data, if the interval of differencing is not constant then first find the interpolation polynomial $P(x)$ approximating $f(x)$ by Lagrange interpolation polynomial or by divided difference interpolation polynomial. Then differentiate $P(x)$ and after that put value of x at which derivative of $f(x)$ is required. Similarly, we can find second or higher order derivatives.

If $f(x)$ is given at equidistant points, our problem is then to compute the derivative either at a grid point or in an interior point.

6.2.1 Derivatives at Interior Points

If the interior point is near the beginning point x_0 and is $x = x_0 + ph$ then $\dfrac{dp}{dx} = \dfrac{1}{h}$.

By Newton's forward interpolation formula

$$y_p = y_0 + \binom{p}{1}\Delta y_0 + \binom{p}{2}\Delta^2 y_0 + \binom{p}{3}\Delta^3 y_0 + \binom{p}{4}\Delta^4 y_0 + \cdots$$

$$= y_0 + p\Delta y_0 + \frac{p^2 - p}{2}\Delta^2 y_0 + \frac{p^3 - 3p^2 + 2p}{6}\Delta^3 y_0 + \frac{p^4 - 6p^3 + 11p^2 - 6p}{24}\Delta^4 y_0 + \cdots$$

$$\therefore \frac{dy}{dx} = \frac{dy}{dp}\frac{dp}{dx} = \frac{1}{h}\left[\Delta y_0 + \frac{2p-1}{2}\Delta^2 y_0 + \frac{3p^2-6p+2}{6}\Delta^3 y_0 + \frac{4p^3-18p^2+22p-6}{24}\Delta^4 y_0 + \cdots\right]$$

From this, $y'(x)$ will be obtained.

Again differentiating it, we can find $y''(x)$.

Similarly if x is near end point then use Newton's backward interpolation formula.

If x is near the midpoint then central difference formula is used and proceeding as above we can find derivative. For Example, when we use Everett formula

$$\begin{aligned} y_p &= \left\{ q y_0 + \binom{q+1}{3}\delta^2 y_0 + \binom{q+2}{5}\delta^4 y_0 + \cdots \right\} \\ &\quad + \left\{ p y_1 + \binom{p+1}{3}\delta^2 y_1 + \binom{p+2}{5}\delta^4 y_1 + \cdots \right\} \\ &= \left\{ q y_0 + \frac{q^3-q}{6}\delta^2 y_0 + \frac{q^5-5q^3+4q}{120}\delta^4 y_0 + \cdots \right\} \\ &\quad + \left\{ p y_1 + \frac{p^3-p}{6}\delta^2 y_1 + \frac{p^5-5p^3+4p}{120}\delta^4 y_1 + \cdots \right\} \end{aligned}$$

Now, using

$$\frac{dy}{dx} = \frac{dy}{dp}\frac{dp}{dx} = \frac{1}{h}\frac{dy}{dp}$$

and

$$\frac{dy}{dx} = \frac{dy}{dq}\frac{dq}{dp}\frac{dp}{dx} = -\frac{1}{h}\frac{dy}{dq} \qquad (\because q = 1-p)$$

we have

$$\begin{aligned} y_p' = \frac{1}{h}\Bigg[&\left\{ -y_0 - \frac{3q^2-1}{6}\delta^2 y_0 - \frac{5q^4-15q^2+4}{120}\delta^4 y_0 + \cdots \right\} \\ &+ \left\{ y_1 + \frac{3p^2-1}{6}\delta^2 y_1 + \frac{5p^4-15p^2+4}{120}\delta^4 y_1 + \cdots \right\} \Bigg] \end{aligned}$$

6.2.2 Derivative at Grid Points

Though the derivative at grid point can also be obtained as for interior points for $p = 0$, but there are simpler formulae to find derivatives at grid points which we shall prove below. The derivative using these formulae will require less computation work. The formulae are

(i) $$y_0' = \frac{1}{h}\left[\mu\delta y_0 - \frac{1}{6}\mu\delta^3 y_0 + \frac{1}{30}\mu\delta^5 y_0 - \cdots\right] \tag{6.1}$$

(ii) $$y_0'' = \frac{1}{h^2}\left[\delta^2 y_0 - \frac{1}{12}\delta^4 y_0 + \frac{1}{90}\delta^6 y_0 - \cdots\right] \tag{6.2}$$

Proof: (i)

$$2\sinh\frac{U}{2}=\delta$$

Differentiate w.r.t. δ

$$\cosh\frac{U}{2}\frac{dU}{d\delta}=1$$

$$\therefore \quad \frac{dU}{d\delta}=\frac{1}{\cosh\frac{U}{2}}=\frac{1}{\sqrt{1+\sinh^2\frac{U}{2}}}=\left(1+\frac{\delta^2}{4}\right)^{-\frac{1}{2}}$$

$$=1-\frac{1}{2}\frac{\delta^2}{4}+\frac{\left(-\frac{1}{2}\right)\left(-\frac{3}{2}\right)}{2!}\frac{\delta^4}{16}+\frac{\left(-\frac{1}{2}\right)\left(-\frac{3}{2}\right)\left(-\frac{5}{2}\right)}{3!}\frac{\delta^6}{64}+\cdots$$

$$=1-\frac{1}{8}\delta^2+\frac{3}{128}\delta^4-\frac{5}{1024}\delta^6+\cdots$$

$\therefore$ integrating, we have

$$U=\delta-\frac{1}{24}\delta^3+\frac{3}{640}\delta^5-\frac{5}{7168}\delta^7+\cdots \tag{6.3}$$

$$=\frac{\mu}{\sqrt{1+\frac{\delta^2}{4}}}\left[\delta-\frac{1}{24}\delta^3+\frac{3}{640}\delta^5-\frac{5}{7168}\delta^7+\cdots\right] \qquad \left(\because \mu=\sqrt{1+\frac{\delta^2}{4}}\right)$$

$$=\mu\delta\left[1-\frac{1}{24}\delta^2+\frac{3}{640}\delta^4-\frac{5}{7168}\delta^6+\cdots\right]\left(1+\frac{\delta^2}{4}\right)^{-\frac{1}{2}}$$

$$=\mu\delta\left[1-\frac{1}{24}\delta^2+\frac{3}{640}\delta^4-\frac{5}{7168}\delta^6+\cdots\right]\left[1-\frac{1}{8}\delta^2+\frac{3}{128}\delta^4-\frac{5}{1024}\delta^6+\cdots\right]$$

$$=\mu\delta\left[1-\left(\frac{1}{8}+\frac{1}{24}\right)\delta^2+\left(\frac{3}{128}+\frac{1}{192}+\frac{3}{640}\right)\delta^4-\left(\frac{5}{1024}+\frac{1}{1024}+\frac{3}{5120}+\frac{5}{7168}\right)\delta^6+\cdots\right]$$

$$=\mu\delta\left[1-\frac{1}{6}\delta^2+\frac{1}{30}\delta^4-\frac{1}{140}\delta^6+\cdots\right]$$

$$\therefore \quad D=\frac{1}{h}\left[\mu\delta-\frac{1}{6}\mu\delta^3+\frac{1}{30}\mu\delta^5-\frac{1}{140}\mu\delta^7+\cdots\right]$$

$$\therefore \quad y_0'=\frac{1}{h}\left[\mu\delta y_0-\frac{1}{6}\mu\delta^3 y_0+\frac{1}{30}\mu\delta^5 y_0-\frac{1}{140}\mu\delta^7 y_0+\cdots\right]$$

(ii) From (6.3)

$$U = \delta - \frac{1}{24}\delta^3 + \frac{3}{640}\delta^5 - \cdots$$

$$\therefore \quad U^2 = \left(\delta - \frac{1}{24}\delta^3 + \frac{3}{640}\delta^5 - \cdots\right)^2$$

$$= \delta^2 - \frac{1}{12}\delta^4 + \left(\frac{3}{320} + \frac{1}{576}\right)\delta^6 - \cdots$$

$$= \delta^2 - \frac{1}{12}\delta^4 + \frac{1}{90}\delta^6 - \cdots$$

$$\therefore \quad y_0'' = \frac{1}{h^2}\left[\delta^2 y_0 - \frac{1}{12}\delta^4 y_0 + \frac{1}{90}\delta^6 y_0 - \cdots\right]$$

These formulae are to be used when the grid point is in the middle.

When the grid point is near the beginning we proceed as follows

$$E = e^{hD} = 1 + \Delta$$

$$\therefore \quad hD = \log(1+\Delta)$$

$$= \Delta - \frac{1}{2}\Delta^2 + \frac{1}{3}\Delta^3 - \frac{1}{4}\Delta^4 + \frac{1}{5}\Delta^5 - \frac{1}{6}\Delta^6 + \cdots$$

$$y_0' = \frac{1}{h}\left[\Delta y_0 - \frac{1}{2}\Delta^2 y_0 + \frac{1}{3}\Delta^3 y_0 - \frac{1}{4}\Delta^4 y_0 + \frac{1}{5}\Delta^5 y_0 - \frac{1}{6}\Delta^6 y_0 + \cdots\right]$$

and

$$h^2D^2 = \left(\Delta - \frac{1}{2}\Delta^2 + \frac{1}{3}\Delta^3 - \frac{1}{4}\Delta^4 + \frac{1}{5}\Delta^5 - \cdots\right)^2$$

$$= \Delta^2 - \Delta^3 + \left(\frac{2}{3} + \frac{1}{4}\right)\Delta^4 - \left(\frac{2}{4} + \frac{2}{6}\right)\Delta^5 + \left(\frac{2}{5} + \frac{2}{8} + \frac{1}{9}\right)\Delta^6 - \cdots$$

$$\therefore \quad y_0'' = \frac{1}{h^2}\left[\Delta^2 y_0 - \Delta^3 y_0 + \frac{11}{12}\Delta^4 y_0 - \frac{5}{6}\Delta^5 + \frac{137}{180}\Delta^6 - \cdots\right]$$

When the grid point is near the end, we proceed as follows

$$E = e^{hD} = (1-\nabla)^{-1}$$

$$\therefore \quad hD = -\log(1-\nabla)$$

$$= \nabla + \frac{1}{2}\nabla^2 + \frac{1}{3}\nabla^3 + \frac{1}{4}\nabla^4 + \frac{1}{5}\nabla^5 + \frac{1}{6}\nabla^6 + \cdots$$

$$\therefore \quad y_n' = \frac{1}{h}\left[\nabla y_n + \frac{1}{2}\nabla^2 y_n + \frac{1}{3}\nabla^3 y_n + \frac{1}{4}\nabla^4 y_n + \frac{1}{5}\nabla^5 y_n + \frac{1}{6}\nabla^6 y_n + \cdots\right]$$

and $$h^2D^2 = \left(\nabla + \frac{1}{2}\nabla^2 + \frac{1}{3}\nabla^3 + \frac{1}{4}\nabla^4 + \frac{1}{5}\nabla^5 + \cdots\right)^2$$

$$= \nabla^2 + \nabla^3 + \left(\frac{2}{3} + \frac{1}{4}\right)\nabla^4 + \left(\frac{1}{2} + \frac{1}{3}\right)\nabla^5 + \left(\frac{2}{5} + \frac{1}{4} + \frac{1}{9}\right)\nabla^6 + \cdots$$

$\therefore$ $$y_n'' = \frac{1}{h^2}\left[\nabla^2 y_n + \nabla^3 y_n + \frac{11}{12}\nabla^4 y_n + \frac{5}{6}\nabla^5 y_n + \frac{137}{180}\nabla^6 y_n + \cdots\right]$$

Now, we shall consider some Examples.

Example 6.1: If $u_0 = 3,\ u_1 = 12,\ u_2 = 81,\ u_3 = 2000,\ u_4 = 100$, calculate Du_0.
Solution: Difference table for given data is

x	u_x	Δu_x	$\Delta^2 u_x$	$\Delta^3 u_x$	$\Delta^4 u_x$
0	3				
		9			
1	12		60		
		69		1790	
2	81		1850		−7459
		1919		−5669	
3	2000		−3819		
		−1900			
4	100				

We have

$$D = \frac{1}{h}\log(1+\Delta)$$

$$= \frac{1}{h}\left(\Delta - \frac{\Delta^2}{2} + \frac{\Delta^3}{3} - \frac{\Delta^4}{4} + \frac{\Delta^5}{5} - \cdots\right)$$

Here, $h = 1$

$\therefore$ $$Du_0 \simeq \Delta u_0 - \frac{\Delta^2 u_0}{2} + \frac{\Delta^3 u_0}{3} - \frac{\Delta^4 u_0}{4}$$

$$= 9 - \frac{60}{2} + \frac{1790}{3} + \frac{7459}{4}$$

$$\simeq 9 - 30 + 596.67 + 1864.75$$

$$= 2440.42$$

Example 6.2: Given the following tabulated values of a function

x	1.0	1.2	1.4	1.6	1.8	2.0	2.2
y	2.7183	3.3201	4.0552	4.9530	6.0496	7.3891	9.0250

Prepare difference table and evaluate y' and y'' at

(i) $x = 1.2$ (ii) $x = 1.6$ (iii) $x = 2.0$ (iv) $x = 2.2$

Solution: Difference table for given data is

x	**y**						
1.0	2.7183						
		0.6018					
1.2	3.3201		0.1333				
		0.7351		0.0294			
1.4	4.0552		0.1627		0.0067		
		0.8978		0.0361		0.0013	
1.6	4.9530		0.1988		0.0080		0.0001
		1.0966		0.0441		0.0014	
1.8	6.0496		0.2429		0.0094		
		1.3395		0.0535			
2.0	7.3891		0.2964				
		1.6359					
2.2	9.0250						

We have

$$h = 0.2$$

For $x = 1.2$

$$D = \frac{1}{h}\log(1+\Delta) = \frac{1}{h}\left[\Delta - \frac{1}{2}\Delta^2 + \frac{1}{3}\Delta^3 - \frac{1}{4}\Delta^4 + \frac{1}{5}\Delta^5 - \cdots\right]$$

$$\therefore \quad y'(1.2) \simeq \frac{1}{0.2}\left[\Delta y(1.2) - \frac{1}{2}\Delta^2 y(1.2) + \frac{1}{3}\Delta^3 y(1.2) - \frac{1}{4}\Delta^4 y(1.2) + \frac{1}{5}\Delta^5 y(1.2)\right]$$

$$= 5\left[0.7351 - \frac{0.1627}{2} + \frac{0.0361}{3} - \frac{0.0080}{4} + \frac{0.0014}{5}\right]$$

$$\therefore \quad y'(1.2) \simeq 3.3203$$

$$D^2 = \frac{1}{h^2}\left[\Delta - \frac{1}{2}\Delta^2 + \frac{1}{3}\Delta^3 - \frac{1}{4}\Delta^4 + \cdots\right]^2$$

$$= \frac{1}{h^2}\left[\Delta^2 - \Delta^3 + \frac{11}{12}\Delta^4 - \frac{5}{6}\Delta^5 + \cdots\right]$$

$$\therefore \quad y''(1.2) \simeq \frac{1}{(0.2)^2}\left[\Delta^2 y(1.2) - \Delta^3 y(1.2) + \frac{11}{12}\Delta^4 y(1.2) - \frac{5}{6}\Delta^5 y(1.2)\right]$$

$$= 25\left[0.1627 - 0.0361 + \frac{(11)(0.0080)}{12} - \frac{5(0.0014)}{6}\right]$$

$$\therefore \quad y''(1.2) \simeq 3.3192$$

For $x = 1.6$

$$D = \frac{1}{h}\left[\mu\delta - \frac{\mu\delta^3}{6} + \frac{\mu\delta^5}{30} - \cdots\right]$$

$$\therefore \quad y'(1.6) \simeq \frac{1}{0.2}\left[\mu\delta y(1.6) - \frac{1}{6}\mu\delta^3 y(1.6) + \frac{1}{30}\mu\delta^5(1.6)\right]$$

$$= 5\left[\frac{1}{2}(0.8978 + 1.0966) - \frac{1}{12}(.0361 + .0441) + \frac{1}{60}(0.0013 + 0.0014)\right]$$

$$= \frac{1}{2}\left[\frac{19.944}{2} - \frac{0.802}{12} + \frac{0.027}{60}\right]$$

$$\therefore \quad y'(1.6) \simeq 4.9528$$

We have

$$D^2 = \frac{1}{h^2}\left[\mu\delta - \frac{\mu\delta^3}{6} + \frac{\mu\delta^5}{30} - \cdots\right]^2$$

$$= \frac{1}{h^2}\mu^2\delta^2\left[1 - \frac{\delta^2}{6} + \frac{\delta^4}{30} - \cdots\right]^2$$

$$= \frac{1}{h^2}\left(1 + \frac{\delta^2}{4}\right)\delta^2\left[1 - \frac{\delta^2}{3} + \left(\frac{1}{36} + \frac{1}{15}\right)\delta^4 + \cdots\right]$$

$$= \frac{1}{h^2}\left(1 + \frac{\delta^2}{4}\right)\left[\delta^2 - \frac{\delta^4}{3} + \frac{17}{180}\delta^4 + \cdots\right]$$

$$= \frac{1}{h^2}\left[\delta^2 - \frac{\delta^4}{12} + \frac{\delta^6}{90} + \cdots\right]$$

$$\therefore \quad y''(1.6) = \frac{1}{(0.2)^2}\left[\delta^2 y(1.6) - \frac{1}{12}\delta^4 y(1.6) + \frac{1}{90}\delta^6 y(1.6) + \cdots\right]$$

$$\therefore \quad y''(1.6) \simeq \frac{100}{4}\left[0.1988 - \frac{0.0080}{12} + \frac{0.0001}{90}\right]$$

$$\therefore \quad y''(1.6) \simeq 4.9534$$

For $x = 2.0$

$$D = \frac{1}{h}\log(1-\nabla)^{-1} = -\frac{1}{h}\log(1-\nabla)$$

$$= \frac{1}{h}\left[\nabla + \frac{\nabla^2}{2} + \frac{\nabla^3}{3} + \frac{\nabla^4}{4} + \cdots\right]$$

$$\therefore \quad y'(2.0) \simeq \frac{1}{0.2}\left(\nabla y(2.0) + \frac{1}{2}\nabla^2 y(2.0) + \frac{1}{3}\nabla^3 y(2.0) + \frac{1}{4}\nabla^4 y(2.0) + \frac{1}{5}\nabla^5 y(2.0)\right)$$

$$= \frac{10}{2}\left[1.3395 + \frac{0.2429}{2} + \frac{0.0441}{3} + \frac{0.0080}{4} + \frac{0.0013}{5}\right]$$

$$= \frac{1}{2}\left[13.395 + \frac{2.429}{2} + \frac{0.441}{3} + \frac{0.080}{4} + \frac{0.013}{5}\right]$$

$$\therefore \quad y'(2.0) \simeq 7.3896$$

$$D^2 = \frac{1}{h^2}\left[\nabla + \frac{\nabla^2}{2} + \frac{\nabla^3}{3} + \frac{\nabla^4}{4} + \frac{\nabla^5}{5} + \cdots\right]^2$$

$$= \frac{1}{h^2}\left[\nabla^2 + \nabla^3 + \left(\frac{1}{4} + \frac{2}{3}\right)\nabla^4 + \left(\frac{1}{2} + \frac{1}{3}\right)\nabla^5 + \left(\frac{1}{9} + \frac{2}{5} + \frac{1}{4}\right)\nabla^6 + \cdots\right]$$

$$= \frac{1}{h^2}\left[\nabla^2 + \nabla^3 + \frac{11}{12}\nabla^4 + \frac{5}{6}\nabla^5 + \frac{137}{180}\nabla^6 + \cdots\right]$$

$$\therefore \quad y''(2.0) \simeq \frac{1}{h^2}\left[\nabla^2 y(2.0) + \nabla^3 y(2.0) + \frac{11}{12}\nabla^4 y(2.0) + \frac{5}{6}\nabla^5 y(2.0)\right]$$

$$= \frac{100}{4}\left[0.2429 + 0.0441 + \frac{11}{12}(0.0080) + \frac{5}{6}(0.0013)\right]$$

$$\therefore \quad y''(2.0) \simeq 7.3854$$

For $x = 2.2$

$$y'(2.2) \simeq \frac{1}{0.2}\left[\nabla y(2.2) + \frac{1}{2}\nabla^2 y(2.2) + \frac{1}{3}\nabla^3 y(2.2) + \frac{1}{4}\nabla^4 y(2.2) + \frac{1}{5}\nabla^5 y(2.2) + \frac{1}{6}\nabla^6 y(2.2)\right]$$

$$= \frac{10}{2}\left[1.6359 + \frac{1}{2}(0.2964) + \frac{1}{3}(0.0535) + \frac{1}{4}(0.0094) + \frac{1}{5}(0.0014) + \frac{1}{6}(0.0001)\right]$$

$$\therefore \; y'(2.2) \simeq 9.0229$$

$$y''(2.2) \simeq \frac{1}{h^2}\left(\nabla^2 y(2.2) + \nabla^3 y(2.2) + \frac{11}{12}\nabla^4 y(2.2) + \frac{5}{6}\nabla^5 y(2.2) + \frac{137}{180}\nabla^6 y(2.2)\right)$$

$$= \frac{100}{4}\left[0.2964 + 0.0535 + \frac{11}{12}(0.0094) + \frac{5}{6}(0.0014) + \frac{137}{180}(0.0001)\right]$$

$$\therefore \; y''(2.2) \simeq 8.9940$$

Example 6.3: A rod is rotating in a plane about one of its ends. The following table gives the angle θ in radians through which the rod has turned for different values of time t seconds. Find the angular velocity at $t = 0.7$ sec.

t (seconds)	0.0	0.2	0.4	0.6	0.8	1.0
θ (radians)	0.00	0.12	0.48	1.10	2.00	3.20

Also find the angular acceleration at $t = 0.7$ sec.

Solution: Backward difference table for given data is

t	θ	$\nabla\theta$	$\nabla^2\theta$	$\nabla^3\theta$	$\nabla^4\theta$	$\nabla^5\theta$
0.0	0.00					
		0.12				
0.2	0.12		0.24			
		0.36		0.02		
0.4	0.48		0.26		0	
		0.62		0.02		0
0.6	1.10		0.28		0	
		0.90		0.02		
0.8	2.00		0.30			
		1.20				
1.0	3.20					

We shall use backward difference interpolation formula to evaluate $\left(\dfrac{d\theta}{dt}\right)_{t=0.7\,\text{sec.}}$

$$t = 0.7,\ t_0 = 0.8,\ h = 0.2$$

$$\therefore \quad p = \frac{t-t_0}{h} = \frac{0.7-0.8}{0.2} = -0.5$$

$$\theta_p = \theta_0 + p\nabla\theta_0 + \frac{(p+1)p}{2!}\nabla^2\theta_0 + \frac{(p+2)(p+1)p}{3!}\nabla^3\theta_0 + \cdots$$

$$\theta'(t) = \theta_p'\frac{dp}{dt} = \frac{1}{h}\left[\nabla\theta_0 + \frac{(2p+1)}{2}\nabla^2\theta_0 + \frac{3p^2+6p+2}{6}\nabla^3\theta_0 + \cdots\right]$$

$$\therefore \quad \theta'(0.7) \simeq \frac{1}{0.2}\left[\nabla\theta(0.8) + \frac{2p+1}{2}\nabla^2\theta(0.8) + \frac{3p^2+6p+2}{6}\nabla^3\theta(0.8)\right]$$

$$= \frac{10}{2}\left[0.90 + \frac{3(.25)+6(-\cdot 5)+2}{6}(0.02)\right]$$

$$= 5\left[0.90 - \frac{(0.25)(0.02)}{6}\right]$$

$\therefore \qquad \theta'(0.7) \simeq 4.5$ radians/sec.

$$\theta''(t) = \theta''_p\left(\frac{dp}{dt}\right)^2 \simeq \frac{1}{h^2}\left[\nabla^2\theta_0 + (p+1)\nabla^3\theta_0\right]$$

$$\therefore \qquad \theta''(0.7) = \frac{1}{(0.2)^2}\left[\nabla^2\theta(0.8) + (0.5)\nabla^3\theta(0.8)\right]$$

$$= \frac{100}{4}\left[0.28 + (0.5)(.02)\right]$$

$$= 7.25$$

$\therefore \qquad \theta''(0.7) \simeq 7.25$ radians/sec^2.

$\therefore$ angular velocity at $t = 0.7$ sec. $\simeq 4.5$ radians/sec.

angular acceleration at $t = 0.7$ sec. $\simeq 7.25$ radians/sec^2.

Example 6.4: Compute $f'''(15)$ given

x	2	4	9	13	16	21	29
$f(x)$	57	1345	66340	402052	1118209	4287844	21242820

Solution: Divided difference table for given data is

x	$f(x)$	$\Delta f(x)$	$\Delta^2 f(x)$	$\Delta^3 f(x)$	$\Delta^4 f(x)$	$\Delta^5 f(x)$	$\Delta^6 f(x)$
2	57						
		644					
4	1345		1765				
		12999		556			
9	66340		7881		45		
		83928		1186		1	
13	402052		22113		64		0
		238719		2274		1	
16	1118209		49401		89		
		633927		4054			
21	4287844		114265				
		2119372					
29	21242820						

By Newton's divided interpolation formula

$$f(x) \approx f(2) + (x-2) f[2,4] + (x-2)(x-4) f[2,4,9]$$
$$+(x-2)(x-4)(x-9) f[2,4,9,13]$$
$$+(x-2)(x-4)(x-9)(x-13) f[2,4,9,13,16]$$
$$+(x-2)(x-4)(x-9)(x-13)(x-16) f[2,4,9,13,16,21]$$
$$+(x-2)(x-4)(x-9)(x-13)(x-16)(x-21) f[2,4,9,13,16,21,29]$$

$$\therefore \qquad f'''(x) \approx \frac{d^3}{dx^3}\Big[556\left(x^3-15x^2+62x-72\right)+45\left(x^4-28x^3+275x^2-878x+936\right)$$
$$+\left(x^5-44x^4+705x^3-878x^2+4990x-14976\right)\Big]$$
$$= 556(6) + 45(24x-168) + 60x^2 - 1056x + 4230$$

$$\therefore \qquad f'''(15) \approx 3336 + 45(360-168) + 13500 - 15840 + 4230$$
$$= 3336 + 45(192) + 13500 - 15840 + 4230$$
$$= 13866$$

Example 6.5: From the following data, find the maximum or minimum value of y

x	0.60	0.65	0.70	0.75
y	0.6221	0.6155	0.6138	0.6170

Solution: Forward difference table for given data is

x	y	Δy	Δ²y	Δ³y
0.60	0.6221			
		−0.0066		
0.65	0.6155		0.0049	
		−0.0017		0
0.70	0.6138		9	
		0.032		
0.75	0			

$$x_0 = 0.60,\ h = 0.05,\ p = \frac{x - x_0}{h} = \frac{x - 0.60}{0.05} = 20x - 12$$

By Newton's forward difference interpolation formula

$$y_p = y_0 + p\Delta y_0 + \frac{p(p-1)}{2}\Delta^2 y_0 + \cdots$$

$$\therefore \quad y_p \simeq 0.6221 - 0.0066p + \frac{0.0049}{2}\left(p^2 - p\right)$$

$$y'_p = \frac{dy}{dp} = -0.0066 + \frac{\cdot 0049}{2}(2p-1)$$

For maxima or minima

$$\frac{dy}{dx} = \frac{dy}{dp}\frac{dp}{dx} = 20\frac{dy}{dp} = 0$$

$$\therefore \quad -0.0066 + \frac{\cdot 0049}{2}(2p-1) = 0$$

$$\therefore \quad 2p - 1 = \frac{2(0.0066)}{0.0049}$$

$$\therefore \quad p = \frac{1}{2}\left[1 + \frac{2(0.0066)}{0.0049}\right] = 1.8469 \simeq 1.85$$

$$\frac{d^2y}{dx^2} = 20 \cdot \frac{d^2y}{dp^2} \cdot \frac{dp}{dx} = 400\frac{d^2y}{dp^2} = 400(0 \cdot 0049) > 0$$

$\therefore$ y is minimum when $p = 1.85$

Minimum $y \simeq 0.6221 - 0.0066(1.85) + \frac{\cdot 0049}{2}(1.85)(0.85)$

$\therefore$ minimum $y \simeq 0.6137$

EXERCISE 6.1

1. Find the first and second derivatives of the function tabulated below at the point $x = 3.0$

x	3.0	3.2	3.4	3.6	3.8	4.0
$f(x)$	–14.000	–10.032	–5.296	0.256	6.672	14.000

2. Find the first and second derivatives of the function tabulated below at the point $x = 1.05$

x	1.00	1.05	1.10	1.15	1.20	1.25	1.30
$f(x)$	1.00000	1.02470	1.04881	1.07238	1.09544	1.11803	1.14017

3. Find $f'(1)$ using the following data:

x	1.0	1.1	1.2	1.3	1.4
$f(x)$	.2500	.2268	.2066	.1890	.1736

4. Find the first, second and third derivatives of the function tabulated below at the point $x = 1.5$

x	1.5	2.0	2.5	3.0	3.5	4.0
$f(x)$	3.375	7.000	13.625	24.000	38.875	59.000

5. Find the first and second derivatives of the function tabulated below at the point $x = 1.1$

x	1	1.2	1.4	1.6	1.8	2.0
$f(x)$	0.0000	0.1280	0.5440	1.2960	2.4320	4.0000

6. Find the first derivative of the function tabulated below at the point $x = 0.4$

x	0.1	0.2	0.3	0.4
$f(x)$	1.10517	1.22140	1.34986	1.49182

7. Find the first three derivatives of the function tabulated below at the point $x = 2.5$

x	1.5	1.9	2.5	3.2	4.3	5.9
$f(x)$	3.375	6.059	13.625	29.368	73.907	196.579

8. Given the following values of $f(x)$ for certain values of x. Find $f'(4)$.

x	1	2	4	8	10
$f(x)$	0	1	5	21	27

9. Find $f'(0.04)$ from the following table

x	.01	.02	.03	.04	.05	.06
$f(x)$	.1023	.1047	.1071	.1096	.1122	.1148

10. Find the first derivative of the function tabulated below at the point $x = 7.50$

x	7.47	7.48	7.49	7.50	7.51	7.52	7.53
$f(x)$	0.193	0.195	0.198	0.201	0.203	0.206	0.208

11. For the following data, approximate $\left(\frac{dy}{dx}\right)$ at $x = 2$

x	0	2	3
y	2	−2	−1

12. Find $\frac{dy}{dx}$ at $x = 5$ from the following table

x	0	2	3	4	7	9
y	4	26	58	112	466	922

13. Given the following pair of values of x and $f(x)$

x	60	75	90	105	120
$f(x)$	28.2	38.2	43.2	40.9	37.7

Find $f'(93)$.

14. From the following table find x correct to two decimal places, for which y is maximum or minimum and find this value of y.

x	1.2	1.3	1.4	1.5	1.6
y	0.9320	0.9636	0.9855	0.9975	0.9996

15. A slide in a machine moves along a straight rod. Its distance x meter along the rod is given below for various values of time t seconds. Find the velocity and acceleration of the slide at $t = 0.3$ sec.

t	0	0.1	0.2	0.3	0.4	0.5	0.6
x	30.13	31.62	32.87	33.64	33.95	33.81	33.24

16. A rod is rotating in a plane. The table below gives various values of θ in radians through which the rod has turned for different values of time t seconds. Find approximately the angular velocity and angular acceleration of the rod at $t = 0.6$ sec.

t	0	0.2	0.4	0.6	0.8	1.0
θ	0	0.12	0.49	1.12	2.02	3.20

ANSWERS 6.1

1. $f'(3) = f''(3) = 18$
2. $f'(1.05) = 0.48763, f''(1.05) = -0.21433$
3. $f'(1) = -0.2483$
4. $f'(1.5) = 4.750, f''(1.5) = 9.000, f'''(1.5) = 6.000$
5. $f'(1.1) = 0.6300, f''(1.1) = 6.6000$
6. $f'(0.4) = 1.49133$
7. $f'(2.5) = 16.750, f''(2.5) = 15.000, f'''(2.5) = 6.000$
8. $f'(4) = 2.8306$
9. $f'(.04) = 0.2558$
10. $f'(7.50) = 0.235$
11. $\left(\dfrac{dy}{dx}\right)_{x=2} = 0$
12. $\left(\dfrac{dy}{dx}\right)_{x=5} = 98$
13. $f'(93) = -0.03627$

14. The maximum value occurs at $x = 1.58$ and the maximum value is 1.
15. Velocity = 5.33 m/sec., acceleration = −45.59 m/sec^2.
16. Angular velocity = 3.82 radians/sec.
    Angular acceleration = 6.75 radians/sec^2.

6.3 NUMERICAL QUADRATURE

Numerical quadrature is the process of computing the approximate value of a definite integral using a set of numerical values of the integrand. We shall assume that the integrand is regular and the interval of integration is finite. Numerical quadrature is performed by approximating the integrand by an interpolating polynomial and then integrating it within the given limits.

6.3.1 General Quadrature Formula

Here the integrand is approximated by Newton's forward interpolation formula up to nth order differences and it is integrated within the limits x_0 to $x_0 + nh$.

Newton's forward interpolation formula is

$$f(x) = f_0 + p\Delta f_0 + \binom{p}{2}\Delta^2 f_0 + \cdots + \binom{p}{n}\Delta^n f_0$$

where $$x = x_0 + ph$$

$\therefore$ $$dx = h\,dp$$

$\therefore$ $$\int_{x_0}^{x_0+nh} f(x)\,dx \simeq h\int_0^n \left[f_0 + p\Delta f_0 + \binom{p}{2}\Delta^2 f_0 + \cdots + \binom{p}{n}\Delta^n f_0\right]dp$$

$$= h\int_0^n \left[f_0 + p\Delta f_0 + \frac{p^2 - p}{2}\Delta^2 f_0 + \frac{p^3 - 3p^2 + 2p}{3!}\Delta^3 f_0 + \cdots\right]dp$$

$$= h\left[p f_0 + \frac{p^2}{2}\Delta f_0 + \frac{1}{2}\left(\frac{p^3}{3} - \frac{p^2}{2}\right)\Delta^2 f_0 + \frac{1}{3!}\left(\frac{p^4}{4} - p^3 + p^2\right)\Delta^3 f_0 + \cdots\right]_0^n$$

$$= h\left[n f_0 + \frac{n^2}{2}\Delta f_0 + \frac{1}{2}\left(\frac{n^3}{3} - \frac{n^2}{2}\right)\Delta^2 f_0 + \frac{1}{3!}\left(\frac{n^4}{4} - n^3 + n^2\right)\Delta^3 f_0 + \cdots\right]$$

From this general formula, we obtain different quadrature formulae by taking $n = 1, 2, 3\ldots.$

6.3.2 Trapezoidal Rule

Here the integrand is approximated by a linear polynomial and integrated over the interval x_0 to $x_0 + h$.

$\therefore$ $$f(x) \simeq f_0 + p\Delta f_0$$

where $$x = x_0 + ph$$

$\therefore \quad dx = h\,dp$

$$\therefore \quad \int_{x_0}^{x_0+h} f(x)\,dx \simeq h\int_0^1 (f_0 + p\Delta f_0)\,dp = h\left(pf_0 + \frac{p^2}{2}\Delta f_0\right)_0^1$$

$$= h\left(f_0 + \frac{1}{2}\Delta f_0\right) = h\left[f_0 + \frac{1}{2}(f_1 - f_0)\right]$$

$$= \frac{h}{2}[f_0 + f_1]$$

If we are to find $\int_a^b f(x)\,dx$ by trapezoidal rule then we divide interval (a, b) into n equal parts at $a = x_0,\ x_0 + h,\ x_0 + 2h, \ldots, x_0 + nh = b$ so that $h = \frac{b-a}{n}$ and apply trapezoidal rule in intervals $(x_0, x_0 + h),\ (x_0 + h, x_0 + 2h), \ldots, \left(x_0 + \overline{n-1}\,h,\ x_0 + nh\right)$ and add, we shall have

$$\int_a^b f(x)\,dx \simeq \frac{h}{2}\left[(f_0 + f_1) + (f_1 + f_2) + \cdots + (f_{n-1} + f_n)\right]$$

$$= \frac{h}{2}\left[f_0 + 2(f_1 + f_2 + \cdots + f_{n-1}) + f_n\right]$$

6.3.3 Simpson's One-third Rule

Here the integrand is approximated by a quadratic polynomial and integrated over the interval x_0 to $x_0 + 2h$.

$$\therefore \quad f(x) \simeq f_0 + p\Delta f_0 + \frac{p(p-1)}{2}\Delta^2 f_0$$

where $\quad x = x_0 + ph$

$\therefore \quad dx = h\,dp$

$$\therefore \quad \int_{x_0}^{x_0+2h} f(x)\,dx \simeq h\int_0^2 \left(f_0 + p\Delta f_0 + \frac{p^2 - p}{2}\Delta^2 f_0\right)dp$$

$$= h\left[pf_0 + \frac{p^2}{2}\Delta f_0 + \frac{1}{2}\left(\frac{p^3}{3} - \frac{p^2}{2}\right)\Delta^2 f_0\right]_0^2$$

$$= h\left[2f_0 + 2\Delta f_0 + \frac{1}{2}\left(\frac{2}{3}\right)\Delta^2 f_0\right]$$

$$= h\left[2f_0 + 2(f_1 - f_0) + \frac{1}{3}(f_2 - 2f_1 + f_0)\right]$$

$$= \frac{h}{3}(f_0 + 4f_1 + f_2)$$

If we are to find $\int_a^b f(x)\,dx$ by Simpson's one-third rule then we divide interval (a, b) into n equal parts at $a = x_0, x_0 + h, x_0 + 2h, \ldots, x_0 + nh = b$ so that $h = \dfrac{b-a}{n}$ and n is even and apply Simpson's one-third rule in intervals $(x_0, x_0 + 2h), (x_0 + 2h, x_0 + 4h), (x_0 + 4h, x_0 + 6h), \ldots, (x_0 + \overline{n-2}\,h, x_0 + nh)$ and add. We shall have

$$\int_a^b f(x)\,dx \simeq \frac{h}{3}\left[(f_0 + 4f_1 + f_2) + (f_2 + 4f_3 + f_4) + \cdots + (f_{n-2} + 4f_{n-1} + f_n)\right]$$

$$= \frac{h}{3}\left[f_0 + f_n + 4(f_1 + f_3 + f_5 + \cdots + f_{n-1}) + 2(f_2 + f_4 + \cdots + f_{n-2})\right]$$

6.3.4 Simpson's Three-eight Rule

Here the integrand is approximated by a cubic polynomial and integrated over the interval x_0 to $x_0 + 3h$.

$$\therefore \qquad f(x) \simeq f_0 + p\Delta f_0 + \frac{p^2 - p}{2}\Delta^2 f_0 + \frac{p^3 - 3p^2 + 2p}{6}\Delta^3 f_0$$

where $\qquad x = x_0 + ph$

$\therefore \qquad dx = h\,dp$

$$\therefore \qquad \int_{x_0}^{x_0+3h} f(x)\,dx \simeq h\int_0^3\left(f_0 + p\Delta f_0 + \frac{p^2 - p}{2}\Delta^2 f_0 + \frac{p^3 - 3p^2 + 2p}{6}\Delta^3 f_0\right)dp$$

$$= h\left[p\,f_0 + \frac{p^2}{2}\Delta f_0 + \frac{1}{2}\left(\frac{p^3}{3} - \frac{p^2}{2}\right)\Delta^2 f_0 + \frac{1}{6}\left(\frac{p^4}{4} - p^3 + p^2\right)\Delta^3 f_0\right]_0^3$$

$$= h\left[3f_0 + \frac{9}{2}\Delta f_0 + \frac{1}{2}\left(9 - \frac{9}{2}\right)\Delta^2 f_0 + \frac{1}{6}\left(\frac{81}{4} - 27 + 9\right)\Delta^3 f_0\right]$$

$$= h\left[3f_0 + \frac{9}{2}(f_1 - f_0) + \frac{9}{4}(f_2 - 2f_1 + f_0) + \frac{3}{8}(f_3 - 3f_2 + 3f_1 - f_0)\right]$$

$$= \frac{3h}{8}\left[8f_0 + 12(f_1 - f_0) + 6(f_2 - 2f_1 + f_0) + f_3 - 3f_2 + 3f_1 - f_0\right]$$

$$= \frac{3h}{8}\left[f_0 + 3f_1 + 3f_2 + f_3\right]$$

If we are to find $\int_a^b f(x)\,dx$ by Simpson's three-eight rule then we divide (a, b) into n equal parts at $a = x_0, x_0 + h, x_0 + 2h, x_0 + 3h, \ldots, x_0 + nh = b$ where n is multiple of 3, $h = \dfrac{b-a}{n}$ and apply Simpson's three-eight rule in intervals $(x_0, x_0 + 3h)$, $(x_0 + 3h, x_0 + 6h)$, $(x_0 + 6h, x_0 + 9h), \ldots, (x_0 + \overline{n-3}\,h, x_0 + nh)$ and add. We shall have

$$\int_a^b f(x)\,dx = \frac{3h}{8}\left[(f_0+3f_1+3f_2+f_3)+(f_3+3f_4+3f_5+f_6)+\cdots+(f_{n-3}+3f_{n-2}+3f_{n-1}+f_n)\right]$$

$$= \frac{3h}{8}\left[f_0+f_n+3(f_1+f_2+f_4+f_5+\cdots+f_{n-2}+f_{n-1})+2(f_3+f_6+\cdots+f_{n-3})\right]$$

6.3.5 Weddle's Rule

Here the integrand is approximated by a 6th degree polynomial and integrated over the interval x_0 to $x_0 + 6h$.

$$\therefore \quad f(x) \simeq f_0 + p\Delta f_0 + \frac{p^2-p}{2}\Delta^2 f_0 + \frac{p^3-3p^2+2p}{6}\Delta^3 f_0$$
$$+\frac{1}{24}\left(p^4-6p^3+11p^2-6p\right)\Delta^4 f_0 + \frac{1}{120}\left(p^5-10p^4+35p^3-50p^2+24p\right)\Delta^5 f_0$$
$$+\frac{1}{720}\left(p^6-15p^5+85p^4-225p^3+274p^2-120p\right)\Delta^6 f_0$$

where $\quad x = x_0 + ph$

$$\therefore \quad dx = h\,dp$$

$$\therefore \quad \int_{x_0}^{x_0+6h} f(x)\,dx = h\int_0^6 \Bigg[f_0 + p\Delta f_0 + \frac{p^2-p}{2}\Delta^2 f_0 + \frac{p^3-3p^2+2p}{6}\Delta^3 f_0$$
$$+\frac{1}{24}\left(p^4-6p^3+11p^2-6p\right)\Delta^4 f_0$$
$$+\frac{1}{120}\left(p^5-10p^4+35p^3-50p^2+24p\right)\Delta^5 f_0$$
$$+\frac{1}{720}\left(p^6-15p^5+85p^4-225p^3+274p^2-120p\right)\Delta^6 f_0 \Bigg] dp$$

$$= h\Bigg[p f_0 + \frac{1}{2}p^2\Delta f_0 + \frac{1}{2}\left(\frac{p^3}{3}-\frac{p^2}{2}\right)\Delta^2 f_0 + \frac{1}{6}\left(\frac{p^4}{4}-p^3+p^2\right)\Delta^3 f_0$$
$$+\frac{1}{24}\left(\frac{p^5}{5}-\frac{3}{2}p^4+\frac{11}{3}p^3-3p^2\right)\Delta^4 f_0$$
$$+\frac{1}{120}\left(\frac{1}{6}p^6-2p^5+\frac{35}{4}p^4-\frac{50}{3}p^3+12p^2\right)\Delta^5 f_0$$
$$+\frac{1}{720}\left(\frac{1}{7}p^7-\frac{5}{2}p^6+17p^5-\frac{225}{4}p^4+\frac{274}{3}p^3-60p^2\right)\Delta^6 f_0 \Bigg]_0^6$$

$$= h\left[6f_0 + 18\Delta f_0 + 27\Delta^2 f_0 + 24\Delta^3 f_0 + \frac{123}{10}\Delta^4 f_0 + \frac{33}{10}\Delta^5 f_0 + \frac{41}{140}\Delta^6 f_0\right]$$

We replace $\frac{41}{140}\Delta^6 f_0$ by $\frac{3}{10}\Delta^6 f_0$, the error will be negligible

$$\therefore \int_{x_0}^{x_0+6h} f(x)dx = h\Big[6f_0+18(f_1-f_0)+27(f_2-2f_1+f_0)+24(f_3-3f_2+3f_1-f_0)$$

$$+\frac{123}{10}(f_4-4f_3+6f_2-4f_1+f_0)+\frac{33}{10}(f_5-5f_4+10f_3-10f_2+5f_1-f_0)$$

$$+\frac{3}{10}(f_6-6f_5+15f_4-20f_3+15f_2-6f_1+f_0)\Big]$$

$$= h\Bigg[\left(6-18+27-24+\frac{123}{10}-\frac{33}{10}+\frac{3}{10}\right)f_0+\left(18-54+72-\frac{492}{10}+\frac{165}{10}-\frac{18}{10}\right)f_1$$

$$+\left(27-72+\frac{738}{10}-33+\frac{45}{10}\right)f_2+\left(24-\frac{492}{10}+33-6\right)f_3$$

$$+\left(\frac{123}{10}-\frac{165}{10}+\frac{45}{10}\right)f_4+\left(\frac{33}{10}-\frac{18}{10}\right)f_5+\frac{3}{10}f_6\Bigg]$$

$$= \frac{3h}{10}[f_0+5f_1+f_2+6f_3+f_4+5f_5+f_6]$$

If we are to find $\int_a^b f(x)dx$ by Weddle's rule then we divide interval (a, b) into n equal parts at $a = x_0,\ x_0+h,\ x_0+2h,\ldots,x_0+nh=b$ where n is multiple of 6, $h=\frac{b-a}{n}$ and apply Weddle's rule in intervals (x_0, x_0+6h), $(x_0+6h, x_0+12h),\ldots,(x_0+\overline{n-6}\,h,\ x_0+nh)$ and add. We shall have

$$\int_a^b f(x)dx = \frac{3h}{10}\Big[f_0+f_n+5(f_1+f_7+f_{13}+\cdots+f_{n-5})$$

$$+5(f_5+f_{11}+f_{17}+\cdots+f_{n-1})+6(f_3+f_9+f_{15}+\cdots+f_{n-3})$$

$$+(f_2+f_4+f_8+f_{10}+\cdots+f_{n-4}+f_{n-2})+2(f_6+f_{12}+f_{18}+\cdots+f_{n-6})\Big]$$

6.3.6 Cote's Formulas

Here the integrand is approximated by Lagrange's interpolation polynomial passing through the points (x_0, y_0), (x_1, y_1), (x_2, y_2) ,…, (x_n, y_n) where $x_k = x_0 + k\,h$; $k = 0, 1, 2, \ldots, n$. Thus, if $f(x)$ is integrand then

$$f(x) \approx P(x) = \sum_{k=0}^{n} l_k(x)\, y_k$$

where
$$l_k(x) = \frac{(x-x_0)(x-x_1)\cdots(x-x_{k-1})(x-x_{k+1})\cdots(x-x_n)}{(x_k-x_0)(x_k-x_1)\cdots(x_k-x_{k-1})(x_k-x_{k+1})\cdots(x_k-x_n)}$$

If we put $x = x_0 + ph$ and hence $dx = hdp$

we have

$$l_k = \frac{p(p-1)\cdots(p-k+1)(p-k-1)\cdots(p-n)}{k(k-1)\cdots(1)(-1)\cdots(k-n)}$$

and we get

$$\int_{x_0}^{x_n} f(x)\,dx \simeq \int_{x_0}^{x_n} P(x)\,dx = h\int_0^n \sum_{k=0}^{n} l_k y_k\, dp$$

$$= nh\,\frac{1}{n}\sum_{k=0}^{n} y_k \int_0^n l_k\, dp$$

Put $$\frac{1}{n}\int_0^n l_k\, dp = {}^nC_k$$

We can write the integration formula

$$\int_{x_0}^{x_n} P(x)\,dx = nh\sum_{k=0}^{n} {}^nC_k\; y_k = (x_n - x_0)\sum_{k=0}^{n} {}^nC_k\; y_k$$

The numbers ${}^nC_k;\ 0 \le k \le n$ are called Cote's numbers.

Since $l_k = l_{n-k}$ and we have already proved in Remark (5.12)(ii) that $\sum l_k(x) = 1$

∴ Cote's numbers follow the properties

$${}^nC_k = {}^nC_{n-k} \text{ and } \sum_{k=0}^{n} {}^nC_k = 1$$

In Cote's formula, taking $n = 1, 2, 3$ we shall get trapezoidal rule, Simpson's one-third rule and Simpson's three-eight rule.

For $n = 1$

$${}^1C_0 = {}^1C_1 = \int_0^1 l_1\, dp = \int_0^1 p\, dp = \frac{1}{2}\left(p^2\right)_0^1 = \frac{1}{2}$$

∴ Cote's formula of order one is

$$\int_{x_0}^{x_0+h} f(x)\,dx \simeq h\sum_{k=0}^{1} {}^1C_k\; y_k = \frac{h}{2}(y_0 + y_1)$$

which is trapezoidal rule.

For $n = 2$

$${}^2C_0 = {}^2C_2 = \frac{1}{2}\int_0^2 l_2\, dp = \frac{1}{2}\int_0^2 \frac{p(p-1)}{2(1)}\, dp = \frac{1}{4}\left(\frac{p^3}{3} - \frac{p^2}{2}\right)_0^2 = \frac{1}{4}\left(\frac{8}{3} - 2\right) = \frac{1}{6}$$

$$\because \quad \sum_{k=0}^{2} {}^2C_k = 1$$

$$\therefore \quad {}^2C_1 = 1 - \left({}^2C_0 + {}^2C_2\right) = 1 - \frac{2}{6} = \frac{2}{3}$$

∴ Cote's formula of order 2 is

$$\int_{x_0}^{x_0+2h} f(x)\,dx \simeq 2h\sum_{k=0}^{2} {}^2C_k\; y_k = 2h\left[{}^2C_0\, y_0 + {}^2C_1\, y_1 + {}^2C_2\, y_2\right]$$

$$= 2h\left[\frac{1}{6}y_0 + \frac{2}{3}y_1 + \frac{1}{6}y_2\right]$$

$$= \frac{h}{3}\left[y_0 + 4y_1 + y_2\right]$$

which is Simpson's one-third rule.

For $n = 3$

$$^3C_0 = {^3C_3} = \frac{1}{3}\int_0^3 l_3\,dp = \frac{1}{3}\int_0^3 \frac{p(p-1)(p-2)}{(3)(2)(1)}\,dp$$

$$= \frac{1}{18}\int_0^3 \left(p^3 - 3p^2 + 2p\right)dp$$

$$= \frac{1}{18}\left(\frac{p^4}{4} - p^3 + p^2\right)_0^3 = \frac{1}{18}\left(\frac{81}{4} - 27 + 9\right) = \frac{1}{8}$$

Now,

$$^3C_1 = {^3C_2} \text{ and } \sum_{k=0}^{3} {^3C_k} = 1$$

$\therefore$
$$^3C_0 + {^3C_1} + {^3C_2} + {^3C_3} = 1$$

$\Rightarrow$
$$^3C_1 = {^3C_2} = \frac{1}{2}\left(1 - {^3C_0} - {^3C_3}\right) = \frac{1}{2}\left(1 - \frac{2}{8}\right) = \frac{3}{8}$$

$\therefore$ Cote's formula of order 3 is

$$\int_{x_0}^{x_0+3h} f(x)\,dx \simeq 3h\sum_{k=0}^{n} {^3C_k}\, y_k$$

$$= 3h\left[\frac{1}{8}y_0 + \frac{3}{8}y_1 + \frac{3}{8}y_2 + \frac{1}{8}y_3\right]$$

$$= \frac{3h}{8}\left[y_0 + 3y_1 + 3y_2 + y_3\right]$$

which is Simpson's three-eight rule.

6.3.7 Error Term in Quadrature Formulae

We shall be using Taylor's series method to find error in quadrature formulae.

Error Term in Trapezoidal Rule

In trapezoidal rule, we approximate

$$\int_{x_0}^{x_0+h} f(x)\,dx \text{ by } \frac{h}{2}\left[f(x_0) + f(x_0+h)\right]$$

$$\therefore \quad \text{error} = \int_{x_0}^{x_0+h} f(x)\,dx - \frac{h}{2}\left[f(x_0) + f(x_0+h)\right]$$

Let $f(x) = F'(x)$

$$\therefore \quad \text{error} = F(x_0+h)-F(x_0)-\frac{h}{2}\left[f(x_0)+f(x_0+h)\right]$$

$$= F(x_0)+hF'(x_0)+\frac{h^2}{2!}F''(x_0)+\frac{h^3}{3!}F'''(x_0)+\cdots$$

$$-F(x_0)-\frac{h}{2}\left[f(x_0)+f(x_0)+hf'(x_0)+\frac{h^2}{2!}f''(x_0)+\frac{h^3}{3!}f'''(x_0)+\cdots\right]$$

$$= hf(x_0)+\frac{h^2}{2!}f'(x_0)+\frac{h^3}{3!}f''(x_0)+\cdots$$

$$-hf(x_0)-\frac{h^2}{2}f'(x_0)-\frac{h^3}{4}f''(x_0)-\frac{h^4}{12}f'''(x_0)-\cdots \quad (\because F'(x)=f(x))$$

$$= -\frac{h^3}{12}f''(x_0)+\cdots = -\frac{h^3}{12}f''(x_0+\theta h);\ 0<\theta<1$$

If we are to find $\int_a^b f(x)\,dx$ and (a, b) is divided into n subintervals at $a=x_0, x_0+h, \ldots, x_0+nh=b$ then

$$\text{Error} = -\frac{h^3}{12}\left[f''(x_0+\theta_1 h)+f''(x_1+\theta_2 h)+\cdots+f''(x_{n-1}+\theta_n h)\right]$$

where $x_i=x_0+ih$ and $0<\theta_1, \theta_2\ldots,\theta_n<1$

If $\quad M = \left|\max_{x_0<x<x_n} f''(x)\right|$

then $\quad |\text{max error}| \le \dfrac{nh^3}{12}M = \dfrac{(b-a)h^2}{12}M$

this shows that the error is of order two.

Error Term in Simpson's One-third Rule

In Simpson's one-third rule, we approximate

$\int_{x_0}^{x_0+2h} f(x)\,dx$ by $\dfrac{h}{3}\left[f(x_0)+4f(x_0+h)+f(x_0+2h)\right]$

$$\therefore \quad \text{error} = \int_{x_0}^{x_0+2h} f(x)\,dx - \frac{h}{3}\left[f(x_0)+4f(x_0+h)+f(x_0+2h)\right].$$

Let $f(x) = F'(x)$

$$\therefore \quad \text{error} = F(x_0+2h)-F(x_0)-\frac{h}{3}\left[f(x_0)+4f(x_0+h)+f(x_0+2h)\right]$$

$$= \left\{2hF'(x_0)+\frac{(2h)^2}{2!}F''(x_0)+\frac{(2h)^3}{3!}F'''(x_0)+\frac{(2h)^4}{4!}F^{iv}(x_0)+\frac{(2h)^5}{5!}F^{v}(x_0)+\cdots\right\}$$

$$-\frac{h}{3}\left[f(x_0)+4\left\{f(x_0)+hf'(x_0)+\frac{h^2}{2!}f''(x_0)+\frac{h^3}{3!}f'''(x_0)+\frac{h^4}{4!}f^{iv}(x_0)+\cdots\right\}\right.$$
$$\left.+\left\{f(x_0)+2hf'(x_0)+\frac{(2h)^2}{2!}f''(x_0)+\frac{(2h)^3}{3!}f'''(x_0)+\frac{(2h)^4}{4!}f^{iv}(x_0)+\cdots\right\}\right]$$
$$=\left\{2hf(x_0)+2h^2f'(x_0)+\frac{4}{3}h^3f''(x_0)+\frac{2}{3}h^4f'''(x_0)+\frac{4h^5}{15}f^{iv}(x_0)+\cdots\right\}$$
$$-\frac{h}{3}\left[(1+4+1)f(x_0)+6hf'(x_0)+4h^2f''(x_0)+2h^3f'''(x_0)+\frac{5}{6}h^4f^{iv}(x_0)+\cdots\right]$$
$$(\because\ F'(x)=f(x))$$
$$=-\frac{h^5}{90}f^{iv}(x_0)+\cdots$$
$$=-\frac{h^5}{90}f^{iv}(x_0+\theta_1 h);\quad 0<\theta_1<2$$

If we are to find $\int_a^b f(x)\,dx$ and (a, b) is divided into n subintervals where n is even at $a = x_0$, $x_0 + h, \ldots, x_0 + nh = b$ then

$$\text{Error} = -\frac{h^5}{90}\left[f^{iv}(x_0+\theta_1 h)+f^{iv}(x_2+\theta_2 h)+f^{iv}(x_4+\theta_3 h)+\cdots+f^{iv}(x_{n-2}+\theta_{n/2}h)\right]$$

where $x_i = x_0 + ih$ and $0<\theta_1,\theta_2,\ldots,\theta_{n/2}<2$

If $\quad M = \left|\max_{x_0<x<x_n}.\ f^{iv}(x)\right|$

then $|\text{max. error}| \le \dfrac{nh^5}{180}M = \dfrac{(b-a)h^4}{180}M$

Hence, error in Simpson's one-third rule is of order 4.

Error Term in Simpson's Three-Eight Rule

In Simpson's there-eight rule, we approximate

$\int_{x_0}^{x_0+3h} f(x)\,dx$ by $\dfrac{3h}{8}\left[f(x_0)+3f(x_0+h)+3f(x_0+2h)+f(x_0+3h)\right]$

$\therefore \qquad \text{error} = \int_{x_0}^{x_0+3h} f(x)\,dx-\dfrac{3h}{8}\left[f(x_0)+3f(x_0+h)+3f(x_0+2h)+f(x_0+3h)\right]$

Let $f(x) = F'(x)$

$\therefore \qquad \text{error} = F(x_0+3h)-F(x_0)-\dfrac{3h}{8}\left[f(x_0)+3f(x_0+h)+3f(x_0+2h)+f(x_0+3h)\right]$

$$= \left\{ 3hF'(x_0) + \frac{9h^2}{2!}F''(x_0) + \frac{27h^3}{3!}F'''(x_0) \right.$$

$$\left. + \frac{81h^4}{4!}F^{iv}(x_0) + \frac{243h^5}{5!}F^{v}(x_0) + \frac{729h^6}{6!}F^{vi}(x_0) + \cdots \right\}$$

$$-\frac{3h}{8}\left[f(x_0) + 3\left\{ f(x_0) + hf'(x_0) + \frac{h^2}{2!}f''(x_0) + \frac{h^3}{3!}f'''(x_0) + \frac{h^4}{4!}f^{iv}(x_0) + \frac{h^5}{5!}f^{v}(x_0) + \cdots \right\} \right.$$

$$+ 3\left\{ f(x_0) + 2hf'(x_0) + 2h^2 f''(x_0) + \frac{4}{3}h^3 f'''(x_0) + \frac{2}{3}h^4 f^{iv}(x_0) + \frac{4}{15}h^5 f^{v}(x_0) + \cdots \right\}$$

$$\left. + \left\{ f(x_0) + 3hf'(x_0) + \frac{9h^2}{2}f''(x_0) + \frac{9h^3}{2}f'''(x_0) + \frac{27}{8}h^4 f^{iv}(x_0) + \frac{81}{40}h^5 f^{v}(x_0) + \cdots \right\} \right]$$

$$= \left\{ 3hf(x_0) + \frac{9}{2}h^2 f'(x_0) + \frac{9}{2}h^3 f''(x_0) + \frac{27}{8}h^4 f'''(x_0) + \frac{81h^5}{40}f^{iv}(x_0) + \frac{81h^6}{80}f^{v}(x_0) + \cdots \right\}$$

$$-\frac{3h}{8}\left[f(x_0) + 3\left\{ f(x_0) + hf'(x_0) + \frac{h^2}{2}f''(x_0) + \frac{h^3}{6}f'''(x_0) + \frac{h^4}{24}f^{iv}(x_0) + \frac{h^5}{120}f^{v}(x_0) + \cdots \right\} \right.$$

$$+ 3\left\{ f(x_0) + 2hf'(x_0) + 2h^2 f''(x_0) + \frac{4h^3}{3}f'''(x_0) + \frac{2h^4}{3}f^{iv}(x_0) + \frac{4}{15}h^5 f^{v}(x_0) + \cdots \right\}$$

$$\left. + \left\{ f(x_0) + 3hf'(x_0) + \frac{9}{2}h^2 f''(x_0) + \frac{9h^3}{2}f'''(x_0) + \frac{27}{8}h^4 f^{iv}(x_0) + \frac{81}{40}h^5 f^{v}(x_0) + \cdots \right\} \right]$$

$$= -\frac{3}{80}h^5 f^{iv}(x_0) + \cdots$$

$$= -\frac{3}{80}h^5 f^{iv}(x_0 + \theta_1 h); \quad 0 < \theta_1 < 3$$

If we are to find $\int_a^b f(x)\,dx$ and (a, b) is divided into n subintervals where n is multiple of 3 at $a = x_0, x_0 + h, \ldots, x_0 + nh = b$ then

$$\text{Error} = -\frac{3}{80}h^5 \left[f^{iv}(x_0 + \theta_1 h) + f^{iv}(x_3 + \theta_2 h) + f^{iv}(x_6 + \theta_3 h) + \cdots + f^{iv}(x_{n-3} + \theta_{n/3} h) \right]$$

where $x_i = x_0 + ih$ and $0 < \theta_1, \theta_2, \ldots, \theta_{n/3} < 3$

If $\quad M = \left| \max_{x_0 < x < x_n} f^{iv}(x) \right|$

then $|\text{max error}| \le \frac{nh^5}{80}M = \frac{(b-a)}{80}h^4 M$

Hence, error in Simpson's three-eight rule is of order 4.

We observe that the error in Simpson's one-third rule is less as compared to error in Simpson's three-eight rule and also it requires less calculations. Hence Simpson's one-third rule is preferred in comparison to Simpson's three-eight rule. This is the reason that we sometimes call Simpson's one-third rule as Simpson's rule.

Example 6.6: Evaluate $\int_0^1 \frac{dx}{1+x}$ using Simpson's one-third rule by taking seven ordinates and compare it with the actual value.

Solution: $h=\frac{1}{6}$

x	0	$\frac{1}{6}$	$\frac{1}{3}$	$\frac{1}{2}$	$\frac{2}{3}$	$\frac{5}{6}$	1
$y=\frac{1}{1+x}$	1	$\frac{6}{7}$	$\frac{3}{4}$	$\frac{2}{3}$	$\frac{3}{5}$	$\frac{6}{11}$	$\frac{1}{2}$

By Simpson's one-third rule

$$\int_0^1 \frac{dx}{1+x} \simeq \frac{h}{3}\left[y_0+y_6+4(y_1+y_3+y_5)+2(y_2+y_4)\right]$$

$$=\frac{1}{18}\left[1+\frac{1}{2}+4\left(\frac{6}{7}+\frac{2}{3}+\frac{6}{11}\right)+2\left(\frac{3}{4}+\frac{3}{5}\right)\right]$$

$$=\frac{1}{18}\left[1+0.5+4\left(\frac{198+154+126}{231}\right)+2\left(\frac{27}{20}\right)\right]$$

$$=\frac{1}{18}[1+0.5+8.2770563+2.7]$$

$$\simeq 0.693169793$$

Exact value of $\int_0^1 \frac{dx}{1+x} = \left(\log(1+x)\right)_0^1 = \log 2 \simeq 0.693147181$

$\therefore$ up to three decimals values are same.

Example 6.7: Evaluate $\int_0^1 \frac{dx}{1+x^2}$ using

(i) Trapezoidal rule taking $h=\frac{1}{4}$

(ii) Simpson's $\frac{1}{3}$ rule taking $h=\frac{1}{4}$

(iii) Simpson's $\frac{3}{8}$ rule taking $h=\frac{1}{6}$

(iv) Weddle's rule taking $h=\frac{1}{6}$

Solution:

x	0	$\frac{1}{4}$	$\frac{1}{2}$	$\frac{3}{4}$	1
$y=\frac{1}{1+x^2}$	1	$\frac{16}{17}$	$\frac{4}{5}$	$\frac{16}{25}$	$\frac{1}{2}$

(i) By trapezoidal rule taking $h=\frac{1}{4}$

$$\int_0^1 \frac{dx}{1+x^2}=\frac{h}{2}\left[y_0+y_4+2(y_1+y_2+y_3)\right]$$

$$=\frac{1}{8}\left[1+\frac{1}{2}+2\left(\frac{16}{17}+\frac{4}{5}+\frac{16}{25}\right)\right]$$

$$\simeq 0.7828$$

(ii) By Simpson's $\frac{1}{3}$ rule taking $h=\frac{1}{4}$

$$\int_0^1 \frac{dx}{1+x^2}=\frac{h}{3}\left[y_0+y_4+4(y_1+y_3)+2y_2\right]$$

$$=\frac{1}{12}\left[1+\frac{1}{2}+4\left(\frac{16}{17}+\frac{16}{25}\right)+2\left(\frac{4}{5}\right)\right]$$

$$\simeq 0.7854$$

For (iii) and (iv), table is

x	0	$\frac{1}{6}$	$\frac{1}{3}$	$\frac{1}{2}$	$\frac{2}{3}$	$\frac{5}{6}$	1
$y=\frac{1}{1+x^2}$	1	$\frac{36}{37}$	$\frac{9}{10}$	$\frac{4}{5}$	$\frac{9}{13}$	$\frac{36}{61}$	$\frac{1}{2}$

(iii) By Simpson's $\frac{3}{8}$ rule taking $h=\frac{1}{6}$

$$\int_0^1 \frac{dx}{1+x^2}=\frac{3}{8}\times\frac{1}{6}\left[y_0+y_6+3(y_1+y_2+y_4+y_5)+2y_3\right]$$

$$=\frac{1}{16}\left[1+\frac{1}{2}+3\left(\frac{36}{37}+\frac{9}{10}+\frac{9}{13}+\frac{36}{61}\right)+2\left(\frac{4}{5}\right)\right]$$

$$\simeq 0.7854$$

(iv) By Weddle's rule taking $h=\frac{1}{6}$

$$\int_0^1 \frac{dx}{1+x^2} = \frac{3h}{10}[y_0+5y_1+y_2+6y_3+y_4+5y_5+y_6]$$

$$=\frac{1}{20}\left[1+5\left(\frac{36}{37}\right)+\frac{9}{10}+6\left(\frac{4}{5}\right)+\frac{9}{13}+5\left(\frac{36}{61}\right)+\frac{1}{2}\right]$$

$$\simeq 0.7854$$

Example 6.8: Find an approximate value of $\int_0^{\frac{\pi}{2}} \sqrt{\cos\theta}\, d\theta$ using Simpson's $\frac{1}{3}$ rule by dividing the interval into six subintervals.

Solution:

θ	0	$\frac{\pi}{12}$	$\frac{\pi}{6}$	$\frac{\pi}{4}$	$\frac{\pi}{3}$	$\frac{5\pi}{12}$	$\frac{\pi}{2}$
$y=\sqrt{\cos\theta}$	1	0.9828	0.9306	0.8409	0.7071	0.5087	0

Using Simpson's $\frac{1}{3}$ rule with $h=\frac{\pi}{12}$

$$\int_0^{\frac{\pi}{2}} \sqrt{\cos\theta}\, d\theta \simeq \frac{\pi}{36}\left[1+0+4(0.9828+0.8409+0.5087)+2(0.9306+0.7071)\right] \simeq 1.187$$

Example 6.9: The speeds of an electric train at various times after leaving one station until it stops at the next station are given in the following table. Find the distance between two stations.

Speed in mph	0	13	33	$39\frac{1}{2}$	40	40	36	15	0
Time in min	0	$\frac{1}{2}$	1	$1\frac{1}{2}$	2	$2\frac{1}{2}$	3	$3\frac{1}{4}$	$3\frac{1}{2}$

Solution:

Let t be time in hrs. and $v(t)$ be speed in mph.

t	0	$\frac{1}{120}$	$\frac{1}{60}$	$\frac{1}{40}$	$\frac{1}{30}$	$\frac{1}{24}$	$\frac{1}{20}$	$\frac{13}{240}$	$\frac{7}{120}$
$v(t)$	0	13	33	$39\frac{1}{2}$	40	40	36	15	0

Let S be the distance between the two stations. As the length of subintervals for $t=0$ to $t=\frac{1}{20}$ hrs. is $\frac{1}{120}$ hrs. and length of subintervals for $t=\frac{1}{20}$ hrs. to $t=\frac{7}{120}$ hrs. is $\frac{1}{240}$ hrs., so distance travelled for $t=0$ to $t=\frac{1}{20}$ hrs. and for $t=\frac{1}{20}$ hrs. to $t=\frac{7}{120}$ hrs. are to be found separately.

$$\therefore \quad S=\int_0^{7/120} v(t)\,dt=\int_0^{1/20} v(t)\,dt+\int_{1/20}^{7/120} v(t)\,dt$$

By Simpson's $\frac{1}{3}$ rule with $h=\frac{1}{120}$ hrs. for the time 0 to $\frac{1}{20}$ hrs. and $h=\frac{1}{240}$ hrs. for the time $\frac{1}{20}$ hrs. to $\frac{7}{120}$ hrs., we have

$$S=\frac{1}{3(120)}\left[0+36+4\left(13+\frac{79}{2}+40\right)+2(33+40)\right]+\frac{1}{3(240)}\left[36+4(15)+0\right]$$

$$=\frac{23}{15}+\frac{2}{15}=\frac{25}{15}=\frac{5}{3}=1.667 \text{ miles.}$$

Example 6.10: Estimate the length of the arc of the curve $3y = x^3$ form (0, 0) to $\left(1,\frac{1}{3}\right)$ using Simpson's $\frac{1}{3}$ rule by taking $h = 0.125$.

Solution: Equation of curve is $y=\frac{1}{3}x^3$

$$\therefore \quad \frac{dy}{dx}=x^2$$

x	0	0.125	0.250	0.375	0.500	0.625	0.750	0.875	1.000
$\sqrt{1+\left(\frac{dy}{dx}\right)^2}=\sqrt{1+x^4}$	1	1.00012	1.00195	1.00984	1.03078	1.07359	1.14735	1.25944	1.41421

If S be the length of the arc of the curve $y=\frac{1}{3}x^3$ form (0, 0) to $\left(1,\frac{1}{3}\right)$

then $S=\int_0^1\sqrt{1+\left(\frac{dy}{dx}\right)^2}\,dx=\int_0^1\sqrt{1+x^4}\,dx$

By Simpson's $\frac{1}{3}$ rule with $h = 0.125$

$$S=\int_0^1\sqrt{1+x^4}\,dx=\frac{0.125}{3}[1+1.41421+4(1.00012+1.00984+1.07359+1.25944)$$
$$+2(1.00195+1.03078+1.14735)]$$

$$=\frac{0.125}{3}\left[1+1.41421+4(4.34299)+2(3.18008)\right]$$

$$=1.08943 \text{ units}$$

$\therefore$ required length of arc $\simeq 1.0894$ units.

Example 6.11: A curve passes through seven points (1, 2), (1.5, 2.4), (2.0, 2.7), (2.5, 2.8), (3, 3), (3.5, 2.6) and (4, 2.1). Obtain the area bounded by the curve, the x-axis and the ordinates at $x = 1$ and $x = 4$. Also find the volume of the solid of revolution obtained by revolving this area about the axis of x.

Solution:

x	1.0	1.5	2.0	2.5	3.0	3.5	4.0
y	2.0	2.4	2.7	2.8	3.0	2.6	2.1
y^2	4.00	5.76	7.29	7.84	9.00	6.76	4.41

By Simpson's $\frac{1}{3}$ rule with $h = 0.5$

$$\text{Required area} = \int_1^4 y\,dx = \frac{0.5}{3}\left[2.0+2.1+4(2.4+2.8+2.6)+2(2.7+3.0)\right]$$

$$\simeq 7.783 \text{ sq.units}$$

Required volume of solid of revolution

$$= \pi\int_1^4 y^2\,dx$$

$$= \pi.\frac{0.5}{3}\left[4.00+4.41+4(5.76+7.84+6.76)+2(7.29+9.00)\right]$$

$$\simeq 64.10 \text{ cubic units}$$

EXERCISE 6.2

1. Given that $y = \log x$ and

x	4.0	4.2	4.4	4.6	4.8	5.0	5.2
y	1.3863	1.4351	1.4816	1.5261	1.5686	1.6094	1.6487

Evaluate $\int_4^{5.2} \log x\,dx$ by

(a) Trapezoidal rule (b) Simpson's $\frac{1}{3}$ rule

(c) Simpson's $\frac{3}{8}$ rule (d) Weddle's rule

Also compare it with exact value.

2. Calculate $\int_2^{10} \frac{dx}{1+x}$ by dividing the range into 16 equal parts.

3. Evaluate $\int_0^6 \frac{dx}{(1+x^2)}$ using

(a) Trapezoidal rule (b) Simpson's $\frac{1}{3}$ rule

(c) Simpson's $\frac{3}{8}$ rule (d) Weddle's rule

and compare the results with the exact value of the integral.

4. Evaluate $\int_0^1 e^{-x^2} dx$ by taking 11 ordinates using

(a) Trapezoidal rule (b) Simpson's rule

5. Calculate by Simpson's rule an approximate value of $\int_{-3}^{3} x^4 dx$ by taking seven equidistant ordinates. Compare it with the exact value and the value obtained by using the trapezoidal rule.

6. Evaluate $\int_{0.5}^{0.7} x^{1/7} e^{-x} dx$ approximately by using a suitable formula.

7. Calculate $\int_0^{\pi/2} e^{\sin x} dx$ by taking seven equidistant ordinates.

8. Find the value of log 2 from $\int_0^1 \frac{x^2}{1+x^3} dx$ using Simpson's $\frac{1}{3}$ rule by dividing the range into four equal parts. Also find the error.

9. Evaluate $\int_0^4 e^x dx$ by Simpson's rule, using the data $e = 2.72$, $e^2 = 7.39$, $e^3 = 20.09$, $e^4 = 54.60$, and compare it with the actual value.

10. Compute the value of the integral $\int_{0.2}^{1.4} \left(\sin x - \log_e x + e^x\right) dx$ by

(a) Trapezoidal rule (b) Simpson's one-third rule

(c) Simpson's three-eight rule (d) Weddle's rule

using 7 ordinates. After finding the true value of the integral, compare the errors in the four cases.

11. Calculate an approximate value of $\int_0^{\pi/2} \sin x \, dx$ by

(a) Trapezoidal rule (b) Simpson's rule

using 11 ordinates.

12. The table below gives the velocity v of speeding car at time t seconds. Approximate the distance travelled by the car in 12 s.

t (in seconds)	0	2	4	6	8	10	12
v (in m/s)	4	6	16	34	60	94	136

13. Find from the following table the area bounded by the curve $y = f(x)$ and the x-axis from $x = 7.47$ to 7.52

x	7.47	7.48	7.49	7.50	7.51	7.52
$f(x)$	1.93	1.95	1.98	2.01	2.03	2.06

14. A river is 80 m wide. The depth 'd' in metres at a distance x metres from one bank is given by the following table. Calculate the area of the cross section of the river using

    (a) Trapezoidal rule (b) Simpson's $\frac{1}{3}$ rule

x	0	10	20	30	40	50	60	70	80
d	0	4	7	9	12	15	14	8	3

15. Find the volume of solid of revolution formed by rotating about the x-axis the area between the x-axis, the lines $x = 0$ and $x = 1$, and a curve through the points with the following coordinates:

x	0.00	0.25	0.50	0.75	1.00
y	1.0000	0.9896	0.9589	0.9089	0.8415

ANSWERS 6.2

1. (a) 1.82766 (b) 1.8278533
   (c) 1.8278475 (d) 1.827858 Exact value 1.8278474
2. 1.299
3. By taking $h = 1$,
   (a) 1.4108 (b) 1.3662 (c) 1.3571 (d) 1.3734 Exact value 1.4056
4. (a) 0.7462 (b) 0.7468
5. By Simpson's rule 98; Exact value 97.2; By trapezoidal rule 115
6. 0.10207 (taking $h = 0.05$ in Simpson's $\frac{1}{3}$ rule)
7. 3.1044 (by Simpson's $\frac{1}{3}$ rule)
8. By Simpson's $\frac{1}{3}$ rule 0.23108; Exact value 0.23105; Error 0.00003
9. 53.873; Actual value 53.598
10. (a) 4.0715 (b) 4.0522 (c) 4.0530 (d) 4.0514
    True value 4.0509; Errors are in
    (a) 0.0206 (b) 0.0013 (c) 0.0021 (d) 0.0005
11. (a) 0.9979 (b) 1.0000045
12. 552 m
13. 0.100 sq. units
14. (a) 705 m^2 (b) 710 m^2
15. 2.8192 cubic units

6.4 NUMERICAL SOLUTIONS OF ORDINARY DIFFERENTIAL EQUATIONS

The analytical solutions of differential equations can be obtained in a limited way as there is no general method of solving the equations. Quite often the differential equations appearing in physical problems do not belong to the known types of analytic solutions. Such equations are solved by numerical methods. We shall consider the methods to solve first order differential equation $\frac{dy}{dx} = f(x, y),\ y(x_0) = y_0$ and then explain the method to solve second order differential equations and simultaneous first order differential equations.

6.4.1 Taylor-series Method

Let $f(x, y)$ be a function which is differentiable for sufficient number of times.

Consider the initial value problem

$$\frac{dy}{dx} = y' = f(x, y),\ y(x_0) = y_0$$

From $y' = f(x, y)$ we can find

$$y'' = f_x + f_y f$$

$$y''' = f_{xx} + f_{xy} f + f_{yx} f + f_y f_x + \left(f_{yy} f + f_y^2\right) f \text{ and } y^{iv}, y^{v}, \ldots.$$

We find their values at $x = x_0$. Then using the Taylor series expansion about x_0 we can find

$$y(x_0 + h) = y_0 + h y_0' + \frac{h^2}{2!} y_0'' + \frac{h^3}{3!} y_0''' + \cdots + \frac{h^n}{n!} y_0^{(n)}$$

The error term will be $\frac{h^{n+1}}{(n+1)!} y^{(n+1)}(\xi);\ x_0 < \xi < x_0 + h$

The number of terms depends upon the permissible error.

If we are to solve the simultaneous differential equations

$$\frac{dy}{dx} = y' = f(x, y, z)$$

$$\frac{dz}{dx} = z' = \phi(x, y, z)$$

subject to the conditions $y(x_0) = y_0$, $z(x_0) = z_0$ then differentiating these equations successively we find

$y'', z'', y''', z''', \ldots$ in this order and then using

$$y(x_0 + h) = y_0 + h y_0' + \frac{h^2}{2!} y_0'' + \cdots$$

$$z(x_0 + h) = z_0 + h z_0' + \frac{h^2}{2!} z_0'' + \cdots$$

we find $y(x_0+h)$ and $z(x_0+h)$.

If we are to solve second order differential equation

$$\frac{d^2y}{dx^2}=f\left(x,y,\frac{dy}{dx}\right);\quad y(x_0)=y_0,\ \ y'(x_0)=y_0'$$

Then, we suppose that $\dfrac{dy}{dx}=z$ and solve the simultaneous equations

$$\frac{dy}{dx}=z=\phi(x,y,z)$$

$$\frac{d^2y}{dx^2}=\frac{dz}{dx}=f(x,y,z);\ z(x_0)=y_0',\ y(x_0)=y_0$$

Example 6.12: Using Taylor's series find the solution of the differential equation $xy'=x-y$; $y(2)=2$ at $x=2.1$ correct to five places of decimals.

Solution: Differential equation is

$$xy'=x-y\,;\ y(2)=2,\qquad \therefore\ y'(2)=1-\frac{y(2)}{2}=1-\frac{2}{2}=0$$

On differentiating, we have

$$xy''+2y'=1\qquad \therefore\ y''(2)=\frac{1-0}{2}=\frac{1}{2}$$

$$xy'''+3y''=0\qquad \therefore\ y'''(2)=-\frac{3}{4}$$

$$xy^{iv}+4y'''=0\qquad \therefore\ y^{iv}(2)=\frac{3}{2},$$

$$\frac{(.1)^5}{5!}y^{v}(\xi)=\frac{.00001}{120}y^{v}(\xi)<(10)^{-6}$$

$\therefore$ By Taylor series expansion

$$y(x)=y(2)+(x-2)y'(2)+\frac{(x-2)^2}{2!}y''(2)+\frac{(x-2)^3}{3!}y'''(2)+\frac{(x-2)^4}{4!}y^{iv}(2)+\cdots$$

$$\therefore\qquad y(2.1)\approx 2+\frac{(.1)^2}{4}-\frac{(.1)^3}{8}+\frac{(.1)^4}{16}$$

$\therefore\qquad y(2.1)=2.00238$ correct to 5 decimal places.

Example 6.13: Using Taylor-series method solve

$\dfrac{dy}{dx}=x^2-y,\quad y(0)=1$ at $x=0.1$, 0.2, 0.3 and 0.4. Compare the values with exact solution.

Solution: $y'=x^2-y;\ y(0)=1 \qquad \therefore\ y'(0)=0-1=-1$

Differentiating both sides, we have

$$y''+y'=2x \qquad \therefore\ y''(0)=-y'(0)=1$$

$$y'''+y''=2 \qquad \therefore\ y'''(0)=2-y''(0)=1$$

$$y^{iv}=-y''' \qquad \therefore\ y^{iv}(0)=-1$$

$$y^{v}=-y^{iv} \qquad \therefore\ y^{v}(0)=1$$

$$y^{vi}=-y^{v} \qquad \therefore\ y^{vi}(0)=-1$$

Hence, $\quad y^{(n)}(0)=(-1)^{n+1};\ n\geq 3$

$\therefore$ By Taylor series expansion

$$y(x)=y(0)+xy'(0)+\frac{x^2}{2!}y''+\frac{x^3}{3!}y'''(0)+\frac{x^4}{4!}y^{iv}(0)+\frac{x^5}{5!}y^{v}(0)+\frac{x^6}{6!}y^{vi}(0)+\cdots$$

$$\therefore \qquad y(x)\simeq 1-x+\frac{x^2}{2}+\frac{x^3}{6}-\frac{x^4}{24}+\frac{x^5}{120}$$

$$\therefore \qquad y(0.1)\simeq 1-(0.1)+\frac{(0.1)^2}{2}+\frac{(0.1)^3}{6}-\frac{(0.1)^4}{24}=0.9052$$

$$y(0.2)\simeq 1-(0.2)+\frac{(0.2)^2}{2}+\frac{(0.2)^3}{6}-\frac{(0.2)^4}{24}=0.8213$$

$$y(0.3)\simeq 1-(0.3)+\frac{(0.3)^2}{2}+\frac{(0.3)^3}{6}-\frac{(0.3)^4}{24}+\frac{(0.3)^5}{120}=0.7492$$

$$y(0.4)\simeq 1-(0.4)+\frac{(0.4)^2}{2}+\frac{(0.4)^3}{6}-\frac{(0.4)^4}{24}+\frac{(0.4)^5}{120}=0.6897$$

All these results are up to 4 decimal places.

Now,

$$\frac{dy}{dx}+y=x^2$$

is Leibnitz's linear differential equation

$$I.F. = e^{\int dx} = e^x$$

∴ solution is

$$ye^x = \int x^2 e^x dx = \left(x^2 - 2x + 2\right)e^x + c$$

$$\Rightarrow \quad y = x^2 - 2x + 2 + ce^{-x}$$

$$\therefore \quad y(0) = 2 + c = 1 \qquad (\because y(0) = 1)$$

$$\therefore \quad c = -1$$

∴ exact solution is

$$y = x^2 - 2x + 2 - e^{-x}$$

From the exact solution

$$y(0.1) = (0.1)^2 - 2(0.1) + 2 - e^{-0.1} = 0.9052$$

$$y(0.2) = (0.2)^2 - 2(0.2) + 2 - e^{-0.2} = 0.8213$$

$$y(0.3) = (0.3)^2 - 2(0.3) + 2 - e^{-0.3} = 0.7492$$

$$y(0.4) = (0.4)^2 - 2(0.4) + 2 - e^{-0.4} = 0.6897$$

These results are taken up to 4 decimal places.

We obtain that the result obtained are same in both cases.

Example 6.14: Given the initial value problem $u' = t^2 + u^2,\ u(0) = 0$. Determine the first three non-zero terms in the Taylor series for $u(t)$ and hence obtain the value for $u(1)$. Also determine t when the error in $u(t)$ obtained from the first two non-zero terms is to be less than 10^{-6} after rounding.

Solution:

$$u' = t^2 + u^2;\ u(0) = 0 \qquad \therefore\ u'(0) = 0 + u^2(0) = 0$$

$$\therefore\ u'' = 2t + 2uu' \qquad \therefore\ u''(0) = 0$$

$$u''' = 2 + 2u'^2 + 2uu'' \qquad \therefore\ u'''(0) = 2$$

$$u^{iv} = 6u'u'' + 2uu''' \qquad \therefore\ u^{iv}(0) = 0$$

$$u^{v} = 6u''^2 + 8u'u''' + 2uu^{iv} \qquad \therefore\ u^{v}(0) = 0$$

$$u^{vi} = 20u''u''' + 10u'u^{iv} + 2uu^{v} \qquad \therefore\ u^{vi}(0) = 0$$

$$u^{vii} = 20u'''^2 + 30u''u^{iv} + 12u'u^{v} + 2uu^{vi} \qquad \therefore\ u^{vii}(0) = 80$$

$$u^{viii} = 70u'''u^{iv} + 42u''u^{v} + 14u'u^{vi} + 2uu^{vii} \qquad \therefore\ u^{viii}(0) = 0$$

$$u^{ix} = 70u^{iv^2} + 112u'''u^{v} + 56u''u^{vi} + 16u'u^{vii} + 2uu^{viii} \qquad \therefore\ u^{ix}(0) = 0$$

$$u^{x} = 252u^{iv}u^{v} + 168u'''u^{vi} + 72u''u^{vii} + 18u'u^{viii} + 2uu^{ix} \qquad \therefore\ u^{x}(0) = 0$$

$$u^{xi} = 252u^{v^2} + 420u^{iv}u^{vi} + 240u'''u^{vii} + 90u''u^{viii} + 20u'u^{ix} + 2uu^{x}$$

$$\therefore\ u^{xi}(0) = 240(2)(80) = 38400$$

$\therefore$ By Taylor series expansion up to first three non-zero terms

$$u(t) = 2\frac{t^3}{3!} + 80\frac{t^7}{7!} + 38400\frac{t^{11}}{11!} = \frac{t^3}{3} + \frac{t^7}{63} + 2\frac{t^{11}}{2079}$$

$$\therefore \qquad u(1) = \frac{1}{3} + \frac{1}{63} + \frac{2}{2079} \simeq 0.350168$$

Error in $u(t)$ obtained from the first two non-zero terms

$$\simeq \frac{2t^{11}}{2079} < 0.0000005$$

$$\therefore \qquad t^{11} < \frac{(5)(2079)}{2}.10^{-7}$$

$$\therefore \qquad t < \left[\frac{(5)(2079)}{2}.10^{-7}\right]^{\frac{1}{11}}$$

$$\therefore \qquad t \simeq 0.50$$

Example 6.15: Using Taylor series, find the numerical solution of $\frac{dy}{dx} = 2y + 3e^x$ at $x = 0.2$, given that $y(0) = 1$. Compare the result with analytical solution.

Solution:

$$y' = \frac{dy}{dx} = 2y + 3e^x;\ \ y(0) = 1, \qquad \therefore\ y'(0) = 2 + 3 = 5$$

Differentiate n times

$$y^{(n+1)} = 2y^{(n)} + 3e^x;\ \ n = 1, 2, 3\ldots$$

$$\therefore \qquad y''(0) = 2y'(0) + 3 = 2(5) + 3 = 13$$

$$y'''(0) = 2y''(0) + 3 = 2(13) + 3 = 29$$

$$y^{iv}(0) = 2y'''(0) + 3 = 2(29) + 3 = 61$$

$$y^{v}(0) = 2y^{iv}(0) + 3 = 2(61) + 3 = 125$$

∴ By Taylor series expansion

$$y(x) \simeq y(0) + xy'(0) + \frac{x^2}{2!}y''(0) + \frac{x^3}{3!}y'''(0) + \frac{x^4}{4!}y^{iv}(0) + \frac{x^5}{5!}y^{v}(0)$$

$$= 1 + 5x + \frac{13}{2}x^2 + \frac{29}{6}x^3 + \frac{61}{24}x^4 + \frac{25}{24}x^5$$

∴ $$y(0.2) \simeq 1 + 5(0.2) + \frac{13}{2}(0.2)^2 + \frac{29}{6}(0.2)^3 + \frac{61}{24}(0.2)^4 + \frac{25}{24}(0.2)^5$$

∴ $y(0.2) = 2.3031$ to 4 decimals.

Analytical solution

$$\frac{dy}{dx} - 2y = 3e^x$$

is Leibnitz's linear differential equation.

$$I.F. = e^{-2x}$$

Solution is

$$ye^{-2x} = \int 3e^x.e^{-2x}dx = -3e^{-x} + c$$

∴ $$y = -3e^x + ce^{2x}$$

∴ $$y(0) = -3 + c = 1 \qquad (\because y(0) = 1)$$

∴ $$c = 4$$

∴ solution is

$$y(x) = 4e^{2x} - 3e^x$$

∴ $$y(0.2) = 4e^{0.4} - 3e^{0.2} = 2.3031$$

It is up to 4 decimals.

Thus, up to 4 decimals results are same.

Example 6.16: Evaluate by means of Taylor's series expansion, the value of y at $x = 0.1, 0.2$ to four significant figures for the problem

$$y'' - x(y')^2 + y^2 = 0; \quad y(0) = 1, \; y'(0) = 0.$$

Solution: Given differential equation is

$$y'' - x(y')^2 + y^2 = 0; \quad y(0) = 1, \; y'(0) = 0$$

Let $y' = z$

$\therefore \quad y'' = z' = xz^2 - y^2 ; \quad y(0) = 1, z(0) = y'(0) = 0, \qquad \therefore \; y''(0) = z'(0) = 0 - (1)^2 = -1$

$y''' = z'' = z^2 + 2xzz' - 2yy' \qquad \therefore \; y'''(0) = 0$

$y^{iv} = z''' = 4zz' + 2xz'^2 + 2xzz'' - 2y'^2 - 2yy'' \qquad \therefore \; y^{iv}(0) = 2$

$\therefore$ By Taylor's series expansion

$$y(t) = y(0) + ty'(0) + \frac{t^2}{2!} y''(0) + \frac{t^3}{3!} y'''(0) + \frac{t^4}{4!} y^{iv}(0)$$

$$= 1 - \frac{t^2}{2} + \frac{t^4}{12}$$

$$\therefore \quad y(0.1) \simeq 1 - \frac{(0.1)^2}{2} + \frac{(0.1)^4}{12} \simeq 0.9950$$

$$y(0.2) \simeq 1 - \frac{(0.2)^2}{2} + \frac{(0.2)^4}{12} \simeq 0.9801$$

EXERCISE 6.3

1. Solve $\frac{dy}{dx} = x + y$, given $y = 0$ when $x = 1$, up to $x = 1.2$ with $h = 0.1$. Compare the final result with the value of the explicit solution.
2. Solve $\frac{dy}{dx} = x - y^2$, $y(0) = 1$ to find $y(0.1)$ correct to four places of decimal by using Taylor series method.
3. Solve $\frac{dy}{dx} = y^2 + 1$, $y(0) = 0$ in the range $0 \le x \le 1$ writing y as series in power of x and check your result with the exact values
4. Use Taylor's series method to solve the equation $\frac{dy}{dx} = -xy$, $y(0) = 1$.
5. Solve the differential equation $y'' = xy$ for $x = 0.5$ and $x = 1$ by Taylor series method where $y(0) = 0$, $y'(0) = 1$.
6. Using Taylor's series method, obtain the solution of $\frac{dy}{dx} = 3x + y^2$ and $y = 1$, when $x = 0$. Find the value of y for $x = 0.1$, correct to four places of decimals.
7. Solve by Taylor's series method:

   $y' = y - \frac{2x}{y}$, $y(0) = 1$ for $x = 0.1$ and -0.1.
8. Find by Taylor's series method, the value of y at $x = 0.1$ and $x = 0.2$ to five places of decimals from $\frac{dy}{dx} = x^2 y - 1$, $y(0) = 1$
9. Solve the following differential equation by Taylor's series method $y' = 2t + 3y$, $y(0) = 1$ and find $y(0.1)$.
10. Using Taylor series expansion evaluate the integral of $y' - 2y = 3e^x$, $y(0) = 0$ at

    (a) $x = 0.1(0.1)\,0.3$ (b) $x = 1.0;\, 1.1$

11. Find the solution u (0.1) and u (0.2) of the initial value problem $u' = x(1 - 2u^2)$; $u(0) = 1$ using the first three non-zero terms of the Taylor series method.
12. Solve numerically the system of simultaneous equations $\frac{dx}{dt}+2x+3y=0$, $\frac{dy}{dt}+3x+2y=2e^{2t}$ with initial conditions $x = 1, y = 2$ at $t = 0$, for $t = 0.1$ by Taylor series method.
13. Use Taylor's series method to obtain the power series in t for x and y satisfying the differential equations $\frac{dx}{dt}=x+y+t$, $\frac{d^2y}{dt^2}=x-t$ under the initial conditions $x(0) = 0$, $y(0) = 1$, $\left(\frac{dy}{dt}\right)_{t=0}=-1$
14. Find solution of the simultaneous equations $\frac{dx}{dt}=xy+2t$, $\frac{dy}{dt}=2ty+x$ subject to the initial conditions $x(0) = 1, y(0) = -1$ by Taylor series method.
15. Solve $\frac{dy}{dx}=x+z$, $\frac{dz}{dx}=x-y^2$ with $y(0) = 2$, $z(0) = 1$ to get $y(0.1)$, $y(0.2)$, $z(0.1)$ and $z(0.2)$ approximately by Taylor's algorithm.
16. Solve the differential equation $\frac{d^2y}{dx^2}+x\frac{dy}{dx}+y=0$, $y(0)=1$, $y'(0)=0$ to approximate $y(0.1)$ by Taylor's series method.

ANSWERS 6.3

1. $y(1.1) = 0.1103$, $y(1.2) = 0.2428$ and results are same by explicit solution.
2. $y(0.1) = 0.9138$
3.

x	0.2	0.4	0.6	0.8	1
Calculated $y(x)$	0.203	0.423	0.684	1.026	1.521
Exact $y(x)$	0.203	0.423	0.684	1.030	1.557

4. $y(x)=1-\frac{x^2}{2}+\frac{x^4}{8}-\frac{x^6}{48}+\ldots\ldots$
5. $y(0.5)\simeq 0.505$, $y(1)\simeq 1.085$
6. $y(x)=1+x+\frac{5}{2}x^2+2x^3+\frac{9}{4}x^4+\frac{177}{60}x^5+\ldots\ldots; y(0.1)=1.12725$
7. $y(0.1)\simeq 1.0954$, $y(-0.1)\simeq 0.8944$
8. $y(0.1)=0.90031$, $y(0.2)=0.80227$
9. $y(0.1)=1.36094$
10. (a) $y(0.1)\simeq 0.34870$, $y(0.2)\simeq 0.81125$, $y(0.3)\simeq 1.41657$
    (b) $y(1.0)\simeq 13.65000$, $y(1.1)\simeq 17.39683$

11. $u(0.1) \simeq 0.99505,\ u(0.2) \simeq 0.9808$
12. $x(0.1) \simeq 0.3291,\ y(0.1) \simeq 1.6668$
13. $x(t) = t + \frac{t^2}{2!} + \frac{t^3}{3!} + \frac{t^4}{4!} + \frac{2t^5}{5!} + \frac{3t^6}{6!} + \cdots\cdots$

    $y(t) = 1 - t + \frac{t^4}{4!} + \frac{t^5}{5!} + \frac{t^6}{6!} + \cdots\cdots$
14. $x(t) = 1 - t + 2t^2 - \frac{3}{2}t^3 + \frac{19}{12}t^4 + \cdots$ and $y(t) = -1 + t - \frac{3}{2}t^2 + \frac{4}{3}t^3 - \frac{9}{8}t^4 + \cdots$
15. $y(0.1) \simeq 2.0845,\ y(0.2) \simeq 2.1367,\ z(0.1) \simeq 0.5867, z(0.2) \simeq 0.1535$
16. $y(0.1) \simeq 0.9950$

6.4.2 Picard's Method

Consider the differential equation

$$\frac{dy}{dx} = y' = f(x, y);\quad y(x_0) = y_0$$

We have

$$\int_{x_0}^{x} y'(x)\,dx = \int_{x_0}^{x} f(x, y)\,dx$$

or

$$y(x) - y(x_0) = \int_{x_0}^{x} f(x, y)\,dx$$

or

$$y(x) = y(x_0) + \int_{x_0}^{x} f(x, y)\,dx$$

It is an integral equation of the given differential equation. First approximation to $y(x)$ denoted by y_1 is obtained by taking $y = y_0$ in the integral

$$\therefore \quad y_1 = y(x_0) + \int_{x_0}^{x} f(x, y_0)\,dx$$

Then, second approximation is obtained by taking $y = y_1$ in the integral

$$y_2 = y(x_0) + \int_{x_0}^{x} f(x, y_1)\,dx$$

Proceeding in this way

$$y_n = y(x_0) + \int_{x_0}^{x} f(x, y_{n-1})\,dx$$

Thus, iterative formula for $y(x)$ is

$$y_n(x) = y(x_0) + \int_{x_0}^{x} f(x, y_{n-1})\,dx$$

If we are to solve the simultaneous differential equations

$$\frac{dy}{dx} = y' = f(x, y, z)$$

$$\frac{dz}{dx} = z' = \phi(x, y, z); \quad y(x_0) = y_0, \; z(x_0) = z_0$$

Then, from the formulae

$$y_n = y(x_0) + \int_{x_0}^{x} f(x, y_{n-1}, z_{n-1})\, dx$$

$$z_n = z(x_0) + \int_{x_0}^{x} \phi(x, y_{n-1}, z_{n-1})\, dx$$

we obtain $y_1, z_1, y_2, z_2, \ldots, \ldots, y_n, z_n$ in this order and when $y_n \simeq y_{n-1}, \; z_n \simeq z_{n-1}$ up to desired accuracy then iteration is stopped.

If we are to solve the second order differential equation

$$\frac{d^2y}{dx^2} = f\left(x, y, \frac{dy}{dx}\right); \quad y(x_0) = y_0, \; y'(x_0) = y_0'$$

then we suppose $\frac{dy}{dx} = z$ and solve the simultaneous differential equations

$$\frac{dy}{dx} = \phi(x, y, z) = z$$

$$\frac{dz}{dx} = f(x, y, z); \quad y(x_0) = y_0, \; z(x_0) = y_0'$$

Remark 6.1:

(i) Picard's method fails if the function $f(x, y)$ is not successively integrable.

(ii) An extra condition called Lipschitz condition

$$|f(x, y) - f(x, z)| < L|y - z| \text{ where } L \text{ is constant}$$

is required for convergence.

Example 6.17: Employing Picard's method, solve the differential equation $\frac{dy}{dx} = 2x - y$, given that $y = 0.9$ at $x = 0$, for $x = 0.2, 0.4$ and $x = 0.6$. Check the result with exact values.

Solution:

$$\frac{dy}{dx} = 2x - y \,;\, y(0) = 0.9$$

By Picard's method, iterative formula for finding approximations to $y(x)$ is

$$y_n(x) = y(0) + \int_0^x (2x - y_{n-1}(x))\, dx = 0.9 + x^2 - \int_0^x y_{n-1}(x)\, dx$$

where $\quad y_0 = y(0) = 0.9$

Following table gives approximations

n	$y_n(x)$
0	0.9
1	$0.9-0.9x+x^2$
2	$0.9-0.9x+1.45x^2-\frac{x^3}{3}$
3	$0.9-0.9x+1.45\left(x^2-\frac{x^3}{3}\right)+\frac{x^4}{12}$
4	$0.9-0.9x+1.45\left(x^2-\frac{x^3}{3}+\frac{x^4}{12}\right)-\frac{x^5}{60}$
5	$0.9-0.9x+1.45\left(x^2-\frac{x^3}{3}+\frac{x^4}{12}-\frac{x^5}{60}\right)+\frac{x^6}{360}$
6	$0.9-0.9x+1.45\left(x^2-\frac{x^3}{3}+\frac{x^4}{12}-\frac{x^5}{60}+\frac{x^6}{360}\right)-\frac{x^7}{2520}$

From last two approximations

$$y(x)\simeq 0.9-0.9x+1.45\left(x^2-\frac{x^3}{3}+\frac{x^4}{12}-\frac{x^5}{60}\right)$$

$$y(0.2)\simeq 0.9-0.9(0.2)+1.45\left[(0.2)^2-\frac{1}{3}(0.2)^3+\frac{1}{12}(0.2)^4-\frac{1}{60}(0.2)^5\right]=0.7743$$

$$y(0.4)=0.9-0.9(0.4)+1.45\left[(0.4)^2-\frac{(0.4)^3}{3}+\frac{(0.4)^4}{12}-\frac{(0.4)^5}{60}\right]=0.7439$$

$$y(0.6)=0.9-0.9(0.6)+1.45\left[(0.6)^2-\frac{(0.6)^3}{3}+\frac{(0.6)^4}{12}-\frac{(0.6)^5}{60}\right]=0.7914$$

Now, given differential equation is

$$\frac{dy}{dx}+y=2x$$

It is Leibnitz's linear differential equation.

$$I.F.=e^x$$

Solution is

$$ye^x = \int 2xe^x \, dx = 2(x-1)e^x + c$$

$\therefore \quad y = 2x - 2 + ce^{-x}$

$\therefore \quad y(0) = -2 + c = 0.9$

$\therefore \quad c = 2.9$

$\therefore$ particular solution is

$$y = 2x - 2 + 2.9e^{-x}$$

$$\therefore \quad y(0.2) = -1.6 + 2.9e^{-0.2} = 0.7743$$
$$y(0.4) = -1.2 + 2.9e^{-0.4} = 0.7439$$
$$y(0.6) = -0.8 + 2.9e^{-0.6} = 0.7916$$

$y(0.6)$ is different at 4th decimal place and $y(0.2)$ and $y(0.4)$ are correct up to four decimal places.

Example 6.18: Solve $\dfrac{dy}{dx} = 1 + xy$, given that $y = 1$ when $x = 0$, in the interval [0, 0.5] for $h = 0.1$ correct to three decimal places.

Solution:

$$\frac{dy}{dx} = 1 + xy \, ; \, y(0) = 1$$

By Picard's method, iterative formula for finding approximations to $y(x)$ is

$$y_n(x) = y(0) + \int_0^x [1 + xy_{n-1}(x)] \, dx = 1 + x + \int_0^x x \, y_{n-1}(x) \, dx$$

where $\quad y_0(x) = y(0) = 1$

Following table gives approximations

n	$y_n(x)$
0	1
1	$1 + x + \dfrac{x^2}{2}$
2	$1 + x + \dfrac{x^2}{2} + \dfrac{x^3}{3} + \dfrac{x^4}{8}$
3	$1 + x + \dfrac{x^2}{2} + \dfrac{x^3}{3} + \dfrac{x^4}{8} + \dfrac{x^5}{15} + \dfrac{x^6}{48}$
4	$1 + x + \dfrac{x^2}{2} + \dfrac{x^3}{3} + \dfrac{x^4}{8} + \dfrac{x^5}{15} + \dfrac{x^6}{48} + \dfrac{x^7}{105} + \dfrac{x^8}{384}$
5	$1 + x + \dfrac{x^2}{2} + \dfrac{x^3}{3} + \dfrac{x^4}{8} + \dfrac{x^5}{15} + \dfrac{x^6}{48} + \dfrac{x^7}{105} + \dfrac{x^8}{384} + \dfrac{x^9}{945} + \dfrac{x^{10}}{3840}$

Now, for maximum $\dfrac{x^7}{105}$ in [0, 0.5]

we have maximum value = $\dfrac{(0.5)^7}{105} \simeq 0.00007$

$\therefore$ for values correct to three decimal places, from last two approximations we have

$$y(x) = 1 + x + \frac{x^2}{2} + \frac{x^3}{3} + \frac{x^4}{8} + \frac{x^5}{15} + \frac{x^6}{48}$$

$$y(0.1) = 1 + (0.1) + \frac{(0.1)^2}{2} + \frac{(0.1)^3}{3} + \frac{(0.1)^4}{8} + \frac{(0.1)^5}{15} + \frac{(0.1)^6}{48} = 1.105$$

$$y(0.2) = 1 + 0.2 + \frac{(0.2)^2}{2} + \frac{(0.2)^3}{3} + \frac{(0.2)^4}{8} + \frac{(0.2)^5}{15} + \frac{(0.2)^6}{48} = 1.223$$

$$y(0.3) = 1 + 0.3 + \frac{(0.3)^2}{2} + \frac{(0.3)^3}{3} + \frac{(0.3)^4}{8} + \frac{(0.3)^5}{15} + \frac{(0.3)^6}{48} = 1.355$$

$$y(0.4) = 1 + 0.4 + \frac{(0.4)^2}{2} + \frac{(0.4)^3}{3} + \frac{(0.4)^4}{8} + \frac{(0.4)^5}{15} + \frac{(0.4)^6}{48} = 1.505$$

$$y(0.5) = 1 + 0.5 + \frac{(0.5)^2}{2} + \frac{(0.5)^3}{3} + \frac{(0.5)^4}{8} + \frac{(0.5)^5}{15} + \frac{(0.5)^6}{48} = 1.677$$

$\therefore$ up to 3 decimals

$$y(0.1) = 1.105,\ y(0.2) = 1.223,\ y(0.3) = 1.355,\ y(0.4) = 1.505,\ y(0.5) = 1.677$$

Example 6.19: Approximate y and z by using Picard's method for the particular solution of $\dfrac{dy}{dx} = x + z,\ \dfrac{dz}{dx} = x - y^2$ for $x = 0.1$ given that $y = 2$, $z = 1$ where $x = 0$

Solution:

$$\frac{dy}{dx} = x + z,\ \frac{dz}{dx} = x - y^2;\ y(0) = 2,\ z(0) = 1$$

By Picard's method iterative formulae for finding approximations to $y(x)$ and $z(x)$ are

$$y_n(x) = y(0) + \int_0^x \left[x + z_{n-1}(x)\right] dx = 2 + \frac{x^2}{2} + \int_0^x z_{n-1}(x)\, dx$$

$$z_n(x) = z(0) + \int_0^x \left[x - y^2_{n-1}(x)\right] dx = 1 + \frac{x^2}{2} - \int_0^x y^2_{n-1}(x)\, dx$$

where $\quad y_0(x) = y(0) = 2,\ z_0(x) = z(0) = 1$

Following table gives approximations to $y(x)$ and $z(x)$ up to x^4.

n	$y_n(x)=2+\frac{x^2}{2}+\int_0^x z_{n-1}(x)\,dx$	$z_n(x)=1+\frac{x^2}{2}-\int_0^x y^2_{n-1}(x)\,dx$
0	2	1
1	$2+x+\frac{x^2}{2}$	$1+\frac{x^2}{2}-\int_0^x 4dx$ $=1-4x+\frac{x^2}{2}$
2	$2+x-\frac{3}{2}x^2+\frac{x^3}{6}$	$1+\frac{x^2}{2}-\int_0^x\left(4+4x+3x^2+x^3\right)dx$ $=1-4x-\frac{3}{2}x^2-x^3-\frac{x^4}{4}$
3	$2+x-\frac{3}{2}x^2-\frac{x^3}{2}-\frac{x^4}{4}$	$1+\frac{x^2}{2}-\int_0^x\left(4+4x-5x^2-\frac{7}{3}x^3\right)dx$ $=1-4x-\frac{3}{2}x^2+\frac{5}{3}x^3+\frac{7}{12}x^4$
4	$2+x-\frac{3}{2}x^2-\frac{x^3}{2}+\frac{5}{12}x^4$	$1+\frac{x^2}{2}-\int_0^x\left(4+4x-5x^2-5x^3\right)dx$ $=1-4x-\frac{3}{2}x^2+\frac{5}{3}x^3+\frac{5}{4}x^4$
5	$2+x-\frac{3}{2}x^2-\frac{x^3}{2}+\frac{5}{12}x^4$	$1+\frac{x^2}{2}-\int_0^x\left(4+4x-5x^2-5x^3\right)dx$ $=1-4x-\frac{3}{2}x^2+\frac{5}{3}x^3+\frac{5}{4}x^4$

From last two approximations

$$y(x)\simeq 2+x-\frac{3}{2}x^2-\frac{1}{2}x^3+\frac{5}{12}x^4$$

$$z(x)\simeq 1-4x-\frac{3}{2}x^2+\frac{5}{3}x^3+\frac{5}{4}x^4$$

$\therefore$

$$y(0.1)\simeq 2+(0.1)-\frac{3}{2}(0.1)^2-\frac{1}{2}(0.1)^3+\frac{5}{12}(0.1)^4\simeq 2.0845$$

$$z(0.1)\simeq 1-4(0.1)-\frac{3}{2}(0.1)^2+\frac{5}{3}(0.1)^3+\frac{5}{4}(0.1)^4\simeq 0.5868$$

Example 6.20: Use Picard's method to approximate y when $x = 0.1$ given that $\frac{d^2y}{dx^2} - x^2\frac{dy}{dx} - x^4 y = 0$ and $y = 5$ and $\frac{dy}{dx} = 1$ when $x = 0$.

Solution:

$$\frac{d^2y}{dx^2} - x^2\frac{dy}{dx} - x^4 y = 0\ ;\ \ y(0) = 5,\ \ y'(0) = 1$$

Let $\frac{dy}{dx} = z$

$$\therefore \qquad \frac{dz}{dx} = x^2 z + x^4 y\ ;\ \ y(0) = 5,\ \ z(0) = 1$$

By Picard's method iterative formulae for finding approximations to $y(x)$ and $z(x)$ are

$$y_n(x) = y(0) + \int_0^x z_{n-1}(x)\ dx = 5 + \int_0^x z_{n-1}(x)\ dx$$

$$z_n(x) = z(0) + \int_0^x \left[x^2 z_{n-1}(x) + x^4 y_{n-1}(x)\right] dx = 1 + \int_0^x \left[x^2 z_{n-1}(x) + x^4 y_{n-1}(x)\right] dx$$

where $y_0 = y(0) = 5,\ z_0 = z(0) = 1$

Following table gives approximations to $y(x)$ and $z(x)$

n	$y_n(x) = 5 + \int_0^x z_{n-1}(x) dx$	$z_n(x) = 1 + \int_0^x \left[x^2 z_{n-1}(x) + x^4 y_{n-1}(x)\right] dx$
0	5	1
1	$5 + x$	$1 + \frac{x^3}{3} + x^5$
2	$5 + x + \frac{x^4}{12} + \frac{x^6}{6}$	$1 + \frac{x^3}{3} + x^5 + \frac{2}{9}x^6 + \frac{1}{8}x^8$
3	$5 + x + \frac{x^4}{12} + \frac{x^6}{6} + \frac{2}{63}x^7 + \frac{1}{72}x^9$	

From last two approximations

$$y(x) \simeq 5 + x + \frac{x^4}{12} + \frac{x^6}{6}$$

$$\therefore \qquad y(0.1) \simeq 5 + (0.1) + \frac{(0.1)^4}{12} \simeq 5.10001$$

EXERCISE 6.4

1. Using Picard's method of successive approximations, obtain a solution up to the fifth approximation of the equation

   $\frac{dy}{dx} = y + x$ such that $y = 1$ when $x = 0$.

   Check your answer by finding the exact particular solution.

2. Use Picard's method to solve $\frac{dy}{dx} = -xy,\ y(0) = 1$.

3. Use Picard's method to approximate y when $x = 0.1$ given that $y = 1$ when $x = 0$ and $\frac{dy}{dx} = \frac{y-x}{y+x}$.

4. Perform two iterations of Picard's method to find an approximate solution of the initial value problem

   $$\frac{dy}{dx} = x + y^2,\ y(0) = 1$$

5. Use Picard's method to approximate the value of y when $x = 0.1$ given that $y = 1$ when $x = 0$

   $$\frac{dy}{dx} = 3x + y^2$$

6. Using three successive approximations of Picard's method, obtain approximate solution of the different equation

   $$y' = x^2 + y^2$$

   satisfying the initial condition $y(0) = 0$

7. Given the differential equation $\frac{dy}{dx} = \frac{x^2}{1+y^2}$ with the initial condition $y = 0$ when $x = 0$, use Picard's method to obtain y for $x = 0.25, 0.5$ and 1.0 correct to three places of decimals.

8. Approximate y and z at $x = 0.1$, using Picard's method for the solution of the equations $\frac{dy}{dx} = z,\ \frac{dz}{dx} = x^3(y+z)$ given that $y(0) = 1$ and $z(0) = \frac{1}{2}$.

9. Use Picard's method to solve the equations

   $\frac{dx}{dt} = -y,\ \frac{dy}{dt} = x$ given that $x = 1,\ y = 0$ when $t = 0$.

10. Use Picard's method to approximate y when $x = 0.1$ given that $\frac{d^2y}{dx^2} + 2x\frac{dy}{dx} + y = 0$ and $y = 0.5,\ \frac{dy}{dx} = 0.1$ when $x = 0$.

ANSWERS 6.4

1. $y_5 = 1 + x + x^2 + \frac{x^3}{3} + \frac{x^4}{12} + \frac{x^5}{60} + \frac{x^6}{720}$

   exact particular solution is

   $$y = 2e^x - x - 1 = 1 + x + x^2 + \frac{x^3}{3} + \frac{x^4}{12} + \frac{x^5}{60} + \frac{x^6}{360} + \cdots\cdots$$

2. $y \simeq 1 - \frac{x^2}{2} + \frac{x^4}{8} - \frac{x^6}{48} + \frac{x^8}{384} y \simeq 1 - \frac{x^2}{2} + \frac{x^4}{8} - \frac{x^6}{48} + \frac{x^8}{384}$

3. 1.0906
4. $y_2 = 1 + x + \frac{3}{2}x^2 + \frac{2}{3}x^3 + \frac{1}{4}x^4 + \frac{1}{20}x^5$
5. $y(0.1) = 1.127$
6. $y_3 = \frac{x^3}{3} + \frac{x^7}{63} + \frac{2x^{11}}{2079} + \frac{x^{15}}{59535}$
7. $y(0.25) = 0.0052,\ y(0.5) = 0.0416,\ y(1.0) = 0.3218$
8. $y(0.1) = 1.05000,\ z(0.1) = 0.50004$
9. $y \simeq t - \frac{t^3}{6} + \frac{t^5}{120},\ x \simeq 1 - \frac{t^2}{2} + \frac{t^4}{24} - \frac{t^6}{720}$
10. $y(0.1) = 0.5075$

6.4.3 Euler's Method

Consider the differential equation

$$\frac{dy}{dx} = f(x, y),\ y(x_0) = y_0$$

We want to find $y(x_n)$ where $x_n = x_0 + nh$.

This method is based on the assumption that in a small interval, the curve is a straight line passing through the initial point along the tangent at initial point. Thus, if $[x_0, x_1]$ is small interval where $x_1 = x_0 + h$ then we approximate the curve by tangent at (x_0, y_0).

Equation of tangent at (x_0, y_0) is

$$y - y_0 = \left(\frac{dy}{dx}\right)_{(x_0, y_0)} (x - x_0) = (x - x_0) f(x_0, y_0)$$

Thus, the value of y corresponding to $x = x_1$, is $y_1 = y_0 + (x_1 - x_0) f(x_0, y_0) = y_0 + h f(x_0, y_0)$.

Now, in the interval $[x_1, x_2]$, we approximate the curve by straight line passing through (x_1, y_1) along the tangent at (x_1, y_1).

Equation of tangent at (x_1, y_1) is

$$y - y_1 = \left(\frac{dy}{dx}\right)_{(x_1, y_1)} (x - x_1) = (x - x_1) f(x_1, y_1)$$

Thus, the value of y corresponding to $x = x_2$ is $y_2 = y_1 + (x_2 - x_1) f(x_1, y_1) = y_1 + h f(x_1, y_1)$.

Repeating this process n times, we shall have $y_n = y_{n-1} + h f(x_{n-1}, y_{n-1})$.

Geometrically, the actual curve of solution through $P(x_0, y_0)$ is shown dotted in Figure (6.1). In the interval (x_0, x_1), we approximate the curve by tangent at P which meets ordinate at $x = x_1$ in P_1. Now, let $P_1 Q_1$ be curve of solution through P_1. In interval (x_1, x_2), we approximate this curve by tangent at P_1 to curve $P_1 Q_1$ which meets ordinate at $x = x_2$ in P_2. Repeating this process n times, we reach point P_n. Thus, point Q on curve will be approximated by P_n by Euler method.

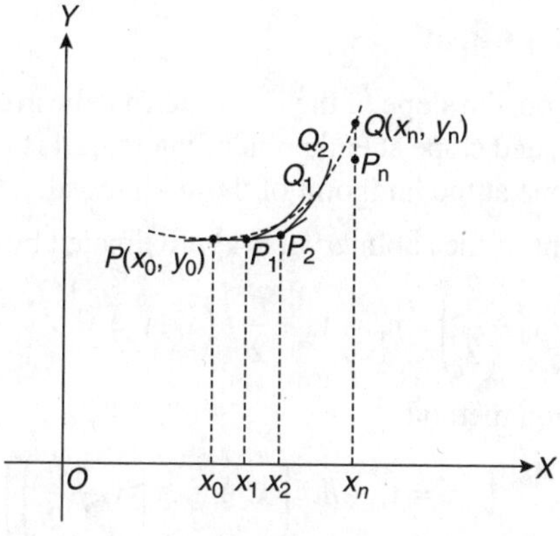

Figure 6.1

If h is not small then error is quite significant. Further, this sequence of lines may also deviate considerably from the curve of solution. Hence, the method is very slow and requires improvement.

6.4.4 Improved Euler's Method

In this method the approximation of y at $x = x_1$ is obtained as in Euler's method. Let it be $y_1^{(p)}$

$$\therefore \quad y_1^{(p)} = y_0 + h f(x_0, y_0)$$

Now, this approximation is improved by approximating the curve of solution in the interval (x_0, x_1) by a straight line passing through (x_0, y_0) but taking the slope of line the average of the slopes at (x_0, y_0) and $\left(x_1, y_1^{(p)}\right)$. Thus, the equation of line becomes

$$y - y_0 = (x - x_0)\left[\frac{1}{2}\left\{f(x_0, y_0) + f\left(x_1, y_1^{(p)}\right)\right\}\right]$$

$$\therefore \quad y_1^{(c)} = y_0 + \frac{h}{2}\left[f(x_0, y_0) + f\left(x_1, y_1^{(p)}\right)\right]$$

Proceeding in this way, we get

$$y_2^{(p)} = y_1^{(c)} + h f\left(x_1, y_1^{(c)}\right)$$

$$y_2^{(c)} = y_1^{(c)} + \frac{h}{2}\left[f\left(x_1, y_1^{(c)}\right) + f\left(x_2, y_2^{(p)}\right)\right]$$

$$\vdots$$

$$y_{n+1}^{(p)} = y_n^{(c)} + h\,f\left(x_n, y_n^{(c)}\right)$$

$$y_{n+1}^{(c)} = y_n^{(c)} + \frac{h}{2}\left[f\left(x_n, y_n^{(c)}\right) + \left(x_{n+1}, y_{n+1}^{(p)}\right)\right]; \quad y_0^{(c)} = y_0 = y(x_0)$$

6.4.5 Modified Euler's Method

In Euler's improvement method, the slope of the line is taken to be average of slope at initial point of sub-interval and approximated slope at end point of interval. It is again modified by taking the slope of the line to be the slope at the midpoint of the subinterval.

The value of y at the midpoint of the subinterval is approximated by Euler method:

$$y\left(x_0 + \frac{h}{2}\right) = y_{\frac{1}{2}} = y_0 + \frac{h}{2} f\left(x_0, y_0\right)$$

and then by Euler modification method

$$y_1 = y_0 + h\,f\left(x_0 + \frac{h}{2}, y\left(x_0 + \frac{h}{2}\right)\right)$$

Thus, the formula becomes

$$y_{n+1} = y_n + hf\left(x_{n+1/2}, y_{n+1/2}\right)$$

where $$y_{n+1/2} \doteq y_n + \frac{h}{2} f\left(x_n, y_n\right)$$

Example 6.21: Solve for y the differential equation $\frac{dy}{dx} = x + y$, $y(0) = 1$ at $x = 0.0\ (0.2)\ 0.8$ using
(i) Euler's method (ii) Picard's method

Also compare with exact values.

Solution:

$$\frac{dy}{dx} = f\left(x, y\right) = x + y; \quad y(0) = 1, \ h = 0.2$$

(i) **By Euler's method**

n	x_n	y_n	$f(x_n, y_n)$	$y_{n+1} = y_n + h\,f(x_n, y_n)$
0	0.0	1.0000	1.0000	1.2000
1	0.2	1.2000	1.4000	1.4800
2	0.4	1.4800	1.8800	1.8560
3	0.6	1.8560	2.4560	2.3472
4	0.8	2.3472		

$\therefore \quad y(0.2) = 1.2000, \ y(0.4) = 1.4800, \ y(0.6) = 1.8560, \ y(0.8) = 2.3472$

(ii) **By Picard's Method**

Iterative formula for approximations of $y(x)$ is

$$y_n(x) = y(0) + \int_0^x (x + y_{n-1}(x))\,dx = 1 + \frac{x^2}{2} + \int_0^x y_{n-1}(x)\,dx$$

where $y_0 = y(0) = 1$

Following table gives approximations

n	$y_n(x)$
0	1
1	$1+x+\frac{x^2}{2}$
2	$1+x+x^2+\frac{x^3}{6}$
3	$1+x+x^2+\frac{x^3}{3}+\frac{x^4}{24}$
4	$1+x+x^2+\frac{x^3}{3}+\frac{x^4}{12}+\frac{x^5}{120}$
5	$1+x+x^2+\frac{x^3}{3}+\frac{x^4}{12}+\frac{x^5}{60}+\frac{x^6}{720}$
6	$1+x+x^2+\frac{x^3}{3}+\frac{x^4}{12}+\frac{x^5}{60}+\frac{x^6}{360}+\frac{x^7}{5040}$

from last two approximations, $y(x)$ is approximated to

$$y(x) \simeq 1 + x + x^2 + \frac{x^3}{3} + \frac{x^4}{12} + \frac{x^5}{60}$$

$$\therefore \quad y(0.2) = 1 + (0.2) + (0.2)^2 + \frac{(0.2)^3}{3} + \frac{(0.2)^4}{12} + \frac{(0.2)^5}{60} \simeq 1.2428$$

$$y(0.4) = 1 + (0.4) + (0.4)^2 + \frac{(0.4)^3}{3} + \frac{(0.4)^4}{12} + \frac{(0.4)^5}{60} \simeq 1.5836$$

$$y(0.6) = 1 + (0.6) + (0.6)^2 + \frac{(0.6)^3}{3} + \frac{(0.6)^4}{12} + \frac{(0.6)^5}{60} \simeq 2.0441$$

$$y(0.8) = 1 + (0.8) + (0.8)^2 + \frac{(0.8)^3}{3} + \frac{(0.8)^4}{12} + \frac{(0.8)^5}{60} \simeq 2.6503$$

Analytically

$$\frac{dy}{dx} - y = x;\; y(0) = 1$$

It is Leibnitz's linear differential equation.

$$I.F. = e^{-x}$$

$\therefore$ solution is

$$ye^{-x} = \int xe^{-x}dx = (-x-1)e^{-x} + c$$

$$\therefore \quad y = -1 - x + ce^{x}$$

$$\therefore \quad y(0) = -1 + c = 1 \qquad (\because\ y(0) = 1)$$

$$\therefore \quad c = 2$$

$\therefore$ particular solution is

$$y = 2e^{x} - 1 - x$$

From this, we have

$$y(0.2) = 2e^{0.2} - 1 - 0.2 \simeq 1.2428$$

$$y(0.4) = 2e^{0.4} - 1 - 0.4 \simeq 1.5836$$

$$y(0.6) = 2e^{0.6} - 1 - 0.6 \simeq 2.0442$$

$$y(0.8) = 2e^{0.8} - 1 - 0.8 \simeq 2.6511$$

By Euler method $y(0.2)$ is correct only up to one decimal and $y(0.4)$, $y(0.6)$, $y(0.8)$ are not correct even up to one decimal. But by Picard's method $y(0.2)$ and $y(0.4)$ are correct up to four decimals, $y(0.6)$ is correct up to three decimals and $y(0.8)$ is correct up to two decimals.

Example 6.22: Solve the differential equation by Euler's method $\dfrac{dy}{dx} = \dfrac{y-x}{y+x}$ with the initial condition $y(0) = 1$ for $x = 0.1$ taking $h = 0.02$.

Solution:

$$\frac{dy}{dx} = f(x,y) = \frac{y-x}{y+x} = 1 - \frac{2x}{y+x}; \quad y(0) = 1, \ h = 0.02$$

n	x_n	y_n	$f(x_n, y_n)$	$y_{n+1} = y_n + h\,f(x_n, y_n)$
0	0.00	1.0000	1.0000	1.0200
1	0.02	1.0200	0.9615	1.0392
2	0.04	1.0392	0.9259	1.0577
3	0.06	1.0577	0.8926	1.0756
4	0.08	1.0756	0.8615	1.0928
5	0.10	1.0928		

$$\therefore \quad y(0.1) \simeq 1.0928$$

Example 6.23: Solve the differential equation $\frac{dy}{dx} = x + y,\ y(0) = 1$ for y (0.1), y (0.2), y (0.3) using modified Euler's method.

Solution:

$$f(x, y) = \frac{dy}{dx} = x + y,\ y(0) = 1,\ h = 0.1$$

n	x_n	y_n	$f(x_n, y_n)$	$y_{n+1/2} = y_n + \frac{h}{2} f(x_n, y_n)$	$f(x_{n+1/2}, y_{n+1/2})$	$y_{n+1} = y_n + hf(x_{n+1/2}, y_{n+1/2})$
0	0	1	1	1.05	1.1	1.11
1	0.1	1.11	1.21	1.1705	1.3205	1.24205
2	0.2	1.24205	1,44205	1.31415	1.56415	1.39846
3	0.3	1.39846				

$\therefore$

$$y(0.1) \simeq 1.11$$

$$y(0.2) \simeq 1.2420$$

$$y(0.3) \simeq 1.3985$$

Example 6.24: Use modified Euler's formula to obtain y (0.2), y (0.4), and y (0.6) correct to three decimal places given that $\frac{dy}{dx} = y - x^2$ with initial condition $y(0) = 1$.

Solution:

$$\frac{dy}{dx} = f(x, y) = y - x^2;\ y(0) = 1$$

We take $h = 0.1$

n	x_n	y_n	$f(x_n, y_n)$	$y_{n+1/2} = y_n + \frac{h}{2} f(x_n, y_n)$	$f(x_{n+1/2}, y_{n+1/2})$	$y_{n+1} = y_n + hf(x_{n+1/2}, y_{n+1/2})$
0	0	1	1	1.05	1.0475	1.1048
1	0.1	1.1048	1.0948	1.1595	1.1370	1.2185
2	0.2	1.2185	1.1785	1.2774	1.2149	1.3400
3	0.3	1.3400	1.2500	1.4025	1.2800	1.4680
4	0.4	1.4680	1.3080	1.5334	1.3309	1.6011
5	0.5	1.6011	1.3511	1.6687	1.3662	1.7377
6	0.6	1.7377				

$\therefore$ $y(0.2) \simeq 1.219,\ y(0.4) \simeq 1.468,\ y(0.6) \simeq 1.738$

Example 6.25: Solve $\dfrac{dy}{dx} = y - \dfrac{2x}{y}$, $y(0) = 1$ in the range $0 \le x \le 0.2$ using

(i) Euler's method (ii) Improved Euler's method (iii) Modified Euler's method.

Solution:

$$\frac{dy}{dx} = f(x, y) = y - \frac{2x}{y}; \; y(0) = 1, \; h = 0.1$$

(i) **By Euler's Method**

x_n	y_n	$f(x_n, y_n)$	$y_{n+1} = y_n + hf(x_n, y_n)$
0	1	1	1.1
0.1	1.1	0.9182	1.1918
0.2	1.1918		

$\therefore \qquad y(0.1) \simeq 1.1, \; y(0.2) \simeq 1.192$

(ii) **By Improved Euler's Method**

x_n	$y_n^{(c)}$	$f(x_n, y_n^{(c)})$	$y_{n+1}^{(p)} = y_n^{(c)} + hf(x_n, y_n^{(c)})$	$f(x_{n+1}, y_{n+1}^{(p)})$	$y_{n+1}^{(c)} = y_n^{(c)} + \frac{h}{2}\left[f(x_n, y_n^{(c)}) + (x_{n+1}, y_{n+1}^{(p)})\right]$
0	1	1	1.1	0.9182	1.0959
0.1	1.0959	0.9139	1.1872	0.8503	1.1841
0.2	1.1841				

$\therefore \qquad y(0.1) \simeq 1.096, \; y(0.2) \simeq 1.184$

(iii) **By Modified Euler's Method**

x_n	y_n	$f(x_n, y_n)$	$y_{n+1/2} = y_n + \frac{h}{2} f(x_n, y_n)$	$f(x_{n+1/2}, y_{n+1/2})$	$y_{n+1} = y_n + hf(x_{n+1/2}, y_{n+1/2})$
0	1	1	1.05	0.9548	1.0955
0.1	1.0955	0.9129	1.1411	0.8782	1.1833
0.2	1.1833				

$\therefore \qquad y(0.1) \simeq 1.096, \; y(0.2) \simeq 1.183$

EXERCISE 6.5

1. Let $\frac{dy}{dx} = x + y$ and $y = 1$ when $x = 0$. Find y when $x = 0.05$, 0.10, 0.15 and 0.20 by Euler's method.

2. Solve by Euler's method $\frac{dy}{dx} = \frac{x-y}{2}$, $y(0) = 1$ over [0, 3] using step size $\frac{1}{2}$.

3. Using Euler's method, find an approximate value of y corresponding to $x = 2$, given that $\frac{dy}{dx} = x + 2y$ and $y = 1$ when $x = 1$.

4. Using Euler's method, solve for y at $x = 0.1$ from $\frac{dy}{dx} = x + y + xy$, $y(0) = 1$ taking step size $h = 0.025$.

5. Use Euler's method to solve numerically the initial value problem $u' = -2tu^2$, $u(0) = 1$ with $h = 0.2$, 0.1 on the interval [0, 1].

6. Apply Euler's modified method to solve $\frac{dy}{dx} = x + 3y$ subject to $y(0) = 1$ to find an approximate value of y when $x = 1$.

7. Solve the equation $\frac{dy}{dx} = x + y^2;\ y(0) = 1$ for y at $x = 0.1$ in two steps using modified Euler's method.

8. Using modified Euler's formula, determine the value of y when $x = 0.1$ given that $y(0) = 1$ and $y' = x^2 + y^2$ by taking $h = 0.1$.

9. Using Euler's modified method, find a solution of the equation $\frac{dy}{dx} = x + \left|\sqrt{y}\right| = f(x, y)$ with initial condition $y = 1$ at $x = 0$ for the range $0 \le x \le 0.6$ in steps of 0.2.

10. Solve $\frac{dy}{dx} = 1 - y$, $y(0) = 0$ in the range $0 \le x \le 0.3$ using
    (i) Euler's method
    (ii) improved Euler's method
    (iii) modified Euler's method

    by choosing $h = 0.1$. Compare the answers with exact solution.

11. Given that $\frac{dy}{dx} = x + y^2$ and $y = 1$ at $x = 0$, find an approximate value of y at 0.5 by improved Euler's method.

ANSWERS 6.5

1. $y(0.05) = 1.05,\ y(0.10) = 1.105,\ y(0.15) = 1.16525,\ y(0.20) = 1.23101$
2. $y(0.5) \simeq 0.7500,\ y(1.0) \simeq 0.6875,\ y(1.5) \simeq 0.7656,$
   $y(2) \simeq 0.9492,\ y(2.5) \simeq 1.2119,\ y(3) \simeq 1.5339$
3. $y(2) \simeq 8.1619$ (taking $h = 0.2$)
4. $y(0.1) \simeq 1.1117$

5. With $h = 0.2,\ u(0.2) \simeq 1,\ u(0.4) \simeq 0.92,\ u(0.6) \simeq 0.7846,\ u(0.8) \simeq 0.6369,\ u(1) \simeq 0.5071$

   With $h = 0.1,\ u(0.1) \simeq 1,\ u(0.2) \simeq 0.98,\ u(0.3) \simeq 0.9416,$

   $u(0.4) \simeq 0.8884,\ u(0.5) \simeq 0.8253,\ u(0.6) \simeq 0.7572,$

   $u(0.7) \simeq 0.6884,\ u(0.8) \simeq 0.6221,\ u(0.9) \simeq 0.5602,\ u(1.0) \simeq 0.5037$

6. $y(1) \simeq 21.082$
7. $y(0.1) \simeq 1.1162$
8. $y(0.1) \simeq 1.1105$
9. $y(0.2) \simeq 1.2298,\ y(0.4) \simeq 1.5230,\ y(0.6) \simeq 1.8828$
10. (i) $y(0.1) \simeq 0.1,\ y(0.2) \simeq 0.19,\ y(0.3) \simeq 0.271$

    (ii) $y(0.1) \simeq 0.095,\ y(0.2) \simeq 0.1810,\ y(0.3) \simeq 0.2588$

    (iii) $y(0.1) \simeq 0.095,\ y(0.2) \simeq 0.1810,\ y(0.3) \simeq 0.2588$

    and exact values are $y(0.1) = 0.0952,\ y(0.2) = 0.1813,\ y(0.3) = 0.2592$
11. $y(0.5) = 2.207$ (taking $h = 0.1$)

6.4.6 Runge's Method

Consider the differential equation $\dfrac{dy}{dx} = f(x, y);\ \ y(x_0) = y_0$

$\therefore$ value of f at $x = x_0$ is $f(x_0, y_0)$.

By modified Euler's method

Value of f at $x = x_0 + \dfrac{h}{2}$ is $f\left(x_0 + \dfrac{h}{2},\ y_0 + \dfrac{h}{2} f(x_0, y_0)\right)$

Now by Euler method at $x = x_0 + h$, y is estimated by

$$y^{(p)} = y_0 + h\, f(x_0, y_0).$$

With this value of y

slope of solution curve at $\left(x_0 + h,\ y^{(p)}\right)$ is $f\left(x_0 + h,\ y^{(p)}\right)$.

Using this, we improve the value of y at $x = x_0 + h$ by $y_0 + hf\left(x_0 + h,\ y^{(p)}\right)$.

Then, value of f at $x = x_0 + h$ is $f\left(x_0 + h,\ y_0 + hf\left(x_0 + h,\ y^{(p)}\right)\right)$.

Now, by Simpson's one-third rule $\left(\text{with } h \text{ replaced by } \frac{h}{2}\right)$

$$\int_{x_0}^{x_0+h} f(x,y)\,dx = \frac{h}{6}\left[f(x_0,y_0)+4f\left(x_0+\frac{h}{2},\, y_0+\frac{h}{2}f(x_0,y_0)\right)\right.$$
$$\left.+ f\left(x_0+h,\, y_0+hf\left(x_0+h,\, y^{(p)}\right)\right)\right]$$

Thus, from differential equation $\frac{dy}{dx} = f(x, y);\ y(x_0) = y_0$

we have $$\int_{x_0}^{x_0+h} \frac{dy}{dx}\,dx = \int_{x_0}^{x_0+h} f(x, y)\,dx$$

$$\Rightarrow \quad y(x_0+h) - y_0 = \frac{h}{6}\left[f(x_0,y_0)+4f\left(x_0+\frac{h}{2},\, y_0+\frac{h}{2}f(x_0,y_0)\right)\right.$$
$$\left.+ f\left(x_0+h,\, y_0+h\,f\left(x_0+h,\, y^{(p)}\right)\right)\right]$$

∴ if we define $$k_1 = hf(x_0, y_0)$$

$$k_2 = h\,f\left(x_0+\frac{h}{2}, y_0+\frac{k_1}{2}\right)$$

$$k' = h\,f(x_0+h,\, y_0+k_1)$$

$$k_3 = h\,f(x_0+h,\, y_0+k')$$

then $$y(x_0+h) = y_0 + \frac{1}{6}(k_1+4k_2+k_3)$$

6.4.7 Runge–Kutta Method

Runge–Kutta method requires the functional values at some points. If solution by Runge–Kutta method agrees with the solution by Taylor's series method up to h^n then it will be called Runge–Kutta method of order n.

First order Runge–Kutta Method

Solution by Euler's method of differential equation $\frac{dy}{dx} = f(x, y);\ y(x_0) = y_0$ is

$$y(x_0+h) = y_0 + hf(x_0, y_0)$$

Also by Taylor series method

$$y(x_0+h) = y_0 + h\,y_0' + \cdots = y_0 + hf(x_0, y_0) + \cdots$$

Thus, solution by Euler's method agrees with solution by Taylor's series solution up to term in h. Hence, Euler method is Runge–Kutta method of the first order.

Second order Runge–Kutta Method

Solution of the differential equation $\frac{dy}{dx} = f(x, y);\ y(x_0) = y_0$ by improved Euler's method is

$$y(x_0 + h) = y_0 + \frac{h}{2}\left[f(x_0, y_0) + f\left(x_0 + h,\ y_1^{(p)}\right)\right]$$

where $$y_1^{(p)} = y_0 + hf(x_0, y_0)$$

$\therefore$
$$y(x_0 + h) = y_0 + \frac{h}{2}\left[f(x_0, y_0) + f\left(x_0 + h,\ y_0 + h\, f(x_0, y_0)\right)\right]$$

$$= y_0 + \frac{h}{2}\left[f(x_0, y_0) + f(x_0, y_0) + h\frac{\partial}{\partial x} f(x_0, y_0) + hf(x_0, y_0)\frac{\partial}{\partial y} f(x_0, y_0) + O\left(h^2\right)\right]$$

$$= y_0 + hf(x_0, y_0) + \frac{h^2}{2}\left[\frac{\partial f}{\partial x} + \frac{\partial f}{\partial y} f\right]_{(x_0, y_0)} + O\left(h^3\right)$$

$$= y_0 + hf(x_0, y_0) + \frac{h^2}{2} f'(x_0, y_0) + O\left(h^3\right)$$

which is same as the solution given by Taylor's series method up to terms in h^2.

Thus, solution by improved Euler's method agrees with solution by Taylor's series solution up to term in h^2. Hence, improved Euler method is Runge–Kutta method of order 2.

Thus, second order Runge–Kutta method can be written as

$$k_1 = hf(x_0, y_0)$$

$$k_2 = hf(x_0 + h,\ y_0 + k_1)$$

$$k = \frac{1}{2}(k_1 + k_2)$$

$$y(x_0 + h) = y_0 + k$$

This is also called as Runge's formula of order 2.

Third Order Runge–Kutta Method

Without giving the proof, we mention that the solution of differential equation $\frac{dy}{dx} = f(x, y);\ y(x_0) = y_0$ by Runge's method is given by

$$y(x_0 + h) = y_0 + \frac{1}{6}(k_1 + 4k_2 + k_3)$$

where $$k_1 = hf(x_0, y_0)$$

$$k_2 = hf\left(x_0 + \frac{h}{2}, y_0 + \frac{k_1}{2}\right)$$

$$k' = hf\left(x_0 + h, y_0 + k_1\right)$$

$$k_3 = hf\left(x_0 + h, y_0 + k'\right)$$

is of third order.

Also, the solution

$$y(x_0 + h) = y_0 + \frac{1}{6}(k_1 + 4k_2 + k_3)$$

where
$$k_1 = hf(x_0, y_0)$$

$$k_2 = hf\left(x_0 + \frac{h}{2}, y_0 + \frac{k_1}{2}\right)$$

$$k_3 = hf\left(x_0 + h,\ y_0 + k_2\right)$$

is of third order. This solution is called third order Runge–Kutta method. It is also known as Kutta's third order rule.

Runge–Kutta Method of 4th Order

The proof of Runge–Kutta method of third or fourth order are beyond the scope of this book. However, we state the fourth order Runge–Kutta method.

If we are to solve the differential equation $\frac{dy}{dx} = f(x, y);\ y(x_0) = y_0$ then solution by Runge–Kutta method of fourth order is

$$y(x_0 + h) = y_0 + \frac{1}{6}(k_1 + 2k_2 + 2k_3 + k_4)$$

where
$$k_1 = hf(x_0, y_0)$$

$$k_2 = hf\left(x_0 + \frac{h}{2}, y_0 + \frac{k_1}{2}\right)$$

$$k_3 = hf\left(x_0 + \frac{h}{2}, y_0 + \frac{k_2}{2}\right)$$

$$k_4 = hf\left(x_0 + h, y_0 + k_3\right)$$

Remark 6.2: When only Runge–Kutta method is mentioned that means Runge–Kutta method of order 4.

Runge–Kutta method can be applied directly to differential equation of second order. Suppose we are to find the solution of

$$\frac{d^2y}{dx^2} = f(x, y, y');\ y(x_0) = y_0,\ y'(x_0) = y_0'$$

we put $$y' = z = \phi(x, y, z)$$

then $$z' = f(x, y, z);\ y(x_0) = y_0,\ z(x_0) = z_0 = y_0'$$

then we find $$k_1 = h\,\phi(x_0, y_0, z_0) = h\,z_0 = h\,y_0'$$

$$l_1 = h\,f(x_0, y_0, z_0)$$

$$k_2 = h\,\phi\left(x_0 + \frac{h}{2}, y_0 + \frac{k_1}{2}, z_0 + \frac{l_1}{2}\right) = h\left(z_0 + \frac{l}{2}\right)$$

$$l_2 = h\,f\left(x_0 + \frac{h}{2}, y_0 + \frac{k_1}{2}, z_0 + \frac{l_1}{2}\right)$$

$$k_3 = h\,\phi\left(x_0 + \frac{h}{2}, y_0 + \frac{k_2}{2}, z_0 + \frac{l_2}{2}\right) = h\left(z_0 + \frac{l_2}{2}\right)$$

$$l_3 = h\,f\left(x_0 + \frac{h}{2}, y_0 + \frac{k_2}{2}, z_0 + \frac{l_2}{2}\right)$$

$$k_4 = h\,\phi(x_0 + h, y_0 + k_3, z_0 + l_3) = h(z_0 + l_3)$$

$$l_4 = h\,f(x_0 + h, y_0 + k_3, z_0 + l_3)$$

Then, $$k = \frac{1}{6}(k_1 + 2k_2 + 2k_3 + k_4)$$

$$l = \frac{1}{6}(l_1 + 2l_2 + 2l_3 + l_4)$$

and $$y(x_0 + h) = y_0 + k$$

$$y'(x_0 + h) = z_0 + l$$

Example 6.26: Apply the fourth order Runge–Kutta method to solve $\frac{dy}{dx} = x^2 + y^2,\ y(0) = 1$. Take step size $h = 0.1$ and determine approximation to $y\,(0.1)$ and $y\,(0.2)$ correct to four decimal places.

Solution:

$$\frac{dy}{dx} = f(x, y) = x^2 + y^2,\ \ y(0) = 1,\ \ h = 0.1$$

If $$y\,(x_n) = y_n$$

then $$k_1 = h\,f(x_n, y_n) = 0.1\left(x_n^2 + y_n^2\right)$$

$$k_2 = h\,f\left(x_n + \frac{h}{2}, y_n + \frac{k_1}{2}\right) = (0.1)\left[\left(x_n + \frac{0.1}{2}\right)^2 + \left(y_n + \frac{k_1}{2}\right)^2\right]$$

$$k_3 = h\,f\left(x_n + \frac{h}{2}, y_n + \frac{k_2}{2}\right) = (0.1)\left[\left(x_n + \frac{0.1}{2}\right)^2 + \left(y_n + \frac{k_2}{2}\right)^2\right]$$

$$k_4 = h\,f(x_n + h, y_n + k_3) = (0.1)\left[(x_n + 0.1)^2 + (y_n + k_3)^2\right]$$

$$k = \frac{1}{6}(k_1 + 2k_2 + 2k_3 + k_4)$$

$$y_{n+1} = y(x_n) + k$$

n	x_n	y_n	k_1	k_2	k_3	k_4	k	$y_{n+1} = y_n + k$
0	0	1	0.1	0.1105	0.11161	0.12457	0.11146	1.11146
1	0.1	1.11146	0.12453	0.14001	0.14184	0.16108	0.14155	1.25301
2	0.2	1.25301						

∴ up to 4 decimals

$$y(0.1) = 1.1115,\quad y(0.2) = 1.2530$$

Example 6.27: Solve the system of differential equations $\frac{dx}{dt} = y - t,\ \frac{dy}{dt} = x + t$ with $x = 1, y = 1$ when $t = 0$, taking $h = 0.1$.

Solution:

$$\frac{dx}{dt} = f(t, x, y) = y - t,\quad \frac{dy}{dt} = \phi(t, x, y) = x + t$$

with $\quad x(0) = 1, y(0) = 1; h = 0.1$

$$k_1 = h\,f(t_0, x_0, y_0) = 0.1(1 - 0) = 0.1$$

$$l_1 = h\,\phi(t_0, x_0, y_0) = 0.1(1 + 0) = 0.1$$

$$k_2 = h\,f\left(t_0 + \frac{h}{2}, x_0 + \frac{k_1}{2}, y_0 + \frac{l_1}{2}\right) = 0.1(1.05 - 0.05) = 0.1$$

$$l_2 = h\,\phi\left(t_0 + \frac{h}{2}, x_0 + \frac{k_1}{2}, y_0 + \frac{l_1}{2}\right) = 0.1(1.05 + 0.05) = 0.11$$

$$k_3 = h\,f\left(t_0 + \frac{h}{2}, x_0 + \frac{k_2}{2}, y_0 + \frac{l_2}{2}\right) = 0.1(1.055 - 0.05) = 0.1005$$

$$l_3 = h\,\phi\left(t_0 + \frac{h}{2}, x_0 + \frac{k_2}{2}, y_0 + \frac{l_2}{2}\right) = 0.1(1.05 + 0.05) = 0.11$$

$$k_4 = h\,f\left(t_0 + h, x_0 + k_3, y_0 + l_3\right) = 0.1(1.11 - 0.1) = 0.101$$

$$l_4 = h\,\phi\left(t_0 + h, x_0 + k_3, y_0 + l_3\right) = 0.1(1.1005 + 0.1) = 0.12005$$

$$k = \frac{1}{6}\left(k_1 + 2k_2 + 2k_3 + k_4\right) = 0.10033$$

$$l = \frac{1}{6}\left(l_1 + 2l_2 + 2l_3 + l_4\right) = 0.11001$$

$\therefore$ $$x(0.1) = x(0) + k = 1.10033$$

$$y(0.1) = y(0) + l = 1.11001$$

$\therefore$ up to four decimals

$$x(0.1) \simeq 1.1003, \quad y(0.1) \simeq 1.1100$$

Example 6.28: In an *L-R-C* circuit, voltage $v(t)$ across the capacitor is given by the equation

$$LC\frac{d^2v}{dt^2} + RC\frac{dv}{dt} + v = 0$$

subject to the conditions $t = 0$, $v = v_0$, $\dfrac{dv}{dt} = 0$.

Taking $h = 0.02$ sec., use Runge–Kutta method to calculate v and $\dfrac{dv}{dt}$ when $t = 0.02$ for the data $v_0 = 10$ volts, $C = 0.1$ farad, $L = 0.5$ henry and $R = 10$ ohms.

Solution:

Using the given data, differential equation is

$$\frac{d^2v}{dt^2} = -\frac{R}{L}\frac{dv}{dt} - \frac{v}{LC} = -20\frac{dv}{dt} - 20v\,;\ v(0) = v_0 = 10, \left(\frac{dv}{dt}\right)_{t=0} = 0$$

Let $$\frac{dv}{dt} = x = f(t, v, x)$$

$\therefore$ $$\frac{dv}{dt} = x = f(t, v, x)$$

and $$\frac{dx}{dt} = \frac{d^2v}{dt^2} = \phi(t, v, x) = -20(v + x);\ v(0) = 10, x(0) = 0, h = 0.02$$

$$k_1 = h\,f\left(t_0, v_0, x_0\right) = (0.02)(0) = 0$$

$$l_1 = h\,\phi(t_0, v_0, x_0) = (0.02)(-200) = -4.0$$

$$k_2 = h\,f\left(t_0 + \frac{h}{2}, v_0 + \frac{k_1}{2}, x_0 + \frac{l_1}{2}\right) = (0.02)(-2.0) = -0.04$$

$$l_2 = h\,\phi\left(t_0 + \frac{h}{2}, v_0 + \frac{k_1}{2}, x_0 + \frac{l_1}{2}\right) = (0.02)\left[-20(10-2)\right] = -3.2$$

$$k_3 = h\,f\left(t_0 + \frac{h}{2}, v_0 + \frac{k_2}{2}, x_0 + \frac{l_2}{2}\right) = (0.02)\left[-1.6\right] = -0.032$$

$$l_3 = h\,\phi\left(t_0 + \frac{h}{2}, v_0 + \frac{k_2}{2}, x_0 + \frac{l_2}{2}\right) = (0.02)\left[-20(10-0.02-1.6)\right] = -3.352$$

$$k_4 = h\,f(t_0 + h, v_0 + k_3, x_0 + l_3) = (0.02)(-3.352) = -0.06704$$

$$l_4 = h\,\phi(t_0 + h, v_0 + k_3, x_0 + l_3) = (0.02)\left[-20(10-0.032-3.352)\right] = -2.6464$$

$$k = \frac{1}{6}(k_1 + 2k_2 + 2k_3 + k_4) = \frac{1}{6}\left[0 + 2(-0.04) + 2(-0.032) - 0.06704\right] = -0.03517$$

$$l = \frac{1}{6}(l_1 + 2l_2 + 2l_3 + l_4) = \frac{1}{6}\left[-4.0 + 2(-3.2) + 2(-3.352) - 2.6464\right] = -3.29173$$

$\therefore$ to four decimals

$$v(0.02) = v(0) + k = 10 - 0.0352 = 9.9648 \text{ volts}$$

$$\left(\frac{dv}{dt}\right)_{t=0.02} = x(0) + l = 0 - 3.2917 = -3.2917 \text{ volts/sec.}$$

Example 6.29: Apply Runge's formula of order 2 to approximate value of y when $x = 1.1$ given $\frac{dy}{dx} = 3x + y^2$ and $y = 1.2$ when $x = 1$.

Solution:

$$\frac{dy}{dx} = 3x + y^2 = f(x, y),\ y(1) = 1.2, h = 0.1$$

$$k_1 = hf(x_0, y_0) = 0.444$$

$$k_2 = hf(x_0 + h, y_0 + k_1) = 0.60027$$

$$k = \frac{1}{2}(k_1 + k_2) = 0.52214$$

$$\therefore \quad y(1.1) = y(1) + k = 1.72214$$

∴ up to four decimal places

$$y(1.1) = 1.7221.$$

Note: Runge's (or Runge–Kutta's) second order method is same as improved Euler's method.

Example 6.30: Use Runge's method to approximate y when $x = 1.6$ given that $y = 0.4$ at $x = 1$ and

$$\frac{dy}{dx} = x - y.$$

Solution:

$$\frac{dy}{dx} = x - y = f(x, y), \ y(1) = 0.4,$$

Taking $h = 0.2$

$$k_1 = hf(x_n, y_n) = 0.2(x_n - y_n)$$

$$k_2 = hf\left(x_n + \frac{h}{2}, y_n + \frac{k_1}{2}\right) = 0.2\left(x_n - y_n + \frac{h}{2} - \frac{k_1}{2}\right)$$

$$k' = hf(x_n + h, y_n + k_1) = 0.2(x_n - y_n + h - k_1)$$

$$k_3 = hf(x_n + h, y_n + k') = 0.2(x_n - y_n + h - k')$$

$$k = \frac{1}{6}(k_1 + 4k_2 + k_3)$$

$$y_{n+1} = y_n + k$$

n	x_n	y_n	k_1	k_2	k'	k_3	k	y_{n+1}
0	1	0.4	0.12	0.128	0.136	0.1328	0.12747	0.52747
1	1.2	0.52747	0.13451	0.14106	0.14760	0.14499	0.14062	0.66809
2	1.4	0.66809	0.14638	0.15174	0.15711	0.15496	0.15138	0.81947
3	1.6	0.81947						

∴ up to four decimals

$$y(1.6) = 0.8195$$

Example 6.31: Apply Runge–Kutta method of third and fourth orders to find an approximate value of y when $x = 0.2$ taking $h = 0.1$ given that

$$\frac{dy}{dx} = x + y, \ y = 1 \text{ when } x = 0.$$

Solution:

$$\frac{dy}{dx} = x + y = f(x, y); \; y(0) = 1, \; h = 0.1$$

By Runge–Kutta method of third order

$$k_1 = hf(x_n, y_n) = 0.1(x_n + y_n)$$

$$k_2 = hf\left(x_n + \frac{h}{2}, y_n + \frac{k_1}{2}\right) = 0.1\left(x_n + y_n + 0.05 + \frac{k_1}{2}\right)$$

$$k_3 = hf(x_n + h, y_n + k_2) = 0.1(x_n + y_n + 0.1 + k_2)$$

$$k = \frac{1}{6}(k_1 + 4k_2 + k_3)$$

$$y_{n+1} = y_n + k$$

n	x_n	y_n	k_1	k_2	k_3	k	y_{n+1}
0	0	1	0.1	0.11	0.121	0.11017	1.11017
1	0.1	1.11017	0.12102	0.13207	0.14422	0.13225	1.24242
2	0.2	1.24242					

∴ up to four decimals

$$y(0.2) = 1.2424$$

(ii) By Runge–Kutta method of fourth order

$$k_1 = hf(x_n, y_n) \quad = 0.1(x_n + y_n)$$

$$k_2 = hf\left(x_n + \frac{h}{2}, y_n + \frac{k_1}{2}\right) = 0.1\left(x_n + y_n + 0.05 + \frac{k_1}{2}\right)$$

$$k_3 = hf\left(x_n + \frac{h}{2}, y_n + \frac{k_2}{2}\right) = 0.1\left(x_n + y_n + 0.05 + \frac{k_2}{2}\right)$$

$$k_4 = hf(x_n + h, y_n + k_3) = 0.1(x_n + y_n + h + k_3)$$

$$k = \frac{1}{6}(k_1 + 2k_2 + 2k_3 + k_4)$$

$$y_{n+1} = y_n + k$$

n	x_n	y_n	k_1	k_2	k_3	k_4	k	y_{n+1}
0	0	1	0.1	0.11	0.1105	0.12105	0.11034	1.11034
1	0.1	1.11034	0.12103	0.13209	0.13264	0.14430	0.13246	1.24280
2	0.2	1.24280						

∴ up to four decimals

$$y(0.2) = 1.2428$$

EXERCISE 6.6

1. Solve the equation $\frac{dy}{dx} = -y$, for values of y at $x = 0.1$ and $x = 0.2$ using Runge–Kutta method of
   (i) order two
   (ii) order three
   (iii) order four.
2. Use Runge's formula (third order) to solve the differential equation $\frac{dy}{dx} = x - y$ for y at $x = 1.1$ subject to $y = 1$ when $x = 1$.
3. Use Runge's method to approximate y when $x = 0.1$ given that $y = 1$ at $x = 0$ and $\frac{dy}{dx} = 3x + y^2$.
4. Apply Runge–Kutta method to find an approximate value of y when $x = 0.2$ in steps of 0.1 if $\frac{dy}{dx} = x + y^2$ given that $y = 1$ when $x = 0$.
5. Use the fourth order Runge–Kutta method to solve the initial value problems $u' = -2tu^2$, $u(0) = 1$ with $h = 0.2$ on the interval [0,0.4].
6. Use the Runge–Kutta fourth order method to find y (0.2) with $h = 0.1$ for the initial value problem
$$\frac{dy}{dx} = \sqrt{x + y},\ y(0) = 1.$$
7. Using Runge–Kutta method of fourth order, solve $\frac{dy}{dx} = \frac{y^2 - x^2}{y^2 + x^2}$ with $y(0) = 1$ at $x = 0.2$ and 0.4.
8. Solve numerically the system of simultaneous equations $\frac{dx}{dt} + 2x + 3y = 0$, $\frac{dy}{dt} + 3x + 2y = 2e^{2t}$ with initial conditions $x = 1, y = 2$ at $t = 0$, for $t = 0.1$ by Runge–Kutta method.
9. Using Runge–Kutta method solve the equations $\frac{dy}{dx} = yz + x$; $\frac{dz}{dx} = xz + y$ given that $y(0) = 1$; $z(0) = -1$ for $y(0.2)$ and $z(0.2)$.
10. Using Runge–Kutta method, solve the equation $y'' = xy'^2 - y^2$ for $x = 0.2$ given that $y = 1, y' = 0$ when $x = 0$.

ANSWERS 6.6

1. (i) $y(0.1) \simeq 0.905,\ y(0.2) \simeq 0.8190$
   (ii) $y(0.1) \simeq 0.9049,\ y(0.2) \simeq 0.8189$
   (iii) $y(0.1) \simeq 0.9048,\ y(0.2) \simeq 0.8187$
2. $y(1.1) \simeq 1.0048$
3. $y(0.1) \simeq 1.12725$
4. $y(0.2) \simeq 1.2736$
5. $u(0.2) \simeq 0.96153,\ u(0.4) \simeq 0.86205$
6. $y(0.2) \simeq 1.2194$
7. $y(0.2) \simeq 1.196,\ y(0.4) \simeq 1.3753$
8. $x(0.1) \simeq 0.3292,\ y(0.1) \simeq 1.6668$
9. $y(0.2) \simeq 0.8522,\ z(0.2) \simeq -0.8341$
10. $y(0.2) \simeq 0.98015$

6.4.8 Milne's Method

First, we prove quadrature formula

$$\int_{x_0}^{x_0+4h} f(x)\,dx = \frac{4h}{3}[2f_1 - f_2 + 2f_3]$$

Let $\quad x = x_0 + ph$

$\therefore \quad dx = h\,dp$

$$\therefore \quad \int_{x_0}^{x_0+4h} f(x)\,dx = h\int_0^4 f(x_0+ph)\,dp = h\int_0^4 E^p f(x_0)\,dp$$

$$= h\int_0^4 (1+\Delta)^p f(x_0)\,dp$$

$$\simeq h\int_0^4 \left[f_0 + p\Delta f_0 + \frac{p(p-1)}{2}\Delta^2 f_0 + \frac{p(p-1)(p-2)}{6}\Delta^3 f_0 \right] dp$$

$$= h\left[p f_0 + \frac{1}{2}p^2\Delta f_0 + \frac{1}{2}\left(\frac{1}{3}p^3 - \frac{1}{2}p^2\right)\Delta^2 f_0 + \frac{1}{6}\left(\frac{1}{4}p^4 - p^3 + p^2\right)\Delta^3 f_0 \right]_0^4$$

$$= h\left[4f_0 + 8(f_1 - f_0) + \frac{20}{3}(f_2 - 2f_1 + f_0) + \frac{8}{3}(f_3 - 3f_2 + 3f_1 - f_0)\right]$$

$$= \frac{4h}{3}[2f_1 - f_2 + 2f_3]$$

Now, suppose we are to find the solution of $\frac{dy}{dx} = f(x, y), f(x_0) = y_0$.

In Milne's method first we find $y(x_0 + h), y(x_0 + 2h), y(x_0 + 3h)$ by Taylor series method or by Picard's method. Let these be y_1, y_2, y_3 respectively.

From these values, we find $f(x_0 + h, y_1) = f_1$, $f(x_0 + 2h, y_2) = f_2$, $f(x_0 + 3h, y_3) = f_3$

Now,
$$\int_{x_0}^{x_0+4h} \frac{dy}{dx}\,dx = \int_{x_0}^{x_0+4h} f\,dx.$$

∴
$$y(x_0 + 4h) = y_0 + \int_{x_0}^{x_0+4h} f\,dx$$

$$y_4 \simeq y_0 + \frac{4h}{3}[2f_1 - f_2 + 2f_3]$$

This is called predictor and we write

$$y_4^{(p)} = y_0 + \frac{4h}{3}[2f_1 - f_2 + 2f_3].$$

Then we use Simpson's one-third rule to improve this value. This is called corrector.

$$y_4^{(c)} = y_2 + \frac{h}{3}\left[f_2 + 4f_3 + f_4^{(p)}\right]$$

where
$$f_4^{(p)} = f\left(x_0 + 4h, y_4^{(p)}\right)$$

If necessary, the process can be repeated by taking this $y_4^{(c)}$ as $y_4^{(p)}$ and then again finding $y_4^{(c)}$ and repetition can be done until values obtained are approximately same as predicted and corrected.

Now, suppose we are to find $y(x_0 + 4h)$ when

$$\frac{d^2y}{dx^2} = f\left(x, y, \frac{dy}{dx}\right);\ y(x_0) = y_0,\ y'(x_0) = y'_0$$

Let
$$\frac{dy}{dx} = z = \phi(x, y, z)$$

∴
$$\frac{dz}{dx} = f(x, y, z); y(x_0) = y_0,\ z(x_0) = y'_0 = z_0$$

find $y(x_0 + h),\ y(x_0 + 2h),\ y(x_0 + 3h)$

and $z(x_0 + h),\ z(x_0 + 2h),\ z(x_0 + 3h)$

by Taylor-series method

and then find $z'(x_0+h)$, $z'(x_0+2h)$, $z'(x_0+3h)$

i.e., f at $x = x_0 + h, x_0 + 2h, x_0 + 3h$

Now by predictor formula find $z^{(p)}(x_0+4h)$ and $y^{(p)}(x_0+4h)$. Then using corrector formula find $z^{(c)}(x_0+4h)$ and then $y^{(c)}(x_0+4h)$.

Example 6.32: Apply Milne's method to find the solution of the differential equation $\dfrac{dy}{dx} = x + y, y(0) = 1$ in the interval [0, 0. 4] in steps of $h = 0.1$. It is given that y (0.1) = 1.1103, $y(0.2) = 1.2428, y(0.3) = 1.3997$

Solution:
$$\frac{dy}{dx} = x + y = f(x, y)$$

$$y(0) = 1, y(0.1) = 1.1103, y(0.2) = 1.2428, y(0.3) = 1.3997;\ h = 0.1$$

n	x	y	f=x+y
0	0	1	1
1	0.1	1.1103	1.2103
2	0.2	1.2428	1.4428
3	0.3	1.3997	1.6997

$$y^{(p)}(0.4) = y(0) + \frac{4h}{3}(2f_1 - f_2 + 2f_3)$$

$$= 1 + \frac{0.4}{3}\left[2(1.2103) - 1.4428 + 2(1.6997)\right] = 1.5836$$

$$f_4^{(p)} = f^{(p)}\left(0.4, y^{(p)}(0.4)\right) = 0.4 + 1.5836 = 1.9836$$

$$y^{(c)}(0.4) = y(0.2) + \frac{h}{3}\left(f_2 + 4f_3 + f_4^{(p)}\right)$$

$$= 1.2428 + \frac{0.1}{3}\left(1.4428 + 4(1.6997) + 1.9836\right)$$

$$= 1.5836$$

which is same as $y^{(p)}(0.4)$

$\therefore$ $y(0.4) = 1.5836$

Example 6.33: Solve the initial value problem $\dfrac{dy}{dx} = 1 + xy^2, y(0) = 1$ for $x = 0.4$ by Milne's method, it is given that y (0.1) = 1.105, y (0.2) = 1.223, y (0.3) = 1.355.

Solution:

$$\frac{dy}{dx} = f(x, y) = 1 + xy^2$$

n	x	y	f = 1 + xy²
0	0	1	1
1	0.1	1.105	1.1221
2	0.2	1.223	1.2991
3	0.3	1.355	1.5508

$$y^{(p)}(0.4) = y(0) + \frac{4h}{3}[2f_1 - f_2 + 2f_3]$$

$$= 1 + \frac{0.4}{3}\left[2(1.1221) - 1.2991 + 2(1.5508)\right] = 1.5396$$

$$f_4^{(p)} = f^{(p)}\left(0.4, y^{(p)}(0.4)\right) = 1 + (0.4)(1.5396)^2 = 1.9481$$

$$y^{(c)}(0.4) = y(0.2) + \frac{h}{3}\left[f_2 + 4f_3 + f_4^{(p)}\right]$$

$$= 1.223 + \frac{0.1}{3}\left[1.2991 + 4(1.5508) + 1.9481\right] = 1.5380$$

$y^{(p)}(0.4)$ and $y^{(c)}(0.4)$ are not equal up to three decimals.

Again, taking $y^{(p)}(0.4)$ as $y^{(c)}(0.4)$ obtained, i.e., 1.5380, we have

$$f_4^{(p)} = f^{(p)}\left(0.4, y^{(p)}(0.4)\right) = 1 + (0.4)(1.5380)^2 = 1.9462$$

$$y^{(c)}(0.4) = y(0.2) + \frac{h}{3}\left[f_2 + 4f_3 + f_4^{(p)}\right]$$

$$= 1.223 + \frac{0.1}{3}\left[1.2991 + 4(1.5508) + 1.9462\right] = 1.5380$$

$\therefore$ up to three decimals, $y^{(p)}(0.4)$ and $y^{(c)}(0.4)$ are equal.

$\therefore \qquad y(0.4) = 1.538$

Example 6.34: Given $\frac{dy}{dx} = \frac{1}{2}\left(1 + x^2\right)y^2$ and $y(0) = 1$, $y(0.1) = 1.06$, $y(0.2) = 1.12$ and $y(0.3) = 1.21$. Evaluate $y(0.4)$ by Milne's predictor–corrector method.

Solution:

$$\frac{dy}{dx} = f(x, y) = \frac{1}{2}\left(1+x^2\right)y^2$$

n	x	y	$f = \frac{1}{2}\left(1+x^2\right)y^2$
0	0	1	0.5
1	0.1	1.06	0.5674
2	0.2	1.12	0.6523
3	0.3	1.21	0.7979

$$y^{(p)}(0.4) = y(0) + \frac{4h}{3}[2f_1 - f_2 + 2f_3]$$

$$= 1 + \frac{0.4}{3}\left[2(0.5674) - 0.6523 + 2(0.7979)\right] = 1.2771$$

$$f_4^{(p)} = f^{(p)}\left(0.4, y^{(p)}(0.4)\right) = \frac{1}{2}\left[1+(0.4)^2\right](1.2771)^2 = 0.9460$$

$$y^{(c)}(0.4) = y(0.2) + \frac{h}{3}\left[f_2 + 4f_3 + f_4^{(p)}\right]$$

$$= 1.12 + \frac{0.1}{3}\left[0.6523 + 4(0.7979) + 0.9460\right] = 1.2797$$

∴ up to two decimals, $y^{(p)}(0.4)$ and $y^{(c)}(0.4)$ are equal.

∴ $\quad y(0.4) = 1.28$

Example 6.35: Solve the differential equation $\frac{dy}{dx} = 2e^x - y$ at $x = 0.5$ using Milne's predictor–corrector method, given that $y(0.1) = 2$.

Solution:

$$\frac{dy}{dx} = 2e^x - y, \ y(0.1) = 2$$

We shall find $y(0.2), y(0.3), y(0.4)$ by Taylor series method.

$$y' = 2e^x - y \qquad \therefore\ y'(0.1) = 2e^{0.1} - 2 = 0.21034$$

$$y'' = 2e^x - y' \qquad \therefore\ y''(0.1) = 2e^{0.1} - 0.21034 = 2.00000$$

$$y^{(n)} = 2e^x - y^{(n-1)};\ n = 2, 3, 4, 5, \ldots$$

$$\therefore \quad y'''(0.1) = 2e^{0.1} - 2 = 0.21034$$

$$y^{iv}(0.1) = 2e^{0.1} - 0.21034 = 2.00000$$

$$y^{v}(0.1) = 2e^{0.1} - 2 = 0.21034$$

$$y^{vi}(0.1) = 2e^{0.1} - 0.21034 = 2.00000$$

By Taylor's series expansion

$$y(x+0.1) = y(0.1) + x\,y'(0.1) + \frac{x^2}{2!}y''(0.1) + \frac{x^3}{3!}y'''(0.1) + \frac{x^4}{4!}y^{iv}(0.1) + \cdots$$

$$\therefore \quad y(x+0.1) \simeq 2\left[1 + \frac{x^2}{2} + \frac{x^4}{24} + \frac{x^6}{720}\right] + 0.21034\left[x + \frac{x^3}{6} + \frac{x^5}{120}\right]$$

$$\therefore \quad y(0.2) = 2\left[1 + \frac{(0.1)^2}{2} + \frac{(0.1)^4}{24} + \frac{(0.1)^6}{720}\right] + 0.21034\left[0.1 + \frac{(0.1)^3}{6} + \frac{(0.1)^5}{120}\right]$$

$$= 2(1.00500) + 0.21034(0.10017) \simeq 2.03107$$

$$y(0.3) = 2\left[1 + \frac{(0.2)^2}{2} + \frac{(0.2)^4}{24} + \frac{(0.2)^6}{720}\right] + 0.21034\left[0.2 + \frac{(0.2)^3}{6} + \frac{(0.2)^5}{120}\right]$$

$$= 2(1.02007) + 0.21034(0.20134) = 2.08249$$

$$y(0.4) = 2\left[1 + \frac{(0.3)^2}{2} + \frac{(0.3)^4}{24} + \frac{(0.3)^6}{720}\right] + 0.21034\left[0.3 + \frac{(0.3)^3}{6} + \frac{(0.3)^5}{120}\right]$$

$$= 2(1.04534) + 0.21034(0.30452) = 2.15473$$

n	x	y	$f = 2e^x - y$
0	0.1	2	0.21034
1	0.2	2.03107	0.41174
2	0.3	2.08249	0.61723
3	0.4	2.15473	0.82892

$$y^{(p)}(0.5) = y(0.1) + \frac{4h}{3}[2f_1 - f_2 + 2f_3]$$

$$= 2 + \frac{0.4}{3}\left[2(0.41174) - 0.61723 + 2(0.82892)\right] = 2.24855$$

$$f_4^{(p)} = f^{(p)}\left(0.5, y^{(p)}(0.5)\right) = 1.04899$$

$$y^{(c)}(0.5) = y(0.3) + \frac{h}{3}\left[f_2 + 4f_3 + f_4^{(p)}\right]$$

$$= 2.08249 + \frac{0.1}{3}\left[0.61723 + 4(0.82892) + 1.04899\right] = 2.248553$$

$\therefore$ up to four decimals, $y^{(p)}(0.5)$ and $y^{(c)}(0.5)$ are equal.

$\therefore \quad y(0.5) = 2.2486$

Example 6.36: Given $y'' + xy' + y = 0, y(0) = 1, y'(0) = 0$ obtain y for $x = 0(0.1)0.3$ by any method. Further continue the solution by Milne's method to calculate $y(0.4)$.

Solution:

Let $\quad y' = z = f(x, y, z)$

$\therefore \quad z' = y'' = -xz - y = \phi(x, y, z)$

where $\quad y(0) = 1, y'(0) = z(0) = 0$

$\therefore \quad z'(0) = -(0)(0) - 1 = -1$

Now, by Picard's method

$$y(x) = y(0) + \int_0^x z(x)\,dx = 1 + \int_0^x z(x)\,dx \tag{1}$$

$$z(x) = z(0) + \int_0^x \left(-x\, z(x) - y(x)\right) dx$$

$$= -x\int_0^x z(x)\,dx + \int_0^x \left[\int_0^x z(x)\,dx\right] dx - \int_0^x y(x)\,dx \quad \text{(integration by parts)}$$

$$= -x\left[y(x) - 1\right] + \int_0^x \left[y(x) - 1\right] dx - \int_0^x y(x)\,dx \quad \text{(from (1))}$$

$$= -xy(x) + x + \int_0^x y(x)\,dx - x - \int_0^x y(x)\,dx$$

$$= -xy(x) \tag{2}$$

$$y' = -xy = f(x, y), \; y(0) = 1$$

$$y(x) = 1 + \int_0^x -x\,y(x)\,dx = 1 - \int_0^x x\,y\,dx$$

$$\therefore \quad y_{n+1} = 1 - \int_0^x xy_n\,dx$$

n	$y_n(x)$
0	1
1	$1-\frac{x^2}{2}$
2	$1-\frac{x^2}{2}+\frac{x^4}{8}$
3	$1-\frac{x^2}{2}+\frac{x^4}{8}-\frac{x^6}{48}$
4	$1-\frac{x^2}{2}+\frac{x^4}{8}-\frac{x^6}{48}+\frac{x^8}{384}$

From last two approximations

$$y(x) \simeq 1-\frac{x^2}{2}+\frac{x^4}{8}-\frac{x^6}{48}$$

From this, we find $y(0.1), y(0.2), y(0.3)$.

n	x	y	$f=-xy$
0	0	1	0
1	0.1	0.99501	−0.09950
2	0.2	0.98020	−0.19604
3	0.3	0.95600	−0.28680

$$y^{(p)}(0.4) = y(0)+\frac{4h}{3}[2f_1 - f_2 + 2f_3]$$

$$= 1+\frac{(0.4)}{3}\left[2(-0.09950)+0.19604+2(-0.28680)\right]$$

$$= 0.92313$$

$$f_4^{(p)} = f^{(p)}\left(0.4, y^{(p)}(0.4)\right) = -(0.4)(0.92313) = -.36925$$

$$\therefore \quad y^{(c)}(0.4) = y(0.2)+\frac{h}{3}\left[f_2 + 4f_3 + f_4^{(p)}\right]$$

$$= 0.98020+\frac{0.1}{3}\left[-0.19604+4(-0.2868)-0.36925\right] = 0.92312$$

$\therefore$ up to 4 decimals,

$$y^{(p)}(0.4) = y^{(c)}(0.4) = 0.9231$$

$$\therefore \quad y(0.4) = 0.9231$$

EXERCISE 6.7

1. Find $y(2)$ if $y(x)$ is the solution of $\frac{dy}{dx}=\frac{1}{2}(x+y)$ assuming $y(0) = 2$, $y(0.5) = 2.636$, $y(1.0)=3.595$ and $y(1.5)=4.968$ by Milne's method.
2. Given $\frac{dy}{dx}=\frac{1}{(x+y)}$, $y(0)=2$, $y(0.2)=2.0933$, $y(0.4)=2.1755$, $y(0.6)=2.2493$, find y (0.8) using Milne's method.
3. Using Milne's predictor–corrector method evaluate the integral of $y'-4y=0$ at $x = 0.4, 0.5$ given that $y(0) = y_0 = 1$, $y_1 = y(0.1) = 1.492$, $y_2 = y(0.2) = 2.226$, $y_3 = y(0.3) = 3.320$. Also compare with exact values.
4. Let $\frac{dy}{dx}=1+y^2$ with y (0) = 0. Compute $y(0.8)$ and $y(1.0)$ by using Milne's method.
5. Using Milne's predictor–corrector method, solve the equation $y'=x-y^2$ in the range $0 \le x \le 1$ given that $y = 0$ at $x = 0$.
6. Use Milne's method to find y (0.3) from $y'=x^2+y^2$, y (0) = 1. Find the initial values y (– 0.1), y (0.1), y (0.2) from the Taylor's series method.
7. Given $\frac{dy}{dx}=x^2(1+y)$ and $y(1) = 1$, $y(1.1) = 1.233$, $y(1.2) = 1.548$, $y(1.3) = 1.979$, evaluate $y(1.4)$ by Milne's predictor–corrector method.
8. Solve $\frac{dy}{dx}=(x+y)y$, $y(0) = 1$ using Milne's predictor–corrector method for $y(0.4)$. The values for $x = 0.1$, 0.2 and 0.3 should be obtained by Runge–Kutta method of fourth order.

ANSWERS 6.7

1. 6.873
2. 2.3164
3.

x	0.4	0.5
Calculated value	4.953	7.388
Exact value	4.953	7.389

4. $y(0.8)\approx 1.0293$, $y(1.0)\approx 1.5554$
5. $y(0.2)\approx 0.0200$, $y(0.4)\approx 0.0795$, $y(0.6)\approx 0.1762$ (by Picard's method)

   $y(0.8)\approx 0.3046$, $y(1)\approx 0.4556$
6. $y(0.3)\approx 1.4397$
7. $y(1.4)\approx 2.575$
8. $y(0.4)\approx 1.83874$

Index

A

Algebraic and transcendental equations, 586
Algebraic equation, 586
Amplitude spectrum, 372
Analytic function, 5
Angle of rotation, 134
Application of Cauchy residue theorem, 106
Applications of Fourier transforms, 416
Applications of Laplace transform, 257
Applications of partial differential equations, 520
Arc, 39

B

Backward difference interpolation polynomial, 696
Backward difference operator, 656
Backward difference table, 662
Backward differences, 662
Bessel functions, 186
Bessel's interpolation formula, 703
Bilinear transformations, 149
Bisection method, 587
Bolzano method, 587
Boundary value problem, 435

C

Cauchy inequality, 53
Cauchy integral formula, 50
Cauchy integral formula for derivatives, 51
Cauchy integral theorem, 45
Cauchy principal value, 114
Cauchy residue theorem, 98
Cauchy-Goursat theorem, 46
Cauchy-Riemann equations, 5–8
Central difference operator, 656
Central difference table, 663
Central differences, 662
Centre of the power series, 67
Change of scale property, 167, 373
Charpit's auxiliary equations, 461
Charpit's method, 460
Circle of convergence, 68
Clairut's equation, 463
Closed curve, 39
Codomain, 1
Coefficient of magnification, 134
Complementary error function, 184
Complementary function, 486–487, 507–509
Complete solution, 435, 486–488
Complex form of Fourier integral, 370
Complex form of Fourier series, 343
Complex Fourier transform, 392
Complex variable, 1
Conformal mapping, 132
Conjugate harmonic function, 22– 23
Conjugate harmonic function in polar form, 25
Conjugate property, 377
Connected domain, 44
Continuity of a function, 4
Continuous curve, 39
Contour, 39
Contour Integrals, 97
Contraction, 136–137
Convergence of Fourier series, 286
Convergence of half range cosine series, 324
Convergence of half range sine series, 324
Convergence of order *m*, 587
Converse of Cauchy integral theorem, 53
Convolution, 221
Convolution theorem, 221, 381
Cote's formulas, 743–745
C-R equations, 5–8
Critical point, 133
Cross ratio, 149–150
Crout's method, 626

D

D'Alembert's method of solving wave equation, 530
Damped vibrations, 169
Daniel Bernoulli, 281

Definition of line integral, 39
Delta function, 200–201
Differentiability, 4
Differentiation of Laplace transform, 172
Differentiation operator, 656
Differentiations of transforms w.r.t. frequency, 378
Diffusivity of string, 528
Dirac-delta function, 199–201
Direct integration method, 444
Direct iteration method, 588
Dirichlet's conditions, 286
Divided differences, 685
Domain, 1
Doolittle method, 625
Double pole, 85

E

Effective value of the function, 343
Electric circuits, 257
Energy of spectrum, 372
Energy theorem, 382
Entire function, 5
Equispaced arguments, 656
Error in quadrature formulae, 745
Error function, 184
Error propagation, 665–666
Error term in Simpson's one-third rule, 746
Error term in Simpson's three-eight rule, 747
Error term in Trapezoidal rule, 745
Errors in numerical computations, 585
Essential singularity, 85
Euler formulae, 283, 285–286
Euler's method, 772
Euler's coefficients, 284
Even function, 286
Expressions for Fourier coefficients, 285
Extrapolation, 585, 683

F

Factorial polynomial, 663
Factorisation method, 623
Faltung theorem, 381
Filtering property of Dirac-delta function, 201
Final value theorem, 227
First order partial differential equation, 434, 448
First–order backward differences, 662
First shifting property, 167
First translation property, 167
Fixed point, 133
Forward difference operator, 656
Forward difference table, 661
Forward differences, 661
Fourier coefficients, 284–286, 353
Fourier cosine integral, 368
Fourier cosine series, 323–324
Fourier cosine transform, 369
Fourier integrals, 367–370
Fourier series, 283
Fourier sine integral, 369
Fourier sine series, 324
Fourier sine transform, 369
Fourier transforms, 371
Fourier transform of integral function, 380
Fourth order Runge-Kutta method, 783
Fractional transformation, 149
Frequency shifting, 373
Function of exponential order, 166
Fundamental period, 281–282
Fundamental theorem of integral calculus, 47

G

Gauss backward difference formula, 703
Gauss elimination method, 615–616
Gauss forward interpolation formula, 701
Gauss backward interpolation formula, 702
Gauss-Jordan method, 616
Gauss-Seidel iteration method, 635
General quadrature formula, 739
General solution, 435
Generalized function, 201
Geometric convergence, 587
Green's theorem, 46
Grid point, 725–726, 728

H

Half range cosine series, 323–324
Half range sine series, 324
Halving method, 587
Harmonic analysis, 352
Harmonic function, 21

Heat equation, 520, 547, 568–569
Heaviside function, 191, 394
Holomorphic, 5
Homogenous and non-homogenous partial differential equations, 434

I

Improved Euler's method, 773
Independence of path, 46
Indirect methods, 634
Infinite series of complex terms, 66
Initial value problem, 435
Initial value theorem, 227
Integral function, 47, 171, 380
Integration of Laplace transform, 172–173
Interpolating polynomial, 683
Interpolation, 585, 683
Interval of differencing, 656, 725
Invariant point, 133, 149
Inverse Fourier transform, 371
Inverse interpolation, 712–714
Inverse Laplace transform, 165, 209
Inverse operator, 489
Inverse transformation, 139
Inversion, 139
Isogonal, 132
Isolated singularities, 85
Iterative methods, 615, 634, 713
Iterative process, 586, 634

J

Jacobi's iterative method, 634
Jordan lemma, 114

K

Kirchoff's laws, 257
Kutta's third order rule, 783

L

Lagrange equation, 448
Lagrange inverse interpolation formula, 712
Lagrange's auxiliary equations, 450
Lagrange's interpolation formula, 683
Lagrange's method, 448, 712
Lagrange's polynomial, 683
Lagrange's subsidiary equations, 450
Laplace equation, 21, 521, 568–570
Laplace transform, 165
Laplace transform of derivative, 170
Laplace transform of integral, 171
Laplace transform of periodic functions, 202
Laplace transform of unit step function, 192
Laplace-Everett's interpolation formula, 704
Laurent's series, 69
Level curves, 22
Limit of a function, 1
Line integral, 39
Linear and non-linear partial differential equations, 434
Linear convergence, 587
Linear operators, 656
Linear transformation, 137
Linearity property, 167, 372
Liouville's theorem, 54
Lipschitz condition, 765
Lower triangular matrix, 623–624
LU decomposition method, 623

M

Magnification, 134, 136–137
Mass-spring system, 271
Mean value operator, 656
Meromorphic function, 85
Method of factorisation, 623–624
Method of false position, 596
Method of separation of variables, 521
Method of tangents, 600
Milne Thomson method, 23–26
Milne's method, 791–793
Milne's predictor-corrector method, 794–795
Mobius transformation, 149
Modification of Newton's iterative formula, 603
Modification of power method, 644
Modified Euler's method, 774
Modulation property of sine and cosine transforms, 375
Modulation theorem, 374
Modulus property, 374
Monge's method, 479–480

Morera's theorem, 53
Multiple curve, 39
Multiple-valued function, 1
Multiply connected domain, 44–45

N

Newton's backward difference interpolation formula, 696
Newton's divided difference interpolation formula, 687
Newton's forward difference interpolation formula, 695
Newton's iteration method, 600
Newton-Raphson method, 600
Non-isolated singularity, 84
Non-linear partial differential equations, 434
*n*th order derivative, 53
Numerical differentiation, 725
Numerical quadrature, 739

O

Odd function, 286
One dimensional wave equation, 520, 527
One-sided Laplace transform, 165
Operator methods, 490, 509
Order of convergence, 595, 597, 602, 605
Ordinary point, 133
Orthogonal system, 22
Orthogonal trajectories, 35–36
Orthogonality of trigonometric system, 282

P

Parabolic, 149, 520
Parametric equation, 39
Parseval's formulae, 341
Parseval's identities, 382
Partial differential equations, 433
Partial pivoting, 616
Particular integral, 486, 489–490, 507, 509
Particular solution, 435, 486
Periodic function, 281
Picard's method, 764–765
Piecewise continuous curve, 39
Piecewise continuous function, 166
Piecewise smooth curve, 39
Pivot element, 615–616
Poisson's integral formula, 54
Polar form of Cauchy-Riemann equations, 7
Pole of order *n*, 85
Power method, 642
Power series, 67–68
Principle of superposition, 487
Problem related to deflection of a loaded beam, 265
Problem related to mechanical systems, 271
Properties of Fourier transforms, 372
Properties of Laplace transforms, 167

Q

Quadratic convergence, 587

R

Radio equations, 560
Radius of convergence, 67
Radius of curvature, 265
Range of function, 1
Rate of convergence, 587
Rectified semiwave function, 203
Reflection, 139
Regula-falsi method, 596
Regular function, 5
Removable isolated singularity, 85
Residue, 90
Residue theorem, 98
Root mean square value, 343
Rotation mapping, 136
Round-off errors, 585
Runge's method, 780
Runge-Kutta method, 781–783

S

Secant method, 594
Second order Runge-Kutta method, 782
Second shifting theorem, 193
Second-order backward differences, 662
Self-reciprocal, 371

Shearing stress, 265
Shifting property, 167, 373
Shifting operator, 656
Signum function, 394
Simple closed curve, 39
Simple curve, 39
Simple pole, 85
Simple zero, 84
Simply connected domain, 44
Simpson's one-third rule, 740
Simpson's three-eight rule, 741
Single-valued function, 1
Singular point, 84
Singular solutions, 435
Singularities of function, 84
    Types of, 84
    Non-isolated singularity, 84–85
    Isolated singularity, 85
Smooth arc, 39
Smooth curve, 39
Spectral density, 372
Spectral representation, 372
Spectrum, 372
Square transformation, 144
Square wave function, 207
Standard transformations, 136
Stirling's formula, 703
System of level curves, 22–23
System of linear equations, 615

T

Taylor series, 68
Taylor's series method, 756
Telegraph equations, 559
Telephone equations, 559
Thermal diffusivity, 547
Third order Runge-Kutta method, 782–783
Transcendental equation, 586
Transform of derivatives, 379
Translation mapping, 136
Transmission line, 558
Trapezoidal rule, 739–740
Triangularisation method, 623
Trigonometric series, 282–283
Triple pole, 85
Triply connected domain, 44–45
Truncation error, 586
Two dimensional wave equation, 576
Two-sided Laplace transform, 165
Type of isolated singularity, 85
Types of errors, 585
    Gross errors, 585
    Round-off errors, 585
    Truncation errors, 586

U

Uniformly convergent series, 67
Unit impulse function, 199–200
Unit step function, 191–192

V

Variable coefficients, 227, 242
Vibrations of a stretched string, 527

W

Wave equation, 520, 527, 576
Weddle's rule, 742

Y

Young's modulus, 265

Z

Zero of an analytic function, 84